COMPREHENSIVE ANALYTICAL CHEMISTRY

ELSEVIER B.V.
Radarweg 29
P.O. Box 211, 1000 AE Amsterdam
The Netherlands

ELSEVIER Inc.
525 B Street, Suite 1900
San Diego, CA 92101-4495
USA

ELSEVIER Ltd
The Boulevard, Langford Lane
Kidlington, Oxford OX5 1GB
UK

ELSEVIER Ltd
84 Theobalds Road
London WC1X 8RR
UK

First edition 2005

Library of Congress Cataloging in Publication Data
A catalogue record is available from the Library of Congress.

British Library Cataloguing in Publication Data
A catalogue record is available from the British Library.

ISBN: 0-444-51718-9
ISSN: 0166-526X

∞ The paper used in this publication meets the requirements of ANSI/NISO Z39.48-1992 (Permanence of Paper). Printed in The Netherlands.

Wilson & Wilson's

COMPREHENSIVE ANALYTICAL CHEMISTRY

Edited by

D. BARCELÓ

Research Professor
Department of Environmental Chemistry
IIQAB-CSIC
Jordi Girona 18-26
08034 Barcelona
Spain

Wilson & Wilson's

COMPREHENSIVE ANALYTICAL CHEMISTRY

VOLUME XLV

ANALYSIS AND DETECTION BY CAPILLARY ELECTROPHORESIS

Edited by

M.L. MARINA

Department of Analytical Chemistry, Faculty of Chemistry
University of Alcalá, 28871 Alcalá de Henares (Madrid), Spain

A. RÍOS

Department of Analytical Chemistry and Food Technology
University of Castilla-La Mancha, 13004 Ciudad Real, Spain

M. VALCÁRCEL

Department of Analytical Chemistry
University of Córdoba, Campus de Rabanales, 14071 Córdoba, Spain

2005

ELSEVIER

AMSTERDAM – BOSTON – HEIDELBERG – LONDON – NEW YORK – OXFORD – PARIS
SAN DIEGO – SAN FRANCISCO – SINGAPORE – SYDNEY – TOKYO

CONTRIBUTORS TO VOLUME XLV

Freek Ariese
Vrije Universiteit Amsterdam, Faculty of Sciences, Department of Analytical Chemistry and Applied Spectroscopy, De Boelelaan 1083, 1081HV Amsterdam, The Netherlands
Antonio Javier Blasco
Department of Analytical Chemistry, Faculty of Chemistry, University of Alcalá, Ctra. Madrid-Barcelona Km. 33.600, 28871 Alcalá de Henares, Madrid, Spain
Hong Chen
Graduate School of Material Science, University of Hyogo, Hyogo, Japan
Alejandro Cifuentes
Department of Food Analysis, Institute of Industrial Fermentations (CSIC), Juan de la Cierva 3, 28806 Madrid, Spain
Antonio L. Crego
Department of Analytical Chemistry, Faculty of Chemistry, University of Alcalá, Ctra. Madrid-Barcelona Km. 33.600, 28871 Alcalá de Henares, Madrid, Spain
Jose Carlos Díez-Masa
Institute of Organic Chemistry (CSIC) Juan de la Cierva, 3, 28006 Madrid, Spain
Dimuthu A. Jayawickrama
Bristol Myres Squibb Company, Pharmaceutical Research Institute, 1 Squibb Drive, New Brunswick, NJ 08903, USA.
Alberto Escarpa
Department of Analytical Chemistry, Faculty of Chemistry, University of Alcalá, Ctra. Madrid-Barcelona Km. 33.600, 28871 Alcalá de Henares, Madrid, Spain
M[a] del Rosario Fernández de la Campa
Department of Physical and Analytical Chemistry, University of Oviedo, Julian Clavería, 8, 33008 Oviedo, Spain
Mercedes de Frutos
Institute of Organic Chemistry (CSIC), Juan de la Cierva, 3, 28006 Madrid, Spain
Maria Àngeles García
Department of Analytical Chemistry, Faculty of Chemistry, University of Alcalá, Ctra. Madrid-Barcelona Km. 33.600, 28871 Alcalá de Henares, Madrid, Spain

Carmen García-Ruiz
Department of Analytical Chemistry, Faculty of Chemistry, University of Alcalá, Ctra. Madrid-Barcelona Km. 33.600, 28871 Alcalá de Henares (Madrid), Spain
Elisa Blanco González
Department of Physical and Analytical Chemistry, University of Oviedo, Julian Clavería, 8, 33008 Oviedo, Spain
Cees Gooijer
Vrije Universiteit Amsterdam, Faculty of Sciences, Department of Analytical Chemistry and Applied Spectroscopy, De Boelelaan 1083, 1081HV Amsterdam, The Netherlands
Kim M. Hanson
Chemistry Department, West Virginia University, Morgantown, WV 26506, USA
Lisa A. Holland
Chemistry Department, West Virginia University, Morgantown, WV 26506, USA
Jonathan V. Sweedler
Department of Chemistry and the Beckman Institute, University of Illinois, Urbana, IL 61801, USA.
Jacobus Kuijt
Vrije Universiteit Amsterdam, Faculty of Sciences, Department of Analytical Chemistry and Applied Spectroscopy, De Boelelaan 1083, 1081HV Amsterdam, The Netherlands
Stephen Kulka
Institute of Chemical Technologies and Analytics, TU Vienna, Getreidemarkt 9, 1060 Vienna, Austria
Pilar Lara-Quintanar
Institute of Organic Chemistry (CSIC), Juan de la Cierva, 3, 28006 Madrid, Spain
Bernhard Lendl
Institute of Chemical Technologies and Analytics, TU Vienna, Getreidemarkt 9, 1060 Vienna, Austria
Gloria Álvarez Llamas
Department of Physical and Analytical Chemistry, University of Oviedo, Julian Clavería, 8, 33008 Oviedo, Spain
Maria Luisa Marina
Department of Analytical Chemistry, Faculty of Chemistry, University of Alcalá, Ctra. Madrid-Barcelona Km. 33.600, 28871 Alcalá de Henares, Madrid, Spain

Contributors to Volume XLV

Alfredo Sanz Medel

Department of Physical and Analytical Chemistry, University of Oviedo, Julian Clavería, 8, 33008 Oviedo, Spain

Theron J. Pappas

Chemistry Department, West Virginia University, Morgantown, WV 26506, USA

Angel Ríos

Department of Analytical Chemistry and Food Technology, University of Castilla-La Mancha, Campus de Ciudad Real, 13004 Ciudad Real, Spain

Bartolomé M. Simonet

Department of Chemistry, University of Girona, Campus de Montilivi, 17071 Girona, Spain

Carolina Simó

Department of Food Analysis, Institute of Industrial Fermentations (CSIC), Juan de la Cierva 3, 28806 Madrid, Spain

Shigeru Terabe

Graduate School of Material Science, University of Hyogo, Hyogo, Japan

Maria Teresa Veledo

Institute of Organic Chemistry (CSIC), Juan de la Cierva, 3, 28006, Madrid, Spain

Miguel Valcárcel

Department of Analytical Chemistry, University of Córdoba, Campus de Rabanales, Annex Building C-3, 14071-Córdoba, Spain

WILSON AND WILSON'S

COMPREHENSIVE ANALYTICAL CHEMISTRY

VOLUMES IN THE SERIES

Vol. IA	Analytical Processes
	Gas Analysis
	Inorganic Qualitative Analysis
	Organic Qualitative Analysis
	Inorganic Gravimetric Analysis
Vol. IB	Inorganic Titrimetric Analysis
	Organic Quantitative Analysis
Vol. IC	Analytical Chemistry of the Elements
Vol. IIA	Electrochemical Analysis
	Electrodeposition
	Potentiometric Titrations
	Conductometric Titrations
	High-Frequency Titrations
Vol. IIB	Liquid Chromatography in Columns
	Gas Chromatography
	Ion Exchangers
	Distillation
Vol. IIC	Paper and Thin Layer Chromatography
	Radiochemical Methods
	Nuclear Magnetic Resonance and Electron Spin Resonance Methods
	X-Ray Spectrometry
Vol. IID	Coulometric Analysis
Vol. III	Elemental Analysis with Minute Sample
	Standards and Standardization
	Separation by Liquid Amalgams
	Vacuum Fusion Analysis of Gases in Metals
	Electroanalysis in Molten Salts
Vol. IV	Instrumentation for Spectroscopy
	Atomic Absorption and Fluorescence Spectroscopy
	Diffuse Reflectance Spectroscopy
Vol. V	Emission Spectroscopy
	Analytical Microwave Spectroscopy
	Analytical Applications of Electron Microscopy
Vol. VI	Analytical Infrared Spectroscopy
Vol. VII	Thermal Methods in Analytical Chemistry
	Substoichiometric Analytical Methods
Vol. VIII	Enzyme Electrodes in Analytical Chemistry
	Molecular Fluorescence Spectroscopy
	Photometric Titrations
	Analytical Applications of Interferometry

Volumes in the Series

Vol. IX	Ultraviolet Photoelectron and Photoion Spectroscopy Auger Electron Spectroscopy Plasma Excitation in Spectrochemical Analysis
Vol. X	Organic Spot Tests Analysis The History of Analytical Chemistry
Vol. XI	The Application of Mathematical Statistics in Analytical Chemistry Mass Spectrometry Ion Selective Electrodes
Vol. XII	Thermal Analysis Part A. Simultaneous Thermoanalytical Examination by Means of the Derivatograph Part B. Biochemical and Clinical Application of Thermometric and Thermal Analysis Part C. Emanation Thermal Analysis and other Radiometric Emanation Methods Part D. Thermophysical Properties of Solids Part E. Pulse Method of Measuring Thermophysical Parameters
Vol. XIII	Analysis of Complex Hydrocarbons Part A. Separation Methods Part B. Group Analysis and Detailed Analysis
Vol. XIV	Ion-Exchangers in Analytical Chemistry
Vol. XV	Methods of Organic Analysis
Vol. XVI	Chemical Microscopy Thermomicroscopy of Organic Compounds
Vol. XVII	Gas and Liquid Analysers
Vol. XVIII	Kinetic Methods in Chemical Analysis Application of Computers in Analytical Chemistry
Vol. XIX	Analytical Visible and Ultraviolet Spectrometry
Vol. XX	Photometric Methods in Inorganic Trace Analysis
Vol. XXI	New Developments in Conductometric and Oscillometric Analysis
Vol. XXII	Titrimetric Analysis in Organic Solvents
Vol. XXIII	Analytical and Biomedical Applications of Ion-Selective Field-Effect Transistors
Vol. XXIV	Energy Dispersive X-Ray Fluorescence Analysis
Vol. XXV	Preconcentration of Trace Elements
Vol. XXVI	Radionuclide X-Ray Fluorecence Analysis
Vol. XXVII	Voltammetry
Vol. XXVIII	Analysis of Substances in the Gaseous Phase
Vol. XXIX	Chemiluminescence Immunoassay
Vol. XXX	Spectrochemical Trace Analysis for Metals and Metalloids
Vol. XXXI	Surfactants in Analytical Chemistry
Vol. XXXII	Environmental Analytical Chemistry
Vol. XXXIII	Elemental Speciation – New Approaches for Trace Element Analysis
Vol. XXXIV	Discrete Sample Introduction Techniques for Inductively Coupled Plasma Mass Spectrometry
Vol. XXXV	Modern Fourier Transform Infrared Spectroscopy
Vol. XXXVI	Chemical Test Methods of Analysis
Vol. XXXVII	Sampling and Sample Preparation for Field and Laboratory
Vol. XXXVIII	Countercurrent Chromatography: The Support-Free Liquid Stationary Phase

Volumes in the Series

Vol. XXXIX	Integrated Analytical Systems
Vol. XL	Analysis and Fate of Surfactants in the Aquatic Environment
Vol. XLI	Sample Preparation for Trace Element Analysis
Vol. XLII	Non-destructive Microanalysis of Cultural Heritage Materials
Vol. XLIII	Chromatographic-mass Spectrometric Food Analysis for Trace Determination of Pesticide Residues
Vol. XLIV	Biosensors and Modern Biospecific Analytical Techniques

Contents

Series Editor's Preface

I am pleased to introduce the new volume on Capillary Electrophoresis, edited by Maria Luisa Marina, Angel Ríos and Miguel Valcárcel. The reason for having such a book in the *Comprehensive Analytical Chemistry* series is quite obvious: capillary electrophoresis is now a consolidated technique that has grown considerably in the last few years, specially in the bio-analytical field, and it is complementary to liquid chromatography in many other areas.

This book contains chapters that cover the fundamentals and general principles, the various separation modes in CE, on-line sample preparation methods and different coupling and detection systems, including UV-Vis, fluorescence, phosphorescence, chemiluminescence, MS, ICP–MS and NMR with a vision of future developments in the area, such as the use of microchips in CE. It will be obvious from the above list of topics that particular attention is given to detection systems, with a comprehensive coverage of detectors coupled to CE.

The book can be used as an academic text, and as a reference book for those working in analytical laboratories who want ready access to information on basic CE through more complex instrumental combinations.

Overall, this book covers an important technique in the field of analytical chemistry. There is no doubt about its usefulness for solving everyday problems in chemical analysis. Indeed, when I took over this series a few years ago I noted that we needed a title on CE. Last year, some Spanish colleagues, under the leadership of a pillar of the analytical chemistry community, Miguel Valcárcel, proposed to edit this title and the book is now in your hands. I would also like to thank all the authors of this book for their time and efforts in preparing this excellent and useful book.

Finally, I would like to take this opportunity to address two issues: *Comprehensive Analytical Chemistry* has recently published a number of books by Spanish authors. It may seem that since the series editor is from Spain, more Spanish authors are publishing in the series. This is partly true, but I would like to add that all book proposals are submitted to three or four referees, and all analytical chemists from anywhere in the world are welcome to submit book proposals. The second comment refers to new developments, and I have good news in this respect. *Comprehensive Analytical Chemistry* will this year become available on Elsevier's ScienceDirect platform, so subscribers will be able to consult the series online. This decision will certainly benefit all of us — the editors, authors and readers.

D. Barceló
Barcelona, Spain

Preface

The real impact of an analytical technique can be assessed using a combination of several quality indicators, such as the number of published scientific articles, books and book chapters on the subject, the availability of instruments in the market and its involvement in official standard methods, among others. From these indicators, we can conclude that, capillary electrophoresis (CE) is already consolidated as a powerful analysis technique that can satisfy a variety of demands for (bio)chemical information in different social and economic areas. CE is now a useful alternative to gas and liquid chromatographies and is increasingly used in routine analytical laboratories for a variety of reasons such as expeditiousness, peak resolution capability, low sample volume required, flexibility based on the variety of CE modes, the feasibility of using homemade approaches, etc. However, we should also be aware of its limitations, the most obvious being the need for an efficient sample pretreatment, the low precision due to the potential lack of representativeness in nanoliter introduction, its lack of sensitivity, which can be attributed to the low concentration of analyte reaching the flow-through detector, the difficulty of extrapolating the results achieved using homemade equipment to commercially available CE instruments used in the laboratory, etc.

The first aim of this book is to show the state of the art of basic CE using modern approaches, beginning with the fundamentals (chapter 1) and separation modes (chapter 2). The importance of preliminary operations in CE is shown in two chapters, one devoted to on-line sample preconcentration techniques (chapter 3) and the other to coupling flow systems for both homemade and commercially available CE instruments (chapter 4). Because continuous detection is one of the aspects of CE that need to be improved, many scientists have devoted much effort to finding better detection alternatives. The second and more extensive part of the book is devoted to this hot topic. Molecular absorption and emission detection modes in CE, which can be considered the conventional ones, are dealt with in chapters 5 and 6, respectively, with attention focussed on the innovations in this context.

The use of detection techniques in CE based on phosphorescence and chemiluminescence, which are also molecular emission spectroscopic techniques, is described in chapter 7. In addition, chapter 11 is devoted to vibrational spectroscopy detection in CE, which can be considered as an emerging alternative. The growing importance of electrochemical detection in CE justifies its inclusion in chapter 8. The so-called hyphenated techniques involving CE are also of great importance; chapter 9, 10 and 12 devoted to mass spectrometric detection, optical emission ICP-MS detection and NMR detection, respectively, constitute an excellent overview of the present and future of detection in CE. Finally, the third part of the book is devoted to recent developments in CE. One chapter is devoted to the emerging and almost consolidated chiral separations in CE (chapter 13), and the other to electrochemical detection in CE as implemented in microchips (chapter 14).

The editors are aware that not all the topics that are of present and future interest have been included in this book. Nevertheless, we provide an almost complete overview of CE that will be useful for scientists and analysts using the technique in their research or as an analytical tool. We wish to express our gratitude to all the prestigious authors involved as well as to Elsevier for the opportunity to publish this work.

The Editors
March 2005

Chapter 1

Fundamentals of capillary electrophoresis

Maria Luisa Marina, Angel Ríos and Miguel Valcárcel

1.1 INTRODUCTION

Capillary electrophoresis (CE) is a relatively recent instrumental separation technique, which has become increasingly important over the last decade. A number of monographs [1–6], reviews dealing with many aspects of CE and a large number of articles about the theory and practice of CE have been published. From a practical point of view, the large variety of available commercial equipment is crucial to the implementation of CE in routine laboratories. In this context, CE is becoming an alternative or complementary technique to the chromatographic ones. In many cases, new analytical possibilities are provided by CE, in addition to the development of new and interesting chromatographic-electrophoretic hybrid modes, mainly represented by electrokinetic chromatography (EKC) and capillary electrochromatography (CEC) [7].

Undoubtedly, practical analytical issues of CE are a present concern of researchers and end users. CE methods need to be appropriately validated for real analytical purposes, as well as be robust enough for assuring the successful transfer to control and routine laboratories. The intrinsic efficiency of CE as a separation technique, together with the recent developments in detection, are two key aspects for its progressive implementation as a routine separation technique. In addition, basic research in this field is providing very interesting contributions in the applied side of CE. Both these aspects of CE (analysis and detection) are the *leitmotif* of this book. In this first chapter, the fundamentals of CE are briefly covered. The basic concepts and instrumentation are considered in the first part of the chapter. Then,

Comprehensive Analytical Chemistry XLV
M.L. Marina, A. Ríos and M. Valcárcel (Eds)
Volume XLV ISSN: 0166-526X DOI: 10.1016/S0166-526X(05)45001-1

an overview of the CE separation modes is given. This topic will be developed in more detail in chapter 2, due to the importance of the CE modes for expanding the analytical scope of the applications associated with the basic CE mode (capillary zone electrophoresis, CZE). The possibility of combining CE with other analytical techniques clearly opens new and promising separation–identification–quantification alternatives. This aspect is shown in section 1.5, but widely expanded through the different chapters of the book. Thus, CE coupled to sample treatment and preparation units (chapter 4), and incorporating additional detectors, in a conventional hyphenated approach (chapters 9, 10 and 12) has been covered. Finally, an overview of the present Field of applications and future Trends of CE is given.

1.2 BASIC CONCEPTS

1.2.1 Electrophoretic mobility and electroosmotic flow

In the presence of a constant electric field, an ionic species in an electrolytic solution contained in a capillary is submitted to an electrostatic force (F), which can be expressed as follows:

$$F = qE \tag{1.1}$$

where q is the charge of the ion and E the strength of the electric field ($E = V/L_{\text{tot}}$,where V is the voltage applied across the capillary and L_{tot} the length of the capillary).

However, opposing this force F is the frictional force (F'), which, according to Stokes's law for spherical particles, can be expressed by the following equation:

$$F' = 6\pi\eta r v_{\text{ep}} \tag{1.2}$$

where η is the viscosity of the electrolytic solution, r the radius of the ion, and v_{ep} its velocity of migration.

When the constant velocity state is reached, F equals F'. Therefore, combining Eqs. (1.1) and (1.2),

$$qE = 6\pi\eta r v_{\text{ep}} \tag{1.3}$$

and the migration velocity of an ion can be expressed as follows:

$$v_{\text{ep}} = \mu_{\text{ep}} E \tag{1.4}$$

where μ_{ep} is the electrophoretic mobility of the ion, which is a function of its q/r ratio and η, according to the following equation:

$$\mu_{ep} = \frac{q}{6\pi\eta r} \tag{1.5}$$

Then, μ_{ep} depends on the ionic species' size and charge, the nature and concentration of the electrolyte, and the temperature. Consequently, each species moves under the influence of an electric field at a specific velocity.

The second driving force in electrophoresis is electroosmosis. electroosmotic flow (EOF) [8] is the liquid flow that originates in the presence of an electric field when an ionic solution is in contact with a charged solid surface. As an example, in a silica solution interphase in contact with an aqueous electrolytic solution, the solid surface has an excess of negative charge due to the ionisation of the surface's silanol groups. A high number of counterions of these anions are on the interphase between the capillary wall and the solution, the origin of the so-called electric double layer which is formed by a stagnant layer adjacent to the capillary wall (Stern layer) and a diffuse layer which is mobile (Gouy–Chapman layer) (see Fig. 1.1). The cationic counterions

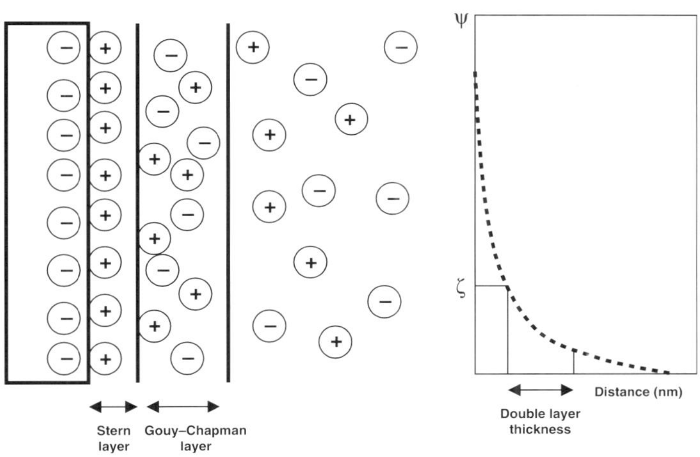

Fig. 1.1. Formation of the double layer and generation of the EOF. ψ, electrical potential; ζ, zeta potential. (Copyright 2000, Agilent Technologies, Inc. Reproduced with permission. Agilent Technologies, Inc. makes no warranty as to the accuracy or completeness of the foregoing material and hereby disclaims any responsibility therefore.)

in the diffuse layer migrate toward the cathode and as these ions are solvated, they drag solvent with them creating the EOF. The linear velocity of the EOF, v_{eo}, depends on the potential across the double layer, the zeta potential (ζ), through the following equation:

$$v_{eo} = \frac{\varepsilon E \zeta}{4\pi\eta} \tag{1.6}$$

ε being the dielectric constant of the solution. Since ε, ζ, and η depend on many properties of solids and liquids, there is a large number of experimental variables that influence the EOF (see sections 1.2.3.4 and 1.3.3).

The extremely small size of the double layer leads to flow at the walls of the capillary, resulting in a flat profile which enables high peak efficiencies in CE with respect to those separation techniques with hydrodynamic flow profiles (pressure-driven techniques). EOF modifies the migration of a species since if it moves in the same direction as EOF, its velocity will increase and if it moves against the EOF its velocity will decrease. This fact is expressed through the concept of apparent mobility (μ_a), defined as the algebraic sum of the electrophoretic mobility and that of the EOF (μ_{eo}):

$$\mu_a = \mu_{ep} \pm \mu_{eo} \tag{1.7}$$

The electrophoretic mobility of an ion can be calculated from μ_a and μ_{eo}, which in turn can be measured using a neutral marker generating a high response in the detector. Neutral molecules move at the same velocity of EOF, which makes possible the simultaneous separation of cationic, neutral, and anionic analytes in CE.

As a consequence, EOF can play an important role in CE separations although in some CE separation modes the existence of the EOF is undesirable (see chapter 2).

1.2.2 Migration time, efficiency, and resolution

The migration time (t_m) of an analyte in CE comprises the time from the injection to the detection of the analyte. This parameter can be calculated from the ratio between the migration distance and velocity as follows:

$$t_m = \frac{l}{v} \tag{1.8}$$

where l is the effective length of the capillary.

Tacking into account Eq. (1.4), t_m can be expressed by the following equation:

$$t_m = \frac{l}{\mu_a E} = \frac{lL_{tot}}{\mu_a V} \tag{1.9}$$

which shows how an increase in the voltage applied across the capillary leads to a decrease in the analysis time.

Peak efficiency in CE is given (as in HPLC) by the number of theoretical plates, N, which can be expressed as follows:

$$N = \left(\frac{l}{\sigma}\right)^2 \tag{1.10}$$

where σ is the standard deviation for a Gaussian peak which measures the dispersion provoked by the differences in solute velocity within the solute zone ($\omega = 4\sigma$, where ω is the baseline peak width).

Since longitudinal diffusion constitutes the main contribution to band broadening in CE, zone dispersion can be expressed as follows [1]:

$$\sigma^2 = 2Dt \tag{1.11}$$

where D is the diffusion coefficient of the analyte and t the analysis time.

Tacking into account Eq. (1.9), the following equation can be derived:

$$\sigma^2 = \frac{2DlL_{tot}}{\mu_a V} \tag{1.12}$$

and substituting into Eq. (1.10),

$$N = \frac{\mu_a lV}{2DL_{tot}} = \frac{\mu_a lE}{2D} \tag{1.13}$$

According to Eq. (1.13), it is possible to increase peak efficiency by increasing the applied voltage if heat dissipation is good and thermal effects are negligible. On the other hand, analytes with low diffusion coefficients and high mobilities give high peak efficiencies.

Following Gidding's approach [9], resolution (R_s) can be defined as follows:

$$R_s = \left(\frac{N^{1/2}}{4}\right)\left(\frac{\Delta v_{ep}}{v_{ep}}\right) \tag{1.14}$$

where $\Delta v_{ep}/v_{ep}$ is the relative velocity difference between the zones of two solutes 1 and 2. Δv_{ep} can be expressed as:

$$\Delta v_{ep} = v_{ep2} - v_{ep1} = \frac{(\mu_{ep2} - \mu_{ep1})V}{L_{tot}} \tag{1.15}$$

where v_{ep} is the average of the velocities corresponding to both analytes:

$$v_{ep} = \frac{(v_{ep1} + v_{ep2})}{2} = \frac{(\mu_{ep} \pm \mu_{eo})V}{L_{tot}} \tag{1.16}$$

and μ_{ep} is the average electrophoretic mobility of the two solutes.

From Eqs. (1.13)–(1.16), and assuming $l \approx L_{tot}$ the following expression for the resolution can be obtained [6]:

$$R_s = 0.177(\mu_{ep2} - \mu_{ep1})\left[\frac{V}{D(\mu_{ep} \pm \mu_{eo})}\right]^{1/2} \tag{1.17}$$

where D is the average diffusion coefficient of the solutes.

Equation (1.17) shows that the resolution between two zones decreases when there is a high EOF in the same direction of the electrophoretic migration although short analysis times will be possible under these conditions. On the contrary, resolution can be increased when EOF is similar but of opposite sign to the electrophoretic mobility. The cost of this high-resolution values is a long analysis time.

Other factors controlling the resolution will be discussed as follows.

1.2.3 Experimental variables affecting the electrophoretic resolution

1.2.3.1 Separation voltage

According to Eq. (1.17), increasing the applied voltage will increase the resolution in CE. However, this equation shows that due to the fact that the resolution is proportional to the square root of the applied voltage, a high increase in the applied voltage is required to enhance the resolution. This increase in the applied voltage is limited by the Joule heating effect. Furthermore, instruments usually have a limitation in the maximum applied voltage (generally 30 kV).

1.2.3.2 Type and dimensions of capillaries

There are several materials that are used to prepare capillaries for CE (see section 1.3.3). This material can have an important effect on the peak efficiency and consequently the resolution. In fact, the nature of the inner capillary wall is a decisive factor, which depends on the material used for preparing the capillary but also on the conditioning treatment. The adsorption of analytes on the inner wall of the capillary may cause important loss in peak efficiency and resolution. Different strategies have been developed in order to minimise this problem (see sections 1.2.3.4 and 1.3.3).

With respect to the dimensions of capillaries, it has been shown that the high surface to volume ratio for capillaries with a small diameter favours a good heat dissipation, which in turn enables the application of high voltages. As a consequence, analysis time is decreased and resolution increased. On the other hand, an increase in the length of the capillary at a constant field strength enables an enhancement in peak efficiency (see Eq. (1.13)) which positively influences the resolution although increasing analysis time (see Eq. (1.9)).

1.2.3.3 Length of the injected zone

The injected sample zone in CE has a finite width whose magnitude can significantly influence the overall band dispersion [2]. The contribution of the sample volume (σ_s^2) to the total peak dispersion can be expressed as follows [10]:

$$\sigma_s^2 = \frac{S^2}{12 v_s^2 \pi^2 (d_c/2)^4} \tag{1.18}$$

where S is the sample volume, v_s the linear velocity of sample injection and d_c the inner diameter of the capillary. It is shown by this equation that a twofold increase in d_c allows a 16-fold increase in the injected sample volume. As a rule, the injection sample zone length should be less than 1–2% of the total length of the capillary [11].

1.2.3.4 Characteristics of the separation buffer (nature, concentration, pH and additives)

The strong dependence of the electrophoretic mobility of an ion as well as of the EOF on the pH of the electrolytic solution makes it necessary to use a buffer in order to keep the pH constant during the separation process. There are a number of desirable characteristics to be fulfilled by the buffer used in CE [12]:

- High buffer capacity in the pH range of choice
- Low response to the detector used
- Low mobility to minimise current generation

- Availability in highly pure reagent grade
- Low temperature coefficient (change in pH per degree celsius).

The nature of the buffer should be chosen taking into account its response to the detection system used, the current intensity generated, and the range of pH in which it can be used (usually between 2 units above its pK_a and 2 units below its pK_a). Table 1.1 groups some of the buffers usually employed in CE together with their pK_a values. Buffers of an organic nature have the advantage of presenting, in general, low conductivity—a property that allows them to be used at higher concentrations than inorganic buffers, which provoke a greater heat generation. However, organic buffers are usually more UV-absorbant than inorganic buffers. Other characteristics that should be taken into account when choosing the buffer solution are that it should not have a

TABLE 1.1

Some of the buffers usually employed in CE together with their pK_a values

Buffer	pK_a
Phosphate	2.12, 7.21, 12.32
Citrate	3.06, 4.74, 5.40
Formate	3.75
Succinate	4.19, 5.57
Acetate	4.75
MES (2-[*N*-morpholino]-ethanesulphonic acid)	6.15
ADA (*N*-[2-acetamido]-2-iminodiacetic acid)	6.60
PIPES (piperazine-*N*,*N*′-bis(ethanesulphonic acid)	6.80
ACES (2-[(2-amino-2-oxoethyl)amino]ethanesulphonic acid)	6.90
MOPSO (3-[*N*-morpholino]-2-hydroxypropanesulphonic acid)	6.90
MOPS (3-[*N*-morpholino]-propanesulphonic acid)	7.20
HEPES (*N*-2-hydroxyethylpiperazine-*N*′-2-ethanesulphonic acid)	7.55
Tricine (*N*-[2-(hydroxymethyl)ethyl]glycine)	8.15
TRIS (tris-(hydroxymethyl)aminomethane)	8.30
Morpholine	8.49
Borate	9.24
CHES (2-[*N*-cyclohexylamino]ethanesulphonic acid)	9.50
CAPS (3-[cyclohexylamino]-1-propane-sulphonic acid)	10.40

Adapted from Ref. [11].

negative effect on the electrophoretic separation and that the solutes should be soluble and stable in that solution. A limited solubility of solutes in aqueous buffers can be solved by adding organic modifiers as additives to the electrolytic solution.

In addition to these considerations, the interaction between solutes and the buffer components can be used in order to increase the separation selectivity or to make possible an electrophoretic separation as occurs in the well-known case of the separation of carbohydrates by complexation with borate buffer.

With respect to the concentration of the buffer, the use of high concentrations in principle allows good peak shapes to be obtained. In fact, the mobilities of analytes are affected by the concentration of the background electrolyte according to the Debye–Hückel–Onsager effect and this can sometimes improve the separation, especially for multivalent analytes [13]. On the other hand, often there is a strong attractive force between positively charged analytes and the negatively charged wall of the silica capillary, resulting in broad tailing sample zones. A high concentration of the background electrolyte can decrease this adhesive behaviour of the sample ions by a competitive adsorption of the co-ions of the electrolyte solution, thus increasing the resolution [13]. However, an increase in the buffer concentration decreases the EOF, which leads to longer analysis times and provokes an increase in conductivity and then in heat generation, which has the negative effects on the separation mentioned above. One possibility to increase buffer concentration without significantly increasing solution conductivity is to choose carefully the salt to be employed. The influence of the anion and the cation of the salt on the conductivity of the solution when the ionic strength and other experimental variables are kept constant has been shown [3,14]. Also, the use of ampholytes has attracted attention for the preparation of low-conductivity background electrolytes in CE. It has been shown that ampholytes can be applied as background electrolytes, but not at a pH close to their pK value. If they are used at pH close to their pK values, the use of ampholytes is not advantageous compared with simple weak bases and acids [15].

Although the nature and concentration of the buffer can considerably influence the separation as stated above, the most important parameter of the separation buffer is its pH value. In fact, since many molecules exhibit acid or basic reactions, if the right pH value is properly chosen, charge differences can be adjusted to achieve a separation [16].

Since the EOF increases with increasing pH of solution, the change in EOF should be taken into account in addition to the change of electrophoretic mobilities of solutes in order to optimize the pH value to achieve the CE separation. As this optimization can be complex, prior testing with standard buffer solutions at different pH has been recommended [16]. Also, buffer systems with pH values close to the pK_a values of the analytes have the best influence on the selectivity, because here the change of charge which is dependent on the pH value is strongest.

If the change in nature, concentration, and pH of the buffer solution is not sufficient to optimize a CE separation, it is possible to use a variety of additives (metal ions, complexing agents, organic modifiers, surfactants, polymers, chiral selectors, etc.) to enhance the separation selectivity. The use of some of these additives can give rise to different CE separation modes such in the case of micellar systems or other pseudophases which give rise to EKC (see chapter 2). Micellar systems and other additives have also been employed to manipulate the magnitude and direction of the EOF and to dynamically coat the inner wall of the capillary, thereby avoiding adsorption of solutes. In fact, cationic surfactants have been used to reverse the EOF and neutral compounds such as hydroxypropylcellulose or polyvinyl alcohols to suppress the EOF. However, control of the EOF by dynamically coating the capillary can give rise to undesirable interactions between additives and solutes. This makes a permanent coating of the inner capillary wall very attractive (see section 1.3.3).

It is interesting to note that buffer preparation and changes of buffer solutions during analysis have been identified as a source of imprecision in CE [17]. Since, as stated above, small changes in buffer pH or composition may result in large changes in migration times, variations introduced when preparing buffer solutions (such as buffer reagents used and pH adjustment, filtration or degassing methods) should be avoided. On the other hand, buffer solutions are not unaffected during CE analysis. In fact, buffer depletion, contamination of the inlet vial or volume reduction change the pH and composition of the buffer. The extent of buffer electrolysis depends on the buffering capacity, buffer concentration, ionic strength, pH, volume of the buffer vials, temperature, current intensity, and analysis time. Then, the use of larger electrolyte reservoirs, the correct choice of the buffer, and the renewal of buffer solutions have been proposed in order to minimise buffer depletion [17].

1.2.3.5 Temperature

Non-thermostated capillaries cause viscosity gradients in the buffer and changes in electroosmotic and electrophoretic mobilities giving rise to perturbed velocity profiles which cause broaden peaks and increased plate heights [8]. For these reasons, CE instruments with a thermostated capillary chamber improve peak efficiency and resolution.

With respect to the convenience of working at low or high temperatures, in the case of thermodynamically controlled processes, the selectivity of the separation decreases when increasing the temperature. It is therefore possible in these cases to enhance the separation selectivity by lowering the temperature. On the other hand, if the selectivity of the system is sufficient, high temperatures can be used to improve peak shape, to short the analysis time (through an increase in the EOF caused by a decrease in the viscosity of the buffer solution) or to provoke some special effects such as conformational changes [12].

1.3 INSTRUMENTATION

Figure 1.2 is a scheme of the Basic instrumentation used in CE. The main components are the following:

- A high-voltage power supply
- Two reservoirs containing an electrolytic solution
- Two electrodes (in general, of platinum) introduced in the reservoirs

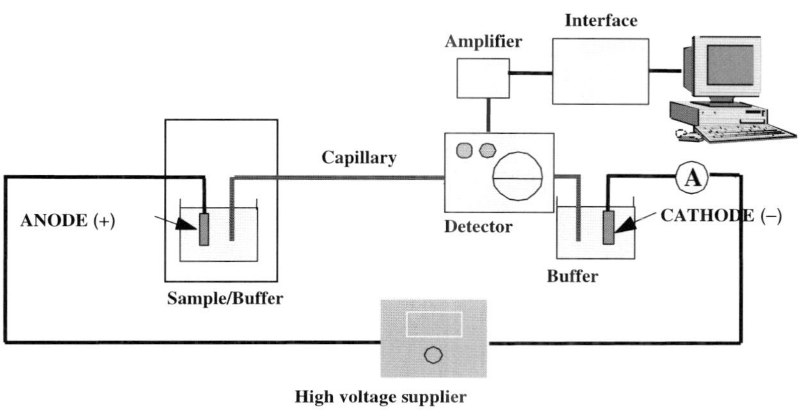

Fig. 1.2. Schematic of a CE instrument.

- A capillary filled with the electrolytic solution
- A sample injection system
- A detection system (on- or off-capillary)
- An amperometer
- A system for control of the instrument and data acquisition and treatment.

Different CE instruments are commercially available but many researchers use home-made instruments due to their ease of construction and higher flexibility. In the simplest working mode in CE, a capillary filled with an electrolytic solution is introduced into two reservoirs, which also contain the electrolytic solution and the electrodes. A high-voltage power supply allows the separation voltage to be applied once the sample has been introduced into one end of the capillary. At the other end of the capillary, detection of solutes is achieved and the detector signal is processed. An amperometer enables measurement of the current intensity through the capillary. The general characteristics of the main components of a CE instrument are described below.

1.3.1 High-voltage power supply

A high-voltage power supply in CE usually allows to apply voltages up to 30 kV. Security mechanisms are generally included to protect the operator from the risk inherent in the use of these high voltages. First of all, it is important to have a high stability in the application of the separation voltages to obtain adequate reproducibility for migration times. Secondly, it is very interesting to also have the possibility to switch polarity. This allows an interchange between the cathode and anode, which enables working with cathodic as well as anodic EOF or without EOF. Finally, other useful characteristics of the high-voltage power supply are: (i) to offer the possibility of working in the constant-voltage or constant-current mode and (ii) to allow programmable changes in the separation voltage, current intensity or field strength during the analysis [12].

1.3.2 Injection systems

As discussed above (see section 1.2.3.3), the size of the injected sample zone can have a great impact on efficiency in CE. Thus, in order to avoid band broadening, only minute sample volumes should be injected in the capillary (usually from 1 to 50 nl depending on the dimensions of the capillary). Electrokinetic and hydrodynamic injection modes are the most used in practice. These are simple and do not usually need any special design, except when they are used for automated injection.

Electrokinetic injection involves replacing the buffer reservoir located at the end of the capillary on which the high voltage is applied by a reservoir containing the sample. Then, a voltage is applied (usually 3–5 times lower than the separation voltage) for a short period of time (generally a few seconds). This brief application of voltage causes a narrow band of sample to migrate into the capillary. The sample reservoir is again replaced by the one containing the buffer. The separation voltage is then applied and the electrophoretic process begins [8].

This technique is simple and effective, causing minimum zone broadening. However, it discriminates among various sample components with different mobilities so that the most mobile species are injected in larger amounts than the less mobile ones. Electromigration injection is also influenced by the sample solution conductivity. For a correct calculation of the amount of sample injected, it is necessary that the sample and buffer solutions possess similiar conductivities [18]. The amount Q of a solute injected in the capillary can be expressed by the following equation [19]:

$$Q = \frac{(\mu_{\mathrm{ep}} \pm \mu_{\mathrm{eo}}) V \pi r'^2 C t_{\mathrm{i}}}{L_{\mathrm{tot}}} \tag{1.19}$$

where r' is the capillary radius, C the concentration of the analyte in the injected sample solution and t_{i} the injection time (time for which voltage is applied to achieve the injection). Equation (1.19) shows how the amount of solute injected can be modified for a given capillary by varying the injection time and the voltage applied during the injection.

In hydrodynamic injection the buffer reservoir is also replaced by the sample vial and it is then elevated to a specific height for a given period of time, creating a difference of height between the levels of liquid in the sample and in the buffer reservoir at the other end of the capillary [8]. This difference in height causes an hydrostatic pressure that allows the sample to enter the capillary. Solute electrophoretic mobility or the composition of sample solution does not influence this kind of injection in CE. The amount Q of solute injected by this technique can be calculated by the following equation [20]:

$$Q = \frac{\rho g \pi r'^4 \Delta h C t_\mathrm{i}}{8\eta L_\mathrm{tot}} \qquad (1.20)$$

where ρ is the density of the sample solution, g the constant for gravitational acceleration, Δh the difference in height between the levels of liquid in the sample and buffer reservoirs, and η the viscosity of the sample solution. Equation (1.20) shows how the amount of solute injected into the capillary can be modified by varying the difference between the reservoir levels and the injection time. Within the hydrodynamic injection mode, pressure and vacuum have also been used. These two injection modes are usually employed with commercial CE instruments.

A study on different approaches to improve precision in CE has shown that the injection process has a very important impact on precision, and it is possible to significantly increase injection repeatability by using automated injection systems [17].

Finally, direct in-line injection has been successfully demonstrated for CE using a commercially available injection valve designed for liquid chromatographic applications. The internal, fluid contacting materials in this valve injector were composed of ceramics and polyetheretherketone. This development was considered to have important implications for the design of completely integrated analysis systems [21].

The small sample size injected is one of the factors contributing to the resulting poor concentration sensitivity that can be observed in CE. For these reasons and as a means to solve this problem, on-capillary sample preconcentration techniques have been developed. These techniques will be discussed in detail in chapter 3. On the other hand, coupling continuous flow systems to CE is an interesting possibility to enhance sensitivity in CE and will be described in chapter 4.

1.3.3 Capillaries

Separation capillaries for CE have been prepared using different materials: glass, fused silica or Teflon [6]. However, the most commonly employed have been those of fused silica due to their following characteristics:

- They have an excellent transparency to UV and visible radiation which allows on-column optical detection in the capillary
- They are easy to fabricate
- They are relatively inexpensive.

Futhermore, they are usually covered by a polymer coating, generally polyimide, which provides flexibility and facilitates their handling. This polyimide coating has to be removed in a small part of the capillary in order to provide an optical detection window. This can be done using a gentle flame or by electric heating which avoids damaging the capillary wall.

Regarding capillary sizes, fused-silica capillaries are commercially available with inner diameters ranging from 2 to 200 μm. Those with 25 to 75 μm inner diameters and, especially 50 μm, are the most commonly employed. As stated above (see section 1.2.3.2), a decrease in the inner diameter of the capillary favours good heat dissipation which allows the application of high voltages thereby increasing peak efficiency and shortening analysis time. The drawbacks of using these small diameter capillaries are a loss in sensitivity when optical detection is used and clogging of the capillary.

With respect to the length of the capillary, it generally ranges from 50 to 100 cm. Short capillaries enable a decrease in the analysis time (see Eq. (1.9)) but there is a practical limit because of the increase in field strength and current intensity when decreasing the length of the capillary.

The most convenient shape of CE capillaries is a cylindrical tube. However, other types of capillaries such as rectangular, Z-shaped, or with a bubble or a multireflection cell have been used in order to enhance the sensitivity in optical detection as will be described in detail in chapter 5. T-shaped capillaries have also been employed for specific applications in CE and allow the collection of sample fractions without having to cut off the applied voltage.

Depending on the CE separation mode, the capillaries will be filled only with an electrolytic solution, or will contain in addition a gel or a stationary phase (open-tubular and packed capillaries) as will be described in chapter 2. As stated in section 1.2.3.2, the interaction of solutes with the inner capillary wall can provoke significant band broadening or even irreversible adsorption. This interaction depends on the nature of the capillary wall and the conditioning treatment, variables that also affect the direction and magnitude of EOF. As a consequence, the control of the state of the inner wall is of paramount importance in CE [22]. In fused-silica capillaries, the dissociation of the silanol groups present in the inner wall of the capillary, which is in contact with an electrolytic solution, leads to the appearance of a negative charge in the inner capillary wall which will be responsible for the attraction of positively charged solutes and the generation of a cathodic EOF.

Modification of the charge of the inner capillary wall can prevent the adsorption of solutes but also change the direction and magnitude of the EOF. Some options are possible:

- Working at low pH ($pH<2$) eliminates the charges of the silica surface by suppressing the dissociation of the silanol groups
- Manipulation of the electric charge of the solutes and the inner wall of the capillary in order to prevent adsorption by the existence of electrostatic repulsion forces
- Addition of chemical reagents which dynamically modify the inner capillary wall and in addition suppress or modify the EOF as stated in section 1.2.3.4.
- Permanent coating of the inner wall of the capillary to avoid the interactions between solutes and the inner wall of the capillary as well as the suppression or modification of the EOF.

As stated in section 1.2.3.4, the modification of the pH or the ionic strength of the electrolyte solution and the addition of reagents can also significantly affect the electrophoretic mobility of the analytes. As a consequence, one of the most attractive possibilities to modify the inner wall of the capillary is permanent coating. Table 1.2 groups some of the coating materials employed.

The extremely high importance that the state of the inner wall of the capillary has on CE separations requires different treatments in order to assure the reproducibility of migration times and the resolution. Three types of treatments are needed:

- A pretreatment before using the capillary for the first time
- A conditioning treatment of the capillary between runs
- A treatment of the capillary for its storage after use.

TABLE 1.2

Some materials used to permanently coat the inner wall of the capillary by chemical bonding

Coating material	Effect on EOF
Linear polyacrylamide	Decrease/elimination
Cross-linked polyacrylamide	Decrease
Polyethylene glycol	Decrease
Aryl pentafluoroepoxydiol	Decrease
Maltose	Decrease
Allylmethylcellulose	Decrease
Hydroxypropylcellulose	Decrease
Poly(vinyl) alcohol	Decrease/elimination
Cross-linked methylcellulose	Decrease/elimination
Ethylene glycol diglycidyl ether	pH independent EOF or elimination
Polyethyleneimine (PEI-18)	pH independent anodic EOF
N-[3-trimethoxysilyl)propyl]ethylenediamine	pH-dependent anodic EOF
Polyethyleneimine 600	pH-switchable cathodic/ anodic EOF
Aminopropyltrichlorosilane	pH-switchable cathodic/ anodic EOF
Ionic polyether	pH-switchable cathodic/ anodic EOF

Adapted from Ref. [12] with permission.

Pretreatments generally include activation of the silica inner surface with sodium hydroxide, and then washing with water and buffer solution. This step prior to the first use of the capillaries is highly recommended to reduce variations between fused-silica capillaries from different batches or manufacturers [17]. Another approach to stabilise the capillary wall is called "voltage conditioning". A short-duration voltage is applied to equilibrate the column. A similar approach is an electroconditioning step, where the capillary is conditioned for 1 h at 30 kV before analysis.

There is also the need to wash the capillary before each run. Washing returns the system to "first run" conditions, leading to less migration time error as the influence of the EOF variation is reduced. The above-mentioned reagents (sodium hydroxide, water, buffer) can usually be employed to achieve the conditioning of the capillary between runs but in this case a short washing routine is recommended in order not to increase analysis time. If there is adsorbed material on the capillary wall (samples containing proteins), an additional step of washing with sodium hydroxide and/or strong acids in the range of 0.1–1 M or sodium dodecyl sulphate (SDS) is used as a between-run washing step [17].

Different strategies have been proposed for the storage of capillaries after use, some of them including washing with sodium hydroxide, distilled water and drawing air through the capillary [23] or washing with methanol [4].

Finally, especially coated capillaries should be washed by following the recommendations of the manufacturer as these coatings can be damaged by washing with sodium hydroxide [12].

1.3.4 Detection systems

The most commonly employed detection systems in CE are those based on optical detection. This type of detection can be achieved on-capillary due to the above-mentioned excellent transparency to UV radiation of fused-silica capillaries. The main drawback of optical detection is its limited sensitivity, which is due to the small dimensions of the CE capillaries, usually with 25–75 μm inner diameters, which limit the volume of sample that can be injected as well as the optical path length.

Optical detection in CE is very commonly achieved by using UV–Vis absorption detectors which are simple, low-cost, easy to use, and have a high versatility. Nowadays, diode-array UV–Vis detectors are usually employed—they have the advantage over conventional UV–Vis absorption of providing peak purity and some structural information.

Other modes for optical detection in CE include thermooptical detection and methods based on the measurement of conventional and laser-induced fluorescence. The use of fluorescence detection allows an increase in the sensitivity with respect to UV–Vis absorption. On the other hand, the use of pre-column, on-column, or post-column derivatisation allows detection of those compounds without native fluorescence. Laser-induced fluorescence detection is one of the most sensitive detection modes in CE. Indirect optical detection in CE is also possible in order to detect non-absorbing or non-fluorescent analytes which expands the application range.

One way to increase sensitivity in optical detection is by using capillaries with larger inner diameters but this is not always possible due to the loss in peak efficiency that results. Another possibility is the use of extended path lengths in the capillaries (see section 1.3.3).

Electrochemical detectors have become increasingly important in CE in recent years. This detection mode is based on a reaction at an electrode surface and, as a consequence, in contrast to optical detection where the response of the detector is dependent on path length, cell volumes can be made very small with no decrease in sensitivity. This means that the use of very narrow capillaries is possible with the above-mentioned advantages (see section 1.3.3).

Conductivity detectors are considered to provide universal detection. They are very useful for detecting non-absorbing ions and can also work in an indirect mode. The combination of CE with suppressed conductivity detection has allowed an increase in the sensitivity obtained with this type of detector.

Mass spectrometry is one of the most attractive and powerful detection techniques in CE due to its sensitivity, universality, and the structural information it provides. Its main drawback is the high cost of mass spectrometers and interfaces.

Other detection methods used in CE are refractive index, radioactivity, novel optical detection technique based on phosphorescence, chemiluminescence, and vibrational spectroscopies (infrared and Raman), or nuclear magnetic resonance (NMR). Many of these have been subject to an intense research effort in order to solve what is considered the main limitation of CE techniques: detection sensitivity. In fact, detection in CE is a challenge due to the high peak efficiencies, small peak volumes, and limited time available to observe analytes [5]. Furthermore, the great number of analytes of very different characteristics and in very complex real matrices that have been analysed by using CE has required the availability of methods to detect low concentrations of these analytes in nanoliter (or even smaller) volumes.

The selection of the detection mode will depend on the properties of the analyte, the sample matrix composition, the required detection limits or the need to obtain qualitative and structural information in addition to quantitative data. Detailed information on the recent advances and achievements is presented in chapters 5–12.

1.4 SEPARATION MODES

The development of different separation modes in CE has expanded the Field of application of these techniques, enabling the analysis of a wide variety of compounds of very different characteristics. A brief description of these CE working modes is presented here since chapter 2 describes in detail the principles and characteristics of these separation modes.

CZE is the simplest separation mode in CE. In CZE, a capillary filled with an electrolytic solution is used. Separation of analytes is based on the differences in their electrophoretic mobilities, which are related to their charge density at a given pH. The direction and velocity of solutes is determined by both electrophoretic mobility and EOF.

EKC was introduced by Terabe's research group and is an analytical separation method which employs the experimental technique of CZE in combination with the principle of chromatography [24]. In addition to the electrolytic solution used in CZE, a major component called the pseudostationary phase is employed. This then satisfies the definition of chromatography where two phases should exist between which the solute is distributed. The electrokinetic phenomenon, including both electrophoresis and electroosmosis, is the means of transporting the pseudostationary phase and solutes inside the capillary. Separation of neutral and charged anaytes is possible based on the existing differences in their distribution coefficients between the pseudostationary phase and the surrounding aqueous phase. A variety of materials can be used as the pseudostationary phase in EKC. Micellar EKC (MEKC) is the term used when micelles are used as the pseudostationary phase [25].

In capillary gel electrophoresis (CGE), analytes are separated according to their charge and size in an electrolytic solution to which a gel-forming medium is added [12]. The gel acts as a molecular sieve to produce a size-based separation. CGE is very useful in the separation of macromolecules such as proteins and nucleic acids because they contain mass-to-charge ratios that do not vary with size [8].

CEC is considered as a hybrid technique between CE and high performance liquid chromatography (HPLC) since the separation of analytes is based on their distribution between mobile and stationary phases, but the mobile phase moves by electroosmosis [7,26]. This enables higher efficiencies than in chromatography due to the characteristic flow-flat profile that exists in CE. Packed and open-tubular capillaries can be employed in CEC.

Capillary isotacophoresis (CITP) is achieved in a discontinuous buffer system. Sample components condense between the so-called leading and terminating electrolytes, producing a steady-state migrating configuration composed of consecutive sample zones. Leading ions have the highest mobility and terminating ions the lowest. In contrast to CZE, where dilution of the sample occurs due to dispersion effects, in CITP very narrow sample zones of high concentration can be obtained. This principle can be used as a preconcentration technique for dilute samples in other CE modes. Cations and anions cannot be simultaneously separated by CITP [1].

In capillary isoelectric focusing (CIEF), amphoteric analytes such as peptides and proteins are separated on the basis of their isoelectric points. Sample ions are mixed with the carrier ampholytes that span the desired pH range, and the mixture is loaded onto the capillary. If an electric field is applied, the ampholytes build up a pH gradient and the sample analytes migrate until they reach a region of pH where they become electrically neutral and therefore stop migrating. Then, zones are focused until a steady-state condition is reached. EOF should be eliminated in CIEF and an additional step is necessary to force the zones pass the detector [1,8].

1.5 HYPHENATED CAPILLARY ELECTROPHORETIC TECHNIQUES

The possibility of combining CE with other equipment and analytical techniques clearly expands the applicability of the electrophoretic separations. In a broad sense, both on-line sample preparation units and additional detectors to the one conventionally included in commercial CE equipment (UV–Vis), can be considered as hyphenated

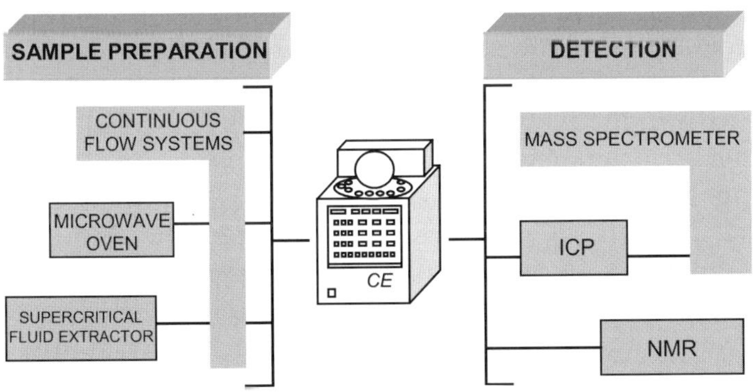

Fig. 1.3. Hyphenated-CE arrangements including both the automatic sample preparation units/equipment and the additional detection alternatives.

arrangements. Fig. 1.3 shows different CE-hyphenated alternatives used to date.

Reviewing the extensive literature on CE, we can conclude that, in many cases, sample preparation issues have scarcely been considered. Some of these publications only demonstrated the CE separation capabilities using synthetic or standard samples. Of course, this is a very limited view that is often not applicable to real situations. Therefore, sample preparation must be systematically considered in CE applications. This is why sample preparation units/equipment coupled on-line to CE can be considered as hyphenated analytical arrangements, integrating the whole analytical process. The different approaches dealing with on-line sample preparation–CE analysis are discussed in detail in chapter 4, as well as the key role played by the interfaces in such arrangements. Three different possibilities are shown in Fig. 1.3, addressed to the sample preparation. The simplest option in this respect is the automatic sample preparation performed in continuous flow systems (CFS) on-line coupled to CE. As has already been pointed out [27–29], the compatibility issues between the continuous and the discrete work of the CFS and the commercial CE equipment, respectively, have to be solved by the appropriate design of the interfaces. In many cases, simple CFS are involved as interface elements. Such is the case when other more sophisticated sample treatment equipment was coupled to CE. For instance, the use of a microwave oven [30], or a supercritical fluid extractor [31] for automating the preparation of complex samples, have been included as interesting possibilities in Fig. 1.3.

Hyphenated techniques were originally associated with the on-line combination of efficient separation techniques with powerful discriminating (identification) techniques. This approach emphasises the detection capabilities of the single separation technique. In this context, CE was firstly combined with mass spectrometry (MS) [32] and, more recently, with inductively coupled plasma-MS(ICP-MS) [33] and NMR [34], demonstrating new and interesting possibilities (Fig. 1.3). All these hyphenated systems (CE–MS, CE–ICP–MS, and CE–NMR) are considered in depth in chapters 9, 10 and 12. The coupling of these powerful detectors to CE provides clear improvements in both the identification and the quantification (by enhancing the low sensitivity of the conventional detector of CE) of a wide variety of analytes. It is evident that these CE-hyphenated arrangements have expanded the applications of CE in the clinical, pharmaceutical, environmental, agricultural and food analysis fields. Again, interfaces are the key elements for the successful combination of techniques working under very different principles, described in the chapters mentioned above.

1.6 FIELD OF APPLICATION AND TRENDS

1.6.1 Role of CE techniques with respect to the analytical potential of other separation techniques

Separation techniques play a decisive role in (bio)chemical analyses. Some of these techniques are commonly used for sample treatments, especially to meet the sensitivity and selectivity requirements of many analytical methods. Such is the case for liquid–solid, solid–liquid and liquid–liquid extraction procedures, gas diffusion, and dialysis, among others. The role of these techniques is, therefore, mainly in preparing the sample for analysis. Other separation techniques (mainly instrumental techniques) are directly related to the determination process itself. In this group, instrumental chromatographic techniques (gas chromatography (GC), HPLC, and supercritical fluid chromatography (SFC)) and CE integrate the continuous separation and detection of the analytes. Therefore, in a first approach, CE has a clearly different objective than that of the non-chromatographic techniques involved in sample treatment.

The information given by CE is similar to that provided by the instrumental chromatographic techniques, although the foundation of the separation process is different. Therefore, the comparison between electrophoretic and chromatographic techniques is of practical analytical interest. Technically, CE is simpler than GC, HPLC and SFC. This simplicity is based on the electrical principle (a physical principle) that produces the separation of the analytes and the movement of the buffer when the EOF is present. In chromatographic techniques, however, mechanical devices (pumps) are responsible for the flow of the mobile phase. In addition, the electrokinetic introduction of the sample in CE is also based on physical principles, in contrast with the mechanical introduction through injection valves in chromatography. This has important analytical advantages because it is possible to introduce very small volumes of samples in CE in a very reproducible manner. A direct practical repercussion derives from this fact: whereas in chromatographic methods an internal standard is normally needed, in electrophoretic methods this is not commonly required. Calibration is thus simplified, as external calibration can be used for the majority of the CE methods.

HPLC and CE have a similar Field of application. The initial tendency to use CE only for separating charged or chargeable compounds was promptly superseded by the appearance of other modes than CZE. Some of these modes include chromatographic principles for the separation of the solutes (even exclusively chromatographic principles), but the mobile phase moves by an electrophoretic effect. Due to the different profiles of the laminar (HPLC) and electroosmotic (CE) flows, CE gives better efficiency and separation resolution than HPLC, in general. This characteristic considerably reduces the time of analysis in CE with respect to HPLC. On the other hand, the important developments in the separation of peptides and proteins in recent years have been mainly obtained using CE. Hence, the role of CE in proteomic studies is clearly more important than that played by chromatographic techniques.

Although miniaturisation has now been introduced in HPLC through μ-HPLC equipment, it does not reach the level of miniaturisation and simplicity of the μ-chips used for CE. This miniaturization is important for the use of portable equipment in field tests, and for *in-situ/in-vivo* measurements. In this way, CE has advantages with respect to HPLC and other chromatographic techniques.

1.6.2 Field of application of CE

In recent years, CE has greatly expanded its Field of application. Indeed, the increasing number of publications show an important number of improvements and developments. On the one hand, CE has been the subject of much basic research devoted to instrumentation and other aspects of this technique. On the other hand, CE has become an interesting alternative in routine analysis, being recognised and established as an acceptable and reliable alternative to traditional analytical methods by many industries, law courts and government regulatory agencies [35].

In the case of small ions, the separation and determination of metal ions and inorganic anions by CE have received considerable attention. Commercial kits are available, the methods are simple and can have analysis times ranging from 2 to 10 min, which are comparable to those obtained by ion chromatography [36]. Also, in elemental speciation analysis, CE has been shown to perform well with application to the separation of metal ions in different oxidation states, metal complexes with inorganic and organic ligands, metalloid oxoanions, organometallic compounds, ionic nonmetal species, etc. [37].

CE is one of the preferred analytical technique in pharmaceutical quality control and clinical chemistry where rapid and accurate determinations are required [16]. Descriptions of CE and specific CE methods have been published in most Pharmacopeias and routine CE methods have been successfully submitted to regulatory authorities, such as the US Food and Drug Administration (FDA) [35]. Application areas include determination of drug content in drug substance and formulations, chiral separations, determination of drug properties (dissociation constants, solubility, etc.), pharmacokinetic studies and clinical monitoring of drug levels. Chiral separations are one of the most interesting applications of CE [38,39]. In fact, due to their especial characteristics, CE-based assays have become very powerful and unique for enantiomeric separation of chiral drugs [40,41]. Clinical determinations of pharmaceuticals in biological fluids by CE offer great possibilities as the direct injection of the sample into the capillary or the injection after minor pre-treatment is in many cases possible [42]. The implementation of CE in this area has been facilitated by the use of different strategies to enhance sensitivity that have been developed recently [35]. Such determinations have been used for residues of drugs in biological fluids [43,44], to indicate the existence of disease states based on the presence of higher or lower levels of specific marker compounds [35], to develop pharmacokinetic applications including drug monitoring and bioavailability studies [40] and for drug metabolites, an area of paramount importance in new fields of biological research, such as metabolomics [41,45]. The use of multicapillary systems allows high sample throughput clinical analysis applications.

Food analysis as a discipline involves the determination of a number of very diverse food components ranging from inorganic ions to proteins. As an added complication, these components can occur at trace levels within a complex sample matrix. CE can meet the requirements in terms of the selectivity and sensitivity that are necessary and the methods are in addition simple and cost-effective [46]. In fact, CE is becoming established as a valuable technique for food analysis and as a powerful tool for food research, with specific applications in the determination of amino acids and biogenic amines, vitamins, carbohydrates and gelling agents, phenolic compounds, food additives, organic acids, inorganic ions, toxins and residues, proteins and peptides, and DNA analysis [35,44,46–48].

Surface modification of capillaries to solve the problem of the adsorption of proteins to the inner wall of the capillary has allowed the development of a wide range of applications to the analysis of serum proteins, glycoproteins, lipoproteins, antibodies, hemoglobins, proteins produced by recombinant DNA technology, and food proteins, and to the study of protein folding [5].

With respect to the application of CE to the analysis, preparation, and physicochemical characterisation of peptides, the enormous potential of the different separation modes in CE has been shown with applications to quality control and purity tests, determination in biomatrices, monitoring of physical and chemical changes and enzymatic conversions, amino acid and sequence analysis, peptide mapping and chiral analysis [49,50].

Environmental monitoring of agrochemicals has also been achieved by CE. The successful determination of herbicides, fungicides, insecticides, and acaricides in water and soils has been reported. Many of these substances are chiral and CE has shown enormous possibilities to perform chiral analysis of these compounds of environmental interest. Applications of CE to trace detection and determination of pesticides in pesticide formulations, environmental samples, and for chiral separations have been described [51]. On the other hand, the determination of numerous compounds such as phenols, surfactants, dyes, polycyclic aromatic hydrocarbons, aromatic and aliphatic amines, aromatic acids, and aromatic sulphonic acids has been achieved in environmental samples by CE. A variety of preconcentration methods has been employed in order to increase sensitivity [35,52].

CE is well suited for the forensic laboratory because of its versatility, small sample requirements, and speed of analysis [53]. In the case of DNA analysis, it should be highlighted that CE has played an important role in this area, as is shown by the fact that the use of capillary arrays and laser-induced fluorescence detection was crucial to the successful completion of the Human Genome project [35].

Most of these applications described in recent years have been possible due to the following developments: the use of sensitive detection systems, the enhancement of sensitivity through the use of different preconcentration techniques or the employment of capillary arrays to increase sample throughput in CE. Chapters 3–12 describe in detail the most recent advances in these areas.

1.6.3 Trends

Despite the above-mentioned achievements, this separation technique is still in a development phase regarding both instrumentation and applications and its usage in the future can be expected to expand. Advances in CE in the near future should increase sensitivity (use of preconcentration techniques, coupling continuous separation techniques to CE, and use of sensitive detection techniques) and sample throughput (capillary arrays) and allow the development of chip-based separations in order to decrease the volume of sample necessary, reagent consumption, and the analysis time.

In the same way that multi-capillary instrumentation has been successfully employed in the Human Genome initative and related protein sequencing, the use of parallel array, multiplexing, is the area where the greatest benefits are foreseen with expanding applications in pharmaceutical, environmental, and clinical analysis [35]. Hyphenating on-line preconcentration with multiplexed CE is highlighted as a promising platform for sensitive and high-throughput analyses of metabolites [45]. The use of selective adsorbents in CE-mass spectrometry for analyte preconcentration and microreactions has been visualised as a powerful three-dimensional tool for multiple chemical and biological applications. The use of affinity adsorbents on the channels of microchips may revolutionise life science research applications because of the possibility of achieving applications in a massive parallel, multichannel, integrated format using the highly specific binding of the antibodies [54].

Non-aqueous solvents, such as methanol and acetonitrile, instead of the conventional aqueous CE electrolytes will be increasingly employed. The wide selection of organic solvents, with their very different physicochemical properties, broadens the scope to manipulate separation selectivity. The lower currents in non-aqueous solvents enable the use of high electric field strengths and wide bore capillaries, the latter in turn allowing larger sample load. In many cases, detection sensitivity can also be enhanced. Other benefits are the analysis of water-insoluble ionisable compounds and the suitability of non-aqueous based electrolytes for the interfacing of CE to mass spectrometry techniques which has given rise to interesting applications [55].

In conclusion, there appears to be a very promising and brilliant future for CE techniques as the number of studies, developments, and applications continue to multiply. CE will allow ultrafast separations (in seconds, not minutes) using microchip technology, and narrower and shorter capillaries will be employed in array format in order to achieve highly efficient and high-throughput analysis. Multiplexing systems will be used not only to sequence DNA but for simultaneous multisample analysis. In addition to its use as an analytical tool, a continued and expanded role of CE in the determination of physicochemical parameters is foreseen. Multidimensional separations employing separation–separation and separation–spectroscopic techniques will be routinely used for the analysis of complex mixtures [53].

REFERENCES

1 S.F.Y. Li, *Capillary Electrophoresis: Principles, Practice and Applications*, Elsevier, Amsterdam, 1992.
2 P.D. Grossman and J.C. Colburn (Eds.), *Capillary Electrophoresis. Theory and Practice*. Academic Press, San Diego, 1992.
3 R. Kuhn and S. Hoffstetter-Kuhn, *Capillary Electrophoresis: Principles and Practice*, Springer-Verlag, Berlin, 1993.
4 J.P. Landers, *Handbook of Capillary Electrophoresis*, CRC Press, Boca Raton, FL, 1994.
5 M.G. Khaledi, *High Performance Capillary Electrophoresis: Theory, Techniques and Applications*, Wiley, New York, 1998.
6 P. Camilleri (Ed.), *Capillary Electrophoresis. Theory and Practice*. CRC Press, Boca Raton, FL, 1998.
7 Z. Deyl and F. Svec, *Capillary Electrochromatography*, Elsevier, Amsterdam, 2001.
8 M.L. Marina and M. Torre, *Talanta*, 41 (1994) 1411–1433.
9 J.C. Giddings, *Sep. Purif. Sci.*, 4 (1969) 181–189.
10 H.T. Rasmussen and H.M. McNair, *J. Chromatogr.*, 516 (1990) 223–231.
11 D. Heiger, *High Peformance Capillary Electrophoresis. An Introduction*, Agilent Technologies, Germany, 2000.
12 B. Chankvetadze, *Capillary Electrophoresis in Chiral Analysis*, Wiley, Chichester, 1997.
13 J.L. Beckers and P. Bocek, *Electrophoresis*, 24 (2003) 518–535.
14 I.Z. Atamna, C.J. Metral, G.M. Muschik and J. Issaq, *J. Liq. Chromatogr.*, 13 (1990) 3201–3210.
15 J.L. Beckers, *Electrophoresis*, 24 (2003) 548–556.
16 H. Wätzig and S. Günter, *Clin. Chem. Lab. Med.*, 41 (2003) 724–738.
17 B.X. Mayer, *J. Chromatogr. A*, 907 (2001) 21–37.
18 X. Huang, M. Gordon and R.N. Zare, *Anal. Chem.*, 60 (1988) 375–377.

19 K.D. Lukacs and J.W. Jorgenson, *J. High Resolut. Chromatogr. Chromatogr. Commun.*, 8 (1985) 407–411.
20 D.J. Rose Jr. and J.W. Jorgenson, *Anal. Chem.*, 60 (1988) 642–648.
21 L.M. Ponton and C.E. Evans, *Anal. Chem.*, 73 (2001) 1974–1978.
22 H. Watzig, S. Kaupp and M. Graf, *Trends Anal. Chem.*, 22 (2003) 588–604.
23 R. Weinberger, *Practical Capillary Electrophoresis*, Academic Press, New York, 1993.
24 S. Terabe, *Trends Anal. Chem.*, 8 (1989) 129–134.
25 J.P. Quirino and S. Terabe, *J. Chromatogr. A*, 856 (1999) 465–482.
26 A.L. Crego, A. González and M.L. Marina, *Crit. Rev. Anal. Chem.*, 26 (1996) 261–304.
27 M. Valcárcel, A. Ríos and L. Arce, *Crit. Rev. Anal. Chem.*, 28 (1998) 63–81.
28 M. Valcárcel, L. Arce and A. Ríos, *J. Chromatogr. A*, 924 (2001) 3–30.
29 M. Valcárcel, M. Gallego and A. Ríos, *Fresenius J. Anal. Chem.*, 362 (1998) 58–66.
30 L. Arce, A. Ríos and M. Valcárcel, *J. Chromatogr. A*, 827 (1998) 113–120.
31 C. Mardones, A. Ríos and M. Valcárcel, *Anal. Chem.*, 72 (2000) 5736–5739.
32 P. Schmitt-Kopplin and M. Frommberger, *Electrophoresis*, 24 (2003) 3837–3867.
33 S.S. Kannamkumarath, K. Wrobel, K. Wrobel, C. B'Hymer and J.A. Caruso, *J. Chromatogr. A*, 975 (2002) 245–266.
34 C.S. Johnson and Q. He, *Adv. Magn. Reson.*, 13 (1989) 131–159.
35 K.D. Altria and D. Elder, *J. Chromatogr. A*, 1032 (2004) 1–14.
36 A.R. Timerbaevtimerbaev, A.R. >, *Electrophoresis*, 23 (2002) 3884–3889.
37 A.R. Timerbaevtimerbaev, A.R. >, *Talanta*, 52 (2000) 573–606.
38 B. Chankvetadze, *Trends Anal. Chem.*, 18 (1999) 485–498.
39 B. Chankvetadze and G. Blaschke, *J. Chromatogr. A*, 906 (2001) 309–363.
40 C.C. Lin, Y.T. Li and S.H. Chen, *Electrophoresis*, 24 (2003) 4106–4115.
41 P.S. Bonato, *Electrophoresis*, 24 (2003) 4078–4094.
42 H. Wätzig, M. Degenhardt and A. Kunkel, *Electrophoresis*, 19 (1998) 2695–2752.
43 M. Hernández, F. Borrull and M. Calull, *Trends Anal. Chem.*, 22 (2003) 416–427.
44 C.L. Flurer, *Electrophoresis*, 24 (2003) 4116–4127.
45 P. Britz-McKibbin and S. Terabe, *J. Chromatogr. A*, 1000 (2003) 917–934.
46 R.A. Frazier and A. Papadopoulou, *Electrophoresis*, 24 (2003) 4095–4105.
47 V. Poinsot, C. Bayle and F. Couderc, *Electrophoresis*, 24 (2003) 4047–4062.
48 J. Lange, K. Thomas and C. Wittmann, *J. Chromatogr. A*, 779 (2002) 229–239.
49 V. Kasicka, *Electrophoresis*, 24 (2003) 4013–4046.

50 G.K.E. Scriba, *Electrophoresis*, 24 (2003) 4063–4077.
51 A.K. Malik and W. Faubel, *Crit. Rev. Anal. Chem.*, 31 (2001) 223–279.
52 Y. Picó, R. Rodríguez and J. Mañes, *Trends Anal. Chem.*, 22 (2003) 133–151.
53 H.J. Issaq, *J. Liq. Chromatogr. Rel. Technol.*, 25 (2002) 1153–1170.
54 N.A. Guzman and R.J. Stubbs, *Electrophoresis*, 22 (2001) 3602–3628.
55 M.L. Riekkola, M. Jussila, S.P. Porras and I.E. Valko, *J. Chromatogr. A*, 892 (2000) 155–170.

Chapter 2

Separation modes in capillary electrophoresis

Maria Ángeles García, Maria Luisa Marina, Angel Ríos and Miguel Valcárcel

2.1 INTRODUCTION

One of the most important characteristics of capillary electrophoresis (CE) is its application to a great variety of analytical problems. This is due to its versatility, which is based mainly on the different separation modes possible within the CE format. These modes are easily accessible simply by changing the composition of the BGE or some experimental conditions of the separation.

Capillary zone electrophoresis (CZE) was the first separation mode developed based on the work achieved by Jorgenson and Lukacs [1], and nowadays continues to be the most commonly employed CE mode. Although in 1983 Hjertén [2] was the first to employ capillaries filled with polyacrylamide gels, it was in 1987 when Cohen and Karger [3] introduced a new CE mode called capillary gel electrophoresis (CGE) for the separation of proteins and nucleic acids. They combined the good resolution of gel electrophoresis with the simple instrumentation of CE. In 1984, Terabe et al. [4] introduced a new separation CE mode called micellar electrokinetic chromatography (MEKC), which enabled the separation of neutral species. The name electrokinetic chromatography (EKC) [5] is nowadays used for all the CE separation modes using a pseudophase other than a micellar system. Another CE mode useful for separating noncharged species is capillary electrochromatography (CEC), which uses a stationary phase as in HPLC together with an applied voltage instead of a pump [6].

Although less commonly employed, capillary isotachophoresis (CITP) and capillary isoelectric focusing (CIEF) are complementary

Comprehensive Analytical Chemistry XLV
M.L. Marina, A. Ríos and M. Valcárcel (Eds)
Volume XLV ISSN: 0166-526X DOI: 10.1016/S0166-526X(05)45002-3

to the other CE modes. The isotachophoresis principle was probably the most commonly employed separation mode before the development of CE due to the possibility of achieving the separation in tubes (250–500 μm) [7]. However, nowadays the isotachophoresis separation principle is more often used as a preconcentration technique [8] than as a separation mode. With respect to CIEF, the first applications of this separation mode in capillaries were achieved by Hjertén and Zhu [9] and Hjertén et al. [10] and it is applied mainly for the separation of proteins and macropeptides on the basis of their different isoelectric points (pI).

Nonaqueous CE (NACE) was first described by Walbroehl and Jorgenson [11] and employed nonaqueous solvents in the separation buffer. The use of this kind of solvent is specially advantageous for separating solute with a very similar electrophoretic behaviour in aqueous solutions, with maximum advantage for apolar analytes.

Figure 2.1 shows the principles of the different separation modes in CE. In some cases, only one separation mode will be useful to achieve a given separation while in other cases the separation will be possible with different separation modes. However, sensitivity and selectivity can be different in each case, and then the most appropriate separation will have to be chosen. In addition, in comparison with HPLC where changing the separation mode implies changing the chromatographic column, in CE we can change the separation mode by using the same capillary (except in CEC separations).

Figure 2.2 shows the degree of utilization of CE techniques over the last 15 years (Fig. 2.2a) as well as their evolution for this period of time (Fig. 2.2b), based on the number of publications that have appeared in

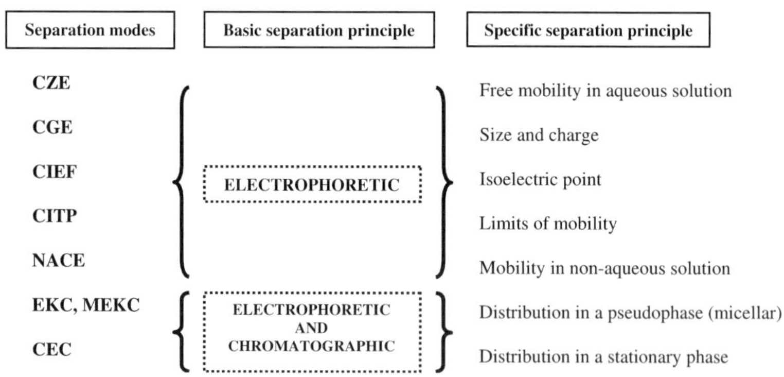

Fig. 2.1. Separation modes and basic and specific separation principles in CE.

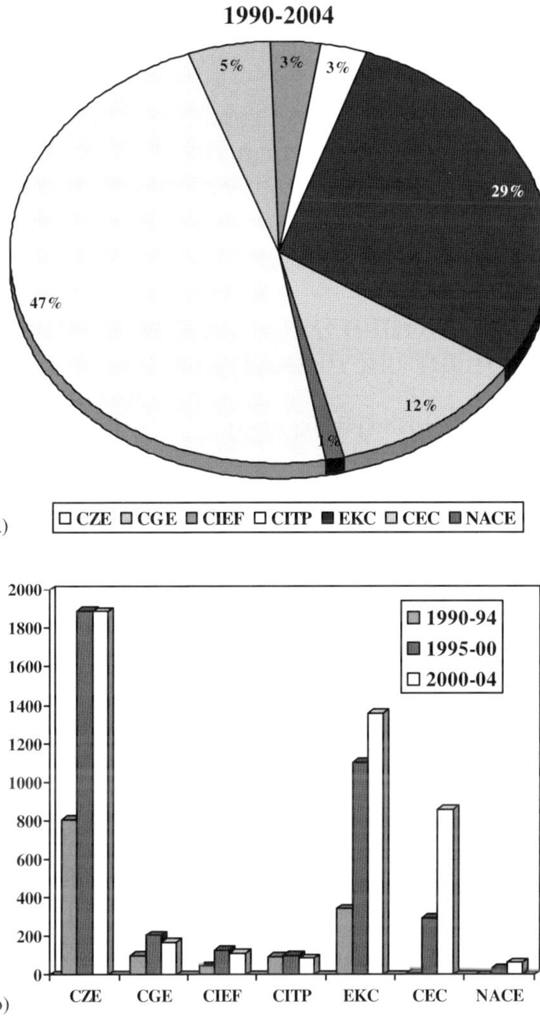

(a)

(b)

Fig. 2.2. (a) Degree of utilization of CE techniques in the last 15 years. (b) Evolution of CE techniques for this period of time. In both cases, the number of publications appeared in the scientific literature has been used.

the scientific literature. It can be observed that CZE, followed by EKC and CEC, have been the most employed CE working modes while CGE, CIEF and CITP have been the least employed modes. On the other hand, the number of publications on EKC, CEC and NACE has continuously increased in the last few years while for the other techniques it has not significantly varied.

For an easy presentation, the different separation modes in CE are categorized here as a function of the separation media (aqueous and nonaqueous) and the separation principle (separation modes based on pure electrophoretic principles and separation modes combining electrophoretic and chromatographic principles). A more detailed description of the characteristics of these separation modes in CE will be presented.

2.2 SEPARATION MODES IN CE BASED ON PURE ELECTROPHORETIC PRINCIPLES

2.2.1 Capillary zone electrophoresis

CZE, also named free CZE (FCZE), is the most commonly employed separation mode in CE due to its versatility, easy of operation and separation power. However, in principle, it has the limitation of being useful only for the separation of charged species (neutral species cannot be separated) [1,12–15].

2.2.1.1 Fundamentals

In CZE the capillary contains only a buffered solution [1,16], and the electrophoretic mobilities of the analytes depend on their charge/mass ratio. In fact, the separation mechanism of the components of the sample is based on these differences (charge/mass ratios), which allow them to have different electrophoretic mobilities, and therefore, different velocities, and it is possible to simultaneously separate cations and anions if the EOF is strong enough to lead anionic substances to the detector located at the cathode (direct polarity).

Figure 2.3 shows the separation principle in CZE. In this case, the sample is injected into the capillary filled with the buffered solution and a separation voltage is then applied. Solutes migrate through the capillary according to their respective migration velocities (v_{ep}), which are related to their electrophoretic mobilities through the following equation:

$$v_{ep} = \mu_{ep}E \tag{2.1}$$

where μ_{ep} is the electrophoretic mobility of the ion, and E the electric field strength. μ_{ep} depends on the charge of the ion (q), its radius (r) and

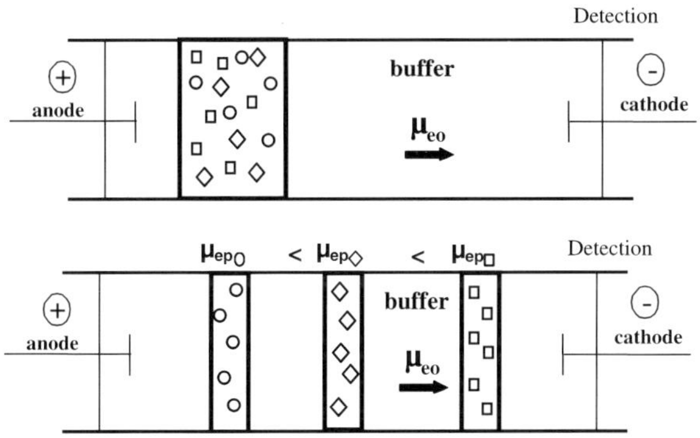

Fig. 2.3. Separation principle in CZE. □, ○ and ◇ are cations.

the viscosity of the solution (η) according to:

$$\mu_{\text{ep}} = \frac{q}{6\pi\eta r} \tag{2.2}$$

If the analyte is present in the solution in different forms which are in dynamic equilibrium (as, for example, in the case of an acid–base equilibrium), the resulting electrophoretic mobility is a combination of the electrophoretic mobility of the possible species in equilibrium (μ_{epi}) pondered by the molar fraction of each species "i" (α_i):

$$\mu_{\text{ep}} = \sum_{i=0}^{n} \alpha_i \mu_{\text{ep}i} \tag{2.3}$$

Electrophoretic mobility controls the separation selectivity in CZE through the q/r ratio, which should be different for the substances to be separated. Analytes with a high charge and low size move under the influence of the electric field more rapidly than those having a low charge and high radius, this difference enabling their separation.

The real electrophoretic mobility does not coincide with that predicted by Eq. (2.2) if EOF (see section 1.2.1) exists.

The total velocity of the species in the presence of EOF can be expressed as follows:

$$v_{\text{tot}} = (\mu_{\text{eo}} \pm \mu_{\text{ep}})E \tag{2.4}$$

where μ_{eo} is the electroosmotic mobility given by:

$$\mu_{\text{eo}} = \frac{\varepsilon\zeta}{4\pi\eta} \tag{2.5}$$

where ε is the dielectric constant, and ζ the named zeta potential which originates in the separation plane of the two layers of the double layer and depends mainly on the nature and amount of ions in the inner surface of the capillary.

From the electroosmotic mobility, the electroosmotic velocity, v_{eo}, can be easily obtained:

$$v_{eo} = \mu_{eo} E \tag{2.6}$$

In CZE, species migrate during the separation due to the combined effect of the two above-mentioned phenomena: the electrophoretic velocity characteristics of each species and electroosmotic velocity, which is equal for all of them [17]. Each species will migrate in the same direction as or against the electroosmotic flow, depending on their charge.

The migration time (t_m), or time in which a solute covers a distance l from the injection to the detection points (effective length of the capillary), can be determined by:

$$t_m = \frac{l}{(\mu_{eo} \pm \mu_{ep})E} \tag{2.7}$$

Equation (2.7) shows that t_m will decrease when increasing the applied electric field. However, the applied electric field is limited by Joule heat generation and the maximum value to be applied will depend on the separation and instrumental conditions employed to achieve the analysis. Equation (2.7) also shows that migration velocities for cations will be higher than that of the EOF since they move in the same direction as the EOF. In fact, cations elute in the first place in order of decreased charge/mass ratio. Neutral molecules move through the capillary at the same velocity as the EOF and elute all together (no separation exists) after the cations. Finally, anions move against the EOF due to their negative charge, but, if the velocity of the EOF is higher than the velocity of each anion of the sample, all of them will be the last to elute at the cathode (after cations and neutral molecules). The elution of anions occurs in order of increased charge/mass ratio as shown in Fig. 2.4.

There are different possibilities to control the EOF, which can be eliminated or even reversed as a function of the separation. If the EOF is eliminated and direct polarity is applied, only cations will migrate through the capillary as they are neutral solutes at the injection point during the analysis and the anions are not detected (they migrate to the anode). On the other hand, the EOF can be reversed (reversed polarity) if negative voltages are applied. In this case, the elution order

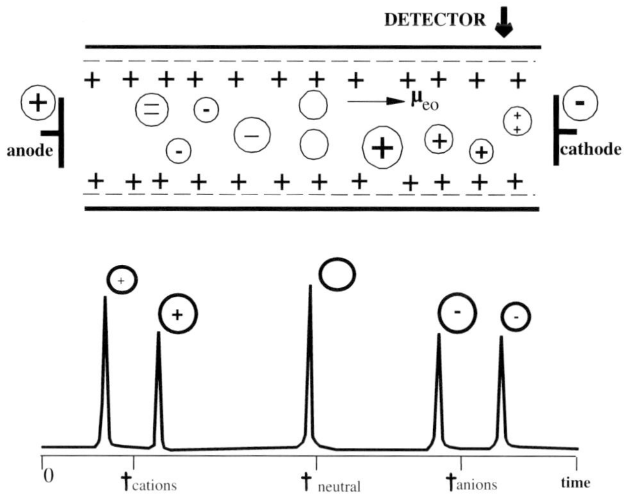

Fig. 2.4. Analyte elution order in CZE.

is also reversed—a very useful strategy for the separation of anions of low mass and high charge.

The term ($\mu_{eo} \pm \mu_{ep}$) is called apparent mobility, μ_a, and can be calculated from experimental data as follows:

$$\mu_a = \frac{lL_{tot}}{t_m V} \qquad (2.8)$$

where L_{tot} is the total length of the capillary and V the applied voltage.

2.2.1.2 Parameters influencing the separation

Electrophoretic mobilities of solutes can be modified by changing the characteristics of the buffer solution—pH, ionic strength and composition—by means of the addition of different types of substances [18,19]. However, these changes can also modify the EOF and then the optimization of the method becomes more difficult. For this reason, in some cases the reduction or elimination of the EOF is recommended, this elimination being also useful to achieve other CE separation modes such as CIEF or CITP [20].

The pH of the electrolytic solution is a key parameter to achieve a separation by CZE since it determines the degree of ionization and, therefore, the relative mobility of the different analytes. As a consequence, a BGE with a good buffering capacity at the working pH should be chosen. Other interesting characteristics of the BGE are low absorptivity at the detection wavelength when using UV-Vis absorbance detection and a low mobility in order to minimize current intensity.

The magnitude of the EOF is another key parameter to be controlled in CZE. Thus, EOF can be too large at high pH values, causing the elution of the analytes before separation, or can be too small at low pH, resulting in unnecessarily long analysis times. The control of the EOF generally requires either modifying the charge of the capillary wall or the viscosity of the BGE since a change in the pH of the BGE also affects the charge of the analyte and therefore its mobility. A decrease in the separation voltage will give rise also to a decrease in the EOF (Eq. (2.6)) but this option has clear drawbacks related to a loss in the separation efficiency and an increase in the analysis time. Finally, the EOF is also affected by changes in the concentration and, therefore, in the ionic strength of the BGE. In fact, an increase in the concentration of the BGE decreases the zeta potential, thereby decreasing the EOF (Eq. (2.5)). Concentrations of BGE usually employed range from 10 to 50 mM, although concentrations higher than 100 mM can also be used in the case of low conductivity buffers.

The use of fused-silica capillaries in CZE may cause the adsorption of analytes with a high positive charge density to the active sites of the charged silica, giving rise to distortion of the electrophoretic peak and to reduction in the separation efficiency. To avoid these adsorption phenomena and to reduce or to eliminate the EOF, additives that are adsorbed to the capillary wall to block the sylanol groups can be introduced in the BGE. Another possibility to avoid the EOF is to use chemically modified capillaries (covalent coating) where the sylanol groups have been derivatized. The coating of the capillaries can be carried out with additives as surfactants, amphoteric salts or linear hydrophilic polymers and can be a complex task. These coated capillaries are nowadays commercially available.

Table 2.1 summarizes the effect of some experimental variables controlling the EOF.

2.2.1.3 Development and optimization of a separation method

Figure 2.5 shows how a CZE separation method can be developed [21]. First of all, it is necessary to know the acid–base properties of the analytes. If these properties are not known, two trials at acid and basic pH values should be carried out, checking the position of the analytes with respect to the EOF. Working in the direct polarity mode, a basic

TABLE 2.1

Influence of different experimental parameters on the EOF

Experimental parameter	Influence	Comment
Electrical field strength (E)	Proportional changes in the EOF	Increasing E increases the Joule effect
Buffer pH	Increasing buffer pH increases the EOF	This is the best option to modify EOF
Ionic strength or buffer concentration	Increasing any of them, decreases zeta potential	It can produce an increase in Joule effect Low ionic strengths can produce an adsorption to the sample
Temperature	Increasing temperature decreases viscosity and increases the EOF	Easy control of temperature is possible due the use of thermostatization systems
Addition of organic modifiers	It modifies zeta potential and viscosity	Generally decreases the EOF and can alter the selectivity

analyte, which is protonated at low pH will migrate before the EOF while an acid analyte at basic pH values (anionic form) will migrate after the EOF. When the pK_a of the analyte is known, its electrophoretic behaviour can be predicted and the most appropriate BGE can be used.

If the sample contains one or more nonionizable analytes, then in the trials carried out at basic as well as at acid pH values, unresolved peaks will be obtained together with the EOF. In this case, the complete separation of the sample by CZE will not be possible and the use of other CE separation mode will be needed, for example, MEKC.

Once the experimental conditions are chosen under which the analytes migrate and can be detected, the optimization of the method is accomplished in terms of peak resolution and analysis time. In order to obtain a sufficient resolution for all the analytes, the choice of the pH of the BGE is a key step since different mobilities will be obtained by modifying this parameter within a pH range in which the analytes

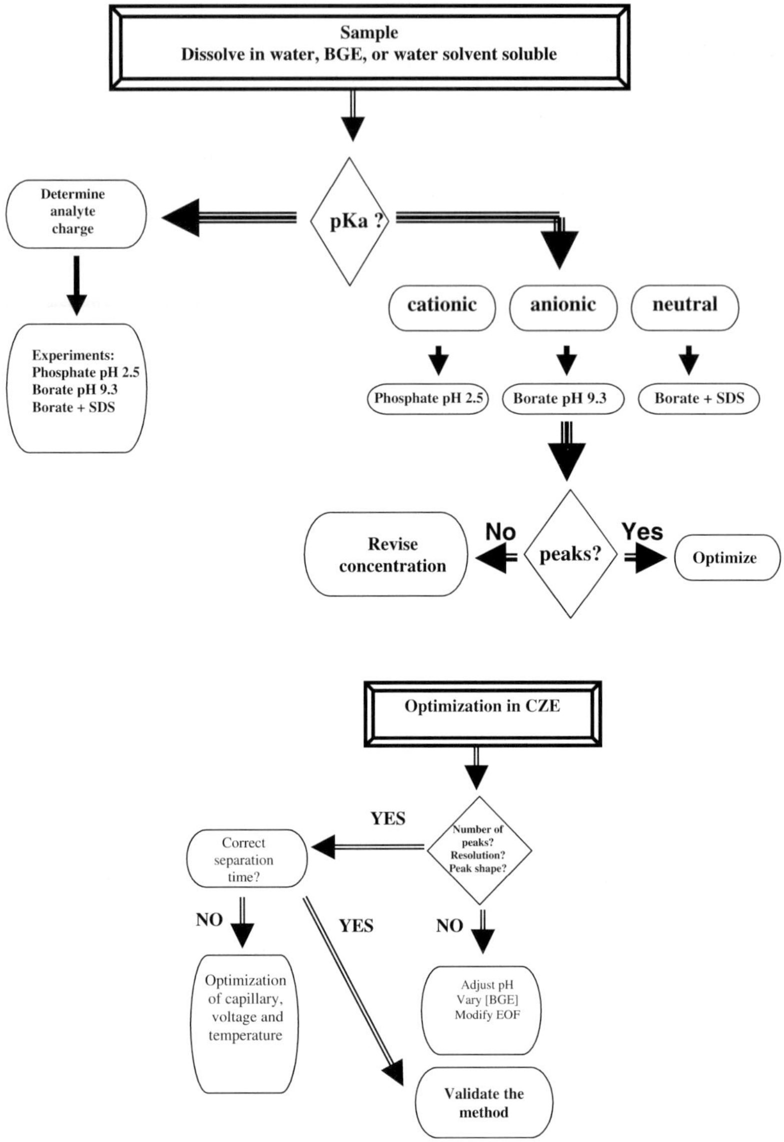

Fig. 2.5. Development and optimization of a CZE analysis method.

remain ionized. The separation of complex mixtures may require the use of different additives with different effects. As an example, a water-miscible organic solvent can be used to modify the viscosity of the medium and a cationic surfactant can be employed to reduce the adsorption of the analytes to the capillary wall. Other additives, such as cyclodextrins, can enhance the separation by selectively including some of the analytes into their cavity. However, when the additive constitutes a pseudophase, as in the case of a cyclodextrin or a micelle, we consider the separation mode to be within the EKC mode.

When the resolution of the analytes is sufficient, analysis time can be shortened by increasing the separation voltage or by decreasing the length of the capillary, although these parameters can also affect the resolution.

Among the main characteristics and limitations of CZE are the following:

- The buffer is homogeneous.
- The applied voltage is constant through the capillary.
- It is useful to separate charged species.
- Moderate reproducibility levels in quantitative analysis (2–5% for real samples).
- Adsorption of analytes on the capillary wall may occur, especially for positively charged compounds (such as basic proteins with $pI > 7$), which may affect the separation.
- Lack of selectivity exists for substances with different molecular weight but equal q/r ratio, as is the case for protein–sodium dodecyl sulphate (SDS) complexes, DNA fragments, polysaccharides or neutral substances that comigrate with the EOF.

2.2.1.4 Applications

CZE is generally employed to separate ionic or ionizable compounds of low molecular weight and sufficient solubility in aqueous solution. However, the enormous amount of additives that can be incorporated into the separation buffer, such as organic modifiers, surfactants at concentrations below their critical micelle concentration (CMC), ion-pair forming reagents, complexing agents, etc. enable to manipulation of the separation selectivity and has expanded the application range of this technique to different disciplines and to a wide number of compounds [22].

Many of the applications of CZE have been developed in the biochemistry area, especially in the separation of peptides and proteins [23–25]. The success of CZE for peptide mapping should be highlighted. In peptide mapping a protein is enzymatically or chemically cleaved into smaller peptide fragments and subsequently separated. The

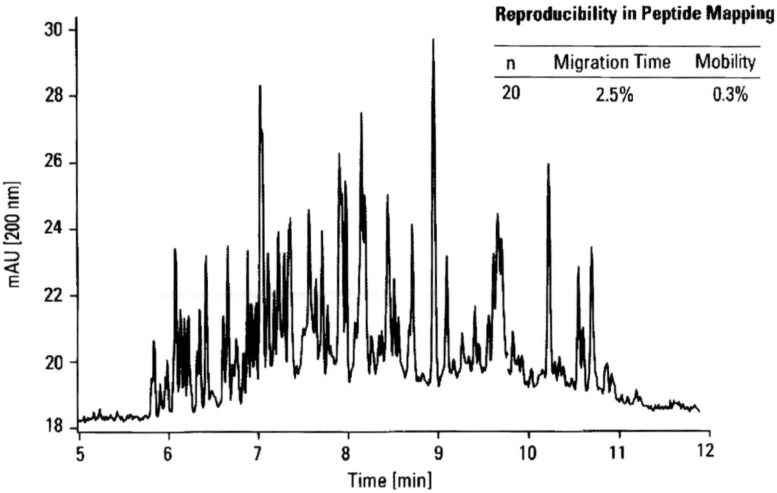

Fig. 2.6. Rapid bovine serum albumin (BSA) peptide mapping. Conditions: buffer (20 mM phosphate, pH 7), sample (tryptic digest of BSA), capillary ($l = 56$ cm, $L_{tot} = 64.5$ cm, 50 µm ID), $E = 350$ V/cm, $T = 35$°C, injection (200 mbar s). Reproduced from [26]. Copyright (2004), Agilent Technologies, Inc., with permission.

analysis is primarily qualitative and is used to detect subtle differences in proteins. As an example, Fig. 2.6 shows a typical peptide map for bovine serum albumin [26]. CZE is also employed as a second-dimension analysis of HPLC-purified peptides.

CZE has been applied with success in the environmental field to the determination of numerous pollutants (herbicides, fungicides, insecticides, accaricides, etc.) and other compounds having ionizable groups, in commercial formulations as well as in different matrices such as waters, soils, foods, etc. [27–30]. CE has enormous potential for the separation of polar and/or nonvolatile analytes. As an example, Fig. 2.7 [31] shows the determination of the residues of fungicides (thiabendazole and procymidone) in three fruits (orange, pear and strawberry) by CZE and MS detection.

On the other hand, CZE has been employed for separating all kind of drugs and their metabolites. In this area of application, the usefulness of this technique to control the purity of drugs or for the determination of their impurities, in toxicological analysis or in pharmacokinetic studies, etc. [32–35] should be highlighted. Figure 2.8 [36] shows the separation at pH 2.5 (Fig. 2.8a) and pH 8.4 (Fig. 2.8b) of a series of

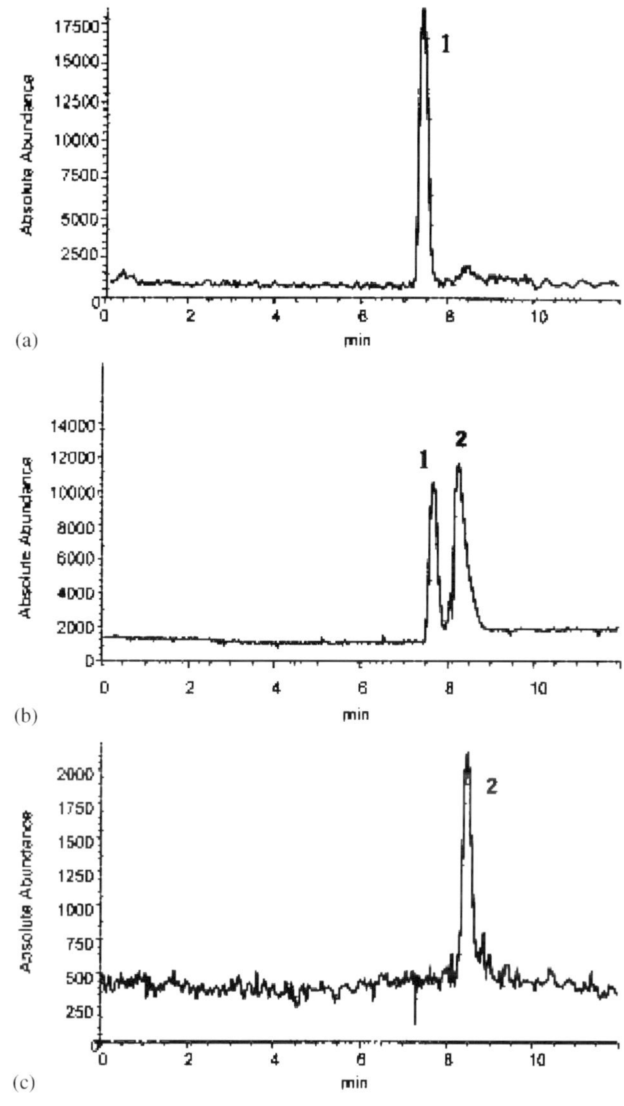

Fig. 2.7. CE–ESI-MS electropherograms in SIM mode of (a) orange that contains thiabendazole at 0.95 mg/kg, (b) pear that contains thiabendazole and procymidone at 0.42 mg/kg and 0.81 mg/kg, respectively, and (c) strawberry sample that contains procymidone at 0.2 mg/kg. Conditions: buffer (12 mM ammonium formate–20 mM formic acid, pH 3.5 with 2% methanol), untreated fused silica capillary (150 cm × 75 μm), $V = 30$ kV, $T = 25$°C, injection (200 mbar s). Reprinted from Ref. [31]. Copyright (2002), with permission from Elsevier.

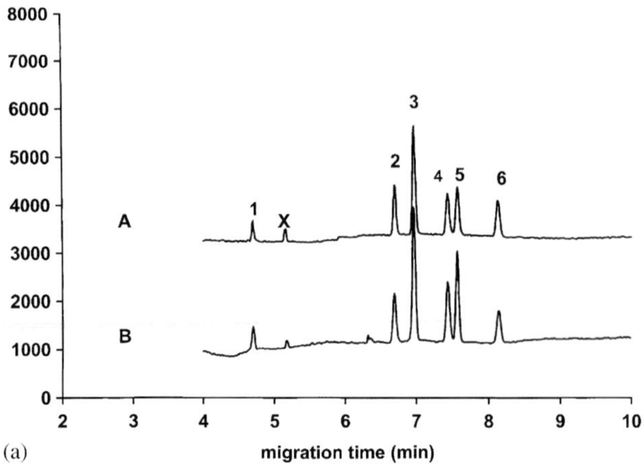

(a)

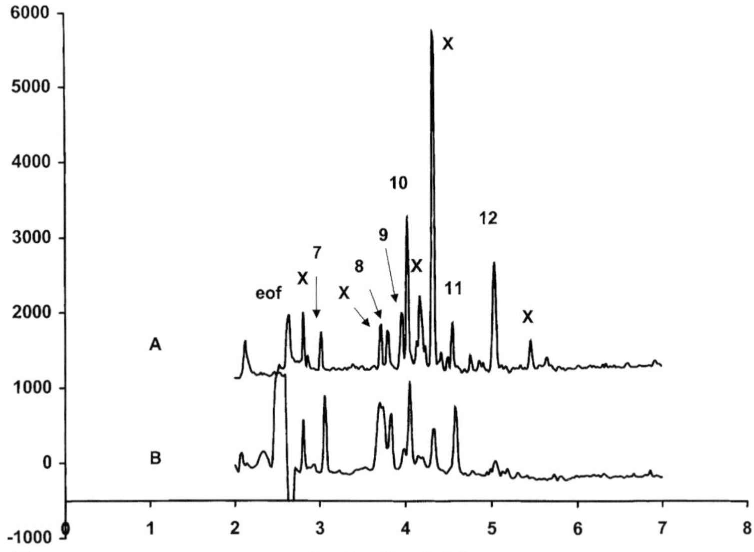

(b)

Fig. 2.8. Electropherograms obtained by CZE of spiked serum extracts (1 μg/ml). (a) 50 mM sodium phosphate buffer, pH 2.5. (b) 50 mM sodium borate buffer, pH 8.4. In both cases the serum extracts are obtained: (A) LLE under basic conditions, (B) SPE. Peaks: basic compounds (1, chloroquine phosphate; 2, mepivacaine; 3, propanolol HCl; 4, papaverine HCl; 5, diazepan; 6, acebutolol); acidic compounds (7, hydrochlorothiazide; 8, piroxicam; 9, ibuprofen; 10, phenobarbital; 11, salicylic acid; 12, chlorothiazide). Other conditions: untreated fused silica capillary (L_{tot} = 57 cm (l = 50 cm), 50 μm ID), V = 30 kV, T = 25°C, UV detection at 214 nm. Reprinted from Ref. [36]. Copyright (2001), with permission from Elsevier.

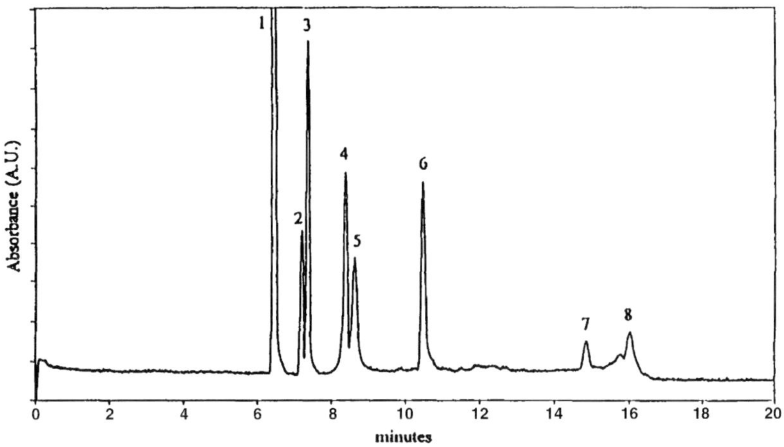

Fig. 2.9. Electropherogram obtained by CZE for the separation of the chloro complexes of osmium (IV), platinum (IV), rhodium (III), palladium (II), gold (III) and ruthenium (III) at 10 μg/ml each. Peaks: 1, contaminant; 2, $OsCl_6^{2-}$; 3, $PtCl_6^{2-}$; 4, $RhCl_6^{3-}$; 5, $PdCl_4^{2-}$; 6, $AuCl_4^{1-}$; 7, $[PtCl_5(H_2O)^{1-}]$; 8, $[RuCl_3OH(H_2O)_2^{-1}]$. Conditions: carrier solution (4 mM H^+–25 mM Cl^-), fused silica capillary (L_{tot} = 60 cm (l = 52.25 cm), 75 μm ID), $V = -10$ kV, sampling time (30 s), UV detection at 214 nm. Reprinted from Ref. [46]. Copyright (1997), with permission from Elsevier.

drugs (acid and basic compounds) in spiked serum extracts obtained by solid-phase extraction (SPE) and liquid–liquid extraction (LLE).

Finally, CZE has been applied to the separation of inorganic and organic ions in different matrices [37–40] and in speciation analysis [41–45]. Figure 2.9 [46] shows as an example the separation of the anionic chloro complexes of osmium (IV), platinum (IV), rhodium (III), palladium (II), gold (III) and ruthenium (III) under acid conditions (pH 2.4).

2.2.2 Capillary gel electrophoresis

CGE combines the good resolution of classical gel electrophoresis with the simple instrumentation used in CZE [3], giving rise to separation efficiencies similar to those obtained by CZE (of the order of 10^7 plates/metre) [47,48].

2.2.2.1 Separation principle

In CGE, a capillary is filled with a gel (with pores of controlled size) that acts as a molecular sieve. A scheme of the separation mechanism is

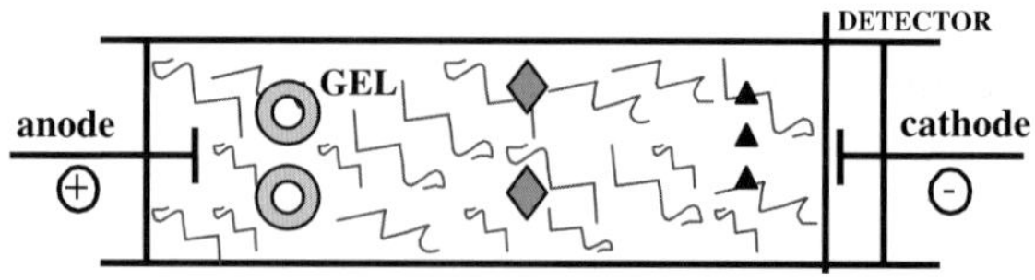

Fig. 2.10. CGE separation principle. ◆ ◎ ▲, substances of different size.

shown in Fig. 2.10. It is based on the different mobilities of the components of a sample through the pores of the gel due to their molecular size (molecular sieving). The substances of small size will traverse the gel matrix easier and will have shorter analysis times than those substances of large size, which will traverse the gel more slowly.

The use of gels in CE confers to the technique some advantages, among which are the following [49]:

- They are anticonvective media (they were introduced originally in classical electrophoresis as anticonvectives).
- They minimize the diffusion of the solute, which decreases band broadening.
- They prevent the adsorption of the analytes on the capillary wall.
- They can eliminate the EOF, which can be a limitation for the separation of neutral molecules that cannot migrate in the absence of EOF.

These gels should have certain characteristics such as stability under different temperature conditions and an appropriate range of pore size.

2.2.2.2 Types of gels

Depending on the sample (characteristics and size of the components to be separated), different types of gels can be employed. Gel consistency may vary depending on its viscosity. Thus, terms such as polymer solution, polymer network, entangled polymer solution, chemical gels, physical gels, etc. are employed to describe these media and even the term "replaceable" gel is used—to describe a gel of low viscosity which can be introduced or eliminated from the capillary [13].

The two types of gels mainly employed in CGE are:

- Chemical gels, which have a high stability and are chemically coated to the capillary wall.
- Physical gels, relatively low viscosity (polymer network solution).

TABLE 2.2

Characteristics of the gels employed in CGE

	Chemical gels	Physical gels
Cross-linked	Yes. Chemically linked to the capillary wall	No. Not attached to the capillary wall
Pore structure	Well defined	Dynamic
Pore size	It cannot be varied after polymerization	It can be varied
Replaceable	Not replaceable, generally high viscosity	Yes, when a relatively low viscosity matrix is used
Temperature	Heat sensitive	Heat insensitive
Efficiency	Higher	Lower

Although the structure of both types of polymers is completely different, the separation mechanism is identical. Table 2.2 shows the most important characteristics of both types of gels.

Chemical gels have a well-defined porous structure, they are rigid, and their pore size can be varied by controlling the polymerization conditions, although after polymerization the pore size does not vary. The first gels of this type to be employed were of cross-linked polyacrylamide. In this case, the porosity of the gel is given by the concentration of the mixture of acrylamide and bisacrylamide (cross-linking reagent) and this porous structure can be modified by varying the proportions of monomer and cross-linking reagent. In fact, it is possible to increase the pore size by decreasing the monomer content to separate molecules of high molecular weight or to increase the concentration of monomers to obtain small pore sizes and to separate molecules of lower molecular weight, such as oligosaccharides. The use of this type of gel presents some problems such as the low resistance and durability of columns, the low reproducibility among different injections and the appearance of bubbles between injections.

Physical gels (polymeric networks) have a dynamic pore structure. They are formed by adding to the separation buffer a hydrophilic polymer (linear polyacrylamide, polyethylenglycol, different alkylcelluloses, polyvinyl alcohol, polyethylene oxide, etc.) [13,15,50,51] that generates a polymeric network due to the interactions among the macromolecules when a certain concentration is reached. Gel viscosity can be selected by varying the concentration and the length of the hydrocarbonaceous chain of the linear polymer [52]. In this case, lower separation efficiencies are usually obtained but the problems mentioned above for chemical gels can be avoided. Furthermore, these gels can be easily replaced in the capillaries by means of vacuum or pressure, they allow the use of hydrodynamic or electrokinetic injection (with chemical gels, electrokinetic injection has to be used due to the rigidity of the gel), high working temperatures can be employed (50–70°C) and very high voltages can be applied (up to 1000 V/cm) without damaging the capillary characteristics [13].

Different semiempirical models have been developed in order to explain the separation mechanism in CGE [53–55].

2.2.2.3 *Influence of experimental parameters on the separation*

An important characteristic of CGE is the high efficiency obtained in DNA separations. In this technique, as in CZE, selectivity can be modified by the addition of an ion-pair forming reagent or other complexing agent such as ethydium bromide for DNA or SDS for proteins. These species can be covalently joined to the gel or simply added to the separation buffer [56].

Separations by CGE are also influenced by other experimental parameters such as the separation voltage, temperature, injection and capillary dimensions [15]. Detection is generally accomplished by UV–Vis absorbance (254 nm for DNA and 214 nm for proteins) or by LIF. It should be noted that as DNA and SDS–protein complexes are negatively charged, their separation in gel-filled capillaries requires a set-up with reversed polarity. On the other hand, the use of voltage gradients (continuous or by steps) increases the separation efficiency and resolution for DNA molecules [57]. Temperature gradients can also be applied; however, high temperatures can be used only when physical gels are employed since temperature changes evoke bubble formation in chemical gels and, therefore, discontinuities in the gel matrix.

2.2.2.4 *Applications*

CGE is applied mainly to the separation of substances possessing a similar charge/mass ratio, but different molecular weights, such as biopolymers, e.g., DNA, proteins (SDS–protein complexes) and carbohydrates.

Thus, CGE has been applied in molecular biology to determine the purity of oligonucleotides, anti-sense gene therapy, DNA sequencing, analysis of PCR products and DNA diagnostics [15,58,59]. Applications in protein chemistry include the separation of native proteins as well as SDS–protein complexes [56,60,61].

Separation of double-stranded DNA restriction fragments, produced by the enzymatic digestion of large DNA molecules, has also been successful. Generally, the more open structure of linear or minimally cross-linked gels are better suited for these large molecules, which can range in size from a few base pairs to more than 106 bp. As an example, Fig. 2.11 shows the separation by CGE of 1-kbp ladder [62]. This sample contains a ladder of 1-kbp DNA extending from 1 to 12 kbp and numerous smaller fragments, which result from the enzymatic nature of the sample preparation. The resolution of fragments from 75 bp to 12 kbp in a single analysis illustrates the wide sample application range of CGE.

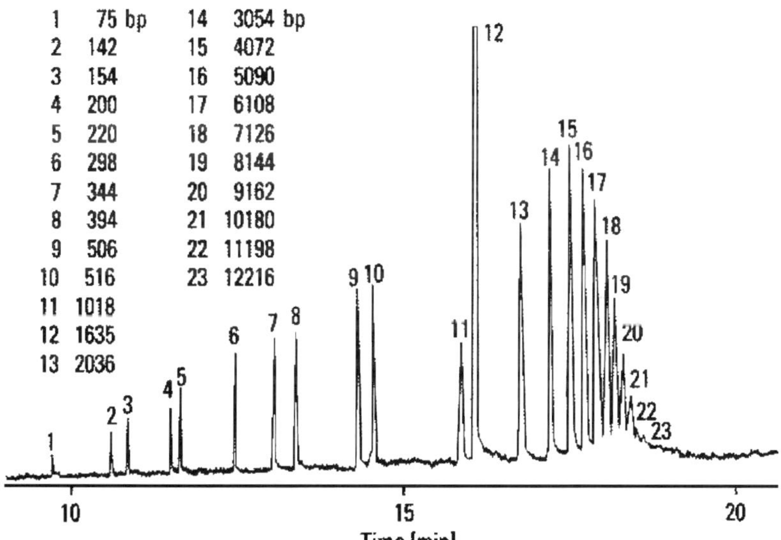

Fig. 2.11. CGE of 1 kbp ladder using minimally crosslinked polyacryl amide. Experimental conditions: Bis-crosslinked polyacrylamide (3% T, 5% C), 100 mM TRIS-borate buffer, pH 8.3, $E = 250$ V/cm, i – 12.5 μA, $L_{tot} = 40$ cm, $l = 30$ cm. 75 μm ID, UV detection at 260 nm, polyacrylamide-coated capillary. Reprinted from Ref. [62]. Copyright (1990), with permission from Elsevier.

2.2.3 Capillary isoelectric focusing

This CE separation mode was introduced by Hjertén and Zhu [9] and Hjertén et al. [10] and it is employed to achieve the separation of peptides, proteins, amino acids and other substances of amphoteric character [63–65]. In CIEF the separation is not based on the difference in electrophoretic mobilities but on the differences in pI of solutes. CIEF enables the separation of proteins with pI differences as low as 0.02 units and, even under optimal conditions lower differences in pI can be distinguished [51].

2.2.3.1 Fundamentals

Separations by CIEF are based on the electrophoretic migration of amphoteric substances in a pH gradient as shown in Fig. 2.12. The pH gradient needed through the capillary is obtained by using a solution of ampholytes (substances of zwitterionic character with different pK_a values (carrier ampholyte). The sample can be added to the carrier electrolyte or injected subsequently. When the electric field is applied, a pH gradient is established within the capillary from the anode (low pH) to the cathode (high pH). Sample components will move until they find a pH value equal to their pI since at this pH they lose their charge and are then stopped (this phenomenon is called focusing). As a consequence, sample analytes will occupy very narrow zones, which confer to this procedure a great efficiency because focusing is opposed to the molecular diffusion of analytes. In fact, if a molecule moves from its zone by molecular diffusion, it becomes charged and then it moves to the anode or to the cathode under the influence of the electric field coming back to its zone. Once the analytes are separated in the capillary, they have to be mobilized and led to the detector, where they can

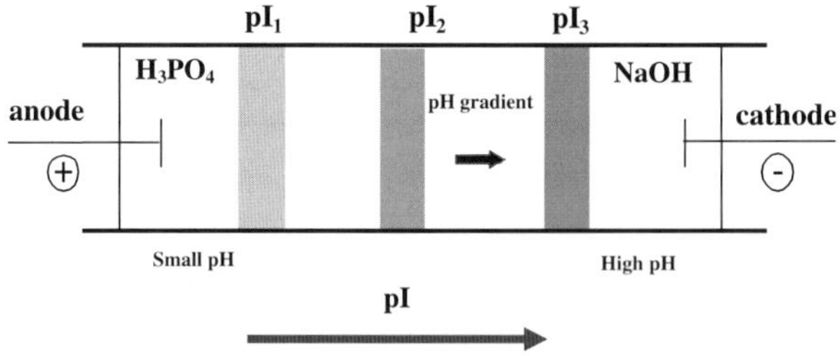

Fig. 2.12. Separation by CIEF based on the electrophoretic migration of amphoteric substances in a pH gradient.

be detected and quantified. Also, it is possible to calculate the pI value for analytes prior to calibration of the system.

The mathematical treatment of the CIEF separation is not simple [66–68]. A model has been proposed that enables the calculation of the minimum difference in pI (ΔpI) that a system can discriminate to obtain a value for the resolution equal to unity. ΔpI depends on the characteristics of the system (pH gradient) and the substances to be separated (variation of the electrophoretic mobilities with pH). The resolution increases when high electric fields are applied, when the variation of the electrophoretic mobilities with the pH is high, and when the diffusion coefficients are low. In addition, resolution is not directly related to the length of the separation capillary, thus, short capillaries can be used to speed up the process.

Regarding the capillary employed, there are basically two ways to perform CIEF [69]: using coated or uncoated capillaries. In the first case, due to suppression of the EOF, the process requires a two-step protocol: a focusing step, followed by mobilisation, in order to force the focused (thus stationary) bands past the detector. In the second case, a partial dynamic coating with soluble polymers, is allows a partial quenching of the EOF, and focusing takes place while the entire pH gradient and train of bands is moving towards the capillary outlet, at the detection point (one-step CIEF method) [63]. In the latter case, EOF is not eliminated but it is controlled during the focusing process.

2.2.3.2 Experimental procedure

The experimental procedure to achieve a separation by CIEF can be carried out in two forms as shown in Fig. 2.13 in one or two steps. In the one-step procedure the focusing is simultaneously achieved together with the mobilization of solute zones and in the two-step procedure the mobilization is carried out after the focusing is finished.

The ampholyte compositions are selected based upon the desired pH range in which analytes are separated. To separate proteins with widely different pI values, or for estimating the pI of an unknown protein, a wide-range ampholyte blend is advisable, e.g., pH 3–10. The final concentration is typically 1–2%. For proteins with similar pI values, the use of narrow-range ampholyte mixtures should be considered to achieve optimum resolution [13]. The capillary is filled generally with the ampholyte mixture and the sample but the injection of the sample can be done separately. In this case, the sample is usually dissolved in the carrier ampholyte and hydrodynamic injection is employed.

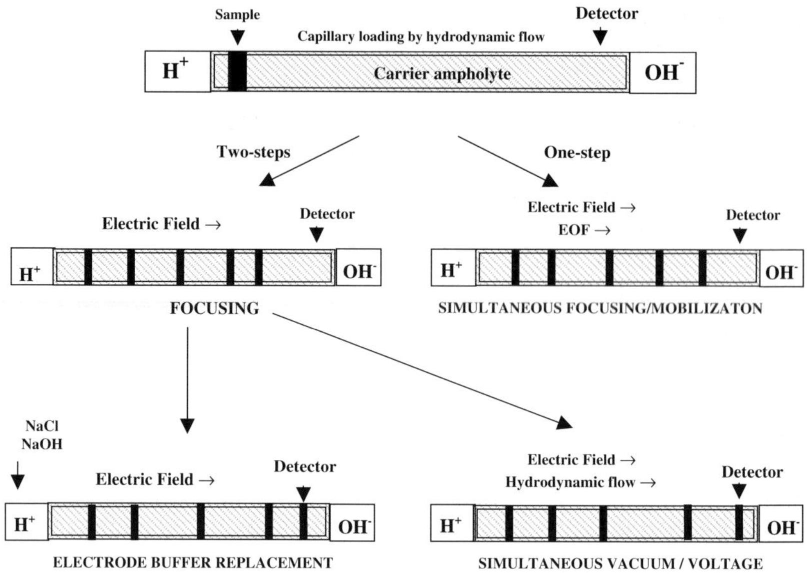

Fig. 2.13. Mobilization schemes in CIEF. Reprinted from Ref. [15]. Copyright (1998), with permission from John Wiley & Sons, Inc.

A basic solution (usually sodium hydroxide) is placed at the cathode and an acid solution (usually phosphoric acid) is placed in the anode. The pH of the acid solution (anode) should be lower than the pH of the most acid ampholyte and the pH of the basic solution (cathode) should be higher than the pH of the most basic ampholyte.

Once the capillary is filled with the carrier ampholyte and the sample, an electric field is applied. A pH gradient is established from the most acid pH at the anode to the most basic pH at the cathode. At the same time, proteins (analytes in general) will migrate within the medium at a region of pH where they are neutral and therefore, they stop or migrate with the EOF (in the one-step process). This region where they stop is the point where pH equals their pI. In addition, the higher is the number of ampholytes in the solution, the narrower will be the zones in which the proteins are focused (more effective separation).

If there is no EOF, analytes focused in the zones should be mobilized in order to be detected.

The different ways to mobilize analytes towards the detector are the following (Fig. 2.13):

- Hydrodynamic flow: pressure is applied at one end of the capillary keeping the electric field constant followed by the focusing effect.
- Electrophoretic mobilization: solution at the anode is replaced with the solution at the cathode or vice versa.
- Saline mobilization: salts are added to the solutions at the anode and the cathode after the separation and the suppression of the electric field. The salt employed should contain a cation different to the proton if it is added to the anode, or an anion different to the hydroxyl if it is added to the cathode. Sodium chloride (NaCl) is usually employed and its concentration generally ranges from 20 to 80 mM. When the electric field is applied again, these salts pass inside the capillary by altering the pH gradient because they compete with protons and hydroxyl ions in each zone. This pH change produces the mobilization of the analytes within the capillary because they recover their electric charge when they encounter pH values different to their pI values. This is the most commonly employed option to mobilize the solutes to the detector.

The mobilization step can produce important problems in CIEF such as uneven separation resolution, poor reproducibility and increased analysis times. An alternative approach is to perform CIEF in uncoated capillaries by using, for example, methyl cellulose as a buffer additive to reduce the EOF and to allow the attainment of steady-state conditions at which protein samples are focused [70,71]. Although this approach is a one-step procedure, it still has limitations associated with the uneven mobilization speeds, long mobilization times for acidic proteins, and incomplete pattern detection at column locations near the capillary end [68].

To avoid some of these problems, some researchers have proposed optical whole-column-imaging detection methods based on moving the separation capillary to the optical detection points; for example, by moving the separation capillary through the detection windows of UV absorbance or fluorescence detectors [68,72]. Although the problem of distorted pH gradient and uneven resolution in the moved CIEF focused zones are solved, the mechanical movement of the capillary increases dynamic noise and analysis time. Also, it is difficult to apply when monitoring a fast CIEF process [67].

Wu and Pawliszyn [73] have demonstrated an ideal approach to CIEF in real-time: whole-column-imaging detection without moving any part of the system. They have developed several types of these systems [68]. In the basic concept of real-time whole-column-imaging detection for CIEF, proteins are concentrated and separated at different pI values in narrow zones during the focusing process. High concentration gradients are created at the boundaries of the separated zones inside the capillary. The concentration gradients will induce refractive index gradients, which in turn cause the deflection of light passed through the gradients. The deflection changes the intensity distribution of transmitted light from the capillary, which is homogeneously illuminated along the capillary axis. Therefore, the focused zones can be imaged by measuring the change in intensity distribution of the transmitted light with a charge-coupled device. If the sample presents optical absorption or fluorescence, the focused sample zones can be imaged simply by means of optical absorption or fluorescence intensity.

2.2.3.3 Applications

The enormous separation possibilities of CIEF have been shown in the resolution of all types of proteins [68,74,75] including proteins with very similar structures such as different types of haemoglobin [76]. CIEF can be applied for routine analysis of protein samples in quality and purity control [77]. However, the applications of CIEF include not only the analysis of proteins but also of mixtures of compounds with different pI values such as different growth hormones [78] or glycomacropeptides [79]. As an example, Fig. 2.14 shows the separation of three haem proteins (cytochrome *c*, myoglobin and peroxidase) by CIEF using chemiluminescence detection [80].

As mentioned above, CIEF can also be applied to the determination of the pI of different proteins. This is carried out by using a mixture of proteins whose pI values are known and by carrying out a prior calibration. This implies plotting pI values as a function of the migration times of proteins. The determination of the migration time of a protein of unknown pI and subsequent interpolation in the prior calibration gives the pI value. The different ways of determining pI points for proteins by CIEF have been reviewed by Righetti [81].

Some drawbacks that are encountered during the experimental development of a separation by CIEF are:

(a) The instability of the polymeric layer employed to eliminate the EOF. After several injections this layer is degraded and the EOF

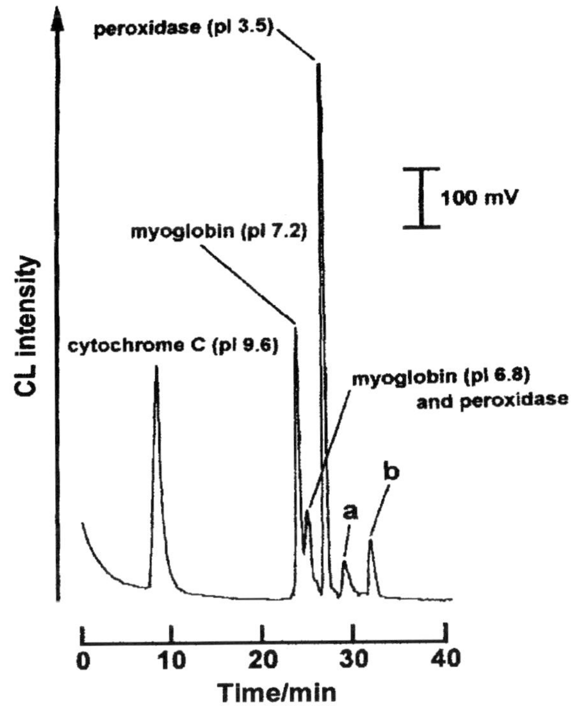

Fig. 2.14. CIEF of cytochrome c and myoglobin. Capillary: 25 cm × 75 μm ID Sample: 2.5×10^{-7} M cytochrome c, 2.5×10^{-6} M myoglobin, 2% pharmalyte 3-10, 0.5% tetramethylethylenediamine, 0.2% hydroxypropylmethylcellulose. Focusing: anolyte, 10 mM H_3PO_4; catholyte, 20 mM NaOH; $V = 10$ kV for 5 min. Mobilization: anolyte, 10 mM H_3PO_4; catholyte, chemiluminescence reagent; $V = 5$ kV. Reprinted from Ref. [80]. Copyright (1999), with permission from Elsevier.

begins to be more important. Then, the pH gradient becomes unstable. This effect can be avoided by using uncoated capillaries, which enables the mobilization of the pH gradient to the detector. In this case, the focusing of the ampholytes to initiate the pH gradient and the focusing of the proteins of the sample take place at the same time that the content of the capillary moves to the detector by the action of the EOF. The value of the EOF and its constancy determine the resolution as well as the reproducibility in migration times.

(b) The possibility that proteins focused under conditions in which their solubility is very low (high concentration and zero net charge) will precipitate [81]. The great variety of additives compatible with CIEF has been reviewed by Conti et al. for native proteins [72]. Nonionic surfactants such as alkylaryl polyether alcohol (Triton X-100) or polyoxyethylene lauryl ether (Brij-35), urea, ethylenglycol, glycerol, etc. can be employed.

(c) The slowness of the overall process.

2.2.4 Capillary isotachophoresis

Although CITP is at present perhaps the least commonly employed CE separation mode, it was the most widely used before the 1980s because it could be carried out in capillaries with relatively large I.D. (250–500 μm) [7].

The name of this separation mode is derived from the words *iso* (equal) and *taco* (velocity) [51]. Two important characteristics of CITP are: it takes place in a discontinuous buffer system and without EOF. This means a very different operation mode with respect to other separation modes such as CZE where a homogeneous buffer is used and where the shape and position of solute zones change continuously. In CITP, cations or anions can be separated, but not simultaneously [82].

2.2.4.1 Separation principle

CITP is based on the use of a terminating electrolyte with an electrophoretic mobility below that of the component of the sample that moves more slowly, and a leading electrolyte with an electrophoretic mobility higher than that of the component which moves more rapidly. Analytes are injected between these two electrolytes in a "sandwich" format. The capillary is filled with the leading electrolyte, the sample is then injected and the terminating electrolyte occupies the outlet vial. On application of an electric field, a potential gradient is initiated through the capillary and the analytes are distributed in adjacent zones as a function of their electrophoretic mobilities, moving within the capillary at the same velocity between the buffers of different mobility. This velocity is given by Eq. (2.9) [83]:

$$v = \mu_{\mathrm{ep1}} E_1 = \mu_{\mathrm{ep2}} E_2 = \mu_{\mathrm{ep3}} E_3 = \cdots = \mu_{\mathrm{ep}i} E_i \qquad (2.9)$$

where $\mu_{\mathrm{ep}i}$ is the electrophoretic mobility of each analyte and E_i the electric field strength in each analyte zone.

The electric field is auto-adjusted in each zone to keep the velocity constant in such a way that in a zone of higher mobility the electric

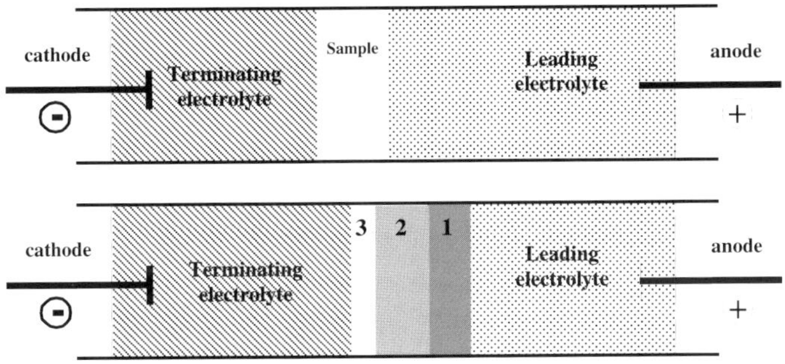

Fig. 2.15. Principle of anionic CITP.

field is lower and vice versa. In fact, the highest electric field corresponds to the zone with the lowest mobility, that is, the zone where the terminating electrolyte is. As CITP is carried out at constant current intensity and the product of this current and the electric field is a measure of the heat generated, temperature will increase in those zones in which the electric field is higher. This means that the terminating electrolyte zone has the highest temperature and justifies the use of thermometric detectors in CITP. If an ion diffuses within an adjacent zone, then its velocity changes and it returns to its zone. This principle is shown in Fig. 2.15 for the separation of a mixture of anions where anion 1 is that of the highest mobility of the sample and anion 3 is that of the lowest mobility.

For the separation of a mixture of anions, the leading electrolyte should contain anions with an electrophoretic mobility higher than that of any anion contained in the sample. Likewise, the terminating electrolyte is chosen so that its anions have an electrophoretic mobility lower than that of any anion of the sample. The same is true for the separation of a mixture of cations.

CITP can be carried out in commercial CE instruments but the EOF has to be eliminated.

Plotting the detector signal in CITP as a function of time, an isotacopherogram is obtained. It is very different to the electropherograms obtained by other CE techniques, which consist of different peaks corresponding to the components of the sample over a baseline. An isotacopherogram consists of a series of steps without a baseline among them. Each step is characterized by its width and height. The height is a qualitative property and can be employed to identify a substance whereas the quantitative analysis is based on the measurement of the

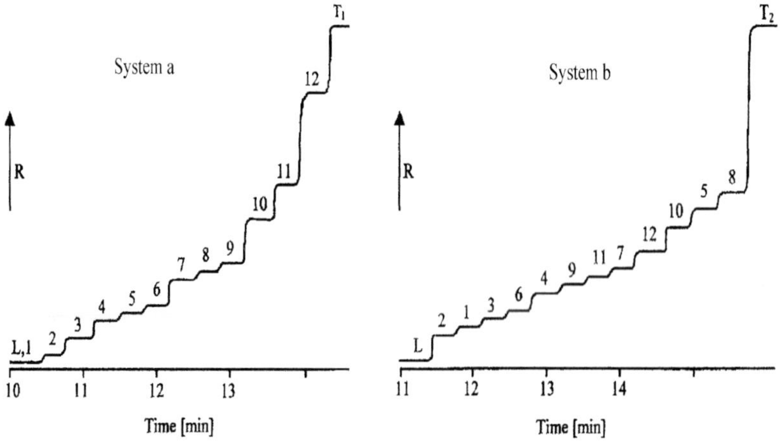

Fig. 2.16. The isotachophoretic separation of sulphate (1), perchlorate (2), oxalate (3), malonate (4), phosphate (5), formate (6), citrate (7), pyroglutamate (8), maleate (9), lactate (10), succinate (11) and acetate (12) (0.1 mmol^{-1} of each) in operating system a (the leading electrolyte (L): 10 mmol^{-1} hydrochloric acid, 0.1% poly(vinylpyrrolidone), β-alanine, 2.9; the terminating electrolyte (T_1): 10 mmol^{-1} nicotinic acid; the driving current 40 μA) and operating system b (the leading electrolyte (L), 10 mmol^{-1} hydrochloric acid, 0.1% poly(vinylpyrrolidone), 5.5 mmol^{-1} 1,3-bis[tris(hydroxymethyl)methylamino]propane; the terminating electrolyte (T_2), 5 mmol^{-1} 2-morpholinoethanesulphonic acid; the driving current 40 μA). R, increasing resistance. Conductimetric detection. Reprinted from Ref. [84]. Copyright (2003), with permission from Elsevier.

stair width. Figure 2.16 [84] shows as an example the isotacopherograms corresponding to a mixture of different anions with two different operating systems. In this case, the CITP operating systems were chosen with the aim of differentiating the acids according to their pK values (system a) and charges (system b).

The current intensity, i, will be the same through the zones of separated substances [51] and its value will be proportional to the product of the electric field strength and the electric conductivity, κ_i, in each zone:

$$I \alpha \kappa_i E_i = \kappa_j E_j = \kappa_k E_k = \cdots = \kappa_T E_T \qquad (2.10)$$

where i, j and k are the different zones. Equation (2.10) explains one of the main applications of ITP, which is sample preconcentration. In fact, when a diluted solution is injected and an electric field is applied, the zone is concentrated in the interphase with the leading electrolyte (see Eqs. (2.9) and (2.10)). This effect can generate enormous trace enrichment and this procedure can be employed also in CZE where a diluted sample can be concentrated since in this case the separation buffer can play the same role as the leading electrolyte in ITP.

The analyte concentrations in each band are adjusted to the concentration of the leading electrolyte ion according to the Kohlraugh regulating function [85]:

$$C_i = C_L \frac{\mu_{\text{ep}i}(\mu_{\text{ep}L} + \mu_{\text{epc}})}{\mu_{\text{ep}L}(\mu_{\text{ep}i} + \mu_{\text{epc}})} \tag{2.11}$$

where C_i is the concentration of species i in an ion band, C_L is the concentration of the leading electrolyte, and $\mu_{\text{ep}i}$, $\mu_{\text{ep}L}$ and μ_{epc}, are the mobilities of species i, the leading electrolyte ion, and the counter ion in the steady state, respectively. C_i then depends on the analyte and leading electrolyte characteristics. CITP will concentrate those ions whose concentration in the sample is lower than their steady-state concentration defined by Eq. (2.11). From this equation it can also be derived that C_i is independent of the original analyte concentration, and then by means of this procedure it is possible to concentrate an analyte several orders of magnitude with a limit at $0.7C_L$.

The recent progress in CITP has been regularly reviewed by Gebauer and Boček [86,87]. Papers reviewed include methodological and instrumental aspects as well as analytical applications.

2.2.4.2 Coupling of CITP with other CE modes

The main limitations in CZE are the low amount of sample injected (normally in the nl range) and, in the case of UV–Vis absorbance detection, the small optical path length which corresponds to the inner diameter of the capillary (usually ranging from 50 to 75 μm). In order to increase CZE sensitivity, one or both of these problems should be solved.

CITP is considered a preconcentration and purification technique [88,89]. It can concentrate samples 10—1000-fold. Concentrated segments can be analysed directly or even coupled by several methods to CZE, using a single or separate capillary for further separation and quantitation [90]. The application of CITP as a preconcentration technique is described in detail in chapter 3.

2.3 SEPARATION MODES IN CE COMBINING ELECTROPHORETIC AND CHROMATOGRAPHIC PRINCIPLES

The modes combining electrophoretic and chromatographic principles in CE include EKC and CEC. EKC is characterized by the use of a pseudostationary phase (PS) in the electrolytic solution and CEC uses a true stationary phase as employed in HPLC.

2.3.1 Electrokinetic chromatography

EKC is a separation mode that was developed to expand the applicability of CE to the separation of neutral compounds. However, nowadays EKC is also employed in the separation of charged compounds, giving rise to separation selectivities that are not possible in CZE [15,91–93].

EKC was defined by Terabe [5] in 1989 as: "an analytical separation method which utilizes the experimental technique of CZE in combination with the principle of chromatography". In EKC a new component, the pseudophase, is added to the separation buffer employed in CZE. This pseudophase gives rise to the existence of a chromatographic phenomenon since the solute is distributed between these two phases (aqueous and pseudostationary phases) [5,94]. In EKC a noncharged solute will migrate either with the velocity of the electroosmotic flow or with the velocity of the PS. As in chromatography, the observed velocity in a solute zone is the mean of two velocities (velocity of the mobile phase and "velocity of the stationary phase") resulting from the partitioning of the solute between these two phases. Consequently, the separation process in EKC can be described in chromatographic terms and the separation of neutral solutes differing in their partitioning coefficients is possible [95].

One of the peculiarities of EKC is the nonexistence of the stationary phase, so the solute zone is also transported (in direction to the detector or in counter-direction) when incorporated into the PS. The ratio of the velocity of the mobile phase (the surrounding medium) to the observed velocity of the PS has a large impact on resolution and peak capacity of the separation system [96–97]. This velocity ratio is called the migration time window and reflects the limited elution window for neutral solutes in EKC in the normal elution mode.

Figure 2.17 shows the separation principle in EKC [94]. It considers a negatively charged PS contained in an aqueous buffer, a

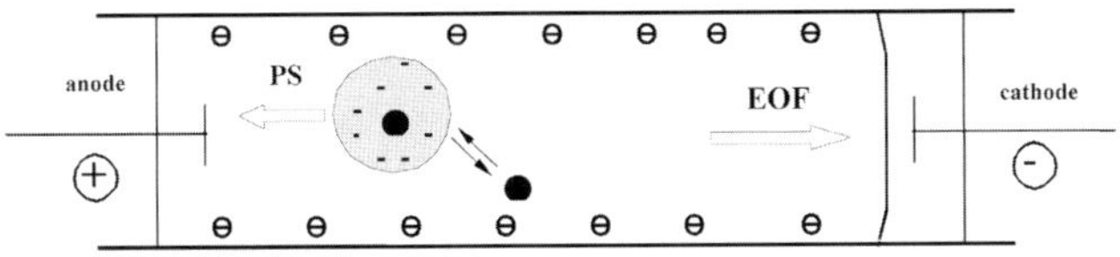

Fig. 2.17. Separation principle in EKC. (PS), anionic pseudostationary phase; (•), neutral analyte. Reprinted from Ref. [94]. Copyright (1999), with permission from Elsevier.

cathode-directed EOF and a neutral solute. The EOF velocity is greater than the PS velocity. The surface of the capillary carries a negative charge due to the dissociation of silanol groups in a fused-silica capillary. The PS velocity is towards the anode, the neutral solute partitions between the PS and the aqueous phase as depicted by the half arrows, and the migration velocity of the solute is towards the cathode due to a higher EOF velocity. Separation of the mixture occurs due to differences in partitioning of analytes between the PS and the surrounding liquid phase. In fact, solutes are distributed in EKC between the pseudophase and the aqueous phase and a distribution coefficient (P) can be defined as:

$$P = \frac{C_{\mathrm{PS}}}{C_{\mathrm{s}}} \tag{2.12}$$

where C_{PS} and C_{s} are the molar concentrations of a solute in the pseudostationary and surrounding phases, respectively. Separation of a mixture occurs due to differences in P values. In fact, the neutral solute acquires an effective electrophoretic velocity, which is the addition of the electrophoretic and electroosmotic velocities, then moves towards the cathode, where the detector is located.

Different working modes are possible in EKC, which can be classified according to the nature of the PS employed or the type of solute migration [94].

- *According to the nature of the pseudophase*: This is the most common classification criterion in EKC. When the pseudophases employed are anionic, cationic, nonionic and zwitterionic surfactants [4,98–104], micelle polymers [105–107], polymer surfactants [105,108] or in situ charged micelles [109–112], then the technique is called MEKC and it is the most widely used EKC mode. Other pseudophases employed in EKC include microemulsions [113–115], cyclodextrins [116–119], macrocyclic and macromolecular phases

[120,121], vesicules [122], resorcarenes [123], dendrimers [105,124,125], polymeric ions [126,127], proteins [128], macrocyclic antibiotics [129,130], polysaccharides and alkaloids [119]. All these EKC modes are usually named by adding the term of the compound employed as pseudophase to the EKC term. For example, when using a cyclodextrin as the PS, the separation mode is named cyclodextrin electrokinetic chromatography (CD-EKC).
- *According to the type of solute migration*: EKC can be classified according to the type of migration of solutes in two modes: normal-migration EKC (NM-EKC) and reversed-migration EKC (RM-EKC). NM-EKC is the most common working mode and is characterized by the fact that the EOF moves more rapidly than the pseudophase [131].

The type of solute migration (normal or reversed) can be achieved by controlling the EOF. As stated above, with fused-silica capillaries a cathodic EOF arises at $pH > 2$. However, when cationic pseudophases are employed, they interact with the capillary wall and the direction of the EOF is reversed. If the pH is lower than 2 or if a coated capillary is used, the EOF can be very low or eliminated and the separation can be carried out with charged pseudophases with normal or reversed solute migration [132].

The migration of neutral and charged solutes in EKC can be described using the same equations except for the calculation of the retention factor [15,92]. The calculation of the retention factor for charged solutes requires measurement of the electrophoretic mobility of these solutes in a CZE system (that is, without the pseudophase). Table 2.3 shows the equations describing the retention factor and the resolution in EKC. In these equations the reciprocal of the term t_0/t_{PS} or t_{PS}/t_0 is used to characterize the ratio of the observable velocities of the two phases in EKC. One widely accepted term for this time ratio is the migration time window [133].

The equations to determine the retention factor of a solute in EKC are more complicated than those corresponding to chromatographic techniques. This is because in EKC the stationary phase is moving continuously, that is, it is a stationary pseudophase. In fact, if this pseudophase were a real stationary phase, which can be achieved by manipulating the EOF [134], then t_{PS} would approach infinity and the above-mentioned equations would be similar to those corresponding to chromatographic techniques. The consequence of the existence of a PS

TABLE 2.3

Equations employed in EKC for the calculation of the retention factor (k) and the resolution (R_s)

EKC mode	Retention factor		Resolution	
NM-EKC and RM-EKC (with EOF)	$k = \frac{t_m - t_0}{t_0\left(1 - \frac{t_m}{t_{PS}}\right)}$	(2.13)	$R_S = \frac{\sqrt{N}}{4}\frac{\alpha - 1}{\alpha}\frac{k_2}{1+k_2}\frac{1 - t_0/t_{PS}}{1+(t_0/t_{PS})k_1}$	(2.14)
RM-EKC (without EOF)	$k = \frac{t_{PS}}{t_m - t_{PS}}$	(2.15)	$R_S = \frac{\sqrt{N}}{4}\frac{\alpha - 1}{\alpha}\frac{k_2}{1+k_2}$	(2.16)

$t_0 = t_{eo}$ for neutral solutes or electrophoretic migration time in the aqueous phase (CZE) for charged solutes.
t_{PS} = migration time of the pseudophase.
t_m = migration time of the solute.
α = separation factor calculated as k_2/k_1 ($k_2 > k_1$). k_1 and k_2 are the retention factors of the analytes 1 and 2, respectively that elute very near.
N = number of theoretical plates.

moving continuously is immediate: while in isocratic chromatography, solutes with increased values of k can be eluted sequentially from the column, in EKC an increase in the retention factor of solutes causes a compression of the peaks near the migration time of the pseudophase, t_{PS}. In fact, some analogies have been found between EKC with micelles as PSs and gradient chromatography [96]. When t_0 is infinite or the electroosmotic flow is completely suppressed, the aqueous phase never comes out from the capillary, and only the pseudophase migrates through the aqueous phase. Thus, electroosmotic flow is not essential in EKC.

Calculation of retention factors requires that t_0 and t_{PS} are known. Generally, to determine the EOF velocity, which is identical to the velocity of the mobile phase, a sample is injected containing a compound (the marker of the velocity of EOF or the mobile phase) that is not retained by the separation carrier or the capillary wall. Several polar substances have been used: formamide, acetone and thiourea [133]. The marker must be detectable by the detector in use. In the case of a UV detector the baseline disturbance caused by a zone of different refractive index than the separation buffer can be used as a signal.

Ahuja et al. [134] have compared the electroosmotic velocities determined by using methanol, acetonitrile, 1-propanol, acetone and tetrahydrofuran, and demonstrated that all these solvents are suited as marker substances with the exception of 1-propanol, because 1-propanol is retained by the separation carrier (in this case a micellar system of SDS). Fuguet et al. [135] investigated for several surfactants the signals produced by different markers and concluded that the suitability of the marker is dependent on the micellar system and that any solvent can be used as marker if t_0 is measured by the first disruption of the baseline.

While the use of a marker seems to be reliable for the determination of the electroosmotic velocity, the determination of the velocity of the separation carrier (t_{PS}) with a marker is prone to systematic error. In the beginning of EKC very nonpolar azo dyes (SUDAN III and SUDAN IV), dodecanophenone or polycyclic aromatic hydrocarbons were employed as marker substances [133]. The inconvenience of working with these nonpolar substances is the difficulty to dissolve them in solvent mixtures compatible with EKC conditions. With aqueous mobile phases (not containing an organic modifier) dodecanophenone is a reliable marker of the velocity of the separation carrier [135]. This parameter can also be evaluated using a homologous series and the iterative procedure proposed by Bushey and Jorgenson [136] and, in general, this method seems to be the most reliable [133–136].

Varying the migration window is, besides regulation of the retention factor, one of the most important parameters for resolution optimization in EKC. The electroosmotic mobility and the electrophoretic mobility of the separation carrier are the key parameters to be controlled [133]. Varying the electroosmotic mobility in fused silica capillaries is very different for anionic separation carriers compared to cationic separation carriers, due to the modification of the capillary wall.

On the other hand, k can be related to the distribution coefficient P of the solute and the phase ratio (volume of the pseudophase/volume of aqueous phase = V_{PS}/V_{aq}) by means of the equation:

$$k = P\left(\frac{V_{PS}}{V_{aq}}\right) \tag{2.17}$$

Since in EKC solutes can interact with the pseudophase to a different extent giving rise to a chromatographic phenomenon—although it is possible that some solutes do not interact with that pseudophase moving instead by an electrophoretic phenomenon—it is preferable to use the term migration time instead of retention time because the first is more general and includes the electrokinetic as well as the chromatographic phenomena.

Of the different EKC modes, MEKC is the most widely employed and will be described in more detail below, followed by other important EKC modes such as those involving the use of cyclodextrins or microemulsions.

2.3.1.1 Micellar electrokinetic chromatography

This EKC mode was introduced by Terabe et al. [4] in 1984, and is characterized by the use of micelles as pseudophase. Some books and interesting reviews on the fundamental characteristics and applications of MEKC have been published [137–143].

Separation principle

Micelles are aggregates that are formed by adding surfactants to the separation buffer at a concentration above their CMC. Surfactants are amphiphilic molecules that comprise a hydrophobic moiety and a polar or ionic head group. They can be recognized by the charge of the head group (as nonionic, anionic, cationic and zwitterionic surfactants) or by the variations in the nature of the hydrophobic moiety (as hydrocarbon, bile salts and fluorocarbon surfactants). When surfactant molecules are at low concentrations in solution they exist as monomers. However, at a given concentration and temperature, they associate spontaneously forming submicroscopic aggregates called micelles. The CMC corresponds to the maximum concentration of surfactant monomers in the solution. Each aggregate or micelle comprises a given number of monomers, which is characteristic of each micellar system and is called the aggregation number.

Ionic surfactants form micelles that move at a different velocity to the EOF. In this case, the separation of neutral analytes is produced by their distribution between the aqueous and the micellar phases. When the micellar phase moves to the detector, the elution range is between the time corresponding to the EOF (t_{eo}) and the migration time of the micelle (t_{PS}) and all analytes will migrate between these two limits depending on their distribution between the two phases. As an example, Fig. 2.18 shows the separation principle of a neutral solute in a MEKC system with anionic micelles. In this case, as the micelle surface is negatively charged, micelles have an electrophoretic mobility towards the anode. However, if a strong cathodic EOF exists, micelles can

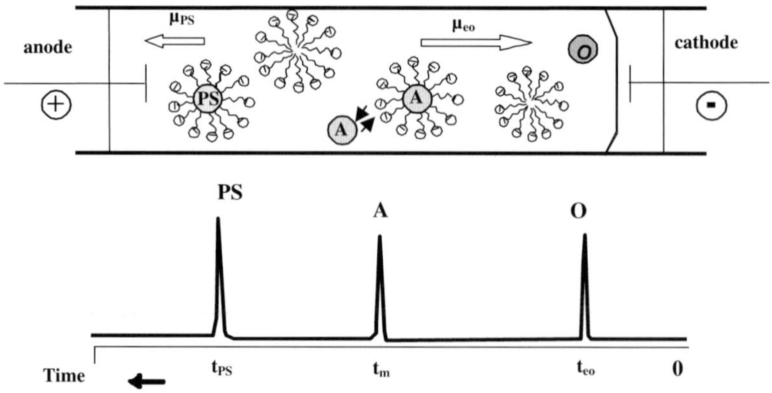

Fig. 2.18. Schematic illustration of the separation principle in MEKC. (PS), micelle; (A), retained compound; (O), nonretained compound.

also move to the cathode at a velocity lower than that of the EOF. As a consequence, the aqueous phase moves rapidly and the micellar pseudophase moves slowly. When a neutral solute is introduced into the system, it distributes between the two phases, migrating at the same velocity as the micelle when it is incorporated in this pseudophase and at the same velocity as the EOF when it does not interact with the micelle. Solute distribution depends on the solute–micelle association constant. As shown in Fig. 2.18, neutral solutes will elute between two limits: the elution time of a solute moving at the EOF velocity (t_{eo}) and the elution time of a solute completely associated with the micelle (micelle migration time, t_{PS}).

When a cationic surfactant is employed, the direction of the EOF will be reversed towards the anode, through the adsorption of the surfactant molecules on the inside wall of the capillary and reversing the surface charges.

The retention factor and resolution in MEKC can be calculated by using the equations included in Table 2.3.

Equation (2.14) (see Table 2.3) shows that the resolution in MEKC can be manipulated by four parameters: separation efficiency (N), separation factor (α), retention factor (k) and migration time window (t_0/t_{PS}). In MEKC higher separation efficiencies always result in a better resolution. Resolution increases proportionally with an increase in the square root of the plate number. Normally, the higher the voltage applied, the higher the plate number that can be attained, unless excessive heating is generated by the Joule effect. Since the diffusion coefficient of the micelle is small, solutes having larger retention factors can yield higher efficiencies. The separation factor, α, is determined by the micellar solubilization process and is influenced by the chemical nature of both the micellar and the surrounding aqueous phases. Its effect on resolution is critical when the resolution is low. This term is discussed in more detail later. The effect of the retention factor on resolution is not derived in a simple manner, as discussed in some articles [137,144]. For a given value of t_0/t_{PS} there is an optimum value of k that maximizes resolution: $k_{opt} = (t_{PS}/t_0)^{1/2}$ [144]. This last term, named column availability, has been discussed by Zhang et al. [145]. It can be easily seen from Eq. (2.14) that the migration time window or column availability is an important factor for improving resolution. Although Eq. (2.14) was derived for the condition in which EOF is larger than the electrophoretic velocity of the micelle, it is valid for other conditions [139].

Efficiency

MEKC can be characterized as a separation method with inherent high efficiency. The number of theoretical plates per capillary (length to the detector 50–70 cm) obtained in MEKC for retarded ($k = 1-5$) neutral solutes under routine conditions is about 200,000–300,000 [142].

Terabe et al. [146] have studied in detail the band-broadening phenomena occurring in the separation capillary during a run. With a suitable injected plug length and adequate on-column detection, extra-column effects can be kept low. The causes of column band-broadening can be ascribed to five mechanisms: longitudinal molecular diffusion, sorption–desorption kinetics, intermicelle mass transfer, radial temperature gradient effect and electrophoretic dispersion of the micelle (heterogeneity in micelle mobilities). At low velocity of the solute zones, longitudinal diffusion is the dominant factor, while at higher velocities sorption–desorption kinetics and heterogeneity in micelle mobilities become significant factors.

In many cases, the plate number observed in MEKC is not as high as that predicted from theory. The main reason is considered to be a too large sample amount injected or some adsorption of the sample onto the inside wall of the capillary [139].

In order to use the high efficiency offered by MEKC, band broadening due to inadequate buffer levels in the inlet and the outlet vials (if not identical) should be avoided. Otherwise a laminar flow independent of the EOF will be induced in the capillary (siphoning effect), which can result in a dramatic decrease in efficiency. The same holds true for band broadening due to solute–wall interactions [142].

Selectivity
A major advantage of MEKC is its flexibility. The composition of the separation electrolyte and, in particular, the PS, can easily be changed by rinsing the capillary. Thus, MEKC is more flexible than a chromatographic technique with immobilized stationary phase. The separation selectivity in MEKC can be controlled through all those factors that also modify the solute–micelle interactions. The most frequent means to modify the separation selectivity is through a change in the surfactant's nature or the use of surfactant mixtures. In addition, due to the dynamic nature of micelles (they are in equilibrium with surfactant monomers), the aggregation of surfactants and the size of micelles are influenced by changes in micellar concentration, pH, temperature, and the addition of modifiers [15,139,143].

Effect of the nature and concentration of the micelle: Since most analytes interact with the micelle on its surface, the hydrophilic or ionic group is generally more important than the hydrophobic one in determining selectivity [96].

Micelles of different structures may give rise to different selectivities. As an example, bile salts, which form helical micelles, can give significantly different selectivity compared with the long alkyl-chain surfactants [147].

A great number of surfactants have been employed as micellar phases in MEKC [148,149]. The most common are anionic with an alkyl chain, especially SDS [15,137]. This fact can be attributed to its high aqueous solubility, low CMC, small UV molar absorptivity, availability and low cost [15]. This surfactant forms micelles with a relative hydrophobicity and gives rise to adequate selectivities for the separation of a great variety of mixtures of solutes with intermediate values for the distribution coefficient. Homologs of SDS are less suitable as surfactants in MEKC.

The cationic surfactants most used in MEKC have an alkylmethylammonium group such as cetyltrimethylammonium bromide (CTAB), tetradecyltrimethylammonium bromide (TTAB) and dodecyltrimethylammonium bromide (DTAB) that show significantly different selectivities to anionic surfactants. The main disadvantage of cationic surfactants is the small migration time windows obtained. The migration window in MEKC with cationic surfactants can be improved by decreasing the EOF while the electrophoretic mobility of the micelles is maintained. This effect can be achieved by addition of certain modifiers such as metal cations [150,151]. CTAB has often been used in CE to modify the capillary surface.

The counterion of an ionic surfactant also affects the selectivity significantly [152,153]. It is recommended that the same ions are employed as the surfactant's counterion and buffer electrolytes. When the counterion of the surfactant is different from the buffer electrolytes, care must be taken to keep the same composition through a series of runs for high reproducibility.

Surfactants with a zero net charge, such as nonionic and zwitterionic surfactants, can be used alone for the separation of charged solutes. These surfactants do not increase the conductivity of the buffer, and can be used at high concentrations while still allowing the use of high voltages and large ID capillaries [148,149]. In addition, both types of surfactants can be employed with an ionic surfactant to form mixed micelles. A mixed micelle consisting of ionic and nonionic surfactants is usually larger than the original ionic micelle and has a lower charge density, and consequently a lower electrophoretic mobility. Thus, a narrower migration time window is obtained and different selectivity can be expected since the surface of the mixed micelle is different to that of the original one [139]. Mixed micelles can increase the selectivity in the case of the separation of complex mixtures of solutes with similar structures.

High-molecular-mass surfactants (HMMS) provide very stable PSs with zero CMC. They can be used for the separation of hydrophobic compounds in the presence of relatively high contents of organic modifiers. Additionally, the use of HMMS is advantageous over conventional micelles for on-line coupling MEKC with electrospray ionization mass spectrometry because the high molecular mass of these surfactants is well beyond the mass range of the spectrometer [154,155].

With respect to the influence of the surfactant concentration, k is related to the surfactant concentration as expressed by the equation[139,143]:

$$k = \frac{v(C_{sf} - CMC)}{1 - v(C_{sf} - CMC)} K \tag{2.18}$$

where v is the molar volume of the surfactant, C_{sf} the total surfactant concentration in the electrolytic solution, and K the solute–micelle association constant per monomer.

At low surfactant concentrations close to CMC values, the denominator in Eq. (2.18) is close to unity and a linear variation is obtained for the retention factor as a function of the total surfactant concentration according to:

$$k = Kv(C_{\mathrm{sf}} - CMC) \tag{2.19}$$

Equation (2.19) shows that solute retention can be controlled through the surfactant concentration in solution.

As an example, Fig. 2.19 [156] illustrates the effect of two different surfactants (SDS and sodium deoxycholate, SDC) on the electrophoretic separation of phosphoamino and amino acids using 3-(2-fluroylquinoline)-2-carboxaldehyde (FQ) as a fluorescent derivatization reagent. As can be seen, SDC can improve the resolution of these derivatives. It can be also observed that an increase in the surfactant concentration leads to an increase in the elution window although in this case the peak shapes were worst for the highest SDC concentration.

Effect of temperature: Temperature affects the separation selectivity because distribution coefficients (P) generally decrease with increasing temperature, causing a decrease in the migration time. In addition, an increase in temperature causes a decrease in the solution viscosity and therefore an increase in the EOF and micelle velocities [157]. Consequently temperature seriously affects the migration time; hence, it is important to maintain the temperature precisely to obtain reproducible results [139].

Effect of pH: Separation selectivity of electrically neutral compounds is not modified by pH changes in the range in which they do not experience acid–base equilibria [158]. However, for ionizable compounds, variations in the buffer pH may produce a change in their dissociation which modifies their charge, electrophoretic mobility and solute–micelle interactions. On the other hand, anionic compounds with similar values of the acid–base dissociation constants (K_{a}) in aqueous solution can have different K_{a} values in micellar solutions. The separation selectivity of ionizable or ionic compounds in MEKC may thus change when varying the buffer pH [158]. As an example, Fig. 2.20 shows the electropherograms corresponding to the separation of a group of benzophenones with a micellar phosphate buffer containing 20 mM SDS at various buffer pH (6.6, 7.0 or 8.0) [159]. Changes in the elution order for some benzophenones are observed, depending on their pK_{a}.

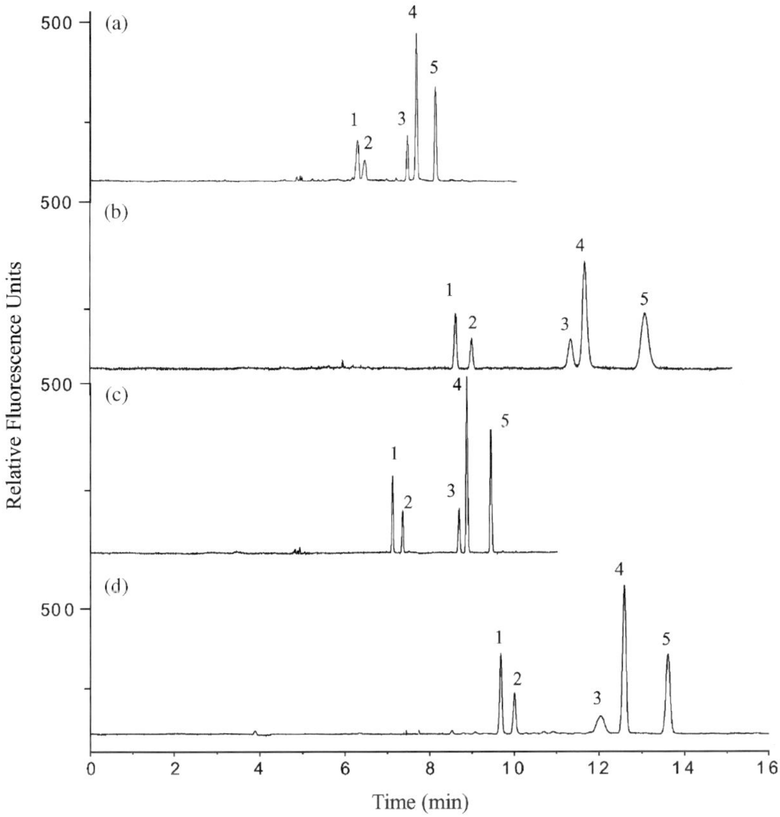

Fig. 2.19. Electropherograms obtained for the separation of five FQ (3-(2-fluroylquinoline)-2-carboxaldehyde) labelled phosphoamino and amino acids in 20 mmol/l borate buffer (pH 9.3) with different surfactants. (a) No surfactant, (b) 10 mmol/l SDS, (c) 20 mmol/l SDC, (d) 40 mmol/l SDC. $V = 25$ kV; injection: 50 mbar for 3 s; Analyte concentrations: 10 µmol/l; peak identification: 1, glutamic acid; 2, aspartic acid; 3, phosphothreonine; 4, phosphotyrosine; 5, phosphoserine. Reprinted from Ref. [156]. Copyright (2004), with permission from Elsevier.

The selection of the buffer pH can also be influenced by the nature of the micellar system employed in MEKC. In fact, it is not possible to employ anionic micelles with high electrophoretic mobilities at low pH because they cannot be eluted with low EOF. As an example, SDS micelles move to the anode at pH < 5 and are not eluted at the cathode under these conditions [158].

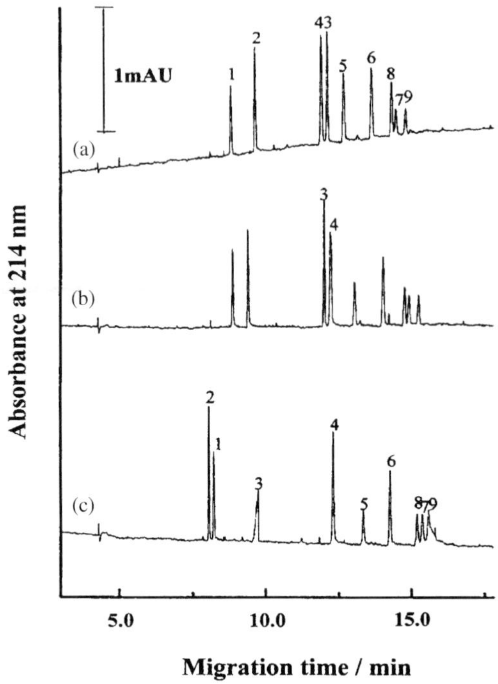

Fig. 2.20. Electropherograms of benzophenones obtained with a micellar phosphate buffer (20 mM) containing 20 mM SDS at varied buffer pH: (a) 6.6; (b) 7.0; (c) 8.0. Capillary: 57 cm × 50 µm, ID; sample concentration: 10 µg/ml; UV detection at 214 nm; other operating conditions: $V = 20$ kV, $T = 25°C$. Peaks identification: 1, 4,4′-dihydroxy-benzophenone; 2, 2,2′,4,4′-tetrahydroxy-benzophenone; 3, 2,4-dihydroxy-benzophenone; 4, 2,2′-dihydroxy-benzophenone; 5, 2-hydroxy-benzophenone; 6, 2,2′-dihydroxy-4-methoxy-benzophenone; 7, 4,4′-dimethoxy-benzophenone; 8, 2-hydroxy-4-methoxy-benzophenone; 9, 2,2′-dihydroxy-4,4′-dimethoxy-benzophenone. Reprinted from Ref. [159]. Copyright (2001), with permission from Elsevier.

Addition of modifiers to the electrolytic solution: The most versatile and effective methods to manipulate the selectivity in MEKC involve the use of additives in the aqueous phase and the selection of suitable surfactants. The additives mainly used in MEKC are cyclodextrins (CDs), organic modifiers, ion-pair reagents and urea.

Cyclodextrins (CD-MEKC): The addition of a cyclodextrin to the buffer solution in MEKC gave rise to cyclodextrin-modified MEKC (CD-MEKC) [160]. The separation principle in CD-MEKC is based on the differential partitioning of compounds among the micellar system, the cyclodextrin and the aqueous phase.

Cyclodextrins are oligosaccharides comprising various D(+)-glucopyranose units. They have the shape of a truncated cone with a relatively hydrophobic, open inner cavity and a hydrophilic outside due to the presence of hydroxyl groups (in 2, 3 and 6 positions of the glucopyranose). α-CD, β-CD and γ-CD are native cyclodextrins, which have six, seven and eight glucopyranose units, respectively. These three CDs have the same depth but different inner diameters, β-CD being the most commonly employed although it is the least soluble in water (its solubility can be increased by adding methanol, ethanol or urea).

Figure 2.21 shows the separation principle in CD-MEKC for a neutral compound using anionic micelles and neutral cyclodextrins. The neutral CD moves at the same velocity as the EOF and we assume that it is not incorporated into the micelle due to its hydrophilic surface. In the presence of the CD the association of a neutral solute with the micelle decreases and the extent to which this association decreases is higher when increasing CD concentration in solution. In fact, the retention factor of the solute can be controlled through the concentrations of the micelle and the CD. On the other hand, solutes migrating at the same velocity as the EOF will give a peak at a migration time t_{eo} while solutes totally incorporated into the micelle that do not interact with the CD will give a peak at a migration time t_{PS}. Solutes interacting with the neutral CD will experience a differential partitioning between the aqueous and micellar phases, showing a peak at migration time t_m. The separation selectivity in these systems can be modified by changing the type and concentration of the cyclodextrin and the surfactant [15].

Organic modifiers: The addition of organic modifiers to the buffer solution in MEKC causes a decrease in the EOF and an increase in the elution range. Furthermore, the solubility of hydrophobic compounds in the aqueous phase is increased and the interactions of the solutes

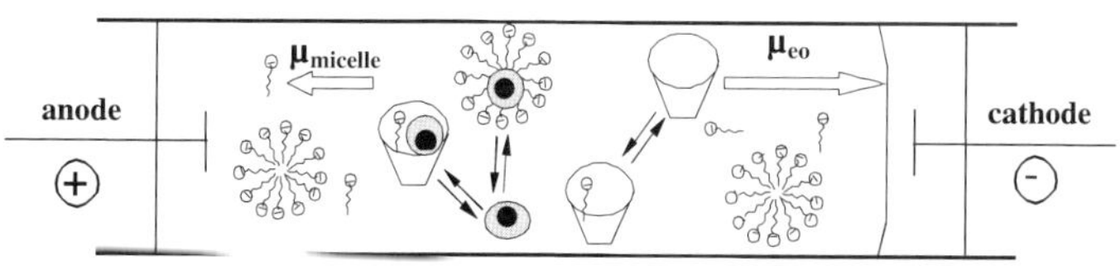

Fig. 2.21. Separation principle in CD-MEKC in the presence of an anionic surfactant and a neutral cyclodextrin. •, neutral compound.

with micelles then decrease [161]. However, in order to maintain the integrity of micelles; high concentrations of organic modifiers cannot be employed in MEKC. Organic modifiers are added usually at percentages below 20–30%, the most commonly used being methanol, *n*-propanol, 2-propanol, butanol, acetonitrile and tetrahydrofuran [139]. As an example, Fig. 2.22 shows the reduction obtained in the EOF when 3% *n*-propanol is added to a 0.050 M 2-(*N*-cyclohexylamino)-ethanesulphonic acid (CHES) buffer, 0.050 M in SDS, for the separation of a mixture of hydrophobic compounds (benzene and naphthalene derivatives and PAHs).

Ion-pair reagents: The use of ion-pair reagents in MEKC causes a remarkable change in the separation characteristics. When a tretraalkylammonium salt is added to an anionic micellar solution,

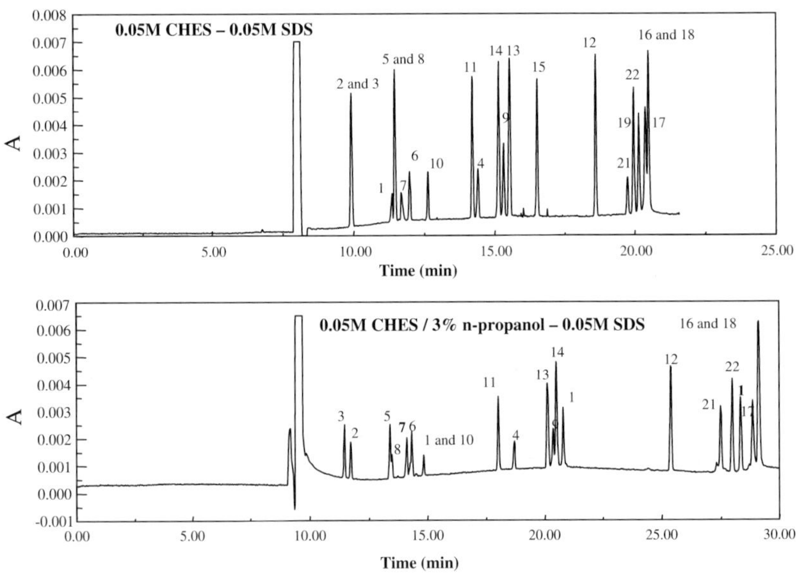

Fig. 2.22. Separation of benzene and naphthalene derivatives and PAHs by MEKC. Capillary, 65 cm (l: 59 cm) × 25 μm ID $V = 15\,kV$. $T = 25°C$. UV detection at 214 nm. Peak identification: 1, benzene; 2, benzilic alcohol; 3, benzamide; 4, toluene; 5, benzonitrile; 6, nitrobenzene; 7, phenol; 8, 2-phenylethanol; 9, chlorobenzene; 10, phenylacetonitrile; 11, 3,5-dimethylphenol; 12, naphthalene; 13, 1-naphthol; 15, 2-naphthol; 16, pyrene; 17, phenanthrene; 18, 2,3-benzofluorene; 19, fluorene; 20, fluoranthene; 21, acenaphthylene; 22, acenaphthene; A=absorbance. Reprinted from Ref. [194]. Copyright (1996), with permission from Elsevier.

anionic analytes form an ion-pair with the ammonium ion; hence, the electrostatic repulsion between the anionic micelle and the anionic analyte is reduced. This ion-pair formation is promoted by an increase in the concentration of the ammonium salt; the higher the concentration of the ammonium salt, the larger the retention factor of the anionic analyte. On the other hand, a cationic analyte competes with the ammonium ion interacting with the anionic micelle, so the migration time of the cation decreases with an increase in ammonium salt concentration [139].

Urea: Urea has been employed as an additive in MEKC to increase the solubility in the aqueous phase of high hydrophobic compounds such as PAHs [162–164] or PCBs [165]. The addition of urea to the micellar solution causes a slight decrease in the EOF, a considerable reduction of the migration velocity of the micelle and a decrease in retention factors [139]. The addition of urea is also effective for improving peak shapes.

Applications

MEKC has been recognized as a useful technique in various analytical fields: drug analysis, analysis of food and beverages, forensic analysis, environmental analysis and bioanalytical separations [142]. In addition, the experimental determination of specific physico-chemical parameters is another interesting application of MEKC, as described below.

The use of MEKC in drug analysis includes purity testing of drugs, assay of drugs, determination of drugs in biological fluids (including the direct injection method), enantiomeric separations and the separation of complex mixtures [166–168]. MEKC with some ionic surfactants allows the direct injection of biological fluids due to the solubilization of proteins. For this application, SDS has been the most employed surfactant in the normal elution mode [168]. Small injection volumes and the possibility of directly injecting body fluids are two features that make MEKC a very interesting technique for clinical analysis [169]. As an example, Fig. 2.23 shows the electropherogram corresponding to the separation of 23 1,4-dihydropyridines carried out with a 0.05 M SDS, 0.08 M CHES–5% *n*-butanol buffer. These compounds are of pharmacological interest owing to their calcium antagonist activity [170].

The fast analysis time, high peak capacity, and extremely small sample requirements of MEKC make this method attractive for forensic [171] as well as environmental analysis [172]. As an example,

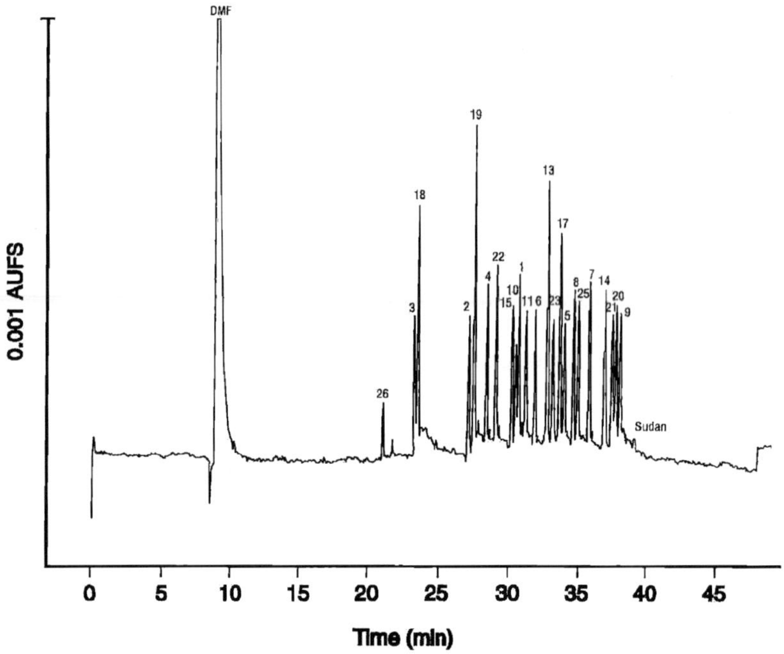

Fig. 2.23. Electropherogram corresponding to the separation of a mixture of 23 1,4-dihydropyridines. Buffer: 0.05 M SDS, 0.08 M CHES – 5% *n*-butanol (pH 10). Injection: 0.02 min at 20 mbar. $T = 31°C$. UV detection at 238 nm. $L_{tot} = 65$ cm, 0.025 mm ID. $V = 15$ kV. Reprinted from Ref. [170]. Copyright (1997), with permission from John Wiley & Sons, Inc.

Fig. 2.24 [173] shows the separation of triazines in a water sample using SDS as micellar system: (a) not spiked and (b) spiked at low level concentration of triazines, near to the limit of detection (0.05 μg/l of each triazine). No triazines were found at quantifiable levels (a). As can be seen, it is possible to determine this group of compounds at the level demanded by the European Union directive (the low maximum permitted per individual pesticide in water destined for human consumption is 0.1 and 0.5 μg/l for the sum of all pesticides).

Typical applications of MEKC are the separation of closely related compounds. CD-MEKC has been frequently used to increase the separation selectivity of structurally related compounds (interacting similarly with anionic micelles and then eluting together) or with a high hydrophobic character (totally incorporated into the anionic micelles and eluting simultaneously at t_{PS}) [160–163,165,174–176].

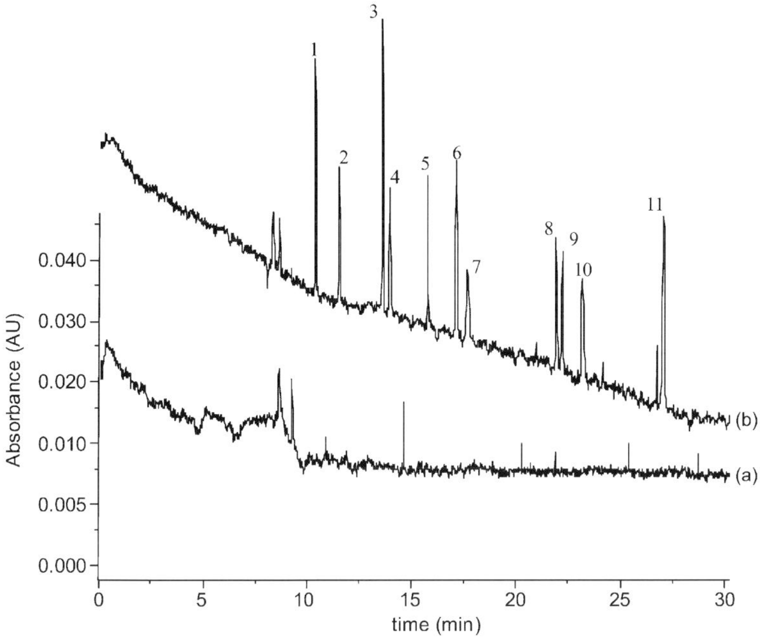

Fig. 2.24. Electropherograms corresponding to the separation of triazines using SDS micellar system in a real water sample: (a) no spiked and (b) spiked at low level concentration of triazines, near to limit detection (0.05 μg/l of each triazine). Buffer solution: 25 mM boric acid, 18 mM phosphate, 24 mM SDS and 5% 1-propanol at pH 9.5. Peaks identification: 1, desethylatrazin-2-hydroxy; 2, simazine; 3, prometon; 4, atrazine; 5, simetryn; 6, ametryn; 7, propazine; 8, prometryn; 9, trietazine; 10, terbutylazine; 11, terbutryn. $L_{tot} = 57$ cm ($l = 50$ cm), 75 μm ID. Injection by pressure, 0.5 psi for 2 s. UV detection at 214 nm. Reprinted from Ref. [173]. Copyright (2004), with permission from Elsevier.

Figure 2.25 shows the separation of 14 PCB congeners (which have a high hydrophobic character) using 50 mM γ-CD (a), 70 mM γ-CD (b), 60 mM γ-CD (c) all in 0.1 M SDS, and using a mixture of 72 mM β-CD and 25 mM γ-CD in 0.09 M SDS (d) [165]. A good separation is obtained using a mixture of β- and γ-CD in the electrophoretic buffer (Fig. 2.25d).

One of the most attractive applications of MEKC is in enantiomeric separations [139]. As a result, MEKC has a wide variety of applications in the pharmaceutical and environmental fields.

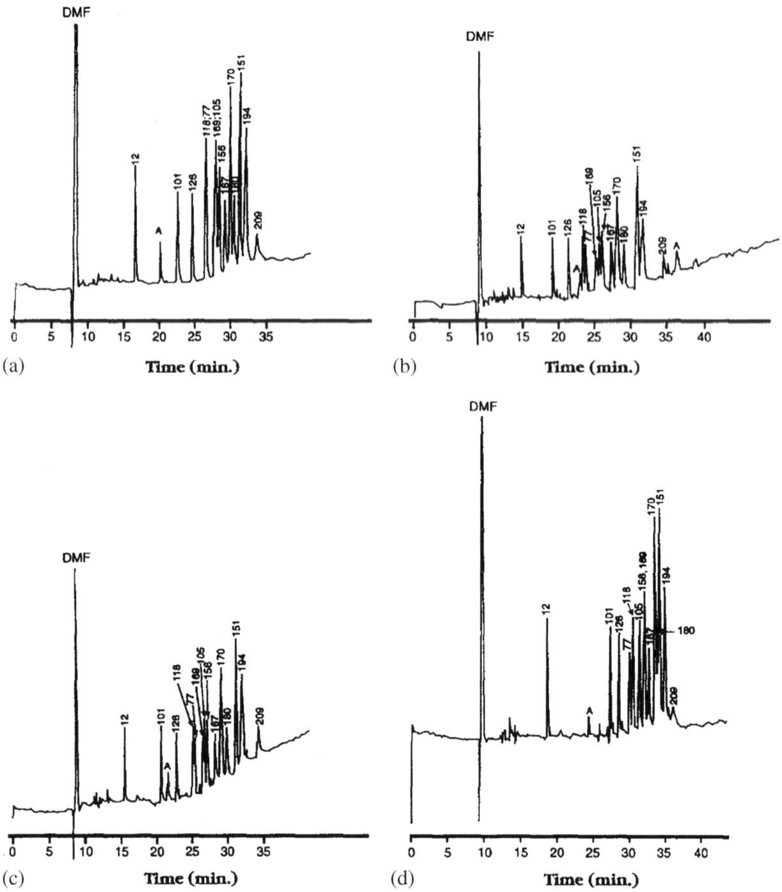

Fig. 2.25. Separation of 14 PCB congeners by CD-MEKC using 50 mM γ-CD (a), 70 mM γ-CD (b), 60 mM γ-CD (c) all in 0.1 M SDS, and using a mixture of 72 mM β-CD and 25 mM γ-CD in 0.09 M SDS (d). Other buffer conditions: 0.08 M CHES (pH 10), 2 M urea. $L_{tot} = 65$ cm (l = 50 cm), 50 μm ID. Injection, 1.2 s, 20 mbar. $V = 15$ kV. $T = 45°$C. UV detection at 240 nm. A = unknown peak. Reprinted from Ref. [165]. Copyright (1997), with permission from Elsevier.

Two main strategies are followed in MEKC to achieve an enantiomeric separation [166]:

- *MEKC with chiral surfactants*: an ionic chiral surfactant is employed as enantioselective PS, or a mixture of a neutral chiral surfactant with an ionic achiral surfactant. Naturally occurring chiral bile salts [166,177–180], digitonins [181] and saponins [182] have been used as well as synthetic chiral surfactants [166] and HMMS [183,184]. Among them, bile salts have been the most widely employed enabling the chiral separation of a great number of compounds [166,177–180]. Bile salts are natural and chiral anionic surfactants, which form helical polar micelles of reversed micelle conformation [138]. Although their solubilization power is less than that of SDS micelles, bile salts can tolerate relatively high concentrations of organic solvents [100]. As an example, Fig. 2.26 shows the chiral separation of PCBs 84 and 95 employing sodium cholate as micellar system [177].

- *CD-MEKC*: several nonderivatized and derivatized CDs are used as chiral selectors and usually achiral ionic surfactants are employed as micellar systems. In fact, CDs can act as chiral PSs. In this case, a chiral separation will be produced when enantiomer–cyclodextrin complexes of different stability are formed for each enantiomer. CD-MEKC has enabled the enantiomeric separation of different chiral compounds such as organochlorine herbicides, polychlorinated biphenyls, neutral fungicides, amino acids, etc. [185–191].

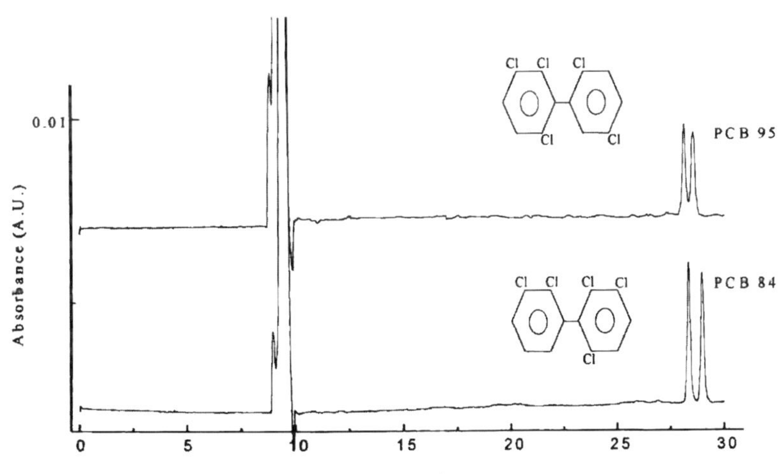

Fig. 2.26. Electropherogram corresponding to the enantioseparation of PCBs 84 and 95. Separation buffer: 50 mM CHES (pH 10), 2 M urea and 150 mM sodium cholate. Injection by pressure, 0.02 min at 20 mbar. UV detection at 235 nm. $L_{tot} = 65$ cm ($l = 50$ cm), 50 μm ID. $V = 15$ kV. Reprinted from Ref. [177]. Copyright (1998), with permission from Wiley-VCH Verlag GmbH.

In addition, MEKC is especially attractive if: (i) very small samples are to be analysed, (ii) peak capacity is needed in conjunction with short run times and (iii) sample clean-up steps that are mandatory for HPLC are to be avoided [142]. Current problems arising from high detection limits (as a consequence of low injection and detection volumes) might be overcome with on-line sample enrichment techniques and more sensitive and selective detection methods [192].

The application of MEKC to the determination of some specific physico-chemical parameters includes, among others, the determination of the solute–micelle association constants and the estimation of the hydrophobicity of organic compounds.

Determination of solute–micelle association constants: Distribution coefficients of solutes in MEKC between the micellar pseudophase and the aqueous phase are related to the solute–micelle association constants (K) through the molar volume of the surfactant (v) as expressed by the Berezin equation [193]

$$K = v(P - 1) \tag{2.20}$$

Knowledge of these solute–micelle association constants is very useful for a better understanding of the separation principle and to predict the retention behaviour of solutes in MEKC [170,194,195].

Solute–micelle association constants per surfactant monomer (K) can be calculated at low surfactant concentrations close to the CMC. In fact, they can be determined from the slope of the straight line obtained from the plot of the variation of the retention factor of a solute as a function of the total surfactant concentration in the electrolytic solution (see Eq. (2.19)). In addition, the errors obtained by MEKC in the determination of these solute–micelle association constants are lower than those obtained by chromatographic techniques such as micellar liquid chromatography (MLC). This is because in MEKC the error in the determination of the solute–micelle association constants is the error associated with the slope of the above-mentioned straight line whereas in MLC that error increases due to the error propagation obtained when calculating a magnitude from a quotient (slope/intercept) [196].

However, MEKC has an important limitation in the determination of the solute–micelle association constants for solutes highly associated with the micelle (with high hydrophobicity), which have migration times close to the micelle migration time. In this case (see Table 2.3), Eq. (2.13) shows that the term ($1{-}t_m/t_{PS}$) becomes close to zero and a large error would be associated with the determination of the retention

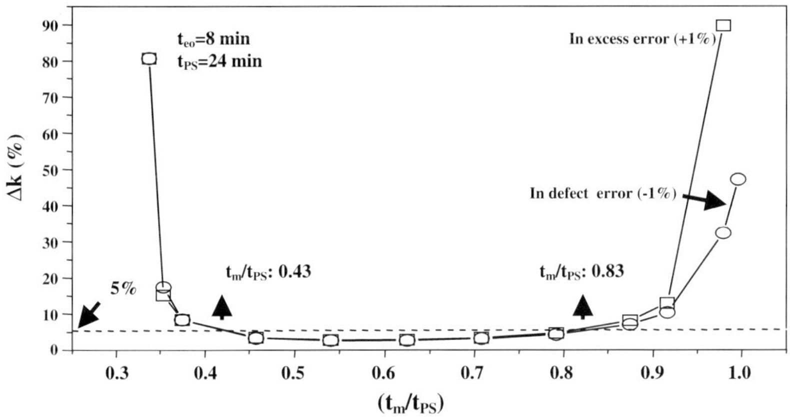

Fig. 2.27. Variation of the estimated error in the determination of the retention factor calculated as a function of t_m/t_{PS} ratio in the hypothetical case in which $t_{eo} = 8$ min and $t_{PS} = 24$ min. (□) With excess error (+ 1%); (○) with defect error (−1%). Reprinted from Ref. [194]. Copyright (1996), with permission from Elsevier.

factor and, therefore, in the determination of the association constant. The same occurs if the migration time of the solute is similar to the migration time of the EOF (t_{eo}). Figure 2.27 shows that when the migration time of solutes is close to t_{eo} or to t_{PS}, the error in the determination of the retention factor exponentially increases and then the variation of k as a function of the total surfactant concentration in solution could not be a straight line for very hydrophobic or very hydrophilic compounds [194]. Consequently, under these conditions, solute–micelle association constants cannot be determined [194]. As an example, Fig. 2.28 shows that the variation of the retention factor for a group of 23 benzene derivatives and PAHs as a function of the total surfactant concentration in solution is linear for the less hydrophobic solutes (Fig. 2.28a) while this linearity is lost for the most hydrophobic solutes (Fig. 2.28b) due to the error in determining their retention factors (t_m values very close to t_{PS}) [194]. However, for other solutes with intermediate hydrophobic character, MEKC has great potential in the determination of the solute–micelle association constants due to its high efficiency and resolution.

Estimation of the hydrophobicity of organic compounds: Hydrophobicity is commonly understood as a measure of the relative tendency of a solute to prefer a nonaqueous rather than an aqueous environment and plays an important role in the biological and physicochemical

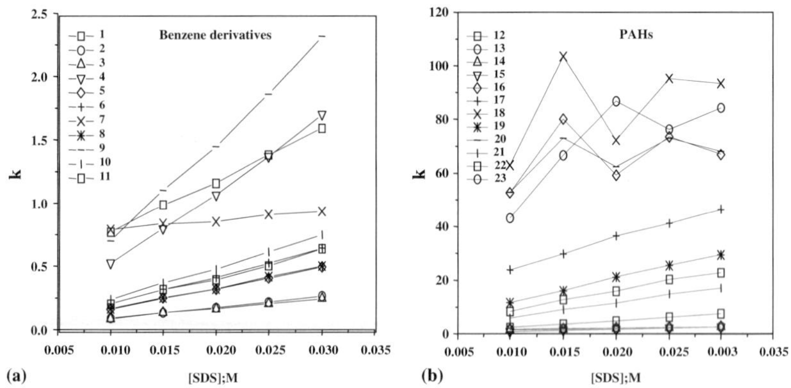

Fig. 2.28. Variation of retention factor as a function of SDS concentration in MEKC. Buffer: CHES 0.05 M/1% n-butanol. $L_{tot} = 65$ cm ($l = 59$ cm), 25 μm ID $V = 15$ kV. $T = 25°$C. UV detection at 214 nm. Solute number as in Fig. 2.22. Peak 23 = anthracene

behaviour of numerous types of organic compounds. This is why the estimation of the hydrophobicity has great importance in disciplines such as drug design and toxicology [197,198].

Traditionally, the logarithm of the octanol–water partition coefficient (log P_{ow}) of a solute has been the most common parameter to measure its hydrophobicity. However, owing to the numerous disadvantages of the standard shake-flask method used for determining this partition coefficient [197,198], several attempts have been made to achieve the indirect estimation of log P_{ow} by other techniques such as chromatography, especially reversed-phase HPLC. This estimation is based on the linear correlation found between the logarithm of the solutes' retention factor in a chromatographic system and their log P_{ow} values [199–201]. The advantages of this technique are that the composition of the mobile phase can be chosen in order to avoid the coelution of interfering peaks with the analyte peak, the small amounts of substances required, the rapidity of the analysis, and the application of the method for log P_{ow} values higher than 4. However, this method has been criticized due to its dependence on the stationary phase and its inadequacy to simulate biomembranes [200]. Since micelles have been considered as simple chemical models for biomembranes [202], MLC and MEKC have been studied as alternative techniques for indirect estimation of hydrophobicity. In the case of MEKC, the existence of only a distribution equilibrium between the micellar and the aqueous phases (no stationary phase exists) can give an ideal system to generate hydrophobicity estimations. In addition, a linear correlation between the values of log k for solutes in MEKC and log P_{ow} can be expected from the Collander equation [203].

MEKC has allowed the indirect estimation of the hydrophobicity of a variety of compounds [164,170,175,195,198,204] although with the above-mentioned limitations in the determination of the retention factors for hydrophobic compounds [194].

As an example, Fig. 2.29 shows the variation of log k with log P_{ow} for a group of 23 aromatic compounds using a 0.05 M CHES buffer with 5% n-butanol and 0.02 M SDS [198]. A good linearity is observed which was improved in the presence of alcohols. In fact, the addition of alcohols to the separation buffer gives rise to media which are more similar to octanol–water media than the purely aqueous micellar systems.

2.3.1.2 Other pseudophases in EKC

The selectivity required for a separation is not always obtainable with micelles. In order to change selectivity, a variety of pseudophases has

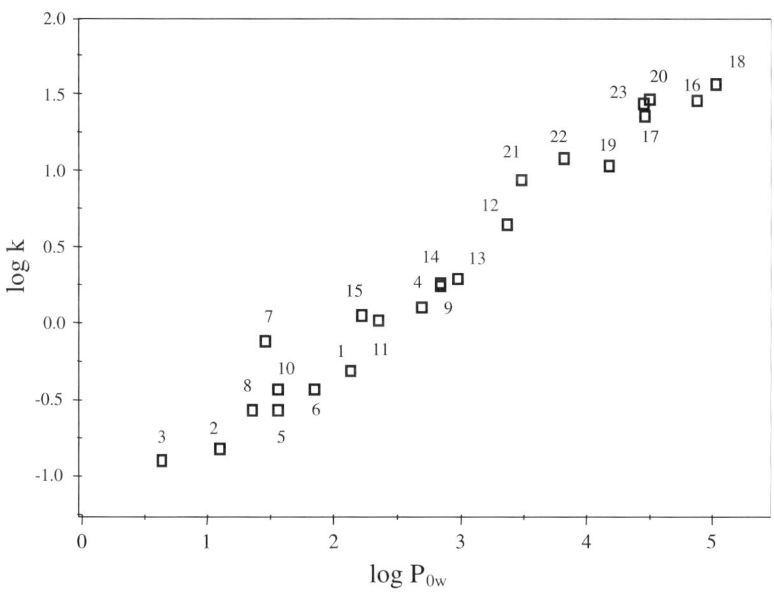

Fig. 2.29. Variation of log k as a function of log P_{ow} in MEKC for a group of 23 aromatic compounds in a 0.05 M CHES–5% n-butanol–0.020 M SDS electrolytic solution. $L_{tot} = 65$ cm ($l = 59$ cm), 25 μm ID $V = 15$ kV, $T = 25$°C, UV detection at 214 nm. Solute numbers as in Fig. 2.22. Peak 23 = anthracene Reprinted from Ref. [198]. Copyright (1996), with permission from Elsevier.

been introduced in EKC: microemulsions [113–115], cyclodextrins [116–119], macrocyclic and macromolecular phases [120,121], vesicles [122], resorcarenes [123], dendrimers [105,124,125], polymeric ions [126,127], proteins [128], macrocyclic antibiotics [129,130], polysaccarides and alkaloids [119]. Due to the great number of pseudophases employed in MEKC and since cyclodextrins and microemulsions have been the most commonly used after the micellar systems, these two EKC modes will be described in more detail.

Cyclodextrin electrokinetic chromatography (CD-EKC)
Cyclodextrins are well-known host compounds for the formation of inclusion complexes with guest compounds that can fit into the CD cavity [5]. Enhancement of the selectivity by the use of cyclodextrins is usually attributable to their ability to selectively include a wide variety of guest organic and inorganic molecules or ions into their hydrophobic cavity. The analyte will be included only if it fits into the CD cavity: inclusion is reflected in a marked increase in the migration time for the analyte with respect to the same electrolytic solution containing no CD. Selective inclusion depends on hydrophobicity, size and hydrogen bonding [15]. Although, natives CDs are nonionic, they can be modified to have ionic groups. In fact, the availability of different CD derivatives has allowed the development of a variety of CD-EKC methods, which are mainly applied for chiral separations [205–208]. In addition, the development and commercialization of new CD derivatives obtained from native CD enables us [206,207]:

- To increase the solubility of CDs.
- To establish different possibilities for solute–CD interaction favouring inclusion complexation.
- To analyse neutral compounds using charged CD derivatives.
- To reverse the migration order of enantiomers to allow solute purity control.

Since chiral separations are the main application of CD-EKC, the following paragraphs focus on this aspect. Figure 2.30 shows the separation principle of a chiral compound by CD-EKC in the presence of an anionic CD [209]. Neutral solutes which do not interact with the CD will move at the same velocity as the EOF, giving a peak at t_{eo}. However, if the chiral solute interacts with the CD, it will move at a different velocity than the EOF. If the EOF velocity is higher than the migration velocity of the anionic CD towards the anode, then the chiral

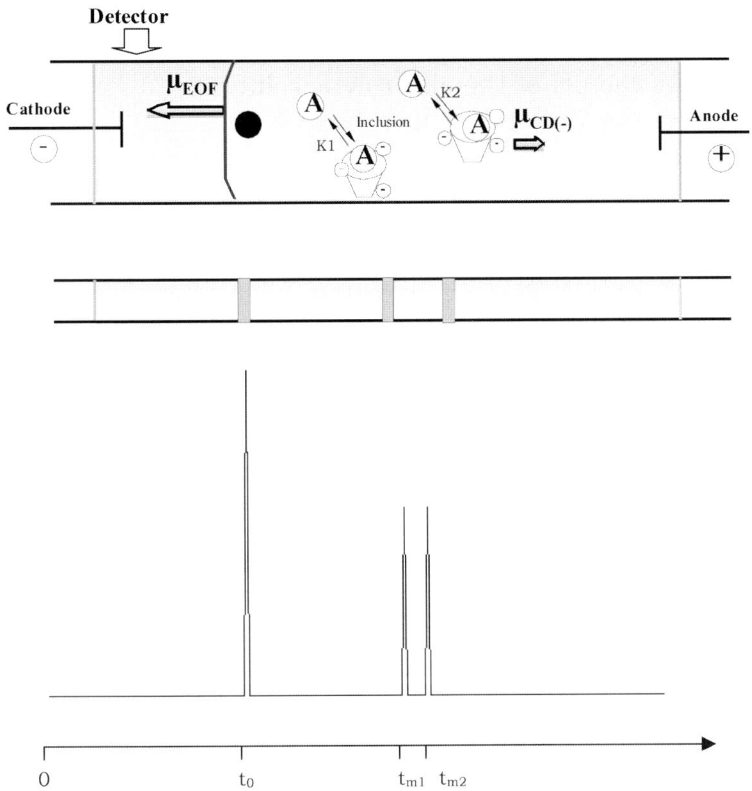

Fig. 2.30. Separation principle of a neutral chiral solute (A) in CD-EKC with an anionic cyclodextrin. (•) Neutral solute not interacting with the CD. From Ref. [209].

solute will elute after the EOF. Two peaks will be observed at different migration times if each enantiomer forms a complex of different stability with the CD. Although analytes interact with the inner cavity of the CD depending on their hydrophobicity, secondary bonds between the analyte and the hydroxyl groups of the CD ring can stabilize the inclusion complex. As primary and secondary hydroxyl groups are linked to asymmetric carbons, these interactions can be stereoselective and thus responsible for a chiral separation [119,206].

The most common ways to control the separation selectivity in CD-EKC are the following:

- To modify the nature and concentration of the CD.
- To use additives in the separation buffer.
- To change the pH of the electrolytic solution.
- To change the working temperature.

With respect to the type and concentration of the CD, selectivity in CD-EKC varies with the size, type of chemical modification, and concentration of CD in the electrolytic solution. Nowadays, a large number of neutral and charged CD derivatives are available for separating neutral and charged compounds by CD-EKC. Chiral resolution generally increases with the CD concentration until an optimum concentration enables maximum resolution. The most commonly employed CDs for chiral separations are described in chapter 13.

Native CDs can include a variety of compounds (particularly if they contain some aromatic ring). Analytes with one aromatic ring bearing a single or no substituent can form inclusion complexes with α-CD, which has the smallest cavity. This group of analytes includes structures with several, unfused rings—at least one of which must be aromatic [208]. Some substances possessing a free, scarcely functionalized ring have been resolved with α-CD [210]. Analytes with a double-substituted aromatic ring, two fused aromatic rings or other six-membered rings fit into the cavity of β-CD [211,212]. Structures with more than two fused rings or containing 5-, 7- or 8-membered heterocycles can also be resolved in this way, provided they possess some aromatic rings [208]. γ-CD appears to be the most suitable CD for resolving analytes with three or four fused rings, or even a single, extensively substituted ring [119]. However, native CDs have not shown the expected benefits because, according to Easton and Lincoln [212], of the inherent symmetry of natural CDs. Increased discrimination can thus be expected from the use of modified CDs with an increased degree of asymmetry. The hydroxyl groups on the rim of the natural CDs can be modified to obtain CDs in variable degrees of substitution [208]. The modification of neutral CDs can lead to neutral or charged CDs.

Neutral modified CD derivatives are obtained by replacing hydroxyl groups with alkyl groups, which increases the solubility and flexibility of the CDs. These two factors help accommodate the guest and increase the stability of the resulting complex [212]. Overall, neutral modified CDs exhibit improved inclusion and enantiodiscrimination capabilities. Neutral CDs have been widely employed in the separation of charged compounds such as drugs and phenoxyacid herbicides [119,208,213–216].

The possibility of resolving neutral and charged analytes fostered the synthesis of charged CDs that provide high resolutions. These charged CDs are subject to several shortcomings, namely [208]:

- Not all charged CDs are commercially available.
- Their presence substantially increases current intensity values.
- The CD–analyte complex can migrate to the cathode or the anode, which entails performing experiments under direct and inverse polarity conditions.
- The high intrinsic mobility of the CD can result in excessive electrodispersion and hence in broad, distorted peaks, and as a result the robustness of the separation (in terms of migration time and resolution) can be diminished.

Anionic CDs have been used for the separation of neutral as well as charged compounds. In the last case, electrostatic interactions play an important role in the formation of solute–CD inclusion complexes [209]. Most applications of anionic CDs are in the separation of basic drugs [212,217–219] but neutral analytes such as terpenes and barbiturates have also been separated [220,221]. Anionic compounds such as phenoxyacid herbicides [222] have been separated with anionic CDs although this case is less favourable due to electrostatic repulsions between solutes and CDs with the same charge. Cationic CDs have been less commonly employed than anionic ones probably due to their lower commercial availability and high cost. They have been employed to separate neutral and charged compounds [221,223,224]. Also, various synthetic amphoteric CDs have been prepared to resolve both cationic and anionic analytes by altering the pH of the electrolytic solution in order to obtain a universal selector [225]. Finally, dual mixtures of CDs enable modification of the separation selectivity [171,226]. Usually, charged and neutral CDs are combined, although mixtures of two neutral CDs have also been employed [178,227–229].

Figure 2.31 shows, as an example, the enantioresolution obtained for isoproterenol with β-CD, its methylated derivatives, and 2-hydroxypropyl-β-cyclodextrin (HP-β-CD) [208]. All modified β-CDs enabled an increase in the chiral resolution of isoproterenol (a substituted aromatic ring) with respect to the use of β-CD, for which there was no baseline separation.

The addition of an alcohol to the electrolytic solution can modify the EOF, the analysis time, the conductivity, the solubility of solutes and

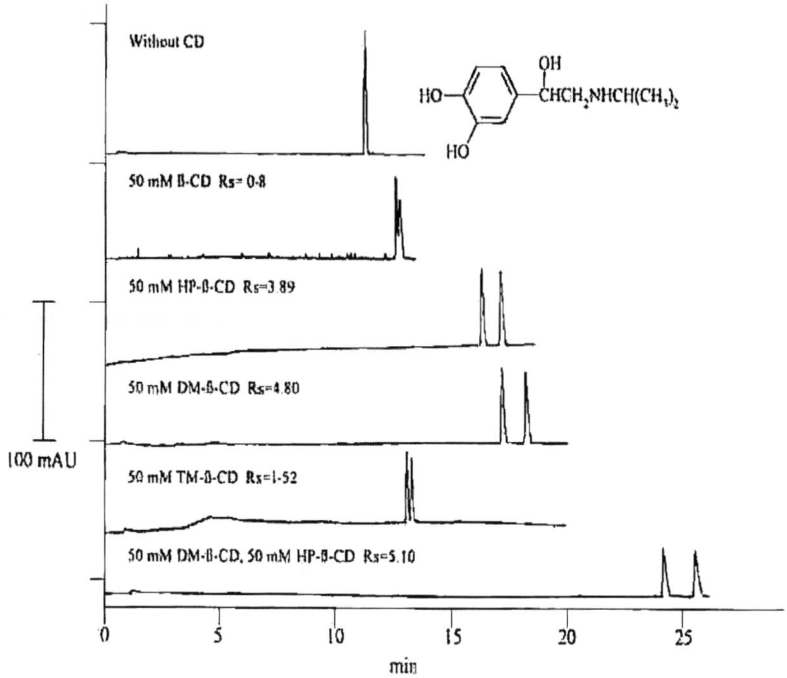

Fig. 2.31. Enantioresolution of isoproterenol with β-CD, its methylated derivatives and HP-β-CD using a 50 mM phosphoric acid/TEA buffer, pH 3 and 30 kV. (β-CD = β cyclodextrin, HP-β-CD = 2-hydroxypropyl-β-cyclodextrin, DM-β-CD = heptakis-(2,6-di-*O*-methyl)-β-cyclodextrin and TM-β-CD = heptakis-(2,3,6-tri-*O*-methyl)-β-cyclodextrin). Reprinted from Ref. [208]. Copyright (2003), with permission from Elsevier.

CDs, and the stability of inclusion complexes. Methanol is the most commonly employed alcohol [228,230,231]. However, other additives such as trietanolamine [232], hydrophilic cellulose derivatives, polyvinyl alcohol, trimethylammonium salts, etc., have also been employed to modify the EOF and avoid the adsorption of cationic analytes to the capillary wall [206,207].

The buffer pH can modify the selectivity through a variation in the EOF, migration times, and charge of the solutes and ionizable CDs [206,207].

Temperature modifies the viscosity of the electrolytic solution and then the mobility of analytes. Furthermore, kinetics and thermodynamics of inclusion equilibria can be modified. However, contradictory results have been observed in relation to the effect that temperature has on the resolution of enantiomers, which can increase or decrease when decreasing the temperature [205–207].

Two main advantages can be highlighted for CD-EKC systems compared to CD-MEKC systems:

- CD-EKC systems are more compatible with organic solvents than CD-MEKC where limitations exist in order to preserve the integrity of micelles, and
- CD-EKC systems are more predecible than CD-MEKC systems since surfactant monomers that may be included in the inner cavity of the CD (interacting with the analytes) do not exist [229].

Microemulsion electrokinetic chromatography (MEEKC)

Microemulsion electrokinetic chromatography (MEEKC) is an electrodriven separation technique, which offers the possibility of highly efficient separations of both charged and neutral solutes covering a wide range of solubilities [115]. The technique uses microemulsions in the buffer to separate solutes based on both their hydrophobicities and electrophoretic mobilities.

Microemulsions [233] are solutions containing dispersed nanometre-sized droplets of immiscible liquid. Typically oil-in-water microemulsions, which are immiscible oil droplets dispersed in aqueous buffer, are used. The oil and water components are totally immiscible as there is a high surface tension between them. The oil droplets are coated with a surfactant to reduce the surface tension between two liquid layers, which allows the formation of the emulsion [234]. The surface tension is further lowered, to approach zero, by the addition of short-chain alcohols, which stabilizes the microemulsion system. When the microemulsion system is unstable it will revert to individual layers of oil and water after a short period of time. The diameter of the oil droplets is below 10 nm. The microemulsion is therefore optically transparent as larger droplets scatter white light [115]. Then, specific combinations of oil, surfactant and co-solvent concentrations are required to form microemulsions.

Microemulsions can be used to achieve chromatographic separations as solutes can partition between the oil droplets and the aqueous buffer phase. Water-insoluble compounds will favour inclusion into the oil droplet rather than into the buffer phase [114]. This situation allows partitioning of the solute between the oil and water phases as in chromatography. The separation basis is similar to that involved in MEKC where ionic surfactant monomers group together to form micelles [113] and the solutes chromatographically interact with the micelles to achieve the separation. However, solutes are more easily able to penetrate the surface of the droplet than the surface of a micelle, which is much more rigid. This allows MEEKC to be applied to a wider range of solutes. MEEKC was introduced in 1991 by Watarai [235] and since then the number of applications has progressively increased.

SDS is the most widely used emulsifier surfactant in MEEKC. The oil droplet is coated with SDS surfactant molecules, making the droplet negatively charged. The alkyl chain of the surfactant penetrates into the oil droplet while the negatively charged hydrophilic sulphate groups reside in the surrounding aqueous phase. High pH buffers, such as phosphate or borate, are generally used. These buffers generate a high electroosmotic flow when the voltage is applied across a capillary filled with the microemulsion. The flow is relatively rapid and is towards the cathode situated near to the detector. The surfactant-coated oil droplets are negatively charged and therefore attempt to migrate towards the anode when the voltage is applied. However, if the EOF is sufficiently strong, the oil droplets will move through the detector [236]. This situation is shown in Fig. 2.32. Highly water-soluble, neutral solutes such as methanol will reside mainly in the aqueous phase and will move rapidly to the detector by the EOF, giving a solvent front. A highly water-insoluble solute such as dodecylbenzene will be strongly retained with an infinitely high retention factor. If a

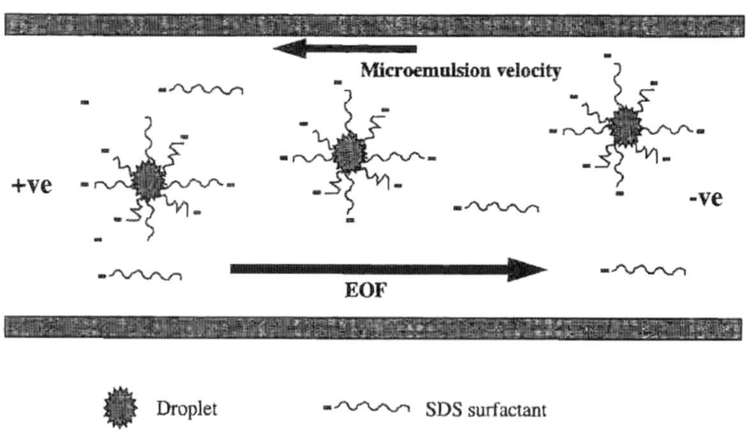

Fig. 2.32. Schematic of MEEKC separation process. Reprinted from Ref. [236]. Copyright (1998), with permission from Elsevier.

moderately soluble solute has a retention factor $k = 1$, then it spends equal amounts of time in both the aqueous phase and the oil droplet. The MEEKC migration time, or retention factor, of a neutral solute can be directly related to the solubility (hydrophobicity) of the solute and thus MEEKC can be used to assess compound solubility with good cross-correlation to other techniques used to measure solubility [237–239].

If a solute is ionized then it will electrophoretically migrate according to its size and number of charges. Repulsion by the negatively charged droplet will occur if the solute is also negatively charged. Conversely, if the solute is positively charged it may have ion-pair type interactions with the negatively charged droplet. The migration times obtained in MEEKC for ionized solutes reflect a combination of both electrophoretically and chromatographic behaviour of the solute ion [239]. The retention factor in MEEKC can be calculated by using equations in Table 2.3. This technique can be equally applied to water-soluble and insoluble compounds and to charged or neutral solutes.

The complexity of the composition of the microemulsions and the MEEKC separation process allows many manipulations to be made during method development in order to achieve a particularly difficult resolution. Selectivity can be modified by the surfactant type and concentration, oil type, addition of organic modifiers, co-surfactant type and concentration, pH, buffer type and concentration, operating temperature and sample solvent [114].

The choice of surfactant has a marked effect on the separation achieved in MEEKC as it affects the oil droplet charge and size, the level and direction of the EOF and the level of any ion-pairing with charged solutes. SDS is an anionic surfactant, which penetrates into the oil droplet. Anionic bile salts have also been used to generate negatively charged droplets with a different selectivity to microemulsions containing SDS surfactant [240]. Cationic surfactants, such as CTAB, produce positively charged droplets and also generate a positively charged surfactant bilayer on the capillary wall, which reverses the EOF direction and a negative polarity voltage is then necessary. In particular, CTAB-based microemulsions have been used to eliminate ion-pair interactions that cationic solutes have with anionic SDS microemulsion droplets [237]. Neutral surfactants are not useful for separating neutral solutes as the neutral droplets migrate at the same velocity as the EOF and the neutral solutes [241]. Mixtures of SDS and neutral surfactants such as Brij 35 have been used in octane microemulsions to separate highly insoluble suntan lotion additives [242] and hydrophobic polymer additives [243]. Neutral surfactants have the advantage that they can be added to the buffer without increasing the operating current. They can also alter the selectivity. The current can also be reduced by selecting a suitable salt of the surfactant [244] and by using a narrow capillary [114]. The low current enables high voltages to be applied, thus achieving fast separations. As an example, Fig. 2.33 illustrates the effect of surfactant type in MEEKC [245]. The selectivity of seven test solutes (caffeine, terbutaline, tropic acid, cinnamic acid, pindolol, hydrocortisone and prednisolon) can be drastically altered as indicated by the change in migration order by replacing SDS (Fig. 2.33a) partly or fully with other surfactants (Fig. 2.33b).

In relation to surfactant concentration [114], higher concentrations of surfactant increase the retention factor of neutral solutes. If a mixture of charged and neutral solutes is analysed, a modification in the surfactant concentration can have an effect on the peak migration order. If a surfactant and the solute have opposite charges, ion-pair interactions occur and migration time increases significantly with surfactant concentration. When both the solute and the surfactant have the same charge, there is an electrostatic repulsion between the droplet and the solute, and the migration time decreases. Increasing surfactant concentration also increases the ionic strength of the buffer, which reduces the EOF and increases analysis time. Finally, higher levels of surfactant reduce surface tension to a greater extent, which generates more stable microemulsions [113,114,236].

The choice of the oil affects the solute partitioning coefficient and has an effect on the selectivity. Generally, octane or heptane has been used to generate oil droplets. Hexane, heptane, and octane have been

---

Fig. 2.33. Electropherograms for seven test solutes (1, caffeine; 2, terbutaline; 3, tropic acid; 4, cinnamic acid; 5, pindolol; 6, hydrocortisone and 7, prednisolon) using four different surfactant systems. (a) 0.8% 5-chloropentane+3.3% SDS+6.6% *n*-butanol+89.3% 10 mM sodium tetraborate (pH 9.2). (b) Same as in (a), except 3.3% SDS is replaced by (A) 1.65% SDS+1.65% Brij 35; (B) 1.65% SDS+1.65% 3-(*N*,*N*-dimethylmyristylammonium) propanesulfonate (MAPS); (C) 1.65% SDS+1.65% polyoxyethylene sorbitan monolaurate (Tween 21) and (D) 3.3% *N*-cetyl-*N*,*N*,*N*-trimethylammonium bromide (CTMA). Other conditions: $L_{tot} = 48.5$ cm ($l = 40$ cm), 50 μm ID, $V = 20$ kV, UV detection at 215 nm. Reprinted from Ref. [245]. Copyright (2000), with permission from Elsevier.

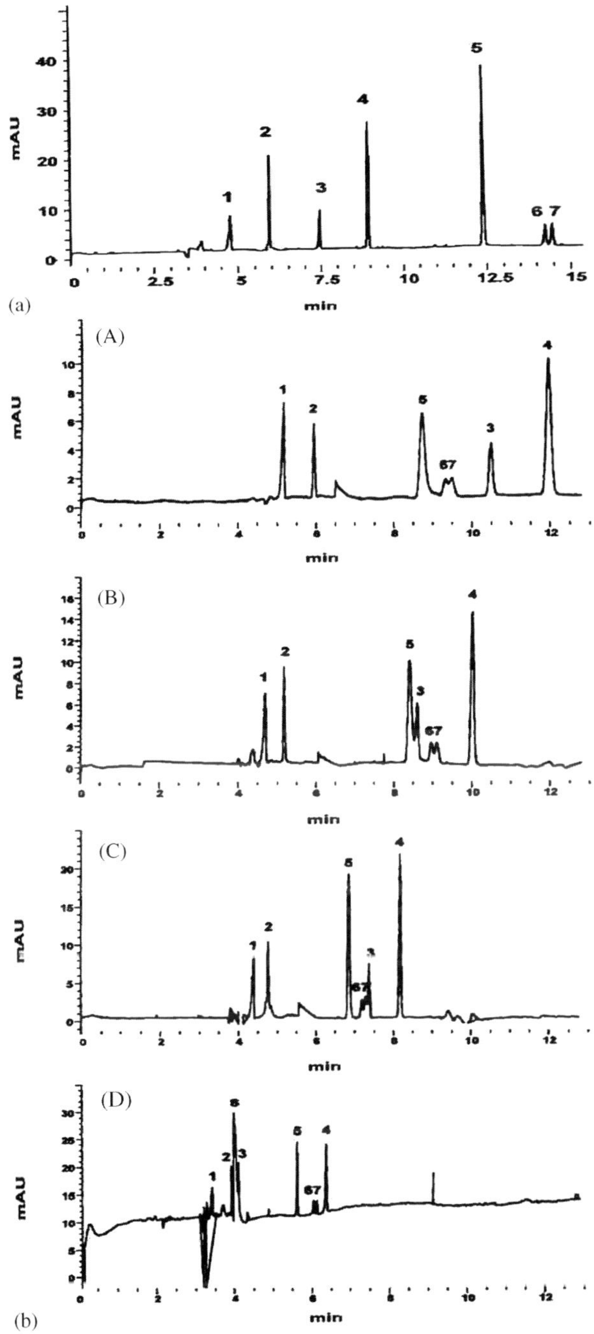
mAU
min
(a)
(A)
(B)
(C)
(D)
(b)
1
2
3
4
5
6 7
67

shown to give similar selectivity and migration times for the separation of a variety of neutral solutes [246]. Other water-immiscible liquids that have been employed are diethyl ether, cyclohexane, chloroform, methylene chloride, amyl alcohol, butyl chloride, ethyl acetate, toluene, diisopropyl ether, butyl acetate, 2-octanone, 1-chloropentane, octan-1-ol and hexan-1-ol [114,245–247]. The evaluation of some of these oils has shown differences in migration order and times [114].

In MEEKC, as in MEKC, organic solvents can be added to reduce retention and improve resolution [237,248], up to a limit as in MEKC. When this limit is exceeded, the microemulsion buffer disintegrates into a cloudy two-phase system, which cannot be used for separation. This is not however the case with propan-2-ol, which can act similarly to a cosurfactant and can be added at very high concentration ($>50\%$, w/w) without disturbing the microemulsion [244]. Butan-1-ol is the most frequently employed cosurfactant. Selectivity of the separation is unaltered by varying the concentration of butan-1-ol [115]. The migration time is altered with varying cosurfactant concentration as it affects the solution viscosity, which in turn affects the EOF rate. The size of the oil droplet increases when increasing cosurfactant concentration, which will affect the charge density on the droplet [114].

The pH of the buffer has a pronounced effect on separation selectivity as it affects solute ionization and the level of the EOF generated. Typically, buffers in the region of pH 7–9 have been used, generating relatively high EOF velocities. Ionized solutes will have different migration properties to neutral solutes. Positively charged compounds will have both partitioning and ion-pair interactions with the negatively charged oil droplet. These compounds will also have an electrophoretic mobility which reduces their MEEKC migration times. Conversely, acidic solutes, which have a negative electrophoretic mobility, will partition into the droplet but will be repelled from the negative droplet. Extreme pH has also been used in MEEKC to specifically suppress solute ionization [239].

Generally, MEEKC can be performed using low ionic strength (5–10 mM) borate or phosphate buffers. They generate relatively low currents and a reasonably fast EOF [114]. High buffer concentrations suppress the EOF and generate high currents, which may limit the level of voltage that can be applied. Zwitterionic buffers (e.g. tris (hydroxymethyl)-aminomethane (TRIS)) have been used to reduce the amount of current produced, which allows higher voltages to be applied and fast separations to be achieved [114,115].

The effect of the sample injection medium in the separation is drastic. Poor separation is obtained if the sample is not dissolved in the microemulsion. This is due to disruption of the microemulsion environment inside the capillary that is adjacent to the sample injection plug. The sample solvent can cause the microemulsion to disintegrate back to unmixed oil and water layers. This is seen as both reduced migration times and retention factors, and peak tailing [114,115].

In conclusion, MEEKC is a versatile EKC technique, which offers the possibility of separating a wide range of compounds of different types. The theory and some applications of MEEKC have been reviewed by Altria et al. [114] and Altria [115]. Poorly water-soluble neutral species such as PAHs, agrochemicals, fat-soluble vitamins, steroids, and fatty acids have been separated by MEEKC but it has been also applied to water-soluble charged solutes such as basic drugs, aromatic acids and proteins. Although oil-in-water microemulsions have generally been employed, Fung-Kee-Fung and Post [249] have demonstrated the use of water-in-oil microemulsions in MEEKC for the separation of very insoluble compounds. MEEKC has also been employed for chiral separations by using a chiral surfactant or a chiral oil [114, 250, 251]. Finally, MEEKC has been shown to be a useful technique for the estimation of physico-chemical parameters such as hydrophobicity of compounds [252–256].

2.3.2 Capillary electrochromatography

2.3.2.1 Fundamentals

CEC combines the advantages of HPLC and CE, such as the high efficiency of CE (movement of solutes by electrical forces) and high selectivity of HPLC (chromatographic interactions). Solutes are separated by the combined action of partitioning between a LC-type stationary phase and mobile phase and, if charged, by the difference in electrophoretic mobility. In CEC, the solutes are transported through the column by the EOF of the solvent and/or by their electrophoretic mobility. In contrast to pressure-driven flow in HPLC, the electrically driven flow velocity is more homogeneous over the cross-section of the column and therefore there is less zone broadening pressure and smaller particles and longer columns than in HPLC can be used [257]. The combined effect of reduced particle size, increased column length and plug flow is the possibility of a resolution domain for CEC similar to that of capillary gas chromatography, and currently not approached by HPLC. Whereas selectivity is most often the basis of separation in HPLC, in CEC it is possible to separate complex mixtures by using the higher available plate number.

Due to the fact that the mobile phase in CEC is impulsed by EOF which has a flat profile, band broadening in CEC is less and thus efficiency is higher than in HPLC where a parabolic profile exists.

Figure 2.34 shows schematically the most significant differences among CE, CEC and HPLC using an open capillary column (OTC) [258]. The separation principle in CE (Fig. 2.34a) is based on the different electrophoretic mobility of solutes (charged species) within a capillary when a voltage is applied and longitudinal diffusion being the main cause of band broadening [259]. The ID of the capillary should be small enough to avoid the band broadening provoked by the radial temperature gradient due to the electric current originated [260]. In CEC (Fig. 2.34b), the separation is based on a distribution of the solute between the stationary and mobile phases combined with electrophoretic forces. These are two main causes for band broadening in CEC:

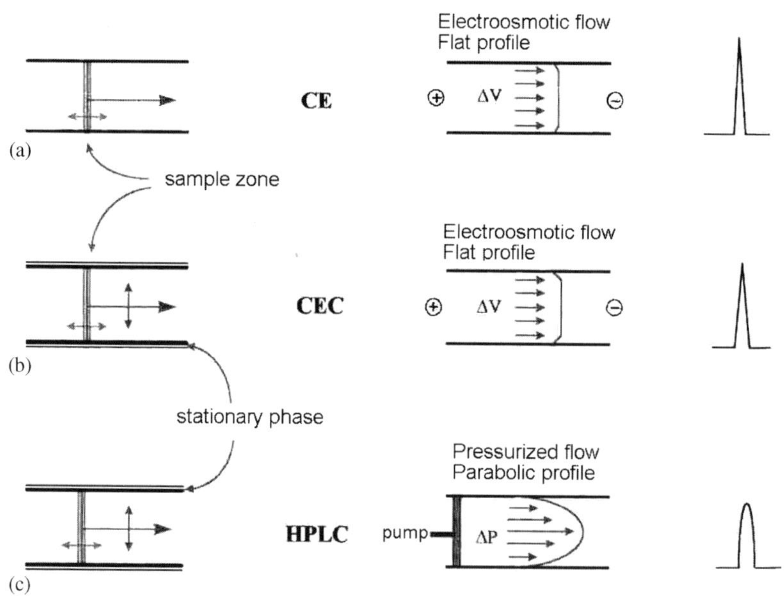

Fig. 2.34. Differences among CE, CEC, and HPLC. Moving mobile phase (→), zone broadening due to axial diffusion (↔), zone broadening due to radial diffusion (↕). Reprinted from Ref. [258]. Copyright (1994), with permission from Wiley-VCH Verlag GmbH.

longitudinal and radial diffusion where kinetics between both phases are important. Radial diffusion determines the ID of the column to be employed, especially when rapid separations are desired. Finally, the separation mechanism in HPLC (Fig. 2.34c), is basically the same as in CEC except for the origin of the flow of the mobile phase: a parabolic flow profile due to a pressure difference in HPLC and a quasi-flat profile due to the application of a voltage in CEC (as in CE).

The history of CEC begin more than 60 years ago when Strain [261] applied an electric field across an adsorption column to demonstrate higher selectivity because of the combination of electrophoretic and chromatographic separation forces. In 1974, Pretorius et al. [6] compared the plate heights between pressure-driven (laminar flow) and electroosmotically driven (plug flow) systems. This investigation was one of the first examples of using the electroosmotic pumping action to move solvent through packed tubes. The viability of CEC in packed capillary columns was demonstrated by Jorgenson and Lukacs in 1981 [1,259]. They used a 10 μm stationary phase of octadecyl silica (ODS) particles packed into a 170 μm ID capillary to separate 2-methylanthracene and perylene, and they obtained good efficiencies but high retention times. Two years later, Stevens and Cortes [262] observed that capillaries packed with particles smaller than 3 μm would have EOF velocities that were too low to be used in CEC due to overlapping of the double layers. Knox and Grant, however, demonstrated first in theory [263] and later in practice [264] that there was no evidence of any reduction in electroosmotic velocities in capillaries packed with particles down to 1.5 μm in size. Reduced plate heights much lower than two are attainable, and consequently columns operating electrically show higher plate numbers than their pressure-driven counterparts, so 100,000 plate separations can be achieved in relatively short time. Later, different groups started experimental work with CEC, and its enormous potential has been illustrated. The theory and practice of CEC has been reviewed in several articles and some books [15,265–272].

Basically, CEC has a similar experimental setup to CE except that the capillary is packed with a stationary phase.

The increasing popularity of CEC is mainly due to the separation power (high efficiency and high selectivity) and other advantages such as the following:

- It offers all the advantages of a miniaturized separation technique: lower solvent consumption, smaller sample volume requirements, increased mass sensitivity and high efficiencies.

- It has the potential arising from retention and selectivity given by a stationary phase (as in HPLC) but with an electrically driven flow which reduces band broadening and noise associated with parabolic pressure flow profiles.
- It enables the use of stationary phases with very low particle diameters (up to 0.5 μm) and since the pressure is not the source of the mobile phase flow, it approach the theoretical limits of HPLC.
- Instrumentation employed in CEC is the same as in CE, with reduced mobile phase flows (nl/min) and injection volumes (pl), and commercial CE instruments can be used.

2.3.2.2 Types of columns employed in CEC

Column packing in CEC plays a dual role: in addition to providing sites for the required interactions, as in HPLC, it is also involved in the generation of EOF. As a result, some packing materials that are excellent for HPLC may have low EOF in CEC, especially those packings with improved properties for the separation of basic solutes [270]. These basic differences between HPLC and CEC have stimulated the development of specific particulate packing with properties tuned for the needs of CEC as well as alternative column technologies. Most reported CEC research utilizes C18 stationary phases designed for liquid chromatography, but an increasing number of new materials—ion-exchange phases, mixed mode phases (reversed phase and ion-exchange phase), organic polymer continuous beds and silica-sol-gel monoliths—are being developed [273–275].

CEC columns are generally classified into three types: packed, open-tubular and continuous-bed or monolithic columns [270].

Most CEC applications are based on the use of packed columns. A packed CEC column is a fused silica capillary (0.05–0.20 mm ID) packed with LC-type packing materials that are confined between two end-frits. Frits are necessary in CEC in order to avoid the stationary phase migrating out of the column. Some practical problems, such as the difficulty in preparing packed columns and operating with them have resulted in the development of new approaches. Among the main problems are frit making, air bubble formation, and Joule heating. The existence of bubbles during the analysis is the most important practical problem since EOF is annulled. This problem takes place in the non-filled section of the capillary. Some parameters (mobile phases, frits, filling procedures, and supplementary pressure) have been investigated in order to obtain stable and reproducible EOF [276]. Bubble formation can be avoided by using columns with very permeable frits, very well degassed mobile phases, working at room temperature, and by applying pressure at the ends of the capillary (2–15 atm). Fragility of the columns should be taken into account during their manipulation since polyimide covering the column is eliminated in the frits and detection window and one can easily break the column. The procedure to fill the columns is crucial for the development of CEC since if the column is badly filled, asymmetric peaks, low efficiencies, and poor resolutions are obtained and bubbles can form.

An OTC is a capillary with a wall-bonded stationary phase that can be a coated polymer, a bonded molecular monolayer, or a synthesized porous layer network. The use of OTCs is an alternative approach that can eliminate the problems encountered in packed columns [277]. With OTCs no frits are required and bubble formation is excluded. The preparation of the stationary phase is paramount for achieving the high performance of these columns in CEC. However, it is very difficult to prepare OTCs with very narrow capillaries (<15 μm ID) and with the appropriate coating materials of sufficient retentive and mass loadability characteristics [266]. The influence of the composition of the mobile phase and of the instrumental parameters (injected volume and applied voltage) on column efficiency in CEC with OTCs of 10 μm ID has been investigated [278,279].

A continuous-bed column is a capillary containing a porous continuous bed that is formed *in situ* in the capillary. Monolithic columns have advantages such as the absence of end-frits and the stability of the column bed but disadvantages such as swelling of organic polymer in solvent and inadequate stability of silica-based monolithic columns under extreme pH conditions [280].

CEC column technologies and a number of applications have been reviewed [277,280–282].

2.3.2.3 *Instrumentation*

CEC can be carried out with basically the same instrumentation used in CE although pressure employed for injection is also used in CEC to avoid bubble formation. For this purpose, pressures from 10 to 12 bar are applied in the mobile-phase reservoirs [283].

The capability of running CEC with commercial CE instruments has stimulated research on this separation technique. Nevertheless, to fully exploit the potential of CEC, a routine CE device can certainly not fulfil all requirements. Two different approaches have helped to overcome this problem. The first was to modify commercial CE instruments for various demands: (i) pressurization of the packed capillary to prevent "air" bubbles (ii) gradient elution capabilities, and (iii) thermostating devices allowing a greater flexibility in column design. A completely different approach is the development of modular laboratory-made instrumentation dedicated to special CEC requirements. In order to increase mobile phase velocity, and thus the speed of analysis, voltages higher than 30 kV were introduced in some of these devices. Gradient elution was achieved by either coupling gradient LC systems or an electroosmotic generation of the changing eluent composition. All of these approaches have been discussed extensively by Steiner and Scherer [284].

When a pressure gradient is applied between both column ends in addition to a voltage gradient, a hybrid between capillary HPLC and CEC results [285]. The chromatographic mode is named pressure assisted CEC. Either CE instruments equipped with additional HPLC pumps or modular laboratory-made devices are suitable for pressure assisted CEC [284].

2.3.2.4 Theory

In CE, the electrical double layer only exists around the walls of the capillary. In CEC, each particle possesses its own electrical double layer. When the electric field is applied, the solvent moves through the capillary around each particle. If the particle diameter becomes too small, the electrical double layers that surround the particles will overlap and the EOF will stop. Also, channels or pores exist within each particle. However, the electrical double layers overlap within these channels, and therefore there is no flow of the mobile phase within a particle, only in between particles [15].

The properties of the EOF in packed capillaries have been discussed [286] and are considered to be the same as in open tubular capillaries, except that the electrical double layer exists on the surface of each silica-based particle in contact with the electrolyte and not on the surface of the silica wall-coated tube. In addition, there are numerous interconnected flow channels between particles of a packed bed. The mean channel diameter is significantly smaller than the particle size (d_p) (about one third dp). It has been theoretically and experimentally demonstrated [264,285,286] that v_{eo} is independent of the channel diameter and should therefore be nearly the same in all channels across a packed bed.

It is important to note that although v_{eo} is not affected by particle size down to 1.5 μm, and for a given material (e.g. silica) it will exhibit the same zeta potential whether in the form of particles or an open tube, the flow will be lower in packed capillaries than in open tubes [266]. This is due to the nonalignment of flow channels with the electric field and the packing porosity [264].

Separation process

Neutral species move in CEC in the same direction as the mobile phase but are retarded with respect to the EOF by their interaction with the stationary phase [269] (see Fig. 2.35a). For a charged solute, three situations are possible depending on the direction and magnitude of the electrophoretic migration with respect to the EOF [287]: (i) co-directional (Fig. 2.35b), where the migration velocities of charged species are always greater than that of the EOF marker; the components emerge before the EOF marker. (ii) counter-directional (Fig. 2.35c), where the EOF velocity is greater than the electrophoretic velocity of charged components, which emerge after the EOF. (iii) counter-directional (Fig. 2.35d), where EOF velocity is lower than the electrophoretic

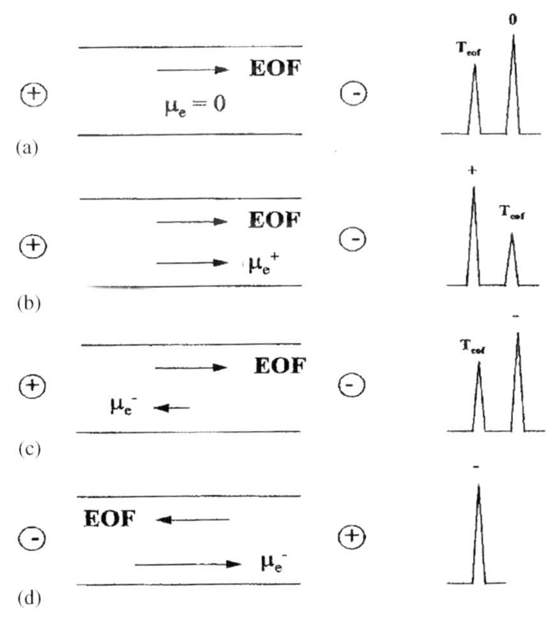

Fig. 2.35. CEC of charged species. Reprinted from Ref. [269]. Copyright (2001), with permission from Elsevier.

velocity, and detection of charged components is only possible with reversed polarity. In this case, the EOF marker is not detected. The sequence in which analytes elute from a CEC column will depend on their migration and their interaction with the stationary phase.

The migration of neutral chromatographically retained compounds in CEC (v_m) is characterized by v_{eo} and the extent of partitioning between two phases (k), according to the following equation:

$$v_m = \frac{v_{eo}}{1+k} \quad (2.21)$$

v_{eo} can be expressed as the electroosmotic mobility (μ_{eo}) of the electrolyte (eluent) using the potential drop (V) over the total length of the column (L_{tot}):

$$v_{eo} = \frac{\mu_{eo}V}{L_{tot}} \quad (2.22)$$

Combining Eqs. (2.21) and (2.22) gives:

$$v_m = \frac{\mu_{eo}}{(1+k)}\frac{V}{L_{tot}} \quad (2.23)$$

The retention time (t_m) of the neutral compound can be calculated, introducing the effective length (injector-to-detector length) of the column (l) and using Eq. (2.23):

$$t_m = \frac{l}{v_m} = \frac{lL_{tot}(1+k)}{\mu_{eo}V} \quad (2.24)$$

As μ_{eo} is related to the time of unretained neutral marker (t_{eo}) by the equation:

$$t_{eo} = \left(\frac{lL_{tot}}{\mu_{eo}V}\right) \quad (2.25)$$

Combining Eqs. (2.24) and (2.25):

$$t_m = t_{eo}(1+k) \quad (2.26)$$

As electropherograms are sufficiently similar to chromatograms, it is possible to think that the degree of retention (k) for the different analytes can be obtained directly from the electrochromatogram as in HPLC by measuring retention times of the unretained neutral solute

(t_{eo}), and of each analyte (t_m):

$$k = \frac{t_m - t_{eo}}{t_{eo}} \tag{2.27}$$

Migration of charged species is more complex than that of neutral species since they move in the mobile phase at a different velocity than that of the EOF. In this case, the migration velocity can be described according to the following equation:

$$v_m = \frac{v_0}{1+k} \tag{2.28}$$

where v_0 is the mean linear velocity of an ionized chromatographically unretained compound, which can be expressed as the sum of electro-osmotic velocity of the electrolyte (v_{eo}) and the effective electrophoretic velocity of the unretained compound (v_{ep}):

$$v_0 = v_{eo} + v_{ep} \tag{2.29}$$

Combining Eqs. (2.28) and (2.29) gives:

$$v_m = \frac{v_{eo} + v_{ep}}{1+k} \tag{2.30}$$

where v_{eo} and v_{ep} can be expressed as a function of μ_{eo} and μ_{ep} using V and L_{tot}:

$$v_{eo} + v_{ep} = (\mu_{eo} + \mu_{ep})\frac{V}{L_{tot}} \tag{2.31}$$

Combining Eqs. (2.30) and (2.31), the following expression for v_m is obtained:

$$v_m = \frac{(\mu_{eo} + \mu_{ep})}{(1+k)}\frac{V}{L_{tot}} \tag{2.32}$$

The retention time of the charged species, t_m, can be calculated by introducing the effective length of the column (l) in Eq. (2.32):

$$t_m = \frac{l}{v_m} = l\left(\frac{L_{tot}(1+k)}{(\mu_{eo} + \mu_{ep})V}\right) \tag{2.33}$$

Combining Eqs. (2.25) and (2.33) the final expression for the retention time of a charged species is obtained:

$$t_{\mathrm{m}} = \left(\frac{(1+k)}{\frac{1}{t_{\mathrm{eo}}} + \frac{\mu_{\mathrm{ep}} V}{l L_{\mathrm{tot}}}} \right) \tag{2.34}$$

Equation (2.34) is more complex than that of corresponding to a neutral species. Contributions to the elution time from electrophoresis and partitioning cannot be ascertained in a single experiment. However, the effective k values should be obtained from Eq. (2.34), including an unretained neutral marker in the sample to establish t_{eo} [266].

Band broadening in CEC

Band broadening in electroseparative techniques is characterized in principle in the same manner as in chromatography, and the dispersive process that can occur is the same, except that the contribution to the theoretical plate height of the phenomena depending on the flow profile is minimized due to the quasi-flat EOF profile.

The observed theoretical plate height (H_{tot}) for any solute in an electrochromatogram is the sum of two contributions [266]:

$$H_{\mathrm{tot}} = H_{\mathrm{col}} + H_{\mathrm{ext\text{-}col}} \tag{2.35}$$

where H_{col} is the equivalent height of a theoretical plate due to band broadening through the column and $H_{\mathrm{ext\text{-}col}}$ is the equivalent height of a theoretical plate due to band broadening out of the column. These two contributions can be studied separately.

H_{col} can be considered the sum of two main phenomena:

$$H_{\mathrm{col}} = H_{\mathrm{ter}} + H_{\mathrm{mig}} \tag{2.36}$$

where H_{ter} is the plate height due to thermal effects through the column and H_{mig} is the plate height due to the solute migration through the column due to the EOF.

Thermal effects are the result of Joule heating which causes a parabolic temperature gradient through the capillary section, resulting in a viscosity gradient in the mobile phase, which in turn can cause solute band broadening. However, these effects have shown to be negligible for capillaries with ID $\leqslant 100\,\mu\mathrm{m}$ [266]. In the case of mobile phases with a high ionic strength, an important efficiency loss can be predicted for capillaries with ID$>$ $75\,\mu\mathrm{m}$ [260].

With respect to band broadening due to solute migration, it depends on the type of column employed [266]. For OTCs, the main process contributing to band broadening is longitudinal molecular diffusion since mass transfer resistance in the mobile phase is considered negligible. In the case of packed capillary columns, in addition to longitudinal molecular diffusion and mass transfer resistance, the inhomogeneities of the packing material should be considered.

On the other hand, column efficiency can be drastically decreased due to inadequate instrumentation. There are two main contributions to band broadening out of the column: the injection (H_{inj}) and the detection (H_{det}):

$$H_{\text{ext-col}} = H_{\text{inj}} + H_{\text{det}} \tag{2.37}$$

However, the extra-column contribution of an on-column detection system can be neglected [288]. With respect to the injection, it should be noted that it is important to employ an injection system capable of minimizing the inlet sample distribution, which depends on the injection profile and the injected sample volume.

2.3.2.5 Mobile-phase considerations

In CEC as in HPLC, the selection of the mobile phase is determined by the type of stationary phase employed. For neutral compounds, the same optimization strategies as in HPLC can be used (variation of the type and concentration of organic modifier) [289]. The biggest difference between CEC and HPLC is that in CEC the mobile phase not only influences retention and selectivity but also EOF and thus the linear velocity of the mobile phase.

EOF generally decreases with the pH for CEC packings due to the decreased ionization of residual silanols at low pH [269].

For a column packed with an ODC silica gel, the most commonly used mobile phase is an aqueous mobile phase with a volume fraction of acetonitrile in the running buffer [15]. Other modifiers that have also been used are methanol, tetrahydrofuran or 2-propanol [290–293].

EOF can be affected by the organic modifier volume fraction, due to changes in the dielectric constant and viscosity of the mobile phase. In general, the ε/η ratio decreases when the organic modifier percentage in the mobile phase increases until a minimum is reached but after this minimum the ε/η ratio increases again. However, modifiers also affect the zeta-potential and thus the charge density of the particle surface in the packed bed [283]. While the mobilities for the methanol system

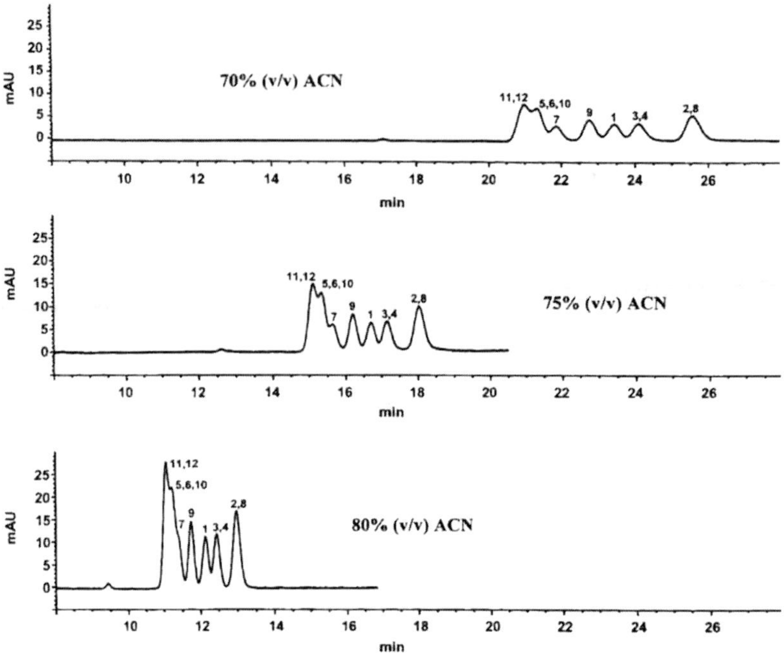

Fig. 2.36. Electrochromatograms showing the effect of acetonitrile volume fraction on the separation of 12 MBAP isomers. Conditions: mobile phase, volume fraction (% (v/v)) of acetonitrile varied/5 mM TRIS buffer, pH 8.0; 25°C; electrokinetic sample injection 5 s, 5 kV, followed by a 3 s, 5 kV run buffer injection; UV detection at 214 nm. Reprinted from Ref. [295]. Copyright (1999), with permission from Elsevier.

show the behaviour expected, the mobilities for the acetonitrile system show a steady increase upon increasing the acetonitrile content [294,295].

The influence that the concentration of organic modifier has on the selectivity of a separation is very similar to HPLC. As an example, Fig. 2.36 shows the effect of the organic modifier content in the mobile phase (acetonitrile) on the separation of 12 methylated benzo[*a*]pyrene (MBAP) isomers by CEC using an ODS column [296]. At 70% (v/v) acetonitrile, the MBAP isomers are retained longer, which results in a wider separation window. The volume fraction of acetonitrile provides good separation selectivity, but is offset by peak broadening and lower detectability. A slightly higher acetonitrile fraction of 75% (v/v) gives a shorter retention of all isomers with increased efficiency, as indicated by sharper peaks. Further increasing the acetonitrile content to 80% (v/v) provided the sharpest peaks, but is offset by a narrow elution window and loss in resolution of several MBAP isomers.

The type of organic modifier used also affects the selectivity of the separation. Ternary mobile phases can be employed to expand the migration time window and to improve resolution and selectivity [296–298].

With respect to the influence of ionic strength, according to theory EOF decreases when increasing buffer concentration due to a reduction in the thickness of the double layer and in the zeta potential. Thus, the ionic strength of the buffer affects the linear velocity and also the plate height [299].

2.3.2.6 Applications

The number of applications in CEC is steadily increasing. It has been applied to the separation of neutral, basic and acidic compounds, inorganic anions and cations, biomolecules (e.g. proteins, peptides, amino acids, nucleotides and nucleosides and carbohydrates), pollutants (e.g. PAHs and pesticides), natural products (e.g. polyphenols, carotenoids, fatty acid esters, retinyl esters), pharmaceuticals (e.g. benzodiazepines, tricyclic antidepressants, steroids, nonsteroidal anti-inflammatory drugs, opiate drugs) and chiral compounds [205,268,270,300]. In the last case, the chiral selector can be either incorporated in the mobile phase or immobilized on the stationary phase [301,302].

As an example illustrating the high efficiencies and short analysis times possible in CEC, Fig. 2.37 shows the baseline separation of six structurally related opiate drugs in 2.5 min [303]. In this case, the addition of 5 mM SDS to the mobile phase was effective in preventing bubble formation as well as for improving the separation efficiency. Efficiencies ranged from 250,000 to 500,000 plates/m.

Figure 2.38 compares HPLC and CEC analyses of a polyphenolic hop extract [270]. As shown in this figure, the elution order was identical for both techniques. In CEC, the separation between the critical pair of isomers, humulone/adhumulone, was satisfactory, although no baseline separation was possible. The pair of isomers lupulone/adlupulone was baseline separated, while this was not the case in HPLC. Retention time repeatability was excellent with RSD values less than 1%, but area repeatability was poor.

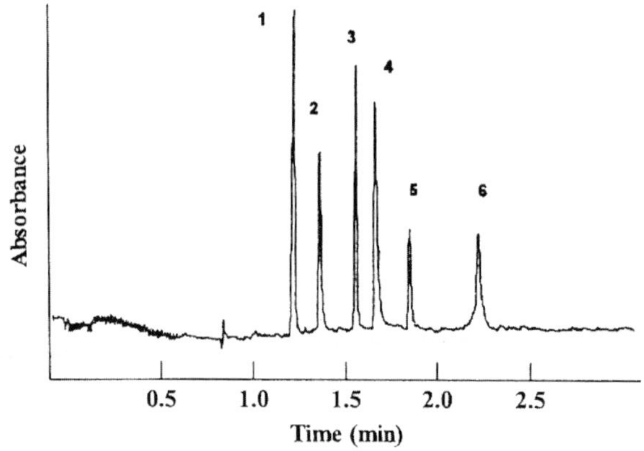

Fig. 2.37. Separation of six opiate drugs. Column, 1.5 μm nonporous ODS (75 μm ID, 150 mm packed length, 280 mm total length); mobile phase, acetonitrile/10 mM TRIS and 5 mM SDS, pH 8.3 (20:80, v/v); $V = 20$ kV; UV detection at 215 nm. Peaks: 1, morphine; 2, hydromorphine; 3, nalorphine; 4, codeine; 5, oxycodone; 6, diacetylmorphine. Reprinted from Ref. [303]. Copyright (2000), with permission from Wiley-VCH Verlag GmbH.

2.4 NONAQUEOUS CAPILLARY ELECTROPHORESIS

2.4.1 Introduction

NACE was first described in 1984 by Walbroehl and Jorgenson [11], who applied CE to the separation of a great number of compounds through the use of nonaqueous solvents as electrolytic solutions. However, it was from 1990 onwards that NACE began to be more generally employed. The potential and achievements of NACE have been documented in some comprehensive reviews [304–308] and in a book chapter [15].

Separations in NACE can be achieved only if the analytes are charged or can interact with a charged additive, producing selectivity in the nonaqueous electrolyte. The proper selection of the organic solvent and additives not only requires the knowledge of their physicochemical properties but also of the analyte characteristics and its interactions with the additives [308,309].

A small change in BGE composition can fine-tune or even drastically alter parameters such as resolution, analysis time, and selectivity [306,307]. The great potential of NACE is in the separation of analytes that are insoluble in water; however, nonaqueous media often also

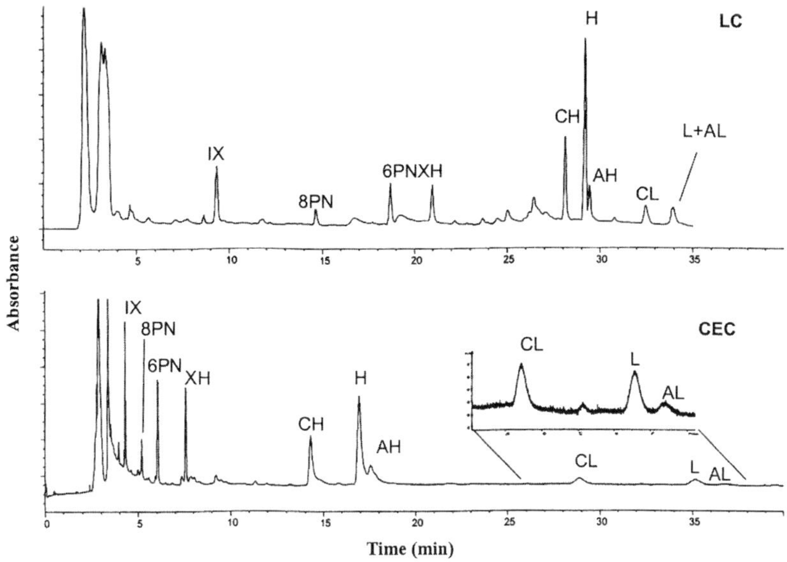

Fig. 2.38. Comparison of LC and CEC analyses of polyphenolic hop extract. LC conditions: Alltima C18 (4.6 mm ID, 250 mm length), mobile phase: 1% formic acid in water (solvent A) and acetonitrile (solvent B), gradient elution flow rate 1 ml/min, UV detection at 280 nm. CEC conditions: column hypersil C18 (100 μm ID, 250 mm packed length, 335 mm total length), mobile phase, acetonitrile /10 mM TRIS+0.5% formic acid (60:40, v/v); 30 kV; UV detection at 280 nm. Peaks: IX, isoxanthohumol; 8PN, 8-prenyl-naringenin; 6PN, 6-prenyl-naringenin; XH, Xanthohumol; CH, cohumulone; H, humulone; AH, adhumulone; CL, colupulone; L, lupulone; AL, adlupulone. Reprinted from Ref. [270]. Copyright (2001), with permission from Wiley-VCH Verlag GmbH.

allow the separation of groups of analytes that exhibit closely similar electrophoretic mobilities in aqueous media. This is because organic solvents offer a separation mechanism based on interactions that can either not take place or are too weak to be measured in aqueous media. In particular, hydrophilic interactions such as hydrogen-bonding and dipole-related and ionic interactions can be exploited in an hydrophobic environment, since these hydrophilic interactions are, in principle, thermodynamically strengthened in nonaqueous media relative to aqueous media. Furthermore, many labile analytes in aqueous media are stable in nonaqueous media, making NACE the method of choice. Another important characteristic is that the low currents present in NACE not only allow the use of higher electrolyte salts' concentrations and higher electric field strengths that provide short analysis times and high separation efficiencies, but also the sample load can be scaled-up employing capillaries with a wider inside diameter. On the other hand, organic solvents offer attractive features for detection systems such as mass spectrometry and electrochemical methods, providing improved sensitivity. Also, the compatibility of the methods needed for pretreatment of real samples and for the final analysis step is improved if, for example, after SPE, concentrated extracts eluted with organic solvents can be directly injected into a NACE system [308]. All of these benefits have been extensively and clearly discussed by Riekkola et al. [305] and Steiner and Hassell [306].

2.4.2 Electrolytic solutions

NACE exploits the vastly different physicochemical properties of organic solvents to control EOF and analyte migration. Table 2.4 groups the physicochemical properties of some solvents usually employed in

TABLE 2.4

Properties of some solvents used in CE at 25°C

Solvent	T_m	T_b	ρ	γ	ε	μ	η	pK_{auto}
MeOH	−97.7	64.5	16.90	22.3	32.66	2.87	0.551	16.91
EtOH	−114.5	78.2	7.89	21.9	24.55	1.66	1.083	19.10
1-PrOH	−126.2	97.1	2.73	23.1	20.45	3.09	1.943	19.40
2-PrOH	−88.0	82.2	6.03	21.2	19.92	1.66	2.044	21.08
1-BuOH	−88.7	117.6	0.82	24.2	17.51	1.75	2.571	20.89
ACN	−43.9	81.6	12.20	28.3	35.94	3.92	0.341	32.20
PC	−55.0	241.7	0.00616	41.4	64.92	4.94	2.530	—
FA	2.5	210.5	0.00882	58.2	109.50	3.37	3.302	16.80
NMF	−3.8	199.0	0.0338	39.5	182.40	3.86	1.650	10.74
DMF	−60.5	153.0	0.53	36.4	36.71	3.82	0.802	23.10
DMSO	18.5	189.0	0.077	43.0	46.45	4.06	1.991	31.80
H_2O	0	100.0	3.17	71.8	78.36	1.85	0.8903	17.51

Units: melting point T_m and boiling point T_b in °C; vapour pressure (ρ) in kPa; surface tension (γ) in mN/m; relative permittivity ε is dimensionless; dipole moment (μ) in D (1 D = 3.33564×10^{-30} Cm); dynamic viscosity (η) in mPa s (numerically equal to cP); autoprotolysis constant pH_{auto} is dimensionless. MeOH, methanol; EtOH, ethanol; PrOH, propanol; BuOH, butanol; ACN, acetronitrile; PC, propylene carbonate; FA, formamide; NMF, *N*-methylformamide; DMF, *N,N*-dimethylformamide; DMSO, dimethyl sulphoxide. Reprinted from Ref. [308]. Copyright (2002), with permission from Wiley-VCH Verlag GmbH.

CE [308]. The solvents used in NACE must meet certain requirements as [308]: (i) they should dissolve the analytes and the electrolytes used in the separation, (ii) they should not be inflammable, toxic or reactive, (iii) they should be available at low cost and as pure substances, (iv) they should preferably be liquids at room temperature for practical reasons, and (v) they should exhibit relatively high relative permittivity (ε, dielectric constant), relatively low viscosity (η) and low vapour pressure (ρ).

Relative permittivity is an important parameter because it describes the strength of the interactions between ions in the solvent. Low dynamic viscosity is preferred because it allows high mobilities of the solvated analyte ions and hence separations in a reasonable time. Also important is the compatibility of the solvent with the detection technique. All solvents listed in Table 2.4 are chemically stable and commercially available with high purity [308].

Methanol and acetonitrile and their mixtures are the most frequently employed organic solvents in NACE [305–307]. The characteristics of acetonitrile such as low viscosity and moderately low polarity make it especially useful. Acetonitrile has been employed in the separation of phenols, alcohols, and carboxylic acids through the formation of heteroconjugates with the undissociated analytes and small inorganic ions [310]. Methanolic electrolyte solutions with their relatively low surface tension and high volatility have frequently been used in electrospray ionization-mass spectrometry, and are the most popular solvent used in NACE [308]. Amide-type solvent, longer chain alcohols (propanol and butanol) and propylene carbonates have also been used [311–315].

The composition of the BGE can be used to control both the direction and the magnitude of the EOF, the magnitude of the EOF being an important parameter for the optimization of the separation efficiency in NACE. The most common buffering systems in organic solvents consist of acids and their ammonium salts, of which acetic acid and ammonium acetate have been most frequently used. Grob and Steiner [316] explored the EOF of organic solvents containing 5 mM ammonium acetate, under basic (addition of triethylamine) and acidic (addition of acetic acid) conditions and showed that the EOF remains high in ACN, DMF and DMSO after addition of triethylamine and it decreases after addition of acetic acid. Conversely, the EOF was unaffected in MeOH, formamide and *N*-methylformamide after addition of acid or base. EOF for MeOH/ACN mixtures showed a different behaviour depending on the concentration of base or acid added. The results indicated that the composition of nonaqueous BGE significantly affects the magnitude and direction of the EOF.

Organic solvents employed in the BGE solution affect the properties of both analytes and electrolyte components by influencing their mobilities and pK_a values [317]. To predict the effect of organic solvents on pK_a values, a differentiation has been made between neutral acids of type HA and cation acids of type HB^+ [308]. Acids of the type HA almost always have higher pK_a values in organic solvents than in water, mainly due to the poor solvation ability of most organic solvents for anions, which are present in the ionization equilibrium of HA. For the second type of acids, HB^+, the effect on the pK_a values is usually much weaker. Another important ion property affected by the solvents is electrophoretic mobility. However, mobility of fully protonated analytes (the actual mobilities) and/or pK_a values can be unknown at the pH of the BGE. Different theoretical studies have appeared in the literature on the shifts of pK_a values from water to organic solvents and the effect of the organic solvent on ionic mobility [318–325].

Porras et al. [318–324] introduced a conventional pH^* scale and have used it to study the basic physicochemical properties of analytes in nonaqueous media. Knowledge of pH^* in nonaqueous BGEs could be used in the determination of actual mobilities and pK_a values [318–320]. The knowledge of these parameters enables optimization of NACE separations and reduces the need to use the complicated solvent mixtures frequently applied in NACE studies [308].

The variety of solvent properties and solvent compositions suitable for NACE offer a good basis for finding good conditions and devising tailor-made solutions for numerous analytical problems [308].

2.4.3 Analyte–additive interactions

The effect of the solvent on the strength of different types of analyte–additive interactions in NACE have been reviewed by Chen et al. [309]. Some of the additives used to modify analyte migration behaviour in NACE are: polyether surfactants such as Brij 35, cyclodextrins, quinine, 2,4,6-triphenylpyrylium, dodecyl trimethyl ammonium ions, tretraalkyl ammonium ions, tretrahexyl ammonium ions, small organic and inorganic ions, tropylium ions, crown ethers, etc. Analyte–additive interactions are generally more complex in NACE than in aqueous systems. Most additives in aqueous CE interact with the analyte

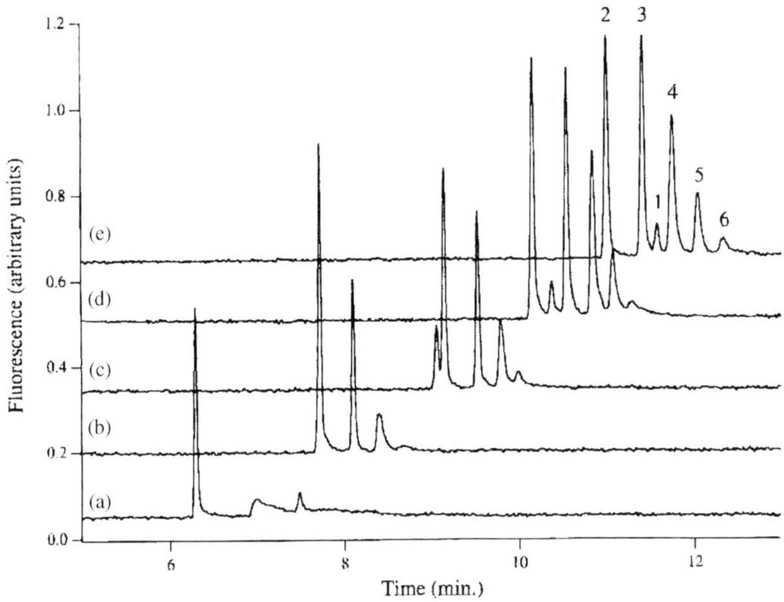

Fig. 2.39. Separation of porphyrin acid monomers utilizing electrostatic ion-dipole interactions. Buffer: 20 mM 3-cyclohexylamino-1-propane sulphonic acid (CAPS) (pH = 10.5) plus (a) 0 mM Brij 35, (b) 5 mM Brij 35, (c) 15 mM Brij 35, (d) 25 mM Brij 35, and (e) 35 mM Brij 35 in methanol. Peaks: 1, mesoporphyrin; 2, coproporphyrin; 3, pentacarboxylporphyrin; 4, hexacarboxylporphyrin; 5, heptacarboxylporphyrin; 6, uroporphyrin. Fluorescence detection, excitation: 488 nm, emission: 620 nm. $V = -30$ kV. Reprinted from Ref. [326]. Copyright (1997), with permission from Wiley-VCH Verlag GmbH.

through hydrophobic effects. Electrostatic and donor–acceptor interactions are generally weak, and with proper choice of conditions their effects can be minimized. Therefore it is possible to focus on the effect of a particular interaction (usually a hydrophobic interaction) in aqueous CE. In NACE, electrostatic and donor–acceptor interactions dominate. A wide range of molecules and ions, as mentioned below, undergo these types of interactions. As an example, Fig. 2.39 shows the separation of porphyrin acid monomers based on a purely electrostatic ion–dipole interaction with the polyether surfactant Brij 35 [326]. In this case, the polyether surfactant forms complexes with the porphyrins through the carboxylic acid functional groups. The equilibrium constants measured for porphyrin monomer-polyether were relatively high for ion–dipole interactions (50–150 M^{-1}). This can be explained by the low dielectric constant of methanol and the multiple binding sites on the porphyrins and the polyether.

Many ions or other buffer additives can be expected to undergo such types of interaction with the analyte. For example, a typical organic anion can be expected to interact with the buffer cations (e.g. sodium or ammonium ions), be involved in acid-base equilibria, as well as interact with any additives present in the buffer. The increase in the strength of the interactions between the analyte and the constituents of the buffer may contribute to the improved separation observed in many non-aqueous systems. This improvement is usually attributed to the auto-protolytic properties of the solvent or differences in the solvation radii of the analytes [304].

2.4.4 Methodological aspects

With aqueous electrolyte solutions, very small internal diameter capillaries have been used to ensure effective dissipation of Joule heat. Less heat is produced in NACE because of the low electrical conductivities of the nonaqueous electrolyte solutions. Organic electrolyte solutions have somewhat lower thermal conductivities than water, which slightly increase the resistance to heat transfer but have only a minor effect on performance. Low heat production leads to enhanced separation efficiency owing to lowered longitudinal diffusion and more uniform temperature distribution inside the capillary. A negative temperature dependence of electrical conductivity was observed for some electrolyte solutions, which may also be advantageous for separation efficiency [327]. In NACE, the capillary diameter can be increased without an excessive rise in the electrolyte solution temperature. The heating power can be further reduced by appropriate choice of the CE solvent, for example by using longer chain alcohols [328].

As nonaqueous solutions have very low conductivity, conventional stacking techniques for preconcentration samples are not easily applied in NACE. The need for preconcentration is nevertheless as great in NACE as in water [308].

As mentioned above the probability that the analytes will adsorb on the capillary wall is reduced in nonaqueous solution, decreasing the need for coating the capillary to suppress analyte–wall interactions. However, it is recognized that the EOF in uncoated capillaries in NACE is often unrepeatable. Belder et al. [329,330] have used different coating materials (ethylene glycol and poly (vinyl) alcohol) and have manipulated the

EOF in these coated capillaries to improve resolution, analysis time, and repeatability in NACE and to optimize the separation of different kinds of compounds. Their studies showed that immobilized ethylene glycol easily interacts with the different cations present in nonaqueous BGEs, leading to an adjustable anodic EOF. Different electrolyte cations (K, Na, Li) favourably affected the direction of the EOF. Also, the magnitude of the anodic EOF depended on the coating thickness. A cathodic EOF was exceptionally observed with lithium ions.

In CE nonaqueous solvents provide higher resistance in the capillary than do conventional aqueous media, at least if the solvent has a much lower relative permittivity than water. Owing to strong ion association, the number of free ions is greatly reduced in such solvent systems. Palonen et al. [311,331] have studied the applicability of different alcohols (methanol, ethanol, 1-propanol, 1-butanol) to NACE separation of cationic and anionic analytes under high electric field strength (600–2000 V/cm). For cationic analytes in a negatively charged, uncoated fused capillary, a clear separation efficiency maximum was located above 30 kV for all the alcohol BGEs studied. The highest separation efficiency was achieved with 1-propanolic BGE with a maximum at 45 kV. The decreases in the separation efficiency with higher voltages were attributed to Joule heating in the case of methanol and ethanol and to analyte adsorption on the capillary wall with 1-propanol and 1-butanol BGE solution [331]. For anionic analytes, the best separation efficiencies were obtained with 1-propanol and the dependency of the efficiency on the separation voltage was linear [311]. 1-Propanol was a particularly suitable BGE solvent for operation at high separation potentials due to the low currents, and it provided reasonably fast and efficient separation despite the long preconditioning periods required. It seems clear that after careful optimization, alcohols with different chain lengths can serve as nonaqueous media under high electric field strengths, expanding the present range of separation parameters [308]. As an example, Fig. 2.40 shows the separation of acidic analytes in propanolic BGE with 15 mM sodium acetate and 15 mM acetic acid at two separation voltages (a) –30 kV and (b) –60 kV [311]. The figure illustrates that 1-propanol enables operation at high voltages with high separation efficiencies.

With respect to detection in NACE, Matysik has published two reviews [332,333] of the different forms of detection (UV, fluorescence, electrochemical and mass spectrometry) employed in NACE, UV detection is the most common. However, care must be taken with the

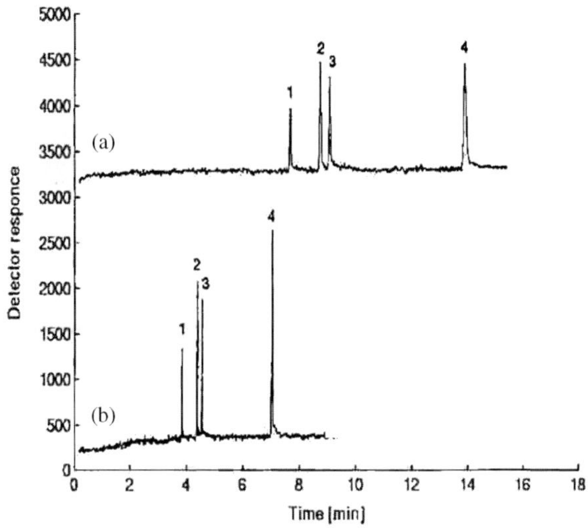

Fig. 2.40. Separation of acidic analytes in propanolic BGE with 15 mM sodium acetate and 15 mM acetic acid. Separation voltage (a) – 30 kV and (b) – 60 kV. Capillary: 50 μm ID, $L_{tot} = 30$ cm, $l = 28.5$ cm, UV detection at 214 nm, analyte concentration 10 μg/ml each, injection: −6 kV for 4 s; air flow thermostatting, 23°C. Peaks: 1, 2,4-dinitrophenol; 2, 3,5-dinitrobenzoic acid; 3, 2,4-dinitrobenzoic acid; 4, 2,6-dichlorobenzoic acid. Reprinted from Ref. [311]. Copyright (2002), with permission from Wiley-VCH Verlag GmbH.

low UV cutoff wavelengths of some organic solvents. Acetonitrile and methanol and their mixtures allow detection at wavelengths as low as 200 nm. In addition, the short optical path length for the light passing through the capillary limits the detection limits. Owing to the low currents in nonaqueous media, the capillary diameter can be increased without an excessive rise in the BGE temperature [326] and, owing to the larger sample load, the overall detection sensitivity can be increased.

2.4.5 Applications

The application of NACE in the pharmaceutical and environmental fields, and increasingly to real samples, has been shown. However, validated methods are needed for routine analysis, and hence more robust experiments are required to confirm the reliability and repeatability of NACE methods [308]. The major advantage of NACE over CZE is that it can be used for the separation of both charged and uncharged analytes.

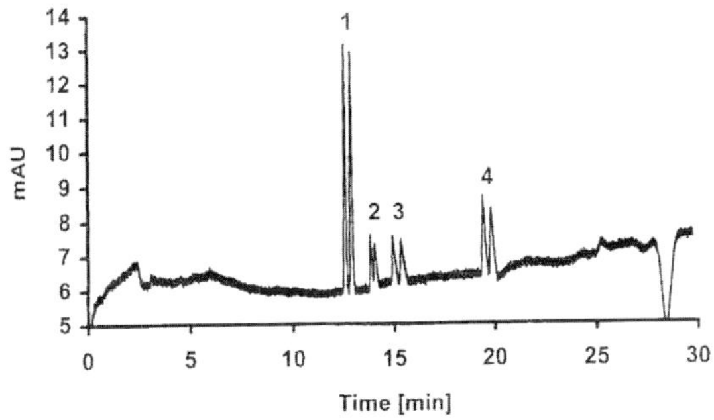

Fig. 2.41. Simultaneous chiral separation of (1) methadone, (2) ephedrine, (3) amphetamine, and (4) metoprolol using a 32 cm (24.5 cm to detector) × 50 μm ID fused silica capillary. Electrolyte solution: 75 mM carboxymethyl-β-cyclodextrin in formamide, injection: 20 kV for 2 s, $T = 40°C$, $V = 15$ kV; UV detection at 254 nm. Reprinted from Ref. [305]. Copyright (2000), with permission from Elsevier.

NACE has also been employed in chiral separations [334]. Many organic solvents facilitate ion-pair formation or ion-dipole interactions between the chiral selectors. This allows the utilization of chiral selectors that cannot be used in conventional aqueous BGEs. Quinine and its derivatives have been used and in this case the ion-pairing interaction between the positively charged quinine and anionic chiral analytes is the major intermolecular interaction responsible for the enantioseparation [308]. Native cyclodextrins and their derivatives have also been used as chiral selectors in NACE. Amide type organic solvents such as formamide are very good solvents for cyclodextrins [305]. Figure 2.41 shows the simultaneous chiral separation of (1) methadone, (2) ephedrine, (3) amphetamine and (4) metoprolol in formamide using carboxymethyl-β-cyclodextrin as chiral selector [305]. No chiral separation of these compounds was achieved in formamide using α-CD, β-CD, dimethyl-β-CD, trimethyl-β-CD or hydroxypropyl-β-CD, nor did chiral separation succeed in *N*-methylformamide.

Nonaqueous solvents have been used in CEC (NACEC) [335]. Nonaqueous studies have mostly included mobile phases consisting of acetonitrile or methanol or their mixtures. Other solvents such as dichloromethane, tetrahydrofuran, *n*-hexane and isopropanol have been added to increase both the solubility of the sample and the elution strength of the mobile phase. NACEC has been applied to enantioseparations using capillaries packed with chiral stationary phases. Comparisons between capillary liquid chromatography and NACEC have established some advantages of NACE for enantioseparations, namely high plate numbers under conditions with similar linear flow rates and better tolerability of high linear flow rates [336,337]. Figure 2.42 illustrates the CEC separation of thalidomide on a vancomycin chiral stationary phase in the polar organic mode (a) and in the reversed-phase mode (b) [338]. Efficiencies and resolution ranged from $N = 190,000$ plates/m and $Rs = 13.8$ in a polar organic mode to $N = 125,000$ plates/m and $Rs = 13.0$ in the reversed mode. Higher efficiencies, resolution, and shorts analysis times are obtained in NACEC compared to aqueous CEC.

2.5 SELECTION OF THE CE MODE AND SEPARATION CONDITIONS

The most appropriate separation modes to analyze different kind of compounds by CE are listed in Table 2.5 [51]. The recommended separation modes are those enabling the best results in the shortest analysis time. While for some species only one or two separation modes are recommended, in other cases many separation modes are possible. As an example, DNA analysis can be carried out by CGE and the separation of small ions by CZE and CITP. However, for small molecules, peptides, and proteins, at least four separation modes can give appropriate results. Finally, chiral separations are possible under very different experimental conditions. The first separation mode cited in each case in Table 2.5 is usually the best for each application.

Table 2.6 summarizes how the different parameters controlling the separation can be optimized assuming that the maximum electric field is applied (velocity, resolution, column charge capacity and sensitivity) [51]. It is also shown how optimizing one of these parameters may affect the other three, how these effects can be compensated and the impact that such compensation has on the separation.

ACKNOWLEDGMENTS

Drs. M. A. García and M. L. Marina thank the Ministry of Science and Technology (Spain) for the research project BQU2003-03638.

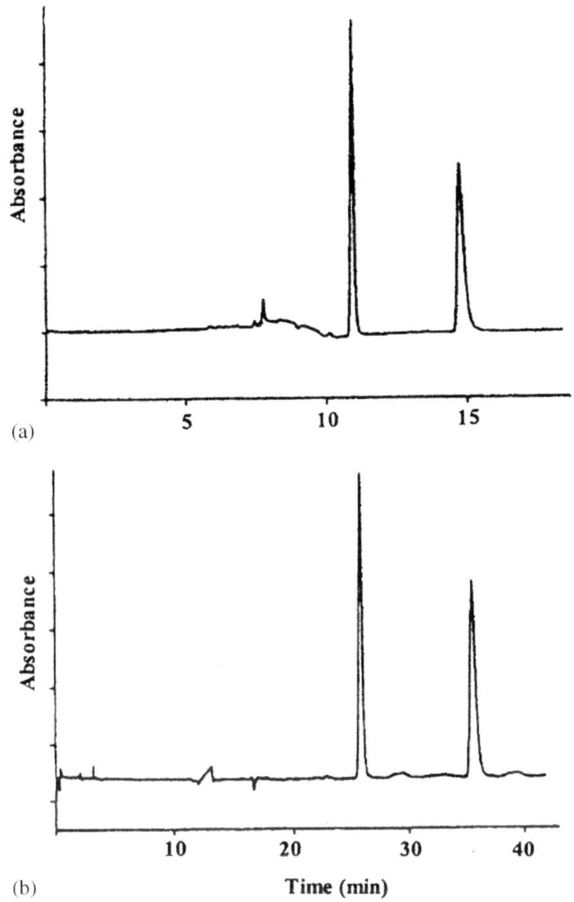

Fig. 2.42. Enantiomeric CEC separation of thalidomide on vancomycin chiral stationary phase (a) in the polar organic or (b) in the reversed-phase mode. Column, 5 μm Chiroviotic V (75 μm ID, 250 mm packed length, 335 mm total length). Conditions: (a) mobile phase, acetonitrile/MeOH/TEA/acetic acid (20:80:0.1:0.1, v/v), 20 kV, UV detection at 200 nm; (b) mobile phase, acetonitrile/0.1% TEA/acetic acid (pH 4) (30:70, v/v), UV detection at 214 nm. Reprinted from Ref. [338]. Copyright (2000), with permission from American Chemical Society.

TABLE 2.5

Selection of the CE mode

Small ions	Small molecules	Hydrophobic compounds	Chiral compounds	Peptides	Proteins	Nucleotides	DNA
CZE	MEKC	CD-EKC	CD-EKC	CZE	CZE	CGE	CGE
CITP	CEC	MEECK	MECK	MEKC	CIEF	MEKC	
	CZE	NACE	CEC	CIEF	CGE		
	CITP		MEECK	CGE	CITP		
			NACE	CITP			

Reprinted from Ref. [51], with permission from Springer-Verlag.

TABLE 2.6

Optimization of the parameters controlling the separation in CE

Parameter to be optimized	Variable to modify	Negative consequence of modification	Compensation to the modification	Disadvantage to the compensation
Velocity	Decrease the capillary length or increase electrical field	Higher Joule effect	Decrease capillary diameter	Lower sensitivity and lower charge capacity
	Decrease the ionic strength			Lower resolution and lower charge capacity
Resolution	Increase capillary length or ionic strength	Lower velocity	None	
	Decrease the amount of injected sample	Lower sensitivity	Increase of capillary diameter	-Higher Joule effect
Charge capacity	Increase capillary diameter or increase ionic strength	Higher Joule effect	Decrease electric field	Lower velocity
	Increase the amount of injected sample	Lower resolution	Increase capillary length	Lower velocity
Sensitivity	Increase capillary diameter	Higher Joule effect	Decrease electric field	Lower velocity
	Increase the amount of injected sample	Lower resolution	Increase ionic strength/increase capillary length	Higher Joule effect/ lower velocity

Reprinted from Ref. [51], with permission from Springer-Verlag.

REFERENCES

1 J.W. Jorgenson and K.D. Lukacs, *Anal. Chem.*, 53 (1981) 1298–1302.
2 S. Hjertén, *J. Chromatogr.*, 270 (1983) 1–6.
3 A.S. Cohen and B.L. Karger, *J. Chromatogr.*, 397 (1987) 409–417.
4 S. Terabe, K. Otsuka, K. Ichikawa, A. Tsuchiya and T. Ando, *Anal. Chem.*, 56 (1984) 111–113.
5 S. Terabe, *Trends Anal. Chem.*, 8 (1989) 129–134.
6 V. Pretorius, B.J. Hopkins and J.D. Schieke, *J. Chromatogr.*, 99 (1974) 23–30.
7 F.M. Evaerts, V. Sustacek and Th.T.E.M. Verheggen, *Isotacophoresis: Theory, Instrumentation and Applications*, Elsevier, Amsterdam, 1976.
8 F. Foret, E. Szoko and B.L. Karger, *J. Microcol. Sep.*, 2 (1990) 229–233.
9 S. Hjertén and M.D. Zhu, *J. Chromatogr.*, 346 (1985) 265–270.
10 S. Hjertén, J.L. Liao and K. Yao, *J. Chromatogr.*, 387 (1987) 127–138.
11 Y. Walbroehl and J.W. Jorgenson, *J. Chromatogr.*, 315 (1984) 135–143.
12 F. Foret, L. Krivankova and P. Boček. In: B.J. Radola (Ed.), *Capillary Zone Electrophoresis*. VCH, Weinheim, 1993.
13 P. Camilleri, *Capillary Electrophoresis: Theory and Practice*, 2nd ed., CRC Press, Boca Raton, Florida, 1998.
14 J.P. Landers, *Handbook of Capillary Electrophoresis*, CRC Press, Boca Raton, Florida, 1993.
15 M.G. Khaledi, *High Performance Capillary Electrophoresis: Theory, Techniques, and Applications*, Wiley, INC., New York, 1998.
16 J.W. Jorgenson, *Trends Anal. Chem.*, 3 (1984) 51–54.
17 J.H. Knox, *J. Chromatogr. A*, 680 (1994) 3–13.
18 M.J. Cugat, F. Borrull and M. Calull, *Trends Anal. Chem.*, 20 (2001) 487–499.
19 C.A. Lucy, *J. Chromatogr. A.*, 850 (1999) 319–337.
20 C. Cruces Blanco, *Electroforesis Capilar*, Servicio de Publicaciones Universidad de Almería, Almería, 1998.
21 I. Valverde Martín, Thesis: *Desarrollo de nuevas meodologías analíticas mediante electroforesis capilar*, Universidad de Barcelona, 2002.
22 K.D. Altria and D. Elder, *J. Chromatogr. A*, 1023 (2004) 1–14.
23 Z.K. Shihabi, *J. Liq. Chromatogr. Rel. Technol.*, 23 (2000) 79–95.
24 A. Pantazaki, M. Taverna and C. Vidal-Madjar, *Anal. Chim. Acta*, 383 (1999) 137–156.
25 W. Voelter, J. Shutz, O.E. Tsitsiloni, A. Weiler, G. Grubler, G. Paulus, S. Stoeva and R. Lechmann, *J. Chromatogr. A*, 807 (1998) 135–149.
26 D.N. Heiger, M. Herold and R. Grimn, *Applications of the HP^{3D} Capillary Electrophoresis System*, Hewlett-Packard GmbH, Germany, 1994.
27 A. Kumar Malik and W. Faubel, *Crit. Rev. Anal. Chem.*, 31 (2001) 223–279.

28 D.T. Eash and R.J. Bushway, *J. Chromatogr. A*, 880 (2000) 281–294.
29 Y. Picó, R. Rodríguez and J. Mañes, *Trends Anal. Chem.*, 22 (2003) 133–151.
30 F. Menzinger, Ph. Schmitt-Kopplin, D. Freitag and A. Kettrup, *J. Chromatogr. A*, 891 (2000) 45–67.
31 R. Rodríguez, Y. Picó, G. Font and J. Mañes, *J. Chromatogr. A*, 949 (2002) 359–366.
32 K.D. Altria, M.A. Nelly and B.J. Clark, *Trends Anal. Chem.*, 17 (1998) 204–214.
33 K.D. Altria, M.A. Nelly and B.J. Clark, *Trends Anal. Chem.*, 17 (1998) 214–216.
34 W.-C. Sung and S.-H. Chem, *Electrophoresis*, 22 (2001) 4244–4248.
35 M. Hernández, F. Borrull and M. Calull, *Trends Anal. Chem.*, 22 (2003) 416–427.
36 C.M. Boone, J.W. Douma, J.P. Franke, R.A. de Zeeuw and K. Ensing, *Forensic Sci. Int.*, 121 (2001) 89–96.
37 J.S. Fritz, *J. Chromatogr. A*, 884 (2000) 261–275.
38 A.H. Harakuwe and P.R. Haddad, *Trends Anal. Chem.*, 20 (2001) 375–385.
39 S.A. Steiner, J. Hooker, J. Dederich, B. Scott and J.S. Fritz, *J. Chromatogr. A*, 1039 (2004) 227–233.
40 S-J- Chen, M.-J. Chen and H.-T. Chang, *J. Chromatogr. A*, 1017 (2003) 215–224.
41 A.R. Timerbaev, *Talanta*, 52 (2000) 573–606.
42 A.R. Timerbaev, *Anal. Chim. Acta*, 4332 (2001) 165–180.
43 K. Pyrzynska, *Talanta*, 55 (2001) 657–667.
44 A.R. Timerbaev, A. Kung and B.K. Keppler, *J. Chromatogr. A*, 945 (2002) 25–44.
45 J-Z Wu and P.C. Ho, *J. Chromatogr. A*, 1026 (2004) 261–270.
46 M.J. Thornton and J.S. Fritz, *J. Chromatogr. A*, 770 (1997) 301–310.
47 A. Guttman, A.S. Cohen, D.N. Heiger and B.L. Karger, *Anal. Chem.*, 62 (1990) 137–141.
48 Y. Baba, T. Matsuura, K. Wakamoto, Y. Morita, Y. Nishitsu and M. Tsuhako, *Anal. Chem.*, 64 (1992) 1221–1225.
49 S.F.Y. Li, *Capillary Electrophoresis: Principles, Practice and Applications*, Elsevier, Amsterdam, 1992.
50 A. Guttman. In: J.P. Landers (Ed.), *Handbook of Capillary Electrophoresis*. CRC Press, Boca Raton, Florida, 1994.
51 M.V. Dabrío et al., *Cromatografía y Electroforesis en Columna*, cap. 8, Springer-Verlag Ibérica, Barcelona, 2000.
52 L. Mitnik, L. Salome, J.L. Viovy and C. Heller, *J. Chromatogr. A.*, 710 (1995) 309–321.
53 O.J. Lumpkin and B.H. Zimm, *Biopolymers*, 21 (1982) 2315–2316.

54 O.J. Lumpkin and P. Dejardin, *Biopolymers*, 24 (1985) 1573–1593.
55 A.E. Barron, H.W. Blanch and D.S. Soane, *Electrophoresis*, 15 (1994) 597–615.
56 D. Heiger, *High Performance Capillary Electrophoresis: An Introduction*, Agilent Technologies, Germany, 2000.
57 A. Guttman, B. Wanders and N. Cooke, *Anal. Chem.*, 64 (1992) 2348–2351.
58 A. Von Broke, T. Freudemann and E. Bayer, *J. Chromatogr. A*, 991 (2003) 129–141.
59 A. Guttman, L. Shi, J. Khandurina and X. Wang, *J. Chromatogr. A*, 1014 (2003) 29–35.
60 H.G. Lee, S. Chang and E. Fritsche, *J. Chromatogr. A*, 947 (2002) 143–149.
61 H.G. Lee and E. Fritsche, *J. Chromatogr. A*, 994 (2003) 213–219.
62 D.N. Heiger, A.S. Cohen and B.L. Karger, *J. Chromatogr. A*, 516 (1990) 33–48.
63 X. Liu, Z. Sosic and S.I. Krull, *J. Chromatogr. A*, 735 (1996) 165–190.
64 R. Rodríguez-Diaz, T. Wehr and M.D. Zhu, *Electrophoresis*, 18 (1997) 2134–2144.
65 P.G. Righetti, C. Gelfi and M. Conti, *J. Chromatogr. B*, 699 (1997) 91–104.
66 J.C. Giddings, *Sep. Sci. Technol.*, 14 (1979) 871–875.
67 Q.L. Mao and J. Pawliszyn, *J. Biochem. Biophys. Methods*, 39 (1999) 93–110.
68 X.-Z. Wu, J. Wu and J. Pawliszyn, *LC-GC*, 19 (2001) 526–545.
69 P.G. Righetti, *J. Chromatogr. A*, 1037 (2004) 491–499.
70 J.R. Mazzeo and I.S. Krull, *Anal. Chem.*, 63 (1991) 2852–2857.
71 J.R. Mazzeo and I.S. Krull, *J. Chromatogr. A*, 606 (1992) 291–296.
72 J. Wu, S.-C Li and A. Watson, *J. Chromatogr. A*, 817 (1998) 163–171.
73 J.Q. Wu and J. Pawliszyn, *Am. Lab.*, 26 (1994) 48–52.
74 A.J.S. Jones, *Adv. Drug Delivery Rev.*, 10 (1993) 29–90.
75 C. Schwer, *Electrophoresis*, 16 (1995) 2121–2126.
76 M. Conti, C. Gelfi, A.B. Bosisio and P.G. Righetti, *Electrophoresis*, 17 (1996) 1590–1596.
77 C. Schwer, *Electrophoresis*, 16 (1995) 2121–2126.
78 K. Shimura and B.L. Karger, *Anal. Chem.*, 66 (1994) 9–15.
79 N.T. Tran, Y. Daali, S. Cherkaoui, M. Taverna, J.R. Neeser and J.-L. Veuthey, *J. Chromatogr. A*, 929 (2001) 151–163.
80 M. Hashimoto, K. Tsukagoshi, R. Nakajima and K. Kondo, *J. Chromatogr. A*, 852 (1999) 597–601.
81 P.G. Righetti, *J. Chromatogr. A*, 1037 (2004) 491–499.
82 P. Boček, M. Deml, P. Gebauer and V. Dolnik, *Analytical Isotachophoresis*, VCH, Weinhein, 1988.

83 Y. Xu, *The Chemical Educator*, 1 (1996) 1–14.
84 J. Sádecká and J. Polonský, *J. Chromatogr. A*, 988 (2003) 161–165.
85 P. Gebauer and P. Boček, *Electrophoresis*, 21 (2000) 3898–3904.
86 P. Gebauer and P. Boček, *Electrophoresis*, 23 (2002) 3858–3864.
87 P. Gebauer and P. Boček, *Electrophoresis*, 21 (2000) 3898–3904.
88 L. Krinvánková, P. Gebauer and P. Boček, *J. Chromatogr. A*, 716 (1995) 35–48.
89 F. Foret, E. Szoco and B.L. Karger, *Electrophoresis*, 14 (1993) 417–428.
90 Z.K. Shihabi, *J. Chromatogr. A*, 902 (2000) 107–117.
91 K. Otsuka, S. Terabe and T. Ando, *J. Chromatogr. A*, 348 (1985) 39–47.
92 P.G. Muijselaar, H.A. Classens and C.A. Cramers, *J. Chromatogr. A*, 765 (1997) 295–306.
93 C. Quang, J.K. Strasters and M.G. Khaledi, *Anal. Chem.*, 66 (1994) 1646–1653.
94 S.P. Quirino and S. Terabe, *J. Chromatogr. A*, 856 (1999) 465–482.
95 U. Pyell, *J. Chromatogr. A*, 1037 (2004) 479–490.
96 S. Terabe, K. Otsuka and T. Ando, *Anal. Chem.*, 57 (1985) 834–841.
97 C.-X. Zhang, Z.P. Sun and D.-K. Ling, *J. Chromatogr. A*, 655 (1993) 309–316.
98 S. Takeda, S. Wikida, M. Yamane, K. Higashi and S. Terabe, *J. Chromatogr. A*, 744 (1996) 135–139.
99 S. Takeda, S. Wikida, M. Yamane, K. Higashi and S. Terabe, *J. Chromatogr. A*, 781 (1997) 11–16.
100 S. Takeda, S. Wikida, M. Yamane, Z. Siroma, K. Higashi and S. Terabe, *J. Chromatogr. A*, 817 (1998) 59–63.
101 N. Matsubara and S. Terabe, *Chromatographia*, 34 (1992) 493–496.
102 K. Koezuka, H. Ozaki, N. Matsubara and S. Terabe, *J. Chromatogr. B*, 689 (1997) 3–11.
103 H.K. Kristensen and S.H. Hansen, *J. Chromatogr.*, 628 (1993) 309–315.
104 C.E. Lin, Y.C. Chem, C.C. Chang and D.Z. Wang, *J. Chromatogr. A*, 775 (1997) 349–357.
105 C.P. Palmer, *J. Chromatogr. A*, 780 (1997) 75–92.
106 K.T. Tellman and C.P. Palmer, *Electrophoresis*, 20 (1999) 152–161.
107 S.A. Shamsi, C. Akbay and I.M. Warner, *Anal. Chem.*, 70 (1998) 3078–3083.
108 S.Y. Yang, J.G. Bumgarner and M.G. Khaledi, *J. High Resolut. Chromatogr.*, 18 (1995) 443–445.
109 J.T. Smith, W. Nashabeth and Z. El Rassi, *Anal. Chem.*, 66 (1994) 1119–1133.
110 J.T. Smith and Z. El Rassi, *J. Microcol. Sep.*, 6 (1994) 127–138.
111 J.T. Smith and Z. El Rassi, *J. Chromatogr. A*, 685 (1994) 131–143.
112 Y. Mechref and Z. El Rassi, *J. Chromatogr. A*, 724 (1996) 285–296.

113 S. Terabe, N. Matsubara, Y. Ishihama and Y. Okada, *J. Chromatogr.*, 608 (1992) 23–29.
114 K.D. Altria, P.-E. Mahuzier and B.J. Clark, *Electrophoresis*, 24 (2003) 315–324.
115 K.D. Altria, *J. Chromatogr. A*, 892 (2000) 171–186.
116 H. Nishi and S. Terabe, *J. Chromatogr. A*, 694 (1995) 245–276.
117 H. Nishi, *J. Chromatogr. A*, 735 (1996) 57–76.
118 R. Vespalec and P. Boček, *Electrophoresis*, 18 (1997) 843–852.
119 G. Gübitz and M.G. Schmid, *J. Chromatogr. A*, 792 (1997) 179–225.
120 S. Terabe, H. Ozaki, K. Otsuka and T. Ando, *J. Chromatogr.*, 332 (1985) 211–217.
121 J.B. Vicent, D.M. Kirby, T.V. Nguyen and G. Vigh, *Anal. Chem.*, 69 (1997) 4419–4428.
122 M. Hong, B.S. Weekly, S.J. Greib and J.P. Foley, *Anal. Chem.*, 70 (1998) 1394–1403.
123 K. Baechmann, A. Bazzanella, K.Y. Han, R. Arnecke, V. Bohmer and W. Vogt, *Anal. Chem.*, 67 (1995) 1722–1726.
124 N. Tanaka, T. Tanigawa, K. Hosoya, K. Kimata, T. Arai and S. Terabe, *Chem. Lett.* (1992) 959–962.
125 N. Tanaka, T. Fukutome, K. Hosoya, K. Kimata and T. Araki, *J. Chromatogr. A*, 716 (1995) 57–67.
126 S. Terabe and T. Isemura, *Anal. Chem.*, 62 (1990) 650–652.
127 N. Tanaka, K. Nakagawa, H. Iwasaki, K. Hosoya, K. Kimata, T. Araki and D.G. Patterson, *J. Chromatogr. A*, 781 (1997) 139–150.
128 D.K. Lloyd, A.F. Aubry and E.D. Lorenzi, *J. Chromatogr. A*, 792 (1997) 349–369.
129 D.W. Armstrong and U.B. Nair, *Electrophoresis*, 18 (1997) 2331–2342.
130 T.J. Ward and A.B. Farris III, *J. Chromatogr. A*, 906 (2001) 73–89.
131 G.M. Janini, G.M. Muschik and H.J. Issaq, *J. Chromatogr. A*, 792 (1997) 125–141.
132 G.M. Janini, G.M. Muschik and H.J. Isaaq, *Electrophoresis*, 17 (1996) 1575–1583.
133 U. Pyell, *J. Chromatogr. A*, 1037 (2004) 479–490.
134 E.S. Ahuja, E.L. Little and J.P. Foley, *J. Liq. Chromatogr.*, 15 (1992) 1099–1113.
135 E. Fuguet, C. Ràfols, E. Bosch and M. Rosés, *Electrophoresis*, 23 (2002) 56–66.
136 M.M. Bushey and J.W. Jorgenson, *Anal. Chem.*, 61 (1989) 491–493.
137 J. Vindevogel and P. Sandra, *Introduction to Micellar Electrokinetic Chromatography*, Hüthig, Heidelberg, 1992.
138 S. Terabe, *Micellar Electrokinetic Chromatography*, Beckman, California, 1993.
139 K. Otsuka and S. Terabe, *Bull. Chem. Soc. Jpn.*, 71 (1998) 2465–2481.

140 P.G. Muijselaar, K. Otsuka and S. Terabe, *J. Chromatogr. A*, 780 (1997) 41–61.
141 M.G. Khaledi, *J. Chromatogr. A*, 780 (1997) 3–40.
142 U. Pyell, *Fresenius J. Anal. Chem.*, 371 (2001) 691–703.
143 K. Otsuka and S. Terabe. In: K. Altria (Ed.), *Micellar Electrokinetic Chromatography in Methods in Molecular Biology*, Vol. 52, Humana Press, Inc., Totowa, NJ, 1996, pp. 125–155 Chapter 12.
144 J.P. Foley, *Anal. Chem.*, 62 (1990) 1302–1308.
145 C.-X. Zhang, Z.-P. Sun and D.-K. Ling, *J. Chromatogr. A*, 655 (1993) 309–316.
146 S. Terabe, K. Otsuka and T. Ando, *Anal. Chem.*, 61 (1989) 251–260.
147 H. Nishi, T. Fukuyama, M. Matsuo and S. Terabe, *J. Chromatogr.*, 513 (1990) 279–295.
148 P.G. Muijeselaar, K. Otsuka and S. Terabe, *J. Chromatogr. A*, 780 (1997) 41–61.
149 M.G. Khaledi, *J. Chromatogr. A*, 780 (1997) 3–40.
150 A. Dworschak and U. Pyell, *J. Chromatogr. A*, 848 (1999) 387–400.
151 A. Dworschak and U. Pyell, *J. Chromatogr. A*, 855 (1999) 669–679.
152 E.S. Ahuja and J.P. Foley, *Anal. Chem.*, 67 (1995) 2315–2324.
153 K.R. Nielsen and J.P. Foley, *J. Microcol. Sep.*, 5 (1993) 347–360.
154 H. Ozaki and S. Terabe, *J. Chromatogr. A*, 794 (1998) 317–325.
155 C. Fujimoto, Y. Fujise and S. Kawaguchi, *J. Chromatogr. A*, 871 (2000) 415–425.
156 X. Liu, L. Ma and Y-T. Lu, *Anal. Chim. Acta*, 512 (2004) 297–304.
157 S. Terabe, T. Katsura, Y. Okada, Y. Ishihama and K. Otsuka, *J. Microcol. Sep.*, 5 (1993) 23–33.
158 M.L. Riekkola, S.K. Wiedmer, I.E. Valkó and H. Sirén, *J. Chromatogr. A*, 792 (1997) 13–35.
159 C.-E. Lin and M-J. Chen, *J. Chromatogr. A*, 923 (2001) 241–248.
160 S. Terabe, Y. Miyashita, O. Shibata, E.R. Barnhart, L.R. Alexander, D.G. Patterson, B.L. Karger, K. Hosoya and N. Tanaka, *J. Chromatogr.*, 516 (1990) 23–31.
161 E. van Hove, R. Szücs and P. Sandra, *J. High Resol. Chromatogr.*, 19 (1996) 674–678.
162 S. Terabe, Y. Miyashita, Y. Ishihama and O. Shibata, *J. Chromatogr.*, 636 (1993) 47–55.
163 B. Jiménez, D.G. Patterson, J. Grainger, Z. Liu, M.J. González and M.L. Marina, *J. Chromatogr. A*, 792 (1997) 411–418.
164 K. Jinno and Y. Sawada, *J. Liq. Chromatogr.*, 18 (1995) 3719–3727.
165 I. Benito, J.M. Saz, M.L. Marina, J. Jiménez-Barbero, M.J. González and J.C. Díez–Masa, *J. Chromatogr. A*, 778 (1997) 77–85.
166 K. Otsuka and S. Terabe, *J. Chromatogr. A*, 875 (2000) 163–178.
167 H. Nishi and S. Terabe, *J. Chromatogr. A*, 735 (1996) 3–27.

168 H. Nishi, *J. Chromatogr. A*, 780 (1997) 243–264.
169 X. Páez, P. Rada and L. Hernández, *J. Chromatogr. B*, 739 (2000) 247–254.
170 I. Benito, M.A. García, J.C. Díez Masa and M.L. Marina, *J. Microcol. Sep.*, 9 (1997) 87–95.
171 I. Lurie, *J. Chromatogr. A*, 780 (1997) 265–284.
172 L. Song, Z. Xu, J. Kang and J. Cheng, *J. Chromatogr. A*, 780 (1997) 297–328.
173 S. Fías, M.J. Sánchez and M.A. Rodríguez, *Anal. Chim. Acta*, 503 (2004) 271–278.
174 R.D. Holland and M.J. Sepaniak, *Anal. Chem.*, 65 (1993) 1140–1146.
175 K. Jinno and Y. Sawada, *J. Cap. Elec.*, 1 (1994) 106–111.
176 U. Krismann and W. Kleiböhmer, *J. Chromatogr. A*, 774 (1997) 193–201.
177 A.L. Crego, M.J. González and M.L. Marina, *Electrophoresis*, 19 (1998) 2113–2118.
178 H. Nishi, *J. High Resol. Chromatogr.*, 18 (1995) 659–664.
179 R.O. Cole, M.J. Sepaniak and W.L. Hinze, *J. High Resol. Chromatogr.*, 13 (1990) 579–582.
180 M. Lin, N. Wu, G.E. Barker, P. Sun, C.W. Hue and R.A. Hartwick, *J. Liq. Chromatogr.*, 16 (1993) 3667–3674.
181 K. Otsuka, M. Kashihara, Y. Kawaguchi, R. Koike, T. Hisamitsu and S. Terabe, *J. Chromatogr. A*, 652 (1993) 253–257.
182 Y. Ishihama and S. Terabe, *J. Liq. Chromatogr.*, 16 (1993) 933–944.
183 J. Wang and I.M. Warner, *Anal. Chem.*, 66 (1994) 3773–3776.
184 S.A. Shamsi and I.M. Warner, *Electrophoresis*, 18 (1997) 853–872.
185 P. Schmitt, A.W. Garrison, D. Freitag and A. Kettrup, *J. Chromatogr. A*, 792 (1997) 419–429.
186 M.L. Marina, I. Benito, J.C. Díez-Masa and M.J. González, *J. Chromatogr. A*, 752 (1996) 265–270.
187 M.L. Marina, I. Benito, J.C. Díez-Masa and M.J. González, *Chromatographia*, 42 (1996) 269–272.
188 A.L. Crego, M.A. García and M.L. Marina, *J. Microcol. Sep.*, 12 (2000) 33–40.
189 R. Furuta and T. Doi, *Electrophoresis*, 15 (1994) 1322–1325.
190 Y. Kurosu, K. Murayama, N. Shindo, Y. Shisa, Y. Satou and N. Ishioka, *J. Chromatogr. A*, 771 (1997) 311–317.
191 H. Wan, A. Engströn and L.G. Blomberg, *J. Chromatogr. A*, 731 (1996) 283–292.
192 J-B. Kim and S. Terabe, *J. Pharm. Biomed. Anal.*, 20 (2003) 1625–1643.
193 I.V. Berezín, K. Martinek and A.Y. Yarsimirskii, *Russ. Chem. Rev. (Engl. Trans)*, 42 (1973) 787–808.
194 M.A. García, J.C. Díez-Masa and M.L. Marina, *J. Chromatogr. A*, 732 (1996) 345–359.

195 C. García-Ruíz, M.A. García and M.L. Marina, *Electrophoresis*, 21 (2000) 2424–2431.
196 M.L. Marina and M.A. García, *J. Chromatogr. A*, 780 (1997) 103–116.
197 D.W. Armstrong and U.B. Nair, *Electrophoresis*, 18 (1997) 2331–2342.
198 M.A. García, J.C. Díez-Masa and M.L. Marina, *J. Chromatogr. A*, 742 (1996) 251–256.
199 R. Kaliszan, *Anal. Chem.*, 64 (1992) 619A–631A.
200 B.J. Herbert and J.G. Dorsey, *Anal. Chem.*, 67 (1995) 744–749.
201 Dummy
202 C. Treiner and A.K. Chattopadhyay, *J. Colloid Interface Sci.*, 109 (1986) 101–108.
203 R. Collander, *Acta Chem. Scand.*, 5 (1951) 774–780.
204 L. Escuder-Gilabert, J.M. Sanchis-Mallols, S. Sagrado, M.J. Medina Hernández and R.M. Villanueva-Camañas, *J. Chromatogr. A*, 823 (1998) 549–559.
205 B. Chankvetadze and G. Blaschke, *J. Chromatogr. A*, 906 (2001) 309–363.
206 S. Fanali, *J. Chromatogr. A*, 792 (1997) 227–267.
207 S. Fanali, *J. Chromatogr. A*, 875 (2000) 89–122.
208 M. Blanco and I. Valverde, *Trends Anal. Chem.*, 22 (2003) 428–439.
209 C. García-Ruíz, Thesis: *Aplicación de la Cromatografía Electrocinética a la separación de compuestos hidrófobos y a la separación quiral de compuestos de interés farmacológico y medioambiental*, Universidad de Alcalá, 2001.
210 B.C. Lin, Y.B. Ji, Y.Y. Chen, U. Epperlein and B. Koppenhoefer, *Chromatographia*, 42 (1996) 106–110.
211 C.Y. Quang and M.G. Khaledi, *J. Chromatogr. A*, 692 (1995) 253–265.
212 C.J. Easton and S.F. Lincoln, *Chem. Soc. Rev.*, 25 (1996) 163–171.
213 C. García-Ruíz and M.L. Marina, *Electrophoresis*, 22 (2001) 3191–3197.
214 D.W. Armstrong, L.W. Chang and S.S.C. Chang, *J. Chromatogr. A*, 793 (1998) 115–134.
215 J.W. Kang, X.M. Zhang, S.M. Zhang and Q.Y. Ou, *Chromatographia*, 50 (1999) 317–320.
216 K. Otsuka, C.J. Smith, J. Grainger, J.R. Barr, D.G. Patterson, N. Tanaka Jr. and S. Terabe, *J. Chromatogr. A*, 817 (1998) 75–81.
217 E. Ekiert, C. García-Ruíz, M.A. García and M.L. Marina, *Electrophoresis*, 24 (2003) 2680–2686.
218 H.Y. Aboul-Enein, M.D. Efstatiade, B. Christian, B. Lin, Y. Ji and Y. Chem, *J. Chromatogr. A*, 20 (1999) 2686–2690.
219 Y.F. Tang, H.L. Wu, S.M. Wu, S.H. Chen and H.S. Kou, *J. Microcol. Sep.* 12 (2000) 366–370.
220 K.H. Gahm, L.W. Chang and D.W. Armstrong, *J. Chromatogr. A*, 759 (1997) 149–155.

221 H. Jakubetz, M. Juza and V. Schurig, *Electrophoresis*, 18 (1997) 897–904.
222 O. Zerbinati, F. Trotta and C. Giovannoli, *J. Chromatogr. A*, 875 (2000) 423–430.
223 A. Bunke and T. Jira, *J. Chromatogr. A*, 798 (1998) 275–280.
224 F. Wang and M.G. Khaledi, *J. Chromatogr. A*, 817 (1998) 121–128.
225 F. Lelièvre, C. Guiet, P. Gareil Y. Bahaddi and H. Galons, *Electrophoresis*, 18 (1997) 891–896.
226 M. Fillet, Ph. Hubert and J. Crommen, *J. Chromatogr. A*, 875 (2000) 123–134.
227 Y. Martín-Biosca, C. García-Ruíz and M.L. Marina, *Electrophoresis*, 22 (2001) 3216–3225.
228 S. Rudaz, J.L. Veuthey, C. Desiderio and S. Fanali, *Chromatographia*, 50 (1999) 369–372.
229 F. Lelièvre, P. Gareil and Y. Bahaddi, *Anal. Chem.*, 69 (1997) 393–401.
230 K.L. Sutton, R.M.C. Sutton, A.M. Scalcup and J.A. Caruso, *Analyst*, 125 (2000) 231–234.
231 B. Chankvetadze, G. Endresz and G. Blaschke, *J. Chromatogr. A*, 700 (1995) 43–49.
232 S.I. Izumoto and H. Hishi, *Electrophoresis*, 20 (1999) 189–197.
233 P. Kumar and K.L. Mittal, *Handbook of Microemulsions Science and Technology*, Marcel Dekker, New York, 1999.
234 F. Sicoli and D. Langevin, *J. Phys. Chem.*, 99 (1995) 14819–14823.
235 H. Watarai, *Chem. Lett.* (1991) 391–394.
236 M.F. Miola, M.J. Snowden and K.D. Altria, *J. Pham. Biomed. Anal.*, 18 (1998) 785–797.
237 Y. Isahihama, Y. Oda, K. Uchikawa and N. Asakawa, *Anal. Chem.*, 67 (1995) 1588–1595.
238 M.H. Abraham, C. Treiner, M. Roses, C. Rafols and Y. Ishihama, *J. Chromatogr. A*, 752 (1996) 243–249.
239 S.J. Gluck, M.H. Benkö, R.K. Haliberg and K.P. Steele, *J. Chromatogr. A*, 744 (1996) 141–146.
240 I. Miksik and Z. Deyl, *J. Chromatogr. A*, 807 (1998) 111–119.
241 H. Watarai and I. Takahashi, *Anal. Commun.*, 35 (1998) 289–292.
242 C.W. Klampfi, T. Leitner and E.F. Hilder, *Electrophoresis*, 23 (2002) 2424–2429.
243 E.F. Hilder, C.W. Klampfi, W. Buchberger and P.R. Haddad, *J. Chromatogr. A*, 922 (2001) 293–302.
244 K.D. Altria, B.J. Clark and P.-E. Mahuzier, *Chromatographia*, 52 (2000) 758–768.
245 S. Pedersen-Bjergaard, C. Gabel-Jensen and S.H. Hansen, *J. Chromatogr. A*, 897 (2000) 375–381.
246 X. Fu and J. Lu, *J. Chromatogr. A*, 735 (1996) 353–356.

247 R.L. Boso, M.S. Bellini, I. Miksik and Z. Deyl, *J. Chromatogr. A*, 709 (1995) 11–19.
248 G. Li, Z. Chen, M. Liu and Z. Hu, *Analyst*, 123 (1998) 1501–1505.
249 C.A. Fung-Kee-Fung and S. Post, *J. Liq. Chromatogr. Rel. Technol.*, 24 (2001) 1133–1151.
250 R. Pascoe and J.P. Foley, *Analyst*, 127 (2002) 710–714.
251 J.H. Aiken and C.W. Huie, *Chromatographia*, 35 (1993) 448–450.
252 Y. Ishihama, Y. Oda and N. Asakawa, *Anal. Chem.*, 68 (1996) 1028–1032.
253 Y. Ishihama, Y. Oda and N. Asakawa, *Anal. Chem.*, 68 (1996) 4281–4284.
254 Y. Mrestani, R.H.H. Neubert and A. Krause, *Pharm. Res.*, 15 (1998) 799–801.
255 S.J. Gluck, M.H. Benkö, R.K. Hallberg and K.P. Steele, *J. Chromatogr. A*, 744 (1996) 141–146.
256 W.L. Klota, M.R. Schure and J.P. Foley, *J. Chromatogr. A*, 930 (2001) 145–154.
257 J.S. Green and J.W. Jorgenson, *J. Chromatogr. A*, 478 (1989) 63–70.
258 J. Vindevogel and P. Sandra, *Electrophoresis*, 15 (1994) 842–847.
259 J.W. Jorgenson and K.D. Lukacs, *J. Chromatogr.*, 218 (1981) 209–216.
260 E. Grushka, R.M. McCormick and J.J. Kirkland, *Anal. Chem.*, 61 (1989) 241–246.
261 H.H. Strain, *J. Am. Chem. Soc.*, 61 (1939) 1292–1293.
262 T.S. Stevens and J.H. Cortes, *Anal. Chem.*, 55 (1983) 1365–1370.
263 J.H. Knox and I.H. Grant, *Chromatographia*, 24 (1987) 135–143.
264 J.H. Knox and I.H. Grant, *Chromatographia*, 32 (1991) 317–323.
265 M.M. Dittman, K. Wienand, F. Bek and G.P. Rozing, *LC–GC Int.*, 13 (1995) 800–807.
266 A.L. Crego, A. González and M.L. Marina, *Crit. Rev. Anal. Chem.*, 26 (1996) 261–304.
267 M.M. Robson, M.G. Cikalo, P. Myers, M.R. Euerby and K.D. Bartie, *J. Microcol. Sep.*, 9 (1997) 357–372.
268 A. Dermaux and P. Sandra, *Electrophoresis*, 20 (1999) 3027–3065.
269 K.D. Bartle and P. Myers, *J. Chromatogr. A*, 916 (2001) 3–23.
270 G. Vanhoenacker, T. Van den Bosch, G. Rozing and P. Sandra, *Electrophoresis*, 22 (2001) 4064–4103.
271 N. Smith, *Capillary Electrochromatography*, Beckman Coulter, Inc., Fullerton, California, 1998.
272 K.D. Bartle and P. Myers (Eds.), *Capillary Electrochromatography*. Royal Society of Chemistry, Cambridge, 2001.
273 M. Pursch and L.C. Sander, *J. Chromatogr. A*, 887 (2000) 313–326.
274 R. Freitag, *J. Chromatogr. A*, 1033 (2004) 267–273.
275 D. Allen and Z. El Rassi, *J. Chromatogr. A*, 1029 (2004) 239–247.
276 B. Behnke, E. Grom and E. Bayer, *J. Chromatogr. A*, 716 (1995) 207–213.

277 K. Jinno and H. Sawada, *Trends Anal. Chem.*, 19 (2000) 664–675.
278 A.L. Crego, J. Martínez and M.L. Marina, *J. Chromatogr. A*, 869 (2000) 329–337.
279 A.L. Crego, J. Martínez and M.L. Marina, *HRC–J. High Resolut. Chromatogr.*, 23 (2000) 373–378.
280 H. Zou, X. Huang, M. Ye and Q. Luo, *J. Chromatogr. A*, 954 (2002) 5–32.
281 Z. Ji, R.E. Majors and E.J. Guthrie, *J. Chromatogr. A*, 842 (1999) 115–142.
282 J.J. Pesek and M.T. Matyska, *J. Chromatogr. A*, 887 (2000) 31–41.
283 M.M. Dittman and G.P. Rozing, *J. Microcol. Sep.*, 5 (1997) 399–408.
284 F. Steiner and B. Scherer, *J. Chromatogr. A*, 887 (2000) 55–83.
285 T. Tsuda, *Anal. Chem.*, 59 (1987) 521–523.
286 J.H. Knox, *Chromatographia*, 26 (1988) 329–335.
287 A.S. Rhathore and Cs. Horváth, *J. Chromatogr. A*, 743 (1996) 231–246.
288 A.L. Crego, J.C. Díez-Masa and M.V. Dabrio, *Anal. Chem.*, 65 (1993) 1615–1621.
289 CEC Guidebook. In: G.A. Ross and G.P. Rozing (Eds.), *Capillary Electrochomatography – Technology and Applications*. Hewlett-Packard, GmbH, Germany, 1999.
290 W. Shuetzner and E. Kenndler, *Anal. Chem.*, 64 (1991) 1991–1995.
291 Y. Guo and L. Colón, *Anal. Chem.*, 67 (1995) 2511–2515.
292 S. Li and D. Lloyd, *Anal. Chem.*, 65 (1995) 3684–3690.
293 A. Banholczer and U. Pyell, *J. Chromatogr. A*, 869 (2000) 363–374.
294 H. Rebscher and U. Pyell, *Chromatographia*, 38 (1994) 737–743.
295 M.G. Cikalo, K.D. Bartle and P. Myers, *J. Chromatogr. A*, 836 (1999) 35–51.
296 D. Norton, J. Zheng and S.A. Shamsi, *J. Chromatogr. A*, 1008 (2003) 205–215.
297 A. Banholczer and U. Pyell, *J. Sep. Sci.*, 24 (2001) 736–742.
298 Z. Liu, H. Zou, M. Le, J. Ni and Y. Zhang, *Electrophoresis*, 20 (1999) 2898–2908.
299 B. Chankvetadze, I. Kartozia, Y. Okamoto and G. Blaschke, *J. Sep. Sci.*, 24 (2001) 635–642.
300 M.C. Breadmore, E.F. Hilder, M. Macka and P.R. Haddad, *Trends Anal. Chem.*, 20 (2001) 355–364.
301 K. Otsuka, C. Mikami and S. Terabe, *J. Chromatogr. A*, 887 (2000) 457–463.
302 S. Fanali, P. Catarcini, G. Blaschke and B. Chankvetadze, *Electrophoresis*, 22 (2001) 3131–3151.
303 J.T. Lim, R.N. Zare, C.G. Bailey, D.J. Rakestraw and C. Yan, *Electrophoresis*, 21 (2000) 737–742.
304 I.E. Valkó, H. Sirén and M.-L. Riekkola, *LC–GC*, 15 (1997) 560–567.

305 M.L. Riekkola, M. Jussila, S.P. Porras and I.E. Valkó, *J. Chromatogr. A*, 892 (2000) 155–170.
306 F. Steiner and M. Hassell, *Electrophoresis*, 21 (2000) 3994–4016.
307 B. Chanckvetadze and G. Blaschke, *Electrophoresis*, 21 (2000) 4159–4178.
308 M.L. Riekkola, *Electrophoresis*, 23 (2002) 3865–3883.
309 M.T. Bowser, A.R. Kranack and D.D.Y. Chen, *Trends Anal. Chem.*, 17 (1998) 424–433.
310 T. Okada, *J. Chromatogr. A*, 771 (1997) 275–284.
311 S. Palonen, M. Jussila, S.P. Porras, T. Hyötyläinen and M.-L. Riekkola, *Electrophoresis*, 23 (2002) 393–399.
312 S.J.O. Varjo, M. Jussila, M. Palonen and M.-L. Riekkola, *Electrophoresis*, 23 (2002) 437–441.
313 J. Muzikár, T. Van der Goor, B. Gaš and E. Kenndler, *Electrophoresis*, 23 (2002) 375–382.
314 J. Muzikár, T. Van der Goor, B. Gaš and E. Kenndler, *Anal. Chem.*, 74 (2002) 428–433.
315 J.M. Jordan, R.L. Moeses, R. Johnson–Watts and D.E. Burton, *J. Chromatogr. A*, 671 (1994) 445–451.
316 M. Grob and F. Steiner, *Electrophoresis*, 23 (2002) 1853–1861.
317 K. Sarmini and E. Kenndler, *J. Biochem. Biophys. Methods*, 38 (1999) 123–137.
318 S.P. Porras, M.-L. Riekkola and E. Kenndler, *J. Chromatogr. A*, 905 (2001) 259–268.
319 S.P. Porras, P. Jyske, M.-L. Riekkola and E. Kenndler, *J. Microcol. Sep.*, 13 (2001) 149–155.
320 S.P. Porras, M.-L. Riekkola and E. Kenndler, *Electrophoresis*, 22 (2001) 3798–3804.
321 S.P. Porras, M.L. Riekkola and E. Kenndler, *Chromatographia*, 53 (2001) 290–294.
322 S.P. Porras, M.L. Riekkola and E. Kenndler, *J. Chromatogr., A*, 924 (2001) 31–42.
323 S.P. Porras, M.L. Riekkola and E. Kenndler, *Electrophoresis*, 23 (2002) 367–374.
324 S.P. Porras and E. Kenndler, *J. Chromatogr. A*, 1037 (2004) 455–465.
325 J. Muzikár, T. van der Goor and E. Kenndle, *Anal. Chem.*, 74 (2002) 434–439.
326 M.T. Browser, E.D. Sternberg and D.D.Y. Chen, *Electrophoresis*, 18 (1997) 82–91.
327 M. Jussila, S. Palonen, S.P. Porras and M.-L. Riekkola, *Electrophoresis*, 21 (2000) 586–592.
328 S.P. Porras, M. Jussila, K. Sinervo and M.-L. Riekkola, *Electrophoresis*, 20 (1999) 2510–2518.
329 D. Belder, H. Husmann and J. Warnke, *Electrophoresis*, 22 (2001) 666–672.

330 D. Belder and J. Warnke, *Langmuir*, 17 (2001) 4962–4966.
331 S. Palonen, M. Jussila, S.P. Porras, T. Hyötyläinen and M.-L. Riekkola, *J. Chromatogr. A*, 916 (2001) 89–99.
332 F.-M. Matysik, *J. Chromatogr. A*, 802 (1998) 349–354.
333 F.-M. Matysik, *Electrophoresis*, 23 (2002) 400–407.
334 F. Wang and M.G. Khaledi, *J. Chromatogr. A*, 875 (2000) 277–293.
335 B. Chankvetadze and G. Blaschke, *Electrophoresis*, 21 (2000) 4159–4178.
336 B. Chankvetadze, I. Kartozia, J. Breitkreuz, Y. Okamoto and G. Blaschke, *Electrophoresis*, 22 (2001) 3327–3334.
337 L. Chankvetadze, I. Kartozia, C. Yamamoto, C. Chankvetadze and G. Blaschke, *Electrophoresis*, 23 (2002) 486–493.
338 C. Karlsson, L. Karlsson, D.W. Armstrong and P.K. Owens, *Anal. Chem.*, 72 (2000) 4394–4401.

Chapter 3

On-line sample preconcentration techniques in capillary electrophoresis

Hong Chen and Shigeru Terabe

3.1 INTRODUCTION

Since capillary electrophoresis (CE) was first demonstrated by Jorgenson and Lukacs in the early 1980s [1], it has been developed into a very powerful analytical technique and may be regarded as one of the most important advancements in analytical chemistry. CE has evolved into a large family of high-resolution separation techniques. Compared with high performance liquid chromatography (HPLC), CE has distinct advantages, such as high speed, high resolution, high efficiency and low running costs. CE has become a successful analytical technique in biological and environmental sciences. Because of its simplicity, flexibility, and cost performance, the most widely accepted detection mode is still UV absorbance. Most commercially available CE equipment uses UV/Vis detection with or without a diode array.

Although CE holds great promise as the method of choice for high resolution of biological samples, it suffers from poor concentration sensitivity, particularly when UV detection is used. In CE, in order to minimize electromigration dispersion caused by convection, the internal diameter of the capillaries has to be less than 100 μm to obtain the high separation efficiency. These small dimensions also require very small injection volumes, of the order of a few nanoliters, so that extra-column effects do not contribute to peak broadening. In normal CE, injection volumes much greater than 1% of the total capillary length result in sample overloading and significant band broadening [2]. Excessive analyte dispersion leads to poorer sensitivity and limited peak resolution. Short optical path lengths (less than 100 μm) and small sample volume (in the range of pg or nl) are two main reasons for the

Comprehensive Analytical Chemistry XLV
M.L. Marina, A. Ríos and M. Valcárcel (Eds)
Volume XLV ISSN: 0166-526X DOI: 10.1016/S0166-526X(05)45003-5

poor concentration sensitivity of CE. In real application, CE's low concentration sensitivity remains a serious hindrance to the analysis of trace levels of analytes. The concentration sensitivity is often much higher than 1 μM for many biomolecules when UV detection is used. Without substantially improved sensitivity, CE can hardly be an acceptable tool for routine use in bioanalysis. Improving sensitivity is essential to promote the use of CE methods. The development of techniques for improving detection sensitivity is currently one of the priority research areas in CE [3].

The approaches to sensitivity enhancement in CE include the following: (i) the use of high-sensitivity detectors, such as laser-induced fluorescence [4], electrochemical detector [5], and mass spectrometry [6]; (ii) the use of chemical reactions, such as chemiluminescence [7] and derivatization [8]; (iii) the use of extended length cells in UV/Vis detectors, such as Z-shaped, bubble cells or multi-reflection detection cells [9,10]; and (iv) sample enrichment via the solid-phase extraction (SPE) method [11] and flow injection analysis coupled CE [12]. However, all of the above-mentioned methods require somewhat complex or expensive apparatuses, which have limited applicability or are time-consuming. These procedures are generally difficult to automate with existing commercial instruments.

On-line sample preconcentration methods based on electrokinetic focusing of large volumes represent one of the easiest ways to enhance sensitivity. These approaches are usually practical and straightforward and involve manipulating the composition of the sample and background solutions together with simple injection procedures using available commercial instrumentation.

In the last two decades, several on-line sample preconcentration techniques in CE have been developed for increasing the amount of sample that can be injected into the capillary. All on-line preconcentration strategies exploit relative differences in the physicochemical properties of the sample matrix and the background electrolyte (BGE). Almost all on-column concentration techniques in CE manipulate the changes in electrophoretic velocity of analytes at the boundary between the sample zone and BGE section. These techniques can be categorized into two methods on the basis of the physical phenomena used to concentrate analytes. One method involves manipulating the electrophoretic velocity of the analyte. The other method partitions the analyte into a stationary or pseudostationary phase to affect its preconcentration.

Two factors affect the preconcentration efficiency in CE. The first of these is the narrowing of analyte bands in the capillary. As the peak width of the analyte decreases, the peak height increases, resulting in a greater signal-to-noise ratio, improving the limits of detection (LOD). The second is the amount of sample that can be loaded onto the column. Because the width of the peaks is significantly narrowed by the stacking procedure, much larger sample volumes may be injected without losing separation efficiency. This results in a greater mass of analyte in the capillary and therefore a higher response at the detector [13]. In general, when on-line preconcentration methods are used in CE, there is a compromise between the degree of sensitivity and the resolution required for complex sample mixtures.

To date, there are four major on-line preconcentration techniques that have been reported in CE: sample stacking, transient isotachophoresis (ITP), sweeping, and dynamic pH junction. Each method relies on a distinct focusing mechanism based on differences in electrolyte properties between sample and BGE zones, such as conductivity (ionic strength), electrolyte co-ion mobility, additive concentration (analyte–additive interactions), and buffer pH, respectively. In this chapter, each sample preconcentration technique is introduced in more detail mainly from the viewpoint of concentration mechanisms.

3.2 SAMPLE STACKING TECHNIQUE

3.2.1 Basic principle of stacking

The sample stacking phenomenon in electrophoresis was probably first suggested by Tiselius and co-workers [14]. In general, the concentration effect of sample stacking relies on the difference in electrophoretic velocity or electric field strength between the high- and low-conductivity zones, which are used as background and sample zones, respectively. The basic principle of sample stacking of anions is summarized in Fig. 3.1. The sample region is a low-conductivity solution while the background region is a high-conductivity solution. When high voltage (V) is applied, the electric field strength (E_1) over the low-conductivity sample zone will be much stronger than the electric field strength (E_2) over the BGE zone. In sample stacking, E_1 can be enhanced several hundred times more than E_2 [15]. Since the velocity of analyte ions is directly proportional to the field strength, the larger the difference in conductivity or resistance between the sample matrix zone and BGE

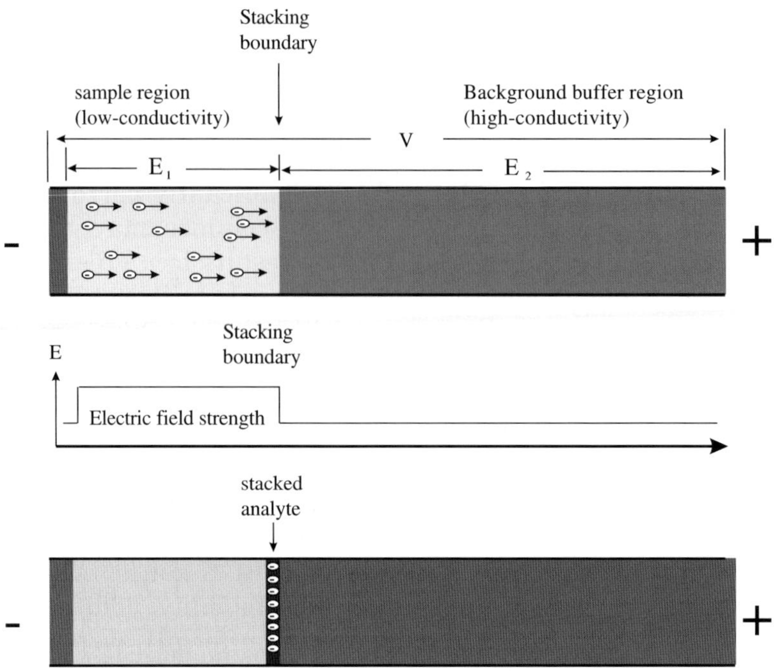

Fig. 3.1. Schematic diagram of sample stacking. (A) Low-concentration and high-resistance sample zone is injected into the capillary filled with BGE, which has high concentration and low resistance. (B) The profile of electric field strength along the capillary. (C) After applying high voltage, analytes are focused across the concentration boundary.

buffer, the faster analytes will migrate through the high-resistance or low-conductivity zone. Therefore, the velocity of the analyte will be higher in the sample zone until it reaches the buffer interface. Once the ions migrate into the region of high-conductivity BGE buffer, they will slow down and experience a lower field strength E_2. The abrupt change in migration velocity of analyte across the concentration boundary results in reduction of sample zone length, thus providing zones with concentrations higher than the original. Sample stacking can provide from 10- to 1000-fold improvement in detector responses, thus expanding the applicability of capillary zone electrophoresis (CZE) to many fields involving trace analysis.

Since sample stacking is defined as the movement of sample ions across the stacking boundary, the narrowing factor (γ) due to sample stacking is correctly defined as equal to the ratio of the electrophoretic velocities of an analyte in the sample and BGE zones as

$$\gamma = E_S/E_{BGS} = v_{BGS}/v_S \quad (3.1)$$

where v is electrophoretic velocity and S and BGS indicate sample zone and background solution, respectively.

In sample stacking, the concentration efficiency is affected by pH, buffer concentration, and composition. Different buffers were shown to amplify the stacking effects for different analytes. Injection of a high-viscosity plug, such as ethylene glycol, prior to the preinjection plug of water, acts as a trap to slow down the electrophoretic velocity of the analytes. Stacking efficiencies were improved by using this procedure [16]. Addition of an organic solvent to the sample matrix also provides an improved signal-to-noise ratio [17].

Sample stacking can be performed in both the hydrodynamic and electrokinetic injection modes. In the simple case of hydrodynamic injection, the samples are dissolved in the low-conductivity matrix, and the resulting solution is introduced into the capillary using pressure or gravity.

3.2.2 Normal stacking mode

Normal stacking mode is the simplest technique for sample stacking. This approach was first proposed in 1979 [18]. It is done by dissolving the sample in a low-conductivity matrix and by injecting the resulting sample solution hydrodynamically. Preconcentration occurs at the interface between the low-conductivity matrix and the BGE. The initial length of injection zone (l_{inj}) in CZE sample stacking with hydrodynamic injection is narrowed by a value equal to the reciprocal of γ:

$$l_{stack} = l_{inj}\,\frac{1}{\gamma} \quad (3.2)$$

where l_{stack} is the length of the analyte zone after sample stacking and γ is given by Eq. (3.1). The resulting concentration after sample stacking of the injected sample can also be predicted by the following equation:

$$C_{stack} = C_{inj}\gamma \quad (3.3)$$

where C_{inj} and C_{stack} are the concentrations of an analyte in the injected sample zone and the resulting stacked zone, respectively. If the conductivity or resistivity of the sample and BGE zones is preserved during application of voltage, γ is equal to the conductivity of the BGE divided by that of the sample. The greater the γ, the higher the sensitivity enhancement that can be expected. In real cases, the value of γ changes right after application of voltage owing to mixing of the sample and BGE zone. Mixing leads to changes in conductivity values and lower γ values. Mixing may be caused by the difference in local electroosmotic velocities between the sample and BGE zones. This is the reason why preconcentration factors determined experimentally were usually smaller than the ones determined by Eq. (3.3).

A limitation in normal stacking mode is that the short optimum sample plug length that can be injected into the capillary without loss of separation efficiency or resolution is very small. The broadening of stacked zones, which can be ascribed to the mixing of low- and high-conductivity zones in the concentration boundary and the laminar flow generated inside the capillary as a result of the mismatch between local and bulk electroosmotic flow (EOF) velocities, are responsible for this. Similar to stacking effects, mixing and laminar flow effects are more significant with higher values of γ. Decrease in resolution and widening of peak widths occur with excessive injection length of the sample. Concentration factors of around 10-fold are easily obtained with normal stacking mode (NSM), improving concentration (LOD) a whole order of magnitude. Concentration factors are ordinarily obtained by dividing the peak height obtained with sample stacking by the peak height obtained with injection. Sensitivity enhancement can be more appropriately evaluated by comparing LOD between the conventional and sample stacking experiments.

3.2.3 Large-volume sample stacking

Large-volume sample stacking (LVSS) is performed by dissolving the sample in water and by hydrodynamically filling 1/3–1/2 of the capillary with the sample solution [19]. When the volume of sample introduced is greater than that found in the optimum normal stacking mode, the sample matrix must be pumped out of the capillary in order to preserve separation efficiency. Pumping may be performed with EOF. The direction of pumping must be opposite to the electrophoretic movement of charged solutes. The velocity of pumping should also be lower than the electrophoretic velocity of the charged solutes. Concentration factors of more than 100 are reported for LVSS, improving LOD by more than two orders of magnitude.

LVSS can be employed with or without polarity switching. The principle of LVSS with polarity switching is shown in Fig. 3.2. Polarity switching is applied to control the direction of EOF. The polarity of voltage for LVSS is chosen according to the charge of the analytes. If anionic analytes are to be stacked, the positive voltage should be applied at the detection end of the capillary. For stacking cationic analyte, the negative voltage should be applied at the outlet end and a cationic surfactant added into BGE to reverse EOF. As a result, the sample matrix will be removed out of the capillary while analytes are stacked at the interface. When the current is monitored to reach approximately 90–99% of its original current (current observed when the capillary is filled with BGE only), the polarity of voltage is returned to separation mode such that the direction of EOF is toward the detector. The normal separation is employed in the usual fashion.

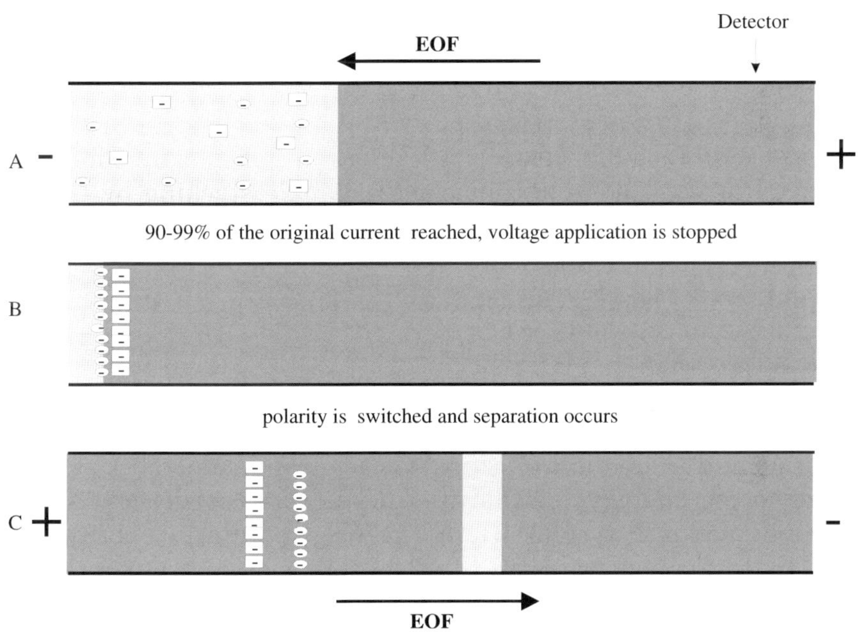

Fig. 3.2. Schematic of the LVSS with polarity switching. (A) Starting situation, injection of sample prepared in a low-conductivity matrix and application of voltage at negative polarity for focusing of zones and removal of sample matrix. (B) Anions are completely focused and most of the sample matrix is removed, voltage is stopped and polarity is reversed. (C) Application of voltage at positive polarity for separation and detection of focused zones.

Under LVSS with polarity switching, the neutrals and ions having different charges from the target analytes in the sample matrix should exit the capillary into the waste reservoir before the polarity returns to normal. However, if the electrophoretic current is not carefully monitored, some analytes may be lost. When using this method, the current should be precisely monitored in order to obtain reproducible results. In real cases, polarity switch is not easy to employ in some commercial CE instrumentations. Since the current should be manually monitored for polarity switching, LVSS with polarity switching is not easy to perform automatically. In order to eliminate these problems, the velocity and direction of EOF should be controlled properly by using LVSS without polarity switching. In this technique, an EOF modifier, such as a cationic surfactant, may be present in the BGE or a low-pH buffer may be used. As the EOF modifier adsorbs on the capillary wall and reverses the direction of the EOF, the polarity-switching step can be eliminated [20]. The limitation of LVSS is that both anions and cations cannot be stacked and separated simultaneously [21].

3.2.4 Electrokinetic injection

Sample stacking with electrokinetic injection was demonstrated in 1991 [22]. The sample is prepared in a low-conductivity matrix and injected by applying voltage. The amount of analyte (N_i) injected into the capillary can be approximated by the following equation:

$$N_i(t) = AC_i\mu_{epi}\gamma E_0 t \tag{3.4}$$

where A is the cross-sectional area of the capillary, C_i the analyte concentration and t the injection time. γ is as given in Eq. (3.1). μ_{epi} is the electrophoretic mobility of analyte i. E_0 is the electric field strength applied by the CE system. Here the electroosmotic velocity is assumed to be much slower than the electrophoretic velocity of the analytes [23].

By using a single electrokinetic injection, only cationic or anionic analytes can be concentrated effectively. Furthermore, there will be a bias effect, i.e., a larger amount of high-mobility ions will be injected and concentrated than lower mobility ions. Compared with hydrodynamic injection, sample stacking with electrokinetic injection can provide larger sensitivity enhancements.

3.2.5 pH-mediated stacking

The normal sample stacking methods, such as field-enhanced sample stacking and LVSS, should be performed with samples either dissolved in water or diluted with a low-conductivity buffer before injection to realize effective on-line sample concentrations. These techniques limit sample solutions to lower ionic concentrations than those of BGE. Moreover, real samples, such as physiological samples, must be significantly diluted before they can be injected. pH-mediated sample stacking is an on-capillary concentration technique that can be employed to overcome these limitations. This approach is used to focus the analyte by titrating the injected sample solution so as to create a low-conductivity region. The pH titration of the CE buffer by acid (utilizing titration of conjugate base of the weakly acid CE buffer with H^+) or base (utilizing titration of the conjugate acid of the weakly basic CE buffer with OH^-) could be used to lower conductivity in the sample zone, resulting in the formation of a high electric field zone [24]. Driven by the high electric field, the ionic analytes are pushed from the titrated zone to the titration boundary. The principle of the acid-stacking mechanism for pH-mediated sample stacking is indicated in Fig. 3.3.

In contrast to the acid-stacking mode, the base-stacking approach cannot be used for CZE under standard conditions. In normal cases, CZE is often operated with the anode at the inlet end, so OH^- ions cannot move into the sample zone to reduce the conductivity of the sample zone. In order to perform base stacking with the cathode at the injection end, the inner wall of the capillary should be coated to suppress EOF, or a cationic surfactant should be added to reverse the direction of EOF [25].

3.2.6 Sample stacking of analytes in micellar electrokinetic chromatography

Micellar electrokinetic chromatography (MEKC) was introduced by Terabe et al. two decades ago [26]. MEKC has developed into a routine separation technique not only for uncharged compounds but also for a large number of ionic compounds to improve selectivity. Although the migration of neutral analytes is not directly affected by enhanced electric field strength, stacking of neutral analytes in MEKC can also be performed by micelles to affect apparent electrophoretic velocities of neutral analytes. The principle of sample stacking in MEKC is similar

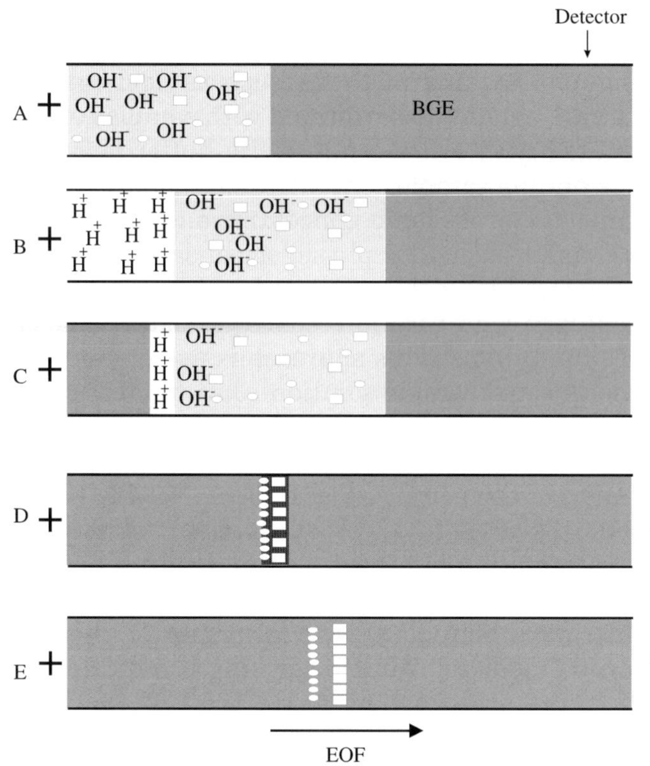

Fig. 3.3. Schematic principle of the acid-stacking approach for pH-mediated stacking. (A) Sample is injected electrokinetically. (B) A zone of strong acid is electrokinetically injected behind the sample zone. (C) Strong acid titrates the sample zone to neutral and the conductivity of sample zone becomes low. (D) The analytes are stacked into narrow bands at the boundary of the titrated zone and the BGE. (E) The analytes are separated by CE.

to that in CZE [27,28]. Sample stacking in MEKC can be grouped on the basis of the type of injection mode, similar to CZE (hydrodynamic or electrokinetic injection stacking mode) [29]. It should be mentioned that the concentration efficiency of sample stacking in MEKC depends on the retention factor of the analytes, because the apparent electrophoretic mobility of the neutral analytes is directly related to the retention factor. The higher the retention factor, the higher the concentration efficiency that can be expected in any stacking modes. Anionic micelles such as sodium dodecyl sulfate (SDS) are assumed in the following techniques unless otherwise mentioned.

3.2.6.1 Hydrodynamic injection modes

Normal stacking mode

The sample prepared in water is injected for a long time after conditioning the capillary with micellar BGS at neutral pH. Micelles from the anodic side of the BGE enter the sample to stack the neutral analytes into the concentration boundary after the application of voltage with the positive polarity at the inlet end. In this method, no injected samples are lost, because the sample matrix is not removed. However, the length of the sample zone is limited by the dispersive effect on the focused bands brought by the local electroosmotic velocity mismatch between the low- and high-conductivity zones. Long injections resulted in broad peaks and consequent loss of resolution [30].

Reversed electrode polarity stacking mode

The samples prepared in water are injected for a much longer time than they are in normal stacking mode. Voltage is then applied at negative polarity. This is to electrokinetically inject micelles from the BGS reservoirs (inlet vial) and to remove sample matrix from the inlet end. Micelles from the inlet vial carry the neutral analytes and stack them into the boundary. The current is monitored carefully, and once the measured current reaches 97–99% of the original value (without sample zone), the polarity is switched from negative to positive. The stacked analytes are separated by MEKC. By this method, a greater concentration factor can be obtained. However, the volume of the sample matrix remaining in the capillary after polarity switching will not be reproducible from run to run [31].

Stacking with reverse migrating micelles

The sample prepared in water is injected after conditioning the capillary with acidic BGS. In acidic solution ($pH < 5$), the electrophoretic velocity of SDS micelles is higher than the electroosmotic velocity in opposite direction. After applying voltage at negative polarity, sample matrix is slowly pumped out by EOF, analytes are stacked with the aid of micelles from the cathodic vial and then analytes are separated in the capillary. An advantage of this approach is that very long injection length can be employed without polarity switching. Because of the removal of the sample matrix, high resolution of peaks, high enhancement in detector responses and high reproducibility can be achieved [32].

Stacking using reverse migrating micelles and water plug
In this method, a water plug is introduced into the capillary first. The sample prepared in an SDS solution or moderately high salt concentration matrix is injected hydrodynamically for a long time. After applying the high voltage, the analytes are stacked at the boundary between the water and BGE zones because the water plug extends the low-conductivity zone and has higher electric field strength than the sample solution zone. During the removal of the sample matrix and water plug by EOF, the analytes are focused in narrow bands and separated by MEKC [33]. The sample solution does not have to be prepared as a low-conductivity solution in this technique.

3.2.6.2 Electrokinetic injection modes

Field-enhanced sample injection (FESI)
A water plug is introduced into the capillary after conditioning the capillary with BGS at neutral pH. The sample prepared in a low-conductivity micellar matrix at neutral pH is then electrokinetically injected (negative polarity) into the capillary. After that, a BGE reservoir replaces the sample reservoir, and the high voltage remains at negative polarity until 97–99% of the original current is reached. Applying high voltage switched at positive polarity, analytes are separated by MEKC. In sample stacking of neutral analytes for MEKC, one of the obstacles in the hydrodynamic injection approach is the limited solubility of hydrophobic solutes in water. This problem can be resolved by using electrokinetic injection modes, because the samples are dissolved in micellar matrices [34].

FESI with reverse migration micelles
The sample is prepared in a low-conductivity micellar matrix of acidic pH. The BGE is also an acidic pH solution. Voltage for electrokinetic injection and separation is at negative polarity. The mode of this approach is similar to FESI. A BGS vial replaces the sample vial when 70–90% of the original current is reached, where the percentage of original current is lower than FESI. Water zones are shorter as well. The method of FESI with reverse migration micelles is suitable for highly hydrophobic compounds [35].

3.2.7 Applications of sample stacking

Many applications of sample stacking have been reported in the fields of biomedical and environmental science, showing their usefulness in real sample analysis. The applications of sample stacking including analytes, sample matrices, LOD and concentration factors, are given in Table 3.1.

However, by using sample stacking, reproducibility of migration times, peak heights, and corrected peak areas remains a problem. Utilization of relative migration time, relative peak height, or relative corrected peak area to a standard can solve this problem. A careful choice of the internal standard is necessary. Diode array (UV/Vis) or mass spectrometric detection can be used for identification of peaks when migration times are very erratic [36].

3.3 ITP PRECONCENTRATION

3.3.1 Principle of ITP

ITP is carried out in a discontinuous electrolyte system, consisting of a leading (L) and a terminating (T) electrolyte as shown in Fig. 3.4 [63]. The capillary and the cathode reservoir are filled with the leading electrolyte, whose cations must have a higher mobility than that of the analyte cations. The leading electrolyte forms the front zone. The anode reservoir is filled with the terminating electrolyte, which must possess a lower mobility than that of the analyte cationic species. The terminating electrolyte forms the rear zone. The sample solution is introduced between T and L zones, and each analyte migrates between L and T zones. Thus, the sample zones are sandwiched between L and T zones. In order to maintain the transport of the constant current through the system, the mixed sample zones are separated further, until each zone contains only one cationic species. When a steady state is reached, all ions migrate at constant velocity. In the sample zone, no BGE is present, and the flow of electricity in the direction of zone migration is maintained only by the analyzed substances [64].

When the steady state has been reached, the different zones will not be broadened further because of the "self-correction" of the zone boundaries: if a cation remains behind a zone with a higher field strength, its migration velocity increases until it reaches its own zone again. If the cation diffuses into a preceding zone where the electric field is lower, its

TABLE 3.1

Application of sample stacking in CE

Analyte	Sample matrix	LOD	SEF	Reference
Normal stacking mode				
Paraquat, diquat and difenzoquat	Water	10 μg/l		[37]
Short-chain organic acids	50% Acetonitrile		17	[38]
Nicotine and it metabolites	Urine	0.55 ng/ml	200	[39]
Phenolic pollutants	Drinking water	0.01 μg/l		[40]
Lactic acid	Food products	0.1 μg/l		[41]
Large-volume sample stacking				
Positively charged analytes	Water or aqueous acetonitrile	ppb	>100	[42]
Weakly acidic compounds	Standards	10^{-8} M	125–170	[43]
Metallothioneins	Eel liver		10	[44]
Plant hormones	Tobacco flowers	0.306 ng/ml	10–600	[45]
Phenols	Standards		500	[46]
Protein		fmol	Q	[47]
Anionic surfactants	Standards	0.002 mg/l	100	[48]
FESI				
Inorganic anions	With water plug	ppb	1000	[49]
Herbicides	Water	0.01 ng/ml	3000	[50]
Opioids	Urine	ppb	1000	[51]
Aromatic sulfonates derivatives	River water	ppb		[52]
Peptide	Proteolytic digestion	200 fmol/l		[53]
Peptides and proteins		nM	3000	[54]

pH-mediated stacking				
DNA sequencing	Purified sequencing sample			[55]
Enantiomers of isoproterenol	Microdialysis samples	0.6 ng/ml		[56]
Coumarin metabolites	Monkey microsomes	10^{-7} M		[57]
Sample stacking in MEKC				
Testosterone, hydrocortisone, cortisone	Pure water	ppb	60	[58]
Neutral analytes	With water plug	ppb	100	[59]
Phenolic compounds	Water	10^{-8} M	~100	[60]
Flavonoids	Chinese medicine	10 ng/ml	40–360	[61]
Multi-residue pesticides	Drinking water and vegetables	0.1 μg/l	3–18	[62]

Note: LOD, limits of detection; SEF, sensitivity enhancement factor.

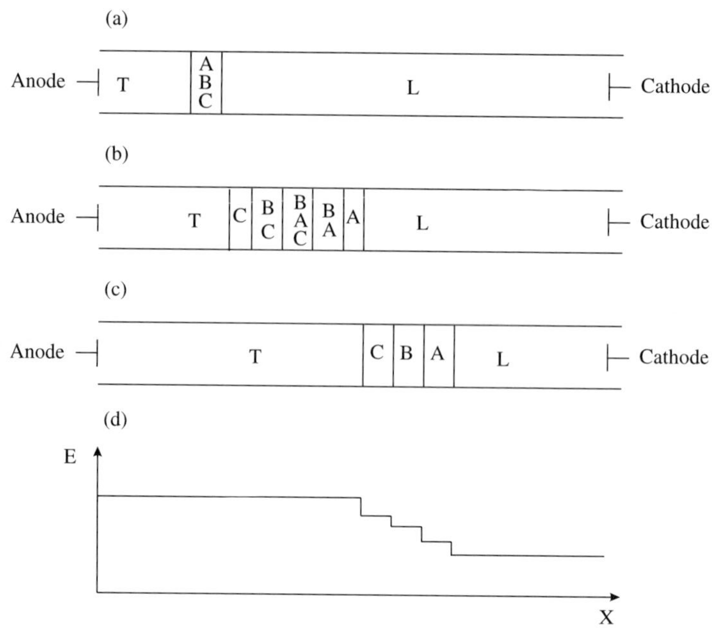

Fig. 3.4. Principle of ITP. (a) Initial state, (b) intermediate state, (c) steady state, (d) the stepwise profile of electric field strength across the separation channel.

velocity will decrease until all sample zones migrate with the same electrophoretic velocity. This is the reason that this electrophoretic mode is expressed by the name "iso-tachos," where "iso" stands for same and "tacho" for speed [64]. The profile of the electric field strength across the separation channel is shown as Fig. 3.4d. In ITP the concentration of each ion is determined by the concentration of L and electrophoretic mobilities of analyte and L as given below.

3.3.2 Normal ITP-zone electrophoresis mode

The use of ITP as an on-line preconcentration technique in CE has already been described since the 1990s [65,66]. Owing to the zone sharpening and concentrating effects, determined by the concentration of the leading ion, diffusion phenomena play a negligible role in the steady state of ITP. ITP is employed as concentration technique for trace components. The combination of these features makes ITP in principle an ideal technique for sample treatment. The degree of the concentration effect in ITP can be derived from the Kohlrausch equation

$$C_L/C_A = [\mu_L/(\mu_L + \mu_C)][(\mu_A + \mu_C)/\mu_A] \tag{3.5}$$

where C_L is the concentration of the leading electrolyte, C_A the analyte concentration and μ the electrophoretic mobility (the subscript C refers to the counter-ion). From this equation it can be seen that the final concentration of the analyte is proportional to the molarity of the leading buffer. Thus, the equation can be written as

$$C_A = C_L K \tag{3.6}$$

where K is a proportionality factor. In the case of similar mobilities, K will have a value of about 1. This equation clearly demonstrates the tremendous concentration potential of ITP. The concentration of the leading ion determines the concentration of the separated ions in the steady state, and thus, the concentration of the compounds can be controlled as required. For example, according to this equation, ITP of an analyte solution (1 μM) in a leading buffer concentration of 5 mM results in a concentration factor of about 5000. As a result of the ITP process, trace compounds are concentrated. Since leading and terminating electrolytes must be chosen carefully, several ITP systems were recommended in some papers [67,68].

For analytical purposes, low concentrations of the leading ion are suitable (0.01–0.005 M), because lower concentrations of the leading ion decrease the total amount of the sample required for the analysis, or decrease the real length of the zone for an equal amount of analyte, hence increasing the sensitivity or accuracy of the analysis. In the analysis of polypeptides, the concentration of the leading ion also determines the ionic strength of a given electrolyte system, which should be selected with respect to the solubility of the analyzed peptides. The concentration effect causes the concentration of polypeptides in the steady state to be relatively high, and their precipitation may take place. For this reason it is also more suitable to select lower concentrations of the leading ion [69].

The counter-ion of the leading electrolyte should be selected so that it has sufficient buffering capacity at the given pH of the leading electrolyte. The concentration of the counter-ion C_C ranges in the interval

$$1.3C_L \leqslant C_C \leqslant 4C_L \tag{3.7}$$

where C_L is the concentration of the leading ion. Otherwise, the roles of counter-ions that form complexes with selected components and that influence the mobility of analytes selectively may increase the separation power of ITP. The use of mixtures of counter-ions in the leading electrolyte increases the buffering capacity over a broader pH interval. In view of the relatively low effective mobility of analytes, extremely slow terminating ions must be selected for their ITP separation. The terminating ions should have a very small specific charge or a high mass-to-charge ratio. The total concentration range of the terminating ion is usually 0.01–0.03 M.

Coupled column uses two capillaries for ITP followed by CZE (Fig. 3.5). Capillary ITP (CITP) is used to focus analytes in the first capillary and the CZE separation then proceeds in the second capillary. In terms of the effective on-column concentration, a 10,000-fold sensitivity enhancement has been reported with this approach [70].

One of the advantages of ITP is that this technique can handle both complex and high salt concentration matrices including blood, plasma, and urine.

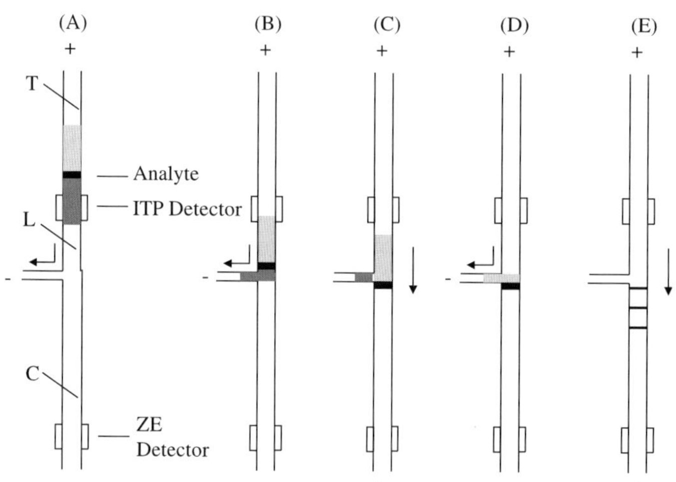

Fig. 3.5. Schematic illustration of the separation phases in combining ITP with CZE in the column-coupling configuration of the separation unit. From [70]. Reprinted with permission from Elsevier. Copyright (1990). (A) ITP separation in the first column (ITP stage); (B) removal of matrix constituents from the separation compartment; (C) transfer of the sample fraction containing the analytes into the second column (ZE stage); (D) removal of matrix constituent from the separation compartment; (E) ZE separation in the second column (ZE stage). L, T, leading and terminating zones, respectively; C, carrier electrolyte.

3.3.3 Transient ITP mode

The on-line coupling of CITP-CZE techniques permits sample preconcentration by at least three orders of magnitude; however, it requires a complicated instrument.

In 1992, Karger's group proposed the technique of on-line transient CITP preconcentration of protein samples where both ITP preconcentration and CZE separation proceed in one capillary on a commercial instrument equipped with a UV detector [71]. In most transient ITP (tITP) cases, two conditions must be fulfilled. First, the injected sample must be supplemented by co-ions with high electrophoretic mobility that can act as leading ions for the ITP migration during the early stages of the separation. The second condition requires that the BGE used for CZE separation contain a co-ion with low electrophoretic mobility that can serve as a terminating ion during the tITP migration. In this method, the analysis is conducted such that both tITP preconcentration and CZE separation proceed in the same capillary. Two basic electrolyte systems can be envisioned as follows: (i) If the BGE is selected with the co-ion having a higher effective mobility than the sample ions, ITP migration is achieved by using a suitable terminating electrolyte behind the sample zone. After the sample is dynamically injected by pressure and after the terminating electrolyte is added, tITP preconcentration of the sample takes place. After on-column tITP preconcentration, the terminating electrolyte in the electrode vessel is replaced with the BGE and separation in the CZE mode, including the migration of a zone of terminating ions behind the sample zones; (ii) on the other hand, if the capillary contains a BGE with a lower effective mobility co-ion, the sample itself must be supplemented by a leading ion (such as NH_4^+, K^+, or Na^+ cationic solutes) to maintain tITP migration. The co-ion in the BGE then serves as a termination ion. Real samples often contain such high-mobility ions in sufficient amount to permit tITP preconcentration. In this case, the capillary is filled with the BGE, and dynamic injection of a large volume of sample dissolved in buffer containing leading ion follows. The inlet end of the capillary is then returned to the BGE reservoir. The leading ions migrate ahead of the sample ions when high voltage is applied. At this point, the sample ion is focused behind the leading ion zone in a narrow band, and keeps moving at constant velocity (as in ITP). The leading ions, however, continue to move through the slower BGE. Consequently, the concentration of leading ions in the leading zone rapidly decreases below the concentration necessary for ITP migration. At this point, the sample is separated as in CZE [72].

The main prerequisite for successful tITP preconcentration is knowledge of the range of solute mobilities in the sample. When the proper BGE is selected and a wall-coated capillary is used to prevent adsorption, transient on-column tITP preconcentration provides reproducible and quantitative results. Furthermore, the sample need not be dissolved in distilled water or dilute buffer and may contain an excess of other ions, which may be useful for stabilization. In order to get insight into the ITP process, the concentration of the leading ion, sample and counter-ion over the course of the ITP experiment were simultaneously determined by using on-line NMR spectroscopy [73].

3.3.4 Applications of ITP preconcentration technique

The combination of tITP and CZE may enhance the performance of analytical CE greatly with respect to both separation power and concentration sensitivity. CZE with tITP preconcentration is applicable both to small ions [74] and to large biomolecules, such as protein and DNA [75]. tITP is one of the critical on-line preconcentration techniques for new types of CE detectors such as NMR [76–78] and time-of-flight mass spectrometry [79]. The former provides high information content for the qualitative identification of unknown analyte, but suffers from poor detection sensitivity.

Detection in microchip electrophoresis needs high sensitivity because the amount of analyte is minute and the detection volume is very small. Although laser-induced fluorescence is the most popular detection method, ITP is a useful and alternative on-line preconcentration technique applied in microchip CE [80,81]. Some applications of CITP are listed in Table 3.2.

3.4 DYNAMIC pH JUNCTION PRECONCENTRATION

3.4.1 Principle of dynamic pH junction

In CE, the pH value of the electrolyte solution is one of the most important parameters of migration velocity. In general, when the ionizable functional groups of the analytes are weak acids or bases, the net charge of the analyte ion is dependent on the degree of ionization given by the pK_a value of the acidic or basic functional group and the pH of

TABLE 3.2

Example for CE analysis of different compounds by using on-line tITP preconcentration

Type	Compound	Sample	Leading ion	Terminating ion	BGE	Detector	LOD	SEF	Reference
Ions	Chlorite	Drinking water	Chloride	Succinic acid	Succinic acid, EACA as counter-ion (pH 3.9)	Conductivity	0.005 mg/l		[82]
	Free sulfite	Wine	Chloride	Tartrate	15 mM succinate, pH 4.0	Conductivity	3 mg/l		[83]
	Iodide	Seawater	Chloride	MES	0.5 M NaCl and mM CTAC	UV (226 nm)	0.2 μg/l		[84]
Organic molecule	Amino acids	Serum	Borate (pH 9.5)	ACES (pH 10.0)	Borate (pH 9.5)	UV	10 nmol/l	200	[85]
	Clenbuterol, salbutamol		Ammonium acetate	β-Alanine	10 mM Ammonium acetate buffer (pH 2.5) containing 40 mM dimethyl-b-cyclodextrin	UV	10^{-6}		[86]
	Flavonoids	Plant leaves or flowers	Chloride	Boric acid	Tris-borate acid (pH 8.3)	UV (254 nm)	21,100 ng ml	1000	[87]
	Phenolic compounds	Red wine	HCl	Boric acid	Tris-boric acid (pH 8.5) containing MOPSO, β-cyclodextrin	UV (254 nm)	30 ng ml		[88]
	Tyramine	Serum sample	Acetate	TEA	Methanol containing 70 mM sodium acetate	UV (280 nm)			[89]
Peptides and proteins	GSH, GSSG	Plasma	Chloride	CAN, borate	300 mM Borate buffer (pH 8.0)	UV (200 nm)	0.3–0.5 μmol/l	15–20	[90]
	Interleukin-3		Triethylamine	β-Alanine	TEA	UV (200 nm)	10^{-8} M	100	[91]
	Protein		Triethylamine	Acetic acid	ε-Aminocaproic acid–acetic acid (pH 4.4)	UV (214 nm)	10^{-8} M		[92]
	Protein		Ammonium acetate	Acetic acid	β-Aminohexanolic acid–acetic acid (pH 4.4)	ESI-MS	10^{-7} M	100	[93]
DNA	DNA restriction fragments		Chloride	Butyrate	Linear polyacrylamide gel buffers	UV (254 nm)			[94]
	DNA fragments		Tris–HCl (pH 8.1)	Glycine–Tris (pH 8.1)	2% HEC	UV			[95]

the solution. Based on this principle, an on-column preconcentration technique in CE, dynamic pH junction, was proposed by Britz-McKibbin and Chen in recent years [96–98].

The principle of dynamic pH junction is visually depicted in Fig. 3.6 for weakly acidic analytes. In this example, the weakly acidic analytes are dissolved in a low pH value sample matrix, whose pH value is lower than p*K* values of the analytes. At this pH, the weakly acidic analyte is neutral. When a large volume of a dilute sample solution is injected into the capillary filled with high pH BGE, there is a pH junction between the sample matrix zone and BGE (Fig. 3.6A). With the application of a high voltage, the original pH junction does not persist, as it is titrated with the hydroxide ions from the BGE. When the sample matrix zone moves along the capillary driven by EOF, dynamic pH junction sweeps across the sample zone until the pH difference is gradually dissipated.

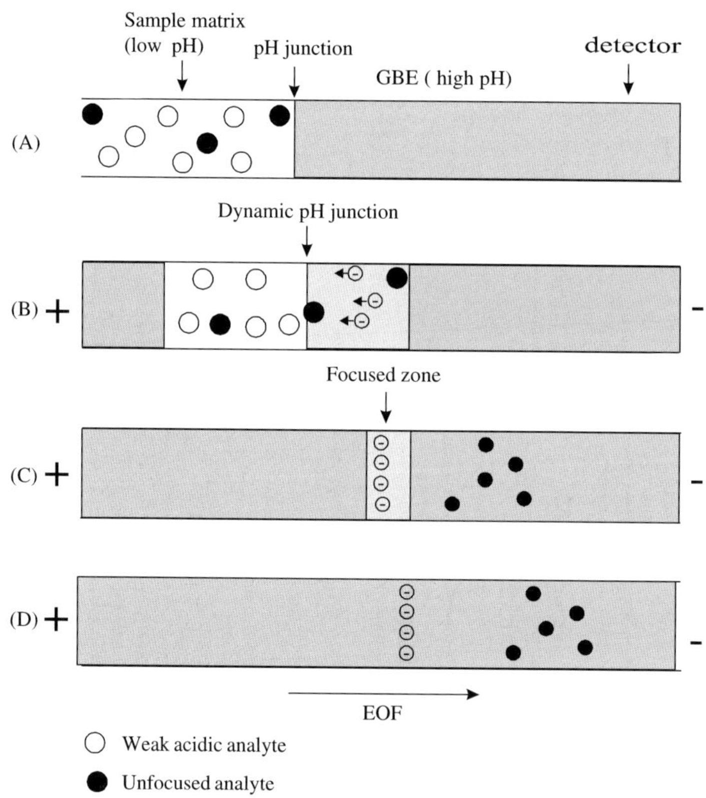

Fig. 3.6. The principle of dynamic pH junction.

In this process, neutral or weakly acidic analytes become negatively charged because of dissociation of proton(s). The migration velocities of weakly acidic analytes become low owing to the large negative mobility they acquire because EOF is strong in the opposite direction. The weakly acidic analytes can therefore be focused into a sharp zone at the boundary between the sample matrix zone and BGE (Fig. 3.6B and C). However, dynamic pH junction is not applicable to neutral or strongly acidic analytes whose mobility is independent of buffer pH. After the sample matrix zone with low pH is titrated by BEG completely, the focused analytes are separated by CZE (Fig. 3.6D). The pH junction range for the focusing of weakly acidic, basic or zwitter-ionic analytes (mobility is pH-dependent) is optimized based on their pK_a. Although the influence of buffer pH, buffer co-ion and ionic strength was examined in some papers, the mechanism of pH junction is not very clear. The progress of dynamic pH junction simulated by a computer is shown in Fig. 3.7. This study helps us to understand the progress of preconcentration of analyte in dynamic pH junction [99].

When the sample matrix zone is introduced into a capillary filled with BGE, a clear pH junction exists between the sample matrix and BGE zone shown in Fig. 3.7A. With application of a high voltage, a steep pH boundary migrates through the sample zone because acetic acid in the sample plug is titrated by borate in the BGE (Fig. 3.7B and C). In this step, the initial sample zone begins to split into two zones (1) stationary and (2) moving zones. Because the pK_a value of the test analyte, *m*-nitrophenol, is between the pH values of sample matrix and BGE, the polarity of the analyte goes from negative to neutral in the boundary between the sample matrix and BGE zone. So the velocity of the analyte slows down at this junction point. As shown in Fig. 3.7C, the analyte is focused when the pH moving boundary migrates through the sample matrix zone. When the migration zone 2 separates from the stationary zone 1, the process of dynamic pH junction is completed (Fig. 3.7D). After this step, the focused peak gradually begins to broaden by longitudinal thermal diffusion (Fig. 3.7D–F). In order to verify analyte focusing by computer simulation, experimental electropherograms are also depicted in Fig. 3.7A–F, which show good agreement with the result of the computer simulations.

In previous dynamic pH junction reports, analytes were dissolved in an acidic or neutral sample matrix and separation was performed in the basic BGE, where borate was used. The velocity of analytes was modified by both pH and borate complexation (for vicinal diol groups) to

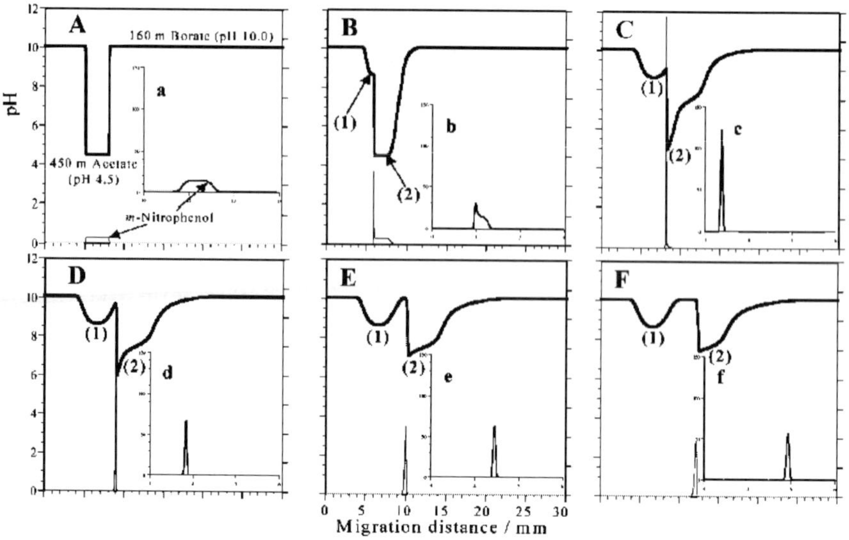

Fig. 3.7. The mechanism of Dynamic pH junction depicted by computer simulation and CE experiment. The profile of the upper line (thick solid line) represents the change in pH, whereas the bottom line (thin solid) shows the concentration profile of the analyte. Insets are experimental electropherograms. Sample matrix, acetic acid buffer (450 mM, pH 4.5); test analyte, *m*-nitrophenol; BGE, borate buffer (160 mM, pH 10.0). In this case, the EOF was assumed to be zero. Reprinted with permission from Ref. [99]. Copyright (2003), American Chemical Society.

induce electrokinetic preconcentration. Recently, it was reported that cationic analytes were preconcentrated by dynamic pH junction in reversed EOF system modified by cetyltrimethylammonium chloride (CTAC). The analytes were prepared in pH 2.0 sample matrix (100 mM phosphate buffer) and the BGE was 250 mM acetate and 2 mM CTAC buffer (pH 4.5) [100].

3.4.2 Applications of dynamic pH junction

Dynamic pH junction is becoming a practical preconcentration technique in CE because it is easily combined with conventional CE. The analytes can be focused effectively in high salt concentration sample matrices. These advantages make this approach useful in a variety of real biological samples. In the beginning report, dynamic pH junction was used to analyze catecholamines in dental anesthetic [101].

Dynamic pH junction is becoming an important method for the trace analysis of the metabolites in real samples such as biological fluids and cellular extract [102]. It has been reported that CZE coupled with dynamic pH junction was used to assay a few metabolites, such as nucleotides [103] and 8-hydroxy-2′-deoxyguanosine (8OHdG), a kind of marker of oxidative DNA damage [104]. In applications of drug analysis, ethynylestradiol derived from a female contraceptive pill extract has been selectively focused and detected, which has resulted in over 100-fold enhancement in concentration sensitivity [105]. Besides small biomolecules, the analysis of trace proteins under acidic conditions based on pH junctions has been demonstrated [106]. The potential for performing high-throughput analyses of purine and sugar metabolites (native, amino and phosphate sugars) was demonstrated by pH junction multiplexed CE (96 capillary array) with UV detection, which would be particularly useful for large sample screening in biological research [107,108].

3.5 SWEEPING

3.5.1 Basic theory of sweeping

The on-column preconcentration techniques based on electrophoretic effects, such as sample stacking, tITP and dynamic pH junction, are able to focus only charged analytes. To separate neutral analytes by CE, it is necessary to provide an electrokinetic vector to the neutral analytes. The pseudostationary phase is used as an additive to generate the electrokinetic vector for neutral analytes to migrate electrophoretically. The pseudostationary phases include not only micelles but also polymers or dendrimers, etc. Partitioning of the analytes between the pseudostationary phase and the surrounding phase promotes the separation of a mixture of analytes. Both neutral and charged analytes can be separated by electrokinetic chromatography (EKC). Neutral analytes are separated by partitioning alone, while charged analytes are separated by partitioning and electrophoresis. The charged and neutral pseudostationary phases can be used in EKC. Only charged analytes can be separated using a neutral pseudostationary phase. There are some on-column preconcentration techniques for focusing neutral analytes in EKC. sample stacking of neutral analytes in EKC is given in an example [109,110].

In 1998, the exceptional narrowing effect of neutral analyte zones in EKC under constant electric field was proposed by Quirino and Terabe [111]. Different from other preconcentration approaches, this special concentration method, which is named sweeping, is based on a physical phenomenon that works well for all analytes with great affinities toward the pseudostationary phase. Sweeping in EKC is defined as the picking and accumulation of analytes by the pseudostationary phase that penetrates the sample zone during application of voltage. The principle of sweeping is depicted in Fig. 3.8.

In this case, the pseudostationary phase is an anionic micelle of SDS. EOF is assumed to be zero by using low pH buffers. The analytes are electrically neutral and prepared in the matrix having same conductivity as that of GBS. However, sample matrix is void of micelle. Figure 3.8A shows a long injection of the sample solution (S) into a capillary previously filled with BGS. When a high voltage is applied, the neutral

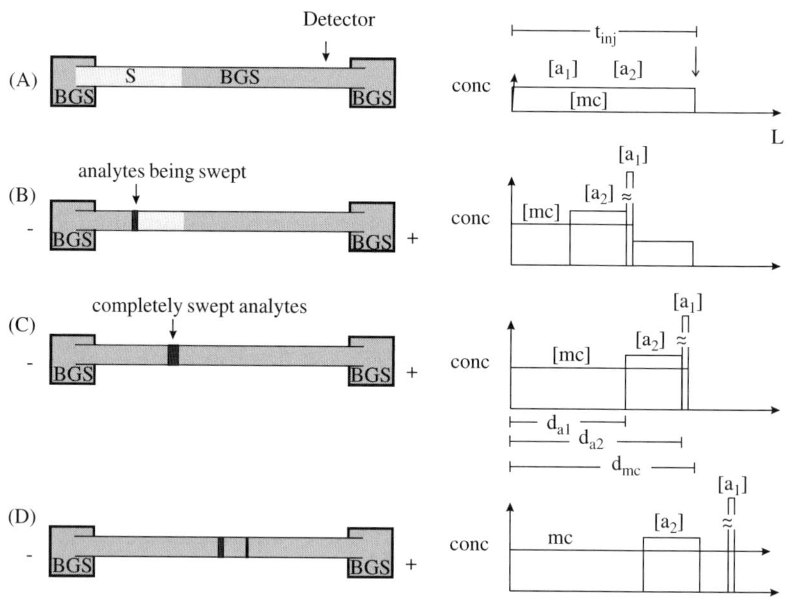

Fig. 3.8. Principle of sweeping. (A) Injection of a sample solution S with length l_{inj} after conditioning the capillary with BGS. The conductivity of the sample matrix is the same as the micellar BGS. (B) Applying a high voltage, micelles from the inlet end enter the sample zone and sweep the analytes into narrower bands whose width depends on the retention factor (in this case, $k_{a1} > k_{a2}$). (C) The sweeping is completed when the micellar BGS fills the whole sample zone.(D) Separation of analytes based on MEKC.

analytes will not migrate by themselves until they are incorporated into the micellar pseudostationary phase. Figure 3.8B shows that the anionic pseudostationary phase enters the capillary and sample zone when a high voltage is applied. As the pseudostationary phase passes through the sample zone, the analytes are picked up and accumulated. After a certain period of time, all analyte molecules are picked and accumulated by the pseudostationary phase as shown in Fig. 3.8C. And then, the focused analytes are separated by MEKC (Fig. 3.8D).

In theory, the length of the swept zone (l_{sweep}) is given as

$$l_{\text{sweep}} = d_{\text{mc}} - d_{\text{a}} \tag{3.8}$$

where d_{mc} is the length of micelles that enter and completely fill the sample zone and d_{a} is the length of analyte picked by pseudostationary phase (see Fig. 3.8C). The length d_{mc} is simply the length of the zone injected (l_{inj}) and d_{a} is l_{inj} multiplied by the retention factor $1/(k+1)$ to incorporate the role of the retention factor. Therefore, the l_{sweep} can be expressed using k and l_{inj} as $l_{\text{sweep}} = l_{\text{inj}} \dfrac{1}{1+k}$ (3.9)

The swept length of the higher retention factor analyte (a_1) is less than that of the lower one (a_2). According to the law of conservation of mass, the concentration of analyte in swept zone (C_{sweep}) is given as

$$C_{\text{sweep}} = C_{\text{inj}}(1 + k) \tag{3.10}$$

where C_{inj} is the concentration of the analyte in the injected zone. Following Eq. (3.10), concentration factor of sweeping ($C_{\text{sweep}}/C_{\text{inj}}$) is basically dependent on the retention factor. This equation also predicts an almost limitless improvement in detection sensitivity for analytes having very high k. It should be noted that the retention factor in Eqs. (3.9) and (3.10) is that in the sample matrix and not in micellar BGS, where additives such as an organic solvent of cyclodextrin derivatives may be used to improve resolution by decreasing k values.

According to Eq. (3.10), both neutral and charged analytes can be concentrated by sweeping in the absence and presence of EOF. The sweeping of neutral or charged solutes is only affected by k and is independent of EOF in the homogeneous electric field system and a fixed l_{inj}. As reported in ref [112], a strong EOF usually causes low concentration efficiency in comparison with low or suppressed EOF. This is probably due to the difference in EOF velocity between the sample and BGS zone, which generates mixing at the boundary, resulting in band broadening. Using a polyacrylamide-coated capillary [113], neutral, positively, and negatively charged analytes were swept in MEKC using cationic surfactants such as tetradecyltrimethylammonium bromide and cetyltrimethylammonium chloride [114]. Ion-pair reagents, such as tetraalkylammonium salts, can improve the sweeping efficiency of anionic analytes [115] with anionic micelles. Since SDS was combined with the zwitter-ionic surfactant can substantial increases in the retention factors of some neutral analytes, such as phenol derivatives which were previously not amenable to sweeping by pure SDS micelles, can be preconcentrated by using anionic–zwitter-ionic mixed micelles [116].

Because of the high concentration efficiency of sweeping, very sharp peaks of analytes can be observed in comparison with the usual injection plug length. By using sweeping microchip Electrophoresis the process of sweeping can be observed directly because the detection position is successively movable along the separation channel [117]. On the basis of this experiment, the profile of the concentration process and the diffusion process in sweeping was elucidated. In the concentration process, the width of the analyte zone was narrowly focused. The focused zone became broad soon after the end of sweeping because of thermal diffusion. The dependence of zone broadening was accurately described by molecular diffusion.

Although EKC with charged pseudostationary phase is the most common separation technique, EKC with neutral pseudostationary phase can also separate many important charged compounds [118,119]. Sweeping can also be performed based on interactions between neutral pseudostationary phase and charged analytes. A theoretical study generates the same equation for sweeping. Unlike sweeping with charged pseudostationary phase, the charged analytes penetrate the neutral pseudostationary phase and are swept into narrow bands [120].

In terms of the principle of sweeping, concentration results from the interaction of analytes and the pseudostationary phase, and includes partitioning, complexation, hydrophobic interaction and chemical derivation with rapid kinetics. The major factor that affects the extent of concentration efficiency is the strength of these interactions involved and the resultant change in analyte velocity. Sweeping is believed to be the most versatile (applicable to both neutral and charged analytes) and one of the most effective ($>$1000-fold increase in sensitivity) online preconcentration techniques in CE. sweeping is a very useful method to analyze real sample because it is less affected by the sample and matrix components. By combining two focusing mechanisms, sweeping and micelle stacking, the analytes in high salt sample solution (the sample is prepared in a matrix two to three times the conductivity of the separation buffer) are able to be preconcentrated and separated successfully [121–126]. One disadvantage of sweeping is that it is not efficient for the preconcentration of weakly interacting analytes with the micelle.

3.5.2 Selective exhaustive injection-sweeping

Selective exhaustive injection-sweeping (SEI-sweeping) is a proper combination of two different on-line preconcentration techniques, sample stacking with electrokinetic injection (FESI) and sweeping [127]. By FESI, as many cationic analytes as possible were selectively introduced to a concentrated stacked zone from a very dilute sample solution. By sweeping, the concentrated stacked zone of cationic analytes was further focused into a sharp band. FESI and sweeping have been shown to provide 1000- and 5000-fold increases in UV detector response, respectively. By this preconcentration approach, the LOD of positively charged analytes can be afforded at a level of parts per trillion (ppt), which is the lowest concentration reported by direct UV detection in CE. Approaching a million-fold improvement in sensitivity is achieved compared with usual injection approaches. The main idea of SEI sweeping is illustrated in Fig. 3.9. In this preconcentration approach, the water plug during the long FESI step helps maintain field enhancement at the tip of the capillary, particularly when the sample contains salt, and may also improve reproducibility. The high-conductivity buffer free of organic solvent improves the total focusing effect. It increases the amount of sample molecules injected and creates a narrower focused zone after the cation selective exhaustive injection (CSEI) step, but does not affect the focusing effect of the sweeping.

A limitation with SEI-sweeping comes from the SEI step, which is sample stacking in CZE in principle. In order to perform sample stacking with electrokinetic injection effectively, the sample has to be prepared in a low-conductivity matrix and only chargeable analytes can be preconcentrated by this approach. In addiction to cationic analytes, anion and acidic compounds, complexes of metal ions, and lysergic acid diethylamide (LSD) were also preconcentrated and separated by applying SEI sweeping [128–131].

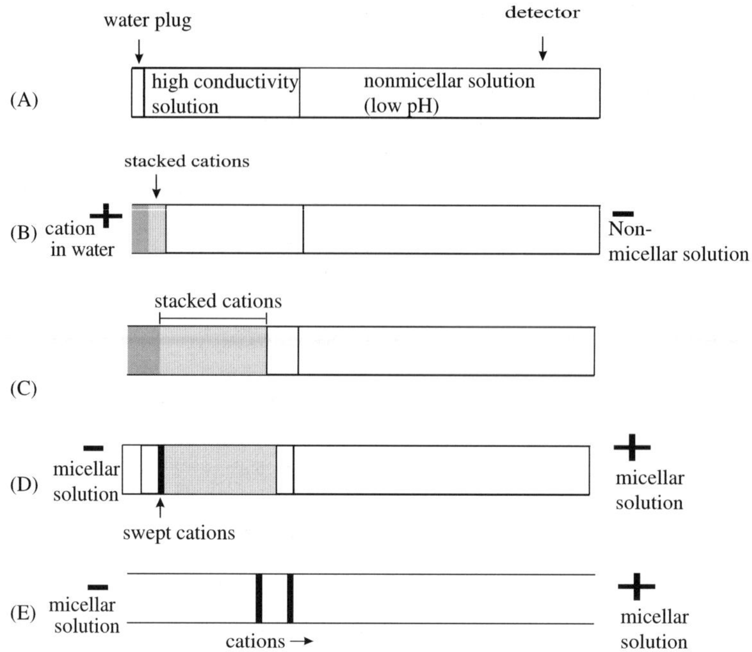

Fig. 3.9. The principle of cation SEI sweeping. (A) Starting situation. The capillary is first conditioned with a nonmicellar BGS, then injection of a high-conductivity solution devoid of organic solvent, followed by injection of a short water plug. (B) Electrokinetic injection for a longer period (e.g., 400 s) at a positive polarity of cationic analytes prepared in a low-conductivity matrix or water, nonmicellar background buffer was used in this step. Cationic analytes focused at the interface between the water zone and high-conductivity buffer void of organic solvent zone. (C) Injection is stopped and micellar solutions are placed at both ends of the capillary. (D) Applying voltage at negative polarity micelles enter into the capillary and sweep the long focused analytes zone to narrower bands. (E) Separation of zones based on MEKC.

3.5.3 Dynamic pH junction sweeping

In real biological research, samples always contain diverse types of analytes, such as intracellular metabolites ranging from anionic, zwitter-ionic, cationic to neutral species. It is difficult to use a single technique of on-column preconcentration to effectively focus some classes of analytes in complex sample mixtures. For instance, the techniques of sweeping and dynamic pH junction are normally effective for focusing either hydrophobic (neutral) or weakly acidic (ionic) analytes, respectively. In order to improve the detection sensitivity of both weakly ionic and neutral (hydrophobic) metabolites simultaneously, the combined dynamic pH junction-sweeping technique was proposed by Britz-McKibbin et al. recently [132].

Dynamic pH junction sweeping is used when the sample has a different buffer pH (dynamic pH junction condition) and is devoid of the pseudostationary phase (sweeping condition) relative to BGE, which contains the pseudostationary phase. The dynamic pH junction sweeping represents a complementary format to optimize the focusing and resolution of diverse classes of metabolites in cases where conventional sweeping or dynamic pH junction is less effective. In general, dynamic pH junction sweeping represents an effective on-line preconcentration technique that can be applied under both strong and suppressed EOF conditions to generate extremely narrow analyte bandwidths.

In Britz-McKibbin's research, using a combined Dynamic pH junction-sweeping technique to analyze flavin metabolites resulted in a 1200-fold improvement in sensitivity relative to conventional injection methods, giving an LOD of about 4.0 pM (10×10^{-14} M) for flavin adenine dinucleotide (FAD) and flavin mononucleotide (FMN). Compared with either sweeping or dynamic pH junction techniques alone, dynamic pH junction sweeping resulted in an over fourfold enhancement in band narrowing [132,133]. Since flavins possess weakly acidic lactam vicinal diols and a hydrophobic isoallaxazine tricyclic ring system, flavins in a large sample plug are focused by different factors, such as buffer pH, borate complexation and SDS interaction [133]. Trace amounts of three flavin derivatives, riboflavin, FMN and FAD, were measured directly from several types of samples, including bacterial cell extracts, recombinant protein and biological fluids.

3.5.4 Applications of Sweeping

Owing to its simplicity and high focusing efficiency, sweeping is becoming an attractive on-line preconcentration technique in CE. It has been applied in environmental and pharmaceutical analysis. In the first report of its performance [111], chiral herbicide spiked at ppt level in lake water was separated and detected by sweeping MEKC. Estrone, β-estradiol and ethynylestradiol spiked in water were extracted using solid-phase extraction and then directly determined by sweeping MEKC. After sweeping, the estrogen-related compounds were detected using a commercial absorbance detection system with an LOD of 0.16–0.30 nM in the original sample [134]. Polycyclic aromatic hydrocarbons, benzo[*a*]pyrene and pyrene, were analyzed by using a combination of sweeping MEKC with liquid–liquid extraction [135]. sweeping MEKC was applied for the analysis of multi-residue pesticides, fungicide and herbicides [136–139]. A kind of endocrine-disrupting compound, bisphenol A, was separated and concentrated on-line by using sweeping [140]. In analysis of real biological sample, a few methods based on sweeping MEKC have been developed for preconcentrating and separating drugs and metabolites. These biological samples include plasma, urine and cell extracts [141–146].

In fields of biological research such as metabolomics, sweeping MEKC has become one of the most important on-line sample preconcentration techniques for the analysis of metabolites [147]. This method was utilized to separate the hydrophobic fractions, while the two-dimension separation approach, which hyphenated micro LC and CE, was employed for the analysis of *Bacillus subtilis* metabolites [148].

REFERENCES

1 J.W. Jorgenson and K.D. Lukacs, *Anal. Chem.*, 53 (1981) 1298.
2 S. Terabe, K. Otsuka and T. Ando, *Anal. Chem.*, 61 (1989) 251–260.
3 B.M. Simonet, A. Ríos and M. Valcárcel, *Trends Anal. Chem.*, 22(10) (2003) 605–614.
4 Y.F. Cheng and N.J. Dovichi, *Science*, 242 (1988) 562–564.
5 Y. Liu, J.A. Vickers and C.S. Henry, *Anal. Chem.*, 76 (2004) 1513–1517.
6 Y. Tanaka, Y. Kishimoto, K. Otsuka and S. Terabe, *J. Chromatogr. A*, 817 (1998) 49–57.
7 Y.-M. Liu and J.-K. Cheng, *J. Chromatogr. A*, 959 (2002) 1–13.
8 T. Fukushima, N. Usui, T. Santa and K. Imai, *J. Pharm. Biomed. Anal.*, 30 (2003) 1655–1687.
9 S.E. Moring, R.T. Reel and R.E.J. van Soest, *Anal. Chem.*, 65 (1993) 3454–3459.
10 T. Wang, J.H. Aiken, C.W. Huie and R.A. Hartwick, *Anal. Chem.*, 63 (1991) 1372–1376.
11 M. Valcarcel, L. Arce and A. l Rýos, *J. Chromatogr. A*, 924 (2001) 3–30.

12 P. Kuban, M. Berg, C. Garcýa and B. Karlberg, *J. Chromatogr. A*, 912 (2001) 163–170.
13 Z.K. Shihabi, *J. Chromatogr. A*, 902 (2000) 107–117.
14 S. Hjerten, S. Jerstedt and A. Tiselius, *Anal. Biochem.*, 11 (1965) 219.
15 R.-L. Chien, *Anal. Chem.*, 63 (1991) 2866–2869.
16 C.-X. Zhang and W. Thomann, *Anal. Chem.*, 70 (1998) 540–548.
17 Z.K. Shihabi, *J. Chromatogr. A*, 853 (1999) 3–9.
18 F. E. P. Mikkers, F. M. Everaerts and Th. P. E. M. Verheggen, *J. Chromatogr. A,* 169 (1979), 11–20.
19 R.-L. Chien and D.S. Burgi, *Anal. Chem.*, 64 (1992) 1046–1050.
20 W.F. Smyth, G.B. Harland, S. McClean, G. McGrath and D. Oxspring, *J. Chromatogr. A*, 772 (1997) 161–169.
21 M. Ablert, L. Debusschere, C. Demesmay and J.L. Rocca, *J. Chromatogr. A*, 757 (1997) 281–289.
22 R.L. Chien and D.S. Burji, *J. Chromatogr.*, 559 (1991) 141–152.
23 R.L. Chien, *Anal. Chem.*, 63 (1991) 2866–2869.
24 Y. Xiong, S.-R. Park and H. Swerdlow, *Anal. Chem.*, 70 (1998) 3605–3611.
25 M.E. Ward, M.R. Smyth, R. O'Kennedy and C.E. Lunte, *J. Pharm. Biomed. Anal.*, 32 (2003) 813–822.
26 S. Terabe, K. Otsuka, K. Ichikawa, A. Tsuchiya and T. Ando, *Anal. Chem.*, 56 (1984) 111–113.
27 J.P. Quirino and S. Terabe, *J. Cap. Elec.*, 4 (1997) 233–245.
28 J.P. Quirino and S. Terabe, *J. Chromatogr. A*, 902 (2000) 119–135.
29 J.-B. Kim and S. Terabe, *J. Pharm. Biomed. Anal.*, 30 (2003) 1625–1643.
30 J.P. Quirino and S. Terabe, *J. Chromatogr.*, 673 (1997) 119–128.
31 J.P. Quirino and S. Terabe, *J. Chromatogr.*, 791 (1997) 255–267.
32 J.P. Quirino and S. Terabe, *Anal. Chem.*, 70 (1998) 149–157.
33 J.P. Quirino and S. Terabe, *J. Chromatogr.*, 686 (1997) 119–128.
34 J.P. Quirino, K. Otsuka and S. Terabe, *J. Chromatogr. B*, 714 (1998) 29–38.
35 J.P. Quirino and S. Terabe, *Anal. Chem.*, 70 (1998) 1893–1901.
36 T. Hirokawa, N. Ikuta and T. Yoshiyama, *J. Chromatogr. A*, 894 (2000) 3–9.
37 O. Nunez, E. Moyano, L. Puignou and M.T. Galceran, *J. Chromatogr. A*, 912 (2001) 353–361.
38 V. Galli and C. Barbas, *Anal. Chim. Acta*, 482 (2003) 37–45.
39 E.E.K. Baidoo, M.R. Clench, R.F. Smith and L.W. Tetler, *J. Chromatogr. B*, 796 (2003) 303–313.
40 I. Rodriguez, M.I. Turnes, M.H. Bollain, M.C. Mejuto and R. Cela, *J. Chromatogr. A*, 778 (1997) 279–288.
41 S. Kodama, A. Yamamoto, A. Matsunaga, T. Soga and K. Minoura, *J. Chromatogr. A*, 875 (2000) 371–377.

42 J.P. Quirino and S. Terabe, *Electrophoresis*, 21 (2000) 355–359.
43 M.-S. Chun and D.-S. Chung, *Anal. Chim. Acta*, 491 (2003) 173–179.
44 G. Álvarez-Llamas, A. Rodriguez-Cea, M.R. Fernández de la Campa and A. Sanz-Medel, *Anal. Chim. Acta*, 486 (2003) 183–190.
45 B.-F. Liu, X.-H. Zhong and Y.-T. Lu, *J. Chromatogr. A*, 945 (2002) 257–265.
46 Y.-R. Chen, M.-C. Tseng, Y.-Z. Chang and G.-R. Her, *Anal. Chem.*, 75 (2003) 503–508.
47 M.-S. Chun, D.-C. Kang, Y.-S. Kim and D.-S. Chung, *Ž. Microchem. J.*, 70 (2001) 247–253.
48 W.-H. Ding and C.-H. Liu, *J. Chromatogr. A*, 929 (2001) 143–150.
49 J.P. Quirino and S. Terabe, *J. Chromatogr. A*, 850 (1999) 339–344.
50 L.- Yan Zhu and H.-K. Lee, *Anal. Chem.*, 73 (2001) 3065–3072.
51 A.B. Wey and W. Thormann, *J. Chromatogr. A*, 924 (2001) 507–518.
52 S. Morales and R. Cela, *Electrophoresis*, 23 (2002) 408–413.
53 S. Locke and D. Figeys, *Anal. Chem.*, 72 (2000) 2684–2689.
54 M.R.N. Monton and S. Terabe, *J. Chromatogr. A*, 1032 (2004) 203–211.
55 Y. Xiong, S-R Park and H Swerdlow, *Anal. Chem.*, 70 (1998) 3605–3611.
56 M.E. Hadwiger, S.R. Torchia, S. Park, M.E. Biggin and C.E. Lunte, *J. Chromatogr. B*, 681 (1996) 241–249.
57 E.M. Ward, M.R. Smyth, R. O'Kennedy and C.E. Lunte, *J. Pharm. Biomed. Anal.*, 32 (2003) 813–822.
58 J.-B. Kim, K. Otsuka and S. Terabe, *J. Chromatogr. A*, 912 (2001) 343–352.
59 J.P. Quirino, K. Otsuka and S. Terabe, *J. Chromatogr. A*, 714 (1998) 29–38.
60 J.P. Quirino and S. Terabe, *J. Chromatogr. A*, 798 (1998) 251–257.
61 S.-F. Wang, Y.-Q. Wu, Y. Ju, X.-G. Chen, W.-J. Zheng and Z.-D. Hu, *J. Chromatogr. A*, 1017 (2003) 27–34.
62 C.L. da Silva, E.C. de Lima and M.F.M. Tavares, *J. Chromatogr. A*, 1014 (2003) 109–116.
63 P. Boček, M. Deml, P. Gebauer and V. Dolník, *Analytical Isotachophoresis*, VCH, 1988.
64 R. Kuhn and S. Hoffstetter-Kuhn, *Capillary Electrophoresis: Principles and Practice*, Springer, Berlin, 1993.
65 D. Kaniansky and J. Marák, *J. Chromatogr.*, 498 (1990) 191–204.
66 F. Evererts, T.P.E.M. Verheggen and F.E.P. Mikkers, *J. Chromatogr.*, 169 (1979) 21–38.
67 L. Křivánková and P. Boček, *J. Chromatogr. B*, 689 (1997) 13–34.
68 J. Pospichal, P. Gebauer and P. Bocek, *Chem. Rev.*, 89 (1989) 419–430.
69 V. Kaŝiĉka and Z. Prusík, *J. Chromatogr.*, 569 (1991) 123–174.
70 D. Kaniansky and J. Marák, *J. Chromatogr.*, 498 (1990) 191–204.
71 F. Foret, E. Szoko and B.L. Karger, *J. Chromatogr.*, 680 (1992) 3–12.

72 F. Foret, E. Szoko and B.L. Karger, *Electrophoresis*, 14 (1993) 417–428.
73 A.M. Wolters, D.A. Jayawickrama, C.K. Larive and J.V. Sweedler, *Anal. Chem.*, 74 (2002) 4191–4197.
74 I. Valaskova and E. Havranek, *J. Chromatogr. A*, 836 (1999) 201–208.
75 M.J. van der Schans, J.L. Beckers, M.C. Molling and F.M. Everaerts, *J. Chromatogr. A*, 717 (1995) 139–147.
76 R.A. Kautz, M.E. Lacey, A.M. Wolters, F. Foret, A.G. Webb, B.L. Karger and J.V. Sweedler, *J. Am. Chem. Soc.*, 123 (2001) 3159–3160.
77 A.M. Wolters, D.A. Jayawickrama, C.K. Larive and J.V. Sweedler, *Anal. Chem.*, 74 (2002) 2306–2313.
78 A.M. Wolters, D.A. Jayawickrama, C.K. Larive and J.V. Sweedler, *Anal. Chem.*, 74 (2002) 4191–4197.
79 Z.D. Peterson, C.R. Bowerbank, D.C. Collins, S.W. Graves and M.L. Lee, *J. Chromatogr. A*, 992 (2003) 169–179.
80 D. Mohan, L. Pasˇa-Tolic, C.D. Masselon, N. Tolic, B. Bogdanov, K.K. Hixson, R.D. Smith and C.S. Lee, *Anal. Chem.*, 75 (2003) 4432–4440.
81 W.N. Vreeland, S.J. Williams, A.E. Barron and A.P. Sassi, *Anal. Chem.*, 75 (2003) 3059–3065.
82 P. Praus, *Talanta*, 62 (2004) 977–982.
83 M. Masár, M. Danková, E. Ölvecká and A. Stachurová, D Kaniansky and Bernd Stanislawski, *J. Chromatogr. A*, 1026 (2004) 31–39.
84 K. Ito, T. Ichihara, H. Zhuo, K. Kumamoto, A.R. Timerbaev and T. Hirokawa, *Anal. Chim. Acta*, 497 (2003) 67–74.
85 D.S. Stegehuis, H. Irth, U.R. Tjaden and J. van der Greef, *J. Chromatogr.*, 538 (1991) 393–402.
86 B. Toussaint, Ph. Hubert, U.R. Tjaden, J. van der Greef and J. Crommen, *J. Chromatogr. A*, 871 (2000) 173–180.
87 M. Urbanek, L. Blechtova, M. Pospýsilova and M. Polasek, *J. Chromatogr. A*, 958 (2002) 261–271.
88 R. Hamoudová, M. Urbánek, M. Pospíšilová and M. Polášek, *J. Chromatogr. A*, 1032 (2004) 281–287.
89 Zakariya K. Shihabi, *Electrophoresis*, 23 (2002) 1628–1632.
90 Y. Kong, N. Zheng, Z.-C. Zhang and R.-Y. Gao, *J. Chromatogr. B*, 795 (2003) 9–15.
91 J. Bergmann, U. Jaehde, M. Mazereeuw, U.R. Tjaden and W. Schunack, *J. Chromatogr. A*, 734 (1996) 381–389.
92 F. Foret, E. Szoko and B.L. Karger, *J. Chromatogr.*, 608 (1992) 3–12.
93 T.J. Thompson, F. Foret, P. Vouros and B.L. Karger, *Anal. Chem.*, 65 (1993) 900–906.
94 M.J. van der Schans, J.L. Beckers, M.C. Molling and F.M. Everaerts, *J. Chromatogr. A*, 717 (1995) 139–147.
95 Z.Q. Xu, T. Hirokawa, T. Nishine and A. Arai, *J. Chromatogr. A*, 990 (2003) 53–61.

96 P. Britz-McKibbin, A.R. Kranack, A. Paprica and D.D.Y. Chen, *Analyst*, 123 (1998) 1461–1463.
97 P. Britz-Mckibbin and D.Y. Chen, *Anal. Chem.*, 72 (2000) 1242–1252.
98 P. Britz-McKibbin, G.M. Bebault and D.D.Y. Chen, *Anal. Chem.*, 72 (2000) 1729–1735.
99 J.B. Kim, P. Britz-McKibbin, T. Hirokawa and S. Terabe, *Anal. Chem.*, 75 (2003) 3986–3993.
100 J.B. Kim, Y. Okamoto and S. Terabe, *J. Chromatogr. A*, 1018 (2003) 251–256.
101 P. Britz-Mckibbin, J. Wong and D.D.Y. Chen, *J. Chromatogr. A*, 853 (1999) 535–540.
102 P. Britz-McKibbin and S. Terabe, *J. Chromatogr. A*, 1000 (2003) 917–934.
103 S.H. Shim, A. Riaz, K. Choi and D.-S. Chung, *Electrophoresis*, 24 (2003) 1603–1611.
104 S.R. Mei, Q.H. Yao, L.S. Cai, J. Xing, G.W. Xu and C.Y. Wu, *Electrophoresis*, 24 (2003) 1411–1415.
105 P. Britz-McKibbin, T. Ichihashi, K. Tsubota, D.D.Y. Chen and S. Terabe, *J. Chromatogr. A*, 1013 (2003) 65–76.
106 S.J. Wang, W.-L. Tseng, Y.-W. Lin and H.-T. Chang, *J. Chromatogr. A*, 979 (2002) 261–270.
107 P. Britz-McKibbin and S. Terabe, *Chromatography*, 23 (2002) 99–100.
108 P. Britz-McKibbin, T. Nishioka and S. Terabe, *Anal. Sci.*, 19 (2003) 99–104.
109 N.J. Munro, J. Palmer, A.M. Stalcup and J.P. Landers, *J. Chromatogr. B*, 731 (1999) 369–381.
110 J. Palmer, D.S. Burgi, N.J. Munro and J.P. Landers, *Anal. Chem.*, 73 (2001) 725–731.
111 J.P. Quirino and S. Terabe, *Science*, 282 (1998) 465–468.
112 J.P. Quirino and S. Terabe, *Anal. Chem.*, 71 (1999) 1638–1644.
113 J.-B. Kim, K. Otsuka and S. Terabe, *J. Chromatogr. A*, 912 (2001) 343–352.
114 J.-B. Kim, J.P. Quirino, K. Otsuka and S. Terabe, *J. Chromatogr. A*, 916 (2001) 123–130.
115 J.-B. Kim, K. Otsuka and S. Terabe, *J. Chromatogr. A*, 979 (2002) 131–136.
116 M.R.N. Monton, K. Otsuka and S. Terabe, *J. Chromatogr. A*, 985 (2003) 435–445.
117 Y. Sera, N. Matsubara, K. Otsuka and S. Terabe, *Electrophoresis*, 22 (2001) 3509–3515.
118 N. Matsubara and S. Terabe, *J. Chromatogr. A*, 680 (1994) 311–315.
119 K. Koezuka, H. Ozaki, N. Matsubara and S. Terabe, *J. Chromatogr. B*, 689 (1997) 3–11.

120 M.R.N. Monton, J.P. Quirino, K. Otsuka and S. Terabe, *J. Chromatogr. A*, 939 (2001) 99–108.
121 J. Palmer, N.J. Munro and J.P. Landers, *Anal. Chem.*, 71 (1999) 1679–1687.
122 J. Palmer and J.P. Landers, *Anal. Chem.*, 72 (2000) 1941–1943.
123 J. Palmer, D.S. Burgi, N.J. Munro and J.P. Landers, *Anal. Chem.*, 73 (2001) 725–731.
124 J.P. Quirino, S. Terabe and P. Bocek, *Anal. Chem.*, 72 (2000) 1934–1940.
125 F. Palmer, *J. Chromatogr. A*, 1036 (2004) 95–100.
126 J.P. Quirino, J.-B. Kim and S. Terabe, *J. Chromatogr. A*, 965 (2002) 357–373.
127 J.P. Quirino and S. Terabe, *Anal. Chem.*, 72 (2000) 1638–1644.
128 J.-B. Kim, K. Otsuka and S. Terabe, *J. Chromatogr. A*, 932 (2001) 129–137.
129 L.-Y. Zhu, C.-H. Tu and H.-K. Lee, *Anal. Chem.*, 74 (2002) 5820–5825.
130 K. Isoo and S. Terabe, *Anal. Chem.*, 75 (2003) 6789–6798.
131 C. Fang, J.-T. Liu and C.-H. Lin, *Talanta*, 58 (2002) 691–699.
132 P. Britz-McKibbin, K. Otsuka and S. Terabe, *Anal. Chem.*, 74 (2002) 3736–3743.
133 P. Britz-McKibbin and S. Terabe, *Chem. Rec.*, 2 (2002) 397–404.
134 H. Harino, S. Tsunoi, T. Sato and M. Tanaka, *Anal. Bioanal. Chem.*, 369 (2001) 546–547.
135 Y. Takagai and S. Igarashi, *Analyst*, 126(5) (2001) 551–552.
136 O. Nunez, J.-B. Kim, E. Moyano, M.T. Galceran and S. Terabe, *J. Chromatogr. A*, 961 (2002) 65–75.
137 C.-E. Lin, Y.-C. Liu, T.-Y. Yang, T.-Z. Wang and C.-C. Yang, *J. Chromatogr. A*, 916 (2001) 239–245.
138 K. Otsuka, M. Matsumura, J.-B. Kim and S. Terabe, *J. Pharm. Biomed. Anal.*, 30 (2003) 1861–1867.
139 C.L. da Silva, E.C. de Lima and M.F.M. Tavares, *J. Chromatogr. A*, 1014 (2003) 109–116.
140 S. Takeda, A. Omura, K. Chayama, H. Tsuji, K. Fukushi, M. Yamane, S. Wakida, S. Tsubota and S. Terabe, *J. Chromatogr. A*, 979 (2002) 425–429.
141 J.P. Quirino, S. Terabe, K. Otsuka, J.B. Vincent and G. Vigh, *J. Chromatogr. A*, 838 (1999) 3–10.
142 R.B. Taylor, R.G. Reid and A.S. Low, *J. Chromatogr. A*, 916 (2001) 201–206.
143 C.-H. Wu, M.-C. Chen, A.-K. Su, P.-Y. Shu, S.-H. Chou and C.-H. Lin, *J. Chromatogr. B*, 785 (2003) 317–325.
144 M.J. Markuszewski, P. Britz-McKibbin, S. Terabe, K. Matsuda and T. Nishioka, *J. Chromatogr. A*, 989 (2003) 293–301.
145 C. Fang, J.-T. Liu and C.-H. Lin, *J. Chromatogr. B*, 775 (2002) 37–47.

146 Y.-Z. Yang, R.I. Boysen, J.I.- Chen Chen, H.H. Keah and M.T.W. Hearn, *J. Chromatogr. A*, 1009 (2003) 3–14.
147 S. Terabe, M.J. Markuszewski, N. Inoue, K. Otsuka and T. Nishioka, *Pure Appl. Chem.*, 73 (2001) 1563–1572.
148 L. Jia, B.-F. Liu, S. Terabe and T. Nishioka, *Anal. Chem.*, 76 (2004) 1419–1428.

Chapter 4

Coupling continuous flow systems to capillary electrophoresis

Bartolomé M. Simonet, Angel Ríos and Miguel Valcárcel

4.1 INTRODUCTION

Since the first publications on capillary electrophoresis (CE) by Mikkers et al. [1] in 1979 and Jorgenson and Lukacs [2] in 1981, this technique has received considerable attention in analytical chemistry. The popularity and acceptance of CE can be explained by its capacity to rapidly determine many analytes at a high resolution with low sample and electrolyte consumption. Despite its good characteristics, several shortcomings have been detected when analysing real samples with complicated matrix composition. For example, owing to the extremely small inner diameter of the capillary, solid particles present in the samples must be removed in order to prevent the capillary from clogging. Another shortcoming is that real samples containing large amounts of proteins and macromolecules cannot be directly injected because these molecules would be adsorbed on the capillary wall. Finally, the use of UV–Vis detectors, used widely because of their simplicity and flexibility, together with the small sample volumes and short light paths (generally limited by the inner diameter of the capillary), results in poor sensitivity, expressed in concentration units in many applications [3]. To solve some of these problems, off-line sample pretreatment steps like filtration, purification or preconcentration must be applied to the CE determination first. As is well known, off-line pretreatments generally lack precision, require manual manipulation and are time-consuming. These generally recognised limitations still affect their broad acceptance in routine applications. Automatic continuous-flow systems (CFSs) offer an elegant and useful alternative to these pretreatment steps.

Comprehensive Analytical Chemistry XLV
M.L. Marina, A. Ríos and M. Valcárcel (Eds)
Volume XLV ISSN: 0166-526X DOI: 10.1016/S0166-526X(05)45004-7

4.2 ROLE OF CFSs

The use of low-pressure CFSs can be considered as one of the most reliable ways of improving the performance of analytical methods through automation, miniaturisation and simplification of the preliminary operations of analytical methodologies. As schematised in Fig. 4.1, the coupling of CFSs to CE is not only limited to sample treatment. It also brings the possibility of implementing automatic calibration procedures and coupling screening devices that allow automatic screening of the positive samples that must be introduced in the CE capillary for its analysis. The total analysis time is the sum of the time required for sample treatment and the time required for the electrophoretic separation. However, it must be taken into account that the use of CFS not only reduces the time required for sample treatment but it also reduces the electrophoretic time. A new sample cannot be introduced in the CE capillary until all the analytes have passed through the detector. This means that the high voltage must be interrupted and the capillary must be rinsed before each run. Kuban et al. [4] proved that it is possible to perform consecutive runs in CE without interrupting the high voltage when using a CFS–CE coupling. This asset has also been shown by Kaljurand et al. [5]. The possibility of making multiple injections into the CE capillary without rinsing before each run has been studied by

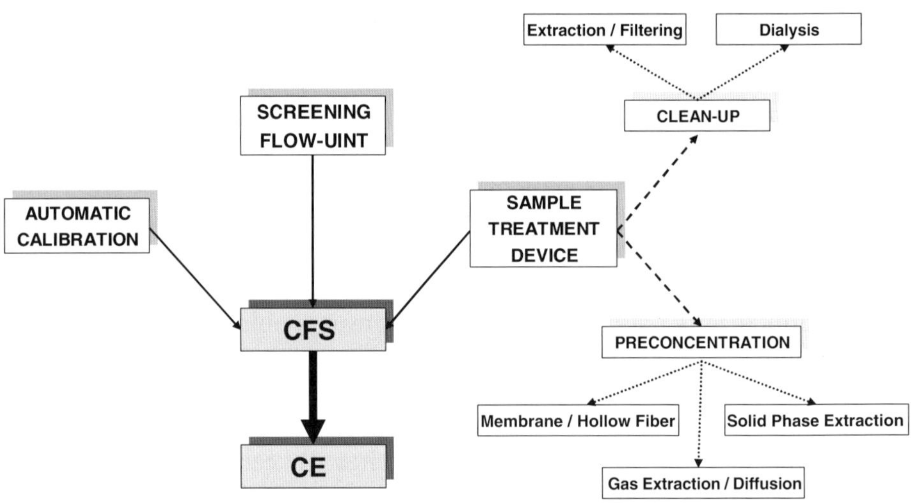

Fig. 4.1. Schematic uses of continuous-flow systems (CFS) coupled to capillary electrophoresis (CE).

Roche et al. [6] demonstrating the possibility of performing 100 sample injections.

The higher the throughput, the more favourable is the influence of the coupling of CFSs to CE on other analytical properties because of the special characteristics of CFSs [7,8]. In fact, they provide higher precision and an increase in sensitivity and selectivity by comparison with the corresponding manual sample pretreatment procedures [3]. Another important point to consider is that the sample volume required for analysis can be greatly reduced when using miniaturised CFS devices. In short, the mentioned approaches for on-line sample pretreatment offer significant advantages over the conventional manual procedures, which are generally tedious. Therefore, it is clear that combining, or hyphenating, CFSs with a high-resolution capacity separation technique like CE is one of the most outstanding advances in chemical measurement processes, which has been fully demonstrated in a wide number of applications.

4.3 INTERFACES FOR COUPLING FLOW SYSTEMS TO CE

The hybridisation of CFSs with CE is characterised by the coupling of a hydrodynamic flow sample treatment with an electroosmotic-flow-driven separation in the capillary system, which requires a special design in interfacing. There have been some technical difficulties in the performance of such a combination. For example, the discontinuity of sample introduction in conventional CE equipment due to the necessity to rinse the CE capillary between analysis, the differences between sample introduction in both techniques (only a few nanolitres are introduced in the CE system) and the electrical separation of the high voltage of the CE equipment from the CFS system. Considering these aspects and that the accessibility and running of homemade CE arrangements and commercial CE equipments are different, the interfaces for coupling CFSs can be separated into two groups: coupling CFS to homemade CE arrangements and coupling CFS to commercial CE equipments. Finally, the coupling of CFS to CE-integrated microchips can also be considered as another group due to its special characteristics and the general trend towards the miniaturisation of electrophoretic systems.

The combination of CFS and CE equipments can be classified according to the degree of integration between these two units. Four levels of CFS sample treatment modules can be coupled with CE, either off-line (manual), at-line (robotic interface), on-line (coupling via a transfer line) or in-line (complete integration between sample preparation and separation system).

4.3.1 CFSs to homemade CE arrangements

The groups of Fang [9] and Karlberg [10] developed split-flow interfaces for coupling CFS to CE. The general scheme of a CFS–CE system configuration with the designs of the two interfaces used in order to couple CFS to CE is shown in Fig. 4.2. Both interfaces are characterised by an electrically grounded low dead volume split-flow cell, in which a small fraction of the sample is electrokinetically introduced into the CE capillary located in the cell when an injected sample zone is transported

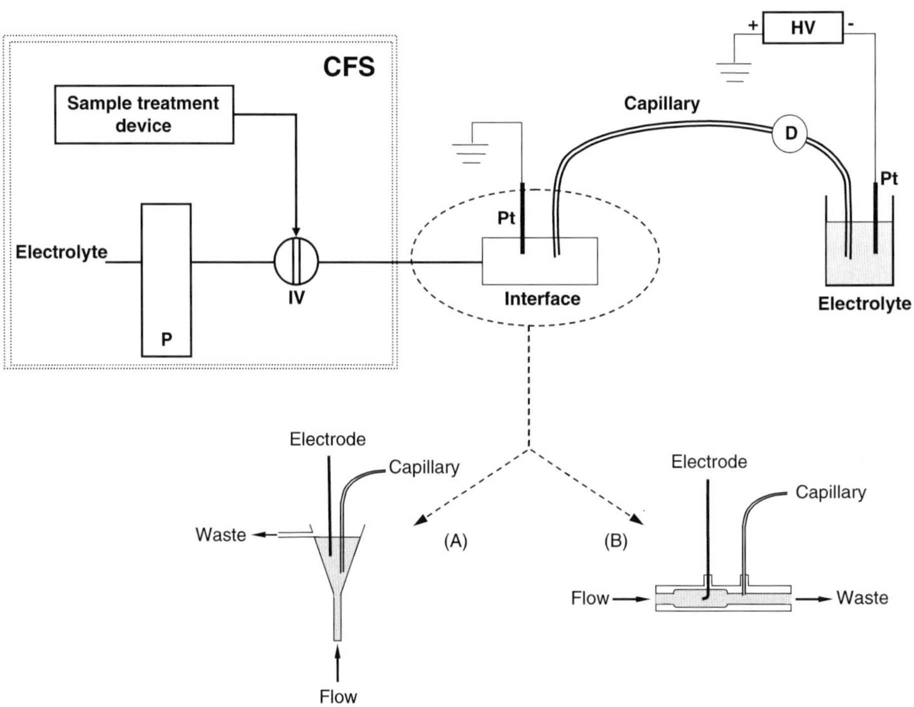

Fig. 4.2. Schematic diagram of CFS–CE system. (A) Split-flow interface in vertical design and (B) split-flow interface in horizontal design. Pt, platinum electrode; P, peristaltic pump; IV, injection valve; D, detector; HV, high voltage. (Split flow cell: adapted from Refs. [9,10].)

past the capillary tip by a carrier electrolyte using a peristaltic or syringe pump. The interface developed by Kuban et al. [11] consisted of a Plexiglas body with a horizontal channel that permited the introduction of the electrolyte carrier and the sample from the CFS device. The CE capillary and the grounded electrode were inserted vertically in the channel as described in Fig. 4.2B. The interface developed by Fang et al. [12] consisted of a vertical, open cell with a conical shape in which the grounded electrode and the CE capillary tip are in contact with the solution from the CFS device (see Fig. 4.2A). The liquid level was maintained constant in the conical cell by applying an outflow rate higher than the inflow rate. In this type of device, it is important to maintain the liquid in the interface at a level identical with that of the capillary outlet in order to avoid hydrodynamic flow in the capillary.

Compared with the conventional methods of sample introduction in CE (both hydrodynamic and electrokinetic injections) the split-flow interface allows an enhancement in sampling and precision. Nevertheless, its use is limited to electrokinetic injection, which can produce biased effects in some cases. In order to perform a hydrodynamic injection, Kuban et al. [13] introduced a valve at the end of the horizontal split-flow interface. Therefore, when the sample plugged in the carrier passed the capillary tip in the interface, the valve was closed and a fraction of the sample was forced into the capillary. During the injection, high voltage was shut off. The main limitation of this methodology is the control of time to open/close the valve and shut off/on the high voltage. At the same time, Pu and Fang [14] used a split-flow cell that made it possible to inject hydrodynamically the sample into the CE capillary using the electroosmotic flow. The interface used a Nafion joint in order to connect the CE capillary to the tube of the CFS system (see Fig. 4.3). In both methods, an electronic time relay system was used to control the switching of the high voltage.

Recently, Ruzicka [15,16] has combined a microsequential injection system with capillary electrophoresis (μSIA–CE). The combination μSIA–CE would apparently be more advantageous over flow injection (FI)–CE because the μSI system, driven by a syringe pump, can pressurize the system to perform hydrodynamic injections of the sample and to refresh the background electrolyte buffer into the capillary. In this way, the advantages of the μSI system for sample dilution, preconcentration and purification are well known. Wu et al. [15] presented a μSI–CE system, interfaced with an integrated Lab-on-Valve (LOV) manifold. The analytical features of the μSI–CE were demonstrated on

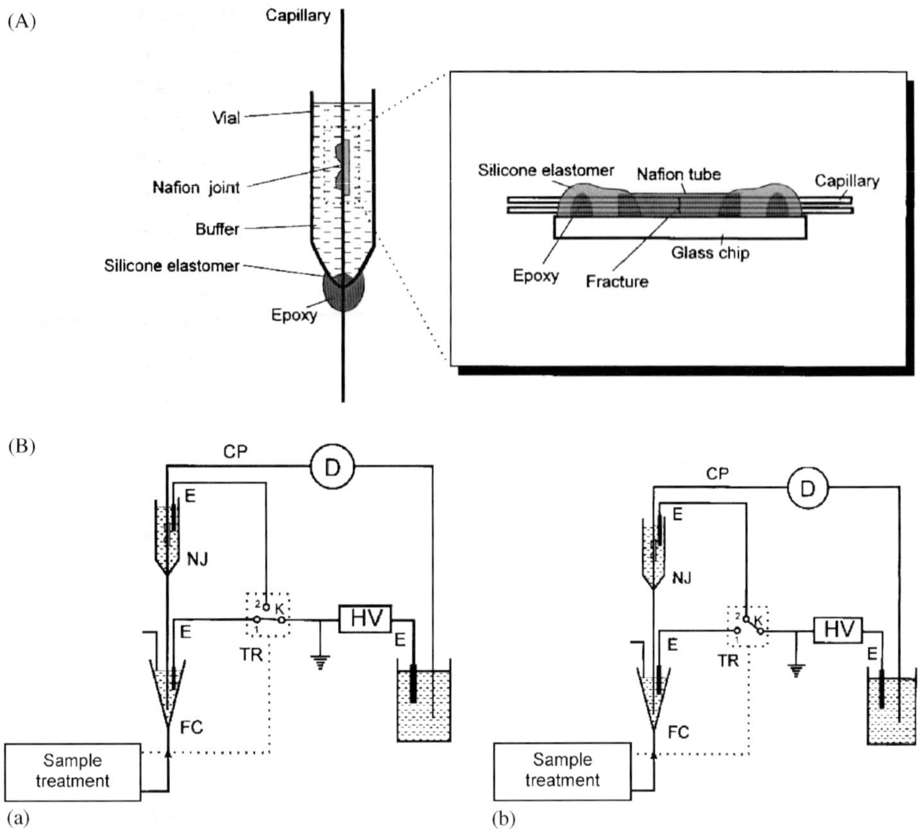

Fig. 4.3. (A) Structure of the Nafion joint. (B) Coupling of CFS to CE with EOF traction injection. (a) Fill position; (b) injection position. FC, split flow cell; NJ, Nafion joint; TR, time relay; HV, high voltage; D, detector; E, platinum electrode; CP, separation capillary; W, waste. (Adapted from Ref. [14].)

assays of samples containing 10 anions. The anions were determined by performing several types of CE sample injections such as hydrodynamic injections, electrokinetic injection and head column field amplification sample stacking injection. The same authors, using the same system (LOV–CE), developed a method for the assays of biomolecules and for peptide derivatisation in situ. The LOV system was used to make the derivatisation with the fluorogenic reagent, and the CE was used for separation. In order to increase sampling frequency, the LOV was used to hydrodynamically flush the remaining sample material out of the capillary [16].

Ponton and Evans [17] demonstrated the possibility of making an in-line valve injection for CE using a commercially available injection valve designed for liquid chromatography (LC). The valve is made of ceramics or polyether-etherketone (PEEK). As a result, the metal injector body is electrically insulated from the high-voltage buffer. The voltage was continuously applied to the CE capillary and was partially interrupted when the valve was switched for injection. The authors proved the in-line injection to be successful. Although this valve was not used in this work as an interface to couple CFS to CE, its use for this purpose is immediate, as it is equivalent to a hydrodynamic sample injection.

4.3.2 Coupling flow systems to commercial CE equipments

The group of Valcárcel [7,18] has developed several systems to couple CFS systems to commercial CE equipment. All of these systems are characterised as at-line approaches. A robotic interface was used in all of the cases.

A typical manifold of coupling CFS to commercial CE is shown in Fig. 4.4. This coupling is based on the combination of mechanical and electronic elements. The mechanical part consists of a custom-made programmable arm and an autosampler included in the commercial CE equipment. The functioning of the coupled elements (CFS, programmable arm, autosampler and electronic system) is controlled by the built-in microprocessor of the instrument via an electronic interface and customised software. As shown in Fig. 4.4, the programmable arm has two different positions with respect to the vial in the autosampler, and it establishes the automated link between the CFS and the capillary end through two injection needles. The injection needles have different lengths, and are both connected to the CFS. The longer one is used to fill the vial and the shorter one to drain it in order to maintain a preset liquid level. While the CFS take place the sample preparation and pumping the prepared sample to the CE vial, the needles are down. When the CFS stops, the needles pass to the upper position, and the CE analysis starts by moving the sample-filled vial to where the capillary end and the electrode are located.

As an alternative to the use of the programmable arm, the replenishment systems available, which are commercial equipments that allow the electrophoretic buffer to be removed from, or changed to, any vial on the carousel, can also be used with some modifications for

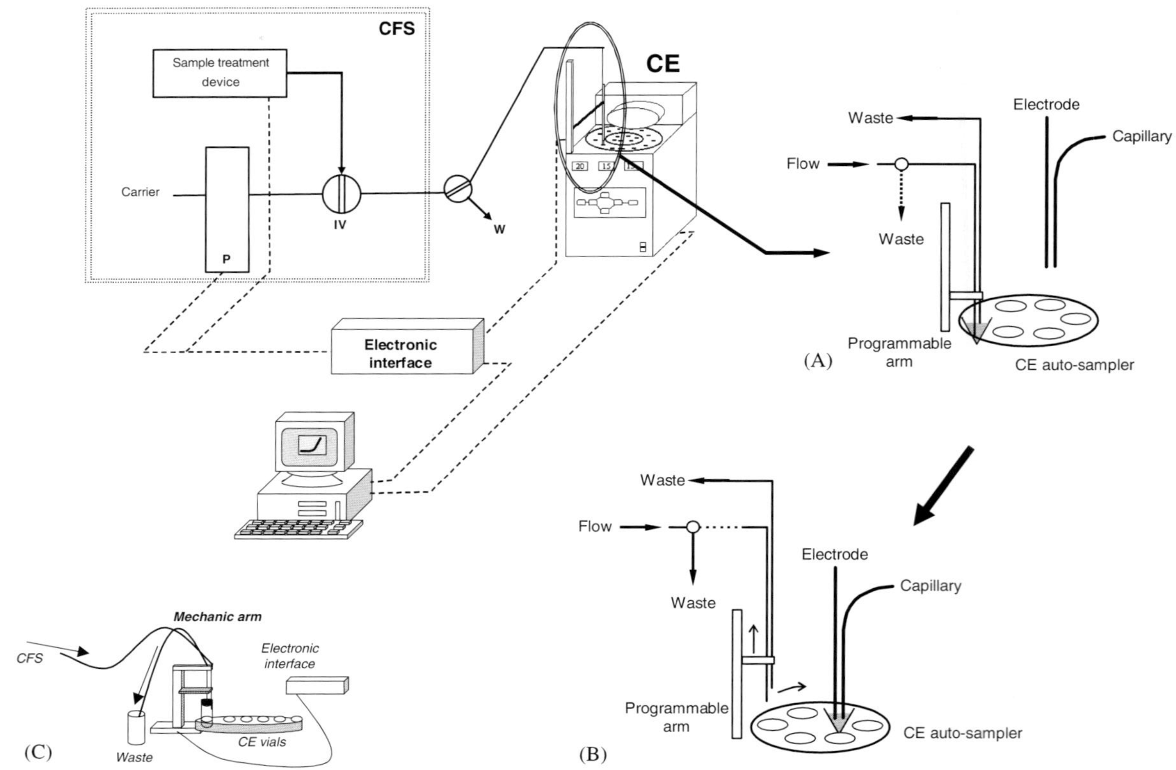

Fig. 4.4. Schematic diagram of the interface used to couple CFS to CE through a programmable arm interface in (A) sample-collection mode and (B) sample-injection mode for electrophoretic analysis. (C) Details of the programmable arm interface. (Adapted from Ref. [3,8].)

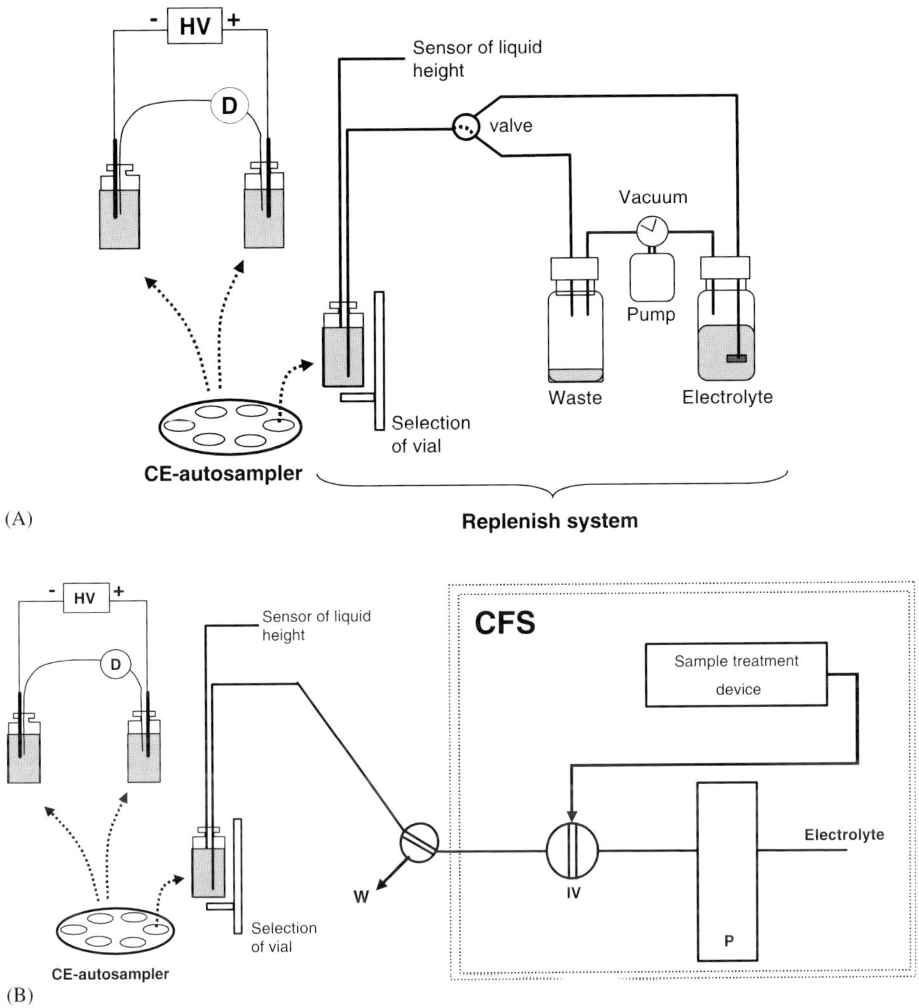

Fig. 4.5. Schematic depiction of the connection of CFS to the CE equipment via the replenishment needle. (A) Normal configuration in the commercial CE and (B) coupling of CFS. (Adapted from Ref. [19].)

coupling CFS devices. To carry out such coupling, Santos et al. [19] disconnected the replenishment needle of the polytetra fluoroethylene (PTFE) tubes in the equipment (connected to the buffer reservoir and the waste, respectively) and replaced them with a PTFE tube leading from the CFS (see Fig. 4.5). Therefore, the CE chemstation was programmed to ensure the on-line introduction of the sample into the vial and its subsequent injection into the capillary.

The same authors used the CFS device without the additional peristaltic or syringe pumps by adequately modifying the replenishment system [20]. The replenishment system includes a needle for insertion into the vials. The needle is connected to a valve that allows vials to be loaded or unloaded by using a vacuum or pressure system. To use this system as a CFS, a minicolumn was inserted between the needle and the valve (see Fig. 4.6) in order to facilitate preconcentration and clean-up of samples. A shortcoming of this approach is that the flow rate is constant and cannot be controlled by the instrument. Therefore, the column dimensions must be adapted in order to ensure an adequate flow rate of samples and eluent.

The split-flow cells were not directly connected to commercial CE equipments because the accessibility to the electrode and CE capillaries is limited. Moreover, the software stations in the commercial equipments are another factor that limits their use. Therefore, the coupling of CFS to CE described in the literature is based on the use of robotic interfaces.

4.3.3 Flow injection–CE integrated microchips

The coupling of CFS to a sequential injection system for the introduction of a microvolume of sample to a micro-fluidic CE system has also been reported [21]. In this work, the CFS was used to deliver microlitres sample by a suitable carrier to a microfluidic split-flow interface (see Fig. 4.7). The work demonstrated the feasibility of using a sequential injection system to couple CE systems on a miniaturised scale, as well as the potential of on-line combining CFS sample treatment on a miniaturised scale with microchip CE-based systems. The system was used to determine derivatised amino acids with isothiocyanate allowing the injection of 1.67 μl into the split-flow interface. The separation performed with the microfluidic CE system was achieved in only 80 s. The coupling of flow injection systems to introduce the sample on a chip platform has also been described. It was achieved by inserting a microdrop in the split-flow interface of the microchip [22–24]. Fu et al. [22] used an H-channel configuration with the separation capillary positioned between two tubular side arms. A falling-drop interface connected to one side arm was used to achieve electrical isolation between the FI and CE systems.

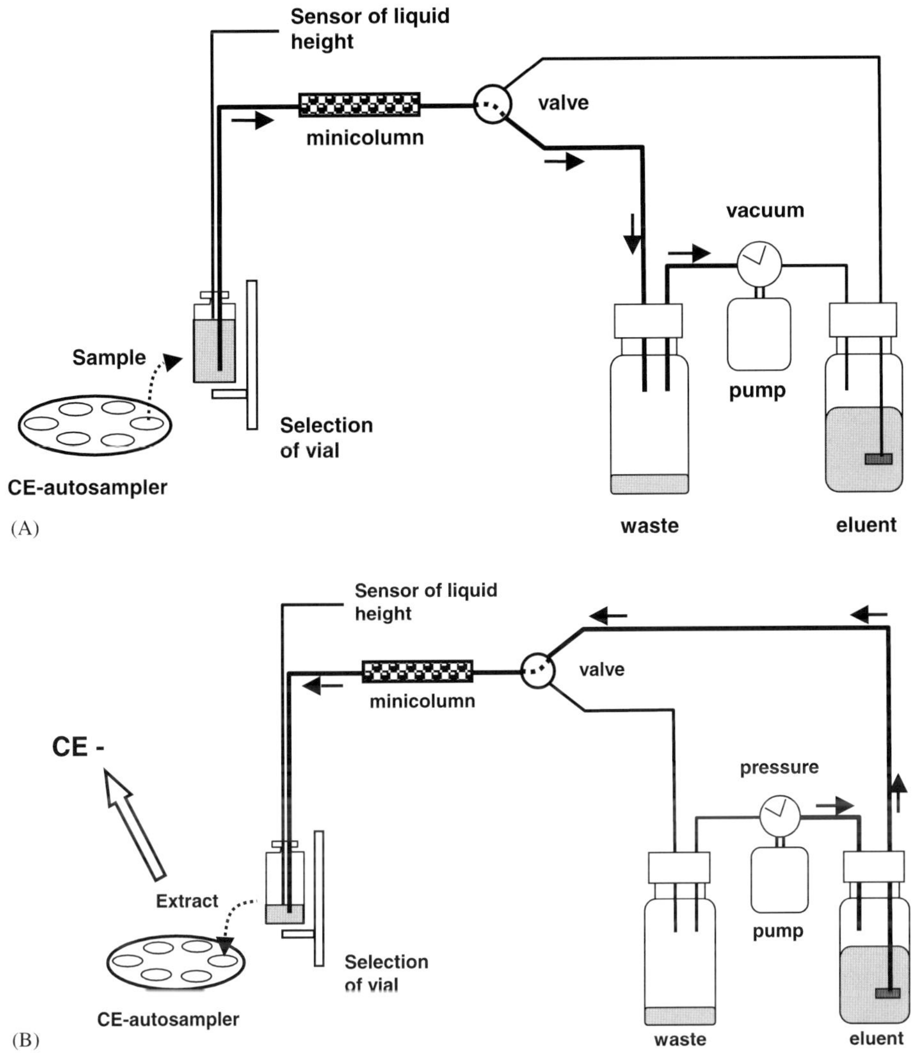

Fig. 4.6. Schematic depiction of the replenishment system used as a clean-up/preconcentration system. (A) Configuration to clean-up and preconcentrate the analytes from the samples and (B) configuration to elute the analytes from the minicolumn for its further CE-analysis. From Ref. [20]. Reprinted with permission from Wiley & Sons, Inc.—Wiley-VCH Verlag GmbH,— Copyright (2004).

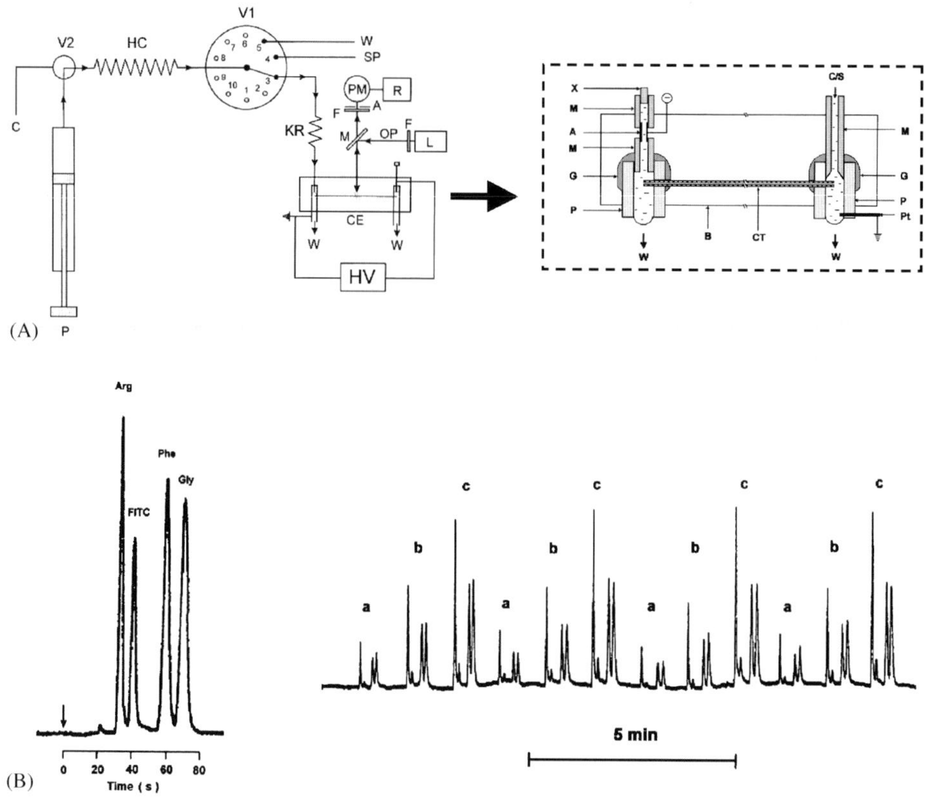

Fig. 4.7. (A) Schematic diagram of the SI-sample introduction microfluidic CE system with split-flow sample introduction interface. P, syringe pump; V1, selection valve; V2, two-way valve; C, carrier; HC, holding coil; KR, knotted reactor; CE, microfluidic CE system; HV; high-voltage power supply; L, laser source; OP, optical fibre; M, dichroic mirror; F, filters; PM, photomultiplier; R; chart record; CT, quartz separation capillary tube (75 μm i.d.); G, epoxy glue; P, tygon tube; M, microline tube; A, platinum tube anode; Pt, platinum grounding electrode; X, glass plug; C/S carrier/sample; W, waste. (B) Typical electropherograms of a mixture of three FITC-labelled amino acids and four cycles of sequentially injected 25, 50 and 100 μM samples. Instrumental conditions: separation length of 25 cm, electric field strength of 250 V/cm, buffer 4 mM sodium tetraborate (pH 9.2), flow rate, 2 ml/min and 3.3 μl sample. Reprinted from Ref. [21], Copyright (1999), with permission from Elsevier.

4.4 SAMPLE INJECTION AND ON-LINE PREPARATION

As can be deduced from the previous section, one of the advantages of coupling CFS to CE is the possibility of making a sample treatment involving clean-up and preconcentration steps. Obviously, the eluent containing the purified sample from the CFS must be compatible with its introduction into the CE equipment.

4.4.1 On-line clean-up and preconcentration

As indicated earlier, the number of routine applications using CE in real samples is limited, one of the main problems being that the analyte—expressed in concentration units—is rather difficult to detect because of the low-volume loadability of the capillary. Two alternative approaches to enhance sensitivity have been described in the literature: (i) the use of on-line capillary-focusing methods based on electrophoretic effects for preconcentration of the analytes and (ii) the use of CFS coupled to CE.

In general, the use of preconcentration methodologies based on electrophoretic effects is limited when real samples are analysed, because preconcentration is based on the differential velocity that the analyte acquires in a medium composed of different electrolytes. Therefore, samples should have a uniform matrix with low conductivity. Small changes in the matrix composition or in the ionic strength of the sample, which can results in important changes in conductivity, resulted a critical aspect for applying such methodologies. These requirements are difficult to achieve when real biomedical or environmental samples are analysed. Thus, analyte clean-up and enrichment are required before CE analysis. For these reasons, preconcentration methodologies based on CFS coupled to CE seem to be more robust for their use in routine analysis. Solid-phase extraction (SPE) is probably the most frequently used methodology to preconcentrate analytes.

Owing to its versatility, SPE has been coupled to CE either at-line, on-line or in-line. Table 4.1 shows some examples, which have been summarised and classified into three groups according to the type of sample analysed: biological samples, environmental samples and food samples. This table does not intend to provide readers a detailed list of all the published papers, but it summarises different examples of the applicability of such methodologies in the analysis of different types of real samples with complex matrices.

TABLE 4.1

Selected examples of coupling CFS involving solid-phase extraction systems to capillary electrophoresis

Samples	Analytes	Stationary phase	LOD	RSD (%)	References
Human urine	Chlorophenols	C18 cartridges	0.08–0.46 μg/ml	206–7.2	[38]
Biological fluids	Nonsteroidal anti-inflammatory drugs	C18 minicolumn	0.88–1.71 μg/ml	2.0–7.7	[34]
Citrus samples	Phenolic constituents	C18 minicolumn	0.16–0.50 mg/l	3.0–6.5	[35]
Human plasma	Pseudoephedrine	C18 minicolumn	12 μg/l	1.2	[36]
Wine	Polyphenols	C18 minicolumn	0.05–0.36 mg/l	3.2–7.1	[37]
Wine	Biogenic amines	C18 minicolumn	0.05–0.1 μg/ml	—	[33]
Water	Pesticides	C18 minicolumn	0.01–0.03 μg/ml	—	[31]
Water	Mercury speciation	C18 minicolumn	0.001–0.002 μg/ml	2.0–2.9	[32]
Food samples	Myo-inositol phosphates	Dowex 1 × 2–200 anion exchanger	11–26 μmol/l	3.9–7.9	[41]
Water	Absorbing ions	Amberlite IRA-410 anion exchanger	10 ng/ml	5	[40]
Water	Inorganic ions	Chelex-100	0.01–0.15 μg/ml	—	[39]
Tap water	Haloacetic acids	LiChrolut EN	2–5 μg/l	5.5–9.8	[73]

SPE has also been incorporated in-line with CE via a microcartridge located in the inlet region of the CE capillary [25,26]. Although the sample volume introduced in the system is only at the nanolitre level, high concentration factors were obtained. The microcartridge is made by introducing sorbing particles into a piece of CE capillary. The sorbing particles are retained in the capillary, using fibre-glass filters as fritters. Finally, this piece is inserted into the CE capillary using tubes and epoxy glue. The final length of the packed bed can be as small as 1–3 mm, so the microcartridge can be easily installed inside the CE cartridge of a commercial instrument. The sample enrichment procedure involves introducing the sample using the pressure capabilities of the CE instrument, and afterwards, desorbing the analytes in a small volume and under optimal electrophoretic conditions (maximal stacking and minimal diffusion due to electroosmotic flow). Using silica-C8 particles with monoclonal antibodies bound on its surface as sorbing material, an useful determination of antrazine in environmental samples has been described, with a 1000-fold amplification of sensitivity [27]. As an alternative, Saito and Jinno [28] performed the extraction in short capillaries, packed lengthwise with filaments of synthetic polymer as the extraction medium. The use of fibre-in tube (FIT)-SPME rather than conventional particle-packed SPE cartridge has the advantage of reducing the pressure drop during extraction and desorption. Therefore, higher volumes of sample can be concentrated using a syringe pump and a commercially available connector (similar to that described by Ponton and Evans [17], see section 4.3.1). Not only does this approach permit the introduction of a high volume of sample, it also allows a connector to be used in order to prevent the introduction of the sample into the CE capillary. This methodology was applied to determine antidepressant drugs in biological samples and achieved preconcentration factors higher than 200 [29]. However, the use of these methodologies is limited due to the difficulty in preparing the capillary with the sorbing particles packed and retained with frits. When the connection valve is omitted, an additional shortcoming is the lifetime of the capillary, specially when samples with complex matrices are analysed. In future research in this field, the use of short open-tubular capillaries or even continuous bed or monolithic capillaries in the inlet region will certainly open interesting new possibilities.

CFS on-line SPE and ion exchange using minicolumn has been demonstrated to be an efficient alternative for preconcentration purposes for a wide variety of analytes and samples [30]. Figure 4.8 shows

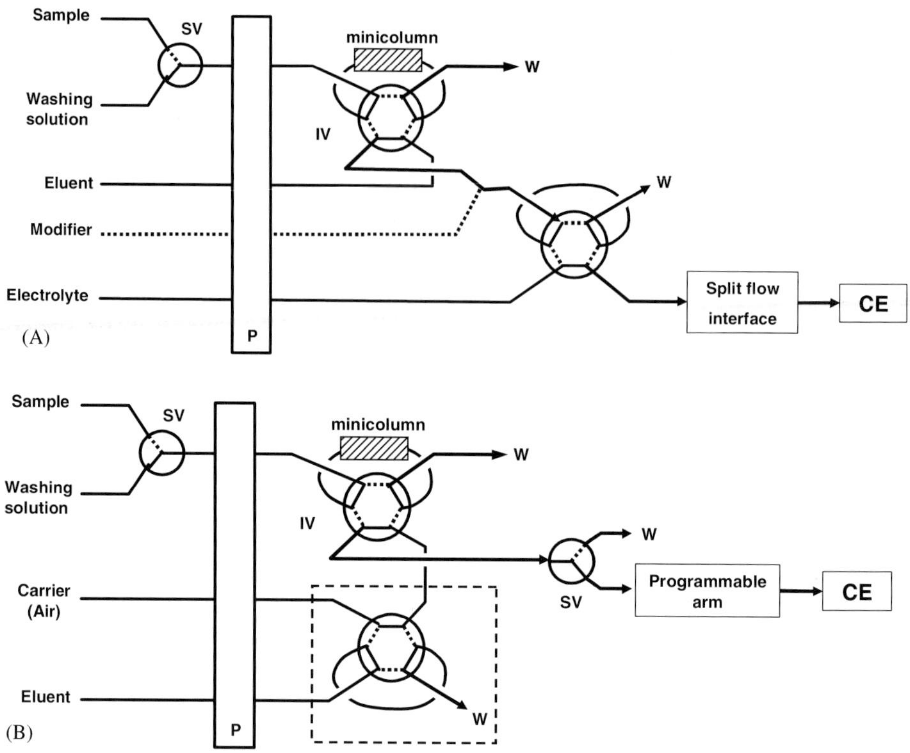

Fig. 4.8. Typical CFS coupled to CE in order to preconcentrate the analyte using SPE when (A) a split-flow interface is used as interface and (B) a programmable arm is used as interface. The manifold depicted in (B) will be also used to couple the CFS through the replenish system.

a typical flow injection manifold used for SPE of the sample as a function of the interface. Figure 4.8A describes the manifold when split-flow interfaces are used, whereas Fig. 4.8B illustrates the situation when robotic interfaces are used. In both cases the solution used to elute the analytes in a minicolumn is compatible with the electrophoretic buffer, or requires a previous modification to its introduction in the CE capillary. Such modification can be easily performed when a robotic interface is used, by collecting the eluent in a vial from the carousel containing the modifier. On the contrary, if split-flow interface is used, a new stream with the modifier must be introduced in the manifold (see Fig. 4.8A). As indicated in the figure, the use of a robotic interface requires the introduction of a controlled volume of sample into the vial. This can be performed by switching the valve or by eluting the sample with a controlled volume of eluent (broken line in Fig. 4.8B), thereby controlling the collection time. The versatility of this manifold has been demonstrated in a large number of applications (see Table 4.1). For example, a C_{18} solid-phase minicolumn can be used to preconcentrate the following substances: pesticides in waters [31]; organo-lead compounds in environmental samples [32]; biogenic amines in wine [33]; non-steroidal anti-inflamatory drugs in biological fluids [34]; phenolic constituents in citrus samples [35]; pseudoephedrine in human plasma [36]; *trans*-resveratrol and other polyphenols in wine [37]; and chlorophenols in human urine [38]. It is important to remark that all of these determinations can be carried out by using the general manifold described in Fig. 4.8, by a single change in the working conditions. In all of these applications, the composition of the solid phase was C18. Other kinds of solid phases used are chelating resins, such as Chelex-100, for the determination of metal ions in water [39]; anion-exchange resins, such as Amberlite IRA-410, for the determination of nitrate, bromide and iodide in water [40]; or Dowex 1×2-200 for the determination of myo-inositol phosphates in food samples [41]. Limits of detection (LOD) achieved with these substances as well as the enrichment factors and precision, are briefly stated in Table 4.1.

4.4.2 Online extraction/filtering

CE capillaries usually have an inner diameter less than 100 μm, therefore, the capillaries are easily clogged by solid particles suspended in sample solutions. For this reason, filtration procedures are always required to analyse liquid samples. Generally, the sample is manually filtered by the operator. Nevertheless, when a CFS is coupled to the CE equipment, the filtration can be done automatically. One example is the work of Santos et al. [19], in which the filtration of wine samples was performed at-line using a simple filter connected to the CFS (see Fig. 4.9). Then, the filtered sample was driven to fill an injection loop. A stream of air was used to transport the filtered sample from the loop of the injection valve to the vial in the CE carousel through the replenishment system (section 4.3.2).

Another example illustrating the usefulness of this approach is the monitoring of a dynamic process. Chen and Fang [42] developed an online filtration–FI–CE system in order to monitor a drug dissolution process. They used a split-flow cell to couple the FI system, which

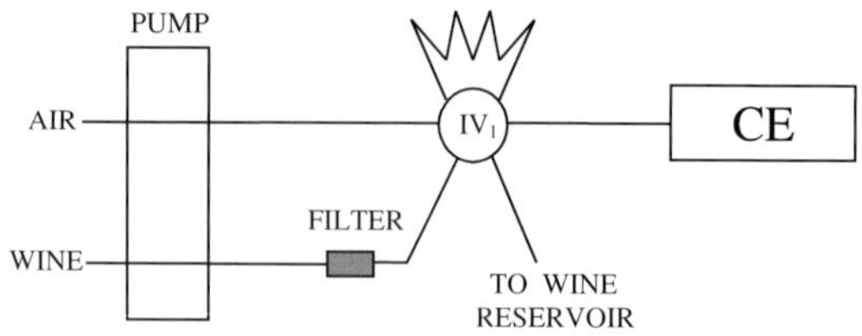

Fig. 4.9. FI manifold used for preliminary filtration of wine samples for the determination of biogenic amines and their insertion into the CE system. IV, injection valve.

involved sample collection from the reactor and filtration to the CE instrument.

Arce et al. [43] proposed a method that permitted the lixiviation of polyphenols present in green tea assisted by microwave radiation. A fraction of the lixiviate solution was filtered on-line and introduced automatically into a vial of the carousel of a commercial CE instrument for further analysis. The same authors used the manifold shown in Fig. 4.10 to develop a methodology for the direct determination of biogenic amines in wine samples. In this case, samples were first filtered by passing them through the microfilters located prior to valve SV1. The filtered samples were then mixed with the buffer and they filled the loop of the injection valve (IV). By switching the IV, the sample plug was driven to the C18 minicolumn located in the second IV, where the amines were preconcentrated. Then, by switching IV2 to its empty position, the elution of the analytes with the HNO_3/MeOH stream selected by valve SV2 took place. The plug containing the preconcentrated analytes went to the CE. Later, valves SV1 and SV2 were switched in order to repeat the process with the next sample. During filtering and mixing with the buffer, MeOH was passed through the C18 minicolumn to flush it, and the analysis of the first sample started in the CE equipment.

4.4.3 On-line dialysis

Dialysis is a powerful sample pretreatment technique used to remove analytes, and it can be coupled with a wide variety of separation techniques. The analytes, which penetrate the membrane, are trapped in the acceptor phase on the other side of the membrane. Membrane characteristics such as porosity, pore size and molecular weight cut-off determine the permeability of the analytes through the membrane.

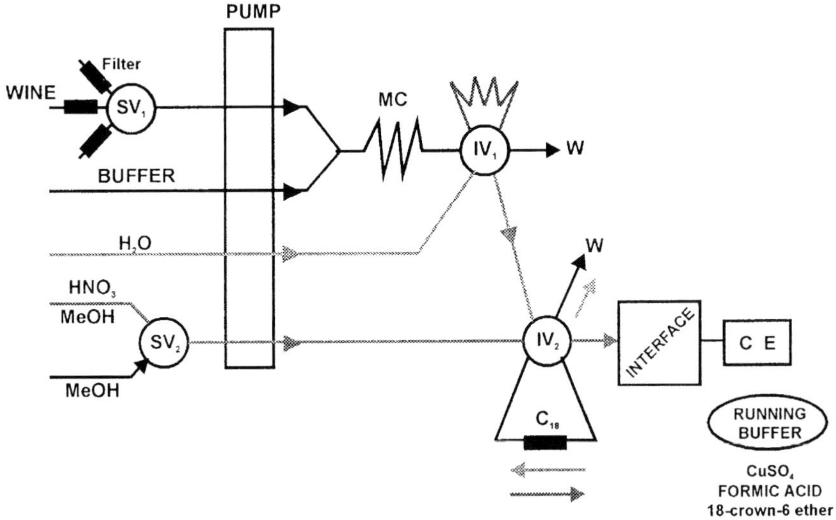

Fig. 4.10. FI manifold used for the determination of biogenic amines in wine samples. The FI manifold involves sample treatment and introducing them in the CE system. (IV; injection valve; SW, switching valve; MC, mixing coil; W, waste). Reprinted from Ref. [33], Copyright (1998), with permission from Elsevier.

Obviously, the transfer process is a consequence of the different concentrations of the analyte on either side of the membrane. However, under normal conditions, a dilution of the analyte will be observed. For this reason, dialysis pretreatment techniques are restricted to situations when the analyte concentrations in the sample are high and when the analytes can easily travel through the membrane. Although the preconcentration factor in some occasions is restricted, this technique is widely used to analyse biological samples because of its ability to separate interfering macromolecular constituents. The main problem posed by macromolecular constituents is their adsorption on the capillary surface: they can produce important changes in the electroosmotic flow. Another practical use of the dialysis unit is the desalination of the sample before its introduction into the CE capillary.

Hogan et al. [44] described an on-line coupling of microdialysis sampling to CE that permits the continuous monitoring of a reaction in vivo. The system allows discrete samples to be taken from the animal and injected into the CE after a microdialysis process. Another example of the utility of microdialysis in monitoring a process in vivo is the work of Lada et al. [45]. In this article, the in vivo monitoring of neurotransmitters glutamate and aspartate using CE with laser fluorescence detection was automatically developed by using a microdialysis device with a flexible loop, microdialysis probe. The probe was made from cellulose fibres (6 kDa cut-off), with 450 μm o.d.(outer diameter) tip diameters and 2-mm tip lengths. The internal volume was only 0.12 μl.

Kuban and Karlberg [46] described the determination of small anions in milk and juice with an on-line dialysis system coupled to CE. The tedious centrifugation procedure, which is time-consuming and it must be inevitably carried out for the CE analysis in milk, can be avoided with the dialysis unit. Using this system, sample pretreatment was not required and no capillary clog was observed even when a large number of determinations were performed.

Veraart et al. [47] developed an automatic system that permitted the determination of acidic drug in urine and serum using on-line dialysis-SPE-CE with UV detection. The system was fully automated. The CE capillary and SPE column were used for over 500 analyses, and the dialysis membrane was replaced after 250 analyses.

In 1990, a rapid system for the determination of aspartate enantiomers in tissue samples by microdialysis on-line coupled to CE was developed by Thompson et al. [48]. The microdialysis probe was directly inserted into the homogenised tissue sample, which generated a continuous sample stream that was filtered and deproteinated.

In general, it can be affirmed that the speed and compatibility of the microdialysis-CE method with automation make this general approach useful for a variety of applications involving high-throughput analysis or sensor-like operations. Other examples of its versatility are summarised in Table 4.2.

4.4.4 On-line membrane and hollow fibre system

In some cases, the use of a suitable impregnated preconcentration membrane that makes it possible to minimise the bed volume of the acceptor solution enhances the sensitivity of the determination. Moreover, by impregnating the membrane with the appropriate solvent, some selectivity can be obtained in the extraction. As an alternative to the use of membranes, the hollow fibre reduces the volume of the sample. In 1997, Palmarsdottir et al. [49] developed a miniaturised hollow fibre device as a liquid membrane support for sample treatment, which was connected to CE on-line and used for the enrichment and

TABLE 4.2

Selected examples of coupling CFS-involving dialysis to capillary electrophoresis

Samples	Analytes	Remarks	LOD	RSD (%)	References
Rat blood	Antineoplastic SR 4233	In vivo microdialysis sampling	—	2.6	[44]
Rat blood	Glutamate and aspartate	In vivo-monitoring microdialysis	200 nM	1–6	[45]
Milk, juice, liquors	Small anions	—	—	1.6–3.3	[46]
Biological samples	Acidic drugs	Dialysis-SPE-CE	0.05–1.0 μg/ml	10	[47]
Tissues samples	Aspartate enantiomers	Microdialysis	10–20 nM	5	[48]
Rat brain tissue	Ascorbic acid	In vivo microdialysis sampling	—	3.8	[74]
Rat caudate nucleus	Primary amines	In vivo microdialysis sampling	20–50 nM	1–9	[75]
Cutaneous animal sample	Nicotina	In vivo microdialysis sampling	1.5 μM	—	[76]

determination of basic drugs in plasma samples. The hollow fibre was impregnated with 6-undecanone. Using this impregnated fibre, the obtained extracts were cleaned. They presented a low ionic strength, making this sample treatment compatible with the capillary zone electrophoresis (CZE) double-stacking procedure. Other examples of the use of liquid membranes are reported in Table 4.3.

4.4.5 On-line gas extraction and gas diffusion

Gas diffusion technique is based on the use of a semi-permeable membrane, similar to that used in dialysis. Unlike dialysis, this technique allows only volatile analytes to pass through the membrane. Therefore, the addition of a reagent into the sample to transform ionic analytes into their respective gaseous species is generally required. The separated analytes are then retained in the acceptor solution by the reconversion of the gaseous species into ions. For example, sulphite or carbonate solutions can be emerged on-line with an acid solution and transformed into sulphur dioxide and carbon dioxide, respectively. The gaseous analytes can penetrate the micropores of a PTFE membrane, and can then be collected in a basic acceptor stream [50].

Lista et al. [51] reported a gas extraction device integrated with a CFS and coupled to CE for the direct determination of trimethylamine and related amines in fish samples. Volatile amines were generated from the solid fish sample by adding sodium hydroxide to the sample and heating it. Volatile amines were aspirated from the gaseous phase 7 min later using a peristaltic pump, and introduced into a vial from the CE carousel containing the acceptor solution. The arrangement allowed the direct determination of the analytes in this type of solid sample.

4.4.6 On-line sample treatment and separations performed in auxiliary equipments

As previously indicated, the implementation of a sample treatment before CE analysis has a great impact on the quality of quantitative and qualitative results. One approach to perform sample treatment is to use CFS coupled to CE, but in some cases the coupling of other separation techniques—supercritical fluid extraction, LC or isotachophoresis, to mention some typical examples—is required.

TABLE 4.3

Selected examples of coupling CFS-involving supported liquid membranes to capillary electrophoresis

Samples	Analytes	Remarks	LOD	RSD (%)	References
Plasma	Basic drugs	Hollow fibre miniaturised	—	5–7	[49]
Haemophilusinfluenzae	Pathogenic lipopolysaccharides	Polymeric membrane	0.9 μM	—	[77]
Plasma	HONO, HNO_3, acetate, formiate	—	9.6–440 μg/l	2.4–6.6	[78]
—	Aliphatic carboxylic acids	—	100 fmol	3	[79]

Mardones et al. [52] described an interface for the automatic coupling of a supercritical fluid extractor with capillary electrophoretic equipment, both commercially available. Cresols and chlorophenols from river water samples or urine samples were extracted from the sample in a C18 cartridge, which was washed, dried and inserted into the SFE sample container. Using the supercritical fluid, analytes were extracted from the C18 and collected in a solid trap packed with diol. After the depressurisation of the SFE, the analytes were eluted from the solid trap with methanol and introduced into the CE equipment via a robotic interface.

The coupling of LC with CE was described for the first time in 1990 by Bushey and Jorgenson [53]. The reversed-phase LC (RPLC) system was used in the first dimension, and eluting fractions were introduced into the CE instrument. The effluent from the chromatographic column fills a loop on a computer-controlled six-port valve. A second pump flushes this loop material over the grounded electrode (anode) tip of a CZE capillary at specific intervals (see Fig. 4.11). As can be seen in the figure, the interfaces are based on a split-flow device constructed by the insertion of the CE capillary and one waste into one of the valve channels.

The group of Jorgenson [54] has developed a new interface for coupling microcolumn high performance liquid chromatography (HPLC) with CZE. The interface is based on the original flow gate design made with plastic. A transverse flow of CZE buffer controls analyte injection onto the CE capillary (Fig. 4.12). The coupling of reversed-phase microchromatography-CZE has been also used by Cabaleiro et al. [55] to separate 4-hydroxybenzoic acid from 3,4,5-trimethoxycinnamic acid. The authors introduce a third electrode in the flow gate interface, which is one of the electrodes from the flow gate interface used to perform electrokinetic injections, and another one to perform electrophoretic separation.

German and Kennedy [56] developed a method in which RPLC was on-line coupled to competitive capillary electrophoresis immunoassay (CEIA) in order to improve sensitivity and to provide a means to detect the multiple species that present cross-reaction with the antibody. This method eliminates two of the most important disadvantages of CEIA, namely, a detection limit that depends on binding constant and the inability to differentiate cross-reactive species. According to the authors, the method may have a variety of applications, particularly in assays of diluted biological samples, Western blot-style analysis and library screening.

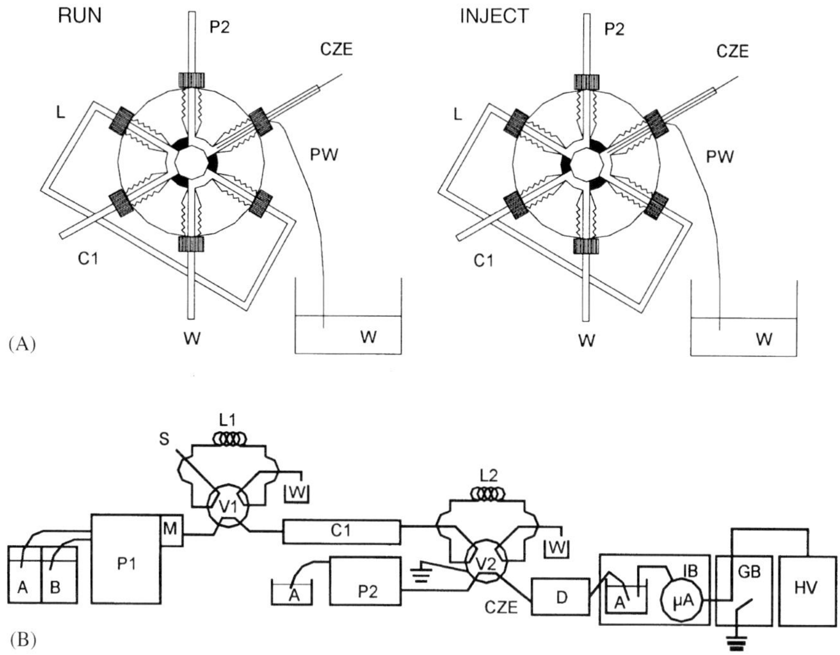

Fig. 4.11. (A) Two configurations of a six-port, computer controlled valve. (B) Schematic of 2-D LC/CZE instrumentation. A and B, buffer; P1, micro-gradient syringe pump; M, mixer; V1, six-port manual injection valve; S, injection syringe; L1, loop; C1, reversed-phase HPLC column; P2, pump 2; L, loop; CZE, CZE-fused silica capillary; PW, paper wick, W, waste; V2, grounded six-port electrically actuated Valco valve; D, fluorescence detector; IB, interlock box; GB, grounding box; HV, high voltage power supply. Reprinted with permission from Ref. [53]. Copyright (1990) American Chemical Society.

Other electrophoretic separation techniques such as isotachophoresis (ITP) can also be used for sample treatment prior to CE analysis. It is well known that ITP is based on the insertion of a sample between two electrolyte solutions, leading and terminating ions, which are, respectively, fast- and slow-mobility ions, at high concentrations. When the electrical field is applied, equilibrium is reached because the electric field is inversely proportional to the ion mobility in each region. Therefore, the potential gradients evolved ensure that all the ions eventually travel in sequence according to ion mobility. The automatic coupling of ITP to CE permits the analytes in the first capillary to be preconcentrated and, afterwards, repeatedly injected into a second capillary of smaller diameter for a more rapid CZE analysis [57].

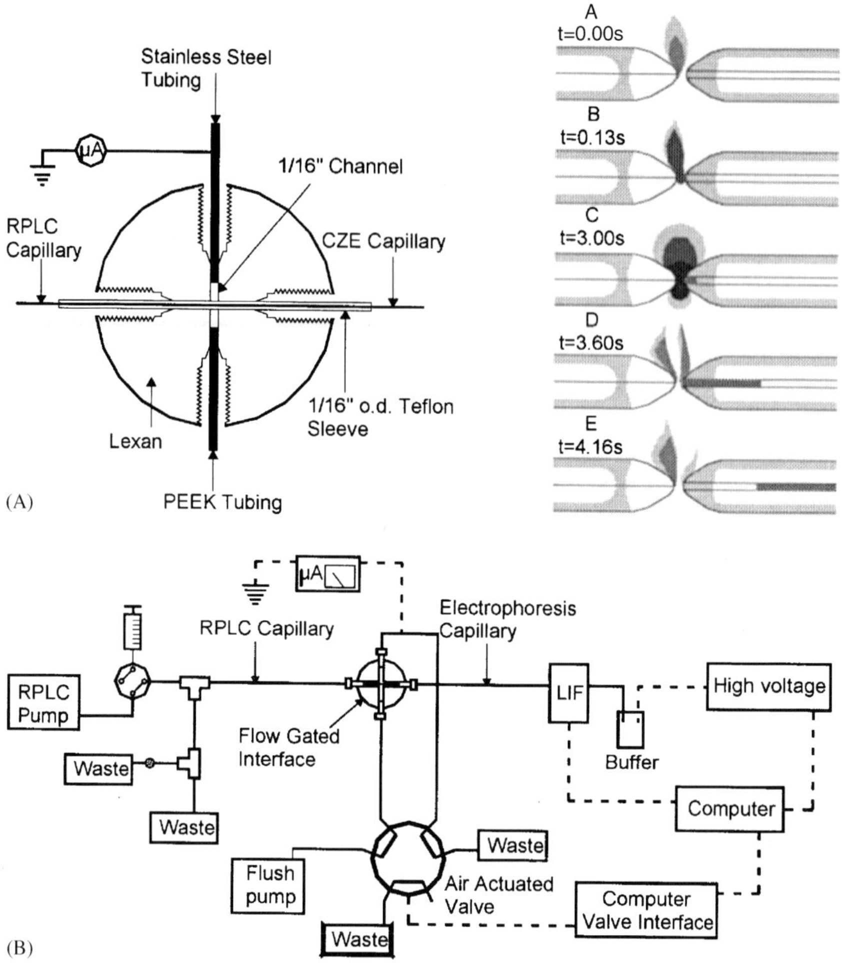

Fig. 4.12. (A) Scheme of the flow gating interface and a frame-grabbed video images of the injection process using the flow gated interface. (B) Schematic of instrumental set-up for 2D micro-RPLC-CZE. Reprinted with permission from Ref. [54]. Copyright (1997), American Chemical Society.

Since only a small portion of the concentrated zones is sequentially injected for CZE separation, overloading was not observed. Moreover, the sensitivity is enhanced because all the concentrated zones are analysed and the result is summed. A schematic representation of the instrumentation for ITP-CZE is described in Fig. 4.13. The four steps during each comprehensive ITP-CZE run are summarised in Fig. 4.14: (i) ITP

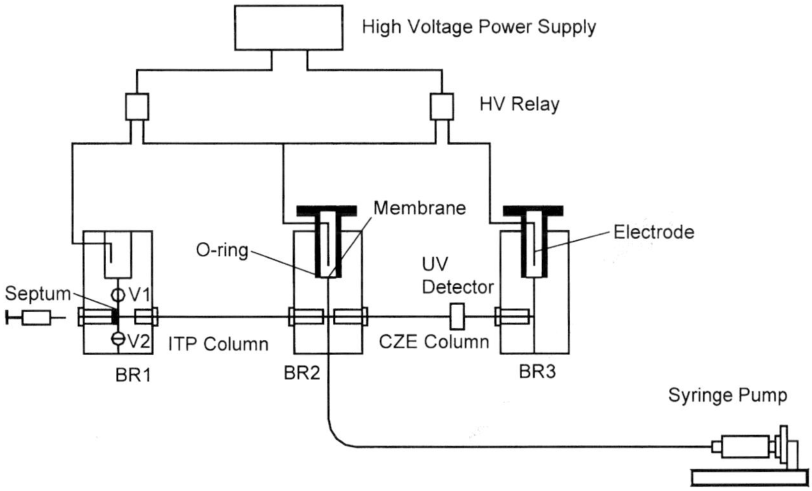

Fig. 4.13. Schematic instrumentation for comprehensive ITP-CZE. BR1, BR2 and BR3, buffer reservoir; V1 and V2, valve. Reprinted with permission from Ref. [57]. Copyright (2000), American Chemical Society.

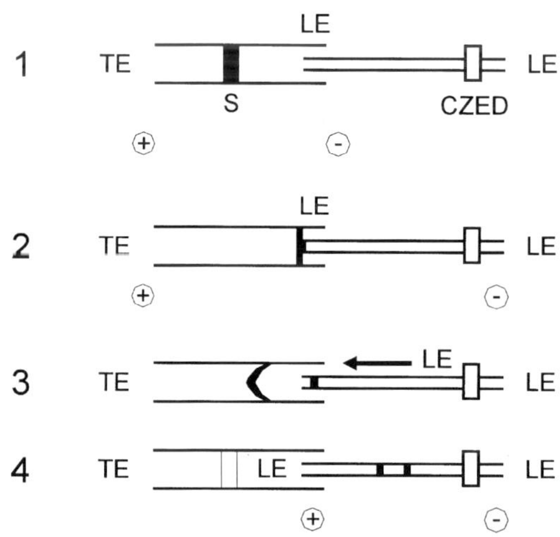

Fig. 4.14. Schematic representation of comprehensive ITP-CZE principles of operation. CZED, CZE detector; TE, terminating electrolyte; LE leading electrolyte; S, sample. Reprinted with permission from Ref. [57]. Copyright (2000), American Chemical Society.

focusing; (ii) CZE injection; (iii) moving the zones back away from the injection end of the CZE column by adding leading electrolyte; and (iv) CZE separation. Comprehensive ITP-CZE involves repetitive runs. One problem in this technique is how to control the CZE injection amount. This problem was solved by correlating the amount injected with the change of current during the injection. This bidimensional configuration has been useful to determine fumaric acid in an apple juice sample with no pretreatment [58]. The on-line coupling of ITP and CZE has been applied to determine a wide variety of analytes and samples, such as flavonoids in methanolic extracts of leaves or flowers [59]; phenolic compounds in plant extracts [60]; inorganic anions in water [61]; orotic acid in urine [62]; separation of enantiomers of 2,4-dinitrophenyl-norleucine and tryptophan in urine [63], etc. Other examples are summarised in Table 4.4.

4.5 AUTOMATIC CALIBRATION IN CE THROUGH CFS

Chemical analytical information is produced as a combination of three experimental activities, namely, sampling, calibration of the equipment involved in the analytical process and, finally, the proper analysis of samples. Major advances in the measuring step have been made during the last decades. Therefore, new developments aimed at simplifying and automating the treatment of the sample and the calibration steps are required. Obviously, both aspects must be controlled because they can dramatically affect the quality of the results. Different approaches have been developed in order to simplify/automate analytical calibration, which involve the often time-consuming preparation of a set of appropriate standards, recording their signals and constructing a calibration graph. A review of flow injection calibration techniques was published in 1988 [64].

Arce et al. [65] described a completely automatic calibration procedure in CE. According to them, the manual calibration operations involved in analytical methods, such as external calibration, standard addition and internal standard, were accomplished with a CFS coupled to a commercial CE. The particular arrangement used to implement the automatic calibration is shown in Fig. 4.15. The computer of the CE equipment controlled the complete system: (i) the electrophoretic equipment in a direct way and (ii) the flow unit and the programmable arm via an electronic interface. The connection between the CE

TABLE 4.4

Selected examples of coupling continuous isotachophoresis systems to capillary electrophoresis

Samples	Analytes	Remarks	LOD	RSD (%)	References
Leaves and flower extracts	Flavonoids	—	100 ng/ml	0.9–2.7	[59]
Plant extracts	Phenolic compounds	—	20–50 ng/ml	0.4–5.2	[60]
Water	Inorganic anions	—	2 μM	3.1–6.8	[61]
Urine	Orotic acid	—	0.2 μM	3–5	[62]
Apple juice	Fumaric acid	—	10 ng/ml	1	[58]
Food products	Lysozyme	—	0.25 μg/ml	3.8	[80]
—	DNA fragments	—	4 mg/l	—	[81]
Pig plasma	Ciprofloxacin, enrofloxavin flumequine	Sample SPE-pre-treatment	50–85 μg/l	4–5	[82]
Serum	Lipoproteins	—	—	0.9–1.1	[83]
Sea water	Nitrite, nitrate	—	0.033–0.063 mg/l	3.9–5.1	[84]
—	Peptides	—	NM	2–12	[85]
Plasma	Peptides	—	2–5 ng/ml	3	[86]

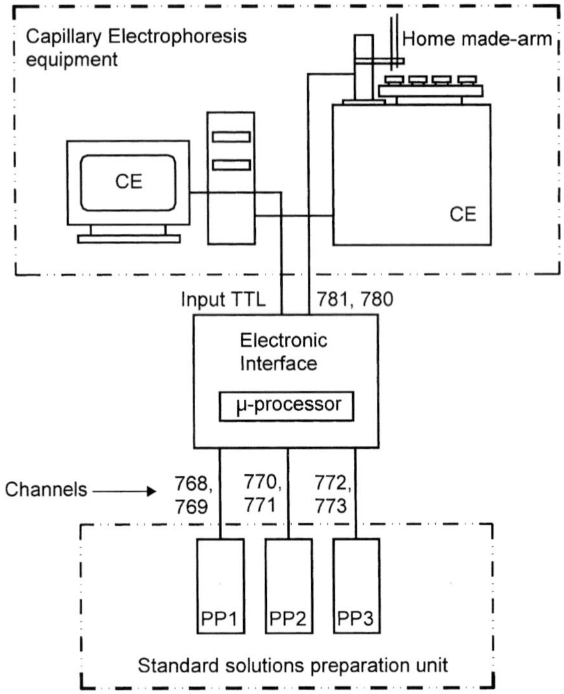

Fig. 4.15. Arrangement used to implement the automatic calibration in a commercial CE equipment. PP, peristaltic pump. From Ref. [65]. Reprinted with permission from Wiley & Sons, Inc.—Wiley-VCH Verlag GmbH—. Copyright (2000).

computer and the electronic interface was made through an INPUT port available in the Beckman® equipment used. Specific output channels of the electronic interface controlled the automatic work of up to three peristaltic pumps (PP) involved in the automatic preparation of the standard solutions. The CFS used to prepare the standard solutions (Fig. 4.16A) also includes the programmable arm that synchronises the introduction of the standard solutions in the CE autosampler. The method developed for external calibration needs two peristaltic pumps, one for the stock solution and the other one for water. The preparation of the different standard solutions (at different concentrations) was carried out automatically by controlling the relative flow rates of pumps P1 and P2, but in such a way that the final flow rate after the mixing point always remained constant. After the merging point, the dilution was homogenised in the mixing coil and injected into the

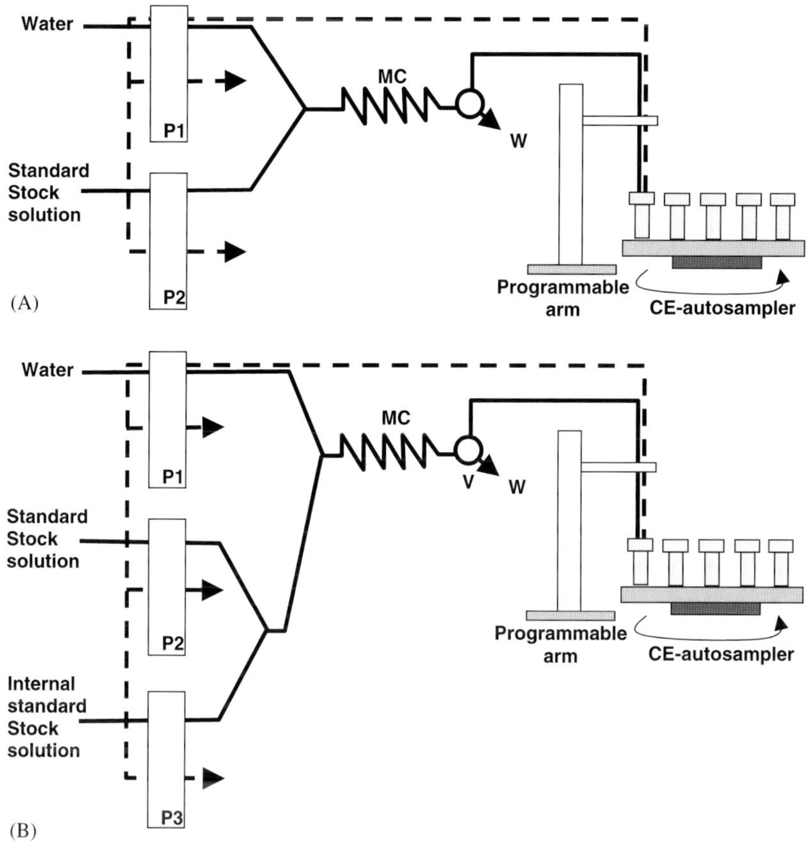

(A) (B)

Fig. 4.16. (A) CFS used to prepare the standard for external calibration. (B) CFS used to prepare the standard for internal calibration. P, peristaltic pump; MC, mixing coil. (Adapted from Ref. [65].)

sample vial via the programmable arm previously described. For calibration using the internal standard method, a third pump was added to the manifold in order to provide the addition of a constant volume of internal standard (Fig. 4.16B). The same manifold was used for the standard addition method.

Karlberg's group [66] described the calibration principles of flow injection analysis-CE with electrokinetic injection. They coupled CFS to CE using a split-flow cell. Best results were obtained with the internal standard method or by applying the conductivity-corrected peak area method. The proposed methods yielded a relative error of prediction of less than 6%. Therefore, they are recommended for the authors for quantitative analysis. The electrokinetic injection has some interesting features, but it has some shortcomings as well. The most important of these is the bias phenomena that can occur when electrokinetic injection is used. The bias arises because the amount of analyte electrokinetically injected is influenced by its effective mobility, which depends on the composition and the presence of other ions in the sample. Another bias problem arises when samples show different conductivities, because the total amount injected is largely affected by the conductivity of the sample. In order to correct the bias problem, the use of an internal standard is critical in a large number of applications.

Another important aspect that must be considered is that the CFS–CE approach for sample introduction markedly differs when home-made or commercial equipment is used. When commercial equipment with robotic interfaces is used, the sample and the standard are homogenised and mixed in a vial that physically replaces the electrolyte vial during injection. However, when home-made CE equipment with split-flow interfaces is used, the electrolyte is always present in the CFS–CE system. Therefore, the mixing of standards with samples occurs during the transport through the flow system. For this reason, the dimensions of the reactor must be critically controlled because a complete mixing of the sample with the standard is desirable. If the mixing is complete, the matrix interferences during injection can be corrected using the internal standard. At the same time, partial mixing of the sample with the electrolyte is unavoidable, but it should be minimised as much as possible.

4.6 SCREENING FLOW UNITS–CE ANALYSIS ARRANGEMENTS

The amount of analytical information provided sometimes exceeds the information required by the client. For these reasons, rapid response analytical systems, that like screening methods, have become of great interest for collecting key analytical information useful for solving problems and to help in taking timely decisions. CFSs are already demonstrating their wide range of possibilities for rapid sample screening [67]. Screening methods are characterised to provide a binary yes/no response related to a specific property of a sample substance or system with the least possible delay. The intrinsic characteristics of CF (namely, automation, miniaturisation, simplification of preliminary operations, low cost, versatility, speed and minimisation of operator errors) make them specially suitable for developing screening methods. The use of screening methods makes sophisticated instruments with a high maintenance cost permanently unnecessary. Thus, the combination of CFS–CE can be used as a screening-confirmation methodology, avoiding the analysis of negative samples by CE, and hence, increasing the throughput in routine laboratories.

Manganiello et al. [68] have developed a new methodology that permits the automatic monitoring of mercury in water (Fig. 4.17). The method combines a screening unit for the total amount of mercury with a CE unit to process the samples containing a detectable level of mercury. Thus, in the manifold connected to the piezoelectric (PZ) flow sensor, conforming the screening unit (Fig. 4.17), a stream of tin(II) chloride was mixed with the sample stream in an acidic medium. Under these conditions, mercury species in the sample were reduced to elemental mercury. When the elemental mercury formed passed through the PZ sensor, an amalgam between Hg and the gold films supported in the quartz crystal was produced, and therefore a change in the

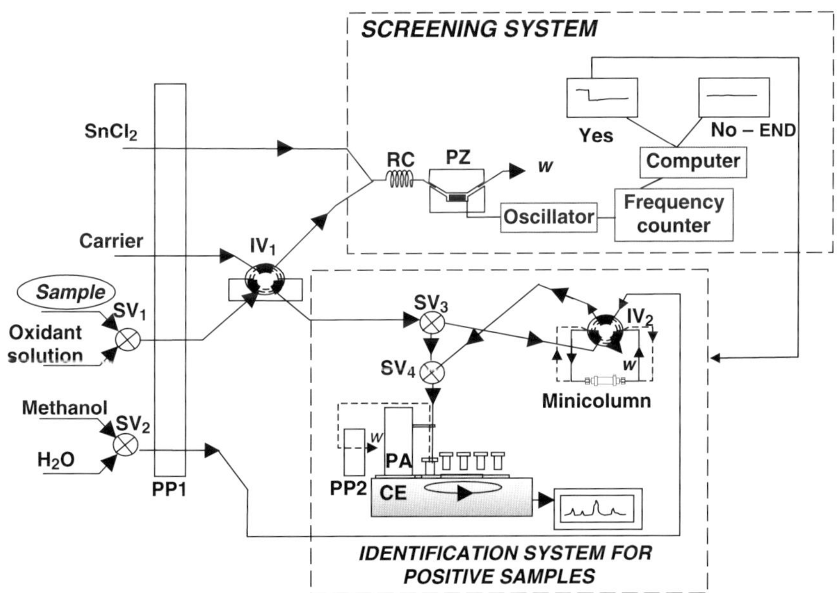

Fig. 4.17. Configuration used for the determination of total mercury of organic and inorganic mercury compounds in water with a sample preconcentration system. The screening system involves the determination of total mercury using a piezoelectric crystal sensor. From Ref. [32]. Reprinted with permission of Wiley & Sons, Inc.—Wiley-VCH Verlag GmbH. Copyright (2002).

frequency of the sensor occurred. This change of frequency is related to the total concentration of mercury species in the sample. Then, a regenerating solution (potassium peroxidisulphate) was used to prepare the sensor for the following sample. This sensor is very selective and sensitive for mercury. The limit of detection is 0.24 ng/ml, and therefore is very suitable as a screening system for total mercury. In order to achieve enough sensitivity for the CE method, which was developed for the separation and quantification of the different mercury species, a preconcentration unit was introduced prior to the CE equipment for concentrating the mercury species of the positive samples. The electrophoretic buffer contained cysteine to form anionic complexes with inorganic mercury and the organomercury compounds, and β-cyclodextrin to allow the separation (in boric acid buffer at pH 8.3). An electropherogram is shown in Fig. 4.18, where the separation of inorganic mercury, diethyl-Hg, methyl-Hg and phenyl-Hg can be seen. The limits of detection for the preconcentration-CE method were low

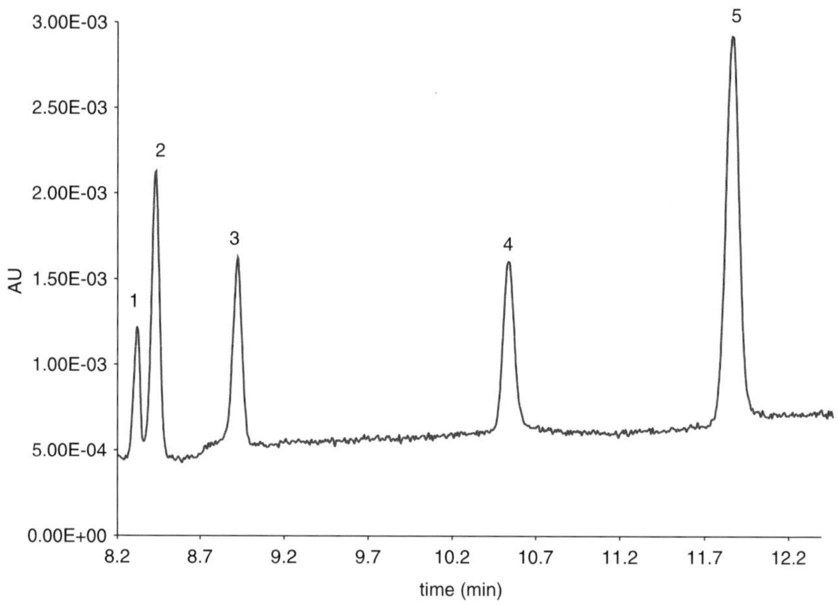

Fig. 4.18. Electropherogram of a standard mixture of different mercury species. Identification peaks: (1) diethyl-Hg; (2) methyl-Hg; (3) phenyl-Hg; (4) cysteine; (5) Hg^{2+}. The concentration of each analyte was 0.15 mg/l of diethyl-Hg; 0.50 mg/l of methyl-Hg, 0.20 mg/l of phenyl-Hg and 0.50 mg/l of Hg^{2+}, respectively. From Ref. [32]. Reprinted with permission from Wiley & Sons, Inc.—Wiley-VCH Verlag GmbH. Copyright (2002).

enough to carry out speciation studies of mercury in water (at ng/ml level in all cases).

Another example of the utility of coupling a screening method to CE is the determination of aflatoxins in feed samples [69]. In this case, a flow system allows the total determination of several compounds by using a fluorimetric screening method. Afterwards, individual analytes are determined in positive samples by micellar electrokinetic capillary chromatography (MECC) analysis prior to their preconcentration by SPE using a C18 mini-column. The screening procedure uses simple equipment and provides a yes/no binary response in a few minutes, thereby avoiding preconcentration and CE analysis in negative samples.

4.7 ANALYTICAL APPLICATIONS BY CFS–CE

When real samples are analysed by CE, a previous sample treatment is commonly a mandatory step. Obviously, depending on the sample matrix and analytes the sample treatment can involve a clean-up or a preconcentration step. In general, it can be affirmed that these sample treatments are simplified by using CFS.

As can be deduced from examples and tables described in previous sections, CFS coupled to CE has been widely used in a large number of applications. Some of them will be described in more detail as selected examples in this section, with the aim of demonstrating the analytical potential of the CFS–CE arrangements. Table 4.5 shows a list with the selected examples.

4.7.1 Biomedical and pharmaceutical analysis

It is well known that in the vast majority of biological fluid assays an important part of the total analysis time is expended on preparing the sample. Owing to the complexity of the matrices, prior to the measuring step, a clean-up or preconcentration step is required. When the instrumental measurement is performed by CE, the pretreatment must be efficient enough to remove interfering solutes, organic salts and particulate matter in order to remove potential interferences. Mardones et al. [34] reported an automatic method for the clean-up, preconcentration, separation and quantification of nonsteroidal anti-inflammatory drugs in biological fluids (human urine and serum) using SPE coupled to CE (Fig. 4.19). In this work, a commercial CE

TABLE 4.5

Selected examples of analytical applications by CFS–CE

Application fields	Analytes	Samples	Sample treatment	References
Biomedical and pharmaceutical analyses	Nonsteroidal anti-inflammatory drugs	Human urine and serum	Solid-phase extraction	[34]
	Chlorophenols	Human urine	Solid-phase extraction	[38]
	Antineoplastic SR 4233	Rat blood	On-line microdialysis	[44]
	D-;L-carnitine	Pharmaceutical formulations	On-line derivatisation	[70]
Environmental analysis	Inorganic ions	Water	Resin exchange	[39]
	Pesticides	Water	Solid-phase extraction	[31]
	Tetracyclines	Water	Solid-phase extraction	[71]
	Chlorophenol and cresol	Water	Supercritical fluid extraction	[52]
Food analysis	Anions	Soil	Extraction/filtering	[72]
	Biogenic amines	Wine	Solid-phase extraction	[33]
	Amines	Fish	Gas extraction	[51]
	Aflotoxins	Feed	Solid-phase extraction	[69]

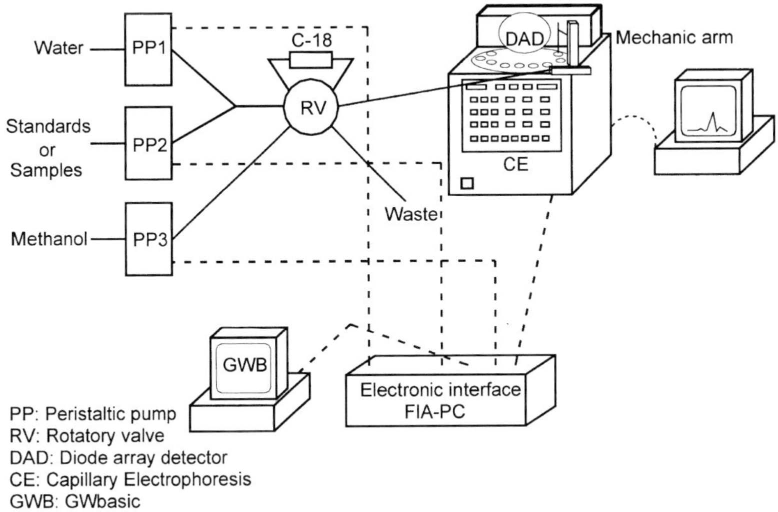

Fig. 4.19. Automatic arrangement used for the determination of nonsteroidal anti-inflammatory drugs in biological fluids. The automatic system involves the on-line solid-phase extraction and capillary electrophoresis. From Ref. [34]. Reprinted with permission from Wiley & Sons, Inc.—Wiley-VCH Verlag GmbH. Copyright (2001).

equipment was used. It was at-line connected to the CFS, which automatically carried out the sample treatment. A programmable arm was used to transport the treated sample from the CFS to the CE equipment. The integrated system was controlled by an electronic interface governed via a programme developed in GWBasic. The clean-up and preconcentration of the drugs, accomplished by using a C18 mini-column, resulted in a 400-fold improvement in the limit of detection. Analytes were directly retained from the sample in the C18 column and then eluted with 75% methanol as eluent. Automatic calibration is also incorporated in this application. Fig. 4.20 shows the electropherograms for the analyses of both spiked urine and serum samples subjected or not to the automatic pretreatment. As can be seen, the direct injection of the spiked samples without pretreatment precluded the determination of the drugs studied. In order to increase the throughput, CFS was developed in order to perform the continuous clean-up and preconcentration of the analytes while the CE system performs the electrophoretic separation.

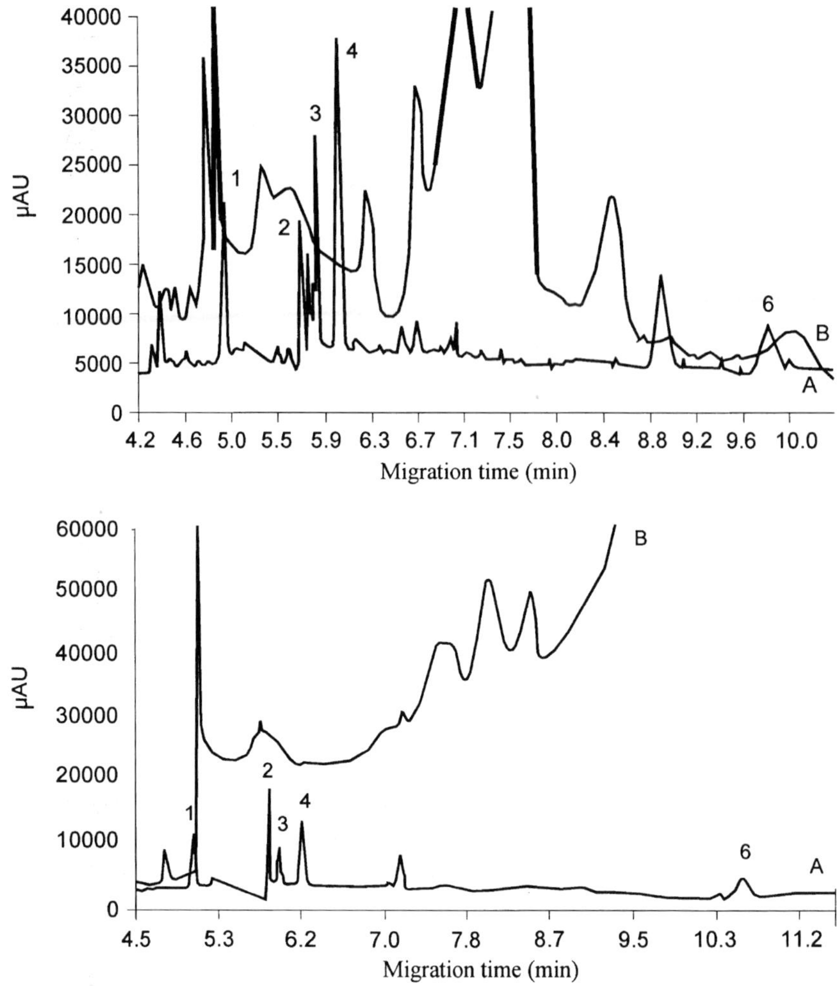

Fig. 4.20. Electropherograms for (top) urine and (bottom) serum samples spiked with 10 and 5 μg/ml concentrations, respectively, of the analytes. (A) Subjected and (B) not subjected to the proposed automatic pre-treatment. Peaks: 1, ibuprofen; 2, fenbufen; 3, tometin; 4, ketoprofen; 6, indomethacin. Electrophoretic conditions: buffer, 20 mM NaH_2PO_4, 20 mM β-cyclodextrin and 50 mM SDS at pH 9.0; potential; 20 kV. From Ref. [34]. Reprinted with permission from Wiley & Sons, Inc.—Wiley-VCH Verlag GmbH. Copyright (2001).

The determination of chlorophenols in human urine is another example of the usefulness of the CFS–CE system [38]. In this case, the compounds were automatically preconcentrated from a urine sample in which the organic matter was previously hydrolysed. The preconcentration of the chlorophenols in the hydrolysed urine was performed using a C18 minicolumn integrated into a CFS coupled on-line to the CE equipment. Figure 4.21 shows the electropherogram for a urine sample, a urine sample spiked with a 50 ng/ml of each chlorophenol and a urine sample hydrolysed and directly injected without pretreatment. As can be seen, the pretreatment is a mandatory step in order to quantify the analytes.

Another example that highlights the importance of this methodology is the monitoring of in vivo reactions. Such studies have been performed via microdialysis, using a CFS system that was coupled on-line to CE equipment. This system takes advantage of microcolumn separations in order to overcome the small sample volume limitation. One example of this type of study is the work of Hogan et al. [44]. The authors studied the pharmacokinetics of the antineoplastic SR 4233 in vivo with male Sprague–Dawley rats. The complete system is shown in Fig. 4.22, whereas Fig. 4.23 corresponds to a typical pharmacokinetic curve obtained with this system. As can be deduced, the on-line microdialysis connected to CE provides a powerful tool to monitor in vivo biochemical reactions. Moreover, one can affirm that this methodology is the only one capable of making such a large number of measurements of multiple species in vivo, mainly because of the sample volume requirements, providing measurements near to those of real time in a conscious of the animal.

Some applications have also been described in the pharmaceutical field, such as the enantiomer separation of D- and L-carnitine by integrating on-line derivatization with CZE [70]. L-carnitine is active as a mitocondrial fatty acid acyltransferase cofactor. On the contrary, D-carnitine has been shown to have a considerable toxic influence on biomedical processes. Nevertheless, since 1960, DL-carnitine has been used as a drug for anorexia, dyspepsia and enterokinesis. So far, in the pharmaceutical formulations, the use of pure L-enantiomer is recommended. Consequently, the separation of D- and L-carnitine is of great interest. In the referred method [70], D- and L-carnitine were derivatised with 9-fluoresymethyl chloroformate in the flow system, working on-line with the CE equipment. The separation was performed using a selective chiral buffer containing 2,6-dimethyl-β-cyclodextrin

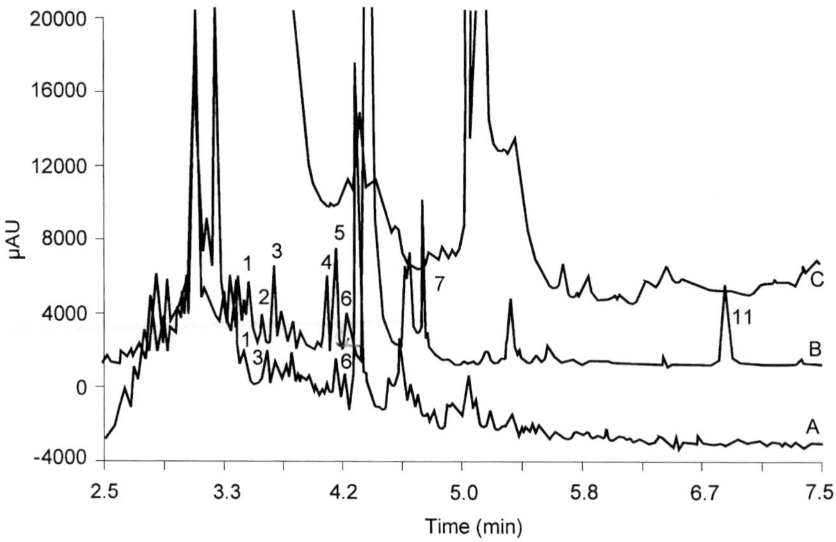

Fig. 4.21. (A) Electropherogram for a hydrolysed urine sample; (B) electropherogram for a hydrolysed urine sample spiked with 0.05 μg/ml of each chlorophenol; (C) electropherogram for the urine sample previously hydrolysed and directly injected without pretreatment. From Ref. [38]. Reprinted with permission from Wiley & Sons, Inc.—Wiley-VCH Verlag GmbH. Copyright (1999).

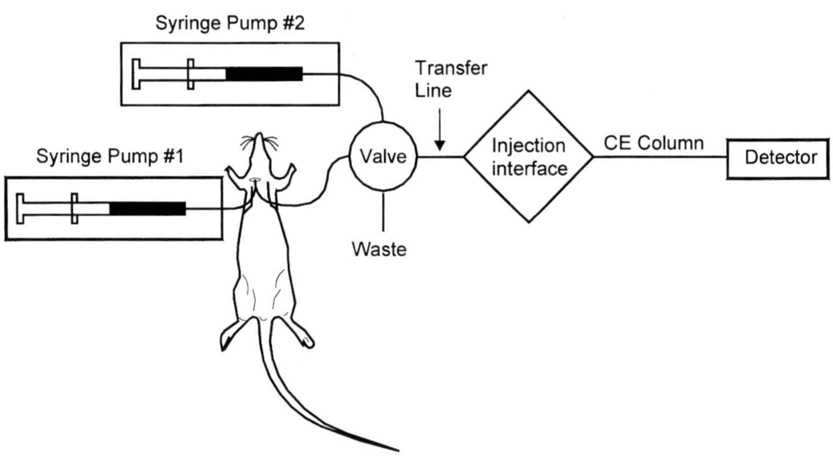

Fig. 4.22. Complete schematic system of the on-line microdialysis-capillary electrophoresis system. Reprinted with permission from Ref. [44]. Copyright (1994), American Chemical Society.

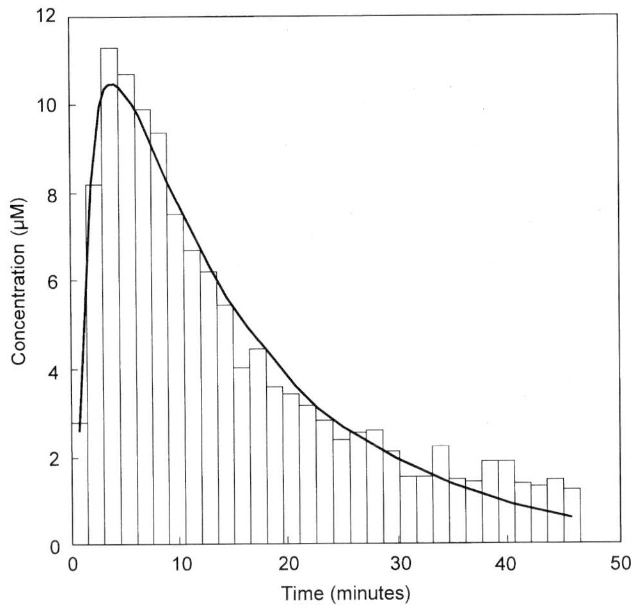

Fig. 4.23. Typical pharmacokinetic curve from a 4 mg/kg ip dose of SR 4233 obtained using the on-line microdialysis–capillary electrophoresis system depicted in Fig. 4.17. Bars represent the microdialysis data while the curve is the bi-exponential fit of the data. Reprinted with permission from Ref. [44]. Copyright (1994), American Chemical Society.

(Fig. 4.24A). The maximal D-: L-carnitine ratio determined was ca. 1:100 (Fig. 4.24B).

4.7.2 Environmental analysis

CE plays an important role in environmental analysis, but the complexity of the samples makes the treatment and conditioning of environmental samples prior to CE analysis necessary. Different applications combining CFS–CE have been already reported in the previous sections for illustrating some automatic sample treatment approaches. From an application point of view, CFS–CE arrangements have proved their suitability for the analysis of both liquid and solid environmental samples. Thus, inorganic anions in water defining their quality (and hence useful for the control of water purification plants) were determined by an on-line coupling of a preconcentration CFS manifold with the CE equipment [39]. Two automated methodologies

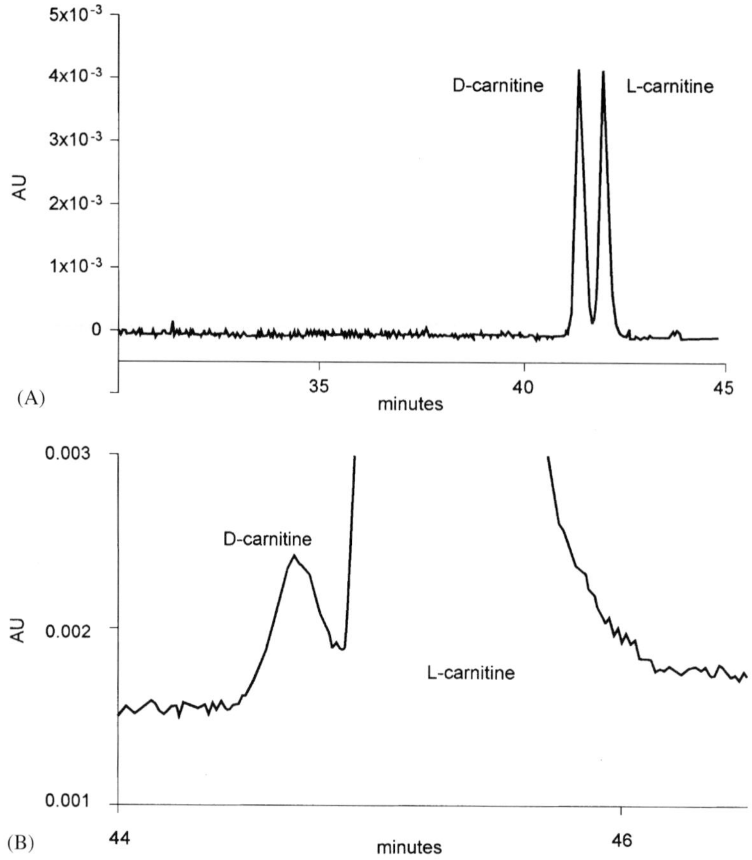

Fig. 4.24. (A) Electropherogram obtained in the separation of D- and L-carnitine isomers using β-cyclodextrine as chiral additive. (B) Electrophoretic separation of a synthetic sample containing a D/L-carnitine ratio of 1:100. From Ref. [70]. Reprinted with permission from Elsevier. Copyright (1999).

were developed for this multiparametric application (anionic and cationic species). The concentration of anions such as chloride, nitrate, nitrite, sulphate and phosphate; cations such as sodium, potassium, ammonium, calcium, magnesium, iron; and heavy metals at trace level (e.g. manganese, aluminium, lead, cadmium) was provided in an integral manner. The determination limits established by legislation in Organisation for Economic Co-operation and Development (OECD) countries were achieved, and the methods were applied to real samples.

Pollutant compounds in water have also been determined by CFS–CE methodologies. For instance, pesticides such as fenuron, simazine, atrazine, carbaryl, ametryn, prometryn and terbutryn were determined by an arrangement, where calibration, preconcentration, elution and introduction of the processed samples were performed in the flow manifold on-line coupled to the CE equipment [31] (Fig. 4.25). At least 15-fold enrichment factors were achieved for every pesticide. The separation was carried out by micellar electrokinetic chromatography (MEKC). Other types of water pollutants, such as tetracylines, were determined in surface waters by using a similar methodology [71].

An interesting combination and powerful analytical tool is the automatic on-line coupling of supercritical fluid extraction (SFE) and CE. SFE is an effective alternative to the extraction of a wide variety of compounds from complex matrices because of the low viscosity and relatively high density of the supercritical fluids (mainly the s-CO_2, which is the most used). On the basis of this approach, chlorophenols and cresols (two families of well-known environmental pollutants) were automatically extracted from river water and biological samples and

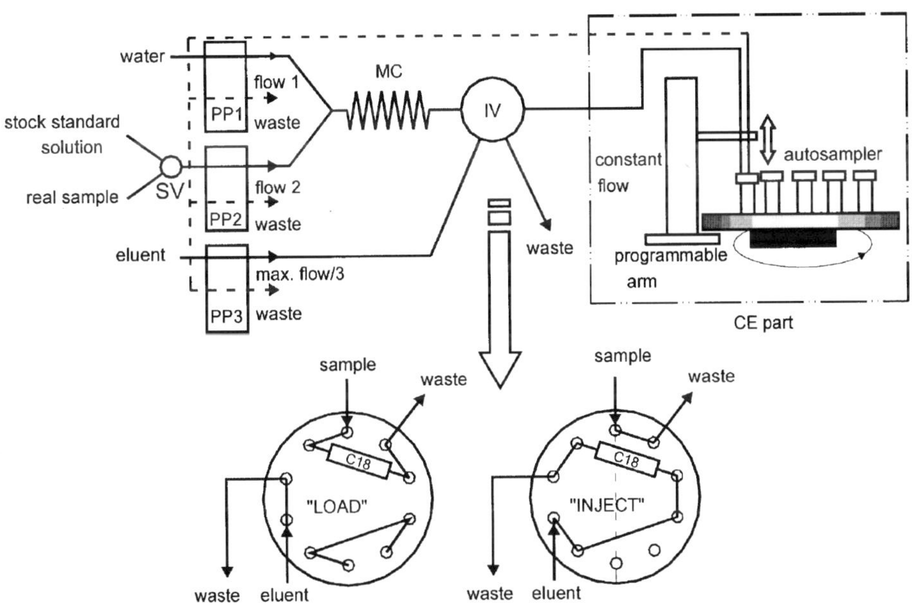

Fig. 4.25. CFS manifold for the preconcentration and analysis of pesticides residues from water samples. From Ref. [31]. Reprinted with permission from Elsevier. Copyright (2000).

directly determined by CE [52]. The previously described electronic and mechanical interfaces (section 4.3.2 and Fig. 4.4) allowed the transfer of the extracted analytes from the SFE equipment to the CE. The liquid samples were previously preconcentrated in C18 mini-cartridges, which were afterwards eluted and transferred into the timbles of the SFE. Phenols extracted by the s-CO_2 were collected into a trap packed with diol material and eluted from the trap at 25°C using 1.5 ml of methanol. This volume was directly fed into the CE vial by the mechanical arm and analysed later.

A more complex situation is presented by the analysis of solid environmental samples. In this case, extraction of the analytes and the proper filtering of the sample are two commonly required steps. This is the case reported by Arce et al. [72] for the direct multiparametric determination of anions in soil samples by integrating on-line automated extraction/filtering with CE. A filtration probe incorporated in the flow system was used in this application, where chloride, sulphate, nitrite and nitrate were determined in soil samples at μg/g level.

4.7.3 Food analysis

The analysis of food samples is another area in which sample treatment is important, because of the complexity of the matrix. Fig. 4.26 shows electropherograms for a red wine sample analysed with and without sample pretreatment. In this work [33], biogenic amines were the analytes subjected to analysis. As can be seen, aside from preconcentration, the clean-up of the sample is crucial in order to determine the biogenic amines. In fact, without the wine sample pre-treatment, no peak can be identified or quantified, owing to matrix interferences.

Solid food samples present a higher complexity, but some applications have been described so far. Thus, a gas extraction sampling device integrated in a CFS was coupled to a CE equipment for the direct determination of trimethylamine and related amines in fish samples [51]. Specifically, trimethylamine is an indicator compound to control the conservation state of the fresh fish. This arrangement allowed the direct introduction and treatment of these solid samples with a high level of automation, as well as the performance of automatic external calibration (Fig. 4.27). The manifold used for preparing the standards and the real samples consisted of four peristaltic pumps, a selection valve, a programmable arm for the sample injection and a heater/stirrer plate. The first peristaltic pump (PP1) was used for water, the

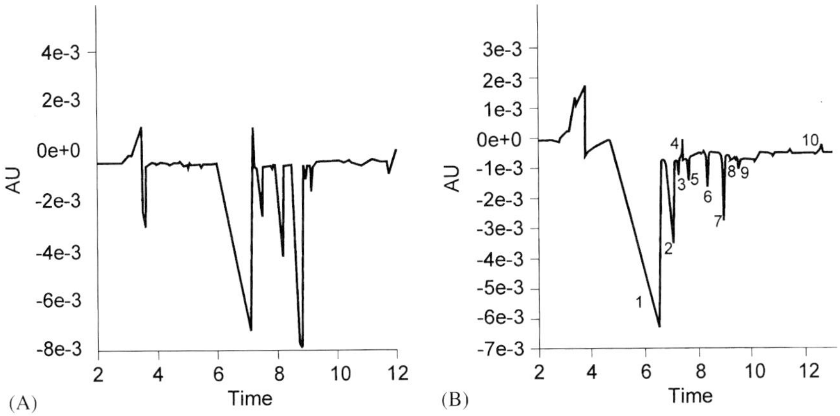

Fig. 4.26. Electropherogram for a red wine sample using a CFS–CE method based on a C18 solid phase extraction pretreatment sample. (A) Wine sample without pretreatment and (B) wine sample after sample clean-up. 1, 2, unknown peaks; 3, putrescine; 4, histamine; 5, cadaverine; 6, 7, unknown peaks; 8, ethanolamine; 9, unknown peak; 10, phenethylamine. Electrophoretic conditions: buffer: 4 mM copper(II) sulphate, formic acid and 18-crown-6; voltage, +15 kV. From Ref. [33]. Reprinted with permission from Elsevier. Copyright (2002).

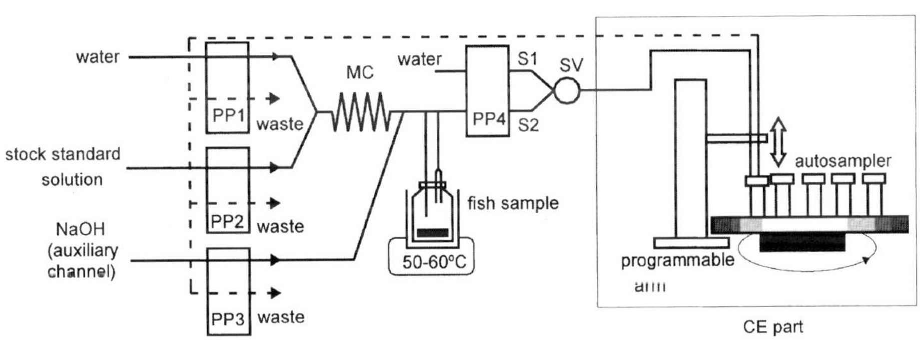

Fig. 4.27. CFS manifold used for the generation of volatile amines from fish samples. From Ref. [51]. Reprinted with permission from Elsevier. Copyright (2001).

second (PP2) for the standard amine solutions, the third (PP3) for NaOH solution and the fourth (PP4) to fill the vial with water and to transport the generated gaseous amines (produced by the action of NaOH over the sample) to the vial. The selection valve allows filling the vial with water (position S1) or gaseous amines (position S2). The functioning of the programmable arm was synchronised with the work of the CFS.

Based on the screening-confirmation approach, a CFS–CE methodology was developed for the rapid control of aflatoxins in feed samples [69]. Figure 4.28 shows the entire arrangement to combine the screening flow system for the binary response with the identification electrophoretic technique. An additional preconcentration unit was also incorporated in order to achieve the required sensitivity to reach the legal limit (0.05 mg/L, as total amount). The screening system was designed to provide a rapid response to the presence or absence of aflatoxins at a threshold concentration level. This system is based on the fluorescence of the oxidised aflatoxins (bromine was used as oxidant). Samples containing concentrations of aflatoxins above 0.1 mg/L were analysed by CE without previous treatment. When the content of aflatoxins was lower than 0.1 mg/L, the samples were preconcentrated before the analysis by CE. After the validation of this methodology, it was applied to different feed samples. Table 4.6 shows the results obtained for five of these samples. As can be seen, the first two samples were negative in the screening method. The third sample was positive

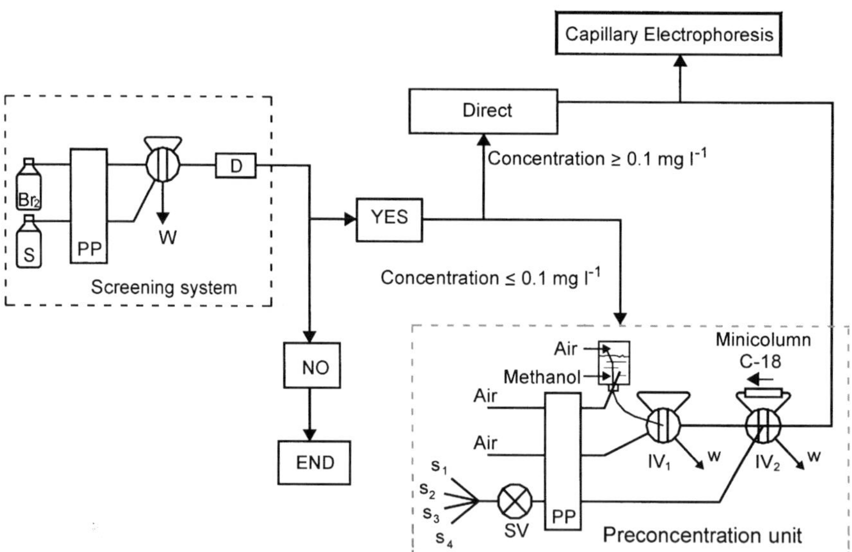

Fig. 4.28. Schematic of the FI systems used for screening and confirmatory analysis for the determination of mycotoxins present in feed samples. From Ref. [69]. Reprinted with permission from Elsevier. Copyright (2002).

TABLE 4.6

Determination of aflatoxins in spiked feed samples using screening system and CE

	Sample				
	1	2	3	4	5
Aflotoxin G1[1]	0	0.025	0	0	0.05
Aflotoxin G2[1]	0	0	0	0	0.05
Aflotoxin B1[1]	0	0	0.04	1	0.05
Aflotoxin B2[1]	0	0	0	0	0.05
Screening results	–	–	+	+	+
Steps			(1+2)	(2)	(1+2)
Screening-CE results			–	+	+

Reproduced from Ref. [69] with permission from Elsevier Science B.V.
[1]Concentration (mg/l) in steps: (1) preconcentration and (2) CE.

in the screening system, being needed its preconcentration and the subsequent analysis by CE. Aflatoxin B1 was found in this sample, but at lower concentration fixed by the legislation. Therefore, at the end, this sample was classified as a negative sample according to the legislation. Samples 4 and 5 were classified as positive samples by the screening system, and then, the results were confirmed by the CE method (without preconcentration for sample 4 and with preconcentration for sample 5).

4.8 FUTURE PROSPECTS

Many practical sample treatment problems can be solved by coupling CFS and CE equipments, thereby improving the overall efficiency and sensitivity of CE methods. Moreover, this system presents intrinsic characteristics—low cost, robustness, and easy to operate and to automate—that make it an excellent analytical tool to develop routine analytical methods.

The miniaturisation and integration of the combined system on micro-fabricated chips may further enhance its performance because of the increased demand for portable equipment suitable for on-site analysis. The capacity and potential of miniaturised systems for chemical analysis has been demonstrated by the development of micromechanized total analysis systems and CE separation systems. This has been accompanied by the development of nanocolumns and filters in microstructures that allow CFS to perform the sample pretreatment in a miniaturised system. The objective is not merely to develop integrated and miniaturised analytical systems, but also to develop analytical systems sufficiently robust to be used by inexpert people for the analysis of complex environmental or biological samples.

In addition to the partly achieved miniaturisation during the last few years, one of the main objectives remains the simplification of the analytical method. In this sense, the coupling of screening methodologies to CE avoids the use of the instruments, when negative samples are detected by the screening method as these samples are not analysed by CE.

ACKNOWLEDGEMENTS

Financial support received from the Spanish Ministry of Education and Science through Grants CTQ2004-01220 and CTQ2004-02362 is gratefully acknowledged.

REFERENCES

1 F.E.P. Mikkers, F.M. Everaerts and T.P.E.M. Verheggen, *J. Chromatogr.*, 169 (1979) 11–20.
2 J.W. Jorgenson and K.D. Lukacs, *Anal. Chem.*, 53 (1981) 1298–1302.
3 B.M. Simonet, A. Ríos and M. Valcárcel, *Trends Anal. Chem.*, 22 (2003) 605–614.
4 P. Kuban, A. Engström, J.C. Olsson, G. Thorsen, R. Tryzell and B. Karlberg, *Anal. Chim. Acta*, 337 (1997) 117–124.
5 M. Kaljurand, M. Ebber and T. Somer, *J. High Resol. Chromatogr.*, 18 (1995) 263–265.
6 M.E. Roche, R.P. Oda, D. Machacek, G.M. Lawson and J.P. Landers, *Anal. Chem.*, 69 (1997) 99–104.
7 M. Valcárcel, M. Gallego and A. Ríos, *Fresenius J. Anal. Chem.*, 362 (1998) 58–66.
8 M. Valcárcel, L. Arce and A. Ríos, *J. Chromatogr. A*, 924 (2001) 3–30.
9 Z.L. Fang, H.W. Chen, Q. Fang and Q.S. Pu, *Anal.Sci.*, 16 (2000) 197–203.
10 P. Kuban and B. Karlberg, *Trends Anal. Chem.*, 17 (1998) 34–41.
11 P. Kuban, A. Engström, J.C. Olsson, G. Thorsén, R. Tryzell and B. Karlberg, *Anal. Chim. Acta*, 337 (1997) 117–124.
12 Z.L. Fang, Z.S. Liu and Q. Shen, *Anal. Chim. Acta*, 346 (1997) 135–143.

13 P. Kuban, R. Pirmohammadi and B. Karlberg, *Anal. Chim. Acta*, 378 (1999) 55–62.
14 Q.S. Pu and Z.K. Fang, *Anal. Chim Acta*, 398 (1999) 65–74.
15 C.H. Wu, L. Scampavia and J. Ruzicka, *The Analyst*, 127 (2002) 898–905.
16 C.H. Wu, L. Scampavia and J. Ruzicka, *The Analyst*, 128 (2003) 1123–1130.
17 L.M. Ponton and C.E. Evans, *Anal. Chem.*, 73 (2001) 1974–1978.
18 M. Valcárcel, A. Ríos and L. Arce, *Crit. Rev. Anal. Chem.*, 28 (1998) 63–91.
19 B. Santos, B.M. Simonet, A. Ríos and M. Valcárcel, *Electrophoresis*, 25 (2004) 3427–3433.
20 B. Santos, B.M. Simonet, A. Ríos and M. Valcárcel, *Electrophoresis*, 25 (2004) 3231–3236.
21 Q. Fang, F.R. Wang, S.L. Wang, S.S. Liu, S.K. Xu and Z.L. Fang, *Anal. Chim. Acta*, 390 (1999) 27–37.
22 C.G. Fu and Z.L. Fang, *Anal. Chim. Acta*, 422 (2000) 71–79.
23 X.J. Huang, S.L. Wang and Z.L. Fang, *Anal. Chim. Acta*, 456 (2002) 167–175.
24 X.J. Huang, Q.S. Pu and Z.L. Fang, *Analyst*, 126 (2001) 281–284.
25 Y.H. Chu, L.Z. Avila, J. Gao and G.M. Whitesides, *Acc. Chem. Res.*, 28 (1995) 461–468.
26 T. Johansson, M. Petersson, J. Johansson and S. Nilsson, *Anal. Chem.*, 71 (1999) 4415–4418.
27 M.K. Little and C.D. Crawley, *Anal. Chim. Acta*, 464 (2002) 25–35.
28 Y. Saito and K. Jinno, *Anal. Bioanal. Chem.*, 373 (2002) 325–331.
29 K. Jinno, M. Kawazoe, Y. Saito, T. Takeichi and M. Hayashida, *Electrophoresis*, 22 (2001) 3785–3790.
30 M. Valcárcel and M.D. Luque de Castro, *Flow Injection Análisis: Principles and Applications*. Ellis Horwood, Chichester, 1987.
31 P. Hinsmann, L. Arce, A. Ríos and M. Valcárcel, *J. Chromatogr. A*, 866 (2000) 137–146.
32 L. Manganiello, L. Arce, A. Ríos and M. Valcárcel, *J. Sep. Sci.*, 25 (2002) 319–327.
33 L. Arce, A. Ríos and M. Valcárcel, *J. Chromatogr. A*, 803 (1998) 249–260.
34 C. Mardones, A. Ríos and M. Valcárcel, *Electrophoresis*, 22 (2001) 484–490.
35 K. Kanitsar, L. Arce, A. Ríos and M. Valcárcel, *Electrophoresis*, 22 (2001) 1553–1560.
36 H.W. Chen and Z.L. Fang, *Anal. Chim. Acta*, 394 (1999) 13–22.
37 L. Arce, M.T. Tena, A. Ríos and M. Valcárcel, *Anal. Chim. Acta*, 359 (1998) 27–38.
38 C. Mardones, A. Ríos and M. Valcárcel, *Electrophoresis*, 20 (1999) 2922–2929.
39 L. Arce, A. Ríos and M. Valcárcel, *J. Chromatogr. A*, 791 (1997) 279–287.

40 L. Arce, P. Kuban, A. Ríos, M. Valcárcel and B. Karlberg, *Anal. Chim. Acta*, 390 (1999) 39–44.
41 B.M. Simonet, A. Ríos, F. Grases and M. Valcárcel, *Electrophoresis*, 24 (2003) 2092–2098.
42 H.W. Chen and Z.L. Fang, *Anal. Chim. Acta*, 376 (1998) 209–220.
43 L. Arce, A. Ríos and M. Valcárcel, *J. Chromatogr. A*, 827 (1998) 113–120.
44 B.L. Hogan, S.M. Lunte, J.F. Stobaugh and C.E. Lunte, *Anal. Chem.*, 66 (1994) 596–602.
45 M.W. Lada, T.W. Vickroy and R.T. Kennedy, *Anal. Chem.*, 69 (1997) 4560–4565.
46 P. Kuban and B. Karlberg, *Anal. Chem.*, 69 (1997) 1169–1173.
47 J.R. Veraart, M.C.E. Groot, C. Gooijer, H. Lingeman, N.H. Vethorts and U.A.T. Brinkman, *Analyst*, 124 (1999) 115–118.
48 J.E. Thompson, T.W. Vickrory and R.T. Kennedy, *Anal. Chem.*, 71 (1999) 2379–2384.
49 S. Palmarsdottir, E. Thordarson, L.E. Edholm, J.A. Jonsson and L. Mathiasson, *Anal. Chem.*, 69 (1997) 1732–1737.
50 P. Kuban and B. Karlberg, *Talanta*, 45 (1998) 477–484.
51 A.G. Lista, L. Arce, A. Ríos and M. Valcárcel, *Anal. Chim. Acta*, 438 (2001) 315–322.
52 C. Mardones, A. Ríos and M. Valcárcel, *Anal. Chem.*, 72 (2000) 5736–5739.
53 M.M. Bushey and J.W. Jorgenson, *Anal. Chem.*, 62 (1990) 978–984.
54 T.F. Hooker and J.W. Jorgenson, *Anal. Chem.*, 69 (1997) 4134–4142.
55 O. Cabaleiro, M. Lores and R. Cela, *Analysis*, 27 (1999) 468–471.
56 I. German and R. Kennedy, *Anal. Chem.*, 72 (2000) 5365–5372.
57 S.J. Chen and M.L. Lee, *Anal. Chem.*, 72 (2000) 816–820.
58 F. Kvasnicka and M. Voldrich, *J. Chromatogr. A*, 891 (2000) 175–181.
59 M. Urbanek, L. Blechtova, M. Pospisilova and M. Polasek, *J. Chromatogr. A*, 958 (2002) 261–271.
60 M. Urbanek, M. Pospisilova and M. Polasek, *Electrophoresis*, 23 (2002) 1045–1052.
61 R. Bodor, V. Madajova, D. Kaniansky, M. Masar, M. Johnck and B. Stanislawski, *J. Chromatogr. A*, 916 (2001) 155–165.
62 M. Dankova, S. Strasik, M. Molnarova, D. Kaniansky and J. Marak, *J. Chromatogr. A*, 916 (2001) 143–153.
63 S. Fanali, C. Desiderio, E. Olvecka, D. Kaniansky and M. Vojtek, *J. High Resol. Chromatogr.*, 23 (2000) 531–538.
64 J.F. Tyson and Z. Fresenius, *Anal. Chem.*, 329 (1988) 663–667.
65 L. Arce, P. Hinsmann, M. Novic, A. Ríos and M. Valcárcel, *Electrophoresis*, 21 (2000) 556–562.
66 P. Kuban, K. Tennberg, R. Tryzell and B. Karlberg, *J. Chromatogr. A*, 808 (1998) 219–227.

67 M. Valcárcel, S. Cárdenas and M. Gallego, *Trends Anal. Chem.*, 21 (2002) 251–258.
68 L. Manganiello, L. Arce, A. Ríos and M. Valcárcel, *J. Sep. Sci.*, 25 (2002) 319–327.
69 R. Peña, M.C. Alcaraz, L. Arce, A. Ríos and M. Valcárcel, *J. Chromatogr. A*, 967 (2002) 303–314.
70 C. Mardones, A. Ríos, M. Valcárcel and R. Cicciarelli, *J. Chromatogr. A*, 849 (1999) 609–616.
71 L. Nozal, L. Arce, B.M. Simonet, A. Ríos and M. Valcárcel, *Anal. Chim. Acta*, 517 (2004) 89–94.
72 L. Arce, A. Ríos and M. Valcárcel, *Fresenius J. Anal. Chem.*, 360 (1998) 697–701.
73 D. Martinez, F. Borrull and M. Calull, *J. Chromatogr. A*, 827 (1998) 105–112.
74 M.W. Lada, G. Schaller, M.H. Carriger, T.W. Vickroy and R.T. Kennedy, *Anal. Chim. Acta*, 307 (1995) 217–225.
75 M.W. Lada and R.T. Kennedy, *Anal. Chem.*, 68 (1996) 2790–2797.
76 J. Zhou, D.M. Heckert, H. Zuo, C.E. Lunte and S.M. Lunte, *Anal. Chem. Acta*, 379 (1999) 307–317.
77 J. Lip, P. Thibault, A. Martin, J.C. Richards, W.W. Wakarchuk and W. van der Wilp, *J. Chromatogr. A*, 817 (1998) 325–336.
78 L. Bao and P.K. Dasgupta, *Anal. Chem.*, 64 (1992) 991–996.
79 B.L. De Backer and L.J. Nagels, *Anal. Chem.*, 68 (1996) 4441–4445.
80 F. Kvasnicka, *Electrophoresis*, 24 (2003) 860–864.
81 Z.Q. Xu, T. Hirokawa, T. Nishine and A. Arai, *J. Chromatogr. A*, 990 (2003) 53–61.
82 M. Hernandez, C. Aguilar, F. Borrull and M. Calull, *J. Chromatogr. B*, 772 (2002) 163–172.
83 K. Inano, S. Tezuka, T. Miida and M. Okada, *Ann. Clin. Biochem.*, 37 (2000) 708–716.
84 K. Fukushi, N. Ishio, M. Sumida, S. Takeda, S.I. Wakida and K. Hiiro, *Electrophoresis*, 21 (2000) 2866–2871.
85 M. Larsson and E.S.M. Lutz, *Electrophoresis*, 21 (2000) 2859–2865.
86 J.C.M. Waterval, H. Krabbe, J. Teeuwsen, A. Bult, H. Lingeman and W.J.M. Underberg, *Electrophoresis*, 20 (1999) 2909–2916.

Chapter 5

UV–Vis absorbance detection in capillary electrophoresis

Antonio L. Crego and Maria Luisa Marina

5.1 INTRODUCTION

The properties of an analyte determine the best technique to be used to achieve its detection in capillary electrophoresis (CE). Figure 5.1 shows how to choose an appropriate detection technique in terms of sensitivity as a function of the characteristics of the analyte. UV–Vis absorbance detection is the first option to be considered, since it is useful for a large number of compounds that contain a chromophore. In CE, it produces detection limits corresponding to a few femtomoles of analyte (at sub-picomole levels), i.e., high mass sensitivity. However, owing to the need of small injection volumes in CE to avoid peak broadening that degrades separation efficiency (pl to nl volumes), such sensitivity is only in the micromolar range, i.e., it appears modest when expressed in terms of concentration (limits of detection [LODs] ranging from 10^{-5} to 10^{-7} M, depending upon the analyte being analyzed). These detection limits are clearly insufficient to solve many analytical problems. Although other detection systems provide low-concentration detection limits (Table 5.1) (see chapters 6 and 8), very often, the choice of the detection system is limited to the equipment available. Therefore, several approaches have been developed to overcome this difficulty. These include solute derivatization or complexation, modified capillary dimensions, and the use of pre-concentration techniques [11]. Some of these approaches will be presented in this chapter.

Table 5.2 shows some examples of LODs obtained for UV–Vis absorbance detection in CE using different pre-concentration strategies. Among them, on-capillary pre-concentration techniques described in

Comprehensive Analytical Chemistry XLV
M.L. Marina, A. Ríos and M. Valcárcel (Eds)
Volume XLV ISSN: 0166-526X DOI: 10.1016/S0166-526X(05)45005-9

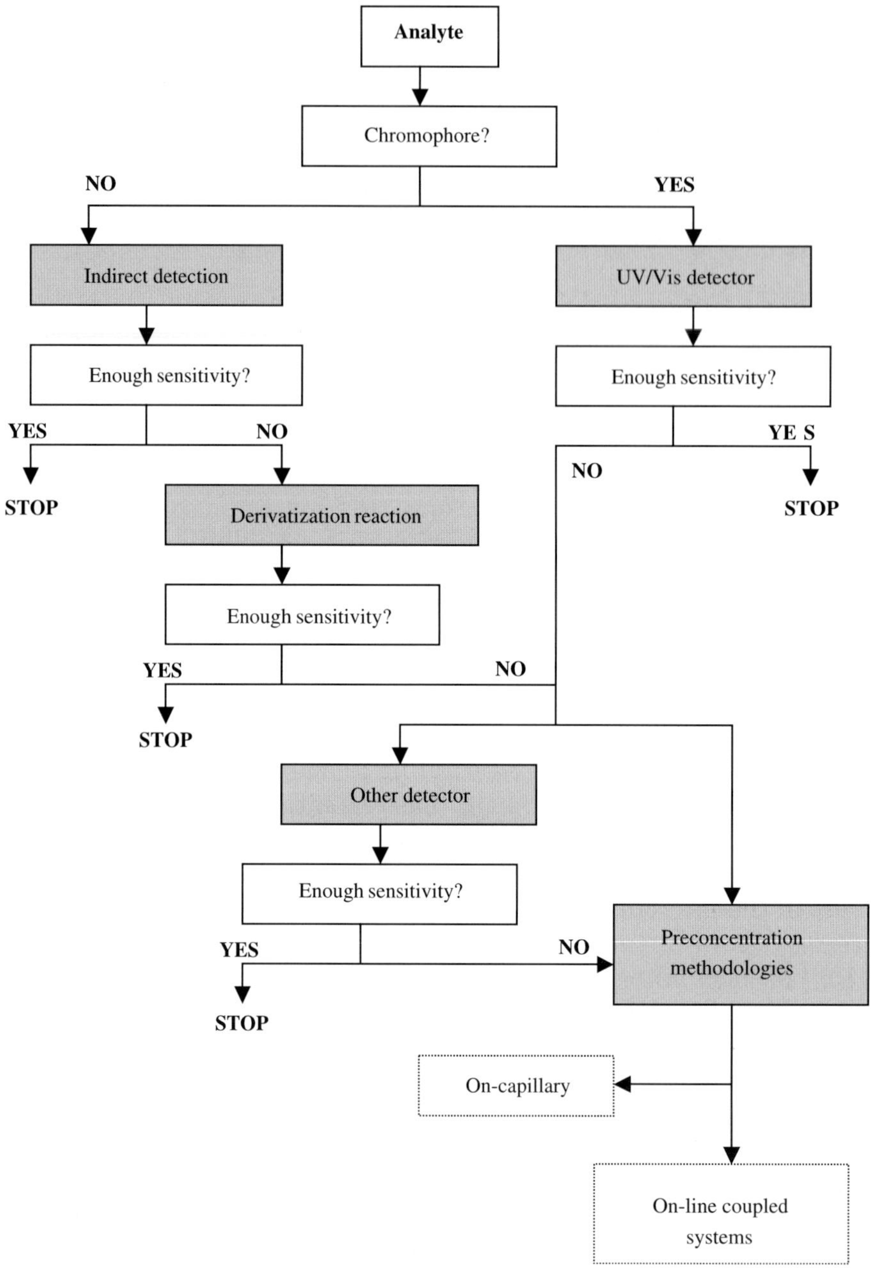

Fig. 5.1. Diagram for the selection of an appropriate detection method in CE in terms of sensitivity as a function of analyte properties. Adapted from Ref. [1], Copyright (2003), with permission from Elsevier.

TABLE 5.1

LODs for different detection techniques in CE (Adapted from ref. [2])

Detection technique	LODs	Reference
Direct absorbance	10^{-5}–10^{-7} M	[3]
Indirect absorbance	10^{-5}–10^{-6} M	[4]
Thermooptical absorbance	10^{-7}–10^{-8} M	[5]
Direct laser-induced fluorescence	10^{-10}–10^{-11} M (native)	[6]
	$< 10^{-13}$ M (chemical derivatization)	[7]
Conductivity	10^{-7}–10^{-8} M	[8]
Potentiometric	10^{-7}–10^{-8} M	[9]
Amperometric	10^{-7}–10^{-8} M	[10]

chapter 3 of this book and the coupling of flow systems to CE described in chapter 4 are the most interesting from the viewpoint of their contribution to the automation of analytical processes. LODs in the 10^{-11} M level can be obtained by direct absorbance detection when using on-line pre-concentration techniques. These results explain why absorbance detectors are by far the most widely used in CE. In fact, in addition to the possibility of achieving the above-mentioned sensitivity by using different pre-concentration strategies, other characteristics, such as the relatively low cost, simplicity, and ease of use of absorbance detectors increase the value of this detection mode.

Other interesting features of this detection mode are as follows:

- It can be used when working in all the separation modes possible in CE, such as CZE, CITP, CIEF, CGE, CEC, and EKC, although some restrictions in the running buffers exist because of their optical properties (inorganic buffers are more useful in this respect).
- It is non-destructive of the analyte, and therefore, other detectors can be connected in order to obtain additional information on the sample or the analyte.
- Qualitative information on peak purity and peak identification can also be obtained from spectra acquired by multiwavelength detector, see section 5.2.2.5.
- Quantitative analysis is easily achieved.
- Although absorbance is not normally considered a universal detection mode, operation at low-UV wavelengths (difficult to use in

TABLE 5.2

LODs obtained for different pre-concentration strategies employed with absorbance detection in CE

Absorbance technique	Pre-concentration system	Analytes	Matrix	LODs	Reference
Direct absorbance	Off-line SPE[a]	Sunlindac	Urine	2.9 μg/l	[12]
	On-line SPE[a]	Terbutaline	Plasma	6×10^{-10} M	[13]
	FESI[b]	Heroin metabolites	Urine	40 μg/l	[14]
	Off-line SPE[a] and FESI[b]	Pesticides	Drinking water	40 ng/l	[15]
	Sweeping	Corticosterone	Plasma	5 μg/l	[16]
	Sweeping	Anthracycline antibiotics	Plasma	10^{-9} M	[17]
	FESI[b]-FI[c]	11 US EPA for priority phenol pollutants	Standards	0.7 μg/l	[18]
	SPE[a]-FI[c]	Nitrite, nitrate, bromide, and iodide	Standards	10 μg/l	[19]
	SPE[a]-FI[c]	Inorganic and organic species of mercury	River water	1.2 μg/l	[20]
	tITP[d]	Iodide	Sea water	0.6 μg/l	[21]
	tITP[d]	Nitrite and nitrate	Sea water	20 μg/l	[22]

	FESI[b]-Water removal and off-line SPE[a]	Phenoxy acid herbicides	River water	10 μg/l	[23]
	CSEI[e]-Sweeping	Laudanosine and naphthylamine	Standards	4 ng/l	[24]
	CSEI[e]-Sweeping	Corticosterone and one of its metabolites	Plasma; urine	3 μg/l	[25]
Complexation prior to UV absorbance	Off-line SPE[a]	Metal ions	Standards	7×10^{-9} M	[26]
	CSEI[e]-Sweeping	Metal ions	Wastewater	2×10^{-11} M	[27]

[a]SPE, solid-phase extraction;
[b]FESI, field-enhanced sample injection (sample stacking with electrokinetic injection);
[c]FI, on-line coupled flow injection;
[d]tITP, transient isotachophoresis;
[e]CSEI-Sweeping, sweeping in conjunction with cation-selective exhaustive injection (sample stacking accompanying electrokinetic injection).

HPLC) or in an indirect mode (see section 5.2.4) enables a near-universal detection of analytes.

Contrary to HPLC, direct UV detection at very low wavelengths such as 185 nm is often feasible in CE. With fused-silica capillaries, detection below 200 nm is possible since fused silica has a UV cut-off of approximately 170 nm. At low-UV wavelengths (180–200 nm), since most organic analytes have some absorbance, detection of molecules without obvious chromophores becomes possible. Some examples of application of low-wavelength absorbance detection in CE include the determination of fermenting acids (185 nm) in very complex matrices such as silage samples [28], the detection of metal ions (185 nm) such as Cr^{3+}, Cu^{2+}, Fe^{3+}, UO_2^{2+}, VO_2^{+}, and VO^{2+} (no complexation was needed) [29], the analysis of organic and inorganic arsenic species (192 nm) and its application to the determination of four aryl arsenic compounds used as additives in animal feed [30], and the determination of morphine (190 nm) at therapeutic concentrations without sample pre-treatment in human plasma [31].

These characteristics of UV–Vis absorbance detection explain why, nowadays, all commercially available CE instruments are provided with a UV–Vis absorbance detector. In fact, many of the analytes separated by CE[1] can be detected by UV–Vis absorbance using any of the different separation modes available.

This chapter presents some of the most significant developments in absorbance detector designs for CE. Additionally, readers are provided with an overview of various approaches and methodologies to improve the use and sensitivity of absorbance detection, including unconventional absorbance detection modes such as thermooptical detectors. The possibilities of UV–Vis absorbance detection in microchip electrophoresis are also considered. In the elaboration of this chapter, relevant information from different chapters of fundamental books on CE [32–36] has been used, together with an important number of reviews, including the five fundamental reviews on CE appeared in *Analytical Chemistry* [37–41], and some recent ones with emphasis on the detection modes in CE [2,11,42–46] have also been published. Likewise, abundant commercial information has been consulted [47–49].

[1]Dyes, small organic and inorganic ions, amino acids, peptides, proteins, sugars, oligosaccharides, carbohydrates, fatty acids, lipids, nucleotides, oligonucleotides, nucleic acids, vitamins, pharmaceutical and forensic drugs, pesticides, pollutants, surfactants, and polymers.

5.2 UV–VIS ABSORBANCE DETECTION IN CE

5.2.1 Basic principles

Any molecule having a chromophore can be detected by UV–Vis absorbance. When absorbing analytes pass through the detection cell, there is a decrease in the intensity of light transmitted to the detector. According to Beer's law, the absorbance (A) is related to the concentration of an analyte (C) through the following equation:

$$A = \varepsilon bC \tag{5.1}$$

where ε is the molar absorptivity and b is the optical path length.

Analytes with molar absorptivities of 10^4 to 10^5 are strong absorbers, and those with molar absorptivities under 10^3 are weak absorbers [47]. According to Eq. (5.1), detection limits are better for analytes that have larger molar absorptivity values. However, the magnitude of the absorbance signal is characteristic of a given substance under a precisely defined set of conditions, such as the chosen wavelength for measuring absorbance, solvent, and temperature. In fact, the pH value, running buffer composition, or degree of ionization are important because these conditions may cause changes in the molar absorptivity value and a shift in the wavelength of maximum absorption [34].

Sensitivity is defined as the slope of the calibration curve (detector signal versus analyte concentration), which depends on the path length and molar absorptivity according to Eq. (5.1). Owing to the much smaller capillary dimensions in CE when compared to column dimensions in HPLC, the detector cell must have proportionally smaller dimensions. This severely limits the optical path length available, and since sensitivity is proportional to path length, this is the factor that mainly limits sensitivity in CE, which decreases as the inner diameter of the capillary is reduced.

On the other hand, assuming good optical and electronic designs, only two factors may influence photometric accuracy and precision: noise and stray light [47]. Absorbance detectors typically have two noise elements. The first element (photon or shot noise) results from the statistical distribution of the photons emitted by a light source. The minimum statistical fluctuation in the number of photons arriving at the photodetector is shot noise, which is proportional to the square root of the intensity of light. Therefore, shot noise can be reduced by increasing the magnitude of light reaching the detector. The second element of noise is inherent to the instrument electronics (detector amplifier, analog-to-digital converter, etc.) and does not depend on the light intensity measured. Noise negatively affects the precision of measurements, and for any single measurement, may introduce errors in accuracy as well.

Stray light is a certain amount of residual light observed by a detector even under conditions of complete absorbance. This background signal limits the linearity of absorbance detectors at high absorbance levels, i.e., stray light causes deviations from Beer's law that limit quantitative analysis. At high analyte concentrations, the deviation usually takes the form of a decreased slope of the calibration curve, which determines a linear dynamic range. Stray light consists primarily of light from the source that does not interact with the analyte and reaches the detector either directly or by reflections. Ideally, in CE, all light should pass through the center of the capillary, not through the wall. However, the stray light becomes more problematic when narrower inner diameter capillaries are used, because more light misses the lumen of the capillary, causing an important limitation in the linear dynamic range [34]. Therefore, in CE the linear dynamic range is significantly lower than that observed in HPLC (generally from 0.4 to 0.7 AU in CE versus 1.2 to 1.5 AU in HPLC) because of the small size and curvature of the capillary [48].

5.2.2 Detector design

UV–Vis absorbance is a very popular detection technique in CE because of its interesting characteristics (described earlier), and its easy adaptation from HPLC detectors by use of the capillary as the detection cell. In fact, absorbance detection must be accomplished on the capillary, i.e., detection occurs across the capillary inner diameter. Thus, in UV–Vis on-capillary detection, a part of the polyimide coating of the capillary is removed to function as a spectrophotometric cell (see Fig. 5.2), as small detection volumes are required to minimize extra-column band broadening[2]. Consequently, since the detection cell is a part of the capillary, there is no zone broadening as a result of the existence of dead-volumes or component mixing (joints, fittings, and connectors are eliminated).

[2]In a first approximation, the limitations on detector volume are similar to those on injection volume, i.e., the detector must function with volumes at nanoliter or sub-nanoliter levels.

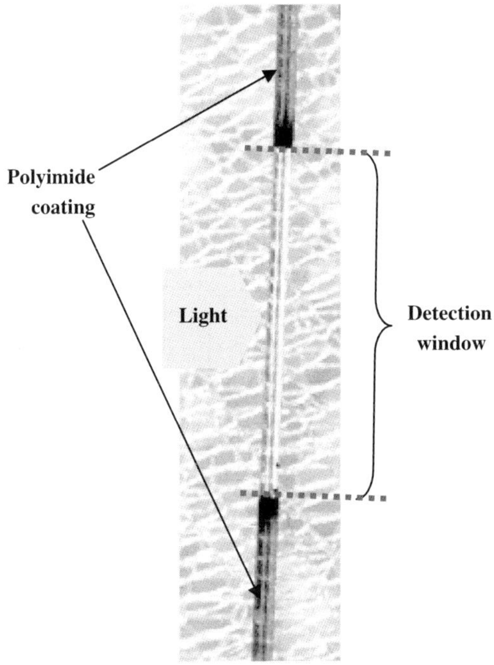

Fig. 5.2. Detection window for on-capillary detection.

The features that a standard UV–Vis detector should have in order to be used as an on-capillary detector in CE are: (i) the ability to secure positioning of the capillary in the light beam in order to avoid vibration of the capillary upon application of high voltages enabling maximum signal-to-noise ratio, and (ii) the ability to focus the light beam onto the inner diameter of the capillary to avoid stray light and non-linear response of the detector. Furthermore, the narrow analyte bands typically obtained in CE require a reduction in the axial length of the detection window to maintain high resolutions.

Figure 5.3 shows one of the first attempts of an optical setup to be used in an on-capillary UV detector for CE [50]. This fixed-wavelength UV absorption detector employed a high-intensity "pen ray" source emitting a radiation that was directly passed through an aperture (a pinhole of 100 μm) before reaching the capillary. The detector also included features to secure positioning of the capillary in the light beam. In addition, the effect of source intensity instabilities was reduced by using a double-beam system as in many commercial systems (where the second optical path serves as a reference to correct source fluctuations

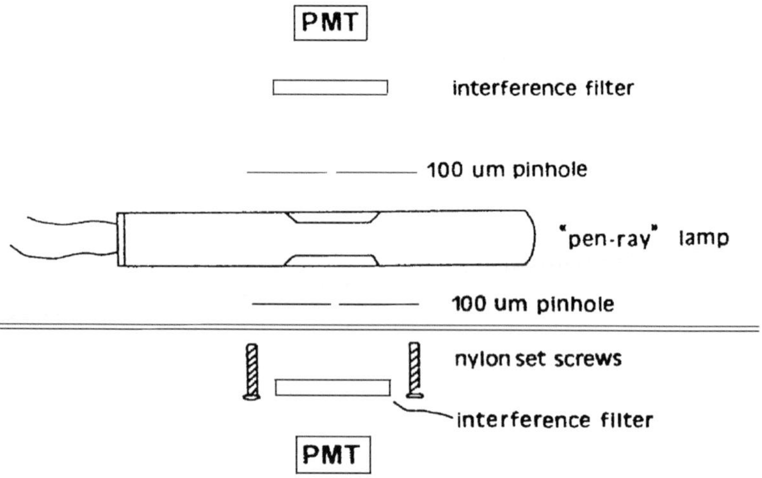

Fig. 5.3. Schematic diagram of the optical layout of a fixed wavelength on-column UV absorption detector. PMT, photomultiplier tube. Reprinted from Ref. [50], Copyright (1984), with permission from Elsevier.

and to decrease the shot noise). However, the reduced dimensions of the aperture used to cut off stray light restricted the amount of light that reached the photomultiplier tube, increasing shot noise. In conclusion, this detector showed poor sensitivity owing to the use of an unfocused light source.

In fact, it is not easy to focus incoherent light through capillaries of inner diameters smaller than 100 μm with sufficient intensity to minimize shot noise. However, spatially incoherent light sources are usually employed in standard UV–Vis detectors (i.e., line and continuous sources). To overcome the above limitations and improve the detector's performance, different designs of UV–Vis absorbance detectors have been optimized. These optimized designs have introduced some changes in different components of a CE absorbance detector, which will be briefly described as follows.

5.2.2.1 Light sources

Since the LOD in absorbance is dictated by the ability to see a small decrease in the light intensity of a high-intensity background, source stability becomes one of the most important criteria of an illumination system. Accordingly, an ideal light source would yield a constant intensity over all the emitted wavelengths with low noise and long-term stability. Unfortunately, such a source does not exist.

Two types of light sources are commonly used in UV–Vis absorbance detectors: lines and continuous sources. The simplest UV light sources are atomic lamps that produce strong emission lines at well-defined wavelengths, e.g., the low-pressure Hg lamp (254 nm), the Cd lamp (229 nm), the Zn lamp (214 nm), or the As lamp (200 nm). A medium-pressure Hg lamp is also available as a line source that has useful emission lines at 254, 280, 313, 365, 405, 436, and 546 nm [34].

With respect to continuous sources, two lamps are usually employed in UV–Vis absorbance detectors. The first source, the deuterium arc lamp, yields a good continuous intensity in the UV region (ranging from 180 to 350 nm) and provides useful intensity in the visible region. Noise from the lamp is often the limiting factor in the instrument's overall noise performance, modern deuterium arc lamps being those having the lowest noise. Over time, the intensity of light from this lamp decreases steadily. Such a lamp typically has a half-life (the time required for the intensity to fall to half of its initial value) of approximately 1000 h. The second source, the tungsten–halogen lamp, yields good intensity over part of the UV spectrum and over the entire visible range (from 280 to 1000 nm). This type of lamp has very low noise and low drift, and typically has a useful life of 10,000 h. Generally, only the deuterium arc lamp is used to measure the UV–Vis range (from 180 to 600 nm), although the light from both types of lamps can be mixed to yield a single broadband source, which permits to obtain an important decrease of the noise in the visible region.

These light sources are incoherent, so it is difficult to focus a significant fraction of the light into the smallest capillaries. The low luminosity and the inefficient coupling of the light source into the capillary result in a decrease in detector performance owing to its low intensity and high stray light [34]. In fact, over the last two decades, developments on laser-based detection techniques for CE have received much attention [51]. There are two main types of lasers: continuous wave (CW) and pulsed lasers. Currently, available lasers are the helium–cadmium (He–Cd) laser, which emits radiation at 325 nm and at 442 nm, the helium–neon (He–Ne) laser, which emits radiation at 633 nm, and the argon-ion laser, which has some lines around 350 and around 475 nm. All of them are CW lasers, which emit their output continuously, generally in the visible range, their intensity being constant over time. Under special conditions, largo-frame argon-ion lasers emit lines as short as 275 nm [52]. Also, intracavity frequency-doubled argon-ion lasers have been introduced commercially, which emit some lines in the deep UV (229, 244, and 257 nm). On the other hand, pulsed lasers typically deliver their optical output as short intense pulses (about 10 ns at a rate of 100 Hz) in the UV region. Krypton-ion (229 and 248 nm) and Nd:YAG (266 nm) lasers are examples of pulsed lasers.

Lasers provide a powerful (between 10 mW and 10 W), directional, and coherent beam of light that can be focused into a small volume (typically 0.1–1 nl). This fact, together with their high spectral radiance (amount of intensity in a certain wavelength interval), enables a reduction in the shot noise. These unique features make a laser an ideal light source for detection purposes in CE [51]. However, absorbance systems based on lasers rarely obtain the minimum value of the shot noise owing to intensity instability. Even the development of a double-beam system to correct source fluctuations that used an argon-ion laser operating at 305 nm for UV detection in CE yielded only a four-fold improvement in the detection limit over that of a commercial detector [53]. As a result, and taking into account the high cost and complexity associated with the use of UV lasers, their limited emission wavelengths, and the small improvement in concentration sensitivity, it is unlikely that laser-based absorbance detectors will become common for commercial CE instruments.

Diode lasers and light-emitting diodes (LEDs) are solid-state light sources that are small, inexpensive, have a stable output (50 times better than a He–Ne laser) and a long lifetime (typically 5×10^4 h or approximately 6 years of continuous service) [54]. A detector based on a diode laser, owing to its small size, can be extremely compact, which is ideal for micro-separation systems. However, this light source has two major disadvantages: (i) it generally operates in the visible and near-infrared region (above 635 nm), although in a few cases continuous emission diode lasers from 390 to 440 nm have been employed [55] (e.g., a violet-emitting laser diode based on gallium nitride is commercially available [56]), and (ii) it provides relatively low optical power (a few mW) [51].

A review about the analytical use of LEDs in the last decade has been published [57]. In this paper, LED-based absorbance measurements and its use in CE, together with other analytical applications, are discussed. The actual emitting chip area of a typical LED is 0.25 mm × 0.25 mm [57]. Therefore, even without focusing optics, the output from an orange LED (605 nm) is good enough to be used in CE, obtaining comparable noise levels for the LED and the cadmium and zinc lamps are comparable [58]. Butler et al. [59] described the construction of a simple absorption detector for CE using a bright blue-green LED as the source (at 525 nm) and optical fibers to couple the capillary to the LED and the diode. They demonstrated the potential of this approach for the direct detection of metal ions and metal ions as complexes of 4-(2-pyridylazo) resorcinol (PAR), and the indirect detection of alkali and alkaline earth metal ions using a background electrolyte (BGE) containing Pyronine G. More recently, Collins and Lu [60], employing a red LED (660 nm) established a LOD of 23 μg/l for U (VI) using Arsenazo III as a pre-capillary complexing ligand. Macka et al. [61] have extensively used LEDs for absorbance detection in CE. In one of their works, they investigated six different LEDs that had emission wavelengths within the visible region (563 to 654 nm), and found that LEDs in general exhibited stable output and markedly lower noise than other light sources, such as mercury, deuterium, and tungsten lamps. They also established attractive detection limits for three metal–PAR complexes using an LED at 568 nm: 9 and 25 μg/l for Nb (V) and Ta (V), respectively [62], and 19 μg/l for V (V) [63]. Recently, they have obtained better LODs with a blue LED (476 nm) than with a deuterium lamp as light source [64].

In contrast to the works in which LEDs operating in the visible and near-infrared region (400–1000 nm) are employed as detector light sources in CE [54], the first use of an LED as a UV light source has been recently reported by King et al. [65]. In this work, a UV LED was investigated as a simple alternative to standard mercury or deuterium lamps for use in indirect detection of inorganic anions using CE with chromate as the BGE. The UV LED used had an emission maximum at 379.5 nm, a wavelength at which chromate absorbs strongly and exhibits a 47% higher molar absorptivity than at 254 nm when using a standard mercury lamp [65]. The results obtained using LED as light source exhibited a performance superior to that of the mercury lamp in terms of noise, sensitivity, and linear range (up to 70% decrease in noise, up to 26% increase in sensitivity, and over 100% increase in the linear range). LODs for common inorganic anions ranged from 3 to 14 μg/l. These excellent results are of great interest, since chromate is by far the most common absorbing probe anion for the indirect detection of anionic species [66–68].

Finally, it can be emphasized that LEDs are ideal sources for integral optical detectors on microchips, as will be described in section 5.4.

5.2.2.2 Optical system

An optical system used for absorbance detection in CE can be set up with the following main elements: the wavelength selector, the focusing optics, the aperture, and the capillary. In addition to these basic components, optical fibers are usually employed.

To isolate the desired wavelength band to illuminate the sample, filters and monochromators have been used. Filters have the advantage of higher throughput, simple optical alignment and design, and low cost. However, monochromators offer the advantage of wavelength selection. For instruments that use fixed-wavelength lamps as a source of radiation, only an appropriate filter is needed for wavelength selection (see Fig. 5.3). However, when a continuous source, such as the deuterium lamp, is used, the line of interest can be isolated either by filters or by a monochromator.

In order to be more generally useful in terms of selectivity, an absorbance detector with the possibility of continuously varying the selected wavelength is desirable. In this case, the monochromator is preferred for wavelength selection over filters since it is more versatile, can be used to scan wavelengths, and produces better resolution. However, the instruments utilizing a monochromator have a drawback: they are usually less sensitive than systems employing filters, which provide more light throughput. Another possibility is to combine a continuous source with a group of filters, as shown in Table 5.3, which groups the general characteristics of the UV–Vis detectors employed in the CE systems supplied by manufacturers of CE equipment at the present time.

The focusing optics and aperture must be carefully designed for a given capillary in order to optimize the linear dynamic range and the sensitivity of the detection system. For a given design that uses a certain aperture (a pinhole or a slit) to block the stray light increasing the light level arriving at the photodetector (e.g., a diode) by increasing the aperture width improves the detection limits by decreasing the baseline noise (see Fig. 5.4a and 5.4b). However, if the aperture used allows stray light to pass through the capillary walls, the linear detection range will be compromised (see Fig. 5.4a and 5.4b). In fact, the aperture width used should be sufficiently small to allow only the inner diameter of the capillary to be illuminated, thereby reducing stray light levels. But if the aperture is too small, excessive background noise can result owing to the decrease in the light throughput. Therefore, the focusing optics should increase the fraction of light passing through the inner

TABLE 5.3

General characteristics of the UV–Vis detectors employed in the CE systems supplied by manufacturers of CE equipments currently

Manufacturer	Model	General characteristics	Web page
Agilent technologies	^{3D}CE	Deuterium lamp multiwavelength selector (monochromator)	www.agilent.com
Beckman coulter	P/ACE^{TM} MDQ DAD	Deuterium lamp multiwavelength selector (monochromator)	www.beckmancoulter.com
	P/ACE^{TM} MDQ UV/Vis	Deuterium lamp fixed-wavelength selector (filters)	
Lumex	CAPEL 103 or 104 models	Mercury lamp for 254 nm fixed-wavelength selector (filters)	www.lumex.ru
	CAPEL-105	Deuterium lamp variable-wavelength selector (monochromator)	
Prince technologies	PrinCE 700 series	Deuterium lamp multiwavelength selector (monochromator)	www.princetechnologies.com

continued

TABLE 5.3 (*continued*)

Manufacturer	Model	General characteristics	Web page
	PrinCE 200, 400, 600 series[a]	Deuterium lamp variable-wavelength selector (monochromator) Deuterium lamp fixed-wavelength selector (filters)	www.princetechnologies.com www.micro-tech.us www.knauer.net www.princetechnologies.com www.knauer.net
Unimicro technologies	TRISEP™ 2010GF/2010 IF	Atomic lamps or tungsten lamp fixed-wavelength selector (filters)	www.unimicrotech.com
	TRISEP™ 2010GV/2010 IV	Deuterium lamp variable-wavelength selector (monochromator)	

[a]Allows the use of any UV–Vis detector.

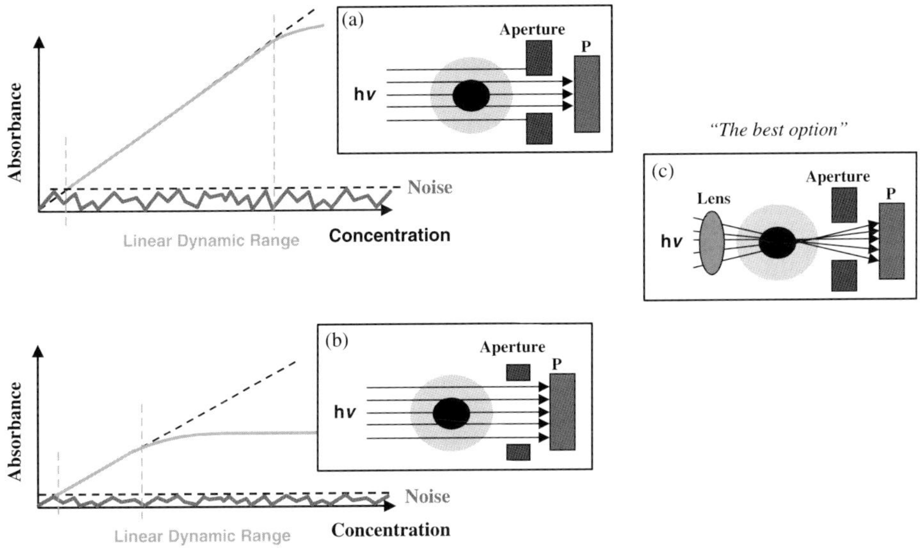

Fig. 5.4. Influence of the light path through the capillary on the linear dynamic range and baseline noise. P: photodetector. Adapted from Ref. [48], Copyright (2000), Agilent Technologies, Inc. Reproduced with permission.

diameter of the capillary, i.e., the optical beam should be tightly focused directly into the capillary to obtain maximum throughput at the aperture and to minimize stray light reaching the detector (see Fig. 5.4c). With this design (aperture together with a lens), it is possible to ensure that most of the light reaching the photodetector passes through the inner diameter of the capillary. Consequently, the linear range and the sensitivity can be increased. The use of ball lenses (small sapphire or quartz spheres) as focusing optics in UV–Vis absorbance detectors can produce systems with very efficient illumination capability if appropriate radius of the ball lenses is chosen. In addition, the performance of these ball lenses depends on the distance between the lens and the capillary, and on the ratio between the inner and outer diameters of the capillary [69].

Optical fibers permit a more flexible coupling of light to the capillary [70,71]. The ability of a fiber to transport light is mainly governed by the diameter of the fiber. Thus, large diameters facilitate the coupling of light into the fiber, but if this diameter is larger than the inner diameter of the capillary, a large fraction of the light passes through the wall of the capillary or outside the capillary, giving rise to stray light. Therefore, the dimensions of the fiber must also be optimized, taking into account the dimensions of the capillary [72], and when the optical fiber is solely employed for the coupling of light through the capillary, its diameter must be kept fairly small in order to avoid stray light. However, reduced dimensions of this fiber restrict the amount of light that reaches the photodetector, and problems with shot noise may arise [73–74], as in the case of the aperture width. Therefore, lenses and fiber optics should be used together to focus the light source onto the capillary and to collect the light and focus it onto the photodetector. For example, Lindberg et al. [71] have described a design that combines a large-diameter source-fiber with focusing optics (see Fig. 5.5). The excellent results obtained with this design (noise about 5×10^{-5} AU and linear dynamic ranges larger than 10^3 for capillaries of 30 to 100 μm I.D. (inner diameter), values similar to those usually obtained with HPLC detectors) demonstrated that the combination of large-core optical fibers with focusing lenses and an aperture that is matched to the dimensions of the capillary lead to a maximized dynamic range and optical throughput, while stray light is minimized. These authors also suggested that the use of a rectangular slit instead of a pinhole decreased the noise of the detector.

Nowadays, all commercial on-capillary absorbance detectors used for CE employ slits with a rectangular window. The dimensions of the slit (detection window) are designed for specific capillary dimensions, with a width slightly narrower than the inner diameter of the capillary, and a length shorter than the solute zone to maintain high resolution. Therefore, since solute zones in CE are typically 2 to 5 mm long, the detection window should be maximally one-third this amount, i.e., in general, values smaller than 1 mm should be used. From the user's viewpoint, easy placement and alignment of the capillary with the detector optics are huge advantages. Thus, these on-capillary detectors are constructed in two subunits that can be press-fitted together,

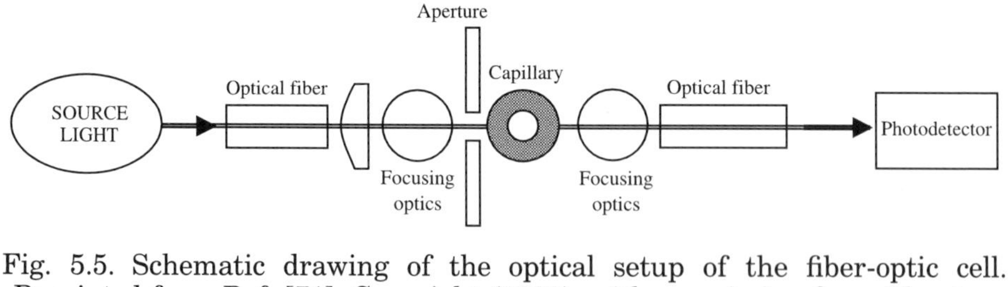

Fig. 5.5. Schematic drawing of the optical setup of the fiber-optic cell. Reprinted from Ref. [71], Copyright (1998), with permission from Elsevier.

thereby enclosing the capillary. In addition, one of the subunits has a V-shaped groove to facilitate a proper alignment of the capillary, avoiding the need for capillary adjustments.

Finally, an effective optical path length is defined in the cylindrical cells commonly used in on-capillary detection for CE owing to its circular cross section. The effective path length would be equal to the capillary inner diameter only if the center of the capillary is illuminated and the ray of a collimated beam can travel through the full length of the inner diameter of the capillary. For beams further away from the central axis, the distance traveled through the capillary will be shorter and finally approaching zero for a ray traveling through the capillary but outside the inner diameter of the capillary. Therefore, assuming parallel light, the effective path length for a circular capillary would be calculated as $(\pi/4)d$, where d is the inner diameter of the capillary, i.e., about 80% of the inner diameter [69]. However, in reality, depending on the width of the aperture defining the illuminated part of the capillary and the quality of detector optics, all those individual path lengths will combine to produce the final effective path length [75]. In fact, important differences in the effective path length were obtained for five commercial detectors (see Table 5.4). The determined effective path length ranged from 49.7 to 64.6 μm for a 75 μm I.D. capillary, i.e., the effective optical path length was always smaller than the inner diameter of the capillary. It was possible to calculate its value from a known probe absorptivity and the measured sensitivity in the linear range of the detector [75]. On the other hand, important differences in the detector linearity were also obtained for the five detectors studied (see Table 5.4). The upper limits of the linear dynamic range (determined as 5% decline in sensitivity) ranged from 0.175 to 1.2 AU. These results clearly show that some instruments have superior optical properties, which can lead to improved results. Thus, as will be shown below, a larger effective path length reflects a better quality of the optical design of the detector, which enables a greater linear dynamic range.

5.2.2.3 Ways for increasing the optical path length

As stated above, there is an important limitation in the sensitivity of absorbance detectors when conventional transmission measurements perpendicular to the capillary are carried out (Fig. 5.6a), since in these cases the sensitivity is proportional to the optical path length that is defined by the inner diameter of the capillary. A small path length,

TABLE 5.4

Upper limits of the linear dynamic range and effective path-length values for five commercial detectors according to ref. [75]

Instrument brand[a]	Detection system	Effective path length (μm)	Linear dynamic range (upper limit in AU)
Agilent technologies ^{3D}CE	Deuterium lamp multiwavelength detector	64.6	1.2
Applied biosystems 270A-HT	Deuterium lamp variable wavelength detector	60.5	0.75
Beckman MDQ PDA	Deuterium lamp multiwavelength detector	54.9	0.55
Beckman MDQ UV	Deuterium lamp fixed wavelength detector	53.6	0.30
Waters CIA	Deuterium lamp fixed wavelength detector	49.7	0.175

[a]A new capillary of 75 μm I.D. × 365 μm O.D. was used for each instrument.

typically 25–100 μm, causes concentration LODs to be 1–2 orders of magnitude lower than in HPLC, where the flow-through cell normally has a 1 cm path length. Although the sensitivity in CE could be increased simply by using wider inner diameter capillaries, the problem is that the increased current and associated Joule heating obtained under these conditions are detrimental for the performance of the separations (see section 1.2). Therefore, various attempts have been made to improve the sensitivity of absorbance measurements in CE. Expansion of the light path by increasing the optical path length has been most actively pursued. Further alternatives, such as axial and multireflection detection, have also been described. All these approaches brought about moderate to significant improvements in detection, but they often impaired separation [76]. These alternatives will be briefly described as follows, although many of them have been too sophisticated to be included in commercial instruments.

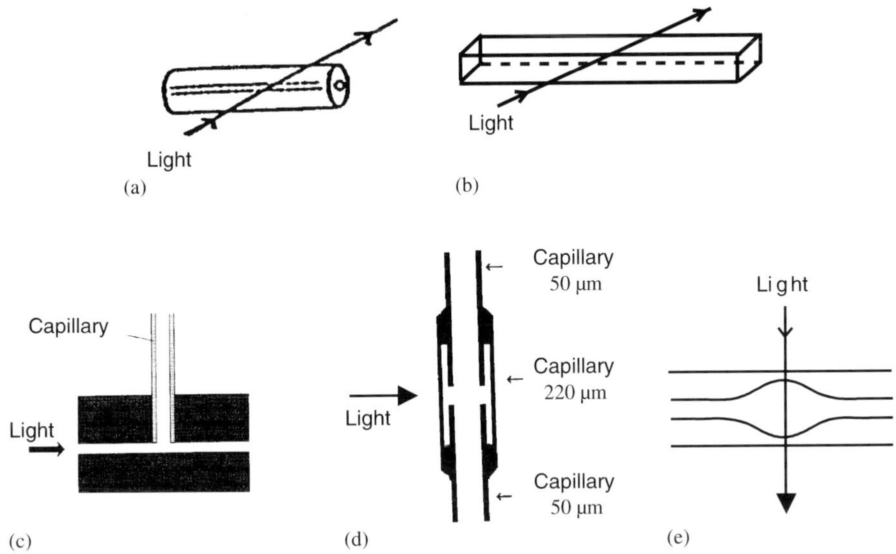

Fig. 5.6. Different detector cells used to increase the optical path length. (a) Conventional capillary, (b) rectangular capillary, (c) post-capillary flow cell, (d) sleeve cell, and (e) "bubble" cell capillary.

Modification of the shape or the diameter of the capillary in the detection zone has been achieved in order to increase the effective path length and to improve detection limits. An alternative approach is to use tubing with a non-circular cross section. Thus, as shown in Fig. 5.6b, a rectangular cross-section (e.g., 50 μm × 1000 μm) capillary can be used [77]. The width of the capillary is much greater than the height (20:1) in order to maximize path length (better absorbance performance) and heat dissipation (higher surface-to-volume ratio) [34]. Despite their advantages, such capillaries are not in widespread use owing to practical difficulties, which include the manufacturing process, limited availability, difficulty of handling (as they are readily breakable), and the need to modify existing instrumentation to accommodate the changed capillary dimensions. In order to avoid such difficulties, capillaries with a square cross section were commercialized in 1998 (Polymicro Technologies, Phoenix, AZ, USA) [78]. However, with a square capillary cell, a smaller increase in the effective path length and higher current intensity than with a rectangular capillary cell are obtained.

Other flow cell shapes focused on improving sensitivity by increasing the optical path length through the capillary have been investigated.

A post-capillary flow cell in which the outlet of the capillary was connected vertically to the middle of a 3-mm microchannel that served as the optical path was also described [79] (see Fig. 5.6c), but a 44% loss in resolution was reported with this design. Finally, on-column designs consisting in connecting a capillary cell with an inner diameter larger than the inner diameter of the separation capillary to increase the optical path length were described and evaluated [80–82]. For example, a simple method for creating a region of extended path length (a sleeve cell) was reported by Djordjevic et al. (see Fig. 5.6d) [82]. An abrupt change in the inner diameter of the capillary in the sleeve cell region (from 50 to 220 μm) did not produce extensive band broadening, and a 3.5-fold increase in corrected peak area was obtained when compared with on-column detection in capillaries of 75 μm I.D. [83].

The "bubble" cell, so called because of its shape (see Fig. 5.6e), is another flow cell shape that has been described [84]. It is made by forming an expanded region, a bubble (three to five times larger than that of the rest of the capillary), directly within the capillary. Although the volume of the bubble is considerably larger than that found in the same section of a non-expanded capillary, separation efficiency and resolution remains nearly unaffected owing to reversible compression of the analyte band in the axial direction (along the capillary) resulting from the temporary decrease in the electric field strength inside the bubble. Concomitant with the axial compression in the bubble, the analyte zone expands radially (across the capillary) to fill the increased area. By a similar process, the solute returns to its original state after it exits the bubble. This phenomenon is illustrated in Fig. 5.7. In addition, there are no adverse effects from Joule heating within the bubble since both the electric field and quantity of heat generated are proportionally lower in the bubble [84]. Therefore, any resolution loss comes from the larger volume of the bubble, but to preserve electrophoretic resolution in the zone compression, the length illuminated in a bubble cell has to be reduced. This fact is illustrated in Fig. 5.7. Thus, whereas the illuminated region of a standard capillary of 50 μm I.D. is defined by a rectangular slit (40 μm × 620 μm), the slit used for the bubble capillary of 50 μm I.D. and 150 μm of bubble is square (145 μm × 145 μm) [84]. This design does not lead to an increase in the signal-to-noise ratio equal to the increase in the optical path length (from three- to five-fold) because a higher baseline noise and a lower light throughput in the detection cell are obtained. Thus, the maximum diameter of the bubble is limited to five times the inner diameter of the

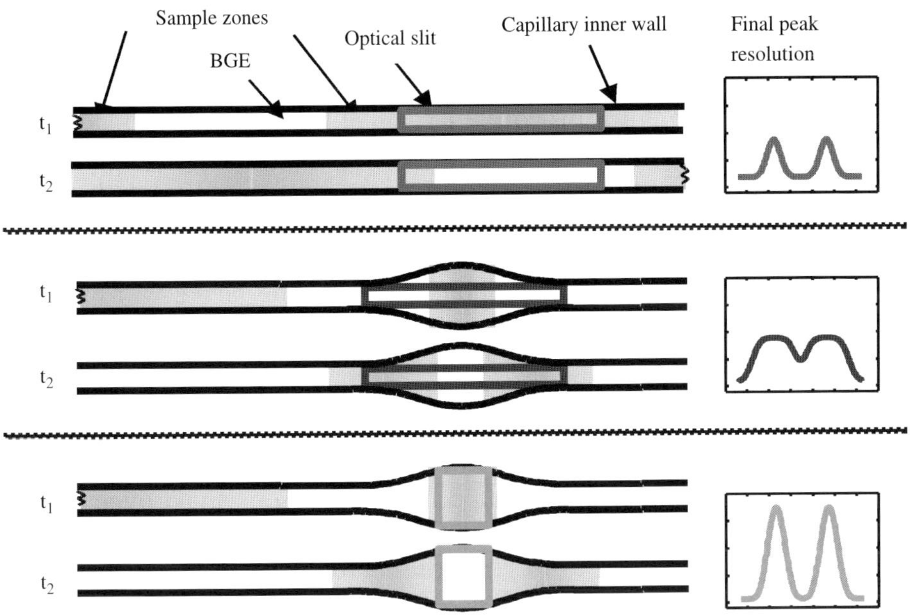

Fig. 5.7. Schematic of bubble cell capillary showing the compression of the analyte band in the axial direction and its resolution according to the length of the detector-viewing region (slit) used. Reprinted from Ref. [84], Copyright (1994), with permission from John Wiley & Sons.

capillary. Larger bubble cells require excessively short illuminated lengths to maintain resolution, which cause elevated baseline noise that compensates the increase obtained in the signal. Nowadays, Agilent Technologies commercializes three types of bubble cell capillaries with different lengths: 25 μm I.D. × 125 μm of bubble (factor 5×, slit: 80 μm × 120 μm); 50 μm I.D. × 150 μm of bubble (factor 3×, slit: 145 μm × 145 μm); and 75 μm I.D. × 200 μm of bubble (factor 2.7×, slit: 150 μm × 200 μm) [85].

Alternatively, a number of researchers have described absorbance measurements along the capillary axis. In this case, if the sample bandwidth is significantly longer than the inner diameter of the capillary, improved absorbance sensitivity can be obtained. For example, a separation that generates one million theoretical plates in a capillary of 50 cm will give rise to a sample bandwidth that occupies approximately a 500-μm section of the capillary. As a result, absorbance measurements made along the band length (axial detection) will generate sensitivity 1 order of magnitude higher than that obtained from absorbance measurements made perpendicular by a capillary of 50 μm I.D. [36].

One approach in axial-direct absorbance measurements uses a Z-shaped cell [86,87]. In this cell, a section of the capillary is bending so that illumination occurs down the bore of a section of the capillary (see Fig. 5.8a), thereby offering greatly increased optical path length. However, band broadening becomes important and separation efficiency can be reduced along with the large gain in sensitivity. If a second band enters the Z-cell before the first band leaves, the peaks will overlap, impairing the separation efficiency. In addition, theoretically predicted increases in sensitivity have not been achieved owing to the increase in the noise levels, which partly offsets the signal gain from the path-length extension, because the Z-cell has a reduced optical throughput as compared with a straight capillary. In practice, a 3000-μm long Z-cell with 75 μm I.D. yields only a 12-fold increase in sensitivity over that of a conventional capillary rather than the theoretical value of 40 (3000/75) [88]. However, the loss in resolution caused by the extended path length was less pronounced than expected (only 14%). It was also shown by the same authors that the gain in a 50-μm capillary was only about four-fold rather than the theoretical value of 60 (3000/50). This

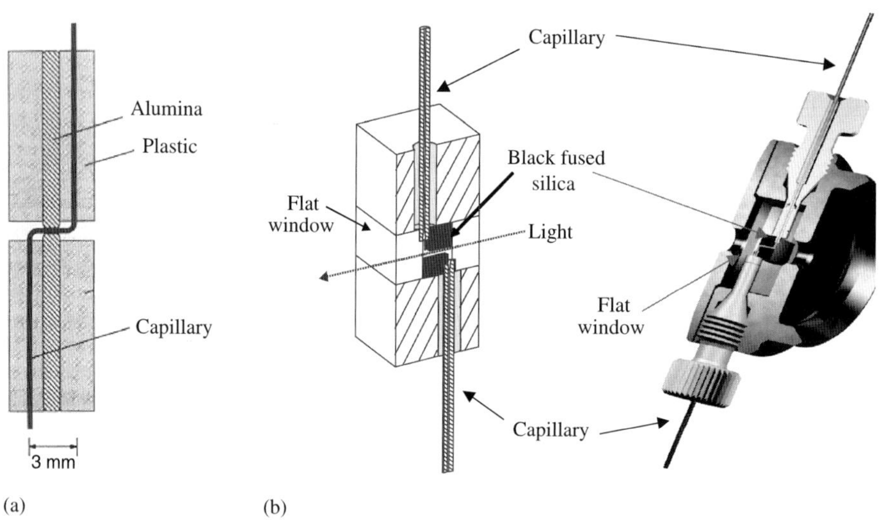

Fig. 5.8. Schematic diagrams of the Z-shaped cells. (a) Reprinted from Ref. [86], Copyright (1991), with permission from Elsevier. (b) Copyright (2004), Agilent Technologies, Inc. Reproduced with permission.

limits the use of the Z-cell in narrower bore capillaries. Nevertheless, in cases where resolution is not a problem and capillaries of 75 μm I.D. can be used, the 75 μm Z-cell can be a useful tool to improve the sensitivity.

Hewlett-Packard [89] introduced a Z-type flow cell that had a path length of 1200 μm, which can be used with 75 μm I.D. capillaries (see Fig. 5.8b). This cell comprises black fused silica and flat windows in order to minimize stray light, which often leads to high detection limits in absorbance measurements. With this Z-cell, Hewlett-Packard reported a ten-fold increase in sensitivity without sacrificing separation efficiency, rather than the theoretical value of 16 (1200/75).

Other approaches to achieve axial absorbance detection have been developed [90–92]. Grant and Steuer described an on-capillary UV detector that extended the effective path length from tens of microns into the millimeter range [90]. A schematic diagram of the axial illumination principle used is shown in Fig. 5.9. Because it would be difficult to directly measure the amount of light absorbed at point A, the addition of a fluorescent marker to the BGE provided a method for indirectly determining the absorbance with a given length of the capillary. For a path length of 2 mm in a capillary of 50 μm I.D., an improvement in sensitivity of 50 times could be obtained compared with the conventional transcolumn illumination method (perpendicular to the capillary), at the expense of 25–30% loss in efficiency [32]. An alternative design of the axial-beam absorption detector for CE has been described by Xi and Yeung [91] and Taylor and Yeung [92]. In this system, the light from a conventional UV source traveled in the capillary by total internal reflection [91]. Consequently, the whole capillary was employed as the detection cell, and the total absorbance represents the sum of all the absorbances of the individual compounds within the

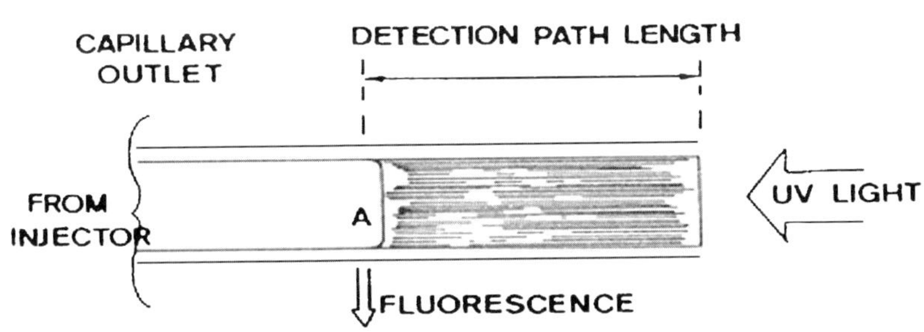

Fig. 5.9. Illustration of the axial illumination principle. The shaded area represents the effective cell volume. Reprinted from Ref. [90] Copyright (1990), with permission from John Wiley & Sons.

capillary [32]. In order to ensure total internal reflection of the light in the capillary, it is necessary to choose adequate BGE and material for the capillary, because the refractive index (RI) of the BGE must be equal to or higher than that of the material of the capillary (e.g., dimethylsulfoxide and fused silica capillary, or aqueous solutions and polytetrafluorethylene (PTFE) capillaries). A seven-fold improvement in sensitivity was obtained in comparison with that obtained by absorbance measurements made perpendicularly to the capillary. On the other hand, when a He–Ne laser was employed as the light source, up to a 15-fold improvement in LODs was reported [92].

Finally, the most promising attempt for improving detection limits in absorbance detection has been accomplished through the use of multireflections and laser-based excitation [93]. The multireflection nanoliter-scale cell is made directly onto the capillary. After removing the polyimide coating of the capillary, a short length of the capillary in the detection region is coated with a layer of metallic silver to form a mirror, as indicated in Fig. 5.10. Two windows located on opposite sides of the capillary, above and below the detection zone, serve as the entrance and exit ports for a He–Ne laser beam. The beam is reflected from the opposite wall and refocused by the curved surface into the center of the capillary. This reflection process is repeated many times before the laser beam exits from the capillary. By adjusting the angle of incidence and the spacing between the entrance and exit windows, it is

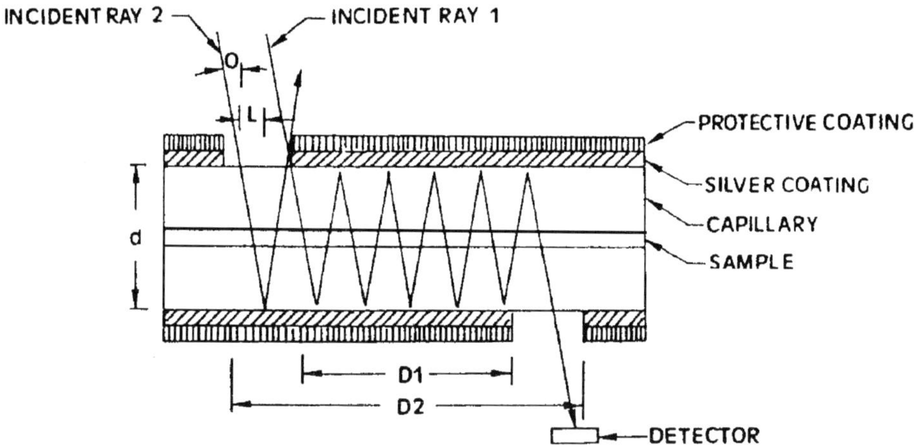

Fig. 5.10. Multireflection nanoliter-scale cell made directly onto the capillary. Reprinted with permission from Ref. [93], Copyright (1991), American Chemical Society.

possible to control the number of reflections produced, which generates a concomitant increase in effective optical path length and proportional increase in sensitivity [36]. For example, using a distance of 1.5 mm between windows, as many as 40 reflections were produced, which permitted a model analyte (a colorant like brilliant green) to become 44 times more sensitive when compared with standard on-capillary UV detection, and to acquire a detection limit of 6.5×10^{-8} M [93]. This detection cell geometry uses a few millimeters of capillary, producing a relatively large probe volume (6.6 nl), but resulting in a loss in separation efficiency. Therefore, the maximum useful spacing between the entrance and exit windows is limited by the bandwidth, as in axial detection. To minimize the loss in efficiency, this distance should be less than a few hundred micrometers. There is another limitation to the number of allowable reflections because attenuation in the beam intensity with each reflection steadily decreases the transmitted intensity of the laser beam, and if there is significant attenuation owing to the imperfect reflections, the loss in beam intensity will result in an increase in shot-noise and degradation in the LOD. Furthermore, owing to the low reflectivity of silver in the UV region of the spectrum, it is not clear whether a significant number of reflections could be used before the attenuation of the beam became significant [36].

Finally, in order for the multireflection absorption cell to be practical for CE, widely tunable laser sources or broadband light must be used for the detection of many analytes. However, lasers with these characteristics are still quite expensive and it does not appear likely that this multireflection cell would be useful with an incoherent light source. Incoherent sources that have broadband emission are characterized by large beam divergence. As a result, the beam tends to spread as it passes through the capillary, making it difficult to focus tightly. Because of optical constraints associated with wavelength dependency involved with refraction and reflection of light, it is hard to achieve total internal reflection within the sample cell [43]. Therefore, the limited wavelength tunability of the laser has resulted in few applications of multireflection detection in CE.

5.2.2.4 *Data acquisition*

The high efficiencies achieved in CE result in very narrow solute zones and sharp peaks, i.e., it is not uncommon to encounter peaks that have widths of only a few seconds or less. This means that data collection rates must be faster than in HPLC to adequately describe these peaks, and the response time of the detector should be low enough to respond to the slope of the peak, otherwise peak broadening can result. Response time describes how fast the detector follows an instantaneous change of the absorbance in the flow cell. Peak width is associated with the detector response time. Thus, the smaller the response time, the more accurately the shorter peaks can be measured. Finally, acquisition rate is associated with peak width and response time. To describe a peak profile adequately, it needs between 15 and 25 data points over the width of the entire peak. Therefore, to accurately measure a peak of 3.0 s, for instance, the associated acquisition rate must be at least 5 Hz [47].

Table 5.5 shows the correspondence between some values of the response time, peak width, and data acquisition rate. Therefore, as peak widths of 3 s or less are common in CE, response times between 0.2 and 1.0 s are usually sufficient. Larger response times reduce noise, because a larger number of the data points are averaged (noise is inversely proportional to the square root of the number of the data points); but since the peak widths increase, the data acquisition rates should be adjusted to the peak widths of the specific analysis.

On the other hand, it is possible to improve absorbance sensitivity by reducing the level and increasing the stability of the background signal from the detectors. This signal results from different sources, such as impurities in the BGE and electrical pick-up in the detector wiring circuitry. A way to reduce the background signal and to increase its stability is to digitize the data and to use mathematical algorithms to smooth out the background noise and amplify the signal. An example of this approach has been described by Saeed et al. [94], who analyzed real samples by CE. They employed a signal enhancement processor (the system "ID/10" developed by Thomas Swan [95]) to significantly

TABLE 5.5

Values of data acquisition rates associated with response times and peak widths

Response time (s)	Peak width	Data acquisition rate (Hz)
0.2	0.6 s = 0.01 min	20
0.5	1.2 s = 0.02 min	10
1.0	3.0 s = 0.05 min	5
2.0	6.0 s = 0.10 min	2.5
4.0	12.0 s = 0.20 min	1.25

improve the signal-to-noise ratio. In addition, the reduction in the baseline noise also extended the linear dynamic range of the measurements.

5.2.2.5 Diode-array detector

The above-mentioned selection of a particular wavelength of light to be passed through the capillary is called "forward optics" design (see Fig. 5.11a). As already stated, polychromatic light from the source is focused on the entrance slit of a monochromator, which selectively transmits a narrow band of light. This light passes through the capillary to the detector (one diode). This is the most frequently used design in variable-wavelength detectors. However, a multiwavelength detector, more well-known as diode-array detector (DAD), employs the "reverse optics" design (see Fig. 5.11b). In this design, the polychromatic light from the source first passes through the capillary, and is then focused on the entrance slit of the monochromator. The diffraction grating of the monochromator disperses the light onto a diode-array (consisting of numerous diodes), after which each diode measures a narrow-band of the spectrum. The bandwidth of light detected by a diode is related to the size of the monochromator entrance slit and to the size of the diode. Each diode performs the same function as the exit

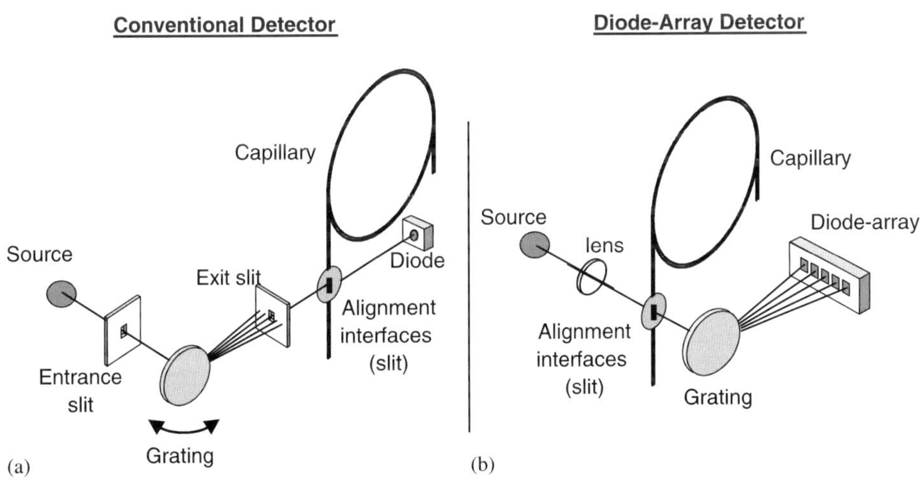

Fig. 5.11. Schematic of (a) a "forward optics" design used in a variable-wavelength detector versus (b) the "reverse optics" design used in a DAD. Copyright (2004), Agilent Technologies, Inc. Adapted and reproduced with permission.

slit of a monochromator, and the number of individual diodes in the array often determines the wavelength resolution of the detector [47].

The used of DADs can greatly simplify the development of a new analysis method by CE. In this case, there is usually little or no information regarding the most adequate detection conditions, and in particular, optimum wavelength to detect analytes is not known. Using a variable-wavelength detector, the sample must be injected repeatedly, and the detector wavelength must be changed each time to make sure that all of the solutes are detected. However, with a DAD, a whole wavelength range can be selected, and, in a single analysis, all solutes absorbing within this range will be detected. Once all peaks have been detected, the DAD can be used to determine the wavelength at which maximum absorbance for the analytes is obtained. Appropriate software can calculate the absorbance maximum automatically. Alternatively, the data can be represented in three-dimensional form, either as absorbance measured as a function of analysis time and wavelength (see Fig. 5.12a) or as an isoabsorbance plot that displays wavelength, analysis time, and absorption intensity (see Fig. 5.12b). The isoabsorbance plot is especially useful to determine the most adequate wavelength to analyze complex mixtures [47].

The first description of a DAD for on-capillary UV detection in CE dates from 1989 [96]. The main advantage of employing DADs is that multiwavelength spectral information can be obtained. It is important to note that DADs can yield detection limits, sensitivity, and linear dynamic ranges that equal or exceed those of variable wavelength detectors. The quantitative and qualitative possibilities of the DAD will be briefly described as follows. For more information about the particular features of DAD in CE, readers are recommended to consult a good monograph, such as ref. [84].

Optimization of sensitivity

A recurring problem in UV–Vis absorbance detection is low sensitivity, which may be due to low concentration of the analyte in the sample or to very weak absorptivity of the analyte. At low absorbances, reducing the noise level directly improves the sensitivity. Two measurement parameters can be employed in a DAD to reduce noise level: wavelength averaging and internal referencing [47].

With respect to wavelength averaging, conventional quantitative measurements would be performed at the absorption maximum, but if data at all wavelengths are available, additional values on either side of

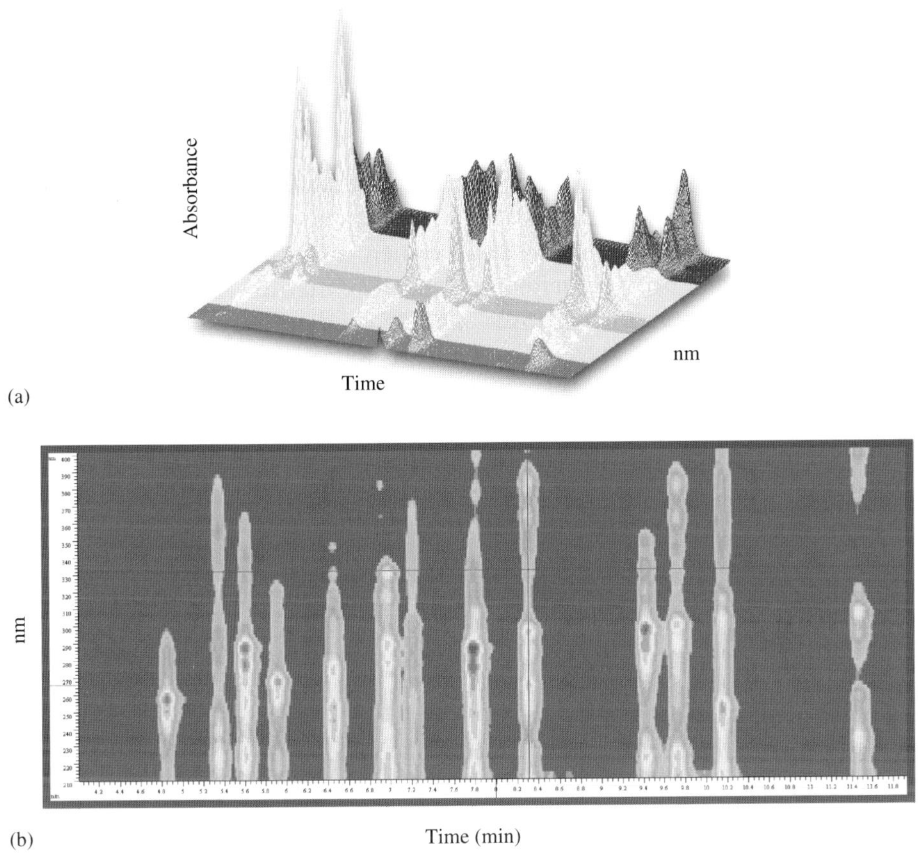

Fig. 5.12. Three-dimensional representations obtained with a DAD. Copyright (2004), Agilent Technologies, Inc. Reproduced with permission.

the maximum can be averaged. As a consequence, the noise is reduced the reduction in the noise being equivalent to the square root of the number of data points. However, as more wavelengths are averaged, the average absorbance is reduced, which has a negative effect on the signal. In conclusion, for a given spectrum, a certain bandwidth will yield an optimum signal-to-noise ratio. Generally, the optimum bandwidth is equal to the natural bandwidth[3] of the absorbing analyte, and typical values range from 20 to 40 nm [47].

[3]The natural bandwidth is the width of the sample absorption band at half the absorption maximum.

Internal referencing can be also used to improve sensitivity at low absorbance levels. This method uses a single reference wavelength to correct baseline drifts induced by lamp drift and fluctuations, and temperature fluctuations. In general, the best choice is a reference wavelength as close as possible to the analytical wavelength, but with no significant absorbance from the analyte inside a broad bandwidth (in the range 50 to 100 nm). The absorbance at the reference wavelength is subtracted from the absorbance measured at the analytical wavelength. Thus, any interference that is constant at all wavelengths is corrected, such as, for example, spikes produced by particles in suspension [47].

Qualitative information

UV–Vis absorbance detection seems to give limited qualitative information on analytes because UV–Vis absorption spectra are characterized by broad bands and can be very similar for structurally similar compounds. However, if DADs are used in conjunction with sophisticated mathematical algorithms, these detectors provide some qualitative information allowing:

- To check the purity of the peak in order to confirm if there is overlapping of peaks.
- To confirm peak assignment during method development.
- To aid the identification of unknown compounds.

One of the most difficult tasks in a separation is determining whether each individual peak corresponds to a single component or if it consists of more than one solute. In single-wavelength detection, this is traditionally done by careful inspection of peak shape. However, skewed peaks are not necessarily impure; in CE, non-Gaussian peaks result from mass overloading or electrodispersion. Multiwavelength absorbance detection can aid in peak purity assessments [34]. The most common method involves acquiring spectra obtained at the upslope, apex, and downslope of a single peak. By normalizing and overlaying the spectra, the plots can be compared, and when the spectra match over all wavelengths, the peak can be considered pure [47]. Another method involves acquiring several spectra during the peak's migration. After normalizing the spectra, a plot of purity factor versus migration time is calculated in conjunction with sophisticated mathematical algorithms. If any purity values exceed the calculated threshold limit, heterogeneity is indicated graphically (see Fig. 5.13).

It is important to highlight that spectral analysis by itself cannot establish peak purity; it can only increase confidence that a peak is pure. This is because the UV–Vis spectra of two comigrating species can be too similar to be differentiated, or an impurity may not be present in a high enough concentration to be detected [34].

During method development, different analytical conditions are applied to a sample to determine the best set of separation parameters; therefore, peaks can change both migration time and order. In single-wavelength detection, tracking the peak under the changing conditions requires sample spiking or rerunning of standards. With multiwavelength detection, the spectral features of the peaks allow compounds to be tracked as the peaks change position. One of the electropherograms obtained during a method development run can be saved and stored in the computer as a reference spectrum. A second method development run can then be made, and the spectra of a particular peak from the second run can be compared to each peak in the reference spectra until a match is found. Thus, peak tracking between different runs is accomplished by spectral matching, which is a numerical interpretation of the similarity of spectra. Peak tracking in this way does not require

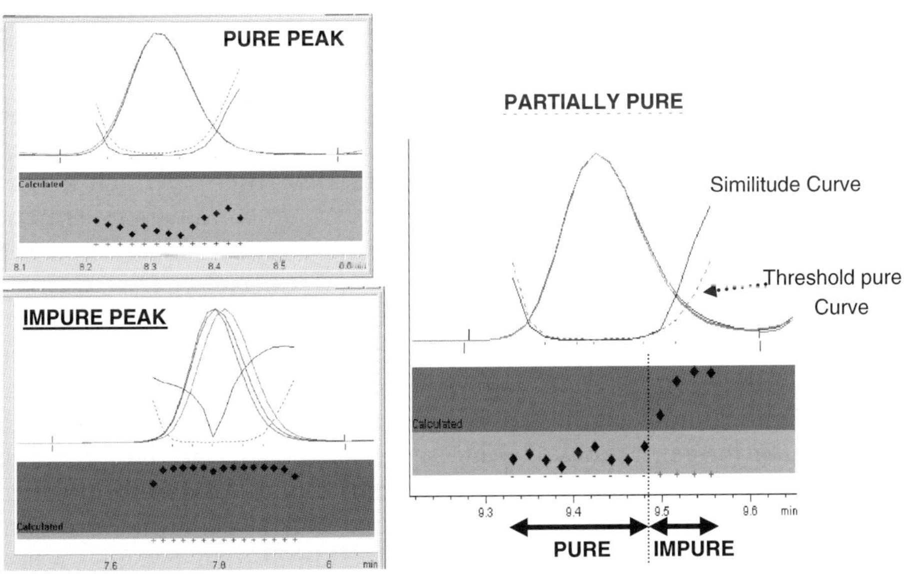

Fig. 5.13. Schematic of peak purity method according to purity factor. Copyright (2004), Agilent Technologies, Inc. Adapted and reproduced with permission.

any prior knowledge about the analyte or the use of pure standards. A complication is that changing the separation media can have a profound effect on spectral band shape, and thus, on the ability to track a peak.

Finally, with respect to the possibility of identifying analytes using single-wavelength detection, separated components can only be identified on the basis of their migration times, which can be problematic in complex sample matrices. The presence of interferents in the matrix can cause migration time shifts, loss of resolution, or comigration of peaks relative to runs made with pure standards. Although multiwavelength absorbance spectra do not normally contain sufficient information to enable a priori identification of compounds, spectra of unknown peaks can be matched to those of reference compounds under the same experimental conditions, or to those found in a standard library. However, peak identification is usually performed on peaks that appear to be homogeneous after peak purity analysis [34]. As an example, the simultaneous monitoring of products and intermediates of L-Dopa oxidation was performed using CE. In this work, DAD permitted peak assignment confirmation by assessing their individual absorbance spectra [97].

5.2.2.6 Absorption imaging detection

An ideal approach for detection in CIEF (see chapter 2) is monitoring the solute zones in real-time using whole-capillary-imaging detection (WCID). CIEF is performed in a two-step operation in which a conventional CE instrument, equipped with an on-column single-point detector located at one end of the capillary, is used. In the first step, a solution of sample (proteins or peptides) and ampholytes are focused (separated and concentrated) to stationary sharp zones according to their respective isoelectric point (pI) values in an internally coated capillary (EOF is eliminated by the coating). Then, the on-column single-point detector requires a second step in which the focused proteins or peptides are moved by electrophoretic or hydrodynamic mobilization so that they can pass the detector point to be measured and recorded. WCID eliminates the mobilization step required for single-point detection after the focusing process. Therefore, it provides a fast analysis speed (3–5 min for each sample), and avoids the disadvantages associated with the mobilization process, such as distortion of pH gradient, loss in resolution, and poor reproducibility [98].

Three types of imaging detection systems have been developed by Pawliszyn's research group: RI gradient [99,100], fluorescence [101], and UV–Vis absorption [102]. Among these, absorption imaging detection is the most practical at the present time owing to its applicability for quantitative analysis and its universal characteristics, and was commercialized in 1998 (Convergent Bioscience Ltd., Toronto, Ontario, Canada). Figure 5.14 illustrates the basic concept of WCID, in which the capillary is homogeneously illuminated along its axis. If samples absorb UV–Vis radiation, the focused sample zones can be imaged simply by measuring the optical absorption, i.e., a decrease in the intensity of the transmitted light is measured. Finally, a suitable optical arrangement, including a set of lenses and filters, between the capillary and the charge-coupled device (CCD) camera, should assemble to achieve the maximum detection sensitivity [103]. All of these first advances in this imaged CIEF technique as well as its applications have been reviewed [98,104].

Other recent progresses in instrumental development as well as in methodology of imaged CIEF have been described. Thus, a whole-capillary-imaging CIEF instrument was simplified and miniaturized by the use of an LED as the light source [105]. The focusing of two pI markers and myoglobin, selected as model samples, was obtained in a

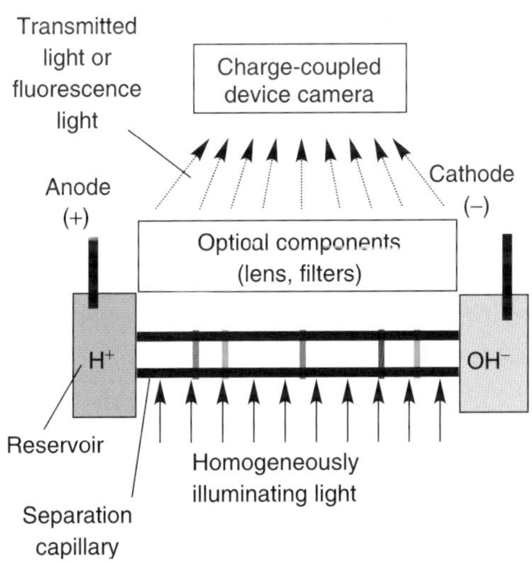

Fig. 5.14. Illustration of the basic concept of the whole-capillary-imaging detection. Reprinted with permission from Ref. [103], Copyright (2001), Advanstar Communications.

capillary of only 1.2 cm. On the other hand, a capillary isoelectric focusing whole- capillary-imaging detection (CIEF-WCID) method has recently been used to determine the pI of norovirus virus-like particles (VLPs). The VLPs were produced from noroviruses that represented the two genogroups. Using the imaged CIEF-WCID technique, separation was accomplished using a short (4–5 cm), internally coated capillary (100 μm I.D.) and a whole-capillary absorption-imaging detector operating at 280 nm. The VLP peaks were shown to be reproducibly resolved. CIEF-WCID shows great promise for norovirus detection in public health, clinical, and food safety applications, as CIEF-WCID overcomes several limitations of the currently used genetic and immunological methods [106].

However, although the WCID system was developed for CIEF, this detection system has also found applications in other CE modes or studies [103]. Researchers have also reported part-capillary absorption-imaging detection with either a diode [107,108] or CCD [109] array detector for CE in order to decrease the absorbance detection limits.

In CE, as noted before, analytes are normally detected at a single point near the output end of the capillary. However, Culbertson and Jorgenson [107,108] have shown the benefits of using a DAD to image a section of a capillary where each of the diodes in the array is treated as a separate detector. The light source in these works was a single-wavelength 254 nm mercury pen-ray lamp. In a first attempt [107], when the average of all electropherograms generated from each of the diodes in a diode array was carried out, an increase of 85 times was obtained in the signal. However, when the LODs were compared to those obtained from a commercial single-point detector, only an improvement of about four times was obtained. This was due to a source flicker noise, which interfered with the signal-averaging process at low analyte concentrations [108]. This noise, moreover, was due to inadequate compensation of the spatial and temporal lamp fluctuations that occurred over the course of a run in this single-beam instrument. To better compensate the source lamp fluctuations and drift, a double-beam detector was constructed [108]. The addition of the reference channel significantly reduced the source flicker noise. Thus, the LODs of four nucleic acids for the double-beam instrument were approximately four times better than those obtained using a single-beam detector, but only about six times better than those obtained using a commercial single-point detector.

Finally, a CCD camera can be used to image a section of a separation channel where each camera pixel is considered as a single-point detector. A CCD is essentially a silicon chip that is very sensitive to light. Thus, the use of a CCD camera together with a conventional deuterium light source was studied to establish the performance characteristics of an array detector for both single-wavelength and multiwavelength UV absorbance (see Fig. 5.15) [109]. When this array detector works as a single-wavelength detector, electropherograms corresponding to individual pixels can be averaged for improved signal-to-noise ratios, i.e., an improvement of the sensitivity is obtained by averaging all of the individual signals. On the other hand, for a multiwavelength detector, the CCD camera images analyte peaks in the capillary dimension, together with wavelength-resolved absorbance in the dimension perpendicular to the capillary. This requirement limits the choice of detector to CCDs. In this work, noise contributions from source intensity fluctuations are reduced by using a portion of the CCD image to provide a baseline reference signal, instead of using a double beam, as in the previous work.

5.2.3 Derivatization and complexation procedures

There are some situations in which the separation and quantitation of species that do not have a prominent chromophore are desired. These

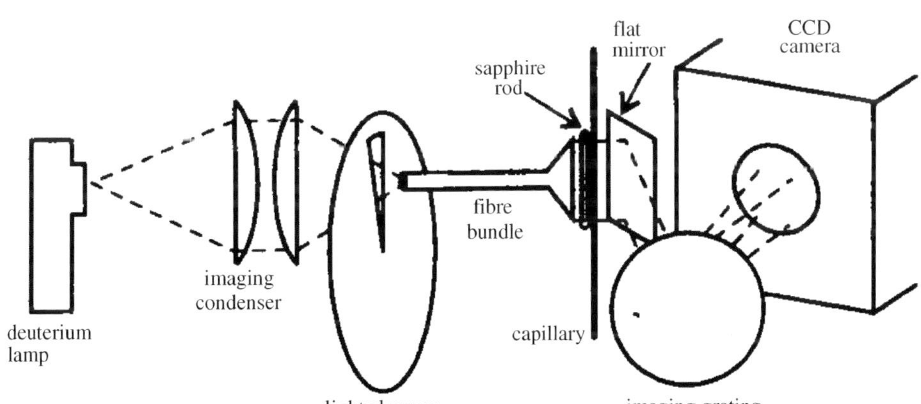

Fig. 5.15. Diagram showing the layout for multiwavelength imaging detection. For single-wavelength detection, the light chopper is effectively replaced by a 10 nm band-pass interference filter. Reprinted with permission from Ref. [109], Copyright (1999), American Chemical Society.

non-absorbing solutes are generally detected in CE using indirect UV detection (see section 5.2.4). However, an alternative to indirect UV detection is the introduction of a chromophore in the molecule of the solute via derivatization or selective complexation in order to make direct UV detection possible. In addition, these strategies (derivatization or complexation) can also be employed to enhance the absorbance detection sensitivity through the conversion of the analytes in products with more favorable detection characteristics, i.e., with increased molar absorptivity.

Chemical derivatization, a common approach in fluorescence detection, can also be employed to achieve UV–Vis absorbance detection. In fact, a recent overview on the use of derivatization protocols in CE analysis has shown that although most derivatization schemes use fluorescence labeling, other possibilities, including UV absorbance, are possible [110]. Various labeling reagents can be used for pre- or in-capillary derivatization in CE.

In a pre-capillary procedure, a derivatization step is employed prior to the analysis to convert the non-absorbing analytes into products that are UV active. Thus, e.g., *benzoyl hydrazine* and *N-(4-aminobenzoyl)-L-glutamic acid* were used as UV-absorbing tags for mono- [111] and oligosaccharides [112], respectively. In both cases, sub-picomole mass detection limits were obtained. Similarly, through chemical derivatization with *dansylhydrazine*, carbonyl compounds in rain water [113] were converted to compounds with higher absorptivity, and LODs in the range of 170 to 300 nM were reached using a UV absorbance Z-cell to improve detection limits. Recently, the analysis of aldehydes (formaldehyde, acetaldehyde, propionaldehyde, and acrolein) in air samples by CE with direct absorbance detection after derivatization with 4-*hydrazinobenzoic acid* (HBA) has been reported [114]. Samples were collected in HBA-impregnated octadecylsilica cartridges. Figure 5.16 shows the separation of formaldehyde and acetaldehyde derivatives in less than 6 min with adequate sensitivity for environmental applications (LODs in the range of 2.7 to 8.8 ng/l).

Derivatization reactions can also be carried out inside the capillary (in-capillary derivatization), which results in a straightforward and automated way of achieving a derivatization. Oguri et al. [115] evaluated two possibilities for sample introduction: on-line and sandwich mode in-capillary derivatization using *1-methoxy-carbonylindolizine-3,5-dicarbaldehyde* (IDA) as labeling reagent of primary amines. The on-line mode consisted of injecting the underivatized

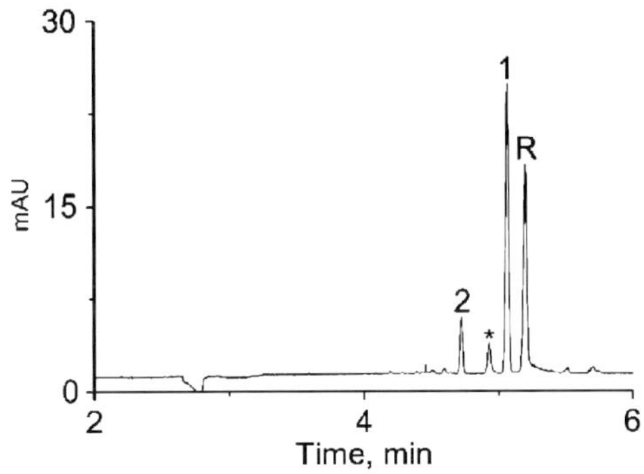

Fig. 5.16. Electropherogram of aldehydes in air samples by CE with direct absorbance detection after pre-capillary derivatization with 4-hydrazinobenzoic acid. Electrolyte: 0.040 M tetraborate buffer (pH 9.3). Voltage, 25 kV. Injection, 20 s at 25 mbar. Detection at 290 nm. (R) excess reagent; (1) formaldehyde; (2) acetaldehyde; *unidentified impurity. Reprinted from Ref. [114], Copyright (2004), with permission from Wiley-VCH Verlag GmbH.

sample by pressure into the capillary filled with run buffer containing IDA, and the sandwich mode consisted of injecting the underivatized sample as a plug between two labeling reagent solution segments. Both modes permitted to obtain LOD of 10^{-6} M for alanine.

In-capillary derivatization procedures may become much more attractive than conventional pre-capillary derivatization in terms of sensitivity and reproducibility. An example of the potential of in-capillary derivatization to enhance the sensitivity of absorbance detection of amino acids in CE has been shown [116]. The method was applied to the determination of amino acids in feed samples, and the results showed that the optimum operation mode consisted of a sandwich mode. Another example was the CE separation of iodine and iodide species followed by selective in-capillary derivatization of iodine to iodide with sulfite, using direct UV detection at 214 nm. The proposed method was applied to the rapid speciation of iodide and iodine in commercially available antiseptics, with LODs of 2×10^{-6} M for I^- and 8×10^{-6} M for I_2 [117].

On the other hand, direct UV detection is often employed in the analysis of metal ions after their complexation with appropriate ligands. The developments in the analysis of metal ions and their complexes by CE, with emphasis on the separation strategies employed and the choice of complexing reagents, were reviewed in 1999 [118]. The advantages and drawbacks of CE compared with HPLC in the analysis of metal ion complexes have also been briefly considered [119]. The speciation of metal ions by CE in different matrices has been reviewed, and various aspects such as sample pre-treatment, metal ion complexation, LODs, and choice of electrolytes have been described as well [120,121].

When using complexation strategies, metal ions are generally chelated pre- or in-capillary with very different complexing ligands to form anionic complexes that are separated by CZE with direct UV detection. Table 5.6 groups the main complexing ligands employed as well as some applications, which have recently been reported for pre- and in-capillary complexation of metal ions. Polyaminecarboxylic acids, 1,10-phenanthroline, thiocyanate, dithiocarbamate, PAR, dithizone sulfonate, or hydroxyisobutyric acid have been employed. As an example, the separation of 5 common metal cations (Fe^{2+}, Co^{2+}, Ni^{2+}, Cu^{2+}, and Zn^{2+}) was achieved in less than 4 min using in-capillary complexation with 1,10-phenanthroline, and the method developed was applied to the determination of Fe(II) and Zn(II) in snow samples (see Fig. 5.17) [133]. Table 5.6 also shows that the most used complexing ligands are the polyaminecarboxylic acids, which are generally employed with a pre-capillary strategy, and the LODs are in the range of μg/l or μM. As an example, Fig. 5.18 shows an excellent separation of 23 metal ions within 10 min using cyclohexane-1,2-diaminetetraacetic acid (CDTA) as complexing reagent and pre-capillary complexation [134]. In this study, LODs ranging from 10^{-7} M for Fe(III) to 4×10^{-6} M for Ca(II) and Hg(II) were obtained.

Complexation strategies in CE are useful tools not only for metal ion analysis, but for other procedures as well. Indeed, enhanced LODs and selectivity for amines, amino acids, and peptides have been achieved by complexing these analytes with Cu(II) ions. Thus, CE with direct UV detection was employed to optimize copper complexation conditions (*biuret* complex formation) for the analysis of neuro-peptides. Pre- and in-capillary strategies were studied, but in-capillary complexation provided significant advantages: an increase in UV absorbance along with peak stacking resulted in a significantly greater response [136]. Finally, another example on the use of complexation strategies in CE is the analysis of non-UV-absorbing polyhydroxylated alkaloids (calystegines) [137]. Figure 5.19 shows the analysis by CE using in-capillary complexation with borate ions of calystegines in plant extracts. In this work, a bubble cell and a low wavelength (191 nm) were employed to further improve method sensitivity. A 35-fold improvement in

TABLE 5.6
Analysis of metal ions by CE using pre- or in-capillary complexation and direct UV-absorbance detection

Analytes	Complexing ligands	Strategy	LODs	Reference
Inorganic and organic species of Hg	Dithizone sulfonate	Pre-capillary	μg/l	[122]
V(IV) and V(V)	EDTA[a]	Pre-capillary	0.1–0.4 mg/l	[123]
	Mo(VI)–P(V) reagent	In-capillary	0.2–0.5 μM	[124]
	EDTA[a]	In-capillary	1–3 μM	[125]
Fe(II) and Fe(III)	1,10-Phenanthroline and CDTA[b]	Pre-capillary	0.06–0.1 mg/l	[126]
Inorganic and organic species of Pb, Se, and Hg	TTHA[c]	Pre-capillary	sub-μg/l	[127]
	NTA[d]			[128]
	EDTA[a]			[129]
Actinides and lanthanides	Hydroxyisobutyric acid	Pre-/in-capillary	μg/l	[130]
Pd(II), Pt(II), and Pt(IV)	Thiocyanate	Pre-capillary	μg/l	[131]
Cu(I), Pb(II), Cd(II), Ni(II), and Hg(II)	Bis(carboxymethyl)dithiocarbamate	Pre-capillary	μg/l	[26]
Cr(III) and Cr(VI)	PDCA[e]	Pre-capillary	2–3 μM	[132]

continued

TABLE 5.6 (*continued*)

Analytes	Complexing ligands	Strategy	LODs	Reference
Fe(II), Co(II), Ni(II), Cu(II), and Zn(II)	1,10-Phenanthroline	In-capillary	1–6 μM	[133]
23 Metal ions	CDTA[b]	Pre-capillary	0.1–4 μM	[134]
Fe(II), Ni(II), and Zn(II)	PAR[f]	Pre-capillary	μg/l	[135]

[a]EDTA, ethylenediaminetetraacetic acid;
[b]CDTA, cyclohexane-1,2-diaminetetraacetic acid;
[c]TTHA, triethylenetetraminehexaacetic acid;
[d]NTA, nitrilotriacetic acid;
[e]PDCA, 2,6-pyridinedicarboxylic acid;
[f]PAR, 4-(2-pyridylazo) resorcinol.

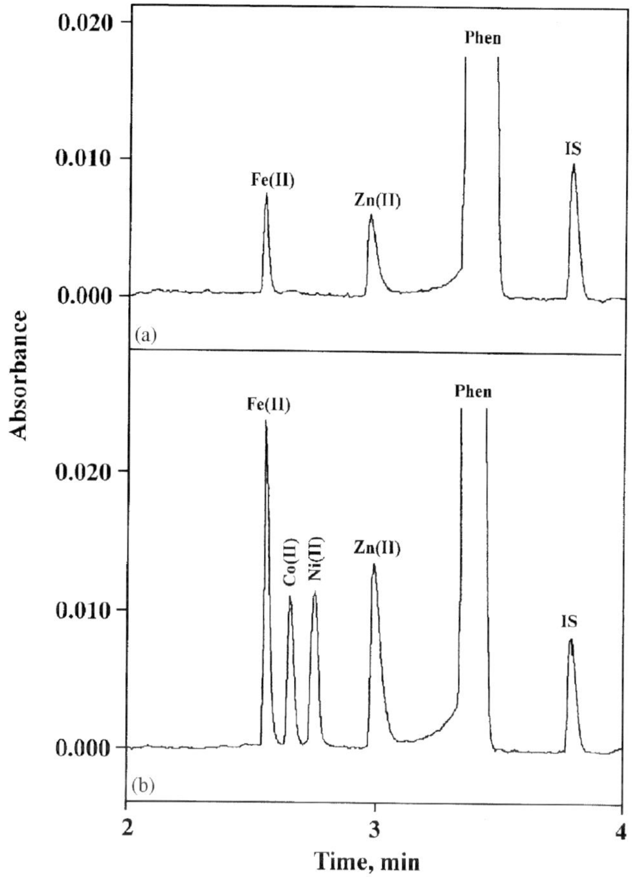

Fig. 5.17. Electropherograms of (a) 1:1 diluted snow sample and (b) the same sample spiked with 1×10^{-6} mol/l of each cation using in-capillary complexation with 1,10-phenanthroline. Injection: (i) hydrodynamic, 2 s, 5×10^{-3} mol/l 1,10-phenanthroline, 70 s electrolyte, and (ii) electromigration, 10 kV for 5 s sample (I.S., benzidine). Separation was carried out in 50 mmol/l glycolic acid electrolyte (pH 6.0 with imidazole) using direct UV detection at 254 nm. Reprinted from Ref. [133], Copyright (2002), with permission from Elsevier.

sensitivity was achieved using the sweeping pre-concentration technique involving borate ions to analyze calystegines. This method was applied to determine miglitol in commercially available pharmaceutical tablets [138].

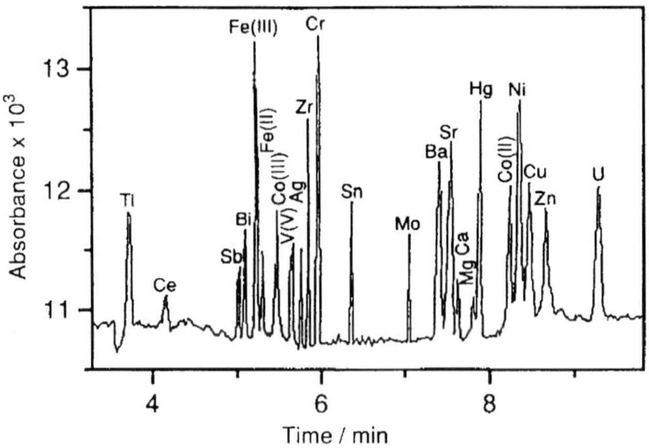

Fig. 5.18. Separation of 23 metal ions (10^{-4}M) using cyclohexane-1,2-diaminetetraacetic acid (CDTA) as complexing reagent and pre-capillary complexation. BGE: 20 mM sodium borate, 1 mM CDTA, and 5% ethylene glycol. Voltage: 12.5 kV. Detection: 214 nm. Reprinted from Ref. [134], Copyright (1996), with permission from Elsevier.

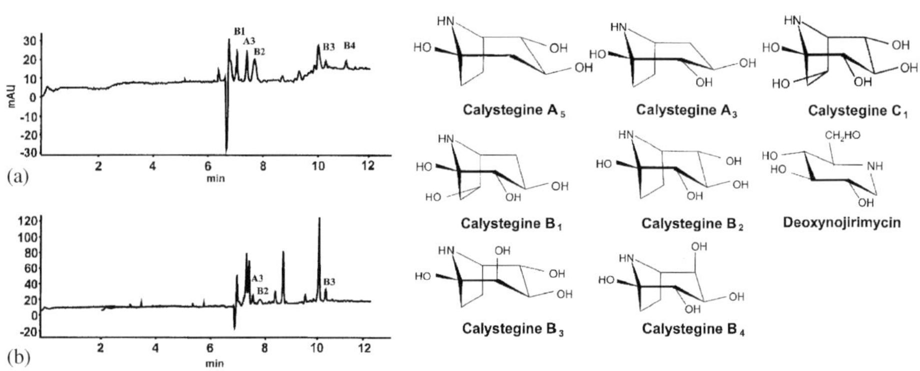

Fig. 5.19. Electropherogram of calystegines (closely related polyhydroxyalkaloids) in plant extracts (a: *Calystegia sepium*, and b: *Atropa belladonna*) using in-capillary complexation with borate ions as complexing reagent. Experimental conditions: 80 mM sodium tetraborate (pH 9.2) at 50°C. Applied voltage 25 kV. The extended path-length capillary was 80.5 cm (72 cm to the detector) ×50 μm I.D. (bubble factor 3). Detection at 191 nm. Sample injection was 5 s at 50 mbar (8 nl injection volume). Reprinted from Ref. [137], Copyright (2000), with permission from Elsevier.

5.2.4 Indirect absorbance detection

This detection mode is accomplished by incorporating an UV-absorbing species (called "probe") into the BGE producing a relatively large background signal that reduces the amount of light reaching the detector. The probe must have the same charge as the analyte. When a non-absorbing analyte migrating through the capillary passes through the detection zone, it will displace the probe and the light that reaches the photodetector increases, resulting in a decrease in the UV signal at the detector. This detection signal depends on the difference between a property of the analyte ion and a property of the probe ion. This displacement process is characterized by the transfer ratio (TR), which represents the number of moles of probe ions displaced by one mole of analyte ions. TR depends on the charge of the analyte ions (z_A) and probe ions (z_P), and also on the electrophoretic mobilities of the analyte ion (μ_A), the probe ion (μ_P), and the counter-ion (μ_X) in the BGE [139]:

$$\mathrm{TR} = \frac{z_A}{z_P} \cdot \frac{\mu_P}{\mu_A} \cdot \left(\frac{\mu_A + \mu_X}{\mu_P + \mu_X} \right) \tag{5.2}$$

In UV–Vis absorbance detection for CE, the change of the absorbance (A) during the elution of an analyte from the capillary is related to the molar absorptivities (ε) of the analyte and the UV-absorbing ion in the BGE, and the term describing TR, according to Beer's law [42]:

$$\Delta A = (\varepsilon_A - \mathrm{TR} \cdot \varepsilon_P) b \cdot C_A \tag{5.3}$$

where b is the optical path length and C_A the concentration of the analyte. For direct UV–Vis detection, ε_P should be zero (UV–Vis transparent carrier electrolytes include phosphate or borate buffers), and the above equation is simplified (see Eq. (5.1) in section 5.2.1). However, for indirect UV–Vis detection, a wavelength should be chosen where ε_A is zero and ε_P is high, in order to observe a decrease in absorbance (negative peak), which is proportional to the amount of analyte present.

In indirect UV detection, the LOD for a non-absorbing analyte is given by the following equation [140,141]:

$$C_{\mathrm{LOD}} = \frac{C_P}{\mathrm{TR} \cdot D_r} = \frac{N_{\mathrm{BL}}}{\mathrm{TR} \cdot \varepsilon \cdot b} \tag{5.4}$$

where C_{LOD} is the concentration limit of detection of the analyte, C_P is the concentration of the probe, D_r is the dynamic reverse (i.e., the ratio of the background absorbance to the noise), N_{BL} is the baseline noise, ε is the molar absorptivity of the probe, and b is the optical path length. According to Eq. (5.4) C_{LOD} can be optimized by reducing C_P or by increasing D_r. However, for indirect absorbance detection, D_r is also related to C_P, so decreasing the probe concentration will not necessarily improve the LOD, because D_r is simultaneously reduced [142]. Consequently, minimizing C_{LOD} often involves maximizing D_r by either reducing the noise or increasing the BGE absorbance. Increasing the molar absorptivity of the probe increases the D_r, and simultaneously reduces the necessary C_P, resulting in a lower LOD. Noise is usually associated with instrumental limitations, so the best way to increase sensitivity is to use probes with high molar absorptivity [143].

In addition to the molar absorptivity of the probe, its mobility is another crucial aspect for the success of the analysis of any species. Thus, the mobility of the probe must be as close as possible to the mobility of the analytes, otherwise electromigration dispersion will occur and unacceptable peak shapes will result. Analytes that migrate faster than the probe have a fronted peak profile, while analytes that migrate more slowly than the probe give tailed peaks. The best peak shapes are obtained when mobilities of the probe and the analyte are exactly matched [143]. Most BGEs designed for indirect absorbance detection usually contain only a single probe because under these conditions the displacement process is relatively simple and well understood; however, when the sample contains analytes having a wide range of mobilities, two or more probes may be used to increase the useful mobility range of the BGE. In this case, fast analytes will displace the fast probe ion and slow analytes the slow probe ion, thereby maintaining good peak shapes throughout the whole electropherogram. Nevertheless, care should be taken when using multiple probe electrolytes, since BGEs containing n probes will lead to $n-1$ system peaks [144]. Analytes that migrate near or on system peaks can be distorted, disappear, or reverse their direction, thereby losing all quantitative properties. Thus, the above approach is only successful if all analytes of interest do not migrate near or on a system peak. A typical example is given in Fig. 5.20, which shows the presence of two system peaks due to the three probes used in the separation of a mixture of anions.

Finally, two other aspects can also be important for the success of the separation: the use of an EOF modifier, if required, and the type of buffer. Thus, when counter-electroosmotic separations are not possible (that is, ions in the sample have higher mobilities than that of the

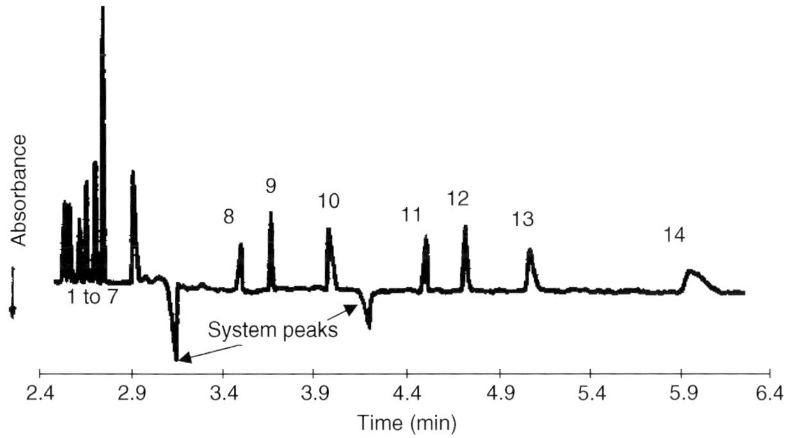

Fig. 5.20. Electropherogram obtained with three probes (5 mM chromic acid, 5 mM phthalic acid, 10 mM benzoic acid) in 60 mM diethanolamine with 0.5 mM tetradecyltrimethylammonium hydroxide (pH 9.2). Indirect detection at 254 nm. Peaks: (1) bromide, (2) chloride, (3) iodide, (4) nitrite, (5) nitrate, (6) sulfate, (7) chlorate, (8) phosphate, (9) carbonate, (10) ethanesulfonate, (11) butanesulfonate, (12) pentanesulfonate, (13) hexanesulfonate, (14) heptanesulfonate. Reprinted with permission from Ref. [144], Copyright (1999), American Chemical Society.

EOF), surfactants are frequently added to the BGE in order to reverse the direction of the EOF and thereby enable the analysis to be conducted in a co-electroosmotic mode with the analytes and the EOF moving toward the anode [143,145]. On the other hand, buffering of BGEs is essential for reproducibility and rugged separations because subjecting unbuffered electrolytes to pH changes occurring in the electrolyte reservoirs leads to variable EOF values and changes in migration times of analytes [143].

Indirect absorbance detection was first used in CE by Hjerten et al. in 1987 [146] to achieve the analysis of anions. Based on this early work, the development of indirect absorbance detection as a universal detection method has led to the establishment of CE as a viable alternative to ion chromatography for the simultaneous determination of inorganic ions [147,148]. However, in spite of this success, when quantitative analysis is required, CE has failed to replace ion chromatography as the method of choice for inorganic ions. The reasons for this apparent failing have been reviewed [66]. Nevertheless, as shown by the substantial number of published articles, many scientists are still contributing to this field [147]. Thus, the popularity of this technique in the analysis of small organic and inorganic anions is evident. These solutes generally have no chromophores and are detected in CE using indirect UV detection. This subject is covered until 1999 by two extensive reviews [139,149], and more recently by another three [66–68]. Several of the published works focus on the evaluation of the appropriate BGE for indirect UV detection. Small inorganic anions and simple organic acids have high negative mobilities, which oppose the typical EOF direction of an uncoated capillary. This counter-migration could result in long migration times and highly diffused peaks. Therefore, capillaries are commonly coated with cationic surfactants to reverse the EOF direction; the application of a negative voltage results in both the reversal of the direction of the EOF and solute migration toward the detector, producing fast and efficient separations. Typically a cationic surfactant such as tetradecyl-trimethylammonium bromide (TTAB) or cetyltrimethylammonium bromide (CTAB) is added to the electrolyte [145]. EOF reversal for the determination of inorganic anions with hydro-organic media has been reported recently [150].

With respect to the probes employed in CE for indirect detection of anions, it should be noted that the most common absorbing probe anion used has been chromate. This can be ascribed, at least in part, to the favorable light absorption properties of chromate in a wide range of light spectra. In addition, the effective mobility of chromate offers a good match with the effective mobilities of many inorganic anions so that their electromigration dispersion can be tolerable in many situations (very different concentrations of analytes in the injected sample). Figure 5.21 shows an excellent example of this type of separation, in which 30 anions are separated in only 3 min [151]. Another example of the use of chromate as probe is shown in Fig. 5.22, where an indirect absorbance detection method was applied to the quantitative determination of free oxalate in amniotic fluid with an LOD of about 100 μg/l [152]. Almost all the studies carried out using chromate as a probe anion have utilized a BGE pH between 8 and 9. On the other hand, sulfosalicylic acid as absorbing probe and a patented EOF modifier from Waters (OFM-OH) were used to analyze mixtures of sulfur-containing compounds (sulfur speciation) with LODs in the low- to mid-μM range, except for sulfide, for which the LOD was 10^{-5} M [153].

The application of indirect detection to the separation of other major groups of analytes including metal ions [68,154] and carbohydrates [155–157] has also been reviewed. In addition, this technique was also investigated to achieve the analysis of phosphates and phosphonates,

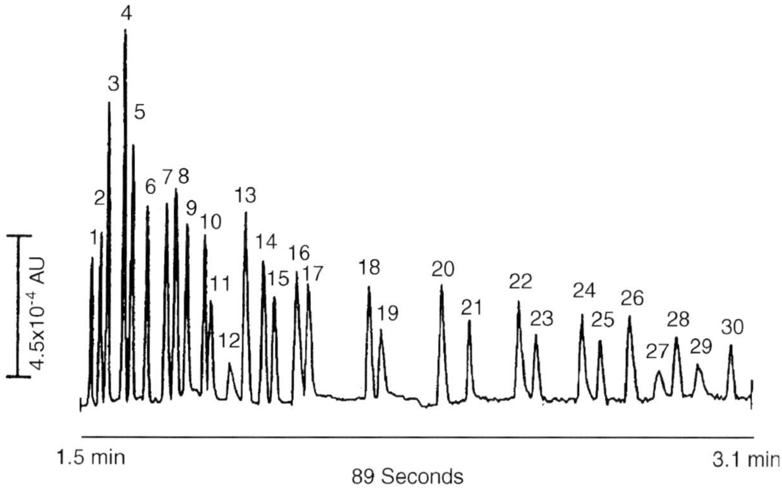

Fig. 5.21. Electropherogram with UV-indirect detection of a mixture of 30 anions (detection at 254 nm) Conditions: BGE 5 mM chromate, 0.5 mM TTAB, pH = 8.0. Compounds: (1) thiosulfate, (2) bromide, (3) chloride, (4) sulfate, (5) nitrite, (6) nitrate, (7) molybdate, (8) azide, (9) tungstate, (10) monofluorophosphate, (11) chlorate, (12) citrate, (13) fluoride, (14) formate, (15) phosphate, (16) phosphite, (17) chorite, (18) galactarate, (19) carbonate, (20) acetate, (21) ethanesulfonate, (22) propionate, (23) propanesulfonate, (24) butyrate, (25) butanesulfonate, (26) valerate, (27) benzoate, (28) L-glutamate, (29) pentanesulfonate, (30) D-glucanate. Reprinted from Ref. [151], Copyright (1991), with permission from Elsevier.

alkylsulfates and alkylsulfonates, surfactants, amino acids, peptides, carboxylic acids, fatty acids, and various other analytes. Suitable probe electrolytes for these analytes include a range of aromatic carboxylates and aromatic sulfonates, and other miscellaneous probes [139]. Table 5.7 groups some of the probes employed in recent applications, indicating whether an EOF modifier is used or not. Figures 5.23–5.25 show some recent examples of these applications. Thus, the performance of pyromellitic probe in comparison with chromate probe in the separation of inorganic selenium species in the presence of selected common anions with hexamethonium hydroxide as EOF modifier is shown in Fig. 5.23 [161]. On the other hand, LODs in the range of 13 to 54 ppb (1×10^{-7} to 6×10^{-7} M) were obtained in the analysis of carboxylic acid metabolites from cell extracts using 2,6-pyridinedicarboxylic acid as a highly UV-absorbing electrolyte probe and CTAB as EOF modifier (see Fig. 5.24) [158]. Finally, a new absorbing electrolyte probe, *p*-aminosalicylic acid, was used for the separation and quantitation of 30

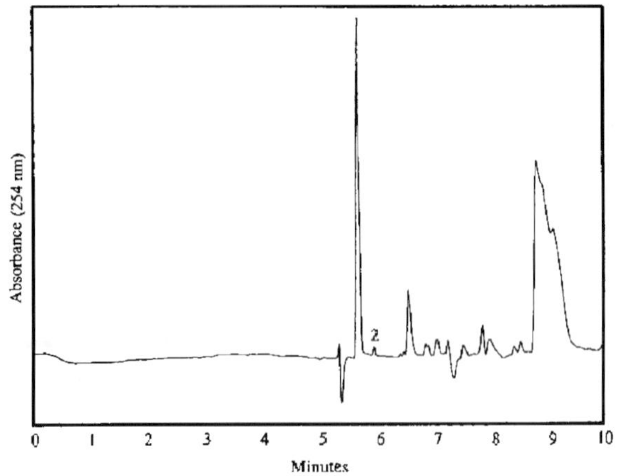

Fig. 5.22. Electropherogram of an amniotic fluid sample. 1:1 (v/v) diluted amniotic fluid sample from a third trimester amniocentesis. Peak identification: (1) sulfate, (2) oxalate. 20°C. Indirect UV-absorbance detection at 254 nm. BGE, 10 mM chromate with 0.5 mM TTAB and 0.1 mM EDTA at pH 8.0. Injection, 10 s (3447 Pa). Applied voltage, −15 kV (≈ 20 μA). Extended light path (200 μm) uncoated fused-silica capillaries (Hewlett-Packard, Wilmington, DE, USA) of 75 μm I.D. ×67 cm (effective length of 60 cm). Reprinted from Ref. [152], Copyright (2000), with permission from Elsevier.

underivatized physiological amino acids and peptides usually present in biological fluids, with LODs ranging from 2 to 20 μM (see Fig. 5.25) [165].

With regard to wavelength values employed, although most of the work with indirect absorbance detection is done at wavelengths in the UV region (between 200 and 300 nm), some highly absorbing dyes like bromocresol green and indigo-tetrasulfonate have also been investigated as electrolyte probes in the visible range. High sensitivity in the determination of small organic and inorganic anions was achieved by utilizing these highly absorbing dyes as probes. LODs were in the sub-μM level. These LODs were an order of magnitude lower than the general detection limit reported for indirect absorbance detection, and were comparable with LODs achieved with indirect fluorescence detection [168]. Optimization of probe concentration using highly absorbing dyes has been reported [169]. In CE with indirect absorbance detection, the concentration of the absorbing probe ion in the background electrolyte should be as high as possible in order to increase the linear dynamic range of the detection method. However, for probes

TABLE 5.7

Applications of CE with indirect absorbance detection

Analytes	Absorbing electrolyte probe	EOF modifier	Reference
Carboxylic acids	2,6-Pyridinedicarboxylic acid	CTAB[a]	[158]
Trans-fatty acids	Sodium dodecylbenzenesulfonate	Brij 35[b]	[159]
Sr and Li	4-Methylbenzylamine		[160]
Inorganic selenium species	Pyromellitic acid	Hexamethonium hydroxide	[161]
Sugar acids	β-Resorcylic acid		[162]
Speciation of sulfur	Pyromellitic acid		[163]
Amino acids and biogenic amines	Sulfonated calixarene		[164]
Amino acids and peptides	p-Aminosalicylic acid		[165]
Phosphates and phosphonates	Phenylphosphonic acid		[166]
Organic acids	8-Hydroxy-7-iodoquinoline sulfonic acid		[167]

[a]CTAB, cetyltrimethylammonium bromide.
[b]Brij 35, polyoxepthylene 23 lauryl ether.

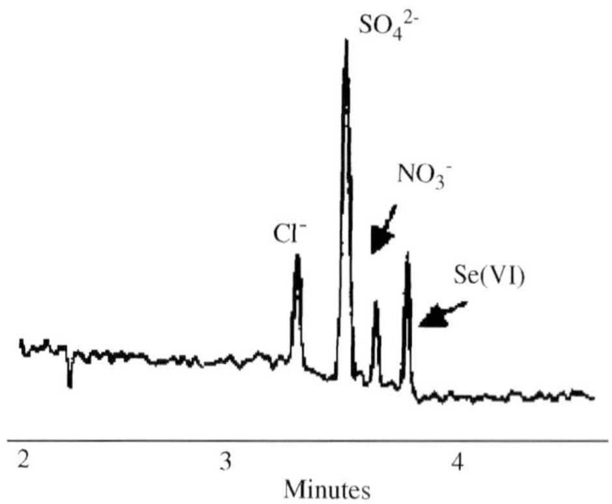

Fig. 5.23. Electropherogram of commercial mineral water "Kinga" spiked with 3 mg/l of Se(VI). BGE: 2.25 mM pyromellitic acid, 3 mM triethanolamine, 1 mM hexamethonium hydroxide, and pH 9. Separations were carried out using 60 cm (50 cm effective length) ×75 μm I.D. Hydrodynamic injection for 10 s. Indirect detection at 254 nm. Reprinted from Ref. [161], Copyright (2003), with permission from Elsevier.

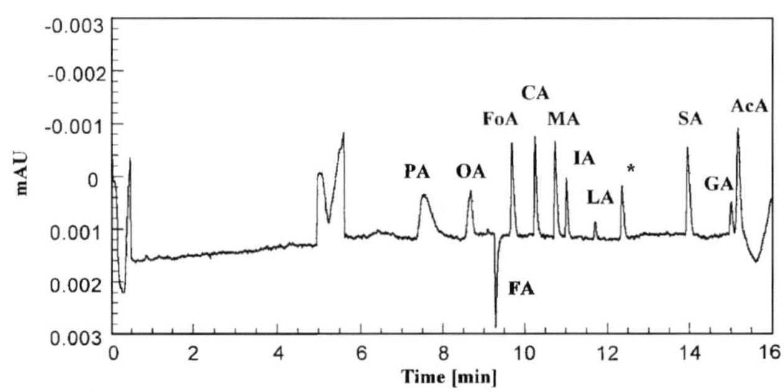

Fig. 5.24. Electropherogram of standard carboxylic acids under the optimized conditions using indirect UV detection at 200 nm. Experimental conditions: BGE, 4 mM 2,6-pyridinedicarboxylic acid, 0.2 mM CTAB (pH 3.5), 10% ethylene glycol and 10% acetonitrile; untreated fused-silica capillary, 75 μm I.D. ×70.6 cm effective length; electrokinetic injection 10 kV 10 s; voltage –25 kV; temperature 15 °C. Samples: PA, pyruvic acid (1.43×10^{-5} M); OA, 2-oxoglutaric acid (8.36×10^{-6} M); FA, fumaric acid (6.04×10^{-6} M); FoA, formic acid (2×10^{-6} M); CA, citric acid (2.64×10^{-6} M); MA, malic acid (3.91×10^{-6} M); IA, isocitric acid (5.27×10^{-6} M); LA, lactic acid (3.28×10^{-6} M); SA, succinic acid (6.77×10^{-6} M); GA, glutamic acid (0.9×10^{-6} M); AcA, acetic acid (7.31×10^{-6} M); *, unidentified peak. Reprinted from Ref. [158], Copyright (2003), with permission from Elsevier.

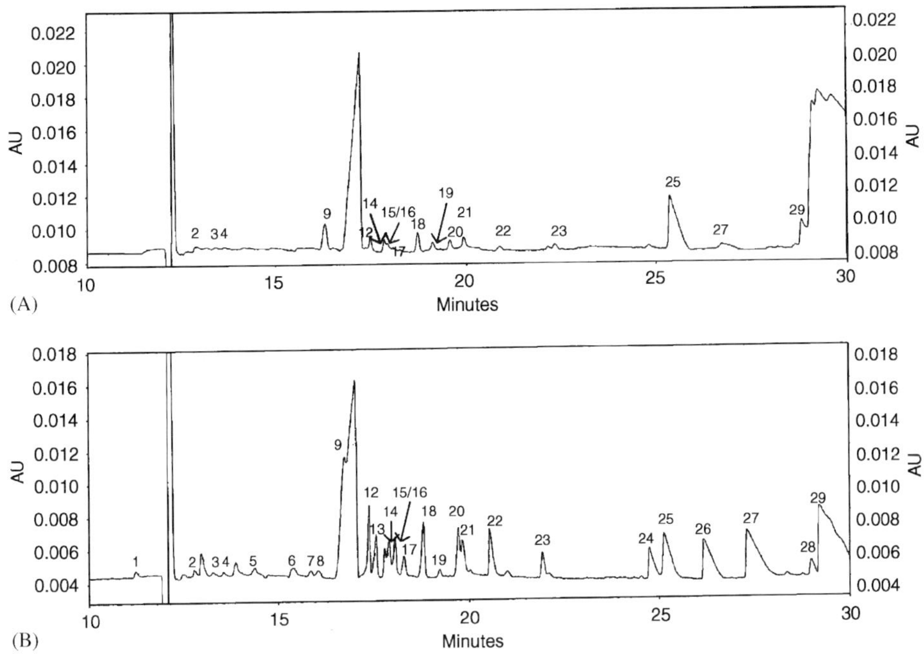

Fig. 5.25. Electropherograms of five-fold diluted deproteinized plasma samples without (A) and spiked with (B) 100 μM standard amino acid/peptide mixture. 8 mM *p*-aminosalicylic acid –2 mM Na_2CO_3 at pH 10.15. Voltage: 15 kV. Temperature: 20°C. Indirect detection at 254 nm. Migration order (peaks) (1) Arg, (2) Lys, (3) Orn, (4) Pro, (5) γ-amino butyric acid, (6) HyLys, (7) αALA, (8) carnosine, (9) Leu/Ile, (10) Trp, (11) Cit, (12) Val, (13) Phe, (14) Ala, (15) His, (16) Met, (17) Gln, (18) Thr, (19) Asn, (20) Gly, (21) Ser, (22) Tyr, (23) hySS, (24) hCyS, (25) CyS(S), (26) oxidized gluthathione hydrate, (27) Glu, (28) reduced gluthathione, (29) Asp. Reprinted from Ref. [165], Copyright (2002), with permission from Elsevier.

with high molar absorptivity values such as dyes, the maximum concentration may be limited by the background absorbance of the electrolyte, which must fall within the linearity range of the detector. Therefore, the linearity of the detector is a crucial parameter when using highly absorbing electrolyte probes [75].

In conclusion, the correct design of the BGE is of paramount importance to ensure the success of the separation by CE with indirect absorbance detection, which requires the consideration of several factors such as the nature of the probe, the relative mobilities of the probe and the analyte, the EOF modifier (if required), and the type of buffer.

Thus, in CE with indirect absorbance detection, the pH and composition of the buffer have to be chosen carefully in order to achieve optimum resolution and sensitivity. Because low buffer concentrations (2–10 mM) are typically used to improve performance, buffering capacity and mass sample overloading become more of a concern, especially when one is analyzing a series of widely varying samples. With the aim of aiding in the development of these separations, Doble et al. [143] have published an excellent discussion about these factors and their importance for designing BGEs for indirect detection of anions by CE. In addition, a comprehensive study of BGEs employed for indirect absorbance detection of anions in CE containing abundant data about analyte and probe mobilities, probe molar absorptivities, pK_a values for probes, and typical BGEs has been compiled by Doble and Haddad [139].

5.3 THERMOOPTICAL DETECTION IN CE

Thermooptical or photothermal techniques are non-traditional absorbance techniques that are based on the generation of heat during the non-radiative relaxation (vibrational relaxation via collisions with the immediately surrounding solvent molecules) after absorption of radiation [2]. The immediate consequence of such a process is an increase in the sample's temperature, which ultimately results in changes in the physical properties of the sample, such as density and RI. Different concepts have been applied to monitor these changes resulting from photothermal processes. This led to the development of various experimental configurations and techniques such as thermal lens spectrometry (TLS), photoacoustic spectrometry (PAS), photothermal deflection spectrometry, photothermal grating, and photothermal shift spectrometry [170]. Among them, TLS and PAS have been employed as detection systems in CE, TLS being especially preferred.

5.3.1 Thermal lens detection

The thermal lens technique is based on absorption spectroscopy, in which a rise in temperature occurs in an illuminated liquid induced by the absorption of small amounts of energy from a laser beam (pump or excitation beam) passing through the liquid. Owing to thermal diffusion, the heat will be drained from the laser beam center to the surrounding liquid. This will result in a cylindrically symmetric temperature distribution, the axis of symmetry extending along the propagation direction of the laser beam. This localized temperature change brings about a gradient of the RI with the same symmetry profile on account of to the dependence of RI on temperature. That is, the RI of the liquid has the lowest value at the beam center and increases to its value at room temperature at the boundary of the beam. This cylindrical RI profile acts as a concave lens, called "thermal lens effect" (TLE), that causes a defocusing of the propagating beam or alters the propagation of a second beam, which is called a probe beam. The TLE is very weak. The temperature change is estimated at the level of a micro-Kelvin per analyte molecule. Therefore, weak signals are recovered by modulating the intensity of the excitation beam and by detecting the synchronous signal with a lock-in amplifier [171].

In order to evaluate the performance of TLS, the signals generated in absorption spectrometry and TLS can be compared. Thus, an enhancement factor of the response compared with Lambert–Beer's law (E) is commonly used in the literature dealing with TLS [172]. The enhancement factor, E, can be expressed through the following equation:

$$E = -\frac{\partial n/\partial T \cdot I_0}{\lambda \cdot \kappa} \tag{5.5}$$

where $\partial n/\partial T$ is the cylindrical RI profile (n is the RI and T the temperature), I_0 is the incident beam power, and κ is the thermal conductivity of the solvent. According to Eq. (5.5), E is proportional to I_0, and therefore, it is advantageous to use a high laser power. However, I_0 should be limited because non-linearity caused by saturation effects and/or photo-decomposition of analyte molecules could take place. On the other hand, organic solvents such as acetonitrile and methanol are more suitable than water, because they show a five-fold higher $\partial n/\partial T$ and three-fold lower κ than water respectively [51]. Thus, the solvent contribution term to the enhancement factor $(\partial n/\partial T)/\kappa$ is a good index for solvent selection. Usually, the index of an organic solvent is more than ten-fold larger than that of an aqueous solution.

Thermal lens spectrometers are generally divided into three groups of instruments: single-beam (SB) instruments, dual-beam (DB) instruments, and pump-probe (PP) instruments. As shown in Fig. 5.26, SB and DB instruments use one laser (in DB instruments, a second beam is obtained by splitting the laser beam before the sample), while the PP instrument requires the use of two lasers: the pump laser and the

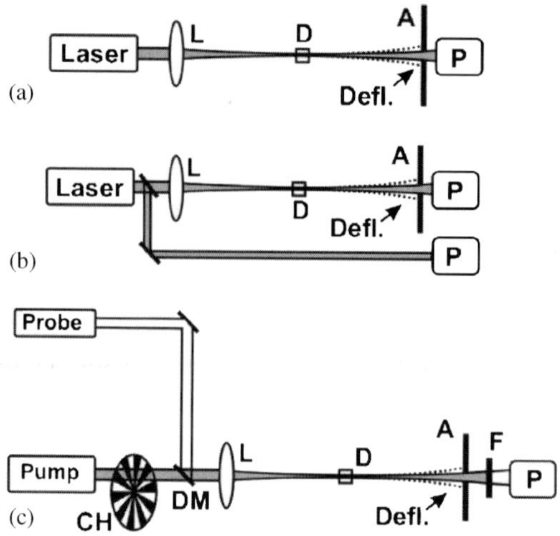

Fig. 5.26. Schematic of TLS detection setups. (a) single-beam; (b) double-beam; (c) pump-probe. Abbreviations: L, focusing lens; D, detector cell; P, diode; CH, chopper; A, aperture; DM, dichroic mirror; F, filter; Defl, beam deflection. Reprinted from Ref. [51], Copyright (2002), with permission from Elsevier.

probe laser. In the latter, the wavelength of the pump (excitation) laser beam is selected to coincide with an absorption band of the analyte, while the wavelength of probe laser beam is chosen so that the sample solution has no absorption.

The development of pump-probe (PP) techniques has led to major improvements in TLS; thus, most TLS instruments operate with this setup (all systems used in CE), as has been reported in a review about laser-based non-fluorescence detection techniques published by Beer et al. [51], which summarizes the basic principles and advances in these detection systems, with an overall evaluation and a discussion of their perspectives.

Pump-probe thermal lens detection (PP-TLD) was developed for CE by Bornhop and Dovichi [173] in 1987. As shown in Fig. 5.27, the pump laser is focused at a right angle to the capillary, and a second laser beam is used to probe the RI of the fluid in the capillary. Both lasers intersect within the capillary defining the detection region (confocal region), and when the analyte passes through the detection region, it absorbs light from the pump laser producing a temperature rise in the capillary. The photothermal effect can be monitored as a change in the intensity of the probe beam by a diode after it passes through a pinhole

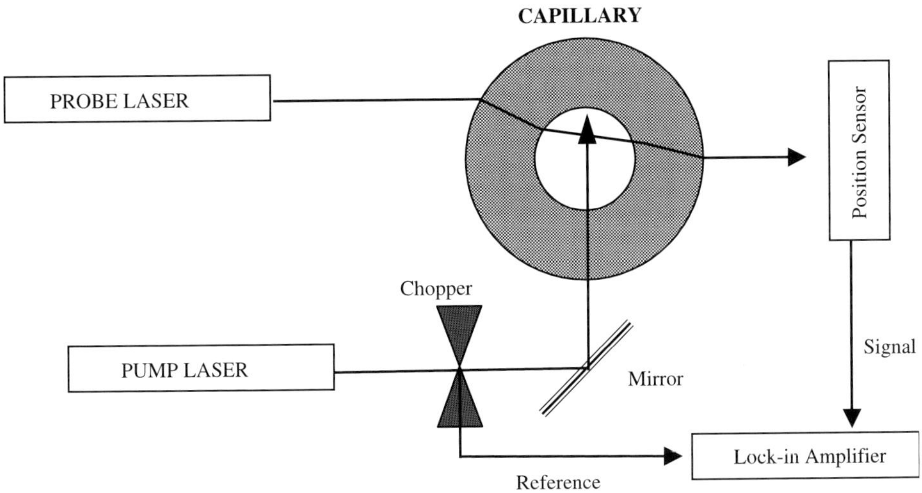

Fig. 5.27. Schematic of pump-probe thermal lens detection for CE.

or aperture. Finally, changes in the probe beam intensity as the analyte passes through the pump beam are proportional to the number of molecules in the confocal region, i.e., proportional to the analyte concentration.

PP-TLD has some interesting advantages [51]. First, it allows the use of a pulsed laser as the pump beam, which enables it to employ high I_0 values, avoiding saturation effects and photo-decomposition of analyte molecules. In addition, since most absorption bands of the analytes are in the UV part of the electromagnetic spectrum and pulsed lasers easily provide output in that wavelength region, pulsed lasers are ideal for the pumping task. Second, lock-in detection—a popular signal-enhancing technique in optical PP setups—can be applied. By amplitude modulation of the pump laser (for instance, by a mechanical chopper), the strength of the thermal lens—and, therefore, the resulting signal—will be modulated as well. By detecting only the modulated signal, the signal-to-noise ratio is improved [174]. In addition, this detection system is hardly affected by the background caused by light scattering because it detects only the light intensity difference synchronized to the modulation of the excitation beam. Therefore, PP-TLD can be applied to turbid samples, unlike absorption and fluorescence methods, which are disturbed by light scattering. However, the main disadvantage of using this detection system is that a second laser is required, and many time-consuming steps to align the laser beams are needed. Generally, an inexpensive low-power CW laser is used to detect the induced lens because there are no restrictions on the wavelength of the probe laser (it should fall within the sensitivity window of the detector and not be absorbed by the solvent). Thus, He–Ne (in almost all reported works) or diode lasers can be used for probing so that the extra costs are marginal. Furthermore, such lasers are very stable, especially the diode lasers, so that their contribution to noise can be ignored.

The use of PP-TLS as a detection technique for the analysis of a variety of compounds separated by CE has been explored by employing different chromophore/laser combinations (see Table 5.8). In one first approach, Bornhop and Dovichi [173] chose a He–Cd laser (at 442 nm) as the pump beam and sunset yellow dye as model analyte. This system was also applied for the detection of dimethylamino-azobenzene sulfonyl (DABSYL) chloride-derivatized amino acids [175]. In this case, a mass detection limit of 200 amol ($\approx$ 300 nM) for DABSYL-glysine was reported. In another work, the detection of DABSYL-methionine at a mass detection limit of 37 amol ($\approx$ 50 nM) was reported using a 130 mW argon-ion laser as the pump source [176,177].

More universal detection would be provided by the use of lasers that operate in the UV region of the spectrum. Bruno et al. [189] reported the first application of UV laser-based thermooptical absorbance detection in CE. They used a frequency-doubled argon-ion laser (at 257 nm) as the pump beam to determine three dansylated amino acids in the 10^{-4} M range. Using the same pump laser, they reported subfemtomole mass detection limits for nucleosides and nucleotides (50 nM for adenosine monophosphate), when the probe beam was guided to the detection region by a holographic optical system [178]. LODs for proteins comparable with those obtained with laser-induced fluorescence detection were also reported (7 nM), and were two orders of magnitude lower than those obtained with absorbance detection [179]. Despite these results, the frequency-doubled argon-ion laser is expensive and not adequate for routine use in the analytical laboratory.

The krypton fluoride (KrF) laser, air-cooled, and less expensive than more conventional UV lasers, was also investigated for the analysis of phenylthiohydantoin (PTH) amino acid derivatives. This laser operates at 248 nm, which matches well the absorbance of the PTH-amino acids. Thus, 20 PTH-amino acids separated by MEKC were detected with LODs of 0.5 fmol (100 fg or 500 nM) for PTH-tryptophan and a linear dynamic range of more than 4 orders of magnitude [5]. This sensitivity was 3 orders of magnitude superior to that of HPLC. Using this system,

TABLE 5.8

Applications of pump-probe thermal lens detection coupled to CE

Analyte	Pump laser (wavelength in nm)	Probe laser (wavelength in nm)	Matrix	LOD	Reference
Sunset yellow dye	He-Cd (442)	HeNe (633)	Standards	150 nM	[173]
Amino acids (DABSYL[a] derivatives)	He-Cd (442)	HeNe (633)	Standards	300 nM	[175]
Amino acids (DABSYL[a] derivatives)	Ar^+ (458)	HeNe (633)	Standards	50 nM	[176, 177]
Nucleosides and nucleotides	Ar^+ (257)	Laser diode (633)	Standards	50 nM	[178]
Proteins	Ar^+ (257)	Laser diode (633)	Standards	7 nM	[179]
Amino acids (PTH[b] derivatives)	KrF (248)	HeNe (633)	Standards	500 nM	[5]
			Peptide sequencing		[180]
Amino acids (DABSYL[a] derivatives)	Ar^+ (458)	HeNe (633)	Standards	200 nM	[181]
Metal ions (after complexation)[c]	Ar^+ (514)	HeNe (633)	Rain water	15 nM	[182]
Nitrophenol pesticides	Ar^+ (257)	HeNe (633)	Standards	23 μg/l	[183]

continued

TABLE 5.8 (*continued*)

Analyte	Pump laser (wavelength in nm)	Probe laser (wavelength in nm)	Matrix	LOD	Reference
Nitroaromatic compounds	Ar^+ (257)	Laser diode (633)	Waste water	60 μg/l	[184]
			Soil		[185]
Pharmaceuticals	Ar^+ (257)	Laser diode (633)	Standards	7 μM	[184]
Antidepressants	KrF (248)	HeNe (633)	Cell culture	100 nM	[186]
Preservatives	KrF (248)	HeNe (633)	Food	400 nM	[187]
Lysine (indirect)	HeNe (633)	HeNe (633)	Standards	5 μM	[188]

[a]DABSYL, 4-dimethylamino-azobenzene sulfonyl chloride;
[b]PTH, phenylthiohydantoin derivatives;
[c]1,10 phenanthroline complexes.

19 PTH-amino acids produced in the Edman degradation reaction for protein sequencing were also deteceted [180].

A novel fiber-optic-modified thermal lens detector was also described [181]. The probe and pump lasers used in TLD were adapted via optical fibers to a commercial injection and high-voltage CE system. The detector performance was demonstrated by monitoring mixtures of (DABSYL) chloride-derivatized amino acids using argon-ion laser at 458 nm (see Fig. 5.28). The miniaturization of the thermal lens detector and the separation of the excitation and probe laser from the sensor head with optical waveguides allowed the combination with typical laboratory CE systems. This system was also used to determine iron in rain water samples using 1,10-phenanthroline as a chromogenic reagent, which forms a stable complex with Fe(II) but not with Fe(III). An LOD of 15 nM was obtained using a 75 μm fused-silica capillary [182].

The usefulness of thermal lens detectors has also been shown in other application areas. Thus, CE with TLDs was used for the analysis of pollutants. Seidel et al. [183] have separated nitrophenol pesticides using MEKC with TLD, and have compared their results with conventional UV detection. The signal-to-noise ratio was significantly improved by the thermal lens detector since an LOD of 23 ppb was obtained, which was 48 times lower than that obtained with a usual absorbance detector at 364 nm. Also, a thermal lens CE detector using a frequency-doubled argon-ion laser (at 257 nm) as pump laser and a laser diode (emission at 633 nm) coupled to an optical fiber as probe laser was used for the analysis by MEKC of nitroaromatic compounds in contaminated water [184] and (non-fluorescent) nitroaromatic explosives in contaminated soils [185]. The results obtained in these works show that this type of detection makes it possible to obtain limits of detection 1–2 orders of magnitude lower than those obtained with UV-absorbance detection (see Fig. 5.29). On the other hand, a pulsed KrF pump laser was also applied to the analysis of preservatives in foods. Optimization of experimental parameters, such as average laser power and separation conditions enabled LODs ten times better than those achievable by on-column UV detection [187]. TLD has also been applied successfully in CE and CEC for the detection of five tricyclic antidepressants [186] and more recently, to the analysis of three pharmaceuticals (tramadol, verapamil, and papaverin) in nanoliter volume samples common in biomedical diagnostics without any pre-concentration step [185].

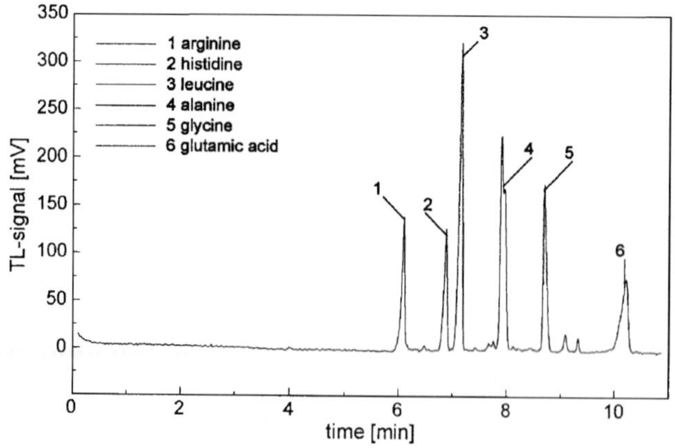

Fig. 5.28. Electropherogram of (DABSYL) chloride-derivatized amino acids using PP-TLD with argon-ion laser (at 458 nm) as pump laser and HeNe (at 633 nm) as probe laser. BGE, 100 mM borate (pH 9.3), 50 mM SDS, 15% methanol, and 1% THF. Applied voltage, 25 kV. Capillary, 50 μm I.D. × 55 cm (35 cm to detection). Reprinted from Ref. [181], Copyright (1998), with permission from Elsevier.

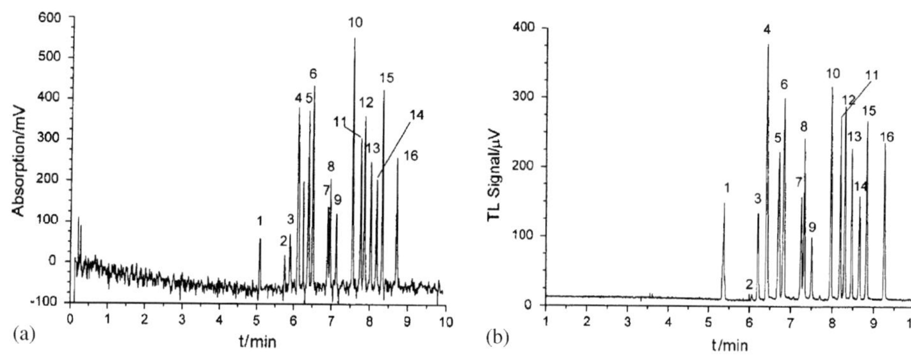

Fig. 5.29. Electropherograms of nitroaromatic compounds: (a) UV-detection at 257 nm; (b) PP-TLD with argon-ion laser (257 nm) and diode laser (633 nm), by MEKC. BGE: 15 mM borate buffer (pH 8.9) with 25 mM SDS. Capillary: 104 (68) cm × 75 μm I.D. Voltage, 30 kV. Injection, vacuum 0.3 s. Compounds: (1) system peak; (2) 1,3,5,7-tetranitro-1,3,5,7-tetraazacyclooctane; (3) 1,3,5-trinitrobenzene; (4) 1-fluoro-2,4-dinitro benzene; (5) 1,3-dinitrobenzene; (6) nitrobenzene; (7) 2,4,6-trinitrotoluene; (8) 2-4-6-*N*-tetranitro-*N*-methylaniline, (9) 2-methyl-3-nitroaniline; (10) 2,4-dinitrotoluene; (11) 2,6-dinitrotoluene; (12) 2-nitrotoluene; (13) 4-nitrotoluene; (14) 2,3-dinitrotoluene; (15) 2-amino-4,6-dinitrotoluene; (16) 4-amino-2,6-dinitrotoluene. Reprinted with permission from Ref. [184], Copyright (2002), American Chemical Society.

Finally, Ren et al. [188] described for the first time indirect TLD using He–Ne beams for the pump (20 mW) and probe (2 mW) beams. Methylene blue was the background absorber utilized in this work, in which LODs slightly poorer than conventional absorbance detection (5 μM for lysine) were reported.

In conclusion, PP-TLS techniques have several advantages as detection systems in CE including very small probe volumes and high sensitivity. The detection volume is determined by the diameter of the capillary in combination with the point of interaction of the two laser beams. The sensitivity is proportional to the pump laser power, and the signal is independent of the capillary diameter. Therefore, unlike typical absorbance measurements, sensitivity is not lost owing to short detection path lengths, and higher power lasers are useful for detecting weakly absorbing or low concentration analytes.

TLS techniques also have disadvantages. Tedious alignment is required for PP-TLD to function, and it is relatively expensive to configure. In addition, the solute must have absorbance features that correspond to the wavelength of the pump laser beam, and κ and $\partial n/\partial T$ values for the BGE must be appreciable in order to produce a highly sensitive measurement. With respect to this last aspect, it should be indicated that, in contrast to HPLC, organic modifiers are not generally added to the aqueous BGE used in CE, so that the BGE solutions have less favorable κ and $\partial n/\partial T$ values (as noted earlier) and no impressive enhancements are expected [51]. Thus, although E values of 10–1000 have been reported, the value of 1000 was quite exceptional, being obtained when the BGE contained 50% acetonitrile [176]. As it is shown in Table 5.8, no real-life samples (standard solutions) have mainly been analyzed using CE-TLS, derivatized amino acids being the most studied analytes.

5.3.2 Photoacoustic detection

Photoacoustic detection (PAD) is based (like TLS) on a temperature change induced by the absorption of radiation. However, in PAD this change is monitored by measuring the accompanying effect on the pressure, which is detected by a piezoelectric transducer (PZT) or a sensitive microphone and further amplified and processed [51,190]. The pump laser used in PAD can be a CW as well as a pulsed laser. The CW laser is usually modulated by means of a mechanical chopper. It is not easy to predict whether a CW or a pulsed laser will give better

LODs in PAD. However, it is known that the quality of the transducers largely determines the sensitivity in PAD. In addition, the transducer should be positioned as close as possible to the focus of the exciting laser beam in order to create a strong signal [51].

The Sawada research group coupled PAD for the first time to CE in 1990 using capillary vibration induced by laser (CVL) [191]. In this technique (see Fig. 5.30a), the separation capillary is supported during the analysis and a tension is applied to the tube by a hanging weight. Then, when an intensity-modulated laser beam irradiates the capillary, analyte molecules absorb radiation and periodically generate heat; hence, pressure waves are formed. These pressure waves cause the capillary to vibrate like a string. Finally, the mechanical vibration of the capillary was detected by measuring the deflection of a second (probe) laser beam (Fig. 5.30a). In this first study, it was possible to measure in an optical sampling volume (detection volume) of 100 pl up to 6 fg, corresponding to 13 amol, of sunset yellow dye used as model analyte [191], which corresponds to the measurement of 1.5×10^{-5} AU. This LOD is 2–3 orders lower than those obtained by the conventional fluorimetric and spectrophotometric methods, which have a sampling volume of about 1 μl [192,193], and the same order as that of the PP-TLS method [173].

To confirm CVL detection in CE and to study the technique further, a vitamin (riboflavin) and two amino acids (phenylalanine and tryptophan) were used as model analytes [194]. In both cases, an argon-ion laser was used as excitation beam, but with riboflavin, the 476 nm argon lasing line was used as visible excitation, whereas with the underivatized amino acids, the 257 nm argon lasing line (obtained by doubling the argon lasing line of 514 nm) was used as UV excitation. The results obtained are summarized in Table 5.9. These results were the first reported detection of femtomole levels of underivatized amino acids based on direct optical absorption.

In a subsequent study [195], the generation mechanism of the capillary vibration was the same as before, and the capillary vibrated like a string between the two supports, one of which was a PZT (Fig. 5.30b). Thus, in this case, mechanical vibration of the capillary was directly detected by the PZT, which allowed the measuring of changes as small as 8.5×10^{-7} AU. Considering the optical sampling volume estimated to be 10 pl, the PZT detection of sunset yellow dye proved to be at least 1–2 orders more sensitive than the use of the probe laser to detect the signal (see Table 5.9).

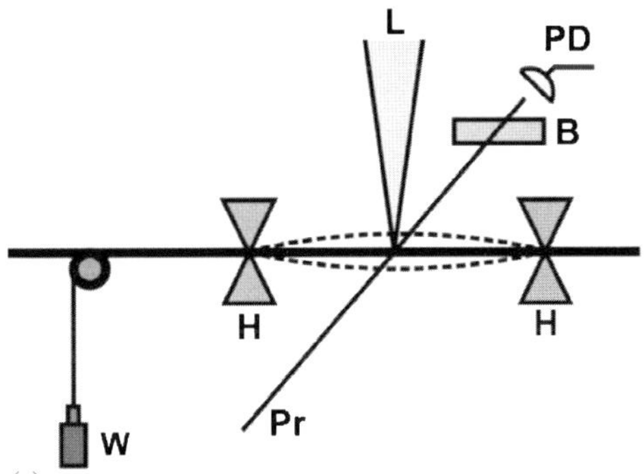

(a)

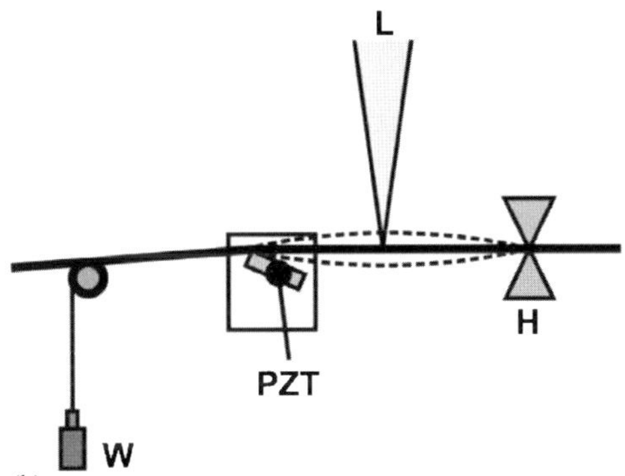

(b)

Fig. 5.30. A schematic drawing of CVL using (a) a second (probe) laser, (b) a piezoelectric transducer (PZT) for detection. Abbreviations: L, pump laser beam; PD, diode; B, blade; H, holders; Pr, probe laser beam; W, weight. Reprinted from Ref. [51], Copyright (2002), with permission from Elsevier.

On the other hand, as shown in Table 5.9, the pump laser used in CVL is generally an intensity-modulated CW laser [191,194–196]. However, a pulsed laser can also be used as a pump laser beam [197] in a CVL system. As shown in Table 5.9, the sensitivity obtained in CVL with a pulsed laser (KrF laser) is comparable to that obtained with a CW laser (argon-ion laser). Thus, to extend the applicability of CVL detection to the UV region (where many biological materials have significant absorption bands), researchers can choose according to their purpose and financial possibilities between a pulsed laser such as the KrF laser, or a CW laser, such as the frequency-doubled argon-ion laser, which is more expensive and less adequate for routine use in the analytical laboratory. In addition, for miniaturization of the detector, a pulsed laser, such as an excimer laser (KrF laser), is more desirable

TABLE 5.9

Applications of photoacoustic detection coupled to CE

Analyte	Pump laser (wavelength in nm)	Signal detector	I.D. capillary (μm)	Detection volume[a] (pl)	LOD (mass)	LOD[b] (conc.)	Reference
Sunset yellow dye	Ar^+ (488)	HeNe (633 nm)	50	100	13 amol (6 fg)	130 nM	[191]
Riboflavin	Ar^+ (476)	HeNe (633 nm)	30	21	80 amol (30 fg)	186 nM	[194]
Phenylalanine	Ar^+ (257)				110 fmol (18 pg)	5 mM	[194]
Tryptophan	Ar^+ (257)				15 fmol (3 pg)	0.7 mM	[194]
Sunset yellow dye	Ar^+ (488)	PZT[c]	50	10	70 zmol (3 ag)	7 nM	[195]
Amino acids (DABSYL[d] derivatives)	Ar^+ (488)	HeNe (633 nm)	30	1	<8 amol[e]	<8 μM[e]	[196]
Phenylalanine	KrF (248)	HeNe (633 nm)	20	1	5 fmol (0.8 pg)	5 mM	[197]
Tryptophan					0.5 fmol (0.1 pg)	0.5 mM	[197]

The analyzed samples were standard solutions.
[a]The detection volume (optical sampling volume) is defined as the capillary tube volume inside the focal area of the excitation beam, which is given by the inner diameter of the capillary tube and the focal length of the lens [191].
[b]LOD in terms of concentration according to LOD in terms of mass and the detection volume.
[c]PZT, piezoelectric transducer.
[d]DABSYL, 4-dimethylamino-azobenzene sulfonyl chloride.
[e]This amount is not LOD.

than a frequency-doubled argon-ion laser (CW laser). Odake et al. [197] showed that a pulsed KrF laser operating at 248 nm is applicable as a CVL excitation source with a stable baseline, and a series of underivatized amino acids were successfully analyzed at the femtomole level using this technique in CE (see Fig. 5.31). The sensitivity of this pulsed laser-induced CVL is at least two orders of magnitude superior than that of the commercially available UV-absorbance detector.

Finally, it is important to emphasize that the number of publications dealing with PAD in CE (see Table 5.9) is much smaller than that dealing with TLD (see Table 5.8). All reported results about this technique refer to standard solutions, and have been published by the same research group.

5.4 UV–VIS ABSORBANCE DETECTION IN MICROCHIP ELECTROPHORESIS

As mentioned above, absorbance is the most widely used detection method in CE. However, detection via absorbance is not commonly used in microchip electrophoresis where absorbance yields only limited sensitivity owing to the small channel depths (i.e., short optical path lengths). Furthermore, rather expensive quartz chips have to be used for high UV transparency. Nevertheless, Schwarz and Hauser [44]

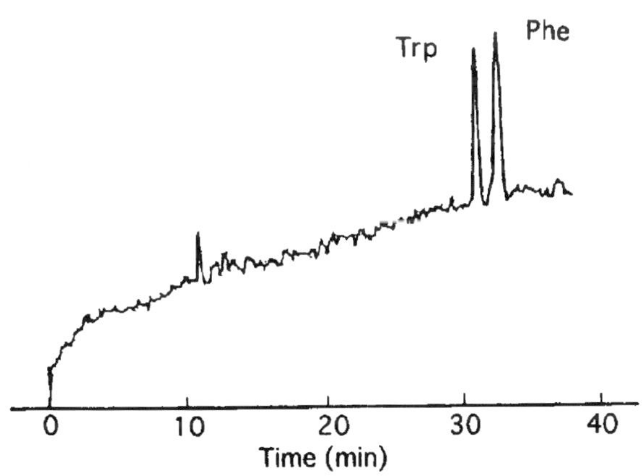

Fig. 5.31. Electropherogram of underivatized amino acids analyzed by CE with CVL detection. Injected amounts were 0.3 pmol for tryptophan and 3.0 pmol for phenylalanine. Applied voltage, 15 kV. Capillary 75 cm × 20 μm I.D. Reprinted with permission from Ref. [197], Copyright (1997), American Chemical Society.

included a paragraph devoted to absorbance detection in an excellent review on microfabricated analytical devices. Another comprehensive review about the analysis of inorganic and small organic ions on microchip platforms has also been published recently [198]. In this article, the application of absorbance detection was discussed together with other detection methods, and the potential utilization of these devices for miniaturized analytical systems was described. Examples of this application area are the determination of phosphate [199], nitrite [200], nitrate [201], and ammonia [202] by means of the adaptation of classical and well-established colorimetric methods. In the same way, several metal ions were separated and detected as their complexes with two different metallochromic complexation ligands, PAR [203,204] and arsenate III [204]. The separation by MEKC on a glass microchip of 7 transition metal ions (Co^{2+}, V^{3+}, Ni^{2+}, Cu^{2+}, Fe^{2+}, Mn^{2+}, and Cd^{2+}) was achievable in less than 70 s (see Fig. 5.32). Detection of the PAR metal chelates was achieved using a green LED and a miniature photomultiplier tube (see Fig. 5.33), obtaining LODs that ranged from 400 ppb for Ni^{2+} to 1.2 ppm for Mn^{2+} [203]. In the case of the arsenate III metal chelates, a red LED was used as light source [204].

In order to solve the problem of the very short path length, several approaches have been described. With traditional absorbance techniques, the last approach to enhance absorbance detection has been the fabrication of rectangular microchannels with reasonable path lengths in UV-transparent materials [205]. UV-absorbance detection of tryptophan was possible vertically through the chip, using pre-aligned optical fibers to guide light to the channel and collect and bring the transmitted light to the detector. Another possible approach to increase the sensitivity of absorbance measurements in microchips is the use of a U-shaped detection cell [206]. The separation of rhodamine 110, 2,7-dichlorofluorescein, fluorescein, and 5-carboxyfluorescein was carried out to demonstrate the utility of this detection cell, in which the waveguides on the device were connected to optical fibers.

Multireflection cells have also been developed to increase the optical path length. Based on an earlier design by Verpoorte et al. [207], a new multireflection cell was microfabricated in glass and tested [208]. Aluminum mirrors were patterned above and below the separation channel, with 30-μm diameter optical entrance and exit apertures for ingoing and outgoing light positioned 200 μm apart. Effective optical path lengths of 50–272 μm were achieved, compared to single-pass path lengths of 10–30 μm, corresponding to sensitivity enhancements of

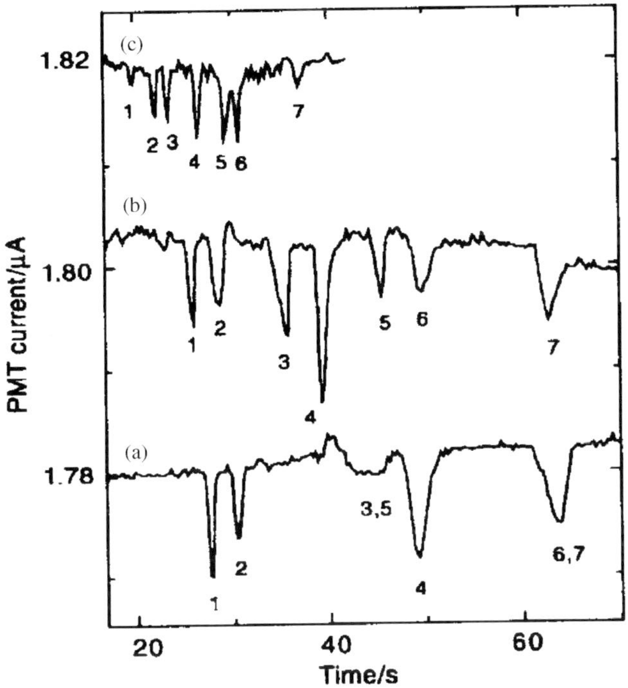

Fig. 5.32. Electropherograms of PAR metal chelates at different pHs on a glass microchip electrophoresis using a green LED. Metal ions: (1) Co^{2+} (1.8 ppm); (2) V^{3+} (3.0 ppm); (3) Ni^{2+} (1.8 ppm); (4) Cu^{2+} (3.0 ppm); (5) Fe^{2+} (3.0 ppm); (6) Mn^{2+} (3.0 ppm); and (7) Cd^{2+} (1.8 ppm). BGE, 1 mM PAR in 10 mM phosphate buffer at: (a) pH = 7.0; (b) pH = 7.5; and (c) pH = 8.0. From Ref. [203]. Reproduced by permission of The Royal Society of Chemistry.

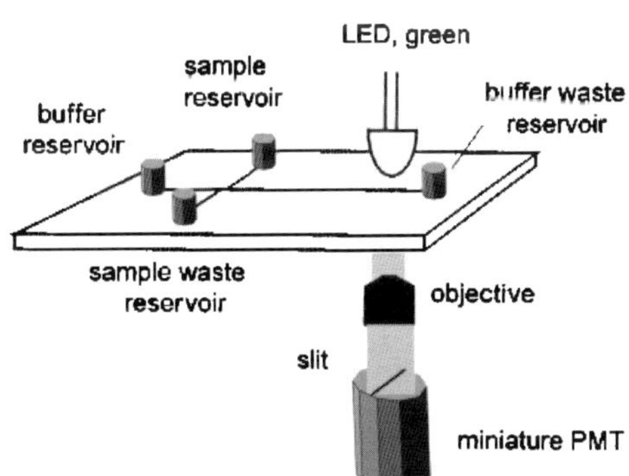

Fig. 5.33. A schematic drawing of microchip electrophoresis instrumental setup for absorbance detection. From Ref. [203]. Reproduced by permission of The Royal Society of Chemistry.

five- to ten-fold over single-pass devices with insignificant contribution from detection to the total band broadening. This result obtained using a He–Ne laser at 633 nm to measure dye-containing solutions showed the potential of absorbance detection for microchip electrophoresis.

A different way to achieve higher sensitivity is the use of non-traditional absorbance techniques such as TLS. The basic fundamentals and applications of this detection technique in conventional CE have already been described (see section 5.3.1). However, to couple a TLS system in microfabricated channel systems, a new development based on the TLE has been described, the thermal-lens microscope (TLM) [209]. A schematic illustration of the TLM is given in Fig. 5.34. Two laser beams (excitation and probe beams) were made coaxial at the dichroic mirror, and were introduced into the optical microscope (Fig. 5.34a). The focus of the probe beam diverges after passing through the thermal lens region (originated by excitation beam), and the degree of the TLE can be obtained by measuring the change in probe beam power after the probe passes through the filter (to eliminate the excitation beam) and the pinhole (Fig. 5.34b). Today, a compact desktop instrument is commercially available (manufactured by the Institute of Microchemical Technology [211]). Recently, a review has been published on integrated chemical systems on microchips showing the future potential of TLM as an ultrasensitive detector for various kinds of chemical and biological systems [171]. However, the majority of the applications of microchip/TLM combinations have been carried out introducing the sample directly into a microchannel. Thus, TLM on microchip was successfully applied to determine model compounds such as a dye (sunset yellow) [210] or a metal ion after complexation (lead porphyrin) [212] with impressive LODs corresponding to yoctomole amounts (10^{-24} mol), which means single-molecule detection. Owing to the extremely small detection volume (in the range of the femtoliters), such sensitivity in terms of concentration detection limits was in nM levels. On the other hand, a thermal lens optical design was optimized for microscopic measurements in microchannels [213]. In this work, the possibilities of fundamental on-chip investigations of chemical reactions were tested on an example of determination of rate constants of formation and dissociation of ferroin at the level of $n \times 10^{-8}$ M, which corresponds to the limit of determination of iron(II). Finally, a TLM detection of CE utilizing microchip technology was developed showing the future possibilities of these systems [214]. Capillaries with an inner diameter of 50 μm were directly connected to a microchannel in a

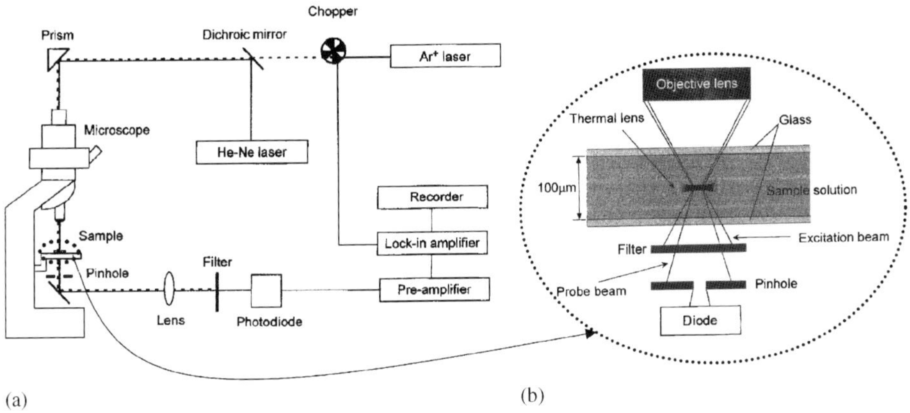

Fig. 5.34. A schematic illustration of the TLM system. (a) Schematic diagram of the experimental setup; (b) schematic drawing of the thermal-lens measurement in a microchannel. Adapted from Ref. [210] with permission of The Japan Society for Analytical Chemistry.

microchip. The LOD of derivatized amino acids determined by CE-TLM was estimated as 2.4×10^{-8} M, which was 100 times lower than that of conventional absorbance detection.

Finally, four groups have developed absorbance detectors based on the whole-channel absorbance imaging detection [105,215–218]. This device allows the direct visualization of the entire separation channel, and therefore, the complete image of processes such as isoelectric focusing (IEF) [105,215] or separation [216–218]. Thus, for example, the feasibility of IEF performed on a quartz chip was demonstrated for the first time via absorption imaging detection at 280 nm [215]. The separation channels were 4 cm long, 100 μm wide, and 10 μm deep, and were coated with linear polyacrylamide to reduce EOF. Low molecular mass pI markers and myoglobin were selected as model samples. The detection limit was about 30 mg/l or 2.4 ng for myoglobin with an optical path length of 10 μm. Nakanishi et al. [216] developed quartz microchips for electrophoresis and a linear imaging UV detector along with the microchip. These microchips have an optical slit, which cut off the stray light in order to improve the sensitivity of UV absorption detection on the chip. In addition, it was also possible to improve the signal-to-noise ratio by repetitive scanning of linear diode array located along the separation channel, and signal averaging during elimination of the potential. Finally, baseline fast chiral separations (in less than 1 min) of 19 basic and acidic compounds have recently been realized on a quartz microchip electrophoresis with a linear imaging UV-detector [218]. The fastest separation could be performed in 2.5 s, which is to date the fastest separation of enantiomers reported. In addition, successful separation of a mixture of three chiral drugs could be performed in a single run in less than 11 s, and utilizing a separation length of only 12 mm. These results show that microchip electrophoresis has great potential for fast chiral analysis and high-throughput screening.

ACKNOWLEDGMENTS

Authors thank the Ministry of Science and Technology (Spain) for the research project BQU2003-03638.

REFERENCES

1 B.M. Simonet, A. Ríos and M. Valcárcel, *Trends Anal. Chem.*, 22 (2003) 605–614.
2 K. Swinney and D.J. Bornhop, *Electrophoresis*, 21 (2000) 1239–1250.
3 J.P. Landers (Ed.), *Handbook of Capillary Electrophoresis*, CRC Press, NY, 1997.
4 M. Mori, W.Z. Hu, J.S. Fritz, H. Tsue, T. Kaneta and S. Tanak, *Anal. Biolanal. Chem.*, 370 (2001) 429–433.
5 K.C. Waldron and N.J. Dovichi, *Anal. Chem.*, 64 (1992) 1396–1399.
6 D.B. Craig, E. Arriaga, J.C.Y. Wong, H. Lu and N.J. Dovichi, *Anal. Chem.*, 70 (1998) 39A–43A.
7 A.T. Timperman, K. Khatib and J.V. Sweedler, *Anal. Chem.*, 67 (1995) 139–144.
8 S. Kar, P.K. Dasgupta, H. Liu and H. Hwang, *Anal. Chem.*, 66 (1994) 2537–2543.
9 A. Nann and E. Pretsch, *J. Chromatogr. A*, 676 (1994) 437–442.
10 H. Lin, D.K. Xu and H.K. Chen, *J. Chromatogr. A*, 760 (1997) 227–233.
11 K.D. Altria and D. Elder, *J. Chromatogr. A*, 1023 (2004) 1–14.
12 P. Zhang, G.W. Xum, J.H. Xiong, Y.F. Zheng, X.Z. Shi, Q. Yang and F.S. Wei, *J. Sep. Sci.*, 26 (2003) 1527–1532.
13 M. Petersson, K.-G. Wahlund and S. Nilsson, *J. Chromatogr. A*, 841 (1999) 249–261.
14 A. Alnajjar and B. McCord, *J. Pharmaceut. Biomed.*, 33 (2003) 463–473.
15 Y.S. Fung and J.L. Mak, *Electrophoresis*, 22 (2001) 2260–2269.
16 C.-H. Wu, M.-C. Chen, A.-K. Su, P.-Y. Shu, S.-H. Chou and C.-H. Lin, *J. Chromatogr. B*, 785 (2003) 317–325.

17 A. Gaveda, J. Sevcik, J. Psotova, P. Vendar, P. Bartak, P. Adamovsky and V. Simanek, *Electrophoresis*, 22 (2001) 2782–2785.
18 P. Kuban, M. Berg, C. Garcia and B. Karlberg, *J. Chromatogr. A*, 912 (2001) 163–170.
19 L. Arce, P. Kuban, A. Rios, M. Varcárcel and B. Karlberg, *Anal. Chim. Acta*, 390 (1999) 39–44.
20 L. Manganiello, L. Arce, A. Rios and M. Varcárcel, *J. Sep. Sci.*, 25 (2002) 319–327.
21 T. Hirokawa, T. Ichihara, K. Ito and A.R. Timerbaev, *Electrophoresis*, 24 (2003) 2328–2334.
22 K. Ito, T. Ichihara, H. Zhuo, K. Kumamoto and A.R. Timerbaev, *Anal. Chim. Acta*, 497 (2003) 67–74.
23 L.Y. Zhu and H.K. Lee, *Anal. Chem.*, 73 (2001) 3065–3072.
24 J.P. Quirino and S. Terabe, *Anal. Chem.*, 72 (2000) 1023–1030.
25 M.-C. Chen, S.-H. Chou and C.-H. Lin, *J. Chromatogr. B*, 801 (2004) 347–353.
26 A. Rudnev, B. Spivakov and A. Timerbaev, *Chromatographia*, 52 (2000) 99–102.
27 K. Isoo and S. Terabe, *Anal. Chem.*, 75 (2003) 6789–6798.
28 W. Buchberger, C.W. Klampfl, F. Eibensteiner and K. Buchgraber, *J. Chromatogr. A*, 766 (1997) 197–203.
29 J. Michelle and J.S. Fritz, *HRC-J. High Res. Chrom.*, 20 (1997) 653–656.
30 B. Sun, M. Macka and P.R. Haddad, *Electrophoresis*, 23 (2002) 2430–2438.
31 S. Emara, I. Darwish, D. Youssef and T. Masujima, *Biomed. Chromatogr.*, 18 (2004) 21–27.
32 S.F.Y. Li, *Capillary Electrophoresis: Principles, Practice and Applications*, Elsevier, Amsterdam, 1992.
33 T.M. Olefirowicz, A.G. Ewing. In:, P.D. Grossman and J.C. Colburn (Eds.), *Capillary Electrophoresis. Detection Methods in Capillary Electrophoresis.* Academic Press, San Diego, 1992.
34 L. Cruz, S.A. Shippy and J.V. Sweedler. In: M.G. Khaledi (Ed.), *High Performance Capillary Electrophoresis. Capillary Electrophoretic Detectors Based on Light.* John Wiley & Sons, New York, 1998.
35 H. Poppe and X. Xu. In: M.G. Khaledi (Ed.), *High Performance Capillary Electrophoresis. Indirect Detection in Capillary Electrophoresis*, John Wiley & Sons, New York, 1998.
36 P. Thibault and N.J. Dovichi. In: P. Camilleri (Ed.), *Capillary Electrophoresis. General Instrumentation and Detection Systems.* CRC Press, Boca Raton, 1998.
37 W.G. Kuhr, *Anal. Chem.*, 62 (1990) 403R–414R.
38 W.G. Kuhr and C.A. Monnig, *Anal. Chem.*, 64 (1992) 389R–407R.
39 C.A. Monnig and R.T. Kennedy, *Anal. Chem.*, 66 (1994) 280R–314R.

40 R.L.St. Claire III, *Anal. Chem.*, 68 (1996) 569R–586R.
41 S.C. Beale, *Anal. Chem.*, 70 (1998) 279R–300R.
42 W.W. Buchberger, *J. Chromatogr. A*, 884 (2000) 3–22.
43 K. Swinney and D.J. Bornhop, *Crit. Rev. Anal. Chem*, 30 (2000) 1–30.
44 M.A. Schwarz and P.C. Hauser, *Lab on a Chip*, 1 (2001) 1–6.
45 F.-M. Matysik, *Electrophoresis*, 23 (2002) 400–407.
46 Y.-M. Liu and J.-K. Cheng, *Electrophoresis*, 24 (2003) 1993–2012.
47 T. Owen, *Fundamentals of UV-Visible Spectroscopy*, Hewlett-Packard, Germany, 1996.
48 D.N. Heiger, *High Performance Capillary Electrophoresis*, Agilent Technologies, Germany, 2000.
49 "*E-seminars*" in http://www.agilent.com/chem/education
50 Y. Walbroehl and J.W. Jorgenson, *J. Chromatogr.*, 315 (1984) 135–143.
51 T. de Beer, N.H. Velthorst, U.A.Th. Brinkman and C. Gooijer, *J. Chromatogr. A*, 971 (2002) 1–35.
52 S.J. Kok, G.Ph. Hoornweg, T. de Ridder, U.A.Th. Brinkman, N.H. Velthorst and C. Gooijer, *J. Chromatogr. A*, 806 (1998) 355–360.
53 Y. Xue and E.S. Yeung, *Appl. Spectrosc.*, 48 (1994) 502–506.
54 A.K. Malik and W. Faubel, *Chem. Soc. Rev.*, 29 (2000) 275–282.
55 S. Nikamura and W. Kaender, *Laser Focus World*, April 1999.
56 http://www.nichia.co.jp
57 P.K. Dasgupta, I.Y. Eom, K.J. Morris and J. Li, *Anal. Chim. Acta*, 500 (2003) 337–364.
58 C.B. Boring and P.K. Dasgupta, *Anal. Chim. Acta*, 342 (1997) 123–132.
59 P.A.G. Butler, B. Mills and P.C. Hauser, *Analyst*, 122 (1997) 949–953.
60 G.E. Collins and Q. Lu, *Anal. Chim. Acta*, 436 (2001) 181–189.
61 M. Macka, P. Andersson and P.R. Haddad, *Electrophoresis*, 17 (1996) 1898–1905.
62 N. Vachirapatama, P. Doble, Z.S. Yu, M. Macka and P.R. Haddad, *Anal. Chim. Acta*, 434 (2001) 301–307.
63 N. Vachirapatama, M. Macka and P.R. Haddad, *Anal. Bioanal. Chem.*, 374 (2002) 1082–1085.
64 C. Johns, M.J. Shaw, M. Macka and P.R. Haddad, *Electrophoresis*, 24 (2003) 557–566.
65 M. King, B. Paull, P.R. Haddad and M. Macka, *Analyst*, 127 (2002) 1564–1567.
66 B. Paull and M. King, *Electrophoresis*, 24 (2003) 1892–1934.
67 V. Galli, A. Garcia, L. Saavedra and C. Barbas, *Electrophoresis*, 24 (2003) 1951–1981.
68 C. Johns, M. Macka and P.R. Haddad, *Electrophoresis*, 24 (2003) 2150–2167.
69 G.J.M. Bruin, G. Stegeman, A.C. van Asten, X. Xu, J.C. Kraak and H. Poppe, *J. Chromatogr.*, 559 (1991) 163–181.

70 F. Foret, D.P. Kirby, P. Vouros and B.L. Karger, *Electrophoresis*, 17 (1996) 1829–1832.
71 P. Lindberg, A. Hanning, T. Lindberg and J. Roeraade, *J. Chromatogr. A*, 809 (1998) 181–189.
72 W. Beck, R. van Hoek and H. Engelhardt, *Electrophoresis*, 14 (1993) 540–546.
73 A.E. Bruno, E. Gassman, N. Periclès and K. Anton, *Anal. Chem.*, 61 (1989) 876–883.
74 A.E. Bruno, R. Jun and H.M. Widmer, *Anal. Chim. Acta*, 234 (1990) 259–262.
75 C. Johns, M. Macka, P.R. Haddad, M. King and B. Paull, *J. Chromatogr. A*, 927 (2001) 237–241.
76 K. Kitagishi and Y. Sato, *Electrophoresis*, 22 (2001) 3395–3400.
77 T. Tsuda, J.V. Sweedler and R.N. Zare, *Anal. Chem.*, 62 (1990) 2149–2152.
78 http://www.polymicro.com/products/products_capillarytubing.htm
79 S. Kim, W. Kim and J.H. Hahn, *J. Chromatogr. A*, 680 (1994) 109–116.
80 S. Liu and P.K. Dasgupta, *Anal. Chim. Acta*, 283 (1993) 747–753.
81 Y. Xue and E.S. Yeung, *Anal. Chem.*, 66 (1994) 3575–3580.
82 N.M. Djordjevic and K. Ryan, *J. Liq. Chromatogr. R. T.*, 19 (1996) 201–206.
83 N.M. Djordjevic, M. Widder and R. Kuhn, *HRC-J. High Res. Chrom.*, 20 (1997) 189–192.
84 D.N. Heiger, P. Kaltenbach and H-J.P. Sievert, *Electrophoresis*, 15 (1994) 1234–1247.
85 Application Note, number 5963–1889E, Agilent Technologies, 1994.
86 J.P. Chervet, R.E.J. van Soest and M. Ursem, *J. Chromatogr.*, 543 (1991) 439–449.
87 P. Jandik and W.R. Jones, *J. Chromatogr.*, 546 (1991) 431–433.
88 S.E. Moring, R.T. Reel and R.E.J. van Soest, *Anal. Chem.*, 65 (1993) 3454–3459.
89 P. Kaltenbach, G. Ross and D.N. Heiger, *High Performance Capillary Electrophoresis*, Anaheim, CA, 1997.
90 I.H. Grant and W. Steuer, *J. Microcolumn Sep.*, 2 (1990) 74–79.
91 X. Xi and E.S. Yeung, *Appl. Spectrosc.*, 45 (1991) 1199–1203.
92 J.A. Taylor and E.S. Yeung, *J. Chromatogr.*, 550 (1991) 831–837.
93 T. Wang, J.H. Aiken, C.W. Huie and R. Hartwick, *Anal. Chem.*, 63 (1991) 1372–1376.
94 M. Saeed, D.H. Craston, M. Depala, J. Reilly, D. Jackson and S.J. Walton, *J. Chromatogr. A*, 836 (1999) 15–24.
95 K.D. Jackson, S.J. Walton and D. Campbell, *J. Autom. Chem.*, 19 (1997) 145–152.

96 S. Kobayashi, T. Ueda and M. Kikumoto, *J. Chromatogr.*, 480 (1989) 179–184.
97 G.M. Robinson and M.R. Smyth, *Analyst*, 122 (1997) 787–802.
98 Q.L. Mao and J. Pawliszyn, *J. Biochem. Bioph. Meth.*, 39 (1999) 93–110.
99 J. Wu and J. Pawliszyn, *Anal. Chem.*, 64 (1992) 224–227.
100 J. Wu and J. Pawliszyn, *Anal. Chem.*, 64 (1992) 2934–2941.
101 J. Wu and J. Pawliszyn, *Electrophoresis*, 16 (1995) 1474–1478.
102 J. Wu and J. Pawliszyn, *Analyst*, 120 (1995) 1567–1571.
103 X.Z. Wu, J. Wu and J. Pawliszyn, *LC-GC N. Am.*, 19 (2001) 526–545.
104 X.H. Fang, C. Tragas, J.Q. Wu, Q.L. Mao and J. Pawliszyn, *Electrophoresis*, 19 (1998) 2290–2295.
105 X.Z. Wu, N.S.K. Sze and J. Pawliszyn, *Electrophoresis*, 22 (2001) 3968–3971.
106 L. Goodridge, C. Goodridge, J.Q. Wu, M. Griffiths and J. Pawliszyn, *Anal. Chem.*, 76 (2004) 48–52.
107 C.T. Culbertson and J. Jorgenson, *Anal. Chem.*, 70 (1998) 2629–2638.
108 C.T. Culbertson and J. Jorgenson, *J. Microcolumn Sep.*, 11 (1999) 652–662.
109 E.T. Bergström, D.M. Goodall, B. Pokric and N.M. Allinson, *Anal. Chem.*, 71 (1999) 4376–4384.
110 W.J.M. Underberg and J.C.M. Waterval, *Electrophoresis*, 23 (2002) 3922–3933.
111 Q. Lin, R. Zhang and G. Liu, *J. Liq. Chromatogr. R. T.*, 20 (1997) 1123–1137.
112 J. Plocek and M.V. Novotny, *J. Chromatogr. A*, 757 (1997) 215–223.
113 A. Mainka and K. Bachmann, *J. Chromatogr. A*, 767 (1997) 241–247.
114 E.A. Pereira, M.O. Rezende and M.F.M. Tavares, *J. Sep. Sci.*, 27 (2004) 28–32.
115 S. Oguri, T. Fujiyoshi and Y. Miki, *Analyst*, 121 (1996) 1683–1688.
116 R.M. Latorre, S. Hernández-Cassou and J. Saurina, *J. Chromatogr. A*, 934 (2001) 105–112.
117 V. Paliulionyte and A. Padarauskas, *Anal. Chim. Acta*, 466 (2002) 133–139.
118 B.F. Liu, L.B. Liu and J.K. Cheng, *J. Chromatogr. A*, 834 (1999) 277–308.
119 V. Pacáková, P. Coufal and K. Stulík, *J. Chromatogr. A*, 834 (1999) 257–275.
120 I. Ali and H.Y. Aboul-Enein, *Crit. Rev. Anal. Chem.*, 32 (2002) 337–350.
121 L. Yan-Ming and C. Jie-Ke, *Electrophoresis*, 24 (2003) 1993–2012.
122 P. Jones and S. Hardy, *J. Chromatogr. A*, 765 (1997) 345–352.
123 J.-F. Jen, M.-H. Wu and T.C. Yang, *Anal. Chim. Acta*, 339 (1997) 251–257.

124 I. Kitazumi, Y. Nakashima and S. Himeno, *J. Chromatogr. A*, 939 (2001) 123–129.
125 Z. Chen and R. Naidu, *Anal. Biolanal. Chem.*, 374 (2002) 520–525.
126 S. Pozdniakova, A. Padarauskas and G. Schwedt, *Anal. Chim. Acta*, 351 (1997) 41–48.
127 W. Liu and H.K. Lee, *Anal. Chem.*, 70 (1998) 2666–2675.
128 W. Liu and H.K. Lee, *J. Chromatogr. A*, 796 (1998) 385–395.
129 W. Liu and H.K. Lee, *Electrophoresis*, 20 (1999) 2475–2483.
130 P. Janos, *Electrophoresis*, 24 (2003) 1982–1992.
131 J. Hamácek and J. Havel, *J. Chromatogr. A*, 834 (1999) 321–327.
132 Z. Chen, R. Naidu and A. Subramanian, *J. Chromatogr. A*, 927 (2001) 219–227.
133 E. Naujalis and A. Padarauskas, *J. Chromatogr. A*, 977 (2002) 135–142.
134 A.R. Timerbaev, O.P. Semenova and J.S. Fritz, *J. Chromatogr. A*, 756 (1996) 300–306.
135 A. Riaz, B. Kim and D.S. Chung, *Electrophoresis*, 24 (2003) 2788–2795.
136 A.J. Gawron and S.M. Lunte, *Electrophoresis*, 21 (2000) 2067–2073.
137 Y. Daali, K. Bekkouche, S. Cherkaoui, P. Christen and J.L. Veuthey, *J. Chromatogr. A*, 903 (2000) 237–244.
138 X. Cahaours, Y. Daali, S. Cherkaoui and J.L. Veuthey, *Chromatographia*, 55 (2002) 211–216.
139 P. Doble and P.R. Haddad, *J. Chromatogr. A*, 834 (1999) 189–212.
140 E.S. Yeung, *Acc. Chem. Res.*, 22 (1989) 125–130.
141 M. Macka and P.R. Haddad, *Electrophoresis*, 18 (1997) 2482–2501.
142 T. Wang and R.A. Hartwick, *J. Chromatogr.*, 607 (1992) 119–125.
143 P. Doble, M. Macka and P.R. Haddad, *Trends Anal. Chem.*, 10 (2000) 10–17.
144 P. Doble and P.R. Haddad, *Anal. Chem.*, 71 (1999) 15–22.
145 V. Pacáková and K. Stulík, *J. Chromatogr. A*, 789 (1997) 169–180.
146 S. Hjerten, K. Elenbring, F. Kilar, J. Liao, A.J. Chen, C.J. Siebert and M.J. Zhu, *J. Chromatogr.*, 403 (1987) 47–61.
147 A.R. Timerbaev, *Electrophoresis*, 23 (2002) 3884–3906.
148 V. Pacáková, P. Coufal, K. Stulik and B. Gas, *Electrophoresis*, 24 (2003) 1883–1891.
149 D. Kaniansky, M. Masar, J. Marak and R. Bodor, *J. Chromatogr. A*, 834 (1999) 133–178.
150 A.G. Diress and C.A. Lucy, *J. Chromatogr. A*, 1027 (2004) 185–191.
151 W.R. Jones and P. Jandik, *J. Chromatogr.*, 546 (1991) 445–458.
152 B.C. Nelson, G.F. Rockwell, T. Campfield, P. O'Grady, R.M. Hernandez and S.A. Wise, *Anal. Chim. Acta*, 410 (2000) 1–10.
153 S. Motellier and M. Descostes, *J. Chromatogr. A*, 907 (2001) 329–335.
154 A.R. Timerbaev and O.A. Shpigun, *Electrophoresis*, 21 (2000) 1093–1103.

155 S. Suzuki and S. Honda, *Electrophoresis*, 19 (1998) 2539–2560.
156 N.K. Karamanos and F. Lamari, *Biomed. Chromatogr.*, 13 (1999) 501–506.
157 Z. El Rassi, *Electrophoresis*, 20 (1999) 3134–3144.
158 M.J. Markuszewski, K. Otsuka, S. Terabe, K. Matsuda and T. Nishioka, *J. Chromatogr. A*, 1010 (2003) 113–121.
159 M.A.L. de Oliveira, V.E.S. Solis, L.A. Gioielli, B. Polakiewicz and M.F.M. Tavares, *Electrophoresis*, 24 (2003) 1641–1647.
160 H. Okamoto, Y. Okamoto, T. Hirokawa and A.R. Timerbaev, *Analyst*, 128 (2003) 1439–1442.
161 M. Dzierzgowska, K. Pyrzyñska and E. Pobozy, *J. Chromatogr. A*, 984 (2003) 291–295.
162 M. De Valme Garcia-Moreno, R.C. Mejias, R.N. Marin and C.G. Barroso, *Eur. Food Res. Technol.*, 215 (2002) 255–259.
163 E. Pobozy, M. Jarczynska and M. Trojanowicz, *Chromatographia*, 56 (2002) 723–728.
164 T.M.H. Choy, L. Jia and C.W. Huie, *J. Sep. Sci.*, 25 (2002) 333–341.
165 G. Zunic, Z. Jelic-Ivanovic, M. Colic and S. Spasic, *J. Chromatogr. B*, 772 (2002) 19–33.
166 J.E. Melanson, B.L.Y. Wong, C.A. Boulet and C.A. Lucy, *J. Chromatogr. A*, 920 (2001) 359–365.
167 M.J. Chen, H.S. Chen, C.Y. Lin and H.T. Chang, *J. Chromatogr. A*, 853 (1999) 171–180.
168 P. Doble, M. Macka and P.R. Haddad, *J. Chromatogr. A*, 804 (1998) 327–336.
169 C. Johns, M. Macka and P.R. Haddad, *Electrophoresis*, 23 (2002) 43–48.
170 M. Franko, *Talanta*, 54 (2001) 1–13.
171 T. Kitamori, M. Tokeshi, A. Hibara and K. Sato, *Anal. Chem.*, 76 (2004) 52A–60A.
172 N.J. Dovichi and J.M. Harris, *Anal. Chem.*, 51 (1979) 728–731.
173 D.J. Bornhop and N.J. Dovichi, *Anal. Chem.*, 59 (1987) 1632–1636.
174 N.J. Dovichi, *CRC Crit. Rev. Anal. Chem.*, 17 (1987) 357–423.
175 M. Yu and N.J. Dovichi, *Microchim. Acta*, 3 (1988) 27–40.
176 M. Yu and N.J. Dovichi, *Anal. Chem.*, 61 (1989) 37–40.
177 M. Yu and N.J. Dovichi, *Appl. Spectrosc.*, 43 (1989) 196–201.
178 B. Krattiger, A.E. Bruno, H.M. Widmer and R. Daendliker, *Anal. Chem.*, 67 (1995) 124–130.
179 J.M. Saz, B. Krattiger, A.E. Bruno, J.C. Diez-Masa and H.M. Widmer, *J. Chromatogr. A*, 699 (1995) 315–322.
180 M. Chen, K.C. Waldron, Y. Zhao and N.J. Dovichi, *Electrophoresis*, 15 (1994) 1290–1294.
181 B.S. Seidel and W. Faubel, *J. Chromatogr. A*, 817 (1998) 223–226.
182 B.S. Seidel and W. Faubel, *Fresen. J. Anal. Chem.*, 360 (1998) 795–797.

183 B.S. Seidel, W. Faubel and H.J. Ache, *J. Biomed. Opt.*, 2 (1997) 326–331.
184 N. Ragozina, S. Heissler, W. Faubel and U. Pyell, *Anal. Chem.*, 74 (2002) 4480–4487.
185 W. Faubel, S. Heissler, U. Pyell and N. Ragozina, *Rev. Sci. Instrum.*, 74 (2003) 491–494.
186 X.-F. Li, C.-S. Liu, P. Roos, E.B. Hansen Jr., C.E. Cerniglia and N.J. Dovichi, *Electrophoresis*, 19 (1998) 3178–3182.
187 K. Waldron and J. Li, *J. Chromatogr. B: Biomed. Appl.*, 683 (1996) 47–54.
188 J. Ren, B. Li, Y. Deng and J. Cheng, *Talanta*, 42 (1995) 1891–1895.
189 A.E. Bruno, A. Paulus and D.J. Bornhop, *Appl. Spectrosc.*, 45 (1991) 462–467.
190 P. Hess and J. Pelzl, *Photoacoustic and Photothermal Phenomena*, Springer, New York, 1987.
191 J. Wu, T. Kitamori and T. Sawada, *Anal. Chem.*, 62 (1990) 1676–1678.
192 T.J. Pang and M.D. Morris, *Anal. Chem.*, 57 (1985) 2153–2155.
193 M.J. Sepaniak, J.D. Vargo, C.N. Kettler and M.P. Maskarinec, *Anal. Chem.*, 56 (1984) 1252–1257.
194 J. Wu, T. Odake, T. Kitamori and T. Sawada, *Anal. Chem.*, 63 (1991) 2216–2218.
195 T. Odake, T. Kitamori and T. Sawada, *Anal. Chem.*, 64 (1992) 2870–2871.
196 T. Odake, T. Kitamori and T. Sawada, *Anal. Chem.*, 67 (1995) 145–148.
197 T. Odake, T. Kitamori and T. Sawada, *Anal. Chem.*, 69 (1997) 2537–2540.
198 H.D. Willauer and G.E. Collins, *Electrophoresis*, 24 (2003) 2193–2207.
199 G.N. Doku and S.J. Haswell, *Anal. Chim. Acta*, 382 (1999) 1–13.
200 G.M. Greenway, S.J. Haswell and P.H. Petsul, *Anal. Chim. Acta*, 387 (1999) 1–10.
201 P.H. Petsul, G.M. Greenway and S.J. Haswell, *Anal. Chim. Acta*, 428 (2001) 155–161.
202 A. Daridon, M. Sequeira, G. Pennarun-Thomas, H. Dirac, J.P. Krog, P. Gravesen, J. Lichtenberg, D. Diamond, E. Verpoorte and N.F. de Rooij, *Sensor. Actuator. B*, 76 (2001) 235–243.
203 Q. Lu and G.E. Collins, *Analyst*, 126 (2001) 429–432.
204 G.E. Collins and Q. Lu, *Sensor. Actuator. B*, 76 (2001) 244–249.
205 L. Ceriotti, K. Weible, N.F. De Rooij and E. Verpoorte, *Microelectronic Engineering*, 67–68 (2003) 865–871.
206 K.B. Mogensen, N.J. Petersen, J. Hubner and J.P. Kutter, *Electrophoresis*, 22 (2001) 3930–3938.
207 E. Verpoorte, A. Manz, H. Lüdi, A.E. Bruno, F. Maystre, B. Krattiger, H.M. Widmer, B.H. van der Schoot and N.F. de Rooij, *Sensor. Actuator. B*, 6 (1992) 66–70.

208 H. Salimi-Moosavi, Y.T. Jiang, L. Lester, G. McKinnon and D.J. Harrison, *Electrophoresis*, 21 (2000) 1291–1299.
209 K. Mawatari, T. Kitamori and T. Sawada, *Anal. Chem.*, 70 (1998) 5037–5041.
210 K. Sato, H. Kawanishi, M. Tokeshi, T. Kitamori and T. Sawada, *Anal. Sci.*, 15 (1999) 525–529.
211 www1.odn.ne.jp/imt
212 M. Tokeshi, M. Uchida, A. Hibara, T. Sawada and T. Kitamori, *Anal. Chem.*, 73 (2001) 2112–2116.
213 M.A. Proskurnin, M.N. Slyadnev, M. Tokeshi and Kitamori, *Anal. Chim. Acta*, 480 (2003) 79–95.
214 K. Uchiyama, A. Hibara, K. Sato, H. Hisamoto, M. Tokeshi and T. Kitamori, *Electrophoresis*, 24 (2003) 179–184.
215 Q. Mao and J. Pawliszyn, *Analyst*, 124 (1999) 637–641.
216 H. Nakanishi, T. Nishimoto, A. Arai, H. Abe, M. Kanai, Y. Fujiyama and T. Yoshida, *Electrophoresis*, 22 (2001) 230–234.
217 Y. Mizukami, D. Rajniak, A. Rajniak and M. Nishimura, *Sensor. Actuator. B*, 81 (2002) 202–209.
218 M. Ludwig, F. Kohler and D. Belder, *Electrophoresis*, 24 (2003) 3233–3238.

Chapter 6

Fluorescence detection in capillary electrophoresis

Maria Teresa Veledo, Pilar Lara-Quintanar, Mercedes de Frutos and Jose Carlos Díez-Masa

6.1 INTRODUCTION

Fluorescence detection has been used since the introduction of analytical chromatographic methods that employ liquid mobile phases [1–3], where it has proved to be a very sensitive detection technique. This also applies to capillary electrophoresis (CE). As will be shown later, to obtain the good sensitivity that fluorescence is able to produce, a laser should be used as light source in microsystems (microcapillaries and microchip channels). In fact, fluorescence detection has become "laser-induced fluorescence" (LIF) monitoring in such separation systems. Using LIF, single-molecule detection has been demonstrated [4,5]. The good sensitivity of LIF detection and the extremely low amount of sample that can be manipulated in CE have allowed the analysis of proteins and other compounds at the level of a single cell, which can lead to new paradigms in physiological studies [6,7]. LIF detection has also allowed the development of fast DNA sequencers based on multicapillary CE, which facilitated the work of the Human Genome Project, and opened up new perspectives in clinical and forensic analysis [8]. Last but not least, fluorescence detection is considered one of the methods of choice for detection in CE analysis carried out by lab-on-a-chip technology (microchips).

6.1.1 Objectives and outline of this chapter

This chapter presents a brief survey of the most important instrumentation aspects and applications of LIF detection in electrophoresis

Comprehensive Analytical Chemistry XLV
M.L. Marina, A. Ríos and M. Valcárcel (Eds)
Volume XLV ISSN: 0166-526X DOI: 10.1016/S0166-526X(05)45006-0

performed in capillaries and microchips. In each section one of the LIF detection topics is discussed, and its basic aspects are presented together with some fundamental references. Since an exhaustive review is beyond the scope of this chapter, we discuss only some application examples in each section and we present an extensive selection in section 6.7.

We start with a brief presentation of the photochemical principles of fluorescence. Then, the main subsystems used for LIF instrumentation are described, including light sources, optical train components, and the main devices used as light measuring detectors. We also present the different configurations employed in LIF detection, such as orthogonal, collinear, and mirror setups, discussing their advantages and limitations. Since only a few substances have native fluorescence at the wavelength of moderate-price lasers, derivatization of the analytes is, in most of the cases, the only way to achieve good sensitivity using LIF detection. Therefore, a section is devoted to the different strategies to carry out derivatization in CE. The fluorescence detection techniques used for monitoring CE separations in microchips, and particularly those that allow the integration of LIF detection in microchips, are reviewed. Although not as sensitive as direct LIF, indirect LIF detection has also been used, and its basic principles and applications are discussed. Finally, a group of LIF techniques with high potential in the near future are presented in section 6.6.

6.1.2 Basic concepts

Fluorescence is the process of spontaneous emission of light from a molecule (or an atom) when it experiences a spontaneous transition from the excited to the fundamental state. When a molecule absorbs light in the electromagnetic spectrum, it passes from the so-called fundamental state to one of the singlet-excited states. From this state, the molecule drops back to the fundamental state by losing energy via two different types of deactivation processes—radiation and radiationless processes. Radiationless processes occur owing to collisions between molecules of the sample and the solvent, causing losses of energy without light emission. Radiation processes are accompanied by the emission of electromagnetic radiation, usually of lower energy (longer wavelength) than that used for excitation. If the emission process takes place in a short period of time after the excitation ($\sim 10^{-9}$ s), the process is called fluorescence emission.

In fluorescence experiments, there are several variables that control the amount of light produced by a given molecule, such as molar absortivity of the analyte at the wavelength of the incident light, quantum yield of the molecule, temperature, nature of the solvent, and amount of oxygen dissolved in the sample [9].

It can be shown that the radiant power of the fluorescence generated by a sample, P_f, is given by

$$P_f = \Phi P_0 abc \varepsilon_x \varepsilon_c \varepsilon_m \varepsilon_{PMT} \tag{6.1}$$

where P_0 is the optical power of the incident light on the capillary; Φ the quantum yield of the molecule (photons emitted/photons absorbed); a, b, and c are the Lambert–Beer law terms; and ε terms are the efficiency of x, the excitation optics; c, the capillary illuminated (detection cell); m, the collection optics; and PMT, the light detection device (typically a photomultiplier), respectively. The equation shows that the signal achieved in fluorescence experiments in CE depends on the power of the incident light at the entrance of the excitation optics, which involves only the light radiated in the direction of capillary where detection takes place, the nature of the molecule (terms Φ and a in the above equation), the optical setup features, and the concentration of the analyte at the detection point. It should be emphasized that if the incident power exceeds a given value, characteristic of each molecule, fluorescence emission is saturated (the power of the emitted light does not increase linearly with P_0 anymore), and even the molecules can experience photodegradation (photobleaching).

6.2 INSTRUMENTAL SETUP

6.2.1 Light sources

To excite the molecule of interest from the fundamental state to the excited state, a light source of the appropriate intensity, wavelength, and spatial characteristics is necessary. Coherent (lasers) and noncoherent (lamps) light sources could be theoretically suitable for this purpose in fluorescence detection for capillary systems; however, the simple calculation that follows demonstrates that it is not so for fluorescence detection in CE (this can be extended to other capillary separation techniques, i.e., capillary HPLC or detection in microchips). To produce fluorescence, light from the source has to be imaged onto the capillary using an optical system. The optical power (P_0) that can be imaged onto the capillary is given by [10]

$$P_0 = E_{t \cdot \Delta\lambda} A_s \Phi_{col} T_f \tag{6.2}$$

where E_t is the spectral radiance of the lamp in units of W/m^2/str at the spectral region $\Delta\lambda$, A_s the illuminated area of the sample (detection window), Φ_{col} the collection efficiency of the illuminating optics, and T_f the transmittance of the filter or monochromator used. Assuming that the spectral radiance of the lamp used is around 20×10^3 W/m^2/str in the range 400–500 nm, the detection area in the capillary is 5×10^{-9} m^2 (a detection window of 100×50 μm), the illuminating optics is able to collect 25% of the light produced by the lamp, and the filter used transmits 60% of the light incident, the power reaching the capillary is around 10–20 μW. Using a moderate-price laser, for instance, an air-cooled Ar-ion laser, and assuming that (i) an output of 1 mW can be obtained at 488 nm with a beam diameter of 2 mm, (ii) a $\times$ 20 microscope with transmission efficiency of 70% is used to focus the laser on approximately the same detection area of the capillary as did the lamp before, and (iii) owing to the monochromaticity of the laser light, neither a spectral filter nor a monochromator is not used, about 200–300 mW can reach the detection point. That is, 10^4 times higher optical power can be obtained using a laser that uses a lamp as light source, which means that fluorescence emission in CE can be more efficiently achieved by using a laser than by using a lamp. Practical experience confirms that lasers are the light source of choice for fluorescence detection in capillary techniques (HPLC, CE, and microchips).

Lasers are used in CE for several different purposes ranging from optical-gated injection [11] to the study of the inner surface of separation capillaries by atomic force microscopy [12]. Of course, the most popular application of lasers in CE is as a light source in several detection techniques, both fluorescent [13,14] and nonfluorescent ones [15].

Lasers produce light in a significantly different way from regular (filament lamp, discharge lamp) light sources. One essential physical phenomenon in laser radiation production is the process of stimulated emission together with spontaneous absorption and spontaneous emission. This leads to the Einstein relations between the rate coefficients of these processes. The creation of a population inversion between the ground state and the metastable excited state of the active medium (the gas, liquid, or solid used to produce the light) causes the optical gain. In most lasers, the active medium is inside an optical resonator to provide a long gain path length. In this way the optical pumping can be produced so that the inversion of population in the media is maintained. Part of the light inside the resonator is drained as a laser beam [16].

Owing to the very process of laser production, laser beams are powerful, directional, and coherent. Thanks to these features, laser beams can be focused into a small spot (typically of a few microns), rendering them very useful for on-capillary detection. Besides, laser light is monochromatic, which allows relatively simple setups (without monochromators or excitation filters) for LIF detection. However, broadly speaking, lasers are expensive (although this is not the case for diode lasers anymore) and have a limited life.

For the sake of simplicity a short review of lasers will be given in this chapter. To know more about the particular features of the different types of lasers, the readers are recommended to consult a good monograph such as Ref. [17]. Lasers can be classified according to different criteria. Depending on the active medium used in the resonator to produce the radiation, lasers can be classified into three categories: gas lasers, solid-state lasers, and semiconductor (diode) lasers. In Fig. 6.1, the output power and wavelengths of various lasers often used in CE are summarized. It is worth mentioning that in recent years, new solid-state lasers and diode lasers with good characteristics (wavelength, power, price, etc.) for LIF detection in CE and microchips have been commercialized. In particular, pulsed diode-pumped solid-state (DPSS)

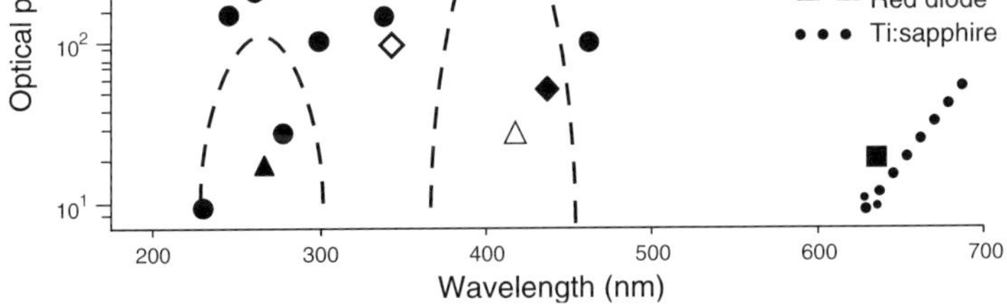

Fig. 6.1. Output power and wavelengths of various laser systems used in LIF detection. Reprinted from Ref. [15]. Copyright (2002), with permission from Elsevier.

UV lasers are solid-state lasers producing very short (<500 ps) high-frequency (100 Hz–2 kHz) pulses of UV radiation (266 nm), which open the way toward a more universal LIF detection in CE analysis [18] because they deliver up to 10 μJ per pulse (average power 20 mW at 2 kHz) at a reasonable price. On the other hand, continuous emission laser diodes in the blue-violet range (400–440 nm) have been commercialized [19,20]. They produce an output of 1–5 mW with a 5 V DC (1 A) power supply. The application of such lasers for LIF detection has been reported using either derivatization of the analyte [21] or indirect detection [22], and they open up new perspectives on portable, miniaturized CE instrumentation.

Finally, light emission diodes (LED), although seldom used in LIF detection, have good characteristics for LIF detection in miniaturized systems. LEDs are solid-state light (noncoherent) sources requiring low driving power (usually 5 V and 20 mA), displaying excellent intensity stability of around $\Delta I/I < 10^{-5}$ (i.e., as good as the current power supply used), and having small size and low price. High-brightness blue LEDs are now available. Most of the benefits of the use of LEDs in chemical sensors have been reported [23]. Light from LEDs can be easily focalized using gradient index (GRIN) lenses, giving rise to the pigtail concept [24] that has been applied to fluorescence detection in CE. The pigtail concept will be presented in more detail in section 6.4.2.1, where the potential application of these systems in microchip-size devices will be considered.

6.2.2 Optical train for LIF

To produce fluorescence in LIF detection, the light from the laser has to be properly focused on the detection window at the outlet end of the capillary, and the fluorescence produced by each analyte has to be imaged on the light measuring device. The careful selection and positioning of the optical components of this train are important in order to collect and transmit as much light as possible from the sample while rejecting as much as possible scatter and Raman light. It should be considered that a very small amount of light (a few photons) is produced by very dilute samples, and that the efficiency of collection optics is a must for good detection sensitivity. Readers wishing to obtain deeper knowledge about the role of each optical component of the optical train are recommended to read a general textbook about optical engineering [25] or some specialized papers [26,27]. We will now discuss briefly the different parts of a LIF detector and the arrangements (orthogonal and collinear) used.

6.2.2.1 Orthogonal arrangement

In this setup, the collection of the fluorescence from the capillary is carried out in a geometric plane placed perpendicular to the plane containing the excitation beam. A typical LIF orthogonal setup is composed of excitation optics, detection cell, and collection optics.

Excitation optics

The laser beam can be directed toward the separation capillary surface using laser quality mirrors. Mirrors with good reflectance at the laser wavelength should be used, although reflectance is not usually a limiting factor because most of the lasers used in LIF detection have more power than necessary to cause fluorescence in the capillary. The main drawback in using mirrors is that owing to the long path of the rays from the laser source to the capillary, small mechanical vibrations of the system can cause an important noise in the detector. In such cases, a good optical table with either active or passive isolating support and a honeycomb tabletop should be used. At least a breadboard on a solid table should be employed. Also, solid optical mounts must be used. Two minor drawbacks in using mirrors are that some skillfulness is necessary for the alignment of the beam with the focusing optics and the capillary, and good opto-mechanical hardware has to be used to maintain this alignment for a long period of time. Most of these limitations can be avoided by using optical fibers to collect the light from the laser tube and to conduct it to the separation capillary [28] or to the focusing optics. Optical fibers with good transmittance in the visible and UV regions of the spectra, connectors, and centering systems can be easily designed and assembled [29].

Since lasers have a beam diameter of 1–2 mm, the beam has to be focused using an aberration corrected lens or microscope objective ($\times 10 - \times 20$) to increase the optical power reaching the detection region of the capillary. When lasers are not focused at the diffraction limit, the focal point of a laser is rather large (laser waist) and easy to image on the capillary surface. In some focusing optics, the output of the lens or of the microscope is coupled to a ball lens that effectively focuses the laser light to the size of the internal diameter of the capillary [30].

Detection cell

One component of an LIF detector that is necessary for achieving good sensitivity is the detection cell. As indicated before, optical power of the light emitted depends on the power of the light arriving to the buffer-sample region, the optical path of the detection cell, and the nature and concentration of the analyte. In order to avoid excessive band broadening, the simplest method is to use a small zone of the separation capillary as detection cell. This on-capillary detection method could yield a separation efficiency of more than 10^6 plates/m and is easy to implement. However, the optical path in on-capillary detection is limited by the ID (inner diameter) of the capillary. Hence it is necessary to maximize the light power arriving inside the capillary while minimizing the background (scatter and Raman radiation) arriving in the detection device. Efficient LIF monitoring using on-capillary detection is difficult. On the one hand, an important amount of light is lost owing to reflection or diffraction on the capillary walls [31]. On the other hand, light scattered at the air–capillary and capillary–buffer interfaces causes a large background signal in fluorescence detection. An obscuration bar perpendicular to the capillary surface, tilting the capillary with respect to the laser beam or at the Brewster's angle while using polarized laser light, can be employed to decrease the background. Unfortunately, all these methods also decrease the fluorescence light reaching the detection device.

An alternative to on-capillary detection of fluorescence is the sheath-flow cuvette. In this postcapillary detection cell [32–34], the capillary end is introduced into a 350–400 μm square-shaped flow chamber constructed from good optical quality quartz. A sheath stream of liquid produced by siphoning enters the cell and surrounds the buffer–analyte stream leaving the capillary. At the low flow rate of the buffer in the electrophoretic capillary, a laminar flow can be easily obtained inside the detection cuvette so that no important broadening of the sample band is produced. Since the sheath-flow cuvette has flat windows for the entrance and the exit of light, almost no diffraction is produced at the air–cell and at the cell–sheath liquid interfaces. Besides, since the separation buffer is used as sheath liquid, no scatter takes place at the interfaces between the sheath and the capillary steams. Consequently, sheath-flow cuvettes produce very low light losses and scatter background. A detection limit reaching single-molecule detection has been reported [5] using sheath-flow cuvettes. This efficient method to eliminate scatter has allowed the development of sensitive LIF detection in multiple capillary instruments such as the DNA sequencers [8], and has accelerated the sequentiation of the human genome [35]. Last but not least, the use of the sheath-flow cuvette is an effective method to carry out postcapillary derivatization without excessive band broadening [36], as will be discussed later.

Collection optics

Only the light produced by the analyte in the detection cell should be collected and measured. One of the most convenient ways to do so is by using a high numerical aperture (NA) microscope. The NA of a lens is proportional to the maximum solid angle of light collected by the lens. The higher the NA, the higher the fraction of light produced by the sample collected by the microscope. Using a microscope with an NA of 0.8 (a typical value for a good quality 50 × microscope) only 20% of the light produced in the detection cell is collected. However, this objective should be placed at only 0.45 mm from the capillary (working distance), which imposes some spatial limitations on the light collection setup [27]. A way to avoid this limitation is to use long working distance microscopes, oil-immersion microscopes, or reflection objectives [37]. Nevertheless, the latter objectives are still rather expensive.

Collection efficiencies close to 50% have been reported [38] using a parabolic mirror and a small scatter mask that block only a small fraction of the fluorescence produced. Optical filters and spatial filters are placed between the light collection device (microscope objective or parabolic mirror) and the light measuring device. Optical filters are used to prevent laser, scatter, and Raman light from reaching the detector while allowing the fluorescence produced by the analyte to go through. Normally color (cutoff) filters are used to avoid laser radiation, and interferometric filters and notch filters are used to select wavelength at the maximum of the fluorescence band. High-transmittance and low-autofluorescence filters should be selected. Spatial filters, such as irises or pinholes, are used to image only the light from the internal diameter of the capillary on the light detection device, and to limit the size of the detection window.

Finally, mechanical hardware (mounting stages, positioning stages, X–Y–Z translating stages, etc.) should be carefully selected in order to design a stable and reproducible piece of equipment.

6.2.2.2 Collinear arrangement

An alternative arrangement to orthogonal setup is the collinear arrangement [39–42]. A dichroic mirror is generally employed in a collinear setup. In this mirror, low-wavelength radiations are reflected while high-wavelength radiations pass through. The laser radiation reflected by the dichroic mirror is imaged on the capillary wall using a microscope objective and the fluorescence emitted by the capillary is collected by the same objective and traverses the dichroic mirror. Afterward, a pinhole and different filters are employed to reject scattered light and to select the detection wavelength. Also, special optical devices containing a ball-lens have been used to obtain a high-efficiency fluorescence collection owing to its very high NA. An inverted fluorescence microscope is very useful in a collinear arrangement. Usually, the photocamera of the microscope is replaced by one of the light measuring devices used in LIF detection.

6.2.3 Light measuring devices for LIF

Once the light from the capillary has been collected by the optics and spatially and optically filtered, it has to be measured. Light measurement involves the conversion of light, or photons, into an electrical signal. This process is based on the photoelectric effect [43].

In LIF detection, very low amount of light has to be usually measured, particularly in those cases in which very diluted analytes have to be detected. Traditionally, photomultiplier tubes (PMTs) have been used, but in recent years, electro-optical technology has improved dramatically, and avalanche photodiodes (APDs) and image detectors (such as charge coupled devices (CCDs) and intensified charge coupled devices (ICCDs)) are frequently used.

In this section a brief description of these measuring elements and their main features are presented.

6.2.3.1 Photomultiplier tubes

PMTs [44] have been and still are the work-horse of the photodetection devices in LIF detection because of their high sensitivity and fast response, although for some applications they are being replaced by the APDs. A typical PMT consists of a photoemissive cathode (photocathode), followed by a focusing electrode, an electron multiplier, and an electron collector (anode) in a vacuum tube (approximately at 10^{-9} bar) (Fig. 6.2).

Photons striking the photocathode generate some amount of electrons by photoelectric effect. These are focused and accelerated by the focusing electrode gaining energy toward an electron multiplier

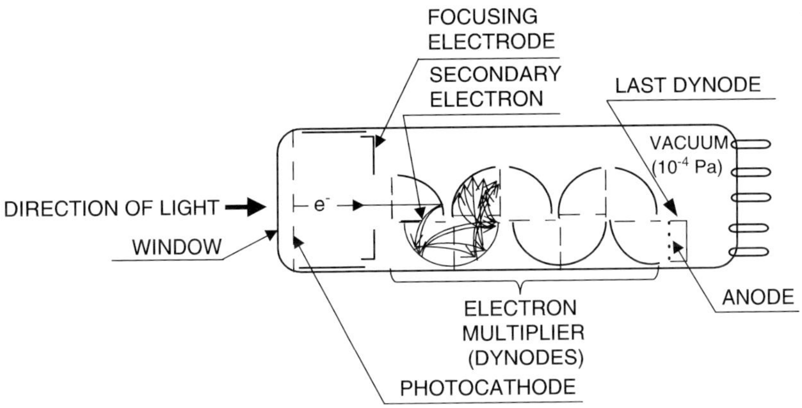

Fig. 6.2. Cross-section of a head-on type PMT.

composed of several (8–19) stages, called dynodes, where a cascade of secondary electrons is generated. In this way one original photoelectron can be amplified greatly (the so-called gain or current amplification). Finally, the electrons are collected in the anode and measured as an electric current between cathode and anode.

Basically, PMTs have two configurations: side-on and head-on. In a side-on configuration, light arrives from one of the sides of the vacuum tube, and in the head-on configuration the light arrives from the front end. In general, side-on PMTs have lower price and are therefore of more general use in light measurements. Head-on configuration presents better spatial uniformity, i.e., the current produced for a given light intensity is more independent of the orientation of the PMT window with respect to the direction of the incident light beam. The configuration used determines the material used for the construction of the photocathode. The photocathode surface is made of different alkali metals (GaAs, InGaAs, SbRbCs, etc.) depending on the spectral response selected. All these materials have very low work function [43]. Most side-on PMTs employ opaque photocathodes that reflect the electrons generated while head-on PMTs use a transparent or semitransparent photocathode that transmits most of the electrons produced. The dynodes of the electron multiplier can be arranged in several ways. In circular-cage type, the most common arrangement for the side-on PMTs, the dynodes are distributed more or less around the photocathode in a compact structure that allows fast response and high gain with a relatively low-voltage polarization. Other types of dynode configurations, such as box-and-grid type and linear-focused type, are often used in head-on PMTs (for a description and discussion of such dynode configurations see Ref. [45]).

The main features to be considered for the choice of a PMT follow. Spectral response is the relationship between photocathode sensitivity and wavelength, and is given by photocathode radiant sensitivity (usually given in mA/W) and the peak wavelength. As mentioned before, spectral response depends on the chemical composition of the photocathode. The gain or current amplification is the ratio of the anode output current to the photoelectric current from the photocathode, and is determined by the collection efficiency of the dynodes and the acceleration between plates by the applied electric field (anode–cathode supplied voltage or bias voltage). Even operating at completely dark state, some amount of current circulates by the PMT circuit (dark current). This current is caused by several effects, such as thermoionic emission of electrons, glass scintillation, and ionization of the residual gases, and it limits the electric noise and, therefore, the detection limit. All these characteristics should be studied carefully when selecting a PMT for LIF detection during the setup design stage.

So far we have considered the linear operation mode of PMTs, where the current produced is proportional to the incident light in the photocathode. However, another effective way to use PMTs in those cases where extremely low light levels are measured is photon counting. If the incident light is so faint that incident photons are temporarily separated from one another, the PMT can produce a burst of current (pulses) for each photon. The number of pulses generated is proportional to the amount of incident light. The photon counting method has the advantage over the linear method of producing a better signal to noise (S/N) ratio in those cases, such as chemiluminescence techniques, in which a very small amount of light is produced by the sample.

6.2.3.2 *Avalanche photodiodes (APDs)*

These photodetectors are solid-state devices that transform the photons into a large number of electrons. In this sense, they operate as PMTs, but the electrons are transported in a doped silicon medium instead of in a vacuum tube [46].

Although APDs have been designed with several different structures, the basic configuration of an APD is composed of two regions, a photon absorption section and an electron gain region. The photon absorption region is a silicon area where the photons are converted into photoelectrons. A small electric field applied to this region drives the photoelectrons to the gain region. In the gain region, a more intense electric field is used to accelerate the electrons, increasing their energy. These electrons of high energy interact with the substrate, producing an increasing number of electrons through impact ionization. In turn, these electrons are accelerated again by the electric field, produce new electrons, and so on. In the whole process, a cascade (avalanche) of electrons is created from a single photon reaching the APD active area.

The APDs, as the PMTs, can operate both in linear mode and in the photon counting mode. In linear mode, low-voltage power supply is required (40–100 V) to polarize the diode, but only moderate gain is achieved (around 100 for APDs versus 10^6 for the PMTs). In the photon counting mode, a voltage higher than the breakdown voltage (~400 V or more) is applied to the APD, and a gain of 10^4–10^8 can be obtained [47].

APDs are advantageous compared to PMTs because they have higher quantum efficiency (>50%) than PMTs (10%), are operated at lower voltage (100 V at maximum in the linear mode) than PMTs (500–1000 V), have small size, and they are relatively inexpensive owing to their mass production.

6.2.3.3 *Imaging detectors: charge coupled devices (CCDs) and intensified charge coupled devices (ICCDs)*

Very often, one single PMT or APD detector is used to measure light in LIF setups. But there are several ways to arrange several detectors in one-dimensional (i.e., photodiode array detectors or linear multianode PMTs) or in two-dimensional arrays (i.e., CCDs, ICCDs, or PMT arrays) to extract physicochemical (spectroscopic or half-life of excited states) information about the analyte, or for more efficient manipulation of the light emitted by the sample, such as in some cases of post-capillary derivatization in CE. In recent years, CCD camera performance and applications have increased substantially, and their price has decreased concomitantly with their widespread use. In this section a brief description of the principles and features of CCD and ICCD detectors will be carried out.

CCD cameras are two-dimensional arrays (columns and rows) of "potential wells," i.e., pixels (*pixel* = the plural for *picture+element*), which retain the electrons generated by photoelectric processes [48] (Fig. 6.3). In CCD functioning, three principles are involved: photodetection, charge storage, and charge transfer, which take place in an

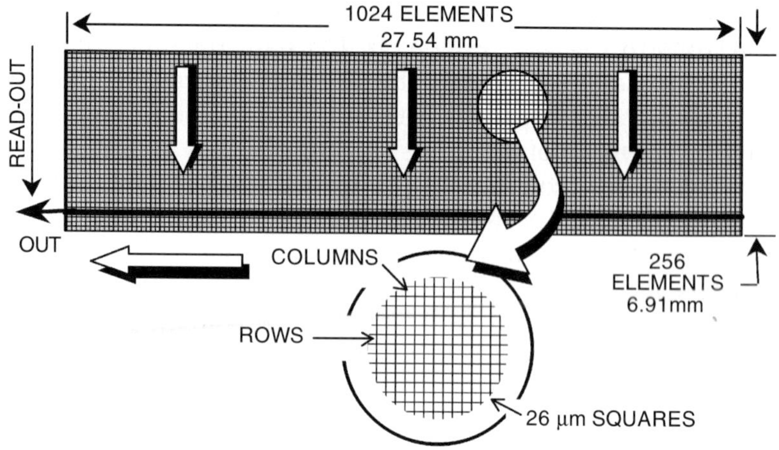

Fig. 6.3. Layout of a CCD sensitive element.

integrated silicon-based optoelectronic component. *Photodetection* is carried out by photodiodes or photo-metal-oxide semiconductors (MOS) capacitors [49] distributed in a bidimensional structure. A MOS is a transparent metal electrode isolated from a semiconducting, silicon-isolating (most often a p-type semiconductor) substrate by a thin film of silicon dioxide. *Charge storage* in MOS is produced in the following way. Since the electrodes are biased positively, the positive carriers (holes) are repelled away from the electrode causing a depletion layer in the semiconductor substrate at a positive potential. Light transmitted by metal electrode and oxide layer generates electrons that recombine with holes to form electron–hole pairs in the p-type silicon, and the pair separates under the electrostatic field. The electrons move toward the oxide layer and are trapped in the depletion layer, which acts as a potential well. The amount of electrical charge accumulated is proportional to the total (integrated) incident radiation during the exposition time. Most CCDs have the so-called three-phase architecture, where three electrodes with their associated charge storage regions compose a pixel. To read out the charge accumulated in each pixel, it has to be transported to the readout register using a *charge transfer* technique (charge coupling) that usually involves the three-phase charge coupled elements of the pixel [50]. By applying a cyclic voltage controlled by the quartz oscillator (clock) to the electrodes, some of the electrons in the vicinity of each electrode will form a discrete packet of charge and move one element toward the readout register for each full clock cycle. The last electrode on the readout row is the input to a transistor amplifier (amplifier output node). The individual pixels that compose a row of the two-dimensional array can be similarly read sequentially (Fig. 6.4), and the information can be stored in the computer for further treatment. Other CCD designs involve two-phase charge-coupled elements instead of the three-phase for the charge transfer process described previously. The two-phase elements are more complicated to build, but they are more efficient during the charge transfer process.

Variables to be considered when choosing a CCD camera for LIF applications include spectral range (range of wavelength at which the camera responds), pixel size, and number of pixels in the active element, S/N ratio that limits sensitivity, dynamic range (the ratio of the saturation charge, given as well capacity, to the detection limit, generally the readout noise), readout rate, maximum spectra rate,

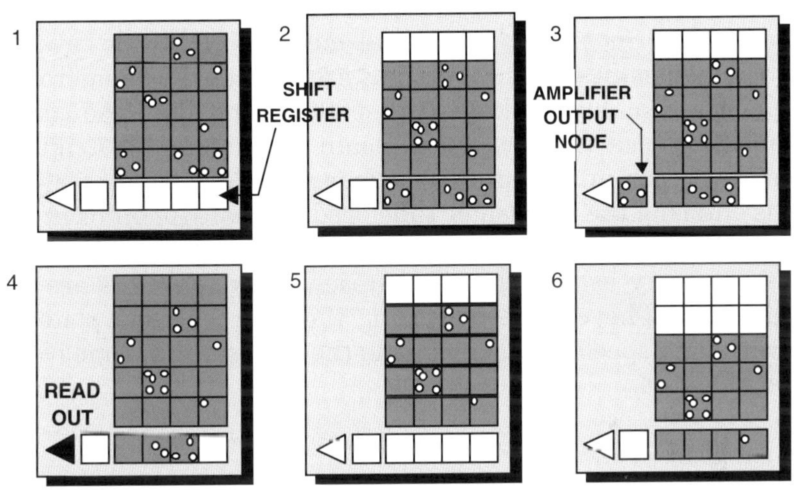

Fig. 6.4. Readout pattern of a two-dimensional CCD element. The element is represented by 20 pixels (composed of four columns and the top five rows of squares), the shift register (the sixth row of pixels), and the amplifier output node. To read out the charge accumulated in each row of pixels (represented by the blank dots), a voltage is applied simultaneously across the electrodes of the columns to move the charge one row down the vertical dimension of the element into the shift register (steps 1–2). In a similar way, these charges are then moved across the shift register toward the amplifier output node (steps 3–4), digitalized, and stored in the computer. Then a new row is read (steps 5–6) by voltage application.

maximum image rate (number of frames read per time unit), and dark current [51]. Although CCD cameras are preferred when spatial resolution is required in light detection, it should be taken into account that, at present, the sensitivity of CCD cameras is about 1–2 orders of magnitude worse than that obtained with PMTs or APDs.

In some LIF applications, measurements of very low amounts of incident light are necessary. Two solutions for such problems can be envisaged using CCD cameras: cooling the CCD array or using a light intensifier tube. One can cool the CCD array using thermoelectric elements or even liquid nitrogen. By doing so, the dark current can be decreased significantly, allowing the integration of the light incident on the array for longer time periods, and increasing the S/N ratio. Cooled CCD cameras are useful in those cases where large dynamic and spectral ranges are required.

A Light Intensifier Tube can be coupled to a CCD camera (ICCD camera) in those cases where faint incident light has to be measured. The intensifier tube first converts photons into electrons using a photocathodic plate (composed of several metallic and oxide layers, generally of doped gallium). The electrons are amplified by a microchannel plate polarized at high voltage (500–6000 V) and then the cascade of electrons produced is transformed again in photons by striking on a fluorescent screen (made out of phosphor particles). In this way the few photons entering the intensifier tube produce many photons at the exit [52], which are collected by the CCD camera. Besides, these ICCD cameras are very useful when time-resolved experiments are used in LIF detection. The voltage polarizing the microchannel plate can be gated with high speed so that events in the nanoseconds time frame can be discriminated.

6.3 DERIVATIZATION METHODS

Derivatization is a modification of analytes, intended to give these compounds more suitable analytical characteristics. Derivatization in CE is oriented to increase the detection sensitivity, associated with the incorporation of fluorescing groups into the analytes.

Two types of derivatization based on the interaction between the analyte and the labeling reagent can be distinguished: covalent and noncovalent labeling reactions [53]. In the first case, a covalent bond is formed, while in the second case, other types of interactions between the analyte and the labeling reagent, such as electrostatic, hydrophobic, or affinity interactions, are included. Amines and the reducing end of (oligo)saccharides are the groups most frequently derivatized covalently, but derivatization through thiols, carbonyl, and carboxyl groups has also been reported.

Derivatization reactions can be also classified depending on when and where the reaction takes place: before introducing the sample in the capillary (pre-capillary), inside the capillary (on-capillary), or after the electrophoretic separation and before the detection point (post-capillary). It is beyond the scope of this chapter to present a complete review of all the papers published on this topic; nevertheless, a selection of the applications of derivatization methods for LIF detection of small molecules are summarized in Table 6.1. We will present the fundamentals, advantages, and limitations of each derivatization technique, illustrated with a few examples. An overview of the reagents and procedures more frequently used for pre-, on-, and post-capillary derivatization in CE is provided in recently published reviews [54–56].

6.3.1 Pre-capillary derivatization

Pre-capillary derivatization can be performed either in the offline mode (manual) or in the online (automated) mode. Offline pre-capillary derivatization is the most employed mode of derivatization because of (i) its simplicity, (ii) the large number of reagents, including fluorescent and fluorogenic ones, that can be employed, (iii) the possibility of removing the excess reagent by evaporation or extraction, (iv) the absence of restrictions with respect to long reaction times or high temperatures, and (v) the possibility of using solvents noncompatible with CE conditions. However, the method is affected by a lack of precision that is concomitant with the fact of its being manual, the fact that it requires large sample volumes, and the long reaction times necessary.

Derivatization is most frequently performed through amine groups. Thus, different compounds, including amino acids [57,58], amines [59,60], proteins [61], and nucleotides [62], have been derivatized through amine residues in various biological matrices, pharmaceutical products, and environmental samples. Capillary zone electrophoresis (CZE) and micellar electrokinetic capillary chromatography (MECC) have been the CE modes mainly employed in these studies. As an example, Fig. 6.5 shows the electrophoretic separation of amino acids

TABLE 6.1

Some applications of CE–LIF for analysis of small molecules

Analyte	Sample	Derivatization	Comments	Reference
Amino acids	Human plasma	Pre-capillary with FQ	Diagnosis of inborn errors of metabolism (IEM)	[57]
Amino acids	Plant cell extracts	Pre-capillary with FITC	LOD in fmol range	[58]
Amino acids	Rats intestinal fluid	On-capillary with OPA/NAC	Separation of enantiomers	[76]
Amino acids	Standards	Post-capillary with OPA	CLOD: 0.83–69 nM. Using sheath-flow cell as reactor	[85]
Amino acids	Cerebral ganglion neurons from *Aplysia californica*	Online with NDA	CLOD: 500 nM. Wavelength-resolved fluorescence detection	[251]
Amino acids and biogenic amines	Brain homogenates	Pre-capillary with FQ	CLOD in low-nanomolar range. 16 biogenic amines and amino acids resolved within 16 min	[60]
Amino acids and amines	In vivo determination in rat brain	Online with OPA/β-mercaptoethanol	CLOD: 13–49 nM. Microdialysis coupled via flow-gate interface to CE–LIF system	[69]
Amino acids and peptides	Standards	Post-capillary with fluorescamine	CLOD: submicromolar range. Using a porous tube as reactor	[252]
Aniline	Real environmental water samples	Pre-capillary with FMOC	CLOD: 10 nM. Solid-state 266 nm laser	[59]
Histamine	Rat peritoneal mast cells	On-capillary with OPA/NAC	LOD: 10 amol. Cell lysis inside the capillary	[75]
Catecholamines	Standards	On-capillary through a photochemical reaction	CLOD in μM range. Multiphoton excitation	[79]

Biogenic amines	Fish, wine, and urine	Pre-capillary with OPA	CLOD: 0.250 μM	[253]
Pteridines	Urine		CLOD under 0.1 nM. Native fluorescence detection using HeCd laser (λ_{exc} = 325 nm)	[254]
Nucleotides	Standards	Pre-capillary with DBD	Synthesis and characterization of the fluorophore probe	[62]
Mono- and disaccharides	Human myometrium and uterine leiomyoma	Pre-capillary with AMAC		[68]
Carragenan polysaccharides	Food products	Pre-capillary with APTS	CLOD ~0.3 μg/ml	[67]
Peptides from the proteolysis of cytochrome *c*	Standards		CLOD in μM range. Native fluorescence detection using frequency-doubled Ti:sapphire laser (λ_{exc} = 445 nm)	[80]
Substance P and metabolites	Rat brain	Post-capillary with NDA/β-mercaptoethanol	CLOD ~100 nM. Using a membrane gap as reactor	[87]
Fatty acids	Standards	Pre-capillary with a polymethine cyanide near-infrared fluorophore label	CLOD: 60 pM. Detection in near-infrared spectral region	[63]
Homocysteine and other thiols	Standards	On-capillary with ABD-F	CLOD: 5 nM for homocysteine	[78]
N-methyl carbamate compounds	Standards	On-capillary with OPA/β-mercaptoethanol	CLOD: 0.5 ppm	[77]

continued

TABLE 6.1 (*continued*)

Analyte	Sample	Derivatization	Comments	Reference
Serotonin, tryptophan, tyrosine	Single metacerebral cell from *Aplysia californica* and single cell release		CLOD: 39 nM serotonin, 45 nM tryptophan. Native fluorescence detection using Ne:Cu laser (λ_{exc} = 248.6 nm)	[255]
Serotonin	Exocytosis from and lysis of single rat peritoneal mast cells		LOD = 1.7 amol. Native fluorescence using Ar^+ laser (λ_{exc} = 275 nm)	[256]
Reduced glutation	Single erythrocyte cell	Pre-capillary with monobromobimane	In packed cells	[257]

Abbreviations: ABD-F, 4-aminosulfonyl-7-fluoro-2,1,3-benzoxadiazole; ACN, acetonitrile; AMAC, 2-aminoacridone; α-Aml, α-amilase; AMP, adenosine-5′-monophosphate; ANS, 1-anilinonaphthalene-8-sulfonate; APTS, trisodium 8-aminopyrene-1,3,6-trisulfonate; 6-AQ, 6-aminoquinoline; AQC, 6-aminoquinolyl-*N*-hydroxysuccinimidyl carbamate; BGE, background electrolyte; BGTPγS, guanosine 5′-*O*-(3-thiotriphosphate) BODIPY FL thioether; BSA, bovine serum albumin; CA, carbonic anhydrase; CAPS, 3-cyclohexylamino-1-propanesulfonic acid; Cat, catalase; CBQCA, 3-(4-carboxybenzoyl)-2-quinoline-carboxaldehyde; α-Chy, α-chymotrypsinogen; CLOD, concentration limit of detection; CMP, cytidine-5′-monophosphate; β-CN, β-casein; Con, conalbumin; CTAB, cetyltrimethylammonium bromide; CTAOH, cetyltrimethylammonium hydroxide; CZE, capillary zone electrophoresis; DBD, 4-(*N*,*N*-dimethylaminosulfonyl)-2,1,3-benzoxadiazole; DETA, diethylentriamine; 2,5-DHB, 2,5-dihydroxybenzoic acid; EtOH, ethanol; FITC, fluorescein isothiocyanate; FMN, flavin mononucleotide; FMOC, 9-fluorenyl-methyl cloroformate; FQ, 3-(2-furoyl)quinoline-2-carboxaldehyde; β-Gal, β-galactosidase; GMP, guanosine-5′-monophosphate; G proteins, guanine-nucleotide-binding proteins; Hb, hemoglobin; HDMB, hexadimethrine bromide; HEPES, *N*-2-hydroxyethyl-piperazine-*N*′-2-ethanesulfonic acid; hGH, human growth hormone; His, histones; HIV, human immunodeficiency virus; HPTS, 8-hydroxypyrene-1,3,6-trisulfonic acid; HSA, human serum albumin; IgA, immunoglobulin A; IgG, immunoglobulin G; α-LA, α-lactalbumin; β-LG, β-lactoglobulin; LOD, limit of detection; Lyz, lysozyme; MECC, micellar electrokinetic capillary chromatography; MeOH, methanol; Met-rhGH, methionyl recombinant human growth hormone; Myo, myoglobin; NAC, *N*-acetylcysteine; NDA, naphthalene-2,3-dicarboxaldehyde; NPY, neuropeptide Y; OPA, *o*-phthaldialdehyde; OVA, ovalbumin; Rib, ribonuclease; TI, trypsin inhibitor; TMP, timidine-5′-monophosphate; TNS, 2-*p*-toluidinonaphthalene-6-sulfonate; Trf, transferrin; Tris, tris(hydroxymethyl)aminomethane; Try, trypsinogen; TTAB, trimethyltetradecylammonium bromide.

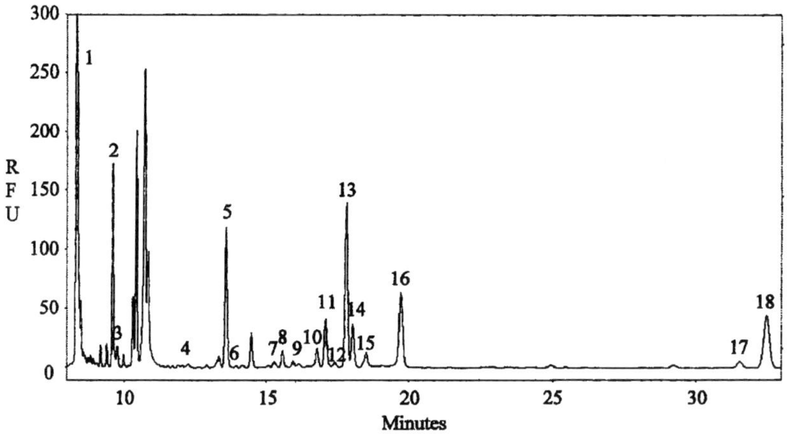

Fig. 6.5. Separation of amino acids from pure phloem sap derivatized with FITC. Separation conditions: 45 mM α-cyclodextrin in 80 mM sodium borate (pH 9.2); voltage, 20 kV. Peak identification: (1) FITC, (2) arginine, (3) lysine, (4) cysteine, (5) leucine/isoleucine, (6) methionine, (7) phenylalanine, (8) threonine, (9) tyrosine, (10) valine, (11) proline, (12) histidine, (13) homoserine, (14) serine, (15) glycine, (16) alanine, (17) glutamate, and (18) aspartate. Peak heights are reported as relative fluorescence units (RFU). Reprinted from Ref. [58]. Copyright (2001), with permission from Elsevier.

from pure phloem sap in which amino acids were pre-capillary (offline) derivatized with FITC. Although most of the studies focus on the derivatization of amine groups, a few papers deal with the derivatization of other functional groups, such as carbonyl [41], carboxyl [63,64], and thiol [65,66] groups, for their use in CE modes.

Mono- and polysaccharides have also been derivatized. Roberts et al. [67] reported the separation and quantitation of carrageenan polysaccharides that are important additives in food products. Trisodium 8-aminopyrene-1,3,6-trisulfonate (APTS) was used as a labeling reagent following an end-labeling procedure with attachment to the reducing end of the carbohydrate chain. Also, mono- and disaccharides were derivatized with 2-aminoacridone (AMAC) after they were released from chondroitin and dermatan sulfates by enzymatic digestion [68].

In the online mode, the derivatization takes place by mixing the analytes with the reagent just before the capillary entrance using a T-junction, and is usually employed in combination with a microdialysis procedure. The online coupling of microdialysis with derivatization allows the handling of submicroliter volumes of sample without any loss of sensitivity. Bowser and Kennedy [69] reported the in vivo determination of neuroactive amino acids and amines using a flow-gated interface, allowing microdialysated fractions to be automatically injected onto the separation capillary.

6.3.2 On-capillary derivatization

In the on-capillary mode, derivatization is performed in the same capillary where the separation takes place without changing the setup of the commercially available CE devices. Several options have been described to do this. Zone-passing derivatization is carried out by applying the separation voltage just after the introduction of the labeling reagent and the sample solutions [70]. In order to carry out the derivatization reaction, the solution with lower electrophoretic mobility must be introduced first into the capillary to allow the solution with higher electrophoretic mobility to pass through it. This strategy is appropriate for fast kinetic processes where high reaction yields are achieved in the few seconds during which analytes and reagent are in contact. In at-inlet derivatization mode, a sample and a reagent solution are introduced at the inlet of the capillary and are allowed to react for a specific time, called mixing time, before applying the separation voltage. In this technique, various sample/reagent introduction modes can be used. For example, Taga and Honda [71] compared tandem and sandwich introduction modes. In the first case, sample and reagent solution are successively introduced into the capillary in the same way as described for zone-passing derivatization. In the sandwich mode, sample solution is introduced between two plugs of reagent solution, and it is not necessary to take into account the sample/reagent relative electrophoretic mobility. In both cases, mixing time controls the mixture between the sample and the reagent solution and, therefore, the derivatization reaction. In throughout-capillary derivatization, the capillary is filled with a running buffer that contains the reagent. When the separation voltage is applied, the components of the sample migrate and mix with the reagent, and the corresponding reaction occurs [72]. A good comparison of all on-capillary derivatization modes discussed above has been reported elsewhere [73].

Generally, the on-capillary derivatization mode is suitable for small sample volumes, because volumes injected onto the capillary are very low (nl), and dilution is reduced to a minimum. Other advantages are short reaction times, the possibility of automation, and the absence of extra devices.

The applicability of on-capillary derivatization for amine groups in very small sample volumes was demonstrated by Lee et al. [74], who derivatized protein extracts from 2.5 cells of HT29 human colon adenocarcinoma line (see Fig. 6.6), and by Oguri et al. [75], who derivatized histamine from mast cells. In both cases, the cells were injected and lysed, and the compounds were on-capillary-derivatized. Labeling reagents were 3-(2-furoyl)quinoline-2-carboxaldehyde (FQ) and a mixture of napthalene-2,3-dicarboxaldehyde (NDA) and *o*-phthaldialdehyde (OPA), respectively. Oguri et al. [76] also studied the degree of amino acid enantiomer absorption from rat intestines with CE separation of on-capillary-derivatized compounds.

A novel method for the analysis of *N*-methyl carbamate pesticides using cetyltrimethylammonium bromide (CTBA) as pseudostationary phase that also catalyzes the degradation of these pesticides was developed by Wu et al. [77]. The same capillary was used subsequently for separation, decomposition, derivatization, and detection purposes. On the other hand, groups different from amine ones, such as thiol groups, have been on-capillary derivatized. Thus, the derivatization of

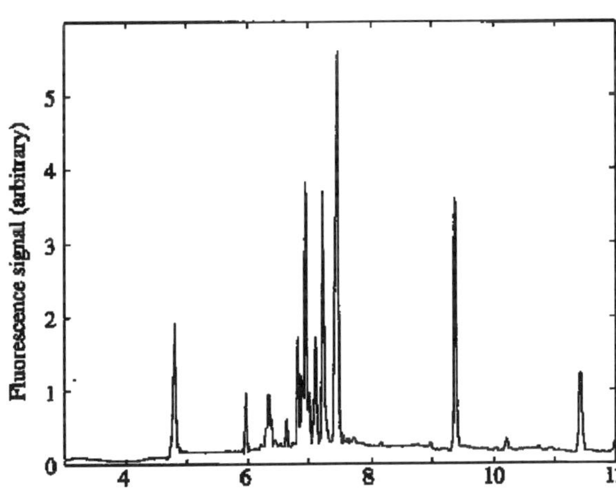

Fig. 6.6. Protein fingerprinting of an extract from an HT29 cell line. Proteins covalently labeled with FQ. Separation conditions: 2.5 mM borax containing 5 mM SDS (pH 9.3); voltage, 400 V/cm. Reprinted with permission from Ref. [74]. Copyright (1998) American Chemical Society.

homocysteine and other thiols with 4-aminosulfonyl-7-fluoro-2,1,3-benzoxadizole (ABD-F) was reported by Kang et al. [78].

On-capillary derivatization can be employed with multiphoton excitation as detection system (see section 6.6.1). Gostkowski et al. [79] described on-capillary derivatization of catecholamines through a photochemistry reaction. The reaction products, that presented fluorescence in the visible spectral range, were detected via two-photon excitation using the near-infrared (NIR) output from a Ti:sapphire laser. Limits of detection were at the μM level. Previously, this group had used the above-mentioned detection system for monitoring the on-capillary proteolysis of cytochrome *c* from different origins. The peptides obtained were detected by two-photon fluorescence with detection limits at the μM level [80].

In the analysis of single cells, on-capillary derivatization methods are the most frequently used. Separation and characterization of amines from individual atrial gland vesicles of *Aplysia californica* have been reported [81]. FQ has been used as labeling reagent to obtain the protein fingerprint of one single cancer cell that was on-capillary lysed and derivatized [82].

6.3.3 Post-capillary derivatization

Post-capillary derivatization [83] has several advantages over the previously mentioned modes. These derivatization modes normally lead to significant changes in the mass-to-charge ratio and hydrophobicity of analytes that, in some cases, become detrimental for the separation. An example of a negative effect of pre-capillary derivatization is described by Craig and Dovichi [84]. A large protein was derivatized using FQ for as little as 1 min. Severe band broadening and several distinct components were visible after the labeling, caused by the binding of FQ to several sites in the molecule. Disadvantages of post-capillary labeling are the negative effects on peak efficiency, sample dilution, incomplete reactions, the requirement of additional equipment, and the higher baseline noise caused by added reagent solution. Band broadening seems to be an inherent problem of post-capillary derivatization, because some turbulence can be expected at the site where the reagent solution is added. Several reactor designs have been explored to minimize this band broadening effect. Efficiency was improved employing a sheath-flow cell as reactor [85,86]. In this case, the sheath liquid containing the labeling reagent entered from the top of the flow cell and surrounded the capillary eluent as it exited the capillary. In this way the original flow profile of the CE buffer was less disturbed compared with other gap reactors.

The use of a porous membrane, positioned as a sleeve over a gap between capillaries, has improved the results of gap reactors, as this membrane allows the reagent to diffuse through it and to mix with the analyte. Kostel and Lunte [87] found that the sample loss for substance P using the membrane reactor was about five times lower than with the standard gap-reactor design.

6.3.4 Protein detection with and without derivatization

The separation of proteins in biological matrices represents an area of tremendous interest in recent years. As proteins are usually present at low concentrations (generally lower than 10^{-6} M) in these matrices, fluorescence is the most employed detection method because of its high sensitivity.

Proteins can be directly detected by fluorescence owing to the intrinsic fluorescence of tryptophan and tyrosine residues that they contain. Thus, the most important problems associated with derivatization reactions, i.e., the formation of multiple reaction products [84] and the difficulties associated with derivatization of proteins at low concentrations (below 10^{-8} M), can be avoided. The formation of multiple reaction products during derivatization reaction is due to the attachment of the labeling reagent to multiple protein sites, since most of protein-labeling agents react through protein amine groups. Thus, 2^n–1 fluorescent reaction products, where n is the number of amine groups on the protein, can be formed. The appearance of different reaction products with different mobility leads to peak broadening of proteins. However, when derivatization takes place at low concentrations, the kinetics controls the reaction and limits the amount of products formed. In addition, derivatization takes time and can generate important modifications in the protein structure.

Optimal excitation and emission wavelengths for tryptophan and tyrosine residues are in the UV region, which limits the type of laser that can be used. UV-operating lasers, such as Nd:YAG (266 nm) [88], frequency-doubled Ar-ionized (275 nm) [89], or KrF excimer (248 nm) [90] lasers, have been used for the analysis of proteins by CE–LIF. However, the systems are expensive and the maintenance needed is costly. This is one of the disadvantages of the use of native fluorescence.

Other drawbacks are the low fluorescence efficiency of tryptophan and tyrosine residues, and the high background fluorescence due to sample or fused-silica capillary impurities. Also, at these wavelengths, the autofluorescence of compounds present in the sample matrix, primarily in biological samples, is a problem since it increases the background fluorescence and can mask the signals of interest. Even deionized water contributes to background fluorescence, since trace organic contaminants are always present. However, by selecting the appropriate pH to maximize the fluorescence efficiency and to avoid band broadening, it is possible to achieve limits of detection of 10^{-10} M [91].

Native fluorescence detection has been used for the analysis of proteins in different single cells, such as erythrocytes, pancreatic, or mast cells [92]. Also, native fluorescence has been applied to detect the separation of isoforms and other microheterogeneities of proteins [88] in order to achieve high-sensitivity separations without modifications in the structure of the protein.

Protein derivatization reduces the background fluorescence compared with native fluorescence detection, employs probes with high fluorescence efficiency, and allows the use of low-cost lasers. Fluorogenic probes are preferred because of the reduction of background signal compared with fluorescent probes. Two types of protein derivatization, covalent or noncovalent labeling, depending on the interaction between protein and probe, have to be considered.

Covalent derivatization has been mainly carried out through amine groups [93,94], but thiol groups [65] have also been used. Numerous works reporting the analysis of proteins by pre-, on-, or post-capillary derivatization have been published. Limits of detection lower than those obtained using native fluorescence were reported. For instance, Pinto et al. [95] obtained a limit of detection of 1×10^{-11} M for conalbumin labeled with FQ.

In non-covalent labeling, different interactions between protein and probe, including hydrophobic, electrostatic, affinity, or a mixture of them, can occur. Several fluorescent hydrophobic probes, such as 1-anilinonaphthalene-8-sulfonate (ANS) and 2-*p*-toluidinonaphthalene-6-sulfonate (TNS) [96,97], Sypro Red [98], Nile Red [99], Indocyanine green [100], or NanoOrange [101], have been used for noncovalent derivatization of proteins. These probes adsorb into the hydrophobic regions of proteins whereupon their fluorescent quantum yield increases substantially. Derivatization employing hydrophobic probes has the advantage over covalent probes that it is faster and the pH reaction control is not as critical.

Also, affinity binding has been widely employed for fluorescent derivatization of proteins. The term affinity includes the idea of a higher or lower degree of selectivity in the binding of two or more molecules. This kind of binding encompasses a wide range of interactions excluding the covalent ones and including electrostatic, van der Waals, hydrophobic, and hydrogen-binding forces. In this section, we will refer to affinity probe capillary electrophoresis (APCE) and related systems. In this technique, the protein is not bound to a fluorescent tag but to a fluorescently tagged ligand, frequently called the probe. Book chapters and reviews devoted to or including aspects related to APCE with fluorescence detection have been recently published [14,102–104].

The protein is made to react with the labeled probe to form a complex. In this way, both the ligand and the complex are fluorescent. CE separation of both peaks allows quantitation of the amount of the protein present in the sample. The increase in the size of the peak corresponding to the complex and the decrease in that from the ligand are related to increasing amounts of the target protein in the sample.

Owing to the well-known affinity between antibodies and antigens (immunoaffinity), in most of the APCE analysis of proteins (antigens), the ligand is a fluorescently labeled antibody. By adopting the names used for traditional immunoassays performed in polystyrene plates, the described method is called a noncompetitive immunoassay performed in CE.

Competitive immunoassays represent a variation on the APCE method in the way it has been described above. In this mode, an analog of the protein is labeled with a fluorescent tag. The protein in the sample and the fluorescent analog compete for binding to a limited amount of the antibody. The larger the concentration of protein in the sample, the larger the concentration of protein–antibody complex and thus the lower the concentration of labeled analog–antibody complex and the larger the amount of free labeled analog. As these two substances are fluorescent, the size of the peak of any of them or the ratio between both can be used to quantitate the amount of protein in the sample [105]. Similarly to what has been described for quantitation of antigens, CE immunoassays can be used for analyzing antibodies.

APCE allows the simultaneous determination of analytes that cross-react with a given antibody. Mono- and di-deamidated and nondeamidated methionyl recombinant human growth hormone (met-rhGH) have been separated by capillary isoelectric focusing (cIEF) into their respective immune complexes with anti-hGH Fab′-labeled with tetramethylrhodamine-iodoacetamide. This label, stable over the pH range applied, allowed in the described system the detection of rhGH with detection limit 5×10^{-12} M [106].

Initially, CE immunoassays were performed by introducing into the capillary column the reagents that were previously mixed offline. Online reaction constitutes an improvement of the method, allowing the separation to be complete in 3 s [105].

As a way for increasing sensitivity and selectivity of CE immunoassays, a reversed-phase liquid chromatography step has been coupled online prior to CE [107]. The method previously demonstrated for the analysis of glucagon was extended for the simultaneous detection of glucagon and neuropeptide Y (NPY) from islets of Langerhans [108]. Analytes preconcentrated onto the capillary LC column are separated by gradient elution to allow their individual detection by using fluorescein-labeled glucagon and NPY in competitive CE assays.

Development of microfluidic devices has allowed the performance of electrophoresis-based immunoassays with fluorescence detection on them. Similar to the evolution followed in capillary columns, the first assays on microchips were performed by introducing into the microsystem the mixture antibody–antigen prepared offline. Introduction of a greater number of channels has made it possible to perform the online mixing of sample, antibody (labeled for noncompetitive formats), and in competitive assays, labeled analog of the analyte. Further modifications and an increase in the number of channels make it possible to perform immunoassays in six channels simultaneously [109]. The multichannel microchip, called SPIDY by the authors, makes use of a scanned LIF detection system. With this system, integration of simultaneous on-chip mixing, reaction, and separation of labeled antibody and complex is possible. For a noncompetitive qualitative assay of ovalbumin (OVA), anti-OVA labeled with Cy-5 has been used. The system has also proved valid for simultaneous quantitative determination of OVA and anti-estradiol using estradiol labeled with fluorescein as probe for this antibody, having a detection limit of 4 nM.

Microchips have been developed to perform continuous monitoring of substances secreted from one live cells [110]. The system to measure insulin secreted from one islet of Langerhans is based on the one previously developed in CE columns [111]. Compared with the previous system, the microchip-based one is easier to assemble and operate. The competitive immunoassay in the microfluidic system makes use of FITC-insulin as a labeled analyte analog, and with a detection limit of 3 nM, allows the resolution of secretory profiles from the islet that show a quick increase of insulin followed by a slower, long-lasting release rate.

For immunoassays based on electrophoresis performed either in capillary columns or in channels, modification of one of the components of the antibody–antigen complex may be needed in order to distinguish the complex from the free components. This was the case, for example, in the assay performed to quantitate human IgA in which a charge-bearing organic molecule (succinic anhydride) had to be introduced into the $F(ab')_2$ fragment of goat antiserum to human IgA labeled with FITC. In this way, the labeled antibody–antigen complex is separated from the labeled antibody and allows detection of IgA by a noncompetitive assay with detection limit lower than 10^{-9} M [112].

One of the drawbacks of CE-based immunoassays is the lack of homogeneity of the antibody, which results in very broad or multiple peaks. The presence of different species in polyclonal antibodies and the fact that each antibody has two paratopes (binding sites for antigen) are the sources of this heterogeneity. Besides, in noncompetitive assays labeling of the antibody does not usually take place at only one position, causing the presence of antibodies with different numbers of tag molecules.

Several approaches to diminish or avoid heterogeneity have been adopted. In general, monoclonal antibodies are preferred to polyclonal ones. Different antibody fragments, such as Fab or Fab′, obtained by the controlled action of enzymes on the whole antibody molecule, or prepared by genetic engineering techniques, e.g., the ScFv (single-chain antibody variable-region) fragment, have been used.

Monoclonal antibody anti-insulin C-terminal pentapeptide was carefully chosen to perform offline and online competitive immunoassays for insulin [105], providing better sensitivity than the Fab fragment previously used in offline systems [113].

Fab′ has been chosen in some instances instead of ScFv, taking into account the lower affinity of the latter fragment [114]. For reasons that are not clear, carbamylation of the antigen, α_1-antitrypsin, was necessary for it to be detected in a reproducible way using the fluorescence-labeled Fab′ fragment of the specific antibody.

On the other hand, strategies to introduce only one molecule of the tag in very well-controlled positions of the antibody have been developed. For example, the thiol group in the hinge region of the Fab′ fragment of the antibody has been used to label the antibody with tetramethylrhodamine-iodoacetamide [106]. Besides the number of tag molecules introduced in an antibody molecule, it is important to control the conditions and the position for labeling, and is preferable to introduce the label far from the paratopes to preserve the affinity for the antigen.

As mentioned above, antibody–antigen recognition is the most widely used affinity system for performing CE assays of proteins with noncovalent fluorescent labeling. However, other biospecific interactions have also proved useful.

Protein G labeled with FITC can be considered a universal fluorescently labeled reagent for immunoassays performed in capillary systems because most antibodies bind to protein G [115]. The system has been proved, for example, to quantitate human IgG_1 in serum using non-labeled anti-human IgG_1.

Aptamers are seen by several authors as an advantageous alternative to antibodies for CE immunoassays [116]. Aptamers are oligonucleotide sequences selected to interact with the target protein. They are produced in vitro by a technique known as SELEX (systematic evolution of ligands by exponential enrichment). As the chemical synthesis can be controlled, the aptamers with the desired properties can be produced. Aptamers fluorescently labeled with 5′-carboxyfluorescein have been used as probes to determine the activity of HIV-1 reverse transcriptase in cell cultures of the HIV-1 virus [117]. The use of a single-stranded DNA-binding protein (SSB) has facilitated the CE separation of the probe from the protein–probe complex when a fluorescein-labeled 15-mer oligonucleotide was employed as an aptamer probe for the analysis of thrombin [118].

Small natural ligands can be also used as affinity probes. This is the case of guanosine 5′-*O*-(3-thiotriphosphate) BODIPY FL thioether (BGTPγS), a fluorescent guanosine 5′-triphosphate (GTP) analog used for the assay of guanine–nucleotide-binding proteins (G proteins). The method, which can be used for kinetics studies and for quantitative analysis, can be applied to mixtures of G proteins and determine multiple analytes at the attomole level in a single assay (see Fig. 6.7) [119].

Interactions between phospholipids and proteins may also be useful. This kind of interaction between bovine serum albumin and phosphatidylserine was monitored by labeling the phospholipid with FQ [120].

A selection of applications of LIF for detection of proteins either without derivatization or using the different strategies described in this section for derivatization are summarized in Table 6.2.

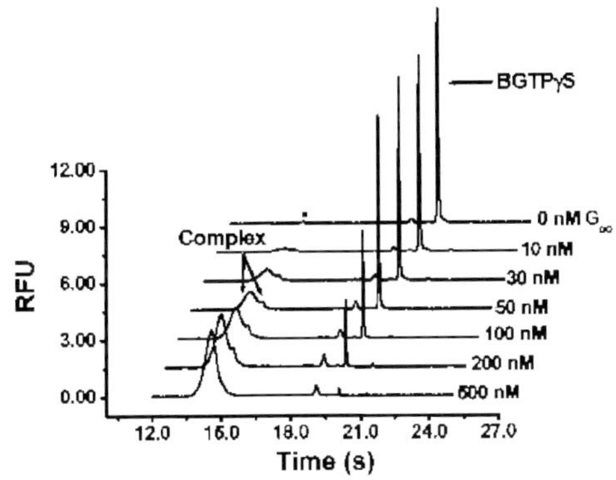

Fig. 6.7. Affinity probe capillary electrophoresis of a G protein using 50 nM BGTPγS as fluorescent affinity probe. The concentration of the G protein was increased from 0 to 500 nM. Separation conditions: 25 mM Tris, 192 mM glycine (pH 8.5); voltage, 5 kV. Peak heights are reported as relative fluorescence units (RFU). Reprinted with permission from Ref. [119]. Copyright (2003), American Chemical Society.

6.4 MINIATURIZED FLUORESCENCE DETECTION FOR MICROCHIPS

During the last 15 years, many efforts have been made to miniaturize analytical instrumentation. In this context, the lab-on-a-chip or micro total analysis (μ-TAS) concept refers to miniaturized devices where all necessary parts and methods to perform a specific analysis have been integrated in the same piece of material, usually glass or polymer. Since its introduction by Manz et al. at the beginning of the 1990s (for a historical perspective see Refs. [121,122]), electrophoresis has been the most frequently used methodology to move liquids inside the microconducts of the chips and to separate analytes (see among other, recent reviews [123–125]). Despite the enormous methodological developments on the use of chips in separation analysis, only a few applications to real-world samples have been published, most of which are summarized in Table 6.3.

The main features that the detection systems using in microchips for electrophoresis have to fulfill are: (i) they have to be sensitive, (ii) they should be miniaturized at the same scale as that of the microchip, and (iii) they should be portable to allow field or point-of-care

TABLE 6.2

Some applications of CE-LIF for analysis of proteins

Analyte	Sample	Derivatization	Comments	Reference
Several proteins	Extracts from HT29 human colon adenocarcinoma cells	Covalent labeling with FQ	CLOD: 0.3 pM for Con. Protein mapping	[74]
Con, OVA, BSA, Myo, CA, α-LA, β-LG	Standards	Covalent labeling with FQ	CLOD: 10 pM for Con	[95]
BSA, OVA, α-Chy	Standards	Covalent labeling with AQC	CLOD: 12–42 pM	[205]
Several proteins	Yeast cells	Covalent labeling with FITC	Two-dimensional separation Protein mapping	[258]
Several proteins	Staphylococcus species	Covalent labeling with FQ	Protein mapping	[259]
Lyz, TI, β-CN, OVA, BSA, Con, β-Gal, Cat, Trf, α-Aml	Human saliva and cerebrospinal fluid		Separation of six isoforms of α-Aml. Native fluorescence detection using Nd:YAG laser (λ_{exc} = 266 nm)	[88]
Phycobiliproteins	Standards		CLOD: 0.25–1.2 μM. Native fluorescence detection using He:Ne laser (λ_{exc} = 594 nm)	[260]
BSA, β-LG A and B, Con	Standards	Non-covalent with ANS and TNS	CLOD: 360–615 nM	[96]
BSA, α-LA, β-LG A and B	Bovine milk	Non-covalent with ANS and TNS	CLOD ranging from 0.06 to 3 μM	[97]

CA, OVA, BSA, Con	Standards	Non-covalent with Sypro Red	CLOD in pM range	[98]
His, OVA, Rib, Try, Lyz, BSA, Cat	Standards	Non-covalent with Nile Red		[99]
Several proteins	Yeast cells	Non-covalent with Alexa Fluor 350 and 660	Protein mapping	[100]
BSA	Standard	Non-covalent with Nano Orange	CLOD: 3 nM	[101]
Hb, Con, BSA, OVA, β-CN	Standards of all. BSA and β-CN in cow milk	Non-covalent with Merocyanine 540 and Sypro Red	CLOD in nM range	[261]
HSA	Human urine and blood cells	Non-covalent with Albumin blue 580	CLOD: 11 nM	[262]
Met-rhGH	Standard	Anti hGH Fab′-labeled with tetramethylrhodamine-iodoacetamide as affinity probe	CLOD: 5 pM	[106]
Glucagon and NPY	Islets of Langerhans	Fluorescein-labeled glucagon and NPY as affinity probes	Online coupling HPLC-CE	[108]
OVA	Standard	Anti-OVA labeled with Cy-5 as affinity probe	Performed in microchip	[109]
Anti-estradiol	Standard	Estradiol labeled with fluorescein as affinity probe	CLOD: 4 nM. Performed in microchip	[109]
Human IgA	Human serum	Succinyl FITC $F(ab')_2$ anti human IgA as affinity probe	CLOD: 0.7 nM	[112]

continued

TABLE 6.2 *(continued)*

Analyte	Sample	Derivatization	Comments	Reference
α_1-Antitrypsin	Human serum	Fluorescence-labeled Fab′ fragment of anti α_1-antitrypsin	Carbamylation of antigen required	[114]
Human IgG_1	Human serum	Protein G labeled with FITC as affinity probe		[115]
HIV-1 reverse transcriptase	Cell cultures of the HIV-1 virus	Aptamer fluorescently labeled with 5′-carboxyfluorescein as affinity probe		[117]
Thrombin	Standard	Fluorescein-labeled 15 mer oligonucleotide as an aptamer affinity probe		[118]
G proteins	Standard	BGTPγS as affinity probe	CLOD: 2 nM	[119]
Protein fingerprint	Single HT29 human colon cancer cell	Covalent labeling with FQ	Separation in CE in submicellar buffer	[263]
Cell cycle-dependent protein fingerprint	Single HT29 human cancer cell	Covalent labeling with FQ	Separation by capillary sieving electrophoresis	[82]
Insulin	Individual insulin-secreting β-cells		LOD = 22 amol. Native fluorescence using Ar^+ laser (λ_{exc} = 275 nm)	[264]
Protein kinase C, protein kinase A, calcium-calmodulin activated kinase II, CDC2 protein kinase	Single rat basophilic leukemia cell and single mouse Swiss 3T3 cell	Fluorescent peptide substrates for each protein kinase were injected in the cell and detected as in vivo reporters	CLOD for the fluorescent peptide 10 nM (assuming 1 pl cell volume)	[265]

For abbreviations see Table 6.1.

TABLE 6.3

Some separations of real samples carried out using microchips and LIF detection

Analyte	Sample	Chip-based method	Comments	Reference
Proteins	Serum	CZE	Glass microchip. One separation channel	[266]
Amino acids	Urine	CZE	Glass microchip. One separation channel. Indirect fluorescence detection using fluorescein as probe	[180]
Amino acids	Urine	MECC	Glass microchip. Cyclic planar microstructure	[267]
Cortisol	Serum	Immunoassay/CZE	Microchip device in a fused silica wafer. One separation channel	[268]
Theophylline	Serum	Immunoassay/CZE	Glass microchip. One separation channel	[269]
Insulin	Islets of Langerhans	Immunoassay/CZE	Glass microchip. Continuous monitoring of hormone secretion	[110]
Immunoglobulin G	Mouse ascites fluids	Immunoassay/CZE	Glass microchip. One separation channel	[270]
Print and film developing agents (CD-3 and CD-4)	Photographic processing solutions	CZE	Glass microchip. Indirect fluorescence detection using fluorescein as probe	[195]

For abbreviations see Table 6.1.

use. UV–Vis absorption, fluorescence emission, and electrochemistry are the analytical techniques most often used to develop microchip detectors. The possiblity of integrating the electrodes in the microchips is an attractive feature for electrochemical detection. However, these detectors cannot compete with fluorescence detectors in terms of sensitivity and limit of detection [126]. That is why fluorescence has been, and still is, the preferred method of detection for microchips.

In this section we discuss first laboratory-made, nonminiaturized setups for on-chip fluorescence detection [127]. This approach includes bulky optic components such as lasers, lenses, microscope objectives, and optical fibers that can require careful alignment, are sensitive to thermal expansions and mechanical vibrations, and use voluminous ancillary instruments that, in comparison with the size of the microchips, can look like Gulliver in the Lilliputian Land of Jonathan Swift's satire. The advances in microoptics and integrated optics based on planar waveguides [128] can allow one to develop fluorescence detection system that are microchip-size and more or less built-in in the microchips. Some of these developments will be also discussed in this section.

6.4.1 On-chip fluorescence detection

For the first developments in microchips, fluorescence was produced using a laser beam focused onto one point of the separation channel [129]. Mirrors, lenses, filters, and PMTs complicate the detection system, which becomes difficult to align, bulky, expensive, and prone to environmental effects (such as thermal expansion and mechanical vibrations) [130]. More user-friendly setups are those in which some optical components used for driving the light in and out of the chip have been replaced by optical fibers. Monomode glass optical fibers are particularly suitable to drive the light from a laser to the detection point on the microchip. A bundle of optical fibers can be used to collect the fluorescent light produced in the chip. The microchip placed in an X–Y–Z translation stage can allow easy optical alignment and optimization of sensitivity [131].

Special setups have been used to render the fluorescence detection easier to carry out in microchips. Under this rubric, confocal microscope and whole-chip detection will be briefly considered.

6.4.1.1 Confocal microscope

Inverted, confocal microscopes render fluorescence detection in microchips easier. The unobstructed X–Y translational stage of the microscope allows the manipulation of the microchips and provides plenty of room for connections of the power supply to the reservoirs placed on the top of the chip [132]. In fact, the few platforms commercialized for microchips so far feature this kind of detection [133,134].

As shown in Fig. 6.8 in the confocal microscope, the exciting light emitted by a source of radiation, such as a laser, is passed through a bandpass filter (optional), reflected by a dichroic mirror, and focused onto the focal plane of the objective where the sample is placed. The fluorescence produced by the specimen is collected by the same objective, passed through the dichroic mirror and focused with tube lens onto a pinhole (confocal aperture) located at the focal point of the lens. Finally, the light, after spectral filtering (through bandpass and

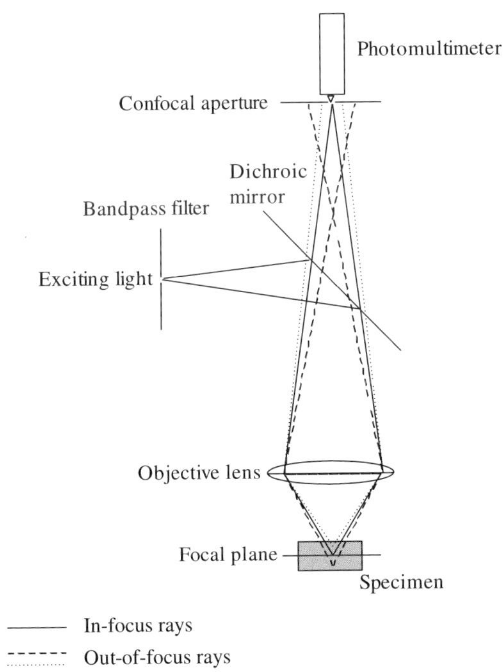

Fig. 6.8. Scheme of confocal microscope. Only the light from the focal plane enters the detector; the broken lines show rays from different depth in the sample.

interferometric or notch filters), is collected by a light measuring device (PMT, APD, or CCD camera), obtaining an analogic signal that can be digitalized and stored in a computer for ulterior analysis [135]. The main features of the confocal microscope are described by Sheppard and Shotton [136]. One of these characteristics, the small depth of field of these optical systems, is essential for sensitive detection in microchips, because only the light from one plane in the interior of the chip channel is probed. This is because the confocal aperture ensures that only fluorescence emitting from the focal plane of the objective is fully detected by the light measuring device. In contrast, fluorescence emissions from regions below and above the focal plane, where most of the scatter light is produced, are rejected and prevented from reaching the detector. In this way, images with good S/N ratios are obtained [40,135,137].

The small sample volumes used in microchip electrophoresis (typically in the nl–pl range) require a very small "detection cell" to avoid postseparation band broadening. This is feasible in confocal microscopy using high NA objectives with long working distance combined with a small pinhole diameter.

Some attention has to be paid to optical parameters that govern the focusing excitation light, fluorescence collection, and band broadening caused by detection. Optimization of power and spot size of the laser beam, careful selection of filter characteristics, optimization of pinhole diameter, and selection of the N.A, working distance, and quality of the objective used for better axial and lateral resolution can improve the detection limit achieved [135,136,138–141].

6.4.1.2 Whole-chip detection

As a general practice in microchip development laboratories, a CCD camera is used to monitor a major part of the chips under the microscope. Usually, CCD cameras are used for qualitative observation to optimize injection [142] or to study displacement of the electrophoretic bands inside the channels [143]. The images are stored in a video recorder or in a computer for ulterior study. Only in a few cases have the images been used to extract quantitative information, either as electrophoretic peaks [144,145] or as diffusion coefficient measurements of the analyte [137].

It should be taken into account that the present sensitivity of CCD cameras for fluorescent detection is about 1–2 orders of magnitude worse than that obtained with PMTs or APDs. However, technology to prepare CCD photosensitive arrays is improving and in the near future CCD cameras could allow enough sensitivity as for LIF detection at any point of the surface of 2D-separation microchips [146].

6.4.2 Miniaturized fluorescence detection

To achieve the full potential of microchip electrophoresis in terms of miniaturization and portability, a substantial reduction in the size of the ancillary equipment used in on-chip fluorescence detection such as laser, focusing and collection optics, and light detection systems should be reached, so that the integration of the whole detection device in the microchips can be carried out. In recent years, advances in solid-state laser, LED, microoptics [128], and detection devices have been achieved. This section reviews some of the concepts that could contribute to downsizing fluorescence detection to microchip size in the near future.

6.4.2.1 Pigtail concept

The pigtail approach for miniaturization of optical detectors was developed by Bruno et al. [24], who borrowed the concept from the telecommunication industry. In pigtail-based devices, light propagates through the optical system with minimal losses due to refractions and reflections of the beam. This effect can be achieved if the number of optical interfaces traversed by the light beam is reduced to a minimum of two. This happens when all the optical components used have a similar refractive index (RI), and they are glued to one another using a refraction index matching (RIM) material. Other main components of the pigtail concept are efficient light sources (such as LEDs or laser diodes), and GRIN lenses [147] for light focusing. In their original paper, Bruno et al. proposed the pigtailing approach for the construction of absorbance, fluorescence, RI, and thermooptical detectors [24].

Figure 6.9 shows the main components of a pigtail-based detector for fluorescence. As light source, it is used a LED where a hole is drilled down into the Plexiglas body and where a GRIN lens provided with an interference filter at the light entrance end [148] is inserted and glued with UV-curing RIM materials. A naked capillary is glued at the exit surface of the GRIN lens (focal point). To collect the light produced, multimode optical fibers are placed at an angle close to 45° in relation to the optical axes of the capillary for minimizing scattered light collected from the inner silica–buffer interface. These elements are fixed

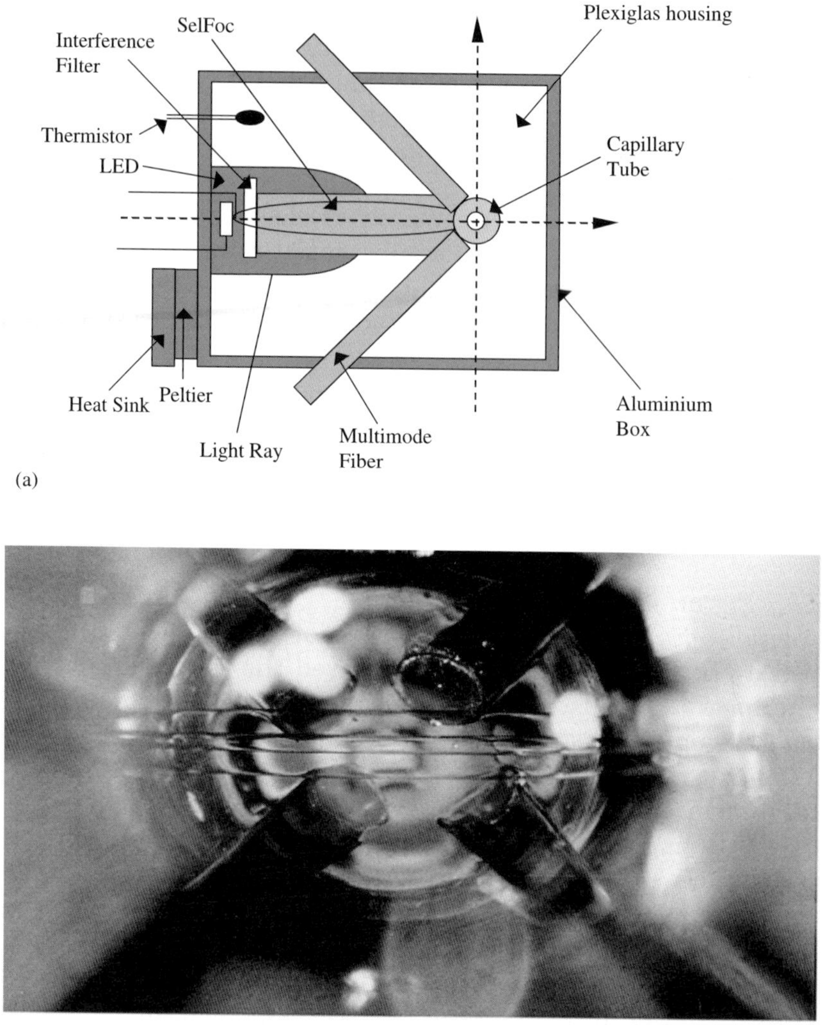

Fig. 6.9. Fluorescence detector based on pigtail concept. (a) Cross-sectional side diagram of the LED-based pigtail capillary fluorescence detector. (b) Picture showing a top view of the detector with the four collecting fibers at an angle of *ca.* 45°. Reprinted from Ref. [24]. Copyright (1994), with permission from Elsevier.

using UV-curing RIM with inspection under the microscope. The whole detector is mounted in an aluminum box and thermostated using a thermoelectric element, platinum sensor, and an external controller.

Pigtail approach provides optical fluorescence detection with minimum optical losses and scattered light collected, with reduced thermal drift, and better stability, which translate into low noise and high sensitivity. All these advantages provide to pigtail concept a lot of potential for making sensitive, miniaturized fluorescence detectors for microchips.

6.4.2.2 Integration of optical elements in microchips

Integrating some of the optical elements inside the chips is not a trivial task. That explains why only a few research groups with technology for and experience in micromachining and microelectronics have assayed to develop some devices in this field. This section briefly reviews some of these developments reported in the literature related to analytical chemistry, thereby obviating the need for a discussion of the papers presented by almost the same research groups in the specialized literature on optics, electronics, or micromachining. For the readers interested in large integration of fluorescence detection in microchips, two recent reviews [127,128] might be of interest.

The most elementary step toward integrating the whole fluorescence detector inside the microchips is to embed optical fibers inside the chip to bring the light in and out of the detection section. Harrison's group [149] demonstrated a procedure for inserting optical fibers into glass substrate close to the detection cell. It involves etching a channel for the fiber on the chip, closing the channel by thermally bonding a top plate, enlarging the channel by etching it with HF solution, and carefully inserting the fiber into the channel. Since the rounded bottom of the channel does not fit the flat shape of the extreme end of the fiber, it has to be glued to the interior wall of the channel with RIM material to avoid light losses due to light scattering. A similar procedure has been used by other researchers to insert optical fibers into glass structures used for flow cytometry [150]. Whitesides' group has demonstrated that optical fibers can be embedded more easily when poly(dimethylsiloxane) (PDMS) is used instead of glass to prepare the chips [151]. A similar approach has been carried out by Tung et al. [152] and Qi et al. [153]. They have described the procedure to incorporate optical fibers for excitation and collection of fluorescence in poly(methyl methacrylate) microchips prepared using hot embossing from nickel-based molds.

One limitation of these approaches is that the whole setup is not very robust, and replacement of the fibers when they break is laborious and time-consuming. One way to avoid this drawback is to develop waveguides monolithically integrated with the microfluidic channels during the preparation of the chip. Hübner et al. [154] used a plasma-enhanced chemical vapor deposition technique to create a silica glass layer from the gas phase on a silicon wafer. The silica of the waveguide is doped with controlled amounts of germanium and nitrogen to render the RI of the guide 1.0–1.5% higher than the surrounding layers, so that the light is driven inside the guide by total internal reflection. Separation channels are then created using photolithography and reactive ion etching. The whole structure is covered by a cladding layer consisting of pure silica glass. By using an external laser diode with a pigtailed fiber glued on top of the separation channel as light source, and a second fiber connected to an interferometric filter and to a photomultiplier as fluorescence collector, an alignment-free, stable, and easily transportable device is developed. The same group has described the application of their device to CE separations [155]. In a similar fashion, Whitesides' group has demonstrated the use of a solid-object printer to create three-dimensional structures in PDMS containing two horizontal channels at 90° to one another, and a vertical channel that allows the insertion of the optical fiber in a precise location. The three-dimensional channel structure allows the fiber to be removed and replaced if necessary [156].

Several research groups have incorporated microlenses in microchip structures to improve the performance of fluorescence detection. In particular, Camou et al. [157] have described the preparation and characterization of PDMS two-dimensional optical lenses integrated in PDMS microchips to render the whole piece of equipment portable. Roulet et al. [158] have described several procedures for integration of PDMS with glass microchips in order to optimize the spectral and spatial separation of the excitation and fluorescent light in the microchips.

One step further toward function unification in one microchip is to integrate the light detection device into the chip. In these cases, an implanted photodiode has been the detector of choice. Webster et al. [159] have developed a monolithic microchip electrophoresis system that features a photodiode created in the silicon substrate. The device incorporates an on-chip thin-film interference filter to prevent the excitation light from reaching the detection diode. Since silicon is a semiconductor material, a limitation of the monolithic silicon devices is that electrical isolation of the high electrical fields used for separation has to be provided. Usually, a layer of polysilicon [155] or aluminum zinc oxide [159] is deposited on the silicon substrate. Whitesides' group [151] has taken integration in PDMS a step further by incorporating with the PDMS chip, besides optical fibers and optical filters, a microavalanche photodiode as light detection element of the fluorescence detector in microchips. In this case, the photodiode was not really implanted in the microchip, but it was fabricated apart and embedded in the PDMS chip during its preparation.

Mathies' group [160] has developed a hydrogenated amorphous Si (a-SiH) photodiode to be coupled (not integrated) to a microchannel separation device. A half-ball lens, a multilayer long-pass optical interference filter with a 250-μm-diameter drilled hole, and an annular a-SiH photodiode were assembled successively on top of one another and supported in an aluminum base used as a platform for attachment of the separation chip. The beam of an Ar-ion laser passed through the annular photodiode, through the hole of the photodiode, and toward the half-ball lens and the separation channel. The fluorescence produced by the sample was collected back, filtered by the interferometric filter, and detected by the a-SiH photodiode. The microchip was glued to the detection platform by using index matching material. The total thickness of the integrated device was only 3.6 mm. This arrangement, combined with vertical cavity surface emitting laser diodes, could facilitate the construction of detection modulus for its application in portable, miniaturized devices. Recently [161], the coupling of an integrated fluorescence detection microdevice with a PDMS microstructure has been described. The integrated detector was obtained by deposition of a CdS thin-film filter on an Si PIN-photodetector using a pulsed laser deposition technique. An (In, Ga)N thin-film blue LED was deposited on top of the CdS filter by using a pixel-to-pixel laser lift-off process from the sapphire substrate where it was grown. The CdS coating acted as a filter of the light produced by the LED to prevent it to reach the photodetector. The light produced by the LED caused the fluorescence of the analyte contained in the PDMS channel mounted above, which was detected by the Si PIN-photodiode. Although promising, the capabilities of this approach in analytical separations have not been demonstrated yet.

6.5 INDIRECT FLUORESCENCE DETECTION

6.5.1 Background concepts: transfer ratio (TR) and dynamic reserve (DR)

Indirect detection is a universal and nondestructive detection method. Therefore, it can be used for detection of compounds without fluorophores, and chemical derivatization is not necessary. Also, the instrumentation required is the same as for direct detection. Indirect detection is based on the displacement of a fluorescing compound, called the probe, present in the background electrolyte (BGE) by the analyte, resulting in a decrease in signal intensity of the detector. In the simplest model, it is assumed that each analyte ion displaces its electrochemical equivalent of probe ions, of the same charge sign, thus maintaining electrical neutrality. The TR is defined as the number of moles of the probe displaced by one mole of analyte ions [162]. As the detector response is proportional to the TR, a higher value of the TR results in a large analyte peak area. On an intuitive level, one would expect displacement on an equivalent-per-equivalent basis so that, for example, the TR between a singly charged solute and a single-charged probe would be expected to be the unity. Consequently, the peak areas for analytes of the same charge and concentration should be the same, allowing the theoretical possibility of using the same calibration curve for different analytes. However, Ackermans et al. [163] showed that a nonlinear relationship existed between peak area and the effective mobilities of the ionic species for an equimolar sample composition. This can be explained by considering the electrophoretic separation mechanism for fully ionized ionic constituents, which can be approximately described by Kohlrausch's regulating function (ϖ) [164]. A consequence of the ϖ function is that TR is dependent on the mobility of the probe and the analyte. Thus, analytes that have a higher mobility than the probe ion result in fronting peaks, and analytes that are less mobile than the probe ion result in tailing peaks because of electromigration dispersion. Therefore, it is advantageous to select a probe ion with mobility as close as possible to the mobility of the pair of substances in the sample that are most difficult to resolve [165]. If two or more ions with the same charge sign as the analyte (co-ions) are used in the BGE, additional considerations, that are discussed in the review of Doble and Haddad [166], have to be taken into account. For example, when the mobility of an analyte was intermediate between the mobilities of two BGE co-ions, the analyte ion would mainly displace the co-ion to which its mobility was closest [167].

Another important parameter is the dynamic reserve, DR, which is defined as the ratio of the background signal to the background noise. In indirect detection, large background signal is required, and the background noise, which in direct detection is practically negligible, can become a serious problem. Thus, the limit of detection in concentration,

C_{LOD}, for indirect fluorescence detection is given by [168]

$$C_{\mathrm{LOD}} = \frac{C_{\mathrm{p}}}{\mathrm{TR} \cdot \mathrm{DR}} \tag{6.3}$$

where C_{p} is the concentration of the fluorescent probe, TR the transfer ratio, and DR the dynamic reserve. Since fluorescence intensity is proportional to the excitation power, excitation source power fluctuations (flicker noise) translate into baseline noise. Increased noise reduces the DR and thereby increases the limit of detection. Thus, indirect fluorescence detection is especially sensitive to excitation source power fluctuations. For this reason, a similar sensitivity or lower sensitivity was obtained in earlier works for indirect fluorescence detection compared to that obtained for indirect UV-absorbance detection (10^{-4}–10^{-5} M) [169]. The use of indirect LIF detection has improved the sensitivity by at least 1 order of magnitude [170] compared with indirect UV absorbance. In spite of this fact, it has been observed that indirect LIF detection has very limited use since it was first reported in 1988 by Kuhr and Yeung [168,171]. The limited use of indirect LIF has been attributed to the high cost associated with lasers, the low number of appropriate probes, and principally, the power instability of conventional gas-based lasers (typically 1%) [172] that are most frequently employed (He–Cd at 325 nm [168,170,171] and 422 nm [173], and Ar-ion at UV region (330–360 nm) [174–177], and 488 nm [178–181]). In some cases different devices for laser power stabilization were employed. Thus, Kuhr and Yeung [168] observed an increase in DR from 299 to 10,493 upon employing laser power stabilization, and Andersson et al. [170] reported a DR of 1005 compared with 80 without stabilization. However, laser power stabilization increases instrument complexity, cost, and loss in laser power (up to 40%).

Diode lasers are an alternative to gas lasers. Diode lasers are especially attractive to LIF [22] owing to their compact size, low cost, long lifetime, and high power stability. Unfortunately, most of diode lasers work in the near-IR region of the electromagnetic spectrum; although recently, diode lasers operating in the blue-violet or in the green region have been commercialized (see section 6.2.1).

6.5.2 Choice of the background electrolyte (BGE): types of electrolyte additives used in indirect fluorescence detection

The choice of the BGE is of paramount importance in indirect detection applications. Besides the requirements imposed by the indirect detection scheme (they will be mentioned below), it is also necessary to take more requirements of the CE separation into account.

The pH in the BGE is dictated by the need to have the analytes ionized; for instance, in the case of sugars [173,180] a pH around 12 is required to obtain sufficient mobilities. To select the type of BGE ions, it is necessary to consider the pH buffering capacity and the possibility of matching the mobility of the most important and most difficult-to-separate analytes. Thus, we will obtain symmetrical and narrow peaks for those compounds.

While selecting the BGE, the electrical conductivity of the concentration selection is important, as it determines power dissipation and the upper limit of the field strength. Thus, concentrations above 0.01 M are not useful [182]. However, low concentrations will result in excessive electromigration dispersion.

Sometimes, special additives are used in the BGE with different aims. Thus, diethylentriamine [183] was used as electroosmotic flow (EOF) suppressor and to reduce capillary wall adsorption. Desbene et al. [184] employed hexadimethrine bromide (HDMB) and tetradecyltrimethylammonium bromide (TTAB) to reverse EOF. Also, zwitterionic compounds are used in the BGE. Thus, Lee and Whang [185] employed *N*-2-hydroxyethylpiperazine-*N*′-2-ethanesulfonic acid (HEPES) in the running buffer containing 6-aminoquinoline (6-AQ) as fluorescent probe. Other compounds, such as hydroxyisobutyric acid [177] and ethylenediaminetetraacetic acid (EDTA) [186], were used as complexing agents to enable the separation of analyte ions that have similar mobilities or to eliminate interferent ions. Also, organic solvents are used in the BGE when nonpolar analytes are separated. Chiu et al. [187] used a mixture of acetonitrile and ethanol in the BGE to separate aliphatic carboxylic acids.

Other considerations imposed by the indirect detection scheme have to be taken into account as were [182]. First, in general, the probe ion should be the unique ion present in the BGE with the same charge sign as the analyte ion. The presence of other nonresponsive ions in the BGE would affect the TR, and therefore, the sensitivity. Second, the probe ion should have a stable response in the detector. Third, the slope of the calibration curve should be large enough to cause a high response in the detector. Fourth, low-conductivity BGEs are preferred because thermal dissipation generates noise. Last, system peaks are best avoided by choosing a BGE with a composition as simple as possible. Additional ions in the BGE, especially with the same charge as the probe ion, could displace the probe ion and generate system peaks that could interfere with the analyte peak.

6.5.3 Direct versus indirect fluorescence detection

Direct LIF detection is at least two orders of magnitude more sensitive than indirect LIF detection, but because most analytes lack fluorescence properties, their derivatization is required in direct methods. Therefore, analytes must be modified in order to reveal fluorescence. Analytes containing highly reactive groups such as amines or thiols have been extensively derivatized, and in these cases, direct LIF detection methods are more frequently employed. In contrast, analytes such as inorganic ions or carboxylic acids are not so easily derivatized, and indirect detection is frequently the best alternative. Also, indirect detection is universal and no chemical manipulation is required.

6.5.4 Applications of indirect fluorescence detection

Traditionally, indirect detection has been applied to the analysis of organic [184,188–190] and more frequently, to inorganic [181,189,191–193] anions. However, the analysis of organic [185] and inorganic [176,177] cations has also been described. Chen et al. [194] reported a simultaneous analysis of cations and anions employing fluorescein as probe and EDTA as complexing agent. The limits of detection found ($S/N = 3$) were around 10 µM (see Fig. 6.10). Amino acids [171], proteins and nucleotides [168], and peptides from tryptic digest [175] were studied by Yeung's group using laser fluorescence and salycilate as fluorescent probe.

Indirect fluorescence detection has also been applied to the analysis of mono- [173] and polysaccharides [178]. Ruiz-Calero et al. [179] described two methods for monosaccharide analysis. The effects of fluorescein (probe) concentration, pH of the BGE, and the influence of the reversal of EOF by addition of HDMB were studied.

Finally, indirect LIF has been used as a detection method on microchips. Thus, Munro et al. [180] developed a method for amino acid analysis on a microchip with an average detection limit of 32.9 µM. This method was applied to the analysis of amino acids in urine samples (see Fig. 6.11). Sirichai et al. [195] presented the detection of both print and

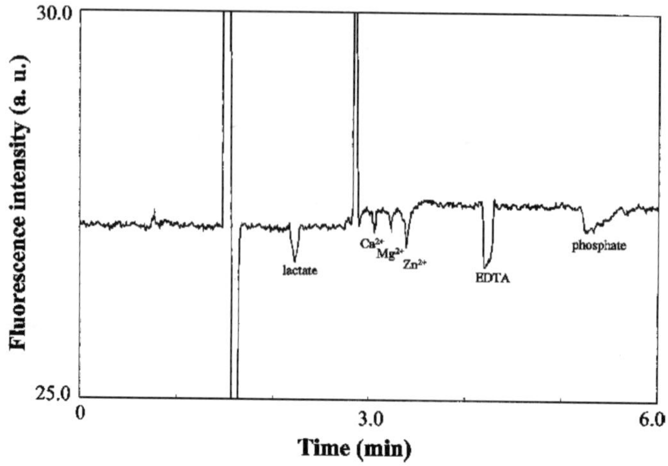

Fig. 6.10. Simultaneous separation of anions and cations in a Chinese herbal soup. Separation conditions: BGE, 5 mM borate containing 5 μM fluorescein and 1 μM sodium chloride (pH 9.2); voltage, 15 kV. Sample diluted 10 times with 5 mM borate containing 10 mM EDTA (pH 9.2). Reprinted from Ref. [194]. Copyright (2003), with permission from Elsevier.

film developing agents (CD-3 and CD-4) in photographic processing solutions using microchips and indirect fluorescence detection employing fluorescein as probe. The limits of detection were 0.17 mM for CD-3 and 0.39 mM for CD-4. Other compounds such as phenols [196] in industrial wastewater, and standards of fatty acids [183] were analyzed employing fluorescein as probe.

In summary, indirect fluorescence detection has been mainly used for the analysis of inorganic ions, but sugars, fatty acids, amino acids, amines, proteins, nucleotides, and other organic ions have also been studied in biological matrices and water. The most employed fluorescent probe has been fluorescein [176,178–180,183,184,188,190–192, 194–199], but other probes such as eosine [181], riboflavin-5-phosphate (FMN) [189], 6-AQ [185], coumarin [173], salycilate [168,171], 8-hydroxypyrene-1,3,6-trisulfonic acid (HPTS) [193], rhodamine 110 [200] 2,5-dihydroxibenzoic acid (2,5-DHB) [201], terbium(III)acetylacetone [202], TNS [203], and merocyanine 540 [187] were also employed. The limits of detection observed were, in general, between 10^{-4} and 10^{-6} M (1–100 ppb). A summary of these and other applications are given in Table 6.4.

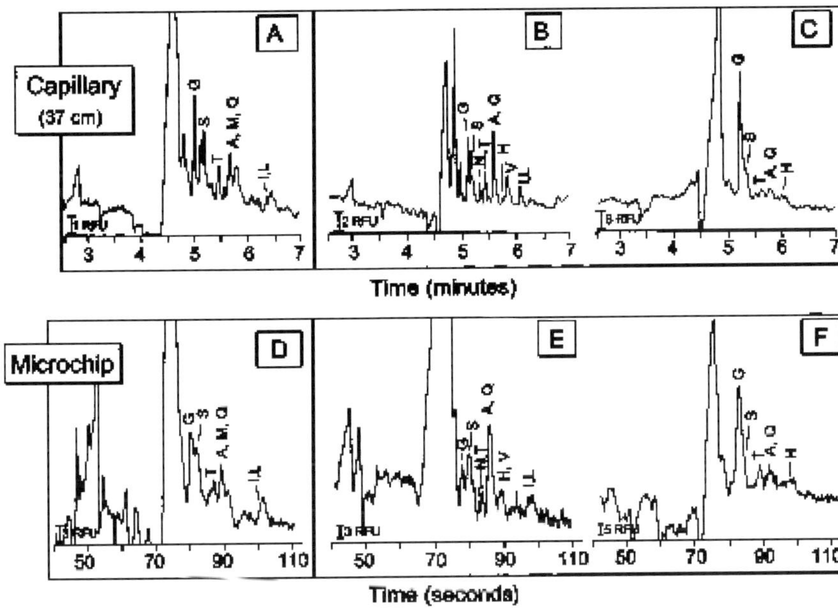

Fig. 6.11. Separation of 19 amino acids in urine samples via capillary and microchip electrophoresis. Buffer: 1.0 mM sodium carbonate, 0.5 mM fluorescein, and 0.2 mM CTAOH at pH 10.3. Capillary conditions: effective length 30 cm, 1 s sample injection, 224 V/cm (reversed polarity) separation voltage, and sample amino acid concentrations of 0.4 mM in 0.5 mM fluorescein. Microchip conditions: effective length 5.5 cm, 15 s injection at 417 V/cm, 183 V/cm (reversed polarity) separation voltage, and sample amino acid concentration of 0.4 mM in 1.0 mM sodium carbonate and 0.2 mM CTAOH. Capillary electrophoresis of (A) the normal specimen, (B) abnormal urine sample 1, and (C) abnormal urine sample 2. Microchip separation of (D) the normal specimen, (E) abnormal urine sample 1, and (F) abnormal urine sample 2. Amino acids named with one-letter notation. Reprinted with permission from Ref. [180]. Copyright (2000), American Chemical Society.

6.6 CONCLUSIONS AND FUTURE TRENDS

As has been discussed in this chapter, LIF detection in capillary and microchip electrophoresis is a very sensitive technique if the molar absorptivity and the quantum efficiency of the analyte at the wavelength of the laser are high enough. Owing to the directional characteristics of laser radiation, sufficient optical power can be focused in a tiny spot with only a few milliwatts of laser output necessary to cause fluorescence. Unfortunately, only a few solutes are able to produce

TABLE 6.4

Some applications of CE with indirect fluorescence detection

Analyte	Sample	Probe	BGE	Comments	Reference
Anions					
Chloride, fluoride, nitrite, nitrate, sulfate, oxalate, citrate	Mineral and spring water	Fluorescein	0.1 M boric acid, 0.02 M sodium tetraborate (pH 8.5), 10^{-5} M TTAB, 8×10^{-5} M fluorescein	CLOD: 10–100 ppb	[184]
Sulfate, nitrite, nitrate, azide, thiocyanate, chlorate	Mineral and tap water	HPTS	50 μM HPTS, 10 mM lysine, 0.19 mM calcium hydroxide (pH 9.9)	CLOD: 0.9–4.7 μM	[193]
Arsenic compounds	Ground water	Fluorescein	1.5 mM fluorescein (pH 9.8)	CLOD: 40–160 ppb	[191]
Arsenic compounds	Mineral water	Fluorescein	2.0 mM sodium hydrogencarbonate, 10^{-7} M fluorescein (pH 9.28)	CLOD: 0.12–0.54 μM	[199]
CN^- and related compounds	Standards	Fluorescein	0.1 mM fluorescein (pH 10.0)	CLOD: ~1 μM	[192]
Bromate, carbonate, chloride, methanesulfonate, nitrate, oxalate, sulfate	Standards	2,5-DHB	1.0 mM 2,5-DHB, 0.5 mM lead acetate	CLOD: 0.2–0.5 μM	[201]
Chromate, hexacyanoferrate(III), nitrite	Standards	Terbium(III)-acetylacetone	0.8 mM acetylacetone, 0.35 mM terbium(III)-chloride, 6 mM sodium chloride (pH 8.3)	CLOD: 0.1 μM	[202]
Bromide, chloride, nitrite, nitrate, sulfate, oxalate, fluoride, succinate, malate, tartrate, citrate, hydrogencarbonate	Mineral and tap water	Eosine	70 mM boric acid, 5 mM sodium tetraborate, 2 mM DETA, 10^{-6} M eosine (pH 8–9)	CLOD: 8–37 ppb	[181]
Cations					
Triorganotin cations	Standards	6-AQ	70 mM HEPES (pH 4), 0.1 mM 6-AQ	CLOD: 8–18 μM	[185]
Na^+ and K^+	Single erythrocyte cell	6-AQ	200 μM CTAB, 500 μM 6-AQ (pH 3.8)		[257]
Anions and cations					
Ba^{2+}, Ca^{2+}, Mg^{2+}, Zn^{2+}, lactate, phosphate	Pond water and Chinese herbal soup	Fluorescein	5 mM borate buffer, 5 μM fluorescein, and 1 μM NaCl (pH 9.2)	CLOD: 3.7–14.6 μM	[194]
Sugars					
Monosaccharides	Standards	Coumarin	1 mM coumarin (pH 11.65)	LOD: 2 fmol	[173]
Monosaccharides and glycosaminoglycans	Standards	Fluorescein	0.0005% (w/v) HDMB, 10^{-6} M fluorescein (pH 12.25)	CLOD: 2.76–50.9 μM	[179]
Polysaccharides (high molecular mass)	Standards	Fluorescein	1 mM fluorescein (pH 11.5)	LOD: pg range	[178]

Amino acids, amines, and peptides					
Amino acids	Standards	Salicylate	1.0 mM sodium salycilate, 200 μM sodium carbonate (pH 9.7)	CLOD: 5 μM	[171]
Biogenic amines	Standards	Rhodamine 110	30 mM phosphate, 3% 2-propanol, 500 μM rhodamine 110 (pH 9.4)	CLOD: ~6 μM. Performed in microchip	[200]
Peptides from tryptic digest	Standards	Salicylate	500 μM salicylate-CAPS (pH 10.9)	LOD: 165 amol	[175]
Carboxylic acids					
Aliphatic carboxylic acids (C_2–C_{18})	Marine lipid concentrate	Merocyanine 540	1.0 mM Tris (pH 7.4), 40% ACN, 30% EtOH, 10 μM merocyanine 540	CLOD: Sub-μM–μM	[187]
Fatty acids (C_6–C_{24})	Standards	Fluorescein	1.0 mM fluorescein containing MeOH and ACN	CLOD: 0.01 mM	[183]
Valproic acid	Serum	Fluorescein	2.5 mM borate–phosphate (pH 8.4), 6 μM fluorescein	CLOD: 900 ppb	[188]
Other compounds					
Nucleotides (AMP, GMP, TMP, CMP) Lyz, iodate, bicarbonate	Standards	Salicylate	250 μM salicylate (pHs 3.5,7.0)	LOD: 70 amol	[168]
Inorganic anions, organic acids, anionic surfactants, and polyphosphates	Standards	FMN	2 mM diethylenetriamine, 10 μM FMN (pH 7.8)	CLOD: 1000–2000 ppb	[189]
Glufosinate, glyphosate, and aminomethylphosphonic acid	Commercial herbicides and ground water	Fluorescein	1.0 mM fluorescein (pH 9.5)	CLOD: μM range	[198]
n-Alkane sulfonates (C_1–C_{10}) and carboxylic acids (C_1–C_{18})	Standards	Fluorescein	5 mM sodium tetraborate (pH 9.2)-EtOH (60:40, v/v), 10^{-5} M fluorescein	CLOD: 700–800 ppb	[190]
Phenols	Industrial wastewaters	Fluorescein	15 mM sodium borate (pH 9.9), 1.0 mM fluorescein	CLOD: 10 ppb	[196]
Dodecyl sulfate	Soap	TNS	0.1 mM TNS, 10 mM sodium hydrogenphosphate, 6.0 mM sodium tetraborate		[203]
Selenite, selenate, selenomethionine, selenocystine, antimonite, antimonate	Standards	Fluorescein	1.2 mM fluorescein (pH 9.5)	CLOD: μM range	[197]

For abbreviations see Table 6.1.

fluorescence at the wavelengths of relatively inexpensive lasers such as Ar-ion, He–Cd, He–Ne, or diode lasers. The most employed method to circumvent this limitation is to derivatize the analyte using pre-, on-, or post-capillary techniques. In LIF practice for real-world samples, sensitivity was not be limited by the detection system, but by the slow kinetics that are characteristic of very dilute solutions of the analyte [204]. A challenging problem is protein detection at trace level. This problem is encountered, among other situations, in host protein detection in biotechnological products, detection of allergenic proteins in hypoallergenic food products, or in proteomics of low-expression proteins. In protein analysis, besides the slow kinetics limitation due to diluted samples, multiple labeling of proteins with the corresponding band broadening or multiple peaks formation for a pure species [84,205] is also an issue. For protein analysis at trace level, on-capillary derivatization or post-capillary derivatization using sheath-flow cuvettes can be of interest in some cases [36,74], particularly now that CCD cameras allow multitrack spectroscopy, that is, spectroscopic analysis of extended light sources. Despite this advantage, the price of the scientific CCD camera is still high and its sensitivity is still low when compared with PMT devices. On the other hand, on-capillary staining techniques that use hydrophobic interactions between protein and dyes can be useful in some cases, particularly with some of the new dyes developed for proteomics [206]. However, the sensitivity achieved with this on-capillary staining strategy is rather poor owing to the background fluorescence of the buffer and the variable uptake of the dye from protein to protein. An alternative methodology to analyte derivatization is indirect fluorescence detection, but the dynamic reserve obtained with most of the dyes used as buffer additives limits the sensitivity of the technique.

In our opinion, fluorescence detection using multiphoton spectroscopy or time-resolved spectroscopy, although a little bit unusual nowadays owing to the price of the instrumentation, can be of interest in the future. Even if fluorescence spectra of most substances are featureless, LIF detection with spatial resolution can be of some help in the identification of some analytes as a complement to mass spectrometry detection (see Chapter 10 of this book). Finally, new dye reagents whose absorption spectra match the light emission of cheap lasers, such as diode lasers, together with good derivatization techniques are still needed, particularly for protein studies. In the following sections, we discuss some of these techniques.

6.6.1 Multiphoton fluorescence detection

Despite its potentialities as detection technique in CE, multiphoton fluorescence spectroscopy has been rarely used mainly because the high-peak power, high-frequency pulsed lasers required for this technique are rather expensive at present [172,207].

The physical principle behind multiphoton fluorescence, illustrated for the case of two-photon excitation (Fig. 6.12), follows [208]. In a single-photon excitation process, the energy of the source photons must be equal to the energy difference between fundamental and excited electronics levels. For the two-photon excitation process, the transition from these two levels can be represented as a sequential process of two transitions: first, from the ground state to an intermediate state called the virtual state, and then from the virtual state to the excited state. This virtual state is a superposition of energy levels that exist as a result of the Heisenberg broadening states. Owing to the extremely short life of the virtual state, a molecule promoted to this state by a first photon must absorb a second photon within ~1 fs (femtosecond, 10^{-15} s). Otherwise, the molecule in the virtual state relaxes back to the ground state, causing light scattering. Obviously, if the molecule returns to the ground state from the excited state, fluorescence is produced by the emission of one photon, but with a much shorter wavelength than that of photons used for excitation. Similarly, multiphoton excitation processes can take place for some molecules using several virtual states as intermediates from the ground to the excited state.

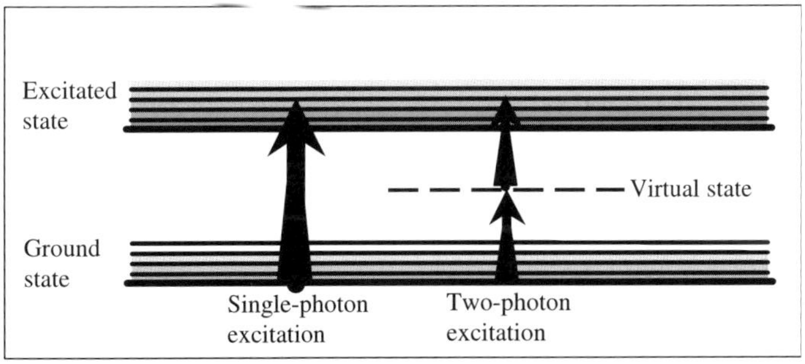

Fig. 6.12. Excitation of molecules of chromophores using in single-photon spectroscopy and two-photon spectroscopy. Reprinted with permission of the author from Ref. [208].

In summary, owing to the low probability of the sequential absorption of several photons by the same molecule and to the very short life of the molecules in the virtual states, multiphoton excitation requires very high power densities and high frequency of pulses. This is now achieved by focusing the beam of an NIR Ti:sapphire mode locked laser in a very small region of the sample by using high NA (NA = 1.3) microscope objectives (spot size around 250 nm, close to the diffraction limit). This laser is able to produce pulses in the range of femtoseconds at frequencies higher than 50 MHz (50×10^6 pulses per second) maintaining a very low average intensity but a very high-peak intensity (for instance, using a $\times 100$ microscope objective a pulse power of $10^{12}\,W\,cm^{-2}$ can be obtained).

The advantages of multiphoton fluorescence techniques over one-photon fluorescence techniques are: (i) because the fluorescence emission is at shorter wavelengths than those of laser emission, the background scatter does not interfere in the fluorescence measurements. (ii) Since laser intensities large enough to generate significant multiphoton excitation are localized to regions within a few micrometers of the focal point, background caused by multiphoton fluorescence by optical components and filters is negligible. Consequently, very high S/N ratios can be achieved. (iii) To increase the probability of exciting molecules in the very small region defined by the tightly focused beam of the laser that requires multiphoton fluorescence, small ID capillaries (around 0.6 μm) should be employed, and an off-capillary detection technique used [209]. With such capillaries, high electrical field intensities can be achieved without thermal overloading. Therefore, fast separations (in the range of a few tens of seconds) can be obtained.

The applicability of multiphoton spectroscopy to CE has been demonstrated using on-capillary detection [210] and post-capillary detection [211,212]. Using this technique a concentration detection limit of 4×10^{-9} M for cumarine 440 [210], limits of detection ranging from 38×10^{-9} M for flavin adenin dinucleotide (FAD) to 110×10^{-9} M for nicotinamide adenine dinucleotide (NADH) [211], and detections limits of 130×10^{-9} M for fluorescamin derivatives of some neurotransmitters such as dopamine, glycine, and glutamate [212] have been obtained. Also, several biological toxins, such as aflatoxins B_1, B_2, G_1, and cholera toxin A-subunit, with detection limits ranging $0.2–0.44 \times 10^{-9}$ M have been analyzed [213].

Finally, the use of very small capillary ID required by multiphoton fluorescence detection, together with their mass sensitivities (at the attomole-level) have allowed the study of the tryptic peptidic maps of several varieties of cytochromes by injecting as few as 2 pl of the sample, digesting it on-capillary, and using two-photon spectroscopy as method [80]. This technique has also been used in the study of the dynamics of photodegradation of serotonin metabolites [214].

6.6.2 Wavelength-resolved LIF detection

Wavelength-resolved fluorescence detection furnishes spectral information to complement migration time, allowing discrimination between analytes. The usefulness of diode array in absorbance detection is similar to that of wavelength-resolved fluorescence detection. The largest impediment to the application of multichannel detectors in fluorescence has been the lack of light detectors furnishing the sensitivity of a single-channel PMT. In fact, until the development of photodiode arrays (PDA) and CCDs and more recently, intensified PDA (IPDA) and intensified CCD (ICCD), the detection limits were poor.

As indicated previously, fluorescence detection systems consist of three major subsystems: excitation, collection, and detection. For multiple-wavelength fluorescence detection, all subsystems are identical to those employed in single wavelength except the detection one because in this case a wavelength selector is required. It can be a series of filters for preselected wavelengths, or a grating or prism for detection over a range of wavelengths [215]. When a short number of filters are used, the output of each filter is sent to a different PMT. If a broader range of wavelengths is of interest, a spectrograph should be used to analyze the light from the capillary. The image of the capillary is focused at the entrance slit of the spectrograph, and the spatially resolved wavelengths at the exit of the spectrograph are recorded by a CCD camera.

Timperman et al. [216] have developed a CE–LIF–CCD system employing an Ar/Kr mixed gas-ion laser. The combination of a spectrograph and a liquid-nitrogen-cooled CCD camera enabled the use of a 500 nm spectral window adjustable in the 300–1000 nm region. A limit of detection of 1.5×10^{-13} M (220 molecules) was obtained for fluorescein. The method was applied to differentiate a mixture of amino acids derivatized with two fluorophores, BODIPY 503/512 C3 and BODIPY 576/589 C3. Capability of the detection system to distinguish multiple fluorophores based on spectral characteristics was realized. Hanning et al. [217] developed an LIF detector based on the principle of total internal reflection (TIR). It was achieved by coating the separation capillary externally with a medium of lower RI than the separation buffer, thereby inducing liquid-core waveguiding so that imaging could occur at the capillary end. Concentration detection limit for fluorescein as good as 6.2×10^{-14} M was obtained. Carson et al. [218] and Fultz et al. [219] reported IPDA (mounted on a spectrograph) wavelength fluorescence detection systems. Detection limits were in the picomolar range.

The application of wavelength-resolved fluorescence detection to the analysis of compounds with native fluorescence allows the detection of different analytes with different spectral profiles without derivatization. Sweedler's group [36,220] described the use of a frequency-doubled Ar–Kr or Ar-ion laser to collect the native fluorescence of tryptophan and tyrosine. Later on, the system developed was improved for single neuron analysis [221], allowing the identification of more than 30 compounds in individual neurons. Limits of detection were between 10^{-7} and 10^{-9} M. Also, Gooijer's group developed an IPDA-based system employing a pulsed KrF excimer dye (Rhodamine 6G) laser with a frequency-doubling crystal to provide tunable UV output [222] or a modified Ar-ion laser to provide 275 nm excitation [223]. In the first case, the system developed was applied to the environmental analysis of naphthalenesulphonates (NS) in river water, achieving detection limits from low- to sub-$\mu g\, l^{-1}$. In the other case, isoflavonoids were determined in red clover using MECC with detection limits between 0.1 and 0.4 μM.

Wavelength-resolved LIF detection has been successfully applied to multifluorophore DNA sequencing [217,218], cellular analysis [220,221], clinical assays [224–228], environmental applications [222], and other biological applications [223]. Also, the information in the fluorescence emission spectrum can yield diagnostic information on the separation. For example, estimation of the pH inside the capillary during a separation has been achieved by use of a pH-sensitive fluorophore, carboxy SNARF-1 [229].

Finally, a CE equipment with wavelength-resolved detection has now been commercialized. The Applied Biosystems (ABI) prism 3100 Genetic Analyzer is an automated instrument designed for analysis of fluorescently labeled DNA fragments separated by capillary gel electrophoresis [230]. A spectrograph–CCD is used to achieve wavelength resolution and to enable differentiation of the different dye-labeled nucleotides.

Identification capabilities of CE have been extended with the coupling of this technique with fluorescence line narrowing (FLN) spectroscopy. In this spectroscopic technique, a capillary cryostat is used to cool the analytes once separated to very low temperatures (4.2 K) using a continuous flow of liquid nitrogen or helium through the cryostat. It makes it possible to obtain that narrow bands for fluorescence spectra of some substances can be achieved instead of the large, featureless band of the regular fluorescence spectra that present the analytes in liquid solutions. In FLN spectroscopy, the well-defined wavelength of a laser light only excites a selection of analyte molecules that possesses the exact energy of the difference between the ground and excited state that matches the wavelength of the laser. If the sample is cooled at low temperature this selection is maintained until the emission takes place, and as a result, a narrowing of the emission bands takes place. Jankowiak's group has demonstrated the applicability of this method to compounds separated by CE. It is beyond the scope of this chapter to describe the capillary cryostat and the optical setup used [231]. The identification possibilities of the technique have been demonstrated in the analysis of the DNA adducts of potent carcinogens such as benzo[*a*]pyrene and dibenzo[*a,l*]pyrene [232].

6.6.3 Time-resolved LIF

Time-resolved detection allows the measuring of fluorescence lifetimes. In order to do so, the fluorophore has to be excited using a pulsed light source whose pulse time duration is much shorter than the lifetime of the fluorescence measured. Once excited, the time difference between excitation and emission is measured for photon counting. As a result, a decay profile from which the fluorophore lifetime can be extracted is constructed.

The measurement of fluorescence lifetimes online in analytical separations has made significant progress over the past few years, primarily due to advances in instrumentation, namely, the development of suitable excitation sources and improvements in detectors appropriate for lifetime measurements [233]. Most time-resolved fluorescence measurements have been performed in the visible region of the spectrum (400–700 nm). This is because of several factors, including the commercial availability of light sources appropriate for time-resolved fluorescence applications in the visible and a wide selection of functional fluorophores that demonstrate excitation/emission in the visible wavelength range. Thus, in one of the first demonstrations of time-resolved fluorescence [234], a nanosecond pulsed nitrogen laser was used to excite the fluorescence, and the emission was monitored at two different times within the decay profile. This process allowed the removal of background photons from the measured signal, significantly improving the signal-to-noise ratio.

First, strike-cameras, and later, ICCD cameras were used as detectors to measure fluorescence lifetimes. Nowadays, single APDs are typically employed as detectors. The associated timing electronics usually consist of a constant fraction discriminator and a time-to-amplitude converter [233].

Diode lasers have been employed with doubling or tripling frequency to produce green or ultraviolet wavelengths. Thus, Basile et al. [235] reported the use of a frequency-tripled diode-pumped Nd:YLF (neodymium doped yttrium lithium fluoride) laser accomplished with a digital oscilloscope for signal processing. Limits of detection were between 10 and 100 pM. Excited-state decays were collected for several enzymatic probes and quinine sulfate, providing lifetimes consistent with those obtained by specialized instruments for lifetime measurements, such as time-correlated, single photon counting.

Recently, new NIR fluorophores have been developed to be excited with NIR diode lasers, such as Ti:sapphire laser. One of the major advantages of working in the NIR spectrum region is that the background level is lower compared with visible region because fewer materials show intrinsic fluorescence in the NIR compared with the visible region [233].

It is important to use fluorophores that possess lifetimes at least as long as the instrument response function, to permit efficient calculation of the lifetime. When multiple fluorophores need to be identified in a single CE run, it is necessary to use fluorophores with sufficiently different lifetimes to allow their differentiation. In fact, the difference in lifetimes required for identification must be larger than the standard deviation in the measurement. In addition, while employing fluorophores that have the same absorption and emission maximum but different lifetimes, only one single detection channel is required (in contrast to wavelength-resolved fluorescence). Lieberwirth et al. [236] and Flanagan et al. [237] have described a set of multiplex dyes that possess these characteristics.

The major application for time-resolved fluorescence in CE has been in the area of DNA sequencing [236–239], in which each nucleotide base is associated with a fluorophore probe, and a lifetime discrimination is made. In this case, fluorophores with similar emission maximum are required. Lifetime discriminations eliminate the problem of cross talk between detection channels associated with the overlapping of emission profiles when different spectral wavelengths are being monitored, such as can occur in wavelength-resolved fluorescence detection.

6.6.4 New approaches to fluorescent labeling of proteins at low concentrations

The interest of detection of proteins at low concentration as well as the difficulties encountered to solve this problem have already been discussed in this chapter. Although several solutions to this problem have been suggested [205,240,241], these are not universal solutions. In this section, we consider briefly two potential methods of protein derivatization that could be of interest in the future—enzymatic tagging and nanoparticle labeling.

Enzymatic reactions, which control most of the biochemical reactions relevant for life, are exquisitely selective, fast, and efficient. Almost every type of chemical reaction involving proteins carried out by living beings, such as oxidation, reduction, moieties addition, and hydrolysis, is governed by one or several enzymes [242]. Banks [243] has suggested the possibility of using transglutaminase for labeling glutamine residues with uncharged fluorophores associated with cadaverine, such as tetramethylrhodamine-cadaverine. Krylov's group [244] has demonstrated the applicability of the reaction of farnesyltranferase to label peptides of the oncogenic form of *Ras* protein with a fluorescent tag. They have also indicated that prenylation of proteins using a similar reaction could be possible.

The development of solid particles with nanometer size has opened up new possibilities in clinical separations [245,246]. Metallic (gold or silver) [247,248] and polystyrene [249] nanoparticles and nanocrystals [250] have demonstrated good sensitivity in laser luminescence detection. The possibility of using this technology in protein derivatization remains to be demonstrated.

6.7 SELECTED APPLICATIONS

Some applications to separations of small molecules (Table 6.1) and proteins (Table 6.2) monitored using LIF detection are summarized. Some separations of real samples carried out using microchips and LIF detection are summarized in Table 6.3. Finally, in Table 6.4, we present some applications of CE separations using indirect LIF detection.

ACKNOWLEDGMENTS

The authors are grateful to CICYT for economic support (Project TIC 2003-0196). M.T.V. and P. L.-Q., thank the Spanish Ministry of Education and Science for FPU pre-doctoral grants.

REFERENCES

1 R.P.W. Scott, Liquid chromatography detectors, *J. Liquid Chromatogr. Library*, 33 (1986).
2 E.S. Yeung, *Detectors for Liquid Chromatography*, Wiley, New York, 1986.
3 R.P.W. Scott, *Chromatographic Detectors*, M. Dekker, New York, 1996, Chromatographic Science Series, Vol. 73.
4 A.J. de Mello, *Lab. Chip.*, 3 (2003) 29N–34N.
5 D.Y. Chen and N.J. Dovichi, *Anal. Chem.*, 68 (1996) 690–696.
6 N.J. Dovichi and S. Hu, *Curr. Op. Chem. Biol.*, 7 (2003) 603–608.
7 B.L. Hogan and E.S. Yeung, *Trends Anal. Chem.*, 12 (1993) 4–9.
8 N.J. Dovichi, *Electrophoresis*, 18 (1997) 21–25.
9 D.A. Skoog, F.J. Holler and T.A. Nimean, *Principles of Instrumental Analysis*, Sounder Golden Sunburst, Orlando, 1998, pp. 356–360.
10 P. Thibault, N.J. Dovichi. In: P. Camilleri (Ed.), *Capillary Electrophoresis. Theory and Practice*, CRC Press, Boca Raton, FL, 1998, Chapter 2.
11 A.W. Moore Jr. and J.W. Jorgenson, *Anal. Chem.*, 65 (1993) 3550–3560.
12 A. Cifuentes, J.C. Diez-Masa, J. Fritz, D. Anselmetti and A.E. Bruno, *Anal. Chem.*, 70 (1998) 3458–3462.
13 Y.W. Lin, T.C. Chiu and H.T. Chang, *J. Chromatogr. B*, 793 (2003) 37–48.
14 S. Hu and N.J. Dovichi, *Anal. Chem.*, 74 (2002) 2833–2850.
15 T. Beer, N.H. Velthorst, U.A.Th. Brikman and C. Gooijer, *J. Chromatogr. A*, 971 (2002) 1–35.
16 F.G. Smith and T.A. King, *Optics and Photonics: An Introduction*, John Wiley, Chichester, 2000 Chapter 15.
17 D.L. Andrews, *Lasers in Chemistry*, Springer, Berlin, 1997.
18 J. Kuijt, C. García-Ruiz, G.J. Stroomberg, M.L. Marina, F. Ariese, U.A.Th. Brinkman and C. Gooijer, *J. Chromatogr. A*, 907 (2001) 291–299.
19 S. Nikamura, W. Kaender, *Laser Focus World*, April 1999.

20 http://www.point-source.com.
21 J.E. Melanson and C.A. Lucy, *Analyst*, 125 (2000) 1049–1052.
22 J.E. Melanson, C.A. Boulet and C.A. Lucy, *Anal. Chem.*, 73 (2001) 1809–1813.
23 A. Holobar, R. Brenes, B.H. Weigl, P. O'Leary, P. Rasport and O.S. Wolfweis, *Anal. Methods Instrum.*, 2 (1995) 92–100.
24 A.E. Bruno, F. Maystre, B. Krattiger, P. Nussbaum and E. Gassman, *Trends Anal. Chem.*, 13 (1994) 190–198.
25 F.T.S. Yu and X. Yang, *Introduction to Optical Engineering*, Cambridge University Press, Cambridge, 1997.
26 S.L. Pentoney Jr. and J.V. Sweedler. In: J.P. Landers (Ed.), *Handbook of Capillary Electrophoresis*, CRC Press, Boca Raton, 1994.
27 S. Wu and N.J. Dovichi, *J. Chromatogr.*, 480 (1989) 141–155.
28 A.E. Bruno, E. Gassmann, N. Pericles and K. Anton, *Anal. Chem.*, 61 (1989) 876–883.
29 E.L. Safford Jr., *The Fiberoptics and Laser Handbook*, Tab Books Inc., Blue Ridge Summit, 1984.
30 G. Nouadje, M. Nertz, Ph. Verdeguer and F. Couderc,, *J. Chromatogr. A*, 717 (1995) 335–343.
31 A.E. Bruno, B. Krattiger, F. Maystre and H.M. Widmer, *Anal. Chem.*, 63 (1991) 2689–2697.
32 L.A. Herzenberg, R.G. Sweet and L.A. Herzenberg, *Sci. Am.*, 254 (1976) 108–117.
33 N.J. Dovichi, J.C. Martin, J.H. Jett and R.A. Keller, *Science*, 219 (1983) 845–847.
34 >Y.F. Cheng, S. Wu, D.Y. Chen and N.J. Dovichi, *Anal. Chem.*, 62 (1990) 496–503.
35 C. Dennis and R. Gallagher, *The Human Genome*, Nature-Palgrave, New York, 2001.
36 K.E. Oldenburg, X.Y. Xi and J.V. Sweedler, *Analyst*, 122 (1997) 1581–1585.
37 F.G. Smith and T.A. King, *Optics and Photonics: An Introduction*, John Wiley, Chichester, 2000, pp. 58–59.
38 S.L. Pentoney Jr., K.D. Konrad and W. Kaye, *Electrophoresis*, 13 (1992) 467–474.
39 L. Hernandez, R. Marquina, J. Escalona and N.A. Guzman, *J. Chromatogr.*, 502 (1990) 247–255.
40 X.C. Huang, M.A. Quesada and R.A. Mathies, *Anal. Chem.*, 64 (1992) 967–972.
41 A. Malek and M.G. Khaledi, *Anal. Biochem.*, 270 (1999) 50–58.
42 S. Bhoopathy and H.T. Karnes, *Biomed. Chromatogr.*, 16 (2002) 1–6.
43 A. Einstein. In: M.H. Shamosed (Ed.), *Great Experiments of Physics*, Dover, New York, 1987, p. 232.

44 D.A. Skoog, F.J. Holler and T.A. Nimean, *Principles of Instrumental Analysis*, Sounder Golden Sunburst, Orlando, 1998, pp. 170–172.
45 *Photomultiplier Tubes*, Hamamatsu Photonics, Toyooka-village, 2002.
46 K.J. Voss, *Methods Enzymol.*, 305 (2000) 53–62.
47 H. Dautet, P. Deschamps, B. Dion, A.D. MacGregor, D. MacSween, R.J. McIntry, C. Trotiier and P.P. Webb, *Appl. Opt.*, 32 (1993) 3894–3900.
48 F.G. Smith and T.A. King, *Optics and Photonics: An Introduction*, John Wiley, Chichester, 2000 Chapter 21.
49 J.M. Martínez-Duart, In: J.M. Abella-Martín, J.J. Jiménez-Lidón and J.M. Martínez-Duart (Eds.), *Optoelectrónica y Comunicación Optica*, Consejo Superior de Investigaciones Científicas, Madrid, 1988, Capítulo 2.
50 G.F. Amelio, *Sci. Am.*, 230 (1974) 23–32.
51 *Multi-channel Detection System and Accessories*, Oriel Instruments, 1998.
52 M. Fischetti, *Sci. Am.*, 289 (2003) 74–75.
53 K.L. Holmes and L.M. Lantz, *Methods Cell Biol.*, 63 (2001) 185–204.
54 H.A. Bardelmeijer, H. Lingeman, C. De Ruiter and W.J.M. Underberg, *J. Chromatogr. A*, 807 (1998) 3–26.
55 J.C.M. Waterval, H. Lingeman, A. Bult and W.J.M. Underberg, *Electrophoresis*, 21 (2000) 4029–4095.
56 W.J.M. Underberg and J.C.M. Waterval, *Electrophoresis*, 23 (2002) 3922–3933.
57 O. Boulat, D.G. McLaren, E.A. Arriaga and D.D.Y. Chen, *J. Chromatogr. B*, 754 (2001) 217–228.
58 K. Artl, S. Brandt and J. Kehr, *J. Chromatogr. A*, 926 (2001) 319–325.
59 P. Chen and M.V. Novotny, *Anal. Chem.*, 69 (1997) 2806–2811.
60 Z. Chen, J. Wu, G.B. Baker, M. Parent and N.J. Dovichi, *J. Chromatogr. A*, 914 (2001) 293–298.
61 M. Horka, T. Willimann, M. Blum, P. Nording, Z. Friedl and K. Slais, *J. Chromatogr. A*, 916 (2001) 65–71.
62 Z.-H. Lan, X. Quian and R.W. Giese, *J. Chromatogr. A*, 831 (1999) 325–330.
63 D.L. Gallaher and M.E. Johnson, *Analyst*, 124 (1999) 1541–1546.
64 D.L. Gallaher Jr. and M.E. Johnson, *Anal. Chem.*, 72 (2000) 2080–2086.
65 C. Chassaing, J. Gonin, C.S. Wilcox and I.W. Wainer, *J. Chromatogr. B*, 735 (1999) 219–227.
66 E. Causse, C. Isaac, P. Malatray, C. Bayle, P. Valdiguie, R. Salvayre and F. Couderc, *J. Chromatogr. A*, 895 (2000) 173–178.
67 M.A. Roberts, H.-J. Zhong, J. Prodolliet and D.M. Goodall, *J. Chromatogr. A*, 817 (1998) 353–366.
68 T.N. Mitropoulou, F. Lamari, A. Syrokou, A. Hjerpe and N.K. Karamanos, *Electrophoresis*, 22 (2001) 2458–2463.
69 M.T. Bowser and R.T. Kennedy, *Electrophoresis*, 22 (2001) 3668–3676.

70 A. Taga, M. Sugimura and S. Honda, *J. Chromatogr. A*, 802 (1998) 243–248.
71 A. Taga and S. Honda, *J. Chromatogr. A*, 742 (1996) 243–250.
72 S. Oguri, K. Yokoi and Y. Motohase, *J. Chromatogr. A*, 787 (1997) 253–260.
73 A. Taga, A. Nishino and S. Honda, *J. Chromatogr. A*, 822 (1998) 271–279.
74 I.H. Lee, D. Pinto, E.A. Arriaga, Z. Zhang and N.J. Dovichi, *Anal. Chem.*, 70 (1998) 4546–4548.
75 S. Oguri, Y. Ohta and C. Suzuki, *J. Chromatogr. B*, 736 (1999) 263–271.
76 S. Oguri, M. Kumazaki, R. Kitou, H. Nonoyama and N. Tooda, *Biochem. Biophys. Acta*, 1472 (1999) 107–114.
77 Y.S. Wu, H.K. Lee and S.F.Y. Li, *Anal. Chem.*, 72 (2000) 1441–1447.
78 S.H. Kang, W. Wei and E.S. Yeung, *J. Chromatogr. B*, 744 (2000) 149–156.
79 M.L. Gostkowski, J. Wei, E. Okerberg and J.B. Shear, *Anal. Biochem.*, 303 (2002) 199–202.
80 E. Okerberg and J.B. Shear, *Anal. Chem.*, 73 (2001) 1610–1613.
81 S.J. Lillard, D.T. Chiu, R.H. Scheller, R.N. Zare, S.E. Rodríguez-Cruz, E.R. Williams, O. Orwar, M. Sandberg and J.A. Lundqvist, *Anal. Chem.*, 70 (1998) 3517–3524.
82 S. Hu, L. Zhang, S. Krylov and N.J. Dovichi, *Anal. Chem.*, 75 (2003) 3495–3501.
83 B. Nickerson and J.W. Jorgenson, *J. Chromatogr.*, 480 (1989) 157–168.
84 D.B. Craig and N.J. Dovichi, *Anal. Chem.*, 70 (1998) 2493–2494.
85 P.G. Coble and A.T. Timperman, *J. Chromatogr. A*, 829 (1998) 309–315.
86 M.L. Ye, S. Hu, W.W.C. Quigley and N.J. Dovichi, *J. Chromatogr. A*, 1022 (2004) 201–206.
87 K.L. Kostel and S.M. Lunte, *J. Chromatogr. B*, 695 (1997) 27–38.
88 W.-L. Tseng, Y.-W. Lin and H.-T. Chang, *Anal. Chem.*, 74 (2002) 4828–4834.
89 T.T. Lee and E.S. Yeung, *J. Chromatogr.*, 595 (1992) 319–325.
90 D.M. Paquette, R. Sing, P.R. Banks and K.R. Waldron, *J. Chromatogr. B*, 714 (1998) 47–57.
91 T.T. Lee, S.J. Lillard and E.S. Yeung, *Electrophoresis*, 14 (1993) 429–438.
92 E.S. Yeung, *J. Chromatogr. A*, 830 (1999) 243–262.
93 E. Zahou, H. Jornvall and T. Bergman, *Anal. Biochem.*, 281 (2000) 115–122.
94 D.P. Richards, C. Stathakis, R. Polakowski, H. Ahmadzadeh and N.J. Dovichi, *J. Chromatogr. A*, 853 (1999) 21–25.
95 D.M. Pinto, E.A. Arriaga, D. Craig, J. Angelova, N. Sharma, H. Ahmadzadeh and N.J. Dovichi, *Anal. Chem.*, 69 (1997) 3015–3021.

96 D.F. Swaile and M.J. Sepaniak, *J. Liq. Chromatogr. & Rel. Technol.*, 14 (1991) 869–893.
97 I. Benito, M.L. Marina, J.M. Sanz and J.C. Diez-Masa, *J. Chromatogr. A*, 841 (1999) 105–114.
98 M.D. Harvey, D. Bandilla and P.R. Banks, *Electrophoresis*, 19 (1998) 2169–2174.
99 J.R. Daban, S. Bartolome and M. Samso, *Anal. Biochem.*, 199 (1991) 169–174.
100 B.L. Legendre Jr. and S.A. Soper, *Appl. Spec.*, 50 (1996) 1196–1202.
101 M.D. Harvey, V. Bablekis, P.R. Banks and C.D. Skinner, *J. Chromatogr. B*, 754 (2001) 345–356.
102 S.N. Krylov and N.J. Dovichi, *Anal. Chem.*, 72 (2000) 111R–118R.
103 M. de Frutos, E. Molina and A. Puerta. In: K.M. Gooding and F.E. Regnier (Eds.), *HPLC of Biological Macromolecules*, Marcel Dekker, New York, 2002.
104 W.S.B. Yeung, G.A. Luo, Q.G. Wang and J.P. Ou, *J. Chromatogr. B*, 797 (2003) 217–228.
105 L. Tao and R.T. Kennedy, *Anal. Chem.*, 68 (1996) 3899–3906.
106 K. Shimura and B.L. Karger, *Anal. Chem.*, 66 (1994) 9–15.
107 I. German and R.T. Kennedy, *Anal. Chem.*, 72 (2000) 5365–5372.
108 I. German, M.G. Roper, S.P. Kaira, E. Rhinehart and R.T. Kennedy, *Electrophoresis*, 22 (2001) 3659–3667.
109 S.B. Cheng, C.D. Skinner, J. Taylor, S. Attiya, W.E. Lee, G. Picelli and D.J. Harrison, *Anal. Chem.*, 73 (2001) 1472–1479.
110 M.G. Roper, J.G. Shackman, G.M. Dahlgren and R.T. Kennedy, *Anal. Chem.*, 75 (2003) 4711–4717.
111 L. Tao, C.A. Aspinwall and R.T. Kennedy, *Electrophoresis*, 19 (1998) 403–408.
112 F.-T.A. Chen, *J. Chromatogr. A*, 680 (1994) 419–423.
113 N.M. Schultz, L. Huang and R.T. Kennedy, *Anal. Chem.*, 67 (1995) 924–929.
114 K. Shimura, M. Hoshino, K.-I. Kamiya, K. Katoh, S. Hisada, H. Matsumoto and K.-I. Kasai, *Electrophoresis*, 23 (2002) 909–917.
115 O.-W. Reif, R. Lausch, T. Scheper and R. Freitag, *Anal. Chem.*, 66 (1994) 4027–4033.
116 S.D. Jayasena, *Clin. Chem.*, 45 (1999) 1628–1650.
117 V. Pavski and X.C. Le, *Anal. Chem.*, 73 (2001) 6070–6076.
118 M. Berezovski and S.N. Krylov, *J. Am. Chem. Soc.*, 125 (2003) 13451–13454.
119 E.E. Jameson, J.M. Cunliffe, R.R. Neubig, R.K. Sunahara and R.T. Kennedy, *Anal. Chem.*, 75 (2003) 4297–4304.
120 S. Hu, L. Zhang and N.J. Dovichi, *J. Chromatogr. A*, 924 (2001) 369–375.

121 D.R. Reyes, D. Iossifidis, P.A. Auroux and A. Manz, *Anal. Chem.*, 74 (2002) 2623–2636.
122 P.A. Auroux, D. Iossifidis, D.R. Reyes and A. Manz, *Anal. Chem.*, 74 (2002) 2637–2652.
123 A.J. Tüdős, G.A.J. Besselink and R.B.M. Schasfoort, *Lab Chip*, 1 (2001) 83–95.
124 E. Verpoorte, *Electrophoresis*, 23 (2002) 677–712.
125 T. Chován and A. Guttman, *Trends Biotechnol.*, 20 (2002) 116–122.
126 J. Wang, *Talanta*, 56 (2002) 223–231.
127 D. Erickson and D. Li, *Anal. Chim. Acta*, 507 (2004) 11–26.
128 E. Verpoorte, *Lab Chip*, 3 (2003) 42N–52N.
129 D.J. Harrison, K. Fluri, K. Seiler, Z. Fan, C. Effenhauser and A. Manz, *Science*, 261 (1993) 895–897.
130 B. He, N. Tait and F.E. Regnier, *Anal. Chem.*, 70 (1998) 3790–3797.
131 A. Manz, D.J. Harrison, E. Verpoorte and H.M. Widmer, *Adv. Chromatogr.*, 33 (1993) 1–66.
132 C.S. Effenhauser, G.J.M. Bruin, A. Paulus and M. Ehrat, *Anal. Chem.*, 69 (1997) 3451–3457.
133 http://www.micralync.com/product_services/toolkit.htm.
134 http://www.chem.agilent.com/Scripts/PCol.asp?lPage=50.
135 G. Ocvirk, T. Tang and D.J. Harrison, *Analyst*, 123 (1998) 1429–1434.
136 C.J.R. Sheppard and D.M. Shotton, *Confocal Laser Scanning Microscopy*, BIOS Scientific Publishers, Oxford, 1997.
137 S. Etho, T. Fujimura, R. Hattatiri and Y. Kuroki, *Jpn. J. Appl. Phys.*, 42 (2003) 4093–4097.
138 L. Hernandez, J. Escalona, N. Joshi and N. Guzman, *J. Chromatogr.*, 559 (1991) 183–196.
139 G. Jiang, S. Attiya, G. Ocvirk, W.E. Lee and D.J. Harrison, *Biosens. & Bioelectron.*, 14 (2000) 861–869.
140 F. Dang, L. Zhang, H. Hagiwara, Y. Mishina and Y. Baba, *Electrophoresis*, 24 (2003) 714–721.
141 A. Lundqvist, D.T. Chiu and O. Orwar, *Electrophoresis*, 24 (2003) 1737–1744.
142 S.C. Jacobson, R. Hergenröder, L.B. Koutny, R.J. Warmack and J.M. Ramsey, *Anal. Chem.*, 66 (1994) 1107–1113.
143 Y. Xu, C.X. Zhang, D. Janasek and A. Manz, *Lab Chip*, 3 (2003) 224–227.
144 M. Vazquez, G. McKinley, L. Mitnik, S. Desmarais, P. Matsudaira and D. Ehrlich, *Anal. Chem.*, 74 (2002) 1952–1961.
145 C.X. Zhang and A. Manz, *Anal. Chem.*, 75 (2003) 5759–5766.
146 Y. Li, J.S. Buch, F. Rosenberger, D.L. DeVoe and C.S. Lee, *Anal. Chem.*, 76 (2004) 742–748.
147 *The Practical Application of Light*, Melles Griot, 1999, pp. 15.16–15.19.

148 E. Schmidt, *Laser Optoelektron.*, 28 (1996) 50–51.
149 Z. Liang, N. Chiem, G. Ocvirk, T. Tang, K. Fluri and J.D. Harrison, *Anal. Chem.*, 68 (1996) 1040–1046.
150 L.M. Fu, R.J. Yang, C.H. Lin, Y.J. Pan and G.B. Lee, *Anal. Chim. Acta*, 507 (2004) 163–169.
151 M.L. Chabinyc, D.T. Chiu, J.C. McDonald, A.D. Stroock, J.F. Christian, A.M. Karger and G.M. Whitesides, *Anal. Chem.*, 73 (2001) 4491–4498.
152 Y.C. Tung, M. Zhang, C.T. Lin, K. Kurabayashi and S.J. Skerlos, *Sens. Actuators B*, 98 (2004) 356–367.
153 S. Qi, X. Liu, S. Ford, J. Barrows, G. Thomas, K. Kelly, A. McCandless, K. Lian, J. Goettert and S. Soper, *Lab Chip*, 2 (2002) 88–95.
154 J. Hübner, K.B. Mogensen, A.M. Jorgensen, P. Friis, P. Telleman and J.P. Kutter, *Rev. Sci. Instrum.*, 72 (2001) 229–233.
155 K.B. Mogensen, N.J. Petersen, J. Hübner and J.P. Kutter, *Electrophoresis*, 22 (2001) 3930–3938.
156 J.C. McDonald, M.L. Chabinyc, S. Metallo, J.R. Anderson, A.D. Stroock and G.M. Whitesides, *Anal. Chem.*, 74 (2002) 1537–1545.
157 S. Camou, H. Fujita and T. Fujii, *Lab Chip*, 3 (2003) 40–45.
158 J.C. Roulet, R. Völken, H.P. Herzig, E. Verpoorte, N.F. de Rooij and R. Dändliker, *Anal. Chem.*, 74 (2002) 3400–3407.
159 J.R. Webster, M.A. Burns, D.T. Burke and C.H. Mastrangelo, *Anal. Chem.*, 73 (2001) 1622–1626.
160 T. Kamei, B.M. Paegel, J.R. Scherer, A.M. Skelly, R.A. Street and R.A. Mathies, *Anal. Chem.*, 75 (2003) 5300–5305.
161 J.A. Chediak, Z. Luo, J. Seo, N. Cheung, L.P. Lee and T.D. Sands, *Sens. Actuators A*, 111 (2004) 1–7.
162 S.M. Cousins, P.R. Haddad and W. Buchberger, *J. Chromatogr. A*, 671 (1994) 397–402.
163 M.T. Ackermans, F.M. Everaerts and J.L. Beckers, *J. Chromatogr.*, 549 (1991) 345–355.
164 R. Kuhn and S. Hoffstetter-Kuhn, *Capillary Electrophoresis: Principles and Practice*, Springer-Verlag, New York, 1993, p. 50.
165 F. Foret, L. Krivankova and P. Bocek. In: B.J. Radola (Ed.), *Capillary Zone Electrophoresis*, VCH, New York, 1993, p. 176.
166 P. Doble and P.R. Haddad, *J. Chromatogr. A*, 834 (1999) 189–212.
167 T. Wang and R.A. Hartwick, *J. Chromatogr. A*, 589 (1992) 307–313.
168 W.G. Kuhr and E.S. Yeung, *Anal. Chem.*, 60 (1988) 2642–2646.
169 M. Albin, P.D. Grossman and S.E. Moring, *Anal. Chem.*, 65 (1993) 489A–497A.
170 P.E. Andersson, W.D. Pfeffer and L.G. Blomberg, *J. Chromatogr. A*, 699 (1995) 323–330.
171 W.G. Kuhr and E.S. Yeung, *Anal. Chem.*, 60 (1988) 1832–1834.

172 C. Gooijer and A.J.G. Mank, *Anal. Chim. Acta*, 400 (1999) 281–295.
173 T.W. Garner and E.S. Yeung, *J. Chromatogr.*, 515 (1990) 639–644.
174 L. Gross and E.S. Yeung, *J. Chromatogr.*, 480 (1989) 169–178.
175 B.L. Hogan and E.S. Yeung, *J. Chromatogr. Sci.*, 28 (1990) 15–18.
176 L. Gross and E.S. Yeung, *Anal. Chem.*, 62 (1990) 427–431.
177 M.N. Church, J.D. Spear, R.E. Russo, G.L. Klunder, P.M. Grant and B.D. Andresen, *Anal. Chem.*, 70 (1998) 2475–2480.
178 M.D. Richmond and E.S. Yeung, *Anal. Biochem.*, 210 (1993) 245–248.
179 V. Ruiz-Calero, L. Puignon and M.T. Galceran, *J. Chromatogr. A*, 873 (2000) 269–282.
180 N.J. Munro, Z. Huang, D.N. Finegold and J.P. Landers, *Anal. Chem.*, 72 (2000) 2765–2773.
181 A.G. Lista, L. Arce, A. Ríos and M. Valcárcel, *J. Chromatogr. A*, 919 (2001) 407–415.
182 H. Poppe and X. Xu. In: M.G. Khaledi (Ed.), *High Performance Capillary Electrophoresis: Theory, Techniques and Applications*, Wiley-Interscience, New York, 1998.
183 T. Wang, H. Wei and S.F.Y. Li, *Electrophoresis*, 19 (1998) 2187–2192.
184 P.L. Desbene, C. Morin and A. Desbene, *Analusis*, 26 (1998) 107–115.
185 Y.-T. Lee and C.-W. Whang, *J. Chromatogr. A*, 746 (1996) 269–275.
186 P.L. Desbene, C.J. Morin, A.M. Desbene and R.S. Groult, *J. Chromatogr. A*, 689 (1995) 135–148.
187 T.-C. Chiu, M.-F. Huang, C.-C. Huang, M.-M. Hsieh and H.-T. Chang, *Electrophoresis*, 23 (2002) 449–455.
188 L.J. Jin, T. Wang and S.F.Y. Li, *Electrophoresis*, 20 (1999) 1856–1861.
189 S. Shamsi, N.D. Danielson and I.M. Warner, *J. Chromatogr. A*, 835 (1999) 159–168.
190 A.M. Desbene, C.J. Morin, N.L. Mofaddel and R.S. Groult, *J. Chromatogr. A*, 716 (1995) 279–290.
191 Y.-M. Huang and C.-W. Whang, *Electrophoresis*, 19 (1998) 2140–2144.
192 V. Marti, M. Aguilar and E.S. Yeung, *J. Chromatogr. A*, 709 (1995) 367–374.
193 N. Ragozina, M. Putz, W. Faubel and U. Pyell, *Electrophoresis*, 24 (2003) 567–574.
194 S.-J. Chen, M.-J. Chen and H.-T. Chang, *J. Chromatogr. A*, 1017 (2003) 215–224.
195 S. Sirichai and A.J. de Mello, *Electrophoresis*, 22 (2001) 348–354.
196 Y.-C. Chao and C.-W. Whang, *J. Chromatogr. A*, 663 (1994) 229–237.
197 S.Y. Chang and H.-T. Chiang, *Electrophoresis*, 23 (2002) 2913–2917.
198 S.Y. Chang and C.-H. Liao, *J. Chromatogr. A*, 959 (2002) 309–315.
199 P. Zhang, G. Xu, J. Xiong, Y. Zheng, Q. Yang and F. Wei, *J. Sep. Sci.*, 25 (2002) 155–159.
200 N.P. Beard and A.J. de Mello, *Electrophoresis*, 23 (2002) 1722–1730.

201 T. Groh and K. Bachmann, *Electrophoresis*, 13 (1992) 458–461.
202 N.W.F. Nielen, *J. Chromatogr.*, 608 (1992) 85–92.
203 D.F. Swaile, C.L. Copper, M.J. Sepaniak, D.E. Burton and L.L. Powell, *Talanta*, 41 (1994) 1499–1505.
204 A.J.G. Mank and E.S. Yeung, *J. Chromatogr. A*, 708 (1995) 309–321.
205 H. Liu, B.Y. Cho, I.S. Krull and S.A. Cohen, *J. Chromatogr. A*, 927 (2001) 77–89.
206 W.F. Patton, *Electrophoresis*, 21 (2000) 1123–1144.
207 M.J. Sepaniak and C.N. Kettler. In: E.S. Yeung (Ed.), *Detectors for Liquid Chromatography*, John Wiley, New York, 1986, Chapter 5.
208 J.B. Shear, *Anal. Chem.*, 71 (1999) 598A–605A.
209 M.L. Gostkowski, J. Wei and J.B. Shear, *Anal. Biochem.*, 260 (1998) 244–250.
210 J.M. Song, T. Inoue, H. Kawazumi and T. Ogawa, *J. Chromatogr. A*, 765 (1997) 315–319.
211 M.L. Gostkowski, J.B. McDoniel, J. Wei, T.E. Curey and J.B. Shear, *J. Am. Chem. Soc.*, 120 (1998) 18–20.
212 J. Wei, M.L. Gostkowski, M.J. Gordon and J.B. Shear, *Anal. Chem.*, 70 (1998) 3470–3475.
213 J. Wei, E. Okerberg, J. Dunlap, C. Ly and J.B. Shear, *Anal. Chem.*, 72 (2000) 1360–1363.
214 M.J. Gordon and J.B. Shear, *J. Am. Chem. Soc.*, 123 (2001) 1790–1791.
215 X. Zhang, J.N. Stuart and J.V. Sweedler, *Anal. Bioanal. Chem.*, 373 (2002) 332–343.
216 A.T. Timperman, K. Khatib and J.V. Sweedler, *Anal. Chem.*, 67 (1995) 139–144.
217 A. Hanning, P. Lindberg, J. Westberg and J. Roeraade, *Anal. Chem.*, 72 (2000) 3423–3430.
218 S. Carson, A.S. Cohen, A. Belenkii, M.C. Ruiz-Martinez, J. Berka and B.L. Karger, *Anal. Chem.*, 63 (1993) 3219–3226.
219 A. Fultz, T.M. Branch and V. Majidi, *Microchem. J.*, 57 (1997) 231–244.
220 Y.H. Park, X. Zhang, S.S. Rubakhin and J.V. Sweedler, *Anal. Chem.*, 71 (1999) 4997–5002.
221 R.R. Fuller, L.L. Moroz, R. Gillette and J.V. Sweedler, *Neuron*, 20 (1998) 173–181.
222 S.J. Kok, E.M. Kristenson, C. Gooijer, N.H. Velthorst and U.A.T. Brinkman, *J. Chromatogr. A*, 771 (1997) 331–341.
223 M.C. Beekman, H. Lingeman, U.A.T. Brinkman and C. Gooijer, *J. Microcolumn Sep.*, 11 (1999) 347–352.
224 U.B. Soetebeer, M.-O. Schierenberg, H. Schulz, G. Grünefeld, P. Andresen and G. Blaschke, *J. Chromatogr. B*, 745 (2000) 271–278.
225 U.B. Soetebeer, M.-O. Schierenberg, H. Schulz, H. Hempel, P. Andresen and G. Blaschke, *Anal. Chem.*, 73 (2001) 2178–2182.

226 U.B. Soetebeer, M.-O. Schierenberg, J.-G. Moller, H. Schulz, G. Grünefeld, P. Andresen, G. Blaschke and G. Ahr, *J. Chromatogr. A*, 895 (2000) 147–155.
227 S. Zaugg, X. Zhang, J.V. Sweedler and W. Thormann, *J. Chromatogr. B*, 752 (2001) 17–31.
228 J.N. Stuart, X. Zhang, J.A. Jakubowski, E.V. Romanova and J.V. Sweedler, *J. Neurochem.*, 84 (2003) 1358–1366.
229 A. Timperman, S.E. Tracht and J.V. Sweedler, *Anal. Chem.*, 68 (1996) 2693–2698.
230 http://www.appliedbiosystems.com/catalog/.
231 R. Jankowiak, D. Zamzow, W. Ding and G.J. Small, *Anal. Chem.*, 68 (1996) 2549–2553.
232 D. Zamzow, C.H. Lin, G.J. Small and R. Jankowiak, *J. Chromatogr. A*, 781 (1997) 73–80.
233 E. Waddell, S. Lassiter, C.V. Owens Jr. and S.A. Soper, *J. Liq. Chrom. & Rel. Technol.*, 23 (2000) 1139–1158.
234 D.J. Desilets, P.T. Kissinger and F.E. Lytle, *Anal. Chem.*, 59 (1987) 1830–1834.
235 F. Basile, A. Cardamone, K.D. Grinstead Jr., K.J. Miller, F.E. Lytle, A. Caprara, C.D. Clark and J.-M. Heritier, *Appl. Spectrosc.*, 47 (1993) 207–210.
236 U. Lieberwirth, J. Arden-Jacob, K.A. Drexhage, D.P. Herten, R. Muller, N. Neumann, A. Schultz, S. Siebert, G. Sagner, S. Klingel, M. Sauer and J. Wolfrum, *Anal. Chem.*, 70 (1998) 4771–4779.
237 J.H. Flanagan Jr., C.V. Owens, S.E. Romero, E. Waddell, S.H. Kahn, R.P. Hammer and S.A. Soper, *Anal. Chem.*, 70 (1998) 2676–2684.
238 B.L. Legendre Jr., D.C. Williams, S.A. Soper, R. Erdmann, U. Ortmann and J. Enderlein, *Rev. Sci. Instrum.*, 67 (1996) 3984–3989.
239 S.A. Soper, B.L. Legendre Jr. and D.C. Williams, *Anal. Chem.*, 67 (1995) 4358–4365.
240 D. Pinto, E.A. Arriaga, S. Sia, Z. Li and N.J. Dovichi, *Electrophoresis*, 16 (1995) 534–540.
241 M.D. Harvey and P.R. Banks, *J. Chromatogr. B*, 793 (2003) 115–125.
242 J.M. Berg, J.L. Tymoczko and L. Stryer, *Biochemistry*, W.H. Freeman and Company, New York, 2002 Chapters 8–10.
243 P.R. Banks, *Trends Anal. Chem.*, 17 (1998) 612–622.
244 M. Berezovsky, W.P. Li, C.D. Poulter and S.N. Krylov, *Electrophoresis*, 23 (2002) 3398–3403.
245 P.S. Doyle, J. Bibette, A. Bancaud and J.L. Viovy, *Science*, 295 (2002) 2237.
246 M. Gorc and K.K. Sheti, *Intl. Biotech. Lab.*, 1 (2004) 6–10.
247 J. Yguerabide and E.E. Yguerabide, *Anal. Biochem.*, 262 (1998) 137–156.
248 J. Yguerabide and E.E. Yguerabide, *Anal. Biochem.*, 262 (1998) 157–176.

249 C.F. Duffy, A.A. McEathron and E.A. Arriaga, *Electrophoresis*, 23 (2002) 2040–2047.
250 W.C.W. Chan, D.J. Maxwell, X. Gao, R. Bailey, M. Han and S. Nie, *Curr. Op. Biotechnol.*, 7 (2003) 40–46.
251 Y.-M. Liu, T. Moroz and J.V. Sweedler, *Anal. Chem.*, 71 (1999) 28–33.
252 R. Zhu and W.Th. Kok, *J. Chromatogr. A*, 814 (1998) 213–221.
253 K.B. Male and J.H.T. Luong, *J. Chromatogr. A*, 926 (2001) 309–317.
254 F. Han, B.H. Huynh, H. Shi, B. Lin and Y. Ma, *Anal. Chem.*, 71 (1999) 1265–1269.
255 H. Miao, S.S. Rubakhin and J.V. Sweedler, *Anal. Bioanal. Chem.*, 377 (2003) 1007–1013.
256 S.J. Lillard, E.S. Yeung and M.A. McCloskey, *Anal. Chem.*, 68 (1996) 2897–2904.
257 B.L. Hogan and E.S. Yeung, *Anal. Chem.*, 64 (1992) 2841–2845.
258 Y. Mao and X. Zhang, *Electrophoresis*, 24 (2003) 3289–3295.
259 Z. Zhang, E. Carpenter, X. Puyan and N.J. Dovichi, *Electrophoresis*, 22 (2001) 1127–1132.
260 P.J. Viskari and C.L. Colyer, *J. Chromatogr. A*, 972 (2002) 269–276.
261 T.-C. Chiu, Y.-W. Lin, C.-C. Huang, A. Chrambach and H.-T. Chang, *Electrophoresis*, 24 (2003) 1730–1736.
262 W.-L. Tseng, T.-C. Chiu, J.-M. Weng and H.-T. Chang, *J. Liq. Chromatogr. & Rel. Technol.*, 24 (2001) 2971–2982.
263 Z. Zhang, S. Krylov, E.A. Arriaga, R. Polakowski and N.J. Dovichi, *Anal. Chem.*, 72 (2000) 318–322.
264 W. Tong and E.S. Yeung, *J. Chromatogr. B*, 685 (1996) 35–40.
265 G.D. Meredith, C.E. Sims, J.S. Soughayer and N.L. Allbritton, *Nat. Biotechnol.*, 18 (2000) 309–312.
266 C.L. Colyer, S.D. Mangru and D.J. Harrison, *J. Chromatogr. A*, 781 (1997) 271–276.
267 F. von Heeren, E. Verpoorte, A. Manz and W. Thormann, *Anal. Chem.*, 68 (1996) 2044–2053.
268 L.B. Koutny, D. Schmalzing, T.A. Taylor and M. Fuchs, *Anal. Chem.*, 68 (1996) 18–22.
269 N.H. Chiem and D.J. Harrison, *Clin. Chem.*, 44 (1998) 591–598.
270 N. Chiem and D.J. Harrison, *Anal. Chem.*, 69 (1997) 373–378.

Chapter 7

Novel optical detection techniques in CE based on phosphorescence or chemiluminescence

Jacobus Kuijt, Freek Ariese and Cees Gooijer

7.1 INTRODUCTION

Detection of low analyte concentrations in complex biological, pharmaceutical or environmental samples continues to be a challenge. From the point of view of separation, capillary electrophoresis (CE) may well be the method of choice if these analytes are highly polar or ionic in nature (see chapter 1). However, it is well known that CE, while providing efficient and fast separation, is hampered by rather poor concentration limits of detection (LODs) for most optical and non-optical detection methods. As discussed in chapters 5 and 9, UV absorption and mass spectrometry are often limited to LODs in the micromolar range because of the intrinsically short optical pathlength or the very small mass loads, respectively. Electrochemical detection (see chapter 8) may in favorable cases be quite sensitive, but is limited in its application range. Of course, luminescence techniques have the potential to provide much better LODs than UV absorption, since these are in principle zero-background techniques, based on the measurements of a light signal in the dark. As discussed in chapter 6, fluorescence detection is well known in CE, and in particular, if combined with lasers for excitation, LODs down to the picomolar range can be obtained. However, its use is restricted to natively fluorescing species unless an extra derivatization step is carried out. In this chapter, two alternative luminescence methods will be discussed: phosphorescence (in section 7.2) and chemiluminescence (CL) (in section 7.3). These techniques merit consideration for use in daily practice. Phosphorescence detection in particular—at present hardly

Comprehensive Analytical Chemistry XLV
M.L. Marina, A. Ríos and M. Valcárcel (Eds)
Volume XLV ISSN: 0166-526X DOI: 10.1016/S0166-526X(05)45007-2

applied—has shown very encouraging results. Emphasis will be on fundamental aspects, advantages and limitations and applications. Special attention is given to the interfaces that are used to couple these detection modes to CE.

7.2 PHOSPHORESCENCE DETECTION IN CE

7.2.1 Introduction

Although phosphorescence detection has not found as widespread an application in analytical chemistry as fluorescence detection, its use has expanded in the past few decades due to its inherent advantages (sensitivity, selectivity and time-resolved detection) [1,2]. In addition, the scope of phosphorescence detection has broadened significantly: while phosphorescence methods in early studies were restricted to low-temperature or solid phases, many different methods based on room-temperature phosphorescence in the liquid state (RTPL) have evolved in more recent years [1,2]. All three RTPL modes—direct, sensitized and quenched phosphorescence—have been coupled to liquid chromatography (LC). The latter two indirect modes appeared to be most promising in terms of analyte detectability [3,4]. It was anticipated that these RTPL detection modes might be even more useful for enhancing detectability in CE, where poor concentration LODs are an even bigger problem. Therefore, both modes were also coupled to CE in our laboratory [5]. One should realize, however, that the sensitized phosphorescence mode has to compete with absorption detection (both techniques depending on high extinction coefficients), and in many cases also with (laser-induced) fluorescence, which is the most sensitive detection technique in CE. Hence, its use will be restricted to the rather limited class of compounds showing either no or weak fluorescence while being efficient phosphorophores. In contrast, quenched RTPL can be used for various analyte classes that do not show phosphorescence themselves, but are able to quench the phosphorescence emitted by another molecule. These analytes will include compounds with poorly absorbing chromophoric groups, which cannot be detected sensitively by UV absorption. The quenched phosphorescence detection mode may well be a sensitive alternative for such analytes without the need for chemical derivatization. Examples will be presented in section 7.2.4.

7.2.2 Basic features of phosphorescence

A simplified scheme of the photophysical processes relevant to directly excited phosphorescence is presented in Fig. 7.1. A significant fraction of the excited molecules may undergo intersystem crossing (ISC) from the lowest excited singlet level S_1 to the triplet state T_1. From this level, (direct) phosphorescence can be subsequently emitted, provided that this deactivation pathway is efficient, i.e., if it is able to compete with the other deactivation pathways open to the triplet state: deactivation by non-radiative decay (not depicted) and bimolecular quenching by dissolved molecular oxygen or other quenchers. In aqueous solutions containing molecular oxygen in equilibrium with the surrounding air ($\sim 10^{-3}$ M), deactivation by bimolecular quenching is rapid ($k_q \approx 10^9\, M^{-1}\, s^{-1}$; therefore $k_q[O_2] \approx 10^6\, s^{-1}$) compared with the radiative phosphorescence transition ($k_p < 10^2\, s^{-1}$). Therefore, low-temperature frozen matrices or other solid matrices—which slow down or prevent oxygen diffusion—are often a prerequisite for obtaining strong phosphorescence. Alternatively, the concentration of dissolved molecular oxygen can be lowered in order to obtain RTPL, typically requiring oxygen levels of about 10^{-7} M. Deoxygenation is usually achieved by purging with nitrogen gas or by chemical reactions consuming molecular oxygen (e.g., oxidation of sulfite ions [6]). It should be noted, however,

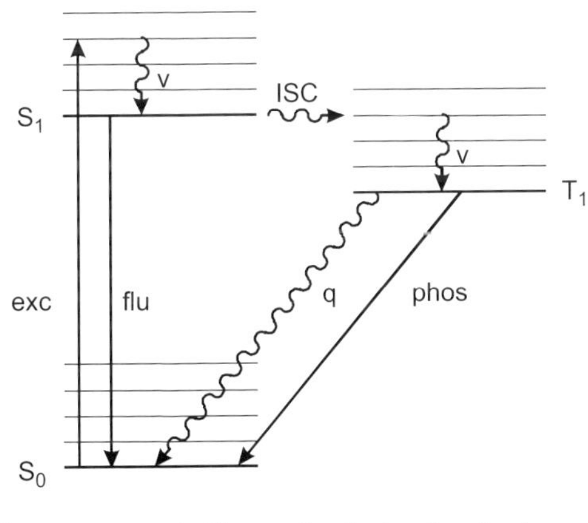

Fig. 7.1. Schematic representation of photophysical processes relevant to directly excited phosphorescence. S_0: singlet ground state, S_1: lowest excited singlet state, T_1: lowest excited triplet state, exc: excitation, flu: fluorescence, v: vibrational relaxation, ISC: intersystem crossing, q: bimolecular quenching, phos: phosphorescence. Reprinted from Ref. [1]. Copyright (2003), with permission from Elsevier.

that under such conditions, RTPL is still an exceptional phenomenon. Only very few compounds that emit strong phosphorescence at 77 K produce strong RTPL. Fortunately, there are some exceptions, the most well-known being biacetyl and brominated naphthalenes. These compounds can be used in indirect RTPL detection in CE, where they are added to the separation buffer. In order to attain the low oxygen levels required, adapted setups are needed for the coupling of phosphorescence detection with CE. These setups will be discussed in section 7.2.3.

7.2.2.1 Sensitized RTPL

As noted above, there are two modes of indirect phosphorescence: sensitized RTPL and quenched RTPL. Sensitized RTPL is obtained when a sensitizer (analyte) molecule is excited and undergoes triplet–triplet energy transfer (ET) to the phosphorophore. Biacetyl has often been used as a phosphorophore in sensitized RTPL, since strong phosphorescence enhancements can be achieved owing to the low absorptivity of biacetyl and the high extinction coefficients provided by suitable analytes [3,7]. In fact, direct excitation of biacetyl (resulting in a phosphorescence signal that is not related to the presence of analytes) is very inefficient and thus hardly contributes to the background.

7.2.2.2 Quenched RTPL

The phenomenon of bimolecular phosphorescence quenching, while posing a problem in the case of molecular oxygen, can be also used in a positive sense, i.e., for detection. In this case, one measures the decrease of a continuous RTPL signal—provided by phosphorophores such as bromonaphthalene derivatives added at relatively high levels to the buffer solution—due to collisions with an analyte (quencher). Quenching is either based on triplet–triplet (ET) from phosphorophore to analyte—the requirements for ET being identical to those for sensitized RTPL [8]—or on photoinduced electron transfer. Photoinduced electron transfer occurs in many molecules since photoexcited molecules are usually much more susceptible to electron transfer than their ground-state counterparts, i.e., excited donor molecules are oxidized more easily ($IP^* < IP$), while excited acceptor molecules are reduced more easily ($EA^* > EA$) [9]. The latter scheme is illustrated in Fig. 7.2; many of the examples that will be discussed below are probably based on this mechanism.

Independent of the specific mechanism involved, the quenching interaction will be governed by the well-known Stern–Volmer equation

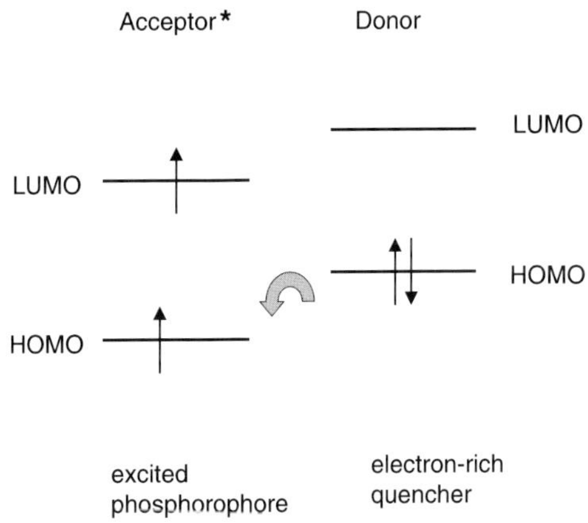

Fig. 7.2. Schematic illustration of reductive electron transfer mechanism. The excited-state phosphorophore (triplet state) is quenched by a ground-state donor molecule, producing two radicals.

for bimolecular quenching of luminescence [8]:

$$\frac{I_0}{I} = 1 + k_q[Q]\tau_0 \qquad (7.1)$$

where I_0 and I are the phosphorescence intensities in the absence and presence of a quencher, respectively, k_q is the rate constant for bimolecular quenching, [Q] the quencher (analyte) concentration and τ_0 the triplet lifetime in the absence of quencher. According to Eq. (7.1), even low concentrations of quenchers will provide efficient quenching if the triplet lifetime, τ_0, and the associated quenching rate constant, k_q, are high enough. In aqueous solutions at room temperature, the maximum value for k_q (attained in case of so-called diffusion-limited quenching) is about $10^{10}\,M^{-1}\,s^{-1}$ [8]. Thus, with a typical triplet lifetime, τ_0, of 10^{-4} s, an analyte concentration as low as 10^{-7} M will cause a measurable 10% reduction in phosphorescence intensity. Obviously, dynamic fluorescence quenching will not provide such a strong response, because singlet lifetimes are about 10^4 times shorter.

7.2.3 Coupling of phosphorescence detection to CE

A central issue in the coupling of phosphorescence detection and CE is the removal of dissolved molecular oxygen from the buffer solution. In contrast to LC phosphorescence, CE phosphorescence does not only concern oxygen diffusion from the surrounding air, but also the oxygen generated at the anode (+), i.e., at the injection side, when using a normal polarity of the separation voltage:

$$2H_2O(l) \rightleftharpoons 4H^+ + O_2(g) + 4e^- \quad (7.2)$$

In all setups used so far, purging with nitrogen gas was chosen for deoxygenation purposes, because this allows for simple and unattended operation, which may not be true for chemical oxygen scavenging.

The setup applied first involved an LC pump, a six-way valve for injection of samples, and an interface (anode) between the LC tubing and the separation capillary through which deoxygenated separation buffer was pumped continuously at 1 ml/min. A minor fraction of the sample (typically 20 nl from a 2 ml sample) was introduced into the separation capillary by timed hydrodynamic injection. Excellent baseline stability was achieved; the influence of electrochemically generated oxygen was minor because of the constant replenishment of separation buffer. Of course, this required rather high volumes of separation buffer, which inhibited the use of expensive buffer additives such as cyclodextrins. Another major disadvantage of this setup was the long deoxygenation time required (~2 h). The setup is discussed in more detail in Ref. [5].

Both drawbacks were resolved by applying in-vial deoxygenation with nitrogen bubbling (see Fig. 7.3), either by using an external sample injection system [10] or by using the injection system of the CE apparatus itself [11]. The nitrogen gas flow was (continuously) monitored by a gas flow meter. To avoid disturbances by nitrogen gas bubbles, the capillary end was positioned well below the nitrogen gas entrance. The startup times were reduced from 2 h to less than 20 min. A nitrogen gas flow of 24 ml/min was required for efficient deoxygenation when using 75 μm ID capillaries. As a result, disturbance of the baseline stability was sometimes observed. Lower nitrogen gas flows (typically 13–18 ml/min) could generally be used with 50 μm ID capillaries, since in that case less oxygen was produced at the anode, and baseline instabilities were completely absent.

7.2.4 Applications of CE–phosphorescence

7.2.4.1 Applications in capillary zone electrophoresis

The first study on CE–phosphorescence was aimed at the detection of naphthalene sulfonates (NS) [5]. Time-resolved detection was provided by a Perkin Elmer LS-40 luminescence detector (equipped with a

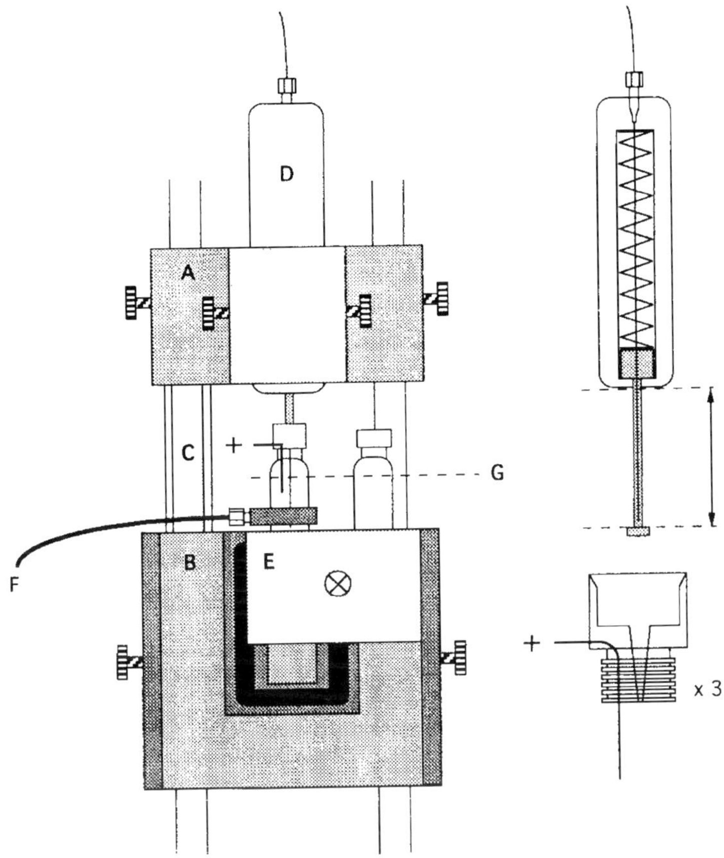

Fig. 7.3. Manually switched buffer deoxygenation/sample injection device. A, B: adjustable blocks; C: spacer; D: cylinder with capillary connection and spring-loaded piston; E: two-position block; F: nitrogen gas inlet; G: liquid level. Reprinted with permission from Ref. [10]. Copyright (2002), American Chemical Society.

pulsed xenon lamp), which was originally intended for use in LC. Rather unexpectedly, this instrument could be used for our detection purposes without changing the optical design: because of the application of time-resolved detection, high background signals due to scatter (or fluorescence) were absent. A 25 mM borate, pH 8.5 buffer with 0.02–0.05 M biacetyl (phosphorophore) was used for separation; with this buffer, a biacetyl triplet lifetime of about 70 μs was obtained. In the sensitized RTPL mode, excitation was performed at 230 nm, the optimum excitation wavelength for NS, while the biacetyl emission was monitored at 510 nm. At 230 nm direct excitation of biacetyl is of minor importance, so the RTPL background is low, as required.

In quenched RTPL detection, the biacetyl excitation maximum at 410 nm was used to generate the required continuous RTPL background, corresponding to I_0 in Eq. (7.1). The resulting electropherograms for both detection modes are presented in Fig. 7.4. Interestingly, the non-substituted NS showed sensitized RTPL, while the substituted NS provided quenching of biacetyl phosphorescence because their triplet (T_1) energies are below that of biacetyl. Under (on-capillary) stacking conditions (hydrodynamic injection) detection limits[1] as low as 10^{-7} M were obtained for both detection modes, in spite of the rather noisy baseline obtained with the quenched RTPL mode (Fig. 7.4, trace b).

Further research was focused on improving the CE–quenched RTPL methodology and on possible applications [12]. Several buffers were tested for compatibility. Borate, succinate, malonate, acetate and phosphate (pH 4.7–8.5) could be used without any problems. NS, nitrophenols, dithiocarbamates, hydroxybenzoic acids and sulfur-containing amino acids were used as the test analytes. For several analytes, the LODs were improved down to 10^{-8} M by sensitizing biacetyl prior to bimolecular quenching, and by using a total emission mirror instead of an emission monochromator. Also, the linear detection range was strongly enhanced. These improvements were due to the 25-fold enhanced signal-to-noise ratio of the continuous RTPL background (I_0) and required no sample stacking.

The quenched RTPL mode was also successfully coupled to CE by using a commercially available Jasco FP-920 fluorescence detector; however, the LODs were ~10-fold higher (~10^{-7} M) [13]. The difference is mainly due to the fact that with the FP-920 detector time-resolved detection is not possible, resulting in ~three-fold lower quenching efficiencies owing to contributions to the background signal from excitation scatter and fluorescence that cannot be quenched. In addition, the signal-to-noise ratio of the baseline was ~three-fold lower.

Sensitive detection of non-derivatized peptides by CE–quenched RTPL was achieved by using a brominated naphthalene sulfonate (BrNS) as the phosphorophore [14]; BrNS provides a strong RTPL background owing to its high extinction coefficient and phosphorescence quantum yield. Moreover, it can be applied over a much wider pH

[1]All LODs in this section: signal equal to 3σ, σ being the root-mean-square-noise of the baseline.

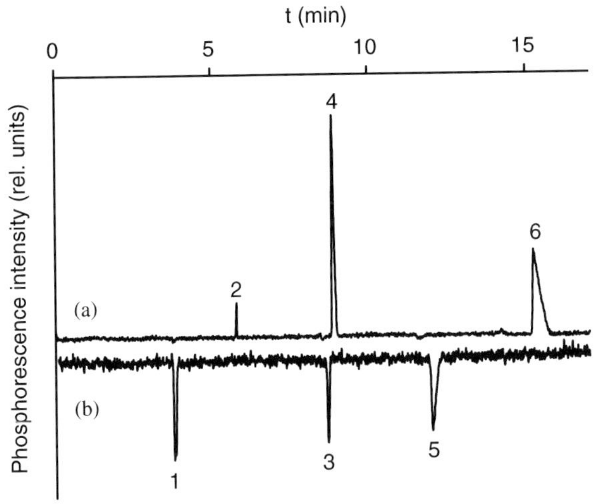

Fig. 7.4. Electropherograms of naphthalene sulfonates, using (a) sensitized RTPL detection, concentration of NS: 10^{-5} M, or (b) quenched RTPL detection, concentration of NS: 2.5×10^{-6} M. Both cases: buffer: 25 mM borate, pH 8.5, containing 0.02 M biacetyl; separation voltage: −30 kV. 1: Oxygen/EOF, 2: 1-naphthalene sulfonate, 3: 4-amino, 5-hydroxy, 2,7-naphthalene disulfonate, 4: 2,6-naphthalene disulfonate, 5: 4,5-dihydroxy, 2,7-naphthalene disulfonate, 6: 1,3,7-naphthalene trisulfonate. Adapted with permission from Ref. [5]. Copyright (1999), American Chemical Society.

range than biacetyl. Interestingly, some of these analytes (e.g., alanine–serine, glycine–asparagine and valine–glycine–glycine) have no suitable chromophoric groups for absorption detection. To enable quenched RTPL detection based on photoinduced electron transfer, a pH of 9.5–10.0 was used (buffer: 50 mM borate, 1 mM BrNS). The probable operative mechanism is illustrated in Fig. 7.2: quenching of the excited-state acceptor (phosphorophore) is possible, provided that the HOMO level of the donor analyte (quencher) is sufficiently high for electron transfer. In the case of amino acids, a relatively high pH is needed because the lone pair of the amino group can function as a donor only if the amino group is not protonated. Electropherograms of some di- and tripeptides are presented in Fig. 7.5. The LODs were quite favorable in comparison with LODs of other detection techniques: they ranged from 2×10^{-8} to 1×10^{-7} M (4–20 μg/l) at moderate injection volumes ($pt = -270$ mbar s).

Finally, the feasibility of laser-based quenched phosphorescence detection in CE was studied with a laboratory-made detection system

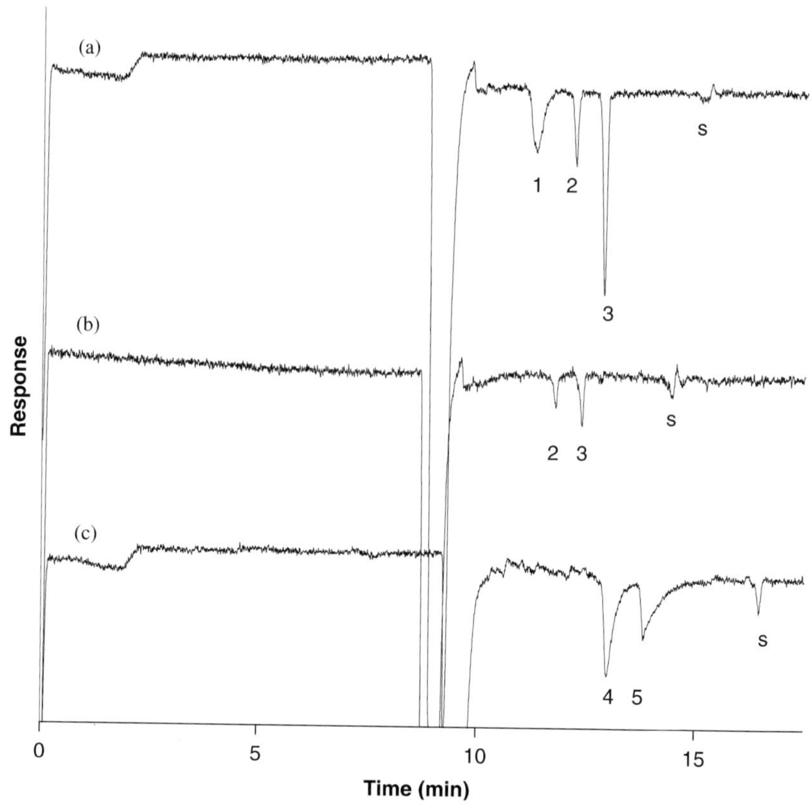

Fig. 7.5. Electropherograms of di- and tripeptides. Buffer: 50 mM borate, 1 mM BrNS. Separation voltage: −15 kV. 1: Alanine-serine, 2: valine-tyrosine-valine, 3: valine-glycine-glycine, 4: glycine-asparagine, 5: glycine-tyrosine, s: system peak. Trace a: pH 9.5; peptide concentration, 100 μg/l. Trace b: pH 9.5; on-capillary stacking ($pt = -630$ mbar s); valine-tyrosine-valine, 20 μg/l; valine-glycine-glycine, 10 μg/l;. Trace c: pH 10; peptide concentrations, 200 μg/l. Reprinted with permission from Ref. [14], Copyright (2001), American Chemical Society.

[15]. Excitation was performed at 266 nm with a small, quadrupled Nd:YAG laser. As expected, an improved signal-to-noise ratio of the RTPL background could be achieved. Unfortunately, this improvement did not result in a reduction in LODs: under laser excitation, the concentration of BrNS molecules in the T_1 state was quite high. Hence, deactivation by triplet–triplet annihilation became strongly competitive with analyte-induced bimolecular quenching.

7.2.4.2 Applications in cyclodextrin-based electrokinetic chromatography

As already noted, the low buffer volume required with the setup of Fig. 7.3 allowed the use of expensive buffer additives, such as (charged) cyclodextrins (CDs), for the separation of neutrals. Based on this, the coupling of cyclodextrin-induced electrokinetic chromatography (CD-EKC) and quenched RTPL was developed in our laboratory [10] and tested for the separation of nitroaromatics. Carboxymethyl-β-cyclodextrin (CM-β-CD) was added at a concentration of 5–12.5 mM to the separation buffer (25 mM borate, 1 mM BrNS, pH 8.5) to induce mobility for the neutral analytes. One expects that the quenching efficiencies would be reduced, since the phosphorophore resides in the aqueous phase while the analytes are inside the CDs for at least part of the time. Nevertheless, the CDs did not inhibit sensitive detection by bimolecular quenching owing to the rather low capacity factors (which are quite common in CD-EKC). Hence, LODs in the 10^{-8} M range were achieved even with 50 μm ID capillaries. The separation could be easily optimized by varying the CM-β-CD concentration in the buffer solution (see Fig. 7.6).

Since CD-EKC has also been applied for the separation of chiral compounds, the question that obviously arises is, can quenched RTPL be utilized in such applications as well? An interesting method for the enantioselective detection of camphorquinone was reported by García-Ruiz et al. [11]. Addition of 10 mM CM-β-CD and 20 mM α-CD to a 25 mM borate, pH 8.5 buffer, containing 1 mM BrNS to provide a constant phosphorescence background, was found optimal for the separation of the camphorquinone enantiomers. The method provided an LOD of $\sim 10^{-7}$ M, which is three orders of magnitude lower than with UV-absorption detection at 200 nm. Because of this, the method could be applied directly—i.e., without involving sample cleanup or analyte preconcentration—to monitor the enantioselective biodegradation of camphorquinone by yeast cells [11]. Figure 7.7 shows the two camphorquinone enantiomers at equal levels in the control experiment (trace a), and the preferential degradation of the (+) enantiomer in the presence of yeast, both outside (trace b) and inside (trace c) the yeast cells.

7.2.5 Concluding remarks on CE–phosphorescence

CE–quenched RTPL yielded promising results: the method provided sensitive detection of several analyte classes down to 10^{-8} M.

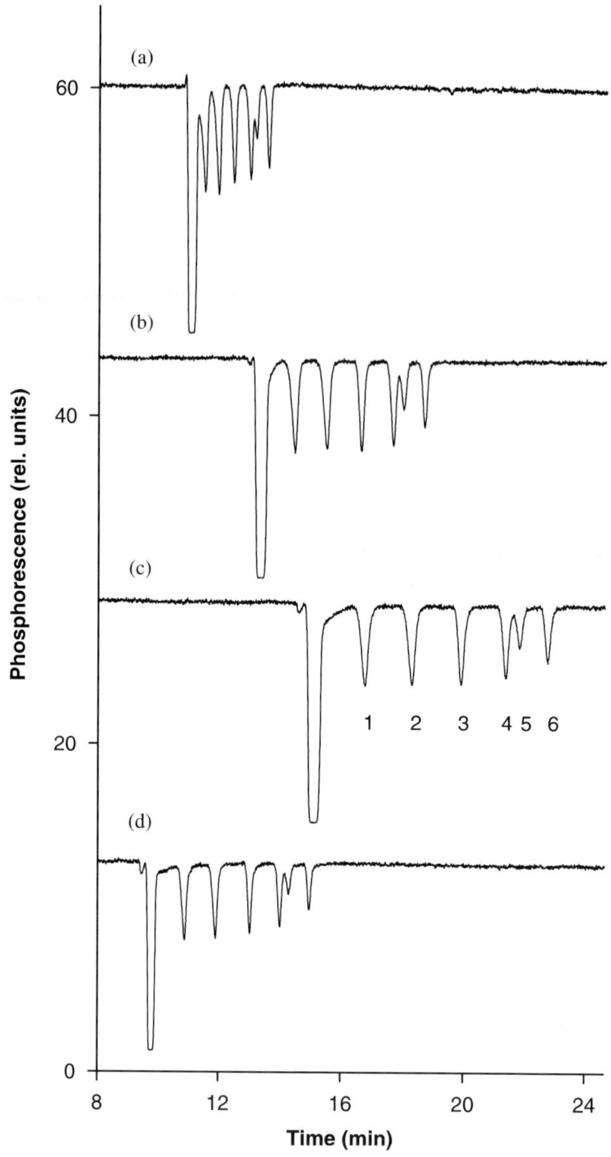

Fig. 7.6. Electropherograms of selected nitromonoaromatic compounds (10^{-6} M) detected with quenched phosphorescence. 1: 1,3-Dinitrobenzene, 2: 2,4-dinitrotoluene, 3: nitrobenzene, 4: 4-nitrotoluene, 5: 2-nitrotoluene, 6: 3-nitrotoluene. Buffer: 25 mM borate, 1 mM BrNS, pH 8.5 with: (a) 5 mM CM-β-CD, (b) 10 mM CM-β-CD, (c) and (d) 12.5 mM CM-β-CD. Separation voltage: -20 kV, except for trace d, -30 kV. Reprinted with permission from Ref. [10]. Copyright (2002), American Chemical Society.

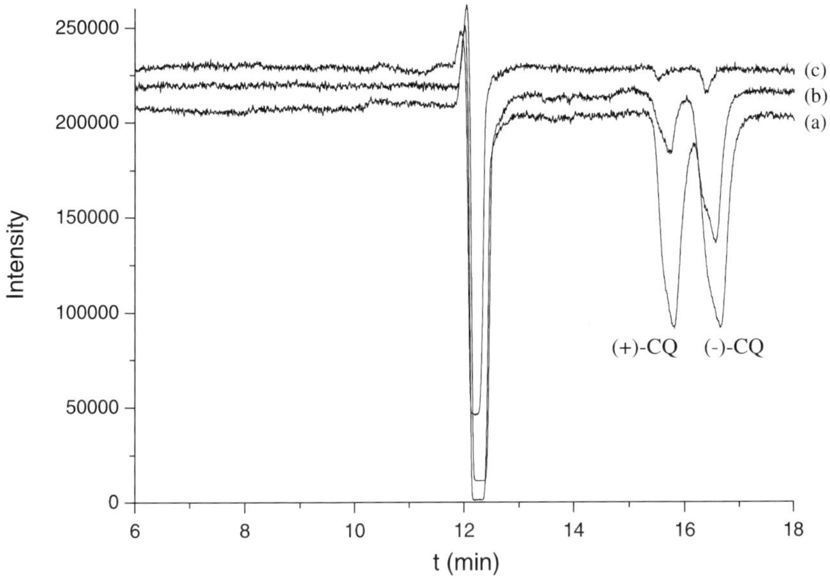

Fig. 7.7. Illustration of chiral CE with quenched phosphorescence detection, used to directly determine the enantioselective degradation of camphorquinone by yeast. (a): racemic mixture from control experiment; (b) yeast medium after 24 h (injected directly after 100 times dilution with methanol); (c): extract from the yeast cells after 24 h. Buffer conditions: 25 mM borate, 10 mM CM-β-CD, 20 mM α-CD, 1 mM BrNS; pH = 9.0. Reprinted with permission from Ref. [11], Copyright (2004), American Chemical Society.

Donkerbroek et al. [16] studied quenched RTPL with a much broader range of analyte classes. Those experiments were carried out under flow conditions at much larger dimensions, but the general conclusions regarding applicability should be valid for CE dimensions as well. Two important CE modes—CZE and (CD)-EKC—were found to be compatible with quenched RTPL detection; several common CE buffers can be applied. CE–quenched RTPL thus provides a viable alternative for optical detection in CE, especially in case of analytes with poorly absorbing chromophores. The technique may even be applicable with ultra narrow capillaries. Suppression of triplet–triplet annihilation will probably be essential for achieving this goal. Further studies should also address another important feature of quenched RTPL, i.e., its inherent selectivity. Undoubtedly, such studies should involve real-life samples in order to establish the practicability of CE–quenched RTPL.

7.3 CHEMILUMINESCENCE DETECTION IN CE

7.3.1 Introduction

Among the different luminescence techniques such as fluorescence and phosphorescence, CL, in biological systems also referred to as bioluminescence, is special in that it requires no light source for excitation. Instead, CL is obtained by means of an exothermic reaction, which produces excited-state products that either give luminescence themselves or are able to excite fluorophores. Generally, the excited-state products contain large amounts of energy, because their structures are strained or contain a peroxide group. In most cases, an oxidation reaction in alkaline solution (using hydrogen peroxide as the oxidant) is involved. A famous example of CL in nature is the firefly bioluminescence reaction, which produces light at an exceptionally high yield of about 90% [17,18]. Although this reaction has been applied in analytical chemistry, the vast majority of CL reactions used for analytical purposes is based on synthetic reactions, i.e., no enzyme systems are involved. In fact, the number of reactions that are used is quite limited: in most cases reactions involve luminol or isoluminol (or its derivatives), peroxyoxalate-type compounds, acridinium esters or permanganate [17]. Apart from these reactions, electrogenerated CL or electrochemiluminescence (ECL) has often been applied [19,20].

7.3.2 Development of CL detection in CE

The last decade has seen a substantial increase in the number of articles dedicated to CL detection in CE. Undoubtedly, the growing interest of analytical researchers in this field is related to the advantages of CL, which are well-known from studies combining CL with techniques such as LC, flow injection analysis (FIA), gas chromatography (GC) and immuno (or hybridization) assays [17,20]. The most notable advantage of CL is its potential for obtaining sensitive detection; absolute detection limits in the fmol–amol range have been obtained. An important factor contributing to this high sensitivity is the fact that no light sources are required for excitation, which is why the CL technique may in many cases be viewed as a true 'dark background' technique. Thus, the noise related to Rayleigh, Raman and other types of scatter will be absent in CL. Furthermore, the technical problem of how to achieve efficient excitation in a CE capillary with an optical pathlength as small as 50–75 μm does not exist, while simple optics can be used for the collection of emission.

In spite of these beneficial features, there are a number of problems and disadvantages in coupling CE and CL, which have presumably delayed the development of CE–CL for quite some years. In almost all CE–CL systems, post-column (post-capillary) addition of one or more CL reagents is required. After mixing of analytes and CL reagents, a transient CL signal is obtained. The mixing can obviously result in additional band broadening, so one of the main attractions of CE may be seriously compromised by using CL detection. As will be seen below, if the CL reactor is designed properly and the CL reactions are fast, this problem can be avoided. It should be noted that this is much more critical in CE than in LC, owing to the high efficiencies normally attained in CE [21]. A second drawback is that the separation conditions (solvent, pH, additives) are not always compatible with the detection conditions, i.e., compromises may be required. This problem can be solved by the addition of post-column reagent.

The different CL reactions applied, as well as the reactors designed for CE–CL, have their specific advantages and limitations, which will be discussed below. First, the CL reactions that have been used most often in CE (and LC) are presented to give an impression of the chemistry involved and possible target analytes (section 7.3.3). After that, the basic features of the setups used most often to couple CE with CL detection are discussed (section 7.3.4). Finally, an overview of the most important applications of CL detection in CE will be given (section 7.3.5).

7.3.3 CL reactions

In LC, four CL reactions have been widely used, i.e., reactions involving luminol (5-amino-2,3-dihydrophthalazine) or isoluminol (6-amino-2, 3-dihydrophthalazine) and its derivatives, acridinium esters, peroxyoxalate-type compounds and the permanganate ion. These reactions have also been applied to CL detection in CE. In addition, the coupling of CE and ECL has been extensively studied in recent years [17,19,20,22–24].

7.3.3.1 Luminol

Luminol, isoluminol and its derivatives have been the subject of many CE studies. Under alkaline conditions, and in the presence of hydrogen peroxide and a catalyst, luminol reacts to give the excited-state product

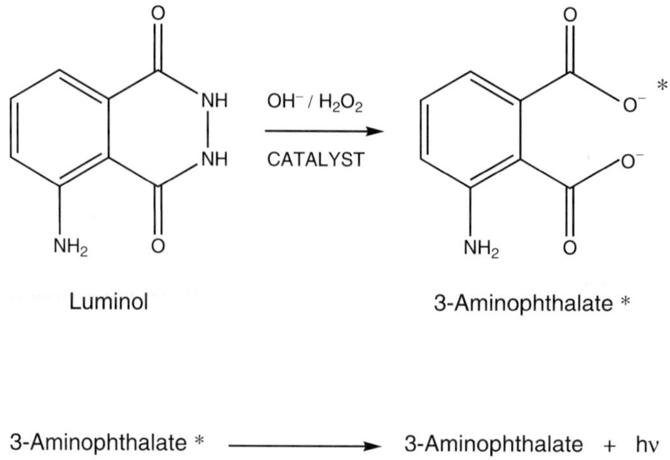

Fig. 7.8. Reaction scheme for luminol-based CL.

3-aminophthalate along with water and nitrogen (see the simplified scheme in Fig. 7.8). 3-Aminophthalate then emits luminescence with a wavelength maximum of 425 nm and an efficiency of about 1%. The lifetime of the transient CL signal obtained after mixing of luminol, hydrogen peroxide and catalyst can be as short as about 10 s, but longer lifetimes up to a minute are also encountered [25,26], depending on the nature of the catalyst, pH and reagent concentrations. A pH of about 10–11 is required, which to some extent limits the pH- range that can be used for the separation buffer, even though highly alkaline reagent solutions can be added post-column.

Several transition metals, such as Fe(II), Fe(III), Co(II), Cu(II), Mn(II), Cr(III) and V(IV), metal-containing compounds and oxidases (for instance, horseradish peroxidase) can be used as catalysts [17,23]. On the basis of this catalytic action, the luminol/hydrogen peroxide/catalyst system can be used for the detection of several analyte classes [17,22,23], including transition metals, metal-containing complexes and proteins (e.g., heme proteins). In addition, organic peroxides, which take over the role of hydrogen peroxide (see Fig. 7.8), can be detected. Last but not least, luminol CL can be applied to detect substrates that can enzymatically produce hydrogen peroxide. An example of the last option is the detection of glucose in presence of glucose oxidase. All these examples are generally referred to as 'direct CL.' Indirect CL is also known: a typical example involves the reduction of CL due to strong complexation of the catalyst metal by certain target analytes.

Isoluminol shows much weaker (~10 × 100-fold less) CL than luminol and therefore seems less appropriate. However, *N*-alkyl isoluminol derivatives are more chemiluminescent than isoluminol itself, while steric hindrance often lowers CL for luminol derivatives. Therefore, isoluminol analogs are used much more often for derivatization [20,24]. Two types of labeling can be distinguished: CL labeling using chemiluminescent tags and chemiluminogenic labeling based on weakly chemiluminescent reagents that become strongly chemiluminescent upon binding; the latter reaction is more selective, but also has a more limited application range. Isoluminol isothiocyanate (ILITC) is appropriate for labeling of amino acids and peptides. *N*-(4-aminobutyl)-*N*-ethylisoluminol (ABEI) has been used for labelling alkylamines, amino acids, peptides and carboxylic acids. The chemiluminogenic labeling agent 6-aminomethylphthalazine-1,4(*2H*,*3H*)-dione (AMP) has been used for derivatization of 5-hydroxyindoles and catecholamines, while 4,5-diaminophthalhydrazide (DPH) was found to be suitable for derivatization of α-keto acids, α-dicarbonyl compounds and aromatic or aliphatic aldehydes [24].

7.3.3.2 *Acridinium ester*

Acridinium ester CL is also obtained in alkaline solutions as the result of a reaction with hydrogen peroxide (see Fig. 7.9). At alkaline pH—obtained in CE by mixing the separation buffer with a strongly alkaline post-column CL reagent solution—hydrogen peroxide is converted into the hydroperoxyl anion (HO_2^-), which attacks the acridinium ring. The excited-state reaction product, *N*-methyl acridone,

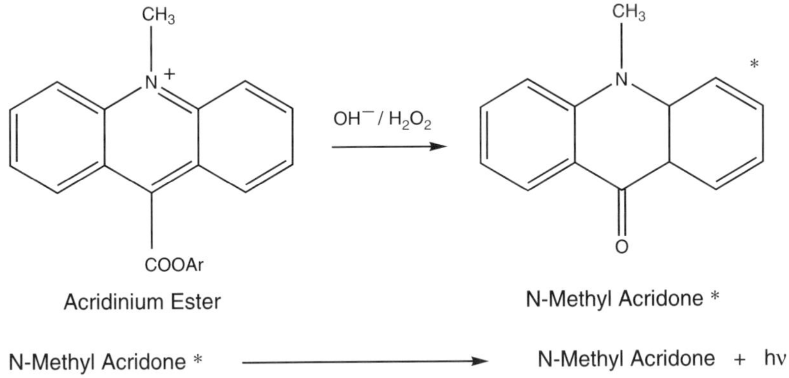

Fig. 7.9. Reaction scheme for acridinium ester-based CL.

emits luminescence with a wavelength maximum of 470 nm [17]. The reaction time is quite short, i.e., only a few seconds [27]. Its use is restricted to separation buffers with $pH < 3$, since acridinium is hydrolyzed at a higher pH. This may be considered to severely restrict its application in CE. Nevertheless, the reaction can be useful for the determination of acridinium-labeled amino acids, peptides and proteins [20], since these analytes are usually separated in CE at pH 2–3. In addition, the CL reaction exhibits a high overall yield of about 10%. In spite of this, according to the literature, this reaction has not been widely applied in CE.

7.3.3.3 Peroxyoxalate

The CL reaction based on the peroxyoxalate/hydrogen peroxide system is depicted in Fig. 7.10. A highly strained reaction product, 1,2-dioxetanedione, is obtained, which may transfer its excess energy in a second step to a suitable acceptor molecule with fluorescent properties. The excited acceptor molecule will then relax to the ground state by emitting fluorescence. Quantum efficiencies range from about 1 to 20% [17]. It should be noted that there has been a lot of discussion in the literature about the nature of the intermediate product molecules. Nevertheless, the reaction scheme depicted in Fig. 7.10 is sufficient to give an indication of the CL reactions involved.

The peroxyoxalate reaction has been widely used in LC, while its application in CE was also studied quite frequently in spite of the disadvantages encountered when using peroxyoxalate-based CL in CE. The most important drawbacks of peroxyoxalate in CE are its instability and limited solubility in aqueous buffers, and its instability under electric fields. In addition, peroxyoxalate has slow reaction kinetics, which reduces the efficiencies obtainable in CE–CL [23]. The most

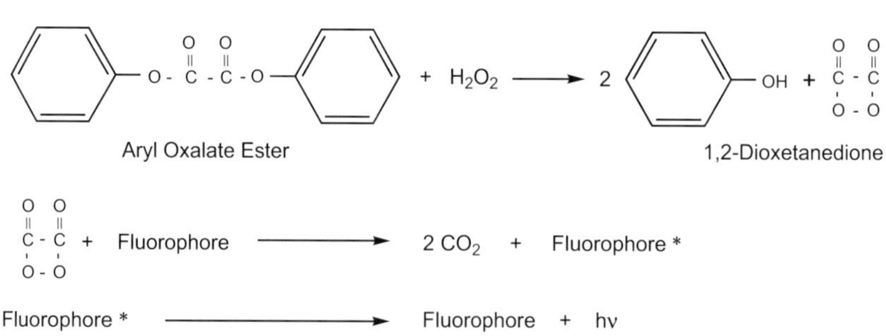

Fig. 7.10. Reaction scheme for peroxyoxalate-based CL.

important application of peroxyoxalate CL is the determination of compounds such as amino acids, proteins, DNA fragments, catecholamines and carboxylic acids, which are derivatized with fluorescent labels such as dansyl chloride, fluorescamine, *o*-phthaldehyde (OPA) and fluorescein isothiocyanate [17,20,22,28].

7.3.3.4 Permanganate

The permanganate (MnO_4^-) ion is a strong oxidant that can be used for direct oxidation of certain target analytes, such as serotonin, catecholamines and morphine derivatives. The reaction takes place post-column under strongly acidic conditions. The oxidized products subsequently emit CL.

7.3.3.5 ECL

ECL has been applied quite broadly in LC and the technique has recently been introduced in CE. In nearly all cases, in situ electrochemical generation of Tris(2,2′-bipyridyl) ruthenium(III) ($Ru(bpy)_3^{3+}$) by oxidation of $Ru(bpy)_3^{2+}$ is involved [22,29]. In situ electrogeneration is necessary because $Ru(bpy)_3^{3+}$ is not stable in aqueous solutions and might react prematurely when added to the separation buffer. The simplified reaction scheme for the detection of reductants is given below [20]:

$Ru(bpy)_3^{2+} \rightarrow Ru(bpy)_3^{3+} + e^-$ electrogeneration of oxidant

$Ru(bpy)_3^{3} + red \rightarrow Ru(bpy)_3^{2+^*} + product$ reaction with reductant

$Ru(bpy)_3^{2+^*} \rightarrow Ru(bpy)_3^{2+} + h\nu$ chemiluminescence

A potential of 1.2 V is required for electrogeneration of $Ru(bpy)_3^{3+}$. Detection by CE–ECL of several reductants, including tertiary, secondary and primary alkyl amines, amino acids, polyamines, nicotineamide–adenine dinucleotide(NADH), peptides and proteins is possible without prior derivatization [20]. An inherent problem of the coupling of CE and ECL is the influence of the electrophoretic currents on the electrode system providing electrogeneration of $Ru(bpy)_3^{3+}$. Generally, decoupling of the electrophoretic system from the detection part is required. Applications of CE–ECL as well as the approaches to decouple both systems from the electrogeneration part will be discussed in detail in section 7.3.5.6.

7.3.4 Post-capillary CL reactors

As was already noted in section 7.3.2, post-column addition of CL reagents is generally required in order to obtain CL. The addition of (all) CL reagents to the separation buffer would obviously result in unwanted CL inside the separation capillary. Instead, the mixing of analytes and CL reagents should take place right outside the separation capillary. The setup for CE–CL should enable detection of as much of the transient CL signal as possible, while maintaining the separation efficiency obtained on-column. The latter can pose a problem since the analytes reacting with the CL reagents outside the column will tend to spread over a larger volume. Four basic designs have been used most often, i.e., the on-column coaxial reactor, two types of the off-column coaxial reactor and the end-column reservoir. These designs are shown in Fig. 7.11. Their basic features, advantages and disadvantages will be discussed below.

7.3.4.1 On-column coaxial reactor

As shown in Fig. 7.11a, the on-column coaxial reactor consists of a separation capillary of which the inlet end is placed in a buffer vial also containing the anode and the outlet end is positioned inside a reaction capillary. The detection window (typically, a few millimeters long) is usually positioned directly behind the outlet of the separation capillary. The outlet end of the reaction capillary is placed in an outlet buffer vial also containing the cathode (ground). CL reagent is mixed with the separated analytes via a three-way connector. In most cases, a syringe pump or a peristaltic pump is used for this purpose. An obvious disadvantage of coaxial reactors is the inevitable spreading of the analytes over a larger volume, which results in band broadening. Thus, it may be advantageous to keep the diameter of the reaction capillary as small as possible. This can be achieved by etching the end of the separation capillary with hydrofluoric acid (HF) in order to decrease its outside diameter [21]. Furthermore, even if larger diameters are used, much of the separation efficiency will be preserved together with high sensitivity, provided that the CL reaction is rapid enough. In that case, high CL reagent flows may be used. If the reactions are too slow, high flows will result in a large decrease in sensitivity since the reaction—and thus light production—will occur largely past the detection window. In this context, it should be realized that the reaction rate ultimately depends on pH, reagent concentrations and, of course, the type of CL reaction

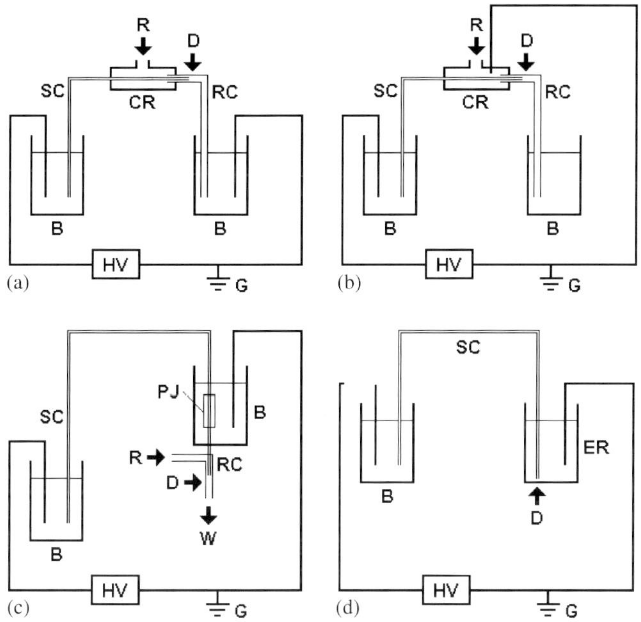

Fig. 7.11. Reactors for CE–CL. (a) On-column coaxial reactor, (b) off-column coaxial reactor, (c) off-column coaxial reactor with porous joint, (d) end-column reservoir. SC: separation capillary, RC: reaction capillary, CR: coaxial reactor, HV: high voltage, B: buffer vial, G: ground, R: CL reagent solution, D: detector, PJ: porous joint, W: waste, ER: end-column reservoir. Adapted from Ref. [22]. Copyright (2000), with permission from Elsevier.

involved. An important problem associated with the on-column coaxial reactor—when utilized in CE—is the generation of bubbles formed upon electrodecomposition of hydrogen peroxide at the anode. This leads to unstable currents through the separation capillary, which will disturb electrophoresis. In addition, the noise of the detected CL signal may increase, resulting in higher LODs.

7.3.4.2 Off-column coaxial reactor

The off-column coaxial reactor design (Fig. 7.11b and c) eliminates the above problems. In fact, the reactor in Fig. 7.11b is not a true off-column reactor, although it is usually denoted as such. Part of the reagent solution is still under the influence of the electric field. Nevertheless, this setup alleviates the above problems, so that higher hydrogen peroxide concentrations can be applied [22]. The coaxial reactors in Fig. 7.11a and b cannot be used for CL detection based on the permanganate anion or other highly mobile, negatively charged CL reagents: under the influence of the electric field, these compounds will back-migrate into the separation capillary, resulting in unwanted in-capillary CL. This problem is solved by the setup of Fig. 7.11c: in this setup, a porous joint is used to provide electrical contact with the cathode (ground). Thus, the reaction capillary is completely decoupled from the electric field over the separation capillary. The details of this setup, especially the construction of porous joints, will be discussed in detail in section 7.3.5.4.

7.3.4.3 End-column reservoir

The last setup discussed here uses an end-column reservoir (see Fig. 7.11d). In fact, this setup may be regarded as an on-column design, since no decoupling of the electrical field is applied. In this case, however, the CL reagent solution is not pumped for mixing with the analyte solution, but rather, the analyte solution flows into the reservoir containing the reagents. A photomultiplier tube (PMT) is mounted on the transparent bottom of the reservoir, while the capillary end is positioned about 1 mm from the bottom. Apart from the problems also associated with the setup in Fig. 7.11a, lower separation efficiencies compared with the coaxial reactor are generally achieved, most probably because of the slower removal of analytes and reagents from the reaction/detection zone. However, the end-column reservoir has the advantage of simple implementation in CE.

More features of the above CE–CL systems will be addressed in section 7.3.5. The setup for CE–ECL, which bears a large resemblance to the setup of Fig. 7.11c, will also be discussed there. Gap reactors and porous tube reactors, which are often used for post-column derivatization in CE with fluorescence detection [21], have not been applied in CE–CL, possibly because their implementation in CE–CL is more difficult. Nevertheless, these reactors may provide high separation efficiencies even with slower CL reactions (e.g., peroxyoxalate-based CL).

7.3.5 Applications of CE– CL

7.3.5.1 Luminol-based CL

Detection based on reactions of luminol (or related compounds) has been used most widely in CE–CL. The analyte classes that may be detected by this type of CL have already been indicated in section 7.3.3.1. Here, some typical examples will be presented.

In one of the first reports on CE–CL, Dadoo et al. [30] used the on-column coaxial reactor (see Fig. 7.11a) to show the feasibility of luminol-based CE–CL. For detection of CL, a parabolic mirror was positioned at the detection window; the reflected light was focused by a lens onto the PMT/photon counting system. In many other cases, the detection system was simply positioned close to the detection window, while no optics were used for the collection of emission. In another study by the same authors [18], luminol, ABEI and ABEI-derivatized amino acids were used as test analytes. In this case, an end-column reservoir (see Fig. 7.11d) was used for mixing of the analytes and a 25 mM ferricyanide (catalyst) solution. Hydrogen peroxide (100 mM) was added to the separation buffer (10 mM sodium phosphate, pH 10.8). An LOD of 20 nM was obtained for luminol. Detection of ATP was also performed, using the firefly luciferin/luciferase CL reaction (see section 7.3.5.5). Perpendicular to the capillary, an optical fiber with a core diameter of 1 mm was positioned at 1 mm from the capillary outlet. Owing to the relatively large detection/reaction zone and rather slow removal of the analyte/reagent mixture from this zone, the separation efficiency was reduced significantly; thus, the resulting number of theoretical plates, N, only ranged from about 10 to 20,000. Nevertheless, this setup has the advantages of simplicity and easy implementation.

A modification of the end-column reservoir system was proposed by Hashimoto et al. [31], who compared two configurations, one with two perpendicular fibers spaced at 1 mm, and one with a fiber opposite to the capillary at 0.3 mm. The latter type was about 100-fold more sensitive than the former, but it should be noted that apart from the configuration, the distances were also different. Using the frontal fiber position, an LOD of 5×10^{-10} M was achieved for luminol. In comparison, the setup of Dadoo et al. [18] allowed for easier optimisation since the optical fiber was placed on a positioner with three degrees of freedom.

Based on the catalytic effect of several transition metals, luminol-based CE–CL may provide in some cases a sensitive, simple and inexpensive alternative to the detection of metal ions with ICP–(inductively coupled plasma) optical emission spectrometry (OES) or even ICP–MS. Ultrasensitive detection of some metal ions was reported, as well as the speciation of some metals with multiple oxidation states [23,32–36]. In all of these studies a coaxial reactor design was used. The pH of the separation buffer was rather low (~3–5), which illustrates that low pH values can be used in luminol-based CE–CL, provided that the highly alkaline CL hydrogen peroxide solution (pH 11.5–12.0) is pumped at a high enough flow rate to provide an adequate pH during detection. In most cases, luminol (~1 mM) was added to the separation buffer [34–36], but it is also possible to add both CL reagents post-column by using two syringe pumps [33].

The catalytic potential of different metals or different oxidation states of the same metals varies extensively, spanning many orders of magnitude. Cr(III), Co(II) and V(IV) are amongst the ions detected most sensitively; concentration LODs of 10,000, 130 and 24 attomolar (aM), respectively, were reported for these ions, [23,32,35,36]. Application of the field-amplified sample injection technique appears to be crucial for obtaining such low LODs: injection of the highly mobile, positively charged analytes is quite selective. In addition, the sample concentration effect is assisted by the low EOF at the applied pH values (pH ~3–5). Going from samples dissolved in the separation buffer to purely aqueous samples, LODs were reported to be lowered by 5–6 orders of magnitude. It should be noted that such a large stacking effect has not been reported earlier in the literature. A detailed explanation of this effect will therefore be required.

Because Cr(III) is a much better catalyst than Cr(VI) (CrO_4^{2-}), and in contrast to Cr(VI), is positively charged, it can be detected much more sensitively. Thus, the difference in LODs was found to be six orders of magnitude, which is rather inconvenient if it is the purpose to determine both oxidation states simultaneously [32]. By applying in-capillary oxidation of Cr(VI), both species could be detected with similar sensitivity (LODs were 0.6 and 8 picomolar (pM) for Cr(III) and Cr(VI), respectively [33]).

An interesting indirect luminol-based CL method was reported by Ren and Huang [34]: on the basis of electrophoretic displacement of Co(II), which was present in the separation buffer at a concentration of only 0.5 μM, 14 lanthanides and some transition metal ions could be detected with LODs in the low nM range; separation was achieved within 3.5 min (see Fig. 7.12). The high sensitivity of the method was due to the low concentration and high catalytic activity of Co(II), and the application of field-amplified sample injection.

In spite of the limitations encountered in the detection of labeled compounds and the drawback of requiring an additional chemical step, several groups have used this approach. Two examples are presented here: Liu and Chang [37] reported a micellar separation of biogenic ABEI-labeled amines (diaminopropane, putrescine, cadaverine and diaminohexane). A 10 mM borate, 80 mM sodium dodecyl sulfate (SDS)

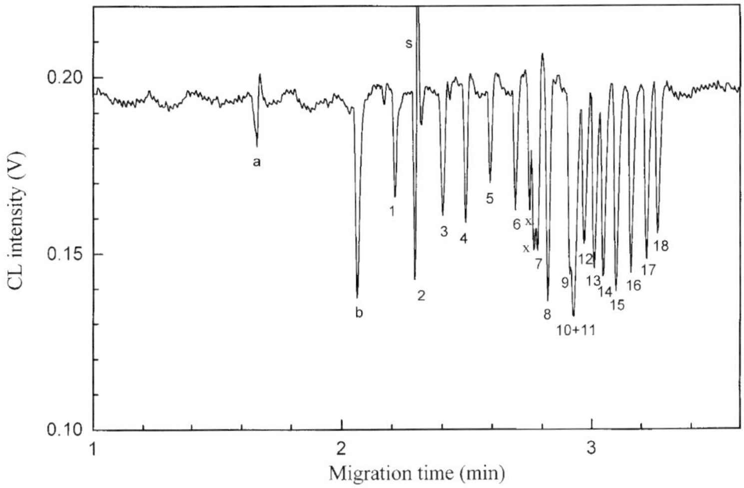

Fig. 7.12. Electropherogram of lanthanides and transition metal ions, detected by indirect luminol-based CL. 1: Mn^{2+}, 2: Cd^{2+}, 3: Ni^{2+}, 4: Pb^{2+}, 5: La^{3+}, 6: Ce^{3+}, 7: Pr^{3+}, 8: Nd^{3+}, 9: Sm^{3+}, 10: Gd^{3+}, 11: Eu^{3+}, 12: Tb^{3+}, 13: Dy^{3+}, 14: Ho^{3+}, 15: Er^{3+}, 16: Tm^{3+}, 17: Yb^{3+}, 18: Lu^{3+}, s: system peak, a: K^+, b: Ca^{2+}, x: unknown. Concentration: 5×10^{-8} M each. Separation buffer: 10 mM sodium acetate, 0.3 mM luminol, 0.5 μM Co^{2+}, adjusted to pH 4.40 with lactic acid. CL reagent solution: 20 mM sodium acetate, 2 mM hydrogen peroxide, pH 11.8. Reprinted with permission from Ref. [34]. Copyright (2001), American Chemical Society.

(pH 9.3) buffer was used for separation. Hydrogen peroxide (100 mM) was added to the buffer, and a strongly alkaline ferricyanide solution was added post-column. LODs ranged from 3.5×10^{-8} to 1.2×10^{-7} M; putrescine and cadaverine were detected in lake water. Using derivatization with luminol, Wang et al. [38] studied CE–CL of seven monosaccharides (see Fig. 7.13); LODs around 10^{-7} M were achieved. The reported plate numbers from this study were very high, i.e., around 200,000. Most likely, these high values are due to the high flow of the CL reagent solution (0.5 μl/s), which may also explain the rather modest LODs for luminol-based CL. A preliminary application of the method to the detection of glucose in rat brain microdialysis samples was reported.

In addition to the indirect luminol CL mode discussed above, an indirect method based on complexation of the catalyst by the analytes was developed by Liao et al. [39], and Liao and Whang [40]. A 15 mM carbonate buffer (pH 10.0) with 5 mM luminol and 25 mM hydrogen

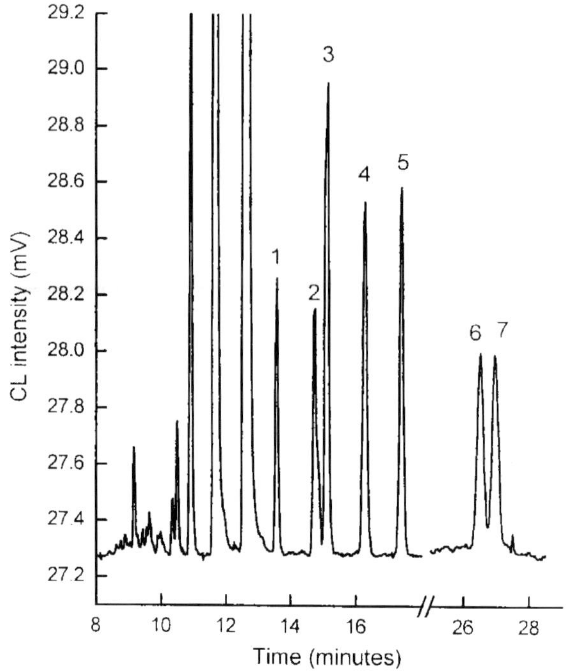

Fig. 7.13. Electropherogram of luminol-derivatized monosaccharides (2 μM each). 1: rhamnose, 2: glucose, 3: arabinose, 4: fucose, 5: galactose, 6: glucuronic acid, 7: galacturonic acid. Separation buffer: 200 mM borate, 100 mM hydrogen peroxide, pH 10.0. CL reagent solution: 25 mM potassium hexacyanoferrate in 3 M NaOH. Reprinted from Ref. [38]. Copyright (2003), with permission from Elsevier.

peroxide was used for separating non-derivatized amino acids and an enkephalin peptide. The catalyst, Cu(II), was added post-column. To prevent precipitation of Cu(II), at high pH, 1 mM tartrate was added to the 0.1 mM $CuSO_4$, 15 mM carbonate buffer solution. In the presence of the amino acids, a strong Cu(II)–amino acid complex was formed. This resulted in negative peaks, since such a complex shows strongly decreased catalytic activity compared with the weak Cu(II)–tartrate complex. LODs of $1–5 \times 10^{-5}$ M were achieved. A disadvantage of this detection mode is the limited linear detection range: in fact, a sigmoidal curve is obtained, which reflects the increased binding of Cu(II) in the presence of increasing concentrations of amino acids.

Based on the above principle, indirect detection of catechol, serotonin and some catecholamines was also achieved [25]. The analyte peaks were reasonably well resolved using a borate buffer of pH 9.5, but significant tailing of the amines was present, and LODs were around 1 μM. The results of Tsai and Whang [25] were about five-fold better than those of Zhang et al. [26], who used Co(II) as a catalyst. The difference may be explained by the much higher complexation constants of the Cu(II)–analyte complexes. Pharmaceutical preparations containing dopamine, norepinephrine and epinephrine, respectively were successfully analyzed [25].

7.3.5.2 Acridinium ester-based CL

In one the earliest reports on CE–CL, Ruberto and Grayeski [27] studied the potential of acridinium ester-based CL. Several acridinium esters were separated using a 50 mM tartrate, 10 mM γ-CD buffer (pH 2.8). An on-column coaxial reactor was used to mix the analytes and CL reagents. An acidic hydrogen peroxide solution and an alkaline phosphate buffer were separately delivered by two syringe pumps in order to stabilize the hydrogen peroxide. A combined CL reagent flow of 11 μl/min (almost 0.2 μl/s) was found suitable for adequate mixing and achieving an alkaline pH; faster flows were reported to show lower CL owing to incomplete reaction in the detection zone. As was already noted, acridinium esters are hydrolyzed at pH $>$ 3, with hydrolysis being almost complete above pH 4. Therefore, after mixing of acridinium and CL reagent solutions, hydrolysis of acridinium may be in strong competition with the CL reaction if the resulting pH is too high; a pH of 8.0 was found to be optimal for the phosphate buffer. Hydrogen peroxide concentrations up to 20 mM could be used without the problems associated with bubble formation. Under optimized conditions, LODs were around 10^{-7} M.

The separation and detection of acridinium-labeled peptides was also reported by the same researchers [41], using a buffer with pH 2.7. The labeling reaction was performed pre-column at pH 8 and was reported to proceed rapidly (within 15 min) and to completion. LODs were in the attomol range.

7.3.5.3 Peroxyoxalate-based CL

In most cases, application of peroxyoxalate-based CE–CL has been aimed at the detection of covalently labeled analytes; in an early study by Hara et al. [42], 'dynamic labeling' was applied, based on the supramolecular complexes of proteins with dyes such as Eosin Y, Rose Bengal and various xanthenes. Tsukagoshi et al. [43] reported the use of peroxyoxalate CL together with dansyl chloride-labeling for the detection of 15 phenols (including chloro- and nitrophenols). An end-column reservoir filled with 2 mM *bis*[2-(3,6,9-trioxadecanyloxycarbonyl)-4-nitrophenyl]oxalate (TDPO), and 50 mM hydrogen peroxide in acetonitrile was used to generate CL. For the separation, a 10 mM phosphate, 2 mM SDS, pH 8.0, 50% acetonitrile buffer was used. The peak efficiencies obtained with this method appeared to be remarkably good for an end-column reservoir system and peroxyoxalate-based CL; no explanation for this was given. The LODs were about 10^{-7} M when using a PMT mounted at the outside of the reservoir, close to the capillary outlet. Using an optical fiber at 0.3 mm from the capillary outlet ~five-fold poorer LODs were obtained, most probably owing to reduced light collection efficiency [18].

The same research group also studied CE–CL of fluorescein isothiocyanate-labeled DNA fragments and proteins [28]. As an alternative to capillary gel electrophoresis, CZE was applied, using a buffer with pH 8.4 that contained 0.5% (w/w) polymer (carboxymethylcellulose). Severely broadened peaks were obtained when using an end-column reservoir setup, which was attributed to slow removal of the analyte/reagent mixture from the reaction/detection zone; obviously, broadening will be most serious for slow CL reactions. Using an on-column coaxial reactor design together with a CL reagent flow rate of 2.5 µl/s or higher (which is extremely high for such a reactor) satisfactory peaks could be obtained. However, the sensitivity was severely compromised by this approach.

A novel and fast co-catalyzed peroxyoxalate reaction for application in CE–CL was described by Kuyper et al. [44]. CL reagents were bis-(2,4,6-trichlorophenyl) oxalate (TCPO), 1,2,2,6,6-pentamethylpiperidine (PMP) and 1,2,4-triazole (triazole). Experiments involving a stopped-flow mixing device showed that the CL signal was reduced to 50% within <2 s. Obviously, such fast reactions will be important in order to obtain high peak efficiencies in CE–CL.

7.3.5.4 Permanganate-based CL

Lee and Whang [45] were the first to apply an off-column coaxial reactor incorporating a porous joint (see Fig. 7.11c) in CE–CL. This type of setup was found necessary to allow the use of permanganate as an oxidant, as discussed before (section 7.3.4.2). The porous joint was constructed as follows: first, the polyimide coating was burned off from a short capillary section close to the end. This part was glued onto a microscope glass for mechanical support after which a small fracture was made in it. Finally, a porous cellulose acetate coating was obtained by carefully applying a few microliters of a 12% (w/v) cellulose acetate solution over the fracture and subsequent curing. To avoid blocking of the fractured section, Forbes et al. [29] temporarily filled the capillary with an electrode of 50 μm in diameter. Similar approaches involved the construction of a Nafion joint [22] or an etched joint assembly [46]. The latter connection, however, was rather fragile and could not be applied above pH 9.

Under acidic conditions, the permanganate ion will generate CL after reaction with suitable reductants. Lee and Whang [45] used permanganate-based CL for the detection of serotonin, catecholamines and catechol. For separation, a 10 mM tetraborate buffer (pH 9.5) was used. A strongly acidic permanganate solution (1 mM $KMnO_4$ in 1 M H_2SO_4) was delivered to the coaxial reaction capillary at 2.5 μl/min. The LODs were quite high ($\sim 10^{-4}$ M). Band broadening due to the porous joint was about 8%, which was significantly smaller than the broadening caused by the interface. There was also a ~5% loss of analytes. The resulting plate numbers were around 8000. As mentioned by the authors, an inherent limitation of this setup is that the EOF should be high enough to transport the analytes past the porous joint. Therefore, low buffer pH will not be beneficial.

Barnett et al. [47] used a completely different approach to solve the problem of permanganate back-migration: the polarity was switched (anode at the capillary outlet) while the EOF was reversed by adding 0.001% (mol/vol) hexadimethrine to the separation buffer (50 mM 6-aminocaproic acid, 80 mM α-CD, pH 4.5). The outlet reservoir contained a solution of 0.5 mM permanganate and 0.5% (mol/vol) sodium polyphosphate at pH 1.9. The concentration of sodium polyphosphate was found to be of critical importance in obtaining high sensitivity. Under optimised conditions, LODs for morphine, oripavine and pseudomorphine ranged from 2.5 to 5×10^{-7} M, which is 10–20-fold better than with UV detection. Remarkably, pumping of the CL reagent in the end-column resulted in strongly enhanced peak efficiencies, as can be seen in Fig. 7.14. This effect seems comparable with the effect of high flow rates in coaxial flow reactors (see sections 7.3.4.1 and 7.3.5.3).

7.3.5.5 Miscellaneous CL reactions

Using a BaO_2 (barium peroxide)-packed detection window, on-column detection was shown to be feasible [48]. This may provide a significant

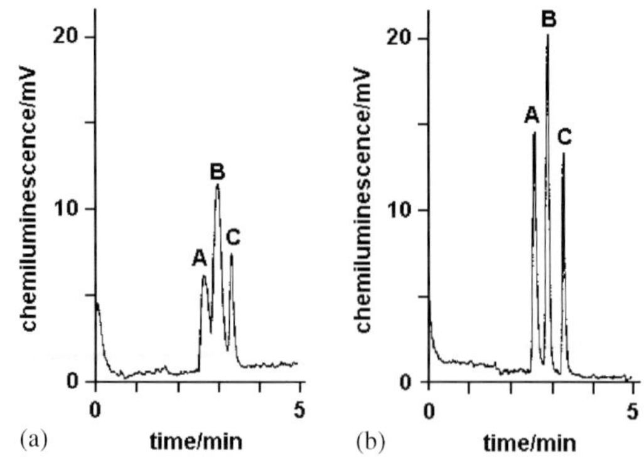

Fig. 7.14. Electropherogram of morphine and related drugs, detected by permanganate-based CL. (a) Without flow, (b) with flow. A: oripavine, B: morphine, C: pseudomorphine. Separation buffer: 50 mM 6-aminocaproic acid, 80 mM α-CD, 0.001% (m/v) hexadimethrine, pH 4.5. CL reagent solution: 0.5 mM permanganate, 0.5% (m/v) sodium polyphosphate, pH 1.9. Adapted from [47], with permission from the Royal Society of Chemistry.

advantage, since the high separation efficiencies of CE may thus be maintained. However, in this study, the plate number, N, was only ~4000 and the stability of the packing was poor. Major improvements over the suggested technique are therefore required.

Of course, in view of the low flow rates in CE, it might be interesting to explore the feasibility in CE of CL detection methods well established in GC. The application of gas-phase nitrogen CL for the CE–CL determination of nitrogen-containing analytes was studied by Mohammad et al. [49] and Sokolowski and Vigh [50]. After having migrated from the capillary, the analytes were nebulized using a pneumatic nebulizer; a sheath fluid was used to maintain electrical contact. Subsequently, they were oxidized at a high temperature to produce nitrogen oxides, which react with ozone to produce chemiluminescent excited-state nitrogen dioxides. Unfortunately, the achievable LODs were only in the sub-mg/l range even when whole-column field-amplified sample stacking was applied.

Dadoo et al. [18] applied firefly luciferin/luciferase bioluminescence for the detection of ATP. The reaction is based on the enzymatic conversion of luciferin into excited-state oxyluciferin in the presence of Mg(II), ATP and molecular oxygen [17]. An LOD of 5 nM was achieved, i.e., approximately three orders of magnitude better than with UV.

7.3.5.6 ECL

Finally, applications of ECL will be discussed. In recent years, many studies have been focused on the implementation of ECL in CE. In a pioneering study by Gilman et al. [51], luminol-based ECL was used to detect *N*-(4-aminobutyl)-*N*-ethylisoluminol–*N*, *N*-disuccinmidylcarbonate (ABEI-DSC)-labeled amines. In all other studies on CE–ECL, however, electrogenerated $Ru(bpy)_3^{3+}$ was applied for oxidation of the target analytes. As noted in section 7.3.3.5, interference from the electrophoretic current may occur when ECL is coupled to CE: the small oxidative currents generated in the detection system (a few μA) may be decreased significantly. Thus, less $Ru(bpy)_3^{3+}$ will be generated, which in turn results in lower ECL intensity. A number of different approaches to tackle this problem have been reported, as will be discussed below.

In the study by Forbes et al. [29], the electrophoretic system was decoupled from the detection part by using a porous joint assembly, as was also used in permanganate-based CL (see sections 7.3.4.2 and 7.3.5.4). The setup is shown in Fig 7.15. It consists of an outlet reservoir with a PMT mounted close to its bottom, downstream of the porous joint. Electrogeneration of $Ru(bpy)_3^{3+}$ was performed by using a

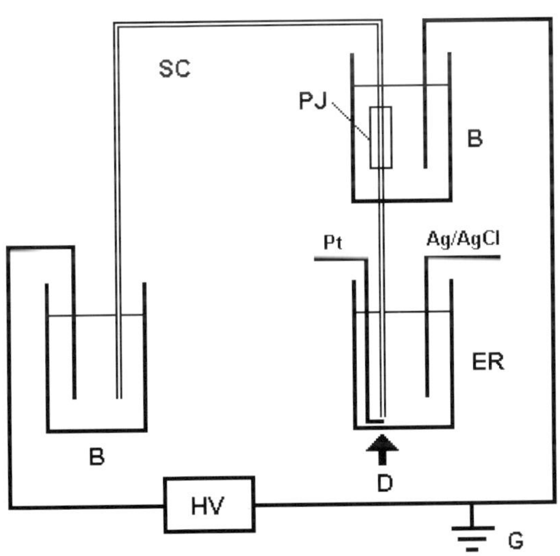

Fig. 7.15. Reactor for CE ECL. SC: separation capillary, HV: high voltage, B: buffer vial, G: ground, D: detector, PJ: porous joint, ER: end-column reservoir, *Pt*: Pt working electrode, Ag/AgCl: Ag/AgCl reference electrode. Adapted from Ref. [22]. Copyright (2000), with permission from Elsevier.

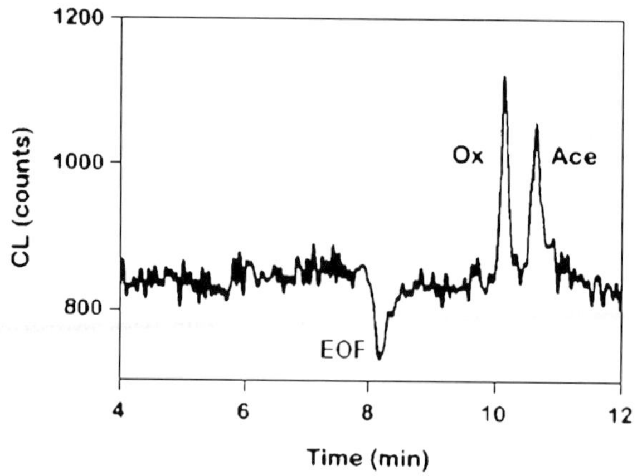

Fig. 7.16. Electropherogram of β-blockers, detected by ECL. Separation buffer: 20 mM borate, 10 mM Triton X-100, 0.5% β-CD, pH 8.8. Ox: oxprenolol, Ace: acebutolol. Concentration: 10 μM each. Adapted from Ref. [29]. Copyright (1997) with permission from Elsevier.

three-electrode potentiostat consisting of an Ag/AgCl reference electrode, an auxiliary electrode and a Pt working electrode inside the capillary. The feasibility of CE–ECL was demonstrated using β-blockers as test analytes (see Fig. 7.16); LODs were in the low mg/l range. As 50–200 μM $Ru(bpy)_3^{2+}$ was added to the separation buffer (20 mM borate, pH 8.8) and the working electrode was inside of the capillary, this is an example of on-column detection. Nevertheless, the plate number was only ~15,000, which might be attributed to the presence of the porous joint. It will also be clear from Fig. 7.16 that a significant background was present, caused by the reaction of $Ru(bpy)_3^{3+}$ with hydroxide ions.

As noted by Dickson et al. [46], addition of $Ru(bpy)_3^{2+}$ to the separation buffer results in equilibration times of several hours; moreover, band broadening due to non-matching mobilities may occur. Therefore, the latter authors used an end-column reservoir containing $Ru(bpy)_3^{2+}$ instead, in combination with electrogeneration of $Ru(bpy)_3^{3+}$ just outside the capillary. The system was tested by performing CE–ECL of two non-derivatized amines, using a 10 mM phosphate, pH 9.0, buffer; LODs in the μM range were achieved. Efficiencies were poor (N~4000), most likely due to the involvement of the end-column reservoir. CE–ECL of non-derivatized amino acids was also studied by Wang and Bobbitt [52], who used a coaxial ECL reactor and achieved sub-μM

LODs. The authors considered the influence of the working electrode diameter and reported a three-fold improvement when the electrode diameter was increased from ~75 to 125 μm.

The interference from the electrophoretic system with the detection may be decreased significantly by reducing the diameter of the separation capillary, since this will strongly lower the generated currents. Of course, the use of low-conductivity buffers will also be beneficial in this respect. This approach was adopted by Liu et al. [53], who performed CE–ECL of four polyamines (putrescine, cadaverine, spermidine and spermine) using capillaries with an ID (inner diameter) as small as 25 μm, and an end-column reservoir approach with at-capillary electrogeneration of $Ru(bpy)_3^{3+}$. Owing to the absence of a porous joint, a phosphate buffer with pH 2 could be used for separation, which prevented tailing of the amines. For the CL reagent solution, a pH of 10–11 was found to be optimal. By using field-amplified sample injection, LODs of 10^{-7}–10^{-8} M were achieved. The method was applied to the detection of polyamines in urine; no interfering compounds appeared to be present.

The same group also studied CE–ECL using 75 μm ID capillaries without a porous joint assembly [54]. At capillary–working electrode distances less than 120 μm, almost no oxidative current and ECL were obtained even if the electrophoretic current was as low as 5 μA. The optimum distance found in this approach was 240 μm, which is significantly larger than in ECL systems with a porous joint (50–70 μm). At the optimum distance, ECL was reported to be strong even with currents as high as 210 μA, indicating that the electrophoretic system was effectively decoupled at this distance. As shown by cyclic voltametry, a higher voltage of about 1.35 V was required to obtain the same current and ECL.

7.3.6 Concluding remarks on CE–CL

During the last decade, several CL reactions, including those based on luminol, acridinium ester, peroxyoxalate and permanganate, have been implemented for detection in CE. In addition, CE–ECL based on reduction of $Ru(bpy)_3^{2+}$ has been developed very recently. Luminol-based CL has been used most often and—because of the variety of target analytes that can be detected with it—appears to be most versatile. CL based on peroxyoxalate or permanganate and electrogenerated CL also offer good perspectives. LODs with the above methods are typically in the 10^{-6}–10^{-8} M range, which is better than LODs usually achieved with UV-absorption detection.

The coupling of CE with CL (and ECL) almost invariably requires post-capillary addition of CL reagents, which has been achieved with a number of different post-capillary reactors. However, the on-column peak efficiencies in CE–(E)CL are usually compromised to some extent, in several cases leading to a serious loss of separation power. The coaxial reactor design generally preserves the peak efficiencies achieved on-column much better than the end-column reservoir type. In addition, successful application of the latter reactor type is much more dependent on fast CL reactions than the former, although peak efficiencies can be improved significantly if the CL reagent solution in the end-column reservoir is continuously replenished. Furthermore, implementation of the end-column reservoir is much simpler than with coaxial reactors, which require tedious construction procedures. Optimal design of the coaxial reactors requires etching of the separation capillary in order to reduce its outside diameter thus allowing insertion into smaller ID reaction capillaries. This is especially important if relatively slow CL reactions are involved. Easy-to-install reactors of the coaxial type are highly desirable and would undoubtedly stimulate a more widespread use of CE–CL. This will probably also apply for the porous joint assembly required in CE–ECL.

Although the number of studies on CE–(E)CL has increased significantly in recent years, the number of applications to real-life samples is rather limited. Of course, an assessment of the practical value of CE–(E)CL in real-life analysis will require efforts in this direction. Such studies should focus on the sensitivity, selectivity, applicability and robustness of the method.

Since no excitation source is required, and scatter problems are virtually absent, rather simple optical designs can be applied. Therefore, CL detection seems very attractive for application in chip-based separations, and this should be one of the main subjects for further research. In fact, some studies in this direction have already been performed [55]. An interesting option, connected with this, may be the use of CE–CL in so-called point-of-care applications, which require inexpensive, disposable analysis tools.

REFERENCES

1 J. Kuijt, F. Ariese, U.A.Th. Brinkman and C. Gooijer, *Anal. Chim. Acta*, 488 (2003) 135–171.

2 R.J. Hurtubise, *Phosphorimetry: Theory, Instrumentation, and Applications*, VCH Publishers, New York, 1990.

3 J.J. Donkerbroek, N.J.R. van Eikema-Hommes, C. Gooijer, N.H. Velthorst and R.W. Frei, *Chromatographia*, 15 (1982) 218–222.

4 J.J. Donkerbroek, N.J.R. van Eikema-Hommes, C. Gooijer, N.H. Velthorst and R.W. Frei, *J. Chromatogr.*, 255 (1983) 581–590.

5 J. Kuijt, U.A.Th. Brinkman and C. Gooijer, *Anal. Chem.*, 71 (1999) 1384–1390.

6 M.E. Díaz-García and A. Sanz-Medel, *Anal. Chem.*, 58 (1986) 1436–1440.

7 J.J. Donkerbroek, C. Gooijer, N.H. Velthorst and R.W. Frei, *Anal. Chem.*, 54 (1982) 891–895.

8 J.A. Barltrop and J.D. Coyle, *Principles of Photochemistry*, Wiley, New York, 1978.

9 G.J. Kavarnos, *Fundamentals of Photoinduced Electron Transfer*, VCH Publishers, New York, 1993.

10 J. Kuijt, D. Arraez Roman, F. Ariese, U.A.Th. Brinkman and C. Gooijer, *Anal. Chem.*, 74 (2002) 5139–5145.

11 C. García-Ruiz, M. Siderius, F. Ariese and C. Gooijer, *Anal. Chem.*, 76 (2004) 399–403.

12 J. Kuijt, U.A.Th. Brinkman and C. Gooijer, *Electrophoresis*, 21 (2000) 1305–1311.

13 J. Kuijt, E. de Rijke, U.A.Th. Brinkman and C. Gooijer, *Anal. Chim. Acta*, 417 (2000) 15–17.

14 J. Kuijt, R. van Teylingen, T. Nijbacker, F. Ariese, U.A.Th. Brinkman and C. Gooijer, *Anal. Chem.*, 73 (2001) 5026–5029.

15 J. Kuijt, F. Ariese, U.A.Th. Brinkman and C. Gooijer, *Electrophoresis*, 24 (2003) 1193–1199.

16 J.J. Donkerbroek, A.C. Veltkamp, C. Gooijer, N.H. Velthorst and R.W. Frei, *Anal. Chem.*, 55 (1983) 1886–1893.

17 D.S. Hage, Chemiluminescent detection in high-performance liquid chromatography. In: G. Patonay (Ed.), *Detection in High-Performance Liquid Chromatography, Newer Methods*. VCH, New York, 1992.

18 R.J. Dadoo, A.G. Seto, L.A. Colón and R.N. Zare, *Anal. Chem.*, 66 (1994) 303–306.

19 S. Kulmala and J. Suomi, *Anal. Chim. Acta*, 500 (2003) 21–69.

20 A.M. García-Campaña, L. Gámiz-Gracia, W.R.G. Baeyens and F. Alés Barrero, *J. Chromatogr. B*, 793 (2003) 49–74.

21 R. Zhu and W.Th. Kok, *J. Pharm. Biomed. Anal.*, 17 (1998) 985–999.

22 X.-J. Huang and Z.-L. Fang, *Anal. Chim. Acta*, 414 (2000) 1–14.

23 Y.-M. Liu and J.-K. Cheng, *J. Chromatogr. A*, 959 (2002) 1–13.

24 M. Yamaguchi, H. Yoshida and H. Nohta, *J. Chromatogr. A*, 950 (2002) 1–19.

25 H.-C. Tsai and C.-W. Whang, *Electrophoresis*, 20 (1999) 2533–2538.

26 Y. Zhang, B. Huang and J.-K. Cheng, *Anal. Chim. Acta*, 363 (1998) 157–163.
27 M.A. Ruberto and M.L. Grayeski, *Anal. Chem.*, 64 (1992) 2758–2762.
28 K. Tsukagoshi, Y. Shikata, R. Nakajima, M. Murata and M. Maeda, *Anal. Sci.*, 18 (2002) 1195–1198.
29 G.A. Forbes, T.A. Nieman and J.V. Sweedler, *Anal. Chim. Acta*, 347 (1997) 289–293.
30 R. Dadoo, L.A. Colón and R.N. Zare, *J. High Resol. Chromatogr.*, 15 (1992) 133–135.
31 M. Hashimoto, K. Tsukagoshi, R. Nakajima and K. Kondo, *J. Chromatogr. A*, 832 (1999) 191–202.
32 Y.-M. Liu and J.-K. Cheng, *Electrophoresis*, 24 (2003) 1993–2012.
33 W.-P. Yang, Z.-J. Zhang and W. Deng, *J. Chromatogr. A*, 1014 (2003) 203–214.
34 J. Ren and X. Huang, *Anal. Chem.*, 73 (2001) 2663–2668.
35 Y.-M. Liu, E.-B. Liu and J.-K. Cheng, *J. Chromatogr. A*, 939 (2001) 91–97.
36 E.-B. Liu, Y.-M. Liu and J.-K. Cheng, *Anal. Chim. Acta*, 456 (2002) 177–181.
37 Y.-M. Liu and J.-K. Cheng, *J. Chromatogr. A*, 1003 (2003) 211–216.
38 X. Wang, Q. Wang, Y. Chen and H. Han, *J. Chromatogr. A*, 992 (2003) 181–191.
39 S.-Y. Liao and Y.-C. Chao, C-W. Whang,, *J. High Resol. Chromatogr.*, 18 (1995) 667–669.
40 S.-Y. Liao and C.-W. Whang, *J. Chromatogr. A*, 736 (1996) 247–254.
41 M.A. Ruberto and M.L. Grayeski, *J. Microcol. Sep.*, 6 (1994) 545–550.
42 T. Hara, J. Yokogi, S. Okamura, S. Kato and R. Nakajima, *J. Chromatogr. A*, 652 (1993) 361–367.
43 K. Tsukagoshi, T. Kameda, M. Yamamoto and R. Nakajima, *J. Chromatogr. A*, 978 (2002) 213–220.
44 C. Kuyper, K. Denham, J. Dickson, J. Murray and R. Milofsky, *Chromatographia*, 53 (2001) 173–178.
45 Y.-T. Lee and C.-W. Whang, *J. Chromatogr. A*, 771 (1997) 379–384.
46 J.A. Dickson, M.M. Ferris and R.E. Milofsky, *J. High Resol. Chromatogr.*, 20 (1997) 643–646.
47 N.W. Barnett, B.J. Hindson and S.W. Lewis, *Analyst*, 125 (2000) 91–95.
48 J.-M. Lin, H. Goto and M. Yamada, *J. Chromatogr. A*, 844 (1999) 341–348.
49 A.A. Mohammad, E. Fujinari, A.O. Okorodudu and J.R. Petersen, *J. Chromatogr. A*, 868 (2000) 121–125.
50 A.D. Sokolowski and G. Vigh, *Electrophoresis*, 22 (2001) 3824–3832.

51 S.D. Gilman, C.E. Silverman and A.G. Ewing, *J. Microcol. Sep.*, 6 (1994) 97–106.
52 X. Wang and D.R. Bobbitt, *Anal. Chim. Acta*, 383 (1999) 213–220.
53 J. Liu, X. Yang and E. Wang, *Electrophoresis*, 24 (2003) 3131–3138.
54 W. Cao, J. Liu, X. Yang and E. Wang, *Electrophoresis*, 23 (2002) 3683–3691.
55 H.D. Willauer and G.E. Collins, *Electrophoresis*, 24 (2003) 2193–2207.

Chapter 8

Electrochemical detection in capillary electrophoresis

Kim M. Hanson, Theron J. Pappas and Lisa A. Holland

8.1 INTRODUCTION

Shortly after literature reports of successful capillary electrophoresis (CE) separations with on-column laser-induced fluorescence (LIF) detection accomplished in 75 and 80 μm inner diameter (ID) glass tubes [1–3], CE was coupled with electrochemical detection (EC) [4,5]. Wallingford and Ewing [4], were the first to report the efficient coupling of a bare fracture decoupler and off-column amperometric detection. These innovations, now in frequent use, were key to the routine implementation of CE–EC. The separation efficiency, speed of analysis, and low sampling requirements of CE are attributes compatible with the wide range of selectivity of EC methods and its excellent limits of detection (LODs). Microelectrodes are frequently coupled to CE systems primarily due to the size compatibility. The ability to reduce electrode dimensions without impairing the signal-to-noise ratio facilitates this coupling. Microelectrodes can be used in resistive solvents, have faster time constants, and are not severely affected by uncompensated resistance. Thus, they provide a means of using two (as opposed to three) electrodes in the EC cell, are better suited for fast voltammetric scanning, and are compatible with non-aqueous CE. Since the report by Wallingford and Ewing [4], the many advantages of coupling CE and EC have led to a wide variety of applications and a large volume of original research publications. A number of reviews on this subject are found in the literature [6–21].

This chapter will cover the fundamentals of electrochemistry that are important to CE. These include modes of operation, electrode materials, electrode geometry, CE geometry, and coupling issues. The

Comprehensive Analytical Chemistry XLV
M.L. Marina, A. Ríos and M. Valcárcel (Eds)
Volume XLV ISSN: 0166-526X DOI: 10.1016/S0166-526X(05)45008-4

content in this chapter is limited to free-zone CE, electrokinetic capillary chromatography, gel electrophoresis, and affinity electrophoresis in conventional fused-silica capillaries. A comprehensive table of research publications from 2002 through mid-2004 organized by analyte class is included. Typical CE–EC applications are described in order to emphasize advantages and considerations necessary for the successful implementation of the methodology. Topics excluded from this chapter are capillary electrochromatography coupled with EC, and CE coupled with electrochemiluminescence. The great increase in chip-based CE warrants a focused discussion of attributes unique to these systems in chapter 14.

8.2 MODES OF EC DETECTION

Electrochemistry is a well-established analytical field that focuses on the effects of shuttling electrons, ion mobility, and membrane potentials. Modes of operation included in this discussion are voltammetry at fixed (amperometry) or scanned potential, conductivity, and potentiometry. Each mode has separate criteria for optimal operation, and often there are different materials, geometries, solvents, statistical parameters, and coupling constraints. Amperometry is by far the most implemented method of EC owing to its excellent LODs. However, other EC modes have various advantages, which will be described in this chapter.

8.2.1 Amperometry

Amperometry is the monitoring of a gain (reduction) or loss (oxidation) of electrons in the presence of a fixed potential, by measuring changes in current. It is highly quantitative, provides excellent LODs (see Table 8.1), good linear range, and reproducibility under well-controlled conditions. Direct amperometric detection is selective for analytes that can undergo oxidation or reduction within the potential window of the electrode — in other words, natively electroactive compounds. The selectivity of this method is tunable, since the applied potential can be changed to discriminate against analytes.

The applicability of amperometry is further increased by employing indirect detection and by derivatization with an electroactive moiety. Derivatization performed prior to separation requires additional sample handling, which necessitates larger working sample volumes, and

TABLE 8.1

CE–EC detection listed by analyte class (2002–2004)

Electrode material	Electrode geometry	LOD	Reference
Amino acids			
PQQ[a] entrapped in PPY[b] matrix on carbon	245 μm disk	23 nM	[99]
Carbon	300 μm disk	0.1 μM	[100]
Carbon fibers @ 6 μm	250 μm toroid	4.5 μM	[39]
Au–Hg	100 μm disk	0.88 μM	[39]
Carbon fibers @ 6 μm	~250 μm disk array	0.3–2 μM	[30]
Steel tubes	2@4 mm cylinder, 2 mm gap	0.2–10 μM	[67]
Silver paint	2@2 mm cylinder, 1 mm gap	9.1–29 μM	[66]
Au–Hg amalgam	3 mm disk	0.5 μM	[101]
Carbon fibers @ 6 μm	1.1×10^{-3} mm^2 area disk[c]	0.17–1.8 μM	[29]
Carbon	300 μm disk	0.15, 0.21 μM	[102]
Carbon fibers @ 6 μm	250 μm disk embedded with 100 μm Au-Hg disk	0.3 μM	[40]
Peptides and proteins			
Au–Hg amalgam	100 μm OD cylinder	1.7 μM	[103]
PQQ[a] entrapped in PPY[b] matrix on carbon	245 μm disk	0.134 μM	[99]
Au–Hg	100 μm disk	5 μM	[39]
Carbon fibers @ 6 μm	~250 μm disk array	4.8 pM	[104]
Carbon fibers @ 6 μm	~250 μm disk array	0.02 nM	[105]
Commercial	Commercial	8, 11 μM	[106]
Carbon fiber	33 μm OD	6 μM[d]	[32]
Commercial	Commercial	0.5–2 μM	[65]
Au–Hg	100 μm disk embedded in electrode array	2.3 μM	[40]
Carbohydrates			
Silver paint	2@2 mm cylinder, 1 mm gap	13–31 μM	[69]
Copper	120 μm disk	0.65–1.8 μM	[107]
Copper	140 μm disk	1.0–4.0 μM	[108]
Copper	100 μm disk	0.62–1.4 μM	[109]
Copper	140 μm disk	0.003–0.007 μM	[110]
Oligonucleotides, purine/ pyrimidine bases			
Carbon fiber	32 μm cylinder	3,~5 nM	[49–51]
Carbon	300 μm disk	1.28–5.02 μM	[111]
Carbon	300 μm disk	0.157–0.767 μM	[112]
Inorganic ions			
Steel tubes	2@4 mm cylinder, 2 mm gap	18–160 nM	[113]

continued

TABLE 8.1 (*continued*)

Electrode material	Electrode geometry	LOD	Reference
Silver paint	2@2 cm, varying gap	0.19–6.6 μM[f]	[114–116]
Steel tubes	2@8 mm cylinder	N.R.	[117]
Aluminum foil	2@2 mm cylinder, varying gap	N.R.	[118]
Aluminum foil	2@10 mm cylinder, varying gap	N.R.	[118]
Steel tubes	2@4 mm, 1 mm gap	31–125 nM	[119]
Steel tubes	2@4 mm, 1 mm gap	55–175 nM[f]	[120]
Steel tubes	2@10 mm, 1 mm gap	1.97–10.08 μM[g]	[121]
Steel tubes	2@4 mm, 1 mm gap	0.89–2.7 μM	[122]
Silver paint	2@7 mm cylinder, 0.4 mm gap	0.156–7.41 μM[f]	[123]
Steel tubes	2@2 mm cylinder, 0.7 mm gap	1–2 μM	[124]
Commercial	Commercial	0.339–0.480 μM	[125]
Commercial	Commercial	0.3 μM	[65]
Steel tubes	2@4 mm, 1 mm gap	20–200 μg/l[h]	[126]
Neurotransmitters			
Platinum	10 μm bands, 10 μm spacing	5 μM	[127]
Carbon fiber	8 μm cylinder	N.R.	[128]
Boron-doped diamond film	50 μm × (~0.3–0.5) mm band	0.019–0.023 μM	[129]
Carbon composite	75 μm	0.03 μM	[130]
Platinum microdisk	25 μm disk	0.46 μM[d]	[131]
Gold	10 μm bands, 10 μm spacing	13 μM dopamine	[97]
Boron doped diamond	Conical 9.8×10^{-4} cm^2 active SA	78 nM	[132]
AchE/ChO modified[e]	200 μm disk	2 μM	[133]
AchE/ChO modified[e]	25 μm cylinder	2 μM	[133,134]
Carbon fiber	5 μm cylinder	N.R.	[135]
Carbon fiber	8 μm cylinder	0.4 μM	[136]
Phenolic compounds			
Carbon	300 μm disk	0.9, 2 μM[f]	[137]
Carbon	350 μm disk	N.R.	[138]
Carbon	300 μm disk	0.2 μM	[139]
Carbon	300 μm disk	0.9–3.0 μM	[140]
Carbon fiber	33 μm disk	0.03 μM	[141]
Carbon	300 μm disk	1.1, 1.7 μM	[142]
Carbon fibers @ 6 μm	~250 μm disk array	0.79, 1.3 μM	[143]
Carbon	300 μm disk	0.14–0.5 μM	[144]
Carbon	300 μm disk	0.49, 0.65 μM	[145]
Carbon	300 μm disk	0.059 μM	[146]
Drugs			
Carbon	300 μm disk	1.7 μM	[142]

continued

TABLE 8.1 (*continued*)

Electrode material	Electrode geometry	LOD	Reference
Carbon fibers @ 6 μm	~250 μm disk array	0.05 μM	[147]
Carbon fibers @ 6 μm	~250 μm disk array	0.5 μM	[148]
Platinum	25 μm disk	0.1 μM[f, h]	[149]
Carbon	300 μm disk	0.2, 0.07 μM	[150]
Carbon fibers @ 6 μm	~250 μm disk array	0.5 μM	[151]
Biomarkers			
Carbon fibers @ 6 μm	250 μm toroid	2.4 μM	[39]
Au–Hg	100 μm disk	0.88, 5 μM	[39]
PQQ[a] entrapped in PPY[b] matrix on carbon	245 μm disk	23 nM	[99]
Carbon fiber	11 μm cylinder	25 nM	[152]
Carbon fiber	7 μm cylinder	20 nM	[153]
Carbon fibers @ 6 μm	~200 μm disk array	0.76 μM	[154]
Carbon fibers @ 6 μm	~250 μm disk array	0.29 U/ml[h]	[155]
Carbon fibers @ 6 μm	~250 μm disk array	0.024 U/ml[h]	[156]
Other			
Gold	50 μm disk	0.1–0.4 μM	[157]
Brass tubes	2@5 mm, 2.5 mm gap	N.R.	[70]
Brass tubes	2@2.5 mm, 1.5 mm gap	N.R.	[70]
Carbon	300 μm disk	0.02 μM	[158]
Carbon fibers @ 6 μm	~250 μm disk array	0.0017 μM	[159]
Carbon	300 μm disk	2.0 μM	[160]
Carbon	500 μm disk	0.9–4 μM	[161]
Carbon	500 μm disk	1.7 μM[f]	[162]

Note: N.R. not reported, SA surface area.
[a]PQQ pyrroloquinoline quinone.
[b]PPY polypyrrole.
[c]Electrode dimensions in text conflict with referenced fabrication.
[d]Reference indicates LOD calculation using S/N = 2.
[e]AchE/ChO acetylcholineesterate/cholineoxidase.
[f]Reported in units of g/l, converted into M for comparison.
[g]Reported in units of ppb, converted into μM for comparison.
[h]Reported LOD conversion into μM not possible.

often makes the analytical method more laborious. On-column and post-column derivatization circumvent these drawbacks, but are more sophisticated in design. Examples of analyte modification resulting in the formation of electroactive products include copper complexation using biuret chemistry [22–24], and derivatization with naphthalene-2,3-dicarboxaldehyde (NDA) [25–31] or *p*-nitrophenol-2,5-dihydroxyphenylacetate bis-tetrahydropyranyl ether (NDTE) [32]. Amperometric reaction detection is another approach to increasing the applicability of EC [33]. In this mode, an electroactive reagent that has the ability to react with analyte is generated. Depletion of the reagent is detected

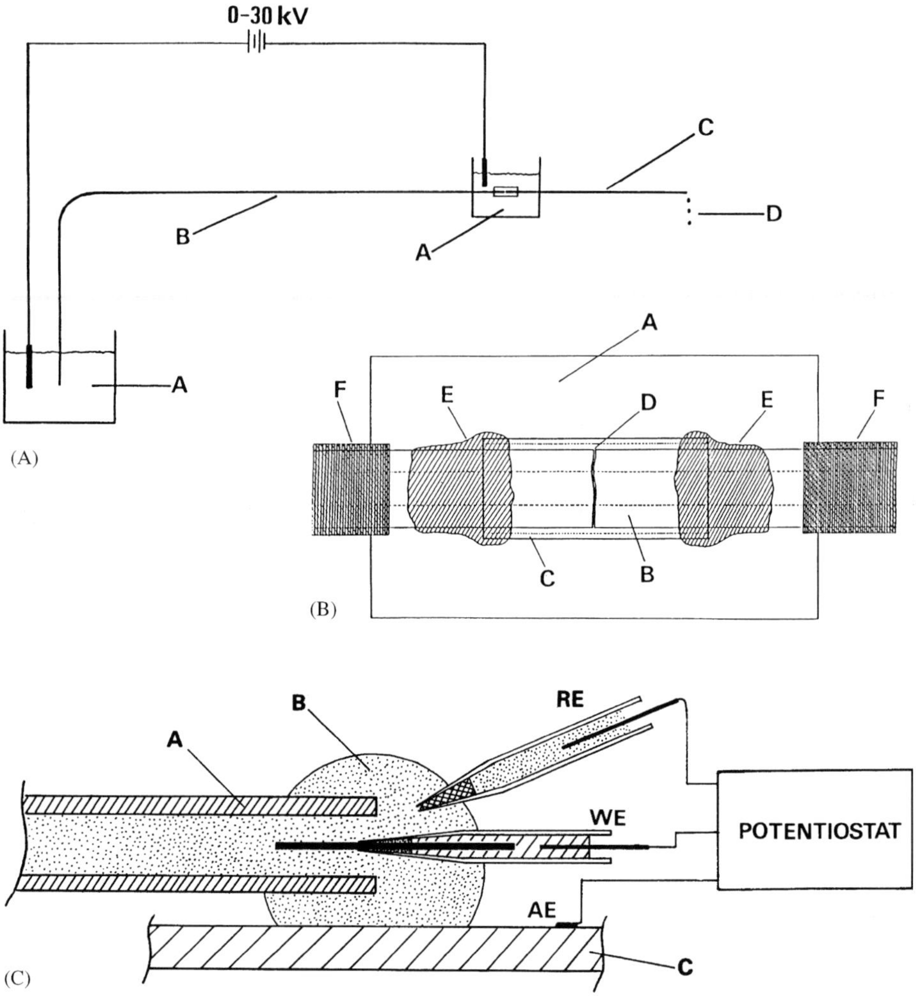

Fig. 8.1. The schematic of the off-column CE–amperometric system first reported by Wallingford and Ewing. The porous glass decoupler is shown in (A) and the arrangement of the EC cell in relation to the CE-capillary connection is shown in (B). Components are labeled as follows: a, microscope slide; b, fused silica capillary; c, porous glass capillary; d, joint; e, epoxy; f, polymer coating; g, eluent drop; h, stainless-steel plate serving as the auxiliary electrode (AE). The working and reference electrodes are labeled WE and RE, respectively. Reprinted with permission from Ref. [4]. Copyright (1987), American Chemical Society.

using amperometry, and the amount of depleted reagent is indicative of analyte concentration.

The geometry of amperometric detectors is reasonably straightforward, and is categorized according to the position of the electrode as well as the presence of current decoupling. Typically, amperometric detection is accomplished with a single working electrode. Electrode configurations are labeled as off-column if they incorporate a physical decoupler (see Fig. 8.1). End-column geometries place the electrode outside the capillary orifice and do not incorporate physical decoupling used in off-column CE-EC systems (see Fig. 8.2). In addition to these single electrode configurations, there are dual electrode configurations in series or parallel, as well as band electrodes. Dual electrodes have been used in series for reversible redox reactions and amperometric reaction detection [33–37]. Parallel configurations for dual electrodes can provide improved qualitative information or selectivity [38–40]. Dual, multiple-band, or multiple-ring electrodes have also been utilized for redox cycling to improve sensitivity [38,41–44]. The most widely reported electrode material is carbon, although gold, platinum, copper, and other materials have been used.

8.2.2 Voltammetry

Voltammetry at scanned potential in CE–EC has employed square [45], cyclic [26,46], staircase [47,48], and sinusoidal [49–51] waveforms. Multistep waveforms that facilitate detection at a constant applied potential, cleaning, and surface activation have also been used in a pulsed EC mode [52–60]. Voltammetry provides additional qualitative information about the nature of the detected analyte on the basis of the applied potential necessary for current production. Thus, voltammetry provides additional selectivity and has been harnessed to detect comigrating analytes with different EC characteristics [46,47]. Electrode geometry is similar to amperometric detection and generally as straightforward. The detection limits of CE–voltammetry methods can be an issue because scanning the applied voltage produces significant contributions from non-Faradic current. Reports of CE–voltammetric detection are limited. Neurotransmitters [45,47], phenolic compounds [26,48], and metal ions [46] have been determined without derivatization by CE voltammetry. Sinusoidal voltammetry in conjunction with redox labels has been applied to oligonucleotide detection for analysis of DNA mutations [50,51] and shows potential for "low

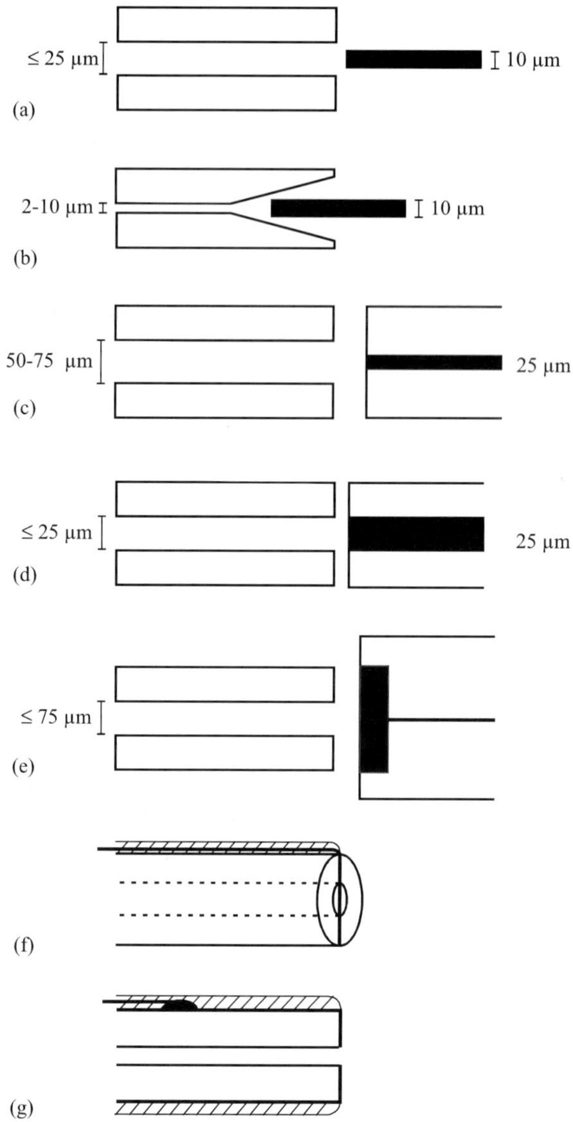

Fig. 8.2. Various capillary and electrode configurations for CE-end-column EC detection. (A) Cylindrical electrode, (B) cylindrical electrode inserted inside a conically etched capillary, (C–E) different disk electrode arrangements, (F) on-capillary wire electrode, (G) on-capillary sputtered electrode. Reprinted from Ref [14]. Copyright (2000), with permission from Wiley–Wiley-VCH Verlag GmbH.

resolution'' sequencing [49]. In one report, adsorption-based processes resulting from modified Osteryoung square-wave voltammetry coupled to CE enabled the detection of amino acids, antibiotics, and illicit drugs [61].

8.2.3 Conductivity

Conductivity is a function of ion mobility, charge, and concentration in the presence of a potential field. Conductimetric detection does not possess the high selectivity of amperometry or voltammetry in CE coupled to conductivity responses are measured on the basis of differences between the conductivity of the background and the analyte. The analyte peak is detectable if it has a significantly higher (or lower) conductivity than the background electrolyte (BGE). However, CE methods development incorporates a means of minimizing differences between sample and BGE mobility to prevent peak asymmetry. Thus, a factor critical in CE-conductivity is the selection of a BGE that balances detection and separation parameters. Early conductivity detectors incorporated dual in-capillary or off-capillary electrodes, and suffered from poor LOD, or electrode geometries that were difficult to fabricate and maintain.

In 1998, the field of CE conductivity reached a great milestone as two separate research groups reported the development of capacitively-coupled contactless conductivity detection (CCD) [62,63]. CCD is based on the transmission and reading of a frequency that responds to changes in resistance between two electrodes acting as capacitors. The electrodes are located on the outside of the capillary, which allows for easier fabrication and eliminates electrode fouling. Much groundwork has been laid in optimization of the design, separation, and detection parameters. Typical electrode arrangements are a few millimeters in length and are separated by gaps of 1–2 mm (see Fig. 8.3). The electrode material is not pertinent, but it is key that the inner diameter of the electrode is well matched in size to the outer diameter of the separation capillary. Decoupling is not necessary for this mode. Historically, CE conductivity has been used for organic and inorganic ion detection (see Table 8.1); but with the recent improvements in performance, design, and BGE selection, conductivity detectors are being used to detect a wider range of biomolecules: proteins, peptides, amino acids, sugars, and long-chain fatty acids [64–70].

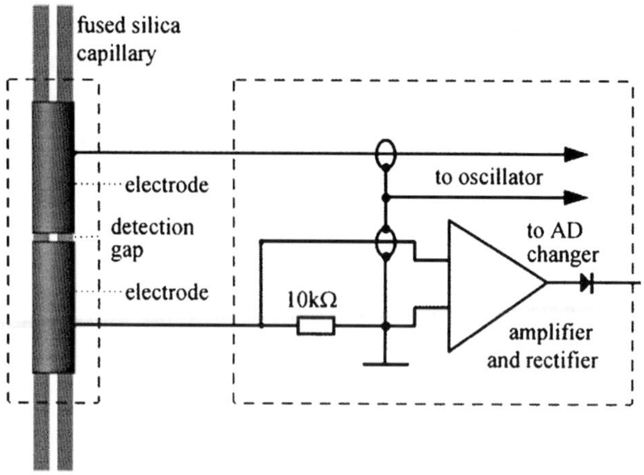

Fig. 8.3. The schematic for the CE–CCD detector as reported by Zemann et al. Reprinted from Zeeman et al. [62]. Copyright (1998), with permission from American Chemical Society.

8.2.4 Potentiometry

Potentiometric detection is based on the measurement of a membrane potential induced by a difference in concentration (activity) of a particular analyte. Fundamental to potentiometry, the response is logarithmic, so the range of quantification is much larger than electrolytic methods of detection. The LODs of potentiometric methods are typically worse than those obtained using electrolytic methods, and the electrodes may not be as durable or easily fabricated. The application has waned since the first report in 1991 [71]. This may be attributable to the lack of commercially available instrumentation or the recent innovation of CE conductivity. Nonetheless, a number of applications have been reported [72–84]. Potentiometric electrodes consist of miniaturized glass-encased membrane electrodes [71–74,85], and polymer-coated wire microelectrodes [75–84].

8.3 DECOUPLING

Decoupling the separation current, which can be as high as 100 μA, from the detection current (nano- to picoamperes) is a concern in electrolytic detection. This has been addressed with physical decoupling, which provides a ground path prior to the detector cell. The physical decoupler reported by Wallingford and Ewing (Fig. 8.1) was fabricated by scoring and fracturing the fused-silica capillary near the cathodic reservoir [4]. The fractured pieces were rejoined inside a porous glass capillary, and this ensemble was affixed to a microscope slide. Electroosmotic and electrophoretic mobility governed the transport of the analyte to the fracture. Beyond this joint, residual flow from electroosmosis in the separation capillary carried analyte to the off-column carbon fiber electrode. In this report, the researchers characterized the effect of distance from the fracture to the detection electrode on separation efficiency. Using a length of 5.7 cm from the fracture to electrode, they obtained separation efficiencies as high as 180,000 theoretical plates.

Physical decouplers may be placed as far as a few centimeters from the end of the capillary. Typically, the cathodic reservoir is grounded and placed at the decoupler. Analyte transport beyond the decoupler is governed by residual flow generated by electroosmosis, and this transport to the electrode must occur in a reasonable time to prevent degradation of the resolution and efficiency generated in the separation. This is generally achieved by maintaining a short distance from the fracture to the electrode by judicious selection of the capillary dimensions. There have been many variations since the first report by Wallingford and Ewing of a porous–glass decoupler [4]. For example, physical decouplers have been formed from bare fractures [86], etched joints [87], or laser-drilled decouplers [88]. Additionally, they may be enhanced with polymeric coatings such as nafion [26] or cellulose [88] to maintain current flow to ground while suppressing the flow of liquid through the gap.

Palladium decouplers, first reported in 1993, offer an additional means of grounding the separation current prior to detection [89]. No physical decoupling (bare fracture, etched porous capillary, etc.) is incorporated in this approach. The current is simply grounded through a palladium joint in contact with the separation capillary. Application of a potential to the CE cathode by direct contact at a joint would normally generate hydrogen gas, which disrupts analyte transport or preturbs the detection electrode downstream from the decoupler. Palladium decouplers adsorb hydrogen and thus serve as a palatable mechanism for isolating separation and detection current.

End-column detection, when implemented appropriately, generally does not normally require current decoupling [90]. Placing the detection electrode a short distance (less than a few hundred microns) from the end of the capillary allows the separation current to travel to ground without significantly hindering the electrode function. Analyte is transported from the capillary orifice to the electrode surface by a combination of residual flow and diffusion. The approach works best when the separation current is kept low. This may be achieved with appropriate selection of the BGE, application of lower separation voltages, and/or maintaining a small inner diameter of the separation capillary. End-column detection has also been facilitated by making the detection end of the capillary porous through etching [91] or through the use of a cast nafion extension [92]. Furthermore, the detection end of the capillary orifice has been physically enlarged through chemical etching, enabling placement of the detection electrode in contact with the capillary without significant interference from the separation current [93]. An additional benefit of this approach is the ease of alignment of the detection electrode with the analyte flow path. Systematic evaluation of parameters relevant to end-column detection has been reported [94–96].

Deviceless decoupling incorporates gold microband electrodes 10 μm wide, 1 mm long, with 10 μm spacing, which are fabricated on a silicon chip [97]. Adjacent bands serve as the working and quasi-reference electrodes, and the band electrode provides precise control over the applied voltage necessary to compensate for the shift in detection potential created by the separation voltage. Another variation of deviceless decoupling has also been facilitated for chip-based CE–EC systems based on the use of a floating potentiostat [98]. This device incorporates optical communication to connect the amplifier that reads the electrode current to the analog-to-digital converter. The voltage necessary to compensate potential shift induced by the separation current is determined experimentally. Although the method has been reported for a chip-based CE–EC system, it is also applicable to conventional CE–EC systems.

8.4 APPLICATIONS

CE–EC analyses have been widespread since the inception of the method. A summary of analyte-specific analyses from 2002 to 2004 is listed in Table 8.1. Although Table 8.1 contains 80 references, it has been restricted to reports relevant to this chapter and to those that provide detailed information about the analytical methodology. Electrode materials, dimensions, and reported LODs are outlined in Table 8.1.

While there is some variation in the detection limits between different EC techniques (conductivity > voltammetry > amperometry), typical EC methods provide detection limits in the low or submicromolar range. For listings of older CE–EC citations, the reader is referred to reviews in the literature [6–21].

The applications of CE–EC are diverse. Analytes are determined in a variety of matrices: foods, plant material, biofluids, cells, tissue, and environmental samples. By far, the most extensive application of CE–EC is in the field of neurochemistry. Separation-based assays comprise another widespread application of CE–EC. Such assays have been used to determine biomarkers and toxins. Physiological activity is also a burgeoning area in CE–EC, both in the determination of enzyme activity, and in the harnessing of enzyme activity for electrocatalytic detection. Finally, following the recent advances in conductivity detection, the number of applications in environmental monitoring is steadily growing.

8.4.1 Neurochemistry

A number of diseases, disorders, and physiological conditions are linked to neurotransmitter dysfunction, and there are many unanswered questions about the physiological mechanisms, function, and implications of neurochemicals [163–166]. Since the first report of CE–EC, the method has been adapted and employed to address a host of questions in the field of neurochemistry. Neurochemical analyses benefit from the applicability of EC detection to natively electroactive neurotransmitters, and the excellent separation efficiency of CE. Neurochemicals determined by CE–EC include catecholamines (epinephrine, norepinephrine, dopamine), histamine, serotonin, acetylcholine, γ-aminobutyric acid (GABA), glutamate [4,5,27,35,37,38,41–45,47, 84,88,91,92,97,127,136,167–197], and neuropeptides [23]. Synthetic precursors to these molecules, including amino acids such as glutamine, tyrosine, phenylalanine, histidine, and tryptophan, are also quantified by CE–EC to better understand function and dysfunction [27,28,30,47,56,61,66,171,174,189,198–202]. The intricacy of neurotransmission is reflected in the variety of models under investigation: single cells, cellular compartments, subcellular fractions, organ-specific in vivo analyses, and the analysis of homogenized whole organisms. The diversity of these physiological models has produced a diversity of methodologies.

Single-cell neurochemical analyses by CE–EC have been applied to intact single snail neuron cells [5,27,169,175,195–197], pheochromocytoma cells [27,128], chromaffin cells [135], lymphocytes [172], and mast cells [193]. Analysis may be accomplished by injecting the entire cell onto a column with the aid of an optical microscope and then lysing the cell prior to the separation [27,128,135,175,196]. The electropherogram shown in Fig. 8.4 was obtained after injecting an intact single pheochromocytoma cell into the separation capillary. Following injection, the cell was lysed, derivatized, and the contents separated by CE [27]. Alternatively, cytoplasmic sampling of single cells has enabled determinations of neurotransmitters in single neurons in the snail *Planorbis corneus* [5,169,195,197]. Recently, synaptosomes, which are subcellular fractions containing high levels of chemical synapses, have been harvested from homogenized rat forebrain for the determination of choline, using chemically modified electrodes containing entrapped enzymes specific for acetylcholine [133,134,192].

While single and subcellular analyses serve as well-controlled systems to elucidate molecular interaction, neurochemical pathways, or

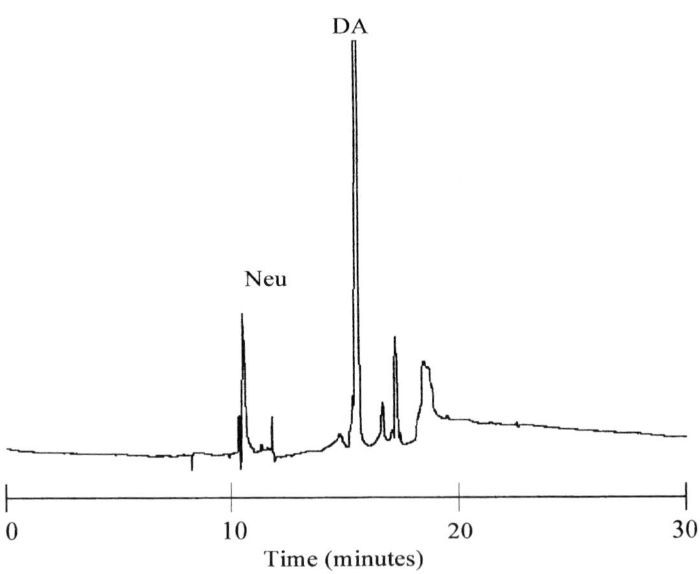

Fig. 8.4. Electropherogram of the contents of a single pheochromocytoma cell. The large peak labeled DA is the response for dopamine. Reprinted from Ref. [27]. Copyright (2000), with permission from Wiley–Wiley-VCH Verlag GmbH.

release mechanisms, there is also a need to study neurochemistry in the context of the whole organism. To address this, microdialysis has been coupled on-line and off-line to CE–EC to provide a means of selective sampling with minimal perturbation of the physiological system [20,25,203,204]. Microdialysis allows any single animal to serve as its own control, and enables both temporal and spatial resolution. Microdialysis–CE–EC has been used to determine neurotransmitters in the brain [25,91,181]. One study incorporated novel end-column decoupling and a rotatable EC detection cell to observe the effect of two drugs (nomifensine and γ–butyrolactone) and electrical stimulation on dopamine release in anesthetized rats. These researchers were able to demonstrate increased levels of dopamine in response to the nomifensine, a dopamine-uptake inhibitor, as well as electrical stimulation and decreased levels of dopamine in response to γ–butyrolactone, an inhibitor of dopaminergic neuron activity [91]. A recent twist to studying neurochemistry in intact organisms is the determination of neurotransmitters in homogenized fruit flies. These studies, which incorporated micellar electrokinetic chromatography coupled to EC to separate monoamine neurotransmitters, demonstrated no detectable dopamine in transgenic flies with inactivated Ddc neurons compared to wild-type flies [205].

8.4.2 Biomarkers

Another application of CE–EC is the determination of biomarkers in clinical or forensic applications. Molecular markers include amino acids, peptides, catecholamines, sugars, xenobiotics, and various metabolites. Analyte has been determined in a variety of biological matrices, including urine [34,38,66,67,99,112,143,150,153,198,199,201,206–214], serum [152,155,156,215–219], plasma [101,106,209,220], airway surface fluid [106,221], saliva [66], and hair [149,222]. Markers of oxidative stress, in particular glutathione [40,103,106,223], have been determined in biological samples using CE–EC. The analyte, 8-hydroxy-deoxyguanosine, a marker of oxidative DNA damage, has been determined using CE–EC [153,211]. These reports revealed elevated levels of 8-hydroxy-deoxyguanosine in individuals who smoke [211] and in cancer patients [153] relative to seemingly healthy individuals. The modified amino acid homocysteine, a precursor to methionine synthesis, has been identified as a risk factor of cardiovascular disease. One CE–EC assay incorporated methodology to determine both free and protein-bound forms of homocysteine [101]. This method, developed for bench-top CE–EC, was duplicated using a chip platform [101]. Finally, forensic analysis of cannabinoids in hair was reported using non-aqueous CE–EC [149]. The approach successfully identified the drug in the hair sample of a known drug user [149].

8.4.3 Physiological activity

Enzymes serve as catalysts enabling metabolism, catabolism, signaling, transport, and mobility. Enzymatic physiological activity is key to investigating mechanisms of physiological function or dysfunction, and their investigation deepens the understanding of the molecular basis of disease. CE–EC methods have been developed to address a wide array of topics within this area. Researchers included copper (II) complexation with CE to facilitate EC of peptides based on the Biuret reaction. The reaction is applicable to peptides between 3 and 18 amino acid residues in length. The method was used to monitor the metabolism of leucine-enkephalin [23]. The potential of employing CE–EC to monitor the activity of angiotensin-converting enzyme in plasma has also been reported with LODs of 200 nM for angiotensin-related peptide fragments [24]. The reaction kinetics of interesting biological systems have also been addressed. For example, the lipoxygenase oxidation of linoleic acid and glucose oxidase-catalyzed oxidation of glucose were monitored using CE–EC [224]. These methods incorporated a sealed reaction chamber, multiplexed injections, and analyte quantification at timed intervals. A representative electropherogram, shown in Fig. 8.5, demonstrates the depletion of oxygen and formation of hydrogen peroxide resulting from the activity of glucose oxidase. Finally, catalytic activity has been harnessed for indirect measurement of enzyme activity with improved detection sensitivity through catalyst-based signal amplification [104,105]. These methods incorporate enzyme substrates that produce electroactive products in the presence of the targeted enzyme. This approach was used to estimate the activity of glucose-6-phosphate dehydrogenase within single erythrocytes [105].

8.4.4 Environmental monitoring

Environmental monitoring by CE–EC has been reported for soil and water samples. The recent growth of CE–CCD has produced an expansion of applications in environmental analyses. A majority of the

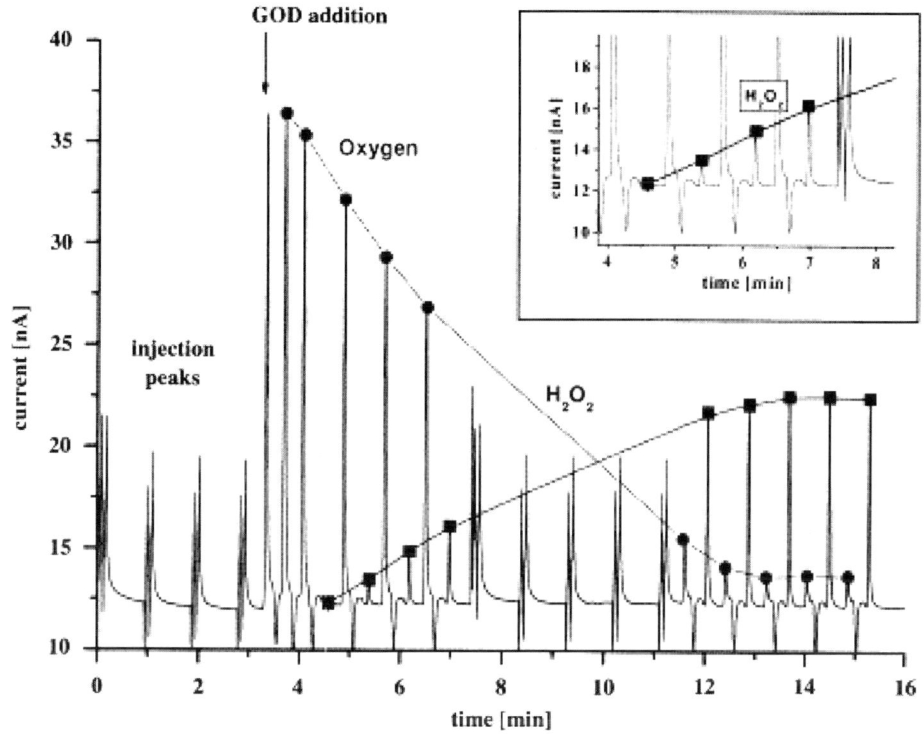

Fig. 8.5. Electropherogram resulting from the CE-amperometric monitoring of the activity of glucose oxidase (GOD). The reaction consumes oxygen and produces hydrogen peroxide. Injections are multiplexed. During the time course of the experiment, the amount of hydrogen peroxide introduced in each injection cycle increase, while the amount of oxygen decreases. Reprinted from Ref. [224]. Copyright (2001) with permission from Elsevier.

work has focused on inorganic ions using conductivity detection [114–116,126,225–229], although amperometry [230,231] and potentiometry [80] have also been employed. Pesticides and herbicides have also been reported using CE–amperometry [161,162]. CE–EC has been used for the environmental monitoring of soil [232,233] as well as for the detection of explosive compounds and degradation products of deadly nerve agents in soil [234,235]. Water samples have been monitored with CE–EC collected from mining or industrial effluent [114,115,225], farm runoff [126], river water [80,226], surface water [115,116,236], rain water [115,116,226,227,229,231], and drinking water [115,116,226,228,237]. In an extensive study, researchers developed fast CE-conductivity to analyze wet-deposition samples in the Sao Paulo region for inorganic ions, collected from May 1997 through March 1998, as well as formic acid and acetic acid collected from February 2000 through October 2000 [229,238].

8.5 CONCLUSIONS AND FUTURE PROSPECTS

Some interesting applications of CE–EC, which point to future directions, are worthy of mention. Separation capillaries with inner diameters of several hundreds of nanometers have been coupled to amperometry for the analysis of neurotransmitters [191,239]. It is foreseeable that such an application could be extended to even further refined subcellular sampling in neurochemistry studies. Patch-clamp detection, shown in Fig. 8.6, has been used in CE–EC for screening of neuroactive compounds and agonists of neuroactive compounds. This mode of detection takes advantage of the selectivity and sensitivity of cellular response in single cells [240–242]. The classical use of CE-conductivity in inorganic ion analyses has been expanded to other exciting applications. CE–CCD has already been exploited for the analysis of museum pieces, providing insight into paint composition, aging, and the condition of valuable works of art [70], and the possibility of protein-binding studies by CE–CCD points to another method for screening binding affinity [64].

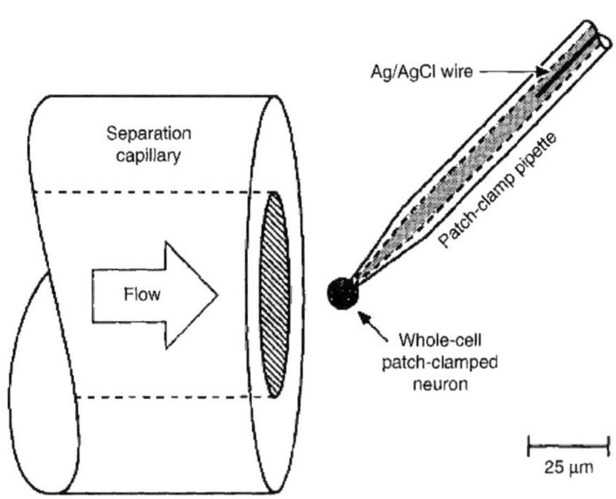

Fig. 8.6. A schematic of a patch clamp electrode coupled with a CE capillary. Reprinted from Ref. [241]. Copyright (1996), with permission from AAAS.

Future applications of CE–EC methods will certainly address improvements in EC detection. New electrode materials, such as boron-doped diamond electrodes, continue to appear in the literature [129,132,243]. Selectivity will also be a topic of future CE–EC reports, either in the area of increased selectivity in complex matrices, or in continued development of universal application seen in advances of CE-conductivity. The constant interest in ease of use, applicability, and portability points to further expansion in CE–EC. Clearly, CE–EC has been exploited for a host of applications. Continued development of innovative and novel CE–EC methodologies can certainly be anticipated in the future.

ACKNOWLEDGMENTS

This material is based upon work supported by the National Science Foundation under Grant No. 0307245. The authors gratefully acknowledge the assistance of A.G. Ewing and F.-M. Matysik for providing high quality files which were reproduced in Figures 1,2, and 4.

REFERENCES

1 J.W. Jorgenson and K.D. Lukacs, *J HRC&CC*, 4 (1981) 230–231.
2 J.W. Jorgenson and K.D. Lukacs, *Anal. Chem.*, 53 (1981) 1298–1302.
3 J.W. Jorgenson and K.D. Lukacs, *J. Chromatogr.*, 218 (1981) 209–216.
4 R.A. Wallingford and A.G. Ewing, *Anal. Chem.*, 59 (1987) 1762–1766.
5 R.A. Wallingford and A.G. Ewing, *Anal. Chem.*, 60 (1988) 1972–1975.
6 T.J. O'Shea. In: S.M. Lunte and D.M. Radzik (Eds.), *Pharmaceutical and Biomedical Applications of Capillary Electrophoresis*, Elsevier, Amsterdam, 1996, pp. 277–306.
7 P.D. Voegel and R.P. Baldwin, *Electrophoresis*, 18 (1997) 2267–2278.
8 C. Haber. In: J.P. Landers (Ed.), *Handbook of Capillary Electrophoresis*, 2nd ed, CRC Press, Boca Raton, FL, 1997, pp. 425–447.
9 L.A. Holland, N.P. Chetwyn, M.D. Perkins and S.M. Lunte, *Pharm. Res.*, 14 (1997) 372–387.
10 L.A. Holland and S.M. Lunte, *Anal. Commun.*, 35 (1998) 1H–4H.
11 T. Kappes and P.C. Hauser, *J. Chromatogr. A*, 834 (1999) 89–101.
12 S. Polesello and S.M. Valsecchi, *J. Chromatogr. A*, 834 (1999) 103–116.
13 R.P. Baldwin, *Electrophoresis*, 21 (2000) 4017–4028.
14 F.-M. Matysik, *Electroanalysis*, 12 (2000) 1349–1355.

15 T. Kappes and P.C. Hauser, *Electroanalysis*, 12 (2000) 165–170.
16 A. Wang and Y. Fang, *Electrophoresis*, 21 (2000) 1281–1290.
17 A.J. Zemann, *Trends Anal. Chem.*, 20 (2001) 346–354.
18 L.A. Holland and A.M. Leigh, *Electrophoresis*, 23 (2002) 3649–3658.
19 J. Tanyanyiwa, S. Leuthardt and P.C. Hauser, *Electrophoresis*, 23 (2002) 3659–3666.
20 S.M. Lunte, R.S. Martin and C.E. Lunte. In: A. Brajter-Toth and J.Q. Chambers (Eds.), *Electroanalytical Methods for Biological Materials*, Marcel Dekker, New York, 2002, pp. 461–490.
21 A.J. Zemann, *Electrophoresis*, 24 (2003) 2125–2137.
22 M. Deacon, T.J. O'Shea, S.M. Lunte and M.R. Smyth, *J. Chromatogr. A*, 652 (1993) 377–383.
23 A.J. Gawron and S.M. Lunte, *Electrophoresis*, 21 (2000) 3205–3211.
24 N.A. Lacher, K.E. Garrison and S.M. Lunte, *Electrophoresis*, 23 (2002) 1577–1584.
25 T.J. O'Shea, P.L. Weber, B.P. Bammel, C.E. Lunte, S.M. Lunte and M.R. Smyth, *J. Chromatogr.*, 608 (1992) 189–195.
26 T.J. O'Shea, R.D. Greenhagen, S.M. Lunte, C.E. Lunte, M.R. Smyth, D.M. Radzik and N. Watanabe, *J. Chromatogr.*, 593 (1992) 305–312.
27 F.D. Swanek, B.B. Anderson and A.G. Ewing, *J. Microcolumn Sep.*, 10 (1998) 185–192.
28 Q. Weng and W. Jin, *Electrophoresis*, 22 (2001) 2797–2803.
29 Q. Dong, W. Jin and J. Shan, *Electrophoresis*, 23 (2002) 559–564.
30 Q. Weng and W. Jin, *Anal. Chim. Acta*, 478 (2003) 199–207.
31 L.-Y. Zhang, Y.-M. Liu, Z.-L. Wang and J.-K. Cheng, *Anal. Chim. Acta*, 508 (2004) 141–145.
32 M.J. Rose, S.M. Lunte, R.G. Carlson and J.F. Stobaugh, *J. Pharm. Biomed. Anal.*, 30 (2003) 1851–1859.
33 L.A. Holland and S.M. Lunte, *Anal. Chem.*, 71 (1999) 407–412.
34 B.L. Lin, L.A. Colón and R.N. Zare, *J. Chromatogr. A*, 680 (1994) 263–270.
35 M. Zhong, J. Zhou, S.M. Lunte, G. Zhao, D.M. Giolando and J.R. Kirchhoff, *Anal. Chem.*, 68 (1996) 203–207.
36 M. Zhong and S.M. Lunte, *Anal. Chem.*, 71 (1999) 251–255.
37 L.A. Holland, N.M. Harmony and S.M. Lunte, *Electroanalysis*, 11 (1999) 327–330.
38 D.-c. Chen, D.-Z. Zhan, C.-W. Cheng, A.-C. Liu and C.-h. Chen, *J. Chromatogr. B*, 750 (2001) 33–39.
39 Q. Weng and W. Jin, *J. Chromatogr. A*, 971 (2002) 217–223.
40 W. Jin, X. Li and N. Gao, *Anal. Chem.*, 75 (2003) 3859–3864.
41 D.-c. Chen, S.-S. Chang and C.-h. Chen, *Anal. Chem.*, 71 (1999) 3200–3205.

42 Z. Liu, O. Niwa, R. Kurita and T. Horiuchi, *J. Chromatogr. A*, 891 (2000) 149–156.
43 O. Niwa, R. Kurita, Z. Liu, T. Horiuchi and K. Torimitsu, *Anal. Chem.*, 72 (2000) 949–955.
44 Z. Liu, O. Niwa, R. Kurita and T. Horiuchi, *Anal. Chem.*, 72 (2000) 1315–1321.
45 G.C. Gerhardt, R.M. Cassidy and A.S. Baranski, *Anal. Chem.*, 70 (1998) 2167–2173.
46 J. Wen, A. Baranski and R. Cassidy, *Anal. Chem.*, 70 (1998) 2504–2509.
47 S.S. Ferris, G. Lou and A.G. Ewing, *J. Microcolumn Sep.*, 6 (1994) 263–268.
48 S. Park, M.J. McGrath, M.R. Smyth, D. Diamond and C.E. Lunte, *Anal. Chem.*, 69 (1997) 2994–3001.
49 S.A. Brazill, P.H. Kim and W.G. Kuhr, *Anal. Chem.*, 73 (2001) 4882–4890.
50 S.A. Brazill and W.G. Kuhr, *Anal. Chem.*, 74 (2002) 3421–3428.
51 S.A. Brazill, N.E. Herbert and W.G. Kuhr, *Electrophoresis*, 24 (2003) 2749–2757.
52 T.J. O'Shea, S.M. Lunte and W.R. LaCourse, *Anal. Chem.*, 65 (1993) 948–951.
53 W. Lu and R.M. Cassidy, *Anal. Chem.*, 65 (1993) 2878–2881.
54 P.L. Weber and S.M. Lunte, *Electrophoresis*, 17 (1996) 302–309.
55 J. Wen and R.M. Cassidy, *Anal. Chem.*, 68 (1996) 1047–1053.
56 W.R. LaCourse and G.S. Owens, *Electrophoresis*, 17 (1996) 310–318.
57 G.S. Owens and W.R. LaCourse, *J. Chromatogr. B*, 695 (1997) 15–25.
58 J. Wen, R.M. Cassidy and A.S. Baranski, *J. Chromatogr. A*, 811 (1998) 181–192.
59 H.-H. Rüttinger and B. Dräger, *J. Chromatogr. A*, 925 (2001) 291–296.
60 X. Sun, X. Yang and E. Wang, *J. Chromatogr. A*, 1005 (2003) 189–195.
61 G.C. Gerhardt, R.M. Cassidy and A.S. Baranski, *Anal. Chem.*, 72 (2000) 908–915.
62 A.J. Zemann, E. Schnell, D. Volgger and G.K. Bonn, *Anal. Chem.*, 70 (1998) 563–567.
63 J.A.F. da Silva and C.L. do Lago, *Anal. Chem.*, 70 (1998) 4339–4343.
64 E.M. Abad-Villar, J. Tanyanyiwa, M.T. Fernández-Abedul, A. Costa-García and P.C. Hauser, *Anal. Chem.*, 76 (2004) 1282–1288.
65 E. Baltussen, R.M. Guijt, G. van der Steen, F. Laugere, S. Baltussen and G.W.K. van Dedem, *Electrophoresis*, 23 (2002) 2888–2893.
66 P. Coufal, J. Zuska, T. van de Goor, V. Smith and B. Gaš, *Electrophoresis*, 24 (2003) 671–677.
67 J. Tanyanyiwa, K. Schweizer and P.C. Hauser, *Electrophoresis*, 24 (2003) 2119–2124.

68 I. Kaml, K. Vcelakova and E. Kenndler, *J. Sep. Sci.*, 27 (2004) 161–166.
69 A.Z. Carvalho, J.A.F. da Silva and C.L. do Lago, *Electrophoresis*, 24 (2003) 2138–2143.
70 I. Surowiec, I. Kaml and E. Kenndler, *J. Chromatogr. A*, 1024 (2004) 245–254.
71 C. Haber, I. Silvestri, S. Röösli and W. Simon, *Chimia*, 45 (1991) 117–121.
72 A. Nann, I. Silvestri and W. Simon, *Anal. Chem.*, 65 (1993) 1662–1667.
73 A. Nann and E. Pretsch, *J. Chromatogr. A*, 676 (1994) 437–442.
74 P.C. Hauser, N.D. Renner and A.P.C. Hong, *Anal. Chim. Acta*, 295 (1994) 181–186.
75 B.L. De Backer and L.J. Nagels, *Biomed. Chromatogr.*, 9 (1995) 257–258.
76 B.L. De Backer and L.J. Nagels, *Anal. Chem.*, 68 (1996) 4441–4445.
77 T. Kappes, P. Schnierle and P.C. Hauser, *Anal. Chim. Acta*, 350 (1997) 141–147.
78 T. Kappes and P.C. Hauser, *Anal. Chem.*, 70 (1998) 2487–2492.
79 P. Schnierle, T. Kappes and P.C. Hauser, *Anal. Chem.*, 70 (1998) 3585–3589.
80 T. Kappes and P.C. Hauser, *Anal. Commun.*, 35 (1998) 325–329.
81 I. Poels and L.J. Nagels, *Anal. Chim. Acta*, 385 (1999) 417–422.
82 I. Poels and L.J. Nagels, *Anal. Chim. Acta*, 401 (1999) 21–27.
83 T. Kappes, P. Schnierle and P.C. Hauser, *Anal. Chim. Acta*, 393 (1999) 77–82.
84 T. Kappes, P. Schnierle and P.C. Hauser, *Electrophoresis*, 21 (2000) 1390–1394.
85 A. Nann and W. Simon, *J. Chromatogr.*, 633 (1993) 207–211.
86 M.C. Linhares and P.T. Kissinger, *Anal. Chem.*, 63 (1991) 2076–2078.
87 S. Hu, Z.-L. Wang, P.-B. Li and J.-K. Cheng, *Anal. Chem.*, 69 (1997) 264–267.
88 D.M. Osbourn and C.E. Lunte, *Anal. Chem.*, 73 (2001) 5961–5964.
89 W.Th. Kok and Y. Sahin, *Anal. Chem.*, 65 (1993) 2497–2501.
90 X. Huang, R.N. Zare, S. Sloss and A.G. Ewing, *Anal. Chem.*, 63 (1991) 189–192.
91 J. Qian, Y. Wu, H. Yang and A.C. Michael, *Anal. Chem.*, 71 (1999) 4486–4492.
92 S. Park and C.E. Lunte, *Anal. Chem.*, 67 (1995) 4366–4370.
93 S. Sloss and A.G. Ewing, *Anal. Chem.*, 65 (1993) 577–581.
94 W. Lu, R.M. Cassidy and A.S. Baranski, *J. Chromatogr.*, 640 (1993) 433–440.
95 W. Lu and R.M. Cassidy, *Anal. Chem.*, 66 (1994) 200–204.
96 F.-M. Matysik, *J. Chromatogr. A*, 742 (1996) 229–234.
97 O. Klett, I. Nischang and L. Nyholm, *Electrophoresis*, 23 (2002) 3678–3682.

98 R.S. Martin, K.L. Ratzlaff, B.H. Huynh and S.M. Lunte, *Anal. Chem.*, 74 (2002) 1136–1143.
99 T. Inoue and J.R. Kirchhoff, *Anal. Chem.*, 74 (2002) 1349–1354.
100 Y. Cao, X. Zhang, Q. Chu, Y. Fang and J. Ye, *Electroanalysis*, 15 (2003) 898–902.
101 S.A. Pasas, N.A. Lacher, M.I. Davies and S.M. Lunte, *Electrophoresis*, 23 (2002) 759–766.
102 Q. Wang, H. Yu, H. Li, F. Ding, P. He and Y. Fang, *Food Chem.*, 83 (2003) 311–317.
103 W. Wang, H. Xin, H. Shao and W. Jin, *J. Chromatogr. B*, 789 (2003) 425–429.
104 Z. He, H. Wang, X. Sun and W. Jin, *Electroanalysis*, 14 (2002) 1674–1678.
105 X. Sun and W. Jin, *Anal. Chem.*, 75 (2003) 6050–6055.
106 K. Govindaraju, V. Govindaraju and D.H. Eidelman, *J. Chromatogr. B*, 788 (2003) 369–376.
107 Q. Wang, H. Yu, J. Zong, P. He and Y. Fang, *J. Pharm. Biomed. Anal.*, 31 (2003) 473–480.
108 Q. Hu, T. Zhou, G. Hu and Y. Fang, *J. Pharm. Biomed. Anal.*, 30 (2002) 1047–1053.
109 Q. Wang, F. Ding, N. Zhu, P. He and Y. Fang, *Biomed. Chromatogr.*, 17 (2003) 483–488.
110 Y. Cao, Y. Wang, X. Chen and J. Ye, *Food Chem.*, 86 (2004) 131–136.
111 G. Chen, X. Han, L. Zhang and J. Ye, *J. Chromatogr. A*, 954 (2002) 267–276.
112 G. Chen, Q. Chu, L. Zhang and J. Ye, *Anal. Chim. Acta*, 457 (2002) 225–233.
113 J. Tanyanyiwa, S. Leuthardt and P.C. Hauser, *J. Chromatogr. A*, 978 (2002) 205–211.
114 P. Kubáň, P. Kubáň and V. Kubáň, *Electrophoresis*, 24 (2003) 1397–1403.
115 P. Kubáň, P. Kubáň and V. Kubáň, *Electrophoresis*, 23 (2002) 3725–3734.
116 P. Kubáň, P. Kubáň and V. Kubáň, *Electrophoresis*, 24 (2003) 1935–1943.
117 M. Macka, J. Hutchinson, A. Zemann, Z. Shusheng and P.R. Haddad, *Electrophoresis*, 24 (2003) 2144–2149.
118 P. Tůma, F. Opekar and K. Štulík, *Electrophoresis*, 23 (2002) 3718–3724.
119 J. Tanyanyiwa and P.C. Hauser, *Electrophoresis*, 23 (2002) 3781–3786.
120 P. Kubáň, P. Kubáň, P.C. Hauser and V. Kubáň, *Electrophoresis*, 25 (2004) 35–42.
121 V. Unterholzner, M. Macka, P.R. Haddad and A. Zemann, *Analyst*, 127 (2002) 715–718.

122 J. Tanyanyiwa, B. Galliker, M.A. Schwarz and P.C. Hauser, *Analyst*, 127 (2002) 214–218.
123 P.S. Vuorinen, M. Jussila, H. Sirén, S. Palonen and M.-L. Riekkola, *J. Chromatogr. A*, 990 (2003) 45–52.
124 J.A.F. da Silva, N. Guzman and C.L. da Lago, *J. Chromatogr. A*, 942 (2002) 249–258.
125 P. Praus, *Talanta*, 62 (2004) 977–982.
126 P. Kubáň, M. Reinhardt, B. Müller and P.C. Hauser, *J. Environ. Monit.*, 6 (2004) 169–174.
127 K.B. Male and J.H.T. Luong, *J. Chromatogr. A*, 1003 (2003) 167–178.
128 L. Zhang, S. Qv, Z. Wang and J. Cheng, *J. Chromatogr. B*, 792 (2003) 381–385.
129 D. Shin, B.V. Sarada, D.A. Tryk, A. Fujishima and J. Wang, *Anal. Chem.*, 75 (2003) 530–534.
130 X. Sun, X. Yang and E. Wang, *J. Chromatogr. A*, 991 (2003) 109–116.
131 F.-M. Matysik, D. Marggraf, P. Gläser and J.A.C. Broekaert, *Electrophoresis*, 23 (2002) 3711–3717.
132 J. Cvačka, V. Quaiserová, J. Park, Y. Show, A. Muck Jr. and G.M. Swain, *Anal. Chem.*, 75 (2003) 2678–2687.
133 T. Inoue, J.R. Kirchhoff and R.A. Hudson, *Anal. Chem.*, 74 (2002) 5321–5326.
134 D.D. Wise, T.V. Barkhimer, P.-A. Brault, J.R. Kirchhoff, W.S. Messer Jr. and R.A. Hudson, *J. Chromatogr. B*, 775 (2002) 49–56.
135 S.W. Suljak, F.D. Swanek, P.F. Gavin and A.G. Ewing, *J. Sep. Sci.*, 26 (2003) 61–68.
136 L. Zhang, W. Huang, Z. Wang and J. Cheng, *Anal. Sci.*, 18 (2002) 1117–1120.
137 Y. Peng, Q. Chu, F. Liu and J. Ye, *J. Agric. Food Chem.*, 52 (2004) 153–156.
138 Q. Wang, F. Ding, H. Li, P. He and Y. Fang, *Electrophoresis*, 24 (2003) 1687–1692.
139 Q. Wang, F. Ding, H. Li, P. He and Y. Fang, *J. Pharm. Biomed. Anal.*, 30 (2003) 1507–1514.
140 Y. Cao, C. Lou, Y. Fang and J. Ye, *J. Chromatogr. A*, 943 (2002) 153–157.
141 X. Sun, C. Gao, W. Cao, X. Yang and E. Wang, *J. Chromatogr. A*, 962 (2002) 117–125.
142 G. Chen, J. Ye, H. Bao and P. Yang, *J. Pharm. Biomed. Anal.*, 29 (2002) 843–850.
143 X. Li, W. Jin and Q. Weng, *Anal. Chim. Acta*, 461 (2002) 123–130.
144 Y. Cao, Q. Chu, Y. Fang and J. Ye, *Anal. Bioanal. Chem.*, 374 (2002) 294–299.
145 G. Chen, L. Zhang, J. Zhao and J. Ye, *Anal. Bioanal. Chem.*, 373 (2002) 169–173.

146 L. Gao, Q. Chu and J. Ye, *Food Chem.*, 78 (2002) 255–260.
147 D. Liu and W. Jin, *J. Chromatogr. B*, 789 (2003) 411–415.
148 D. Liu and W. Jin, *J. Chromatogr. B*, 783 (2003) 509–515.
149 U. Backofen, F.-M. Matysik and C.E. Lunte, *J. Chromatogr. A*, 942 (2002) 259–269.
150 T. Zhou, H. Yu, Q. Hu and Y. Fang, *J. Pharm. Biomed. Anal.*, 30 (2002) 13–19.
151 W. Jin and L. Jiang, *Anal. Chim. Acta*, 461 (2002) 117–121.
152 J.L. Boughton, B.W. Robinson and T.G. Strein, *Electrophoresis*, 23 (2002) 3705–3710.
153 S.-R. Mei, Q.-H. Yao, L.-s. Cai, J. Xing, G.-W. Xu and C.-y. Wu, *Electrophoresis*, 24 (2003) 1411–1415.
154 Q. Dong, R. Dong, M. Jin and W. Jin, *J. Chromatogr. B*, 774 (2002) 121–126.
155 Z. He, N. Gao and W. Jin, *Anal. Chim. Acta*, 497 (2003) 75–81.
156 Z. He, N. Gao and W. Jin, *J. Chromatogr. B*, 784 (2003) 343–350.
157 X. Sun, X. Yang and E. Wang, *J. Chromatogr. A*, 1005 (2003) 189–195.
158 Q. Wang, F. Ding, N. Zhu, H. Li, P. He and Y. Fang, *J. Chromatogr. A*, 1016 (2003) 123–128.
159 M. Jia, Z. He and W. Jin, *J. Chromatogr. A*, 966 (2002) 187–194.
160 Y. Cao, Q. Chu and J. Ye, *Anal. Bioanal. Chem.*, 376 (2003) 691–695.
161 M. Chicharro, A. Zapardiel, E. Bermejo, A. Sánchez and R. González, *Electroanalysis*, 16 (2004) 311–318.
162 M. Chicharro, A. Zapardiel, E. Bermejo and A. Sánchez, *Anal. Chim. Acta*, 469 (2002) 243–252.
163 G.J. Yohrling IV and J.-H.J. Cha. In: G. Bates, P. Harperand and L. Jones (Eds.), *Huntington's Disease*, 3rd ed, Oxford University Press, New York, 2002, pp. 276–308.
164 R.D. Grubbs and D.K. Sundberg. In: P.M. Conn (Ed.), *Neuroscience in Medicine*, 2nd ed, Humana Press, Totowa, NJ, 2003, pp. 457–489.
165 W. Gsell, G. Jungkunz and P. Riederer, *Curr. Pharm. Design*, 10 (2004) 265–293.
166 J. Kelly, *Curr. Med. Chem.: Cent. Nerv. Syst. Agents*, 3 (2003) 311–322.
167 R.A. Wallingford and A.G. Ewing, *Anal. Chem.*, 60 (1988) 258–263.
168 R.A. Wallingford and A.G. Ewing, *Anal. Chem.*, 61 (1989) 98–100.
169 T.M. Olefirowicz and A.G. Ewing, *Anal. Chem.*, 62 (1990) 1872–1876.
170 A.J. Tüdös, M.M.C. Van Dyck, H. Poppe and W.Th. Kok, *Chromatographia*, 37 (1993) 79–85.
171 M.A. Malone, P.L. Weber, M.R. Smyth and S.M. Lunte, *Anal. Chem.*, 66 (1994) 3782–3787.
172 J. Bergquist, A. Tarkowski, R. Ekman and A. Ewing, *Proc. Natl. Acad. Sci. USA.*, 91 (1994) 12912–12916.
173 M.-C. Chen and H.-J. Huang, *Anal. Chem.*, 67 (1995) 4010–4014.

174 F.-M. Matysik and U. Backofen, *Fresen. J. Anal. Chem.*, 356 (1996) 169–172.
175 F.D. Swanek, G. Chen and A.G. Ewing, *Anal. Chem.*, 68 (1996) 3912–3916.
176 A.M. Fermier, M.L. Gostkowski and L.A. Colón, *Anal. Chem.*, 68 (1996) 1661–1664.
177 P.D. Voegel, W. Zhou and R.P. Baldwin, *Anal. Chem.*, 69 (1997) 951–957.
178 H.-y. Chen, A.-m. Yu and D.-k. Xu, *Fresen. J. Anal. Chem.*, 359 (1997) 542–545.
179 S.R. Wallenborg, L. Nyholm and C.E. Lunte, *Anal. Chem.*, 71 (1999) 544–549.
180 B.B. Anderson and A.G. Ewing, *J. Pharm. Biomed. Anal.*, 19 (1999) 15–32.
181 W. Jin, L. Jin, G. Shi and J. Ye, *Anal. Chim. Acta*, 382 (1999) 33–37.
182 R. Kurita, H. Tabei, Z. Liu, T. Horiuchi and O. Niwa, *Sens. Actuators B*, 71 (2000) 82–89.
183 W.R. Everett, C. Bohs and M.I. Davies, *Curr. Sep.*, 19 (2000) 25–28.
184 G. Chen, J.N. Ye and J.S. Cheng, *Chromatographia*, 52 (2000) 137–141.
185 L. Hua and S.N. Tan, *Anal. Chim. Acta*, 403 (2000) 179–186.
186 Q. Weng, F. Xia and W. Jin, *Electroanalysis*, 13 (2001) 1459–1461.
187 S. Wilke, R. Schürz and H. Wang, *Anal. Chem.*, 73 (2001) 1146–1154.
188 G. Chen, J. Cheng and J. Ye, *Fresen. J. Anal. Chem.*, 370 (2001) 930–934.
189 M. Goto, S. Inagaki and Y. Esaka, *Anal. Sci.*, 17 (2001) 1383–1387.
190 W.-C. Yang, A.-M. Yu, X.-D. Yu and H.-Y. Chen, *Electrophoresis*, 22 (2001) 2025–2031.
191 L.A. Woods, T.P. Roddy, T.L. Paxon and A.G. Ewing, *Anal. Chem.*, 73 (2001) 3687–3690.
192 T.V. Barkhimer, J.R. Kirchhoff, R.A. Hudson and W.S. Messer Jr., *Electrophoresis*, 23 (2002) 3699–3704.
193 Q. Weng, F. Xia and W. Jin, *J. Chromatogr. B*, 779 (2002) 347–352.
194 C.-W. Cheng, K.-C. Lee, S.-S. Chang, D.-c. Chen, C.-M. Yu and C.-h. Chen, *Anal. Chem.*, 74 (2002) 3906–3910.
195 T.M. Olefirowicz and A.G. Ewing, *J Neurosci. Methods*, 34 (1990) 11–15.
196 H.K. Kristensen, Y.Y. Lau and A.G. Ewing, *J. Neurosci. Methods*, 51 (1994) 183–188.
197 J.B. Chien, R.A. Wallingford and A.G. Ewing, *J. Neurochem.*, 54 (1990) 633–638.
198 J. Ye and R.P. Baldwin, *Anal. Chem.*, 66 (1994) 2669–2674.
199 J. Zhou and S.M. Lunte, *Electrophoresis*, 16 (1995) 498–503.
200 J. Zhou and S.M. Lunte, *Anal. Chem.*, 67 (1995) 13–18.
201 A.M. Fermier and L.A. Colón, *J. High Resolut. Chromatogr.*, 19 (1996) 613–616.

202 Q. Dong, X. Wang, L. Zhu and W. Jin, *J. Chromatogr. A*, 959 (2002) 269–279.
203 J. Zhou, D.M. Heckert, H. Zuo, C.E. Lunte and S.M. Lunte, *Anal. Chim. Acta*, 379 (1999) 307–317.
204 S.M. Lunte, M.A. Malone, H. Zuo and M.R. Smyth, *Curr. Sep.*, 13 (1994) 75–79.
205 P.J. Ream, S.W. Suljak, A.G. Ewing and K.-A. Han, *Anal. Chem.*, 75 (2003) 3972–3978.
206 J. Zhou, T.J. O'Shea and S.M. Lunte, *J. Chromatogr. A*, 680 (1994) 271–277.
207 T.J. O'Shea and S.M. Lunte, *Anal. Chem.*, 66 (1994) 307–311.
208 X. Huang and W.Th. Kok, *J. Chromatogr. A*, 716 (1995) 347–353.
209 D.-K. Xu, L. Hua, Z.-M. Li and H.-Y. Chen, *J. Chromatogr. B*, 694 (1997) 461–466.
210 T. You, X. Yang and E. Wang, *Analyst*, 123 (1998) 2357–2360.
211 D.J. Weiss and C.E. Lunte, *Electrophoresis*, 21 (2000) 2080–2085.
212 S. Zhang, Z. Yuan, H. Liu, H. Zou, H. Xiong and Y. Wu, *Electrophoresis*, 21 (2000) 2995–2998.
213 W. Jin, Q. Dong, D. Yu and X. Ye, *Electrophoresis*, 21 (2000) 1535–1539.
214 S.S. Zhang, H.X. Liu, Y.J. Wu and C.L. Yu, *Analyst*, 126 (2001) 441–445.
215 X. Huang, T.-K.J. Pang, M.J. Gordon and R.N. Zare, *Anal. Chem.*, 59 (1987) 2747–2749.
216 M. Zhong and S.M. Lunte, *Anal. Commun.*, 35 (1998) 209–212.
217 A. Wang, L. Zhang, S. Zhang and Y. Fang, *J. Pharm. Biomed. Anal.*, 23 (2000) 429–436.
218 Q. Dong, D. Yu, X. Ye and W. Jin, *Electrophoresis*, 22 (2001) 128–133.
219 J.A.F. da Silva, N.L. Ricelli, A.Z. Carvalho and C.L. do Lago, *J. Braz. Chem. Soc.*, 14 (2003) 265–268.
220 D.-K. Xu, L. Hua and H.-Y. Chen, *Anal. Chim. Acta*, 335 (1996) 95–101.
221 K. Govindaraju, E.A. Cowley, D.H. Eidelman and D.K. Lloyd, *Anal. Chem.*, 69 (1997) 2793–2797.
222 J. Mo, W. Liu and S. Mo, *Anal. Commun.*, 35 (1998) 365–367.
223 T.J. O'Shea and S.M. Lunte, *Anal. Chem.*, 65 (1993) 247–250.
224 A. Radschuweit, H.-H. Rüttinger and P. Nuhn, *J. Chromatogr. A*, 937 (2001) 127–134.
225 D. Schlegel, J. Mattusch and R. Wennrich, *Fresen. J. Anal. Chem.*, 354 (1996) 535–539.
226 D. Kaniansky, V. Zelenská and D. Baluchová, *Electrophoresis*, 17 (1996) 1890–1897.
227 S. Valsecchi, G. Tartari and S. Polesello, *J. Chromatogr. A*, 760 (1997) 326–332.
228 R. Bodor, D. Kaniansky and M. Masár, *J. Chromatogr. A*, 916 (2001) 31–40.

229 F.R. Rocha, J.A.F. da Silva, C.L. Lago, A. Fornaro and I.G.R. Gutz, *Atmos. Environ.*, 37 (2003) 105–115.
230 E.P.C. Lai, W. Zhang, X. Trier, A. Georgi, S. Kowalski, S. Kennedy, T. MdMuslim and E. Dabek-Zlotorzynska, *Anal. Chim. Acta*, 364 (1998) 63–74.
231 B. Tenberken, P. Ebert, M. Hartmann, M. Kibler, A. Mainka, T. Prokop, A. Röder and K. Bächmann, *J. Chromatogr. A*, 745 (1996) 209–215.
232 A. Hilmi, J.H.T. Luong and A.-L. Nguyen, *Environ. Sci. Technol.*, 31 (1997) 1794–1800.
233 A. Hilmi and J.H.T. Luong, *Anal. Chem.*, 73 (2001) 2536–2540.
234 A.-E.F. Nassar, S.V. Lucas, W.R. Jones and L.D. Hoffland, *Anal. Chem.*, 70 (1998) 1085–1091.
235 A. Hilmi, J.H.T. Luong and A.-L. Nguyen, *Anal. Chem.*, 71 (1999) 873–878.
236 F. Hissner, J. Mattusch and K. Heinig, *J. Chromatogr. A*, 848 (1999) 503–513.
237 S.R. Wallenborg, S.M. Dorholt, A. Faibushevich and C.E. Lunte, *Electroanalysis*, 11 (1999) 362–366.
238 A. Fornaro and I.G.R. Gutz, *Atmos. Environ.*, 37 (2003) 117–128.
239 L.A. Woods and A.G. Ewing, *Chem. Phys. Chem.*, 4 (2003) 207–211.
240 K. Jardemark, O. Orwar, I. Jacobson, A. Moscho and R.N. Zare, *Anal. Chem.*, 69 (1997) 3427–3434.
241 O. Orwar, K. Jardemark, I. Jacobson, A. Moscho, H.A. Fishman, R.H. Scheller and R.N. Zare, *Science*, 272 (1996) 1779–1782.
242 C. Farre, A. Sjöberg, K. Jardemark, I. Jacobson and O. Orwar, *Anal. Chem.*, 73 (2001) 1228–1233.
243 D. Shin, B.V. Sarada, D.A. Tryk and A. Fujishima, *Chem. Sens.*, 18 (2002) 124–126.

Chapter 9

Mass spectrometry detection in capillary electrophoresis

Carolina Simó and Alejandro Cifuentes

9.1 INTRODUCTION

Capillary electrophoresis–mass spectrometry (CE–MS) has emerged as a powerful analytical tool that combines the separation speed, high resolving power and minimum sample consumption of CE with the selectivity and structural information provided by MS. The potential of CE–MS was first realized and developed by Smith's group in the late 1980 s [1–3] and since then, the number of works devoted to CE–MS has increased greatly [4–10]. In the last 10 years, there have been significant developments in CE–MS instrumentation and applications. The main instrumental advances, apart from the increasingly sophisticated and powerful mass analyzers, have focused on the ionization techniques and methodological procedures developed for direct coupling of CE and MS [11]. Among the interfaces developed for facilitating the transfer of analytes from the liquid phase of CE to the gas phase of MS, electrospray ionization (ESI), and to a lesser extent, inductively coupled plasma (ICP) have been the more frequently applied. However, some other procedures, for example , atmospheric pressure chemical ionization (APCI), and more importantly, matrix-assisted laser desorption/ionization (MALDI) are being actively investigated for direct coupling of CE with MS. Interestingly, the number of applications and types of compounds addressed by CE–MS have also become broader, ranging from small inorganic ions to complex biomolecules such as DNAs or proteins. As a result, CE–MS can be considered as a real alternative procedure to LC–MS for analyzing polar and charged compounds, which are often not very well analyzed by LC when their separation is based on hydrophobic interactions.

Comprehensive Analytical Chemistry XLV
M.L. Marina, A. Ríos and M. Valcárcel (Eds)
Volume XLV ISSN: 0166-526X DOI: 10.1016/S0166-526X(05)45009-6

This chapter will principally focus on the use of CE–ESI–MS. For the use of ICP–MS together with CE, see chapter 10. A review of the current status and instrumentation involved in CE–ESI–MS is given, paying special attention to the compatibility of CE (a liquid-phase technique) and MS (a gas-phase technique), with particular emphasis on the development of methods (including type of buffers, additives, solvents, partial filling techniques (PFTs), etc.) that can be used to overcome the intrinsic incompatibility between the different CE modes and MS. The different strategies to improve sensitivity in CE–MS are also reviewed, and we include a brief discussion about the hyphenation of microchips and MS. Finally, an updated description of the different applications of CE–ESI–MS, such as the analysis of peptides, drugs, proteins, pesticides, nucleic acids, etc., is provided.

9.2 PRINCIPLES AND INSTRUMENTATION

CE, like any other liquid-phase separation technique, requires a suitable interface for coupling it to mass spectrometric detection systems. It has been interfaced to continuous flow–fast atom bombardment (CF–FAB) [12], MALDI (mostly for off-line CE–MS couplings) [13,14], APCI [15,16], ESI, etc. [11]. In spite of the large number of ionization techniques available, the principal interface used for direct coupling of CE to MS has been ESI, since this ionization procedure allows molecules to be transferred directly from the separation capillary to the mass spectrometer. Moreover, ESI allows the efficient analysis of polar, labile and large compounds (typically up to 100 kDa), and is easy to implement, sensitive and has a wide range of applications.

The first problem in coupling CE and MS via an ESI interface is the flow rate compatibility, since CE usually provides flow rates up to 100 nl/min, while the most common interfaces work in the 1–200 μl/min, range. To solve this problem, two different strategies have been carried out: (i) the use of make-up flows and (ii) the use of micro- or nano-ESI devices that are stable and have extremely low flow rates. This has led to the development of different ESI instrumental designs for direct CE–MS coupling to different mass analyzers (*vide infra*).

The second problem in directly coupling CE to ESI–MS is that CE, in its different modes (CZE, MEKC, CGE, etc.), requires the use of aqueous buffers containing nonvolatile compounds, such as borate, phosphate or sodium dodecyl sulfate (SDS), which are noncompatible with

ESI–MS. These compounds significantly increase the chemical noise during the detection, ruining the sensitivity and, under extreme conditions, clogging the system. As a result, different strategies (instrumental and/or methodological) have been developed to overcome this important limitation and they are summarized below.

As described by Severs and Smith [4], the ideal interface between CE and MS should (i) be sensitive, precise and linear in response; (ii) maintain separation efficiency and resolution; (iii) maintain electrical continuity across the separation capillary; (iv) cope with all eluents presented by the CE separation step; and (v) provide efficient ionization from low flow rates for mass analysis. ESI seems to be close to this ideal, although important limitations still have to be overcome regarding its compatibility with many of the substances usually employed in the CE buffers.

9.2.1 Electrospray ionization

Although Dole et al. described the first analytical use of ESI [17] in the late 1960s it was in the 1980s when Fenn et al. were responsible for the major developments of this technique [18,19]. ESI works at atmospheric pressure and is based on a liquid stream flowing through a metal capillary tube that is held at about 2–4 kV potential. The flowing stream becomes charged as it passes through the tube. Evaporation of the solvent leads to highly charged droplets. It is thought that ions are either directly desorbed from the droplets or that the droplets undergo coulombic explosion, leading to smaller droplets that can eventually lead to single ions, or both. A characteristic of ESI is that it produces multiply-charged ions, which has revolutionized the field. As mass analyzers measure the mass/charge ratio (m/z), an increase in the number of charges on a molecule will therefore decrease the measurable m/z of that compound. In this way, it becomes possible to detect very large compounds on a relatively limited mass range mass spectrometer. As indicated by Tomer [8]

"The impact ...[ESI] technique had on mass spectrometry in general and separations/MS can be likened to the effect the meteor impact in the Cretaceous Era had on dinosaurs. The cutting-edge of bioanalytical MS instrumentation was commonly accepted as bigger magnets, bigger instruments. The ability of ESI to form multiply charged ions from large biomolecules allowed these analyses to be performed on relatively low-cost analyzers, such as quadrupoles...."

This ionization technique can work in positive-or negative-ion mode. Thus, in positive ion mode, ESI gives rise to multiply protonated molecules $[M+nH]^{n+}$, where n assumes several integral values giving a series of related ions, or to adduct ions obtained from the attachment of, for example, sodium, lithium or ammonium ions. In negative-ion mode, deprotonated molecules $[M-nH]^{n-}$ are usually observed.

9.2.1.1 Interface designs for CE–ESI–MS

Over the past 15 years, a variety of CE–ESI–MS interfaces have been introduced, as can be deduced from the various reviews recently published on this topic [4,7–11,20–22]. All the developed designs for the CE–ESI–MS interface aim at achieving both a stable spray-current and high efficiency of ionization. With this goal in mind, three principal interface designs have been reported so far: sheathless, coaxial sheath flow and liquid junction devices. Table 9.1 summarizes the properties of these three systems.

Sheathless

This interface design for direct coupling of CE with MS was first reported in 1987 [1], and it used a metal coating on the tip of a 100 μm i.d. internal diameter capillary in order to make the electrical connection between the separation buffer and the ESI needle. This interface design is conceptually the simplest, but possibly the most difficult experimentally. Besides, the durability of the coated tip is usually a few running hours. In order to improve the sensitivity and stability of the coated tip, different procedures have been developed that use conductive coating substances such as silver, gold, graphite, nickel, conductive epoxy adhesives, etc. [1,23–25]. In addition, the tip of the capillary is usually tapered to provide an increased electrical field by etching the tip (e.g., from 360 μm o.d. × 20 μm i.d. to 40 μm o.d. × 3 μm i.d. using 40% hydrofluoric acid solution) before applying the conductive coating, giving rise to the so-called "microspray" or "nanospray." The use of tapered, coated tips (alone or combined with other procedures such as, e.g., first plating the tip with Ni or Ni/Cr alloy before coating with gold) has demonstrated significant gains in sensitivity and sample usage as a result of the more effective ionization taking place with this type of modified capillaries [23,26–28].

Coating the tip of the CE capillary with a conductive substance, as described earlier, is the most common sheathless design for CE–MS [1,23–25]. However, other alternative sheathless designs that can give rise to ionization while maintaining the electrical continuity across the

TABLE 9.1

Summary of characteristics of ESI-interfaces designed to hyphenate CE and MS

Design	Electrical contact with CE	Spray	Advantages	Drawbacks
Sheathless	Metal coating Inserted wire	From capillary tip (0–100 nl/min, nano-ESI)	High sensitivity	Less robust (instable coating, clogging,…) Dependent on EOF
Coaxial sheath liquid	Sheath liquid	Sheath liquid (1–4 µl/min)	Commercially available Stable spray Independent on EOF (buffer)	Low sensitivity
Liquid junction	Liquid gap between CE capillary and ESI needle	Direct from ESI emitter (nano-ESI)	High sensitivity Independent on EOF More robust than sheathless (no coating)	Peak broadening Less robust than coaxial (clogging of tip, contamination)

separation capillary have also been developed. They used, for example, an electrode inserted into the end of the CE capillary [29,30], the CE voltage to drive spray formation [31] or a direct connection of the CE column to a separate spraying tip or stainless-steel liner [32,33]. In this latter design, the separation capillary is placed through the liner (made from a syringe needle) and the electrical connection is maintained by a liquid film that builds up at the outer surface of the tip of the separation capillary [34]. Recently, a sheathless split-flow interface has been introduced, in which the electrical connection to the CE capillary outlet is achieved by diverting part of the CE buffer out of the capillary through an opening near the capillary outlet [35]. In this design, the CE buffer exiting the opening contacts a sheath metal tube, which acts as the CE outlet/ESI shared electrode. If the ESI source uses a metal needle, the voltage contact to the CE buffer can be achieved by simply inserting the outlet of the CE capillary, which contains an opening, into the existing ESI needle. Since the liquid flows out of the capillary through the opening, there is no dead volume associated with this interface. Moreover, bubble formation due to redox reactions of water at the electrode does not affect CE–ESI–MS performance (a typical problem in the electrode-inserted-design mentioned above), because the actual metal/liquid contact occurs outside of the CE capillary. The sensitivity associated with this sheathless CE–MS interface, the ease of fabrication, universality and lack of any dead volume are claimed to make this design a superior CE–ESI–MS interface. The performance of this interface is demonstrated by analyses of a peptide mixture and a protein digest using a variety of capillary dimensions [35]. Some examples of different types of sheathless interfaces developed up to now are given in Figs. 9.1A–C [8].

Although, as indicated above, the sheathless design provides better sensitivity, the tip of the capillary is often quite small and can lead to clogging. Besides, the preparation of these interfaces can be problematic and frequently labor-intensive while the procedures are not very robust. Moreover, other problems, including the high dependence of the ionization efficiency on both the separation buffer used and the electroosmotic flow (EOF) generated inside the separation capillary, have directly led to the development of the coaxial-sheath flow interface.

Coaxial sheath flow

This approach, reported by Smith et al. [3] in 1988 is predominantly used today because of its ease of operation. As mentioned above, standard ESI sources operate optimally with flow rates around 5 µl/min.

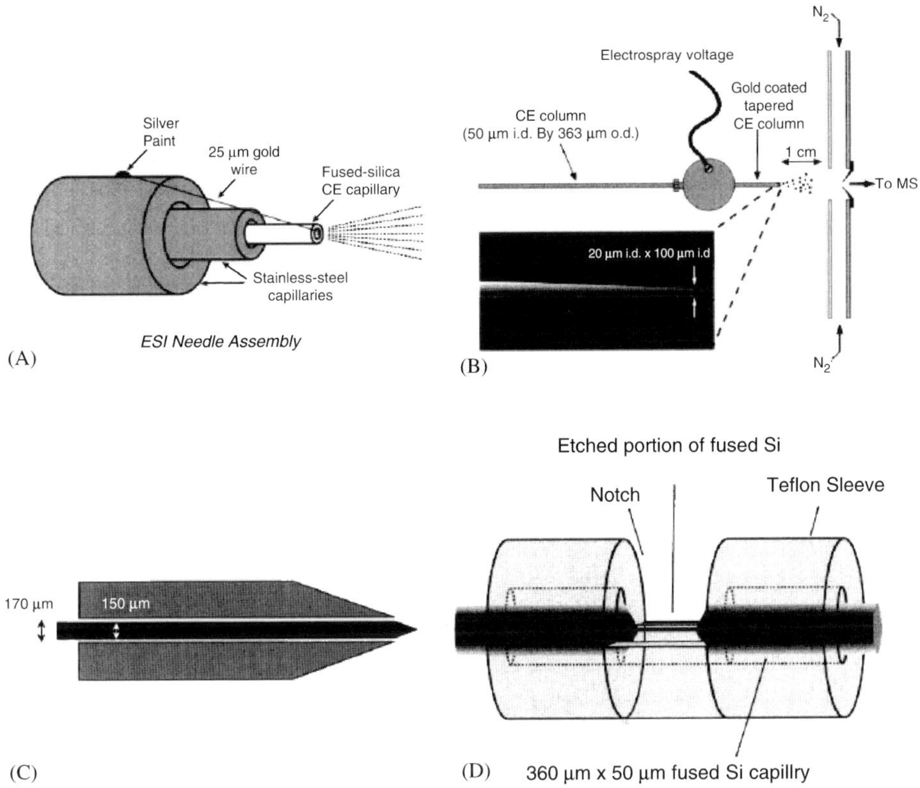

Fig. 9.1. Interface designs for the direct coupling of CE to ESI–MS. (A) Wire in-capillary design. (B) Tapered metallized capillary tip design in which the CE capillary end is tapered and the sputter-coated with gold. (C) Liner interface in which the outlet of the CE capillary was led through a stainless-steel liner the tip of which has been ground to a sharp end and polished. The end of the fused silica was also ground to a sharp point. The gap between the fused silica column and the liner was approximately 10 μm. (D) Porous glass joint design. The center of the fused silica is etched so that the capillary walls are <20 μm thick and is center in the notch of a Teflon sleeve. This joint is immersed in a reservoir containing 1% acetic acid in water. The CE termination/micro-ESI voltage is applied to the acid solution. From Ref. [8]. Copyright (2001), American Chemical Society.

Since CE flow rates are in the low nanoliter per minute range, additional liquid flow needs to be added to the CE flow. The sheath flow is, therefore, required to increase the actual flow rate from the capillary and to provide for a stable electrospray. Moreover, it also allows closing the CE electric circuit since the sheath liquid contacts both the metal sprayer (usually connected to ground) and the buffer flowing out of the CE capillary. In sheath-flow designs (see Fig. 9.2), the sheath liquid is introduced coaxially via a capillary concentric with the separation capillary. Nebulizing gas is usually introduced through a third concentric capillary to favor the evaporation of the solvents coming from both the separation capillary and, mostly, the sheath liquid, thereby improving the nebulization process.

The main advantages of this design are that it is robust and relatively simple to implement. However, it also has some limitations. Thus, the coaxial interface leads to some dilution of the solute bands, reducing the sensitivity. In this sense it is interesting to indicate that at flow rates lower than ~100 nl/min, the ESI source becomes a concentration-sensitive technique; as a result, the dilution of analytes by the sheath liquid taking place in the coaxial interface will decrease the sensitivity [9]. Moreover, some other experimental parameters, such as sheath liquid composition and flow rates, pressure, flow and temperature of the nebulizing gas, capillary dimension and position on the interface and nature of CE electrolytes, can modify both the sensitivity of this interface and influence the separation resolution (generating, e.g., a suction effect that pulls the sample bands and buffer through the capillary) [36,37]. As an example, Fig. 9.3 shows the influence of sheath liquid flow rate on the intensity of the MS signal. Thus, as can be seen, the higher the flow rate, the lower is the intensity of the MS signal owing to solute dilution. Fig. 9.4 shows, on the other hand, the effect of the nebulizing gas pressure on resolution and ion abundance. As can be seen, an increase in the gas pressure can improve the MS signal but, at the same time, can reduce the resolution of the separation (as a consequence of the hydrodynamic

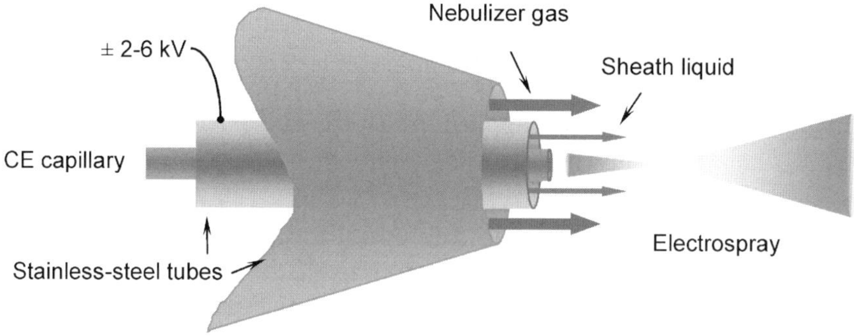

Fig. 9.2. Interface design for a coaxial-sheath flow CE–ESI–MS coupling.

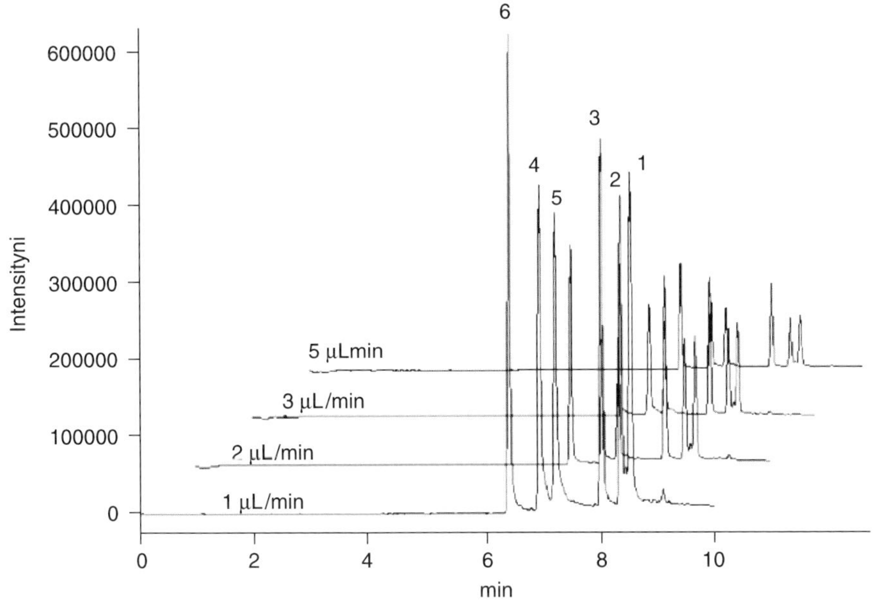

Fig. 9.3. Influence of sheath liquid flow rate on intensity of fluoxetine and its related compounds. CE conditions: Fused silica capillary (80 cm l_t, 50 μm i.d., 360 μm o.d.); running buffer: acetonitrile containing 20 mM ammonium acetate and 1 M acetic acid; pressure injection: 25 mbar for 5 s; separation voltage: 30 kV, temperature: 20°C. ESI–MS conditions: SIM positive ion mode (5 ions), capillary voltage: 4 kV, fragmentor: 50 V, drying gas: N_2, temperature: 150°C, nebulizer pressure: 4 psi. Sheath liquid composition: 0.5% formic acid in water-isopropanol (20:80, v/v). Reprinted from Ref. [38]. Copyright (2002), John Wiley & Sons, Inc.–Wiley-VCH Verlag GmbH.

flow generated inside the separation capillary). Therefore, a compromise should preferably be found for this parameter.

Some preliminary experimental recommendations can be made as follows: (i) the degree to which the CE capillary exits the tube should be about one-third of its outer diameter; (ii) the sheath liquid should be sufficiently conductive to complete the CE circuit and to permit ESI, but not of such a high ionic strength that arcing or discharges are created in the ESI source (usually, the sheath liquid comprises more than 50% of organic solvent with less than 1% of acid or base added, depending on the ionization of the analytes of interest); (iii) the flow rates of both nebulizing gas and sheath liquid should be kept to a minimum to maintain the stability of the electrospray without generating noticeable dilution and syphoning effect; and (iv) a

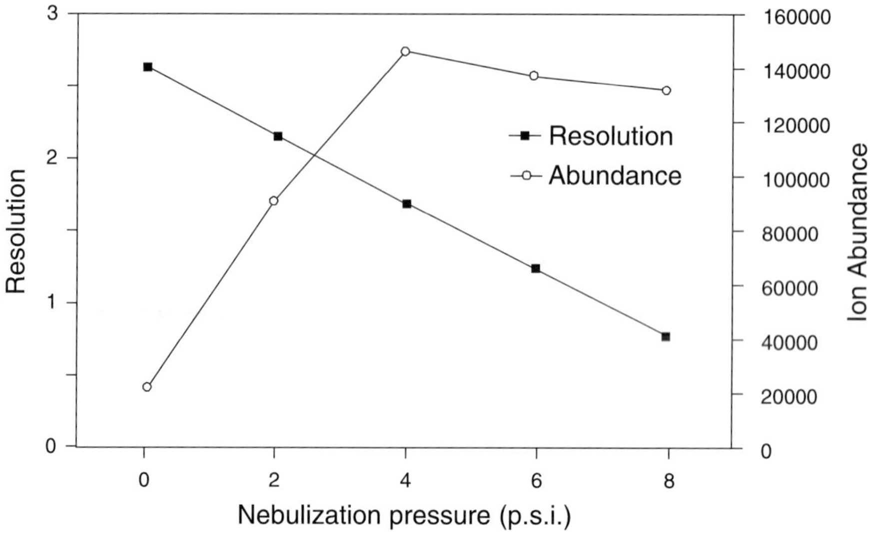

Fig. 9.4. Effect of nebulizing gas pressure on resolution of fluoxetine and its meta-isomer as well as flouxetine ion abundance. Sheath flow: 2 μl/min. Other conditions are as in Fig. 9.3. Reprinted from Ref. [38]. Copyright (2002), with permission from John Wiley & Sons, Inc.–Wiley-VCH Verlag GmbH.

compromise value for the gas pressure (usually around 2–8 psi) should be found. These are very general rules that in some cases will not match the requirements of a specific analysis.

An additional problem that has been observed by Foret et al. [39] is that the use of a sheath flow under determined conditions can affect the separation. This seems to be due to counterions from the liquid sheath migrating into the separation capillary, forming a moving ionic boundary inside the capillary. This effect is experimentally observed as changes in migration order, delays in migration time, and loss of resolution compared to off-line CE. Moreover, these disturbances are more noticeable for columns with low EOFs.

In spite of all these drawbacks, this type of interface design is nowadays the most widely used to hyphenate CE and MS. As mentioned above, it is robust as well as easy to implement and use, and up to now the only ESI device commercially available for CE–ESI–MS.

Liquid junction

The first report describing this interface for CE–MS was published in 1989 [40]. In this design the separation capillary terminates in a reservoir filled with a makeup fluid that is used to close the CE electrical circuit. A transfer capillary (also called the ESI emitter capillary) is positioned opposite the end of the separation capillary with a very narrow gap (ca. 10–25 μm) between the two in such a way that the liquid and analytes from the separation capillary pass to the transfer tube and are then sprayed. A scheme of this interface can be seen in Figs. 9.1D and 9.5.

The advantage of this setup is that the ESI process can be optimized by adding a freely chosen liquid sheath flow, since CE and ESI can be operated independently owing to the partial electrical and physical disconnection of the CE separation from the ESI emitter. The main disadvantages linked to the use of this type of interface have been the difficulty of positioning the transfer capillary in a reproducible way, and potential loss of separation efficiency in passage through the ESI emitter [41,42]. Also, the ESI emitter can originate a counter-hydrodynamic flow in the CE column, and in some cases, pressure (typically 20–40 mbar) may have to be applied at the inlet capillary to counterbalance this effect. Although some new developments improving the original design have been reported [43–47], some of the original difficulties (i.e., peak broadening, and thus, loss of separation efficiency) frequently remain, while the construction of the junction between the CE capillary and the ESI emitter continues to be a difficult task. Moreover, the column washings required to regenerate the CE inner surface frequently contaminate the liquid junction, modifying the overall sensitivity of the system. These inconveniences can probably explain the few new developments on this subject reported in recent years. However, the research on this type of interface seems to be more

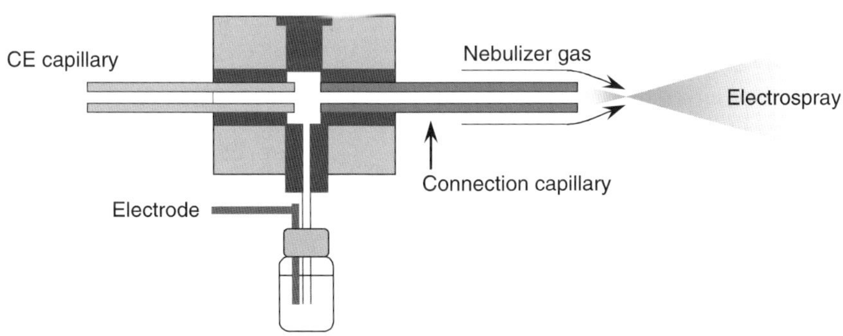

Fig. 9.5. Interface design for a liquid junction CE–ESI–MS coupling.

active for microchip–MS coupling, as discussed in detail in Section 9.2.4 of this chapter.

9.2.2 Mass analyzers

Mass analyzers must provide accurate identification and quantification of solutes with a high degree of sensitivity and selectivity at speeds compatible with the narrow bands generated by CE. Although in principle any type of mass spectrometer can be coupled with CE, up to now, quadrupole (Q) and ion trap (IT) analyzers have predominately been used [7,8]. This has been mostly because these analyzers are relatively less expensive, and impose lower technical difficulties in the introduction of liquids than, for example , Fourier transform–ion cyclotron resonance (FT–ICR) or sector instruments and also have an adequate scan speed for coupling to CE. IT instruments, moreover, offer significant gains in transmission efficiencies since all ions can be stored simultaneously and analyzed sequentially. The new IT analyzers also provide both the possibility of carrying out MS–MS experiments and, as mentioned earlier, adequate scanning speed, making them very compatible with the high separation efficiencies achievable by CE [48–50]. Time-of-flight (TOF) analyzers have seen an impressive development in the last 10 years as a result of the large mass range that can be analyzed when coupled with MALDI sources for biopolymer analysis [51]. The main advantage of the TOF mass spectrometer is that the entire mass spectrum can be acquired with efficient ion transmission and relatively fast duty cycles, which makes it very compatible for coupling to CE [28,30,52–54]. Moreover, the prices of these analyzers are nowadays becoming more affordable.

Combinations of CE with other analyzers for example, Fourier transform–mass spectrometers (FT–MS, mainly FT–ICR analyzers), Q-TOF, or triple quadrupole–mass spectrometers (QqQ—MS) have also been described in the literature [8]. Thus, FT–MS offers excellent sensitivity, high resolution ($>$100,000 with low ppm mass measurement error), and the possibility to carry out MS^n experiments [55]. Besides, the new generation of FT–ICR analyzers with magnetic fields higher than 9 T is capable of sub-attomole sensitivity at nearly 1 s acquisition times [56,57]. Recently, the Hadamard transform–TOF MS (HT–TOFMS) has proved to be a promising detector for any capillary-format separation that can be coupled to MS by electrospray ionization. HT–TOFMS was able to sample peaks having widths in the millisecond range [58]. However, the very high cost of some of these MS equipments (e.g., Q-TOF, FT–MS) keeps them out of reach for most laboratories and, therefore, they cannot be considered as routine instruments.

A comparative summary of the characteristics (e.g., scan speed, resolution, sensitivity, cost, etc.) of the principal analyzers that can be used today with CE is given in Table 9.2.

9.2.3 CE–MS compatibility

The high separation efficiency of CE makes it extremely attractive for the separation of complex mixtures, especially biological or environmental mixture, but the low concentration limits of detection is still a drawback. Unfortunately, many of the compounds used in the CE buffers are not volatile and are not suitable for CE–ESI–MS since they often suppress the ionization of the analyte, yielding poor mass spectral sensitivity or, as mentioned above, can even clog the system. This limitation has to be taken into account when using CE in its different modes, in which substances that are not compatible with ESI–MS frequently have to be used in order to achieve the adequate CE separation of the analytes. Nonetheless, CE–ESI–MS seems well suited for a number of applications.

In this section, the compatibility of the different modes of CE with ESI–MS is studied. Also, some compatible buffers and compounds for use in CE–ESI–MS are identified, and a summary of the different methods that have been reported to increase both the compatibility (e.g., PFT) and sensitivity of CE–MS (e.g., on-line preconcentration methods) is provided.

9.2.3.1 CE separation modes

A revision of the literature dealing with CE-MS shows that till now reported applications involve hyphenation with free solution CE (CZE or FSCE) and, to a lesser extent, capillary electrochromatography (CEC). It can also be observed that coupling between MS and other CE modes such as capillary isoelectric focusing (CIEF), MEKC, capillary gel electrophoresis (CGE) and capillary isotachophoresis (CITP) has been rarely reported. This is mainly due to the intrinsic incompatibility between the ESI source and the nonvolatile compounds used in these latter CE modes.

TABLE 9.2

Comparative characteristics of some mass-spectrometry analysers used in CE–MS

Analyzer	Scan-speed	Sensitivity full scan (SIM)	Resolution	MS–MS	Quantitation	Cost	Applications (roughly)
Q	–	– (+)	0	—	+	Low	Low price mass selective detector
QqQ	–	-– (+)	+	+	++	High	Quantification of known targets in complex matrices
IT	+	+	+	++	+	Medium	Unknown substances in complex matrices
TOF	++	+	++	(+)	+	High	Unknown substances in complex matrices, quantification
FT–MS	+	+	+++	+	+	Very high	Structure elucidation in complex mixtures

Capillary zone electrophoresis-mass spectrometry
To date, CZE is the most widely used CE technique and it is also valid for CZE–MS, probably because CZE is the easiest CE mode to run (i.e., a single buffer is typically used). Therefore, the discussion in Section 9.2.3.2 as well as the applications described below in Section 9.3 apply mostly to this CE mode; if otherwise, the specific mode will be stated.

Micellar electrokinetic chromatography- mass spectrometry
MEKC has rarely been used in combination with MS, mainly owing to both the potential for source contamination by the surfactants, and to suppression effects on the analyte signal due to competition between the charged surfactant and the analyte for the available charges [59].

An example of this deleterious phenomenon can be seen in Fig. 9.6 where the influence of the concentration of SDS (the surfactant most commonly used in MEKC) on the MS signal is shown in two different buffers. Although different approaches have been attempted for combining MEKC with MS, PFT seems to be the most widely used, and will be discussed in detail in Section 9.2.3.3. Some other solutions to combine MEKC with MS have been reported, based on the use of, for example, APCI [61], mixed aqueous–organic buffers containing suitable amounts of zwitterionic detergents and neutral, hydrophilic, coated capillaries to reduce the EOF, minimizing in that way the interference with the ESI signal [62]; the use of high-molecular-weight surfactants whose electrosprayed ions are above the mass range scanned by MS [63]; the use of surfactants that migrate toward the anode with the micellar velocity being controlled by adjustment of the EOF [64]; and elimination of the entrance of the surfactant into the MS by using two capillaries combined through a liquid junction, the first employed for MEKC and the second for CZE analyte transfer to the ESI source [65]. Recently, the use of fluorinated surfactants such as perfluorooctanoic acid and perfluorooctanesulfonic acid have been shown to have some compatibility with ESI–MS without a significant decrease in sensitivity and similar selectivity to that provided by the most common SDS [66,67]. Another approach that has been investigated to make feasible the coupling of MEKC with MS is the direct introduction of nonvolatile buffer and surfactant into the electrospray interface [68,69]. Following this idea, it has been shown that up to 20 mM SDS could be used in the direct coupling of MEKC with ESI–MS while still obtaining useful signals [70]. In other similar work, an on-line MEKC–MS coupling was employed under conventional MEKC conditions with a background

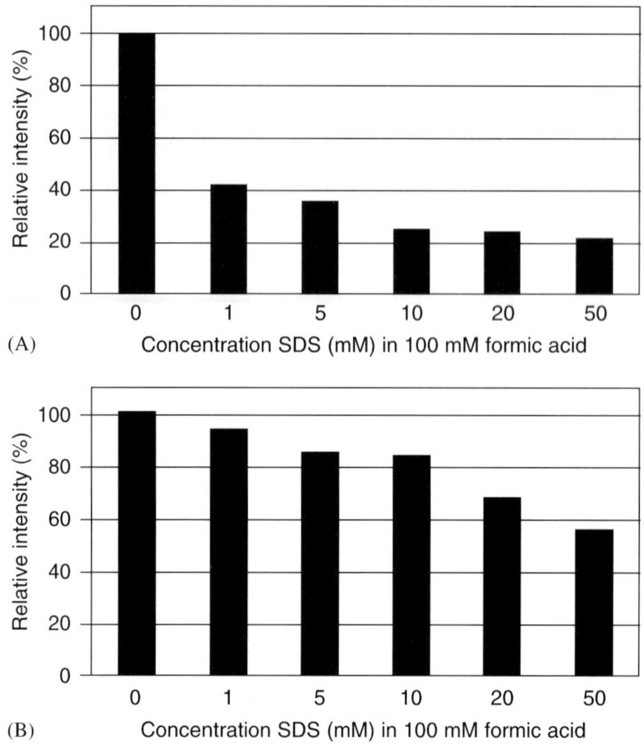

Fig. 9.6. Influence of the SDS concentration on the relative MS signal of 1 μg/ml mebeverine (monitored at *m/z* 430) in (A) 100 mM formic acid at pH 2.4 and (B) 10 mM sodium phosphate at pH 7.5. Reprinted from Ref [60]. Copyright (2003), with permission from Elsevier.

electrolyte containing SDS, sodium phosphate buffer and organic modifier. Although significant suppression of the analyte signal occurs, using extracted-ion chromatograms (obtained in the full scan mode), detection limits better than 1 μg/ml are achieved, as can be seen in Fig. 9.7, where MS detection below 100 ng/ml is possible when SIM is used. Such sensitivity will be quite sufficient for several relevant analytical applications, such as the impurity profiling of drugs [60]. However, in the two later examples, the ESI interface was working in the positive-ion mode, where SDS interference is less severe than in the negative-ion mode. Recently, the development of a low-flow ESI–MS interface (using several hundred nanoliters per minute as sheath liquid flows) have been shown to make the MEKC–MS coupling more compatible [71]. According to the authors, this low-flow interface showed better

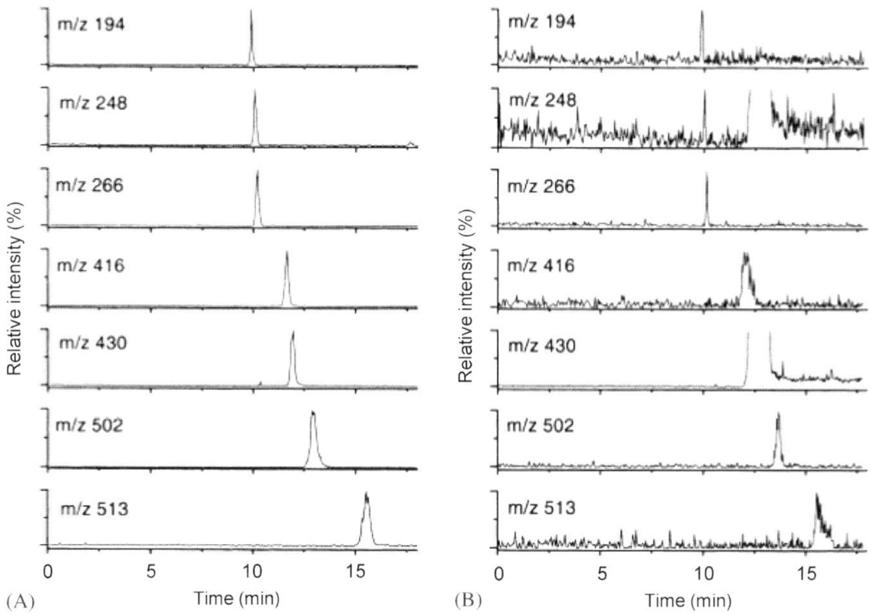

Fig. 9.7. MEKC–MS of a mixture of mebeverine and related compounds. Running buffer: 10 mM sodium phosphate at pH 7.5, 20 mM SDS and 25% acetonitrile. Compound concentrations: (A) 10 μg/ml each, and (B) 1 mg/ml of mebeverine and 1 μg/ml of related compounds. Reprinted from Ref. [60]. Copyright (2003), with permission from Elsevier.

sensitivity than a conventional sheath liquid interface, which was primarily attributed to a low dilution factor and a reduction in the sprayer orifice size. The interface is also more tolerant to the presence of nonvolatile salts. Because of tolerance to the surfactant SDS, this interface can be used to couple MEKC with ESI–MS. The performance of the interface in an MEKC–MS application, using triazines, was significantly better than that obtained with a conventional sheath liquid interface. Also, this interface can be easily used for large-volume sample-stacking applications. [71]. The improved sensitivity was explained by the higher salt tolerance, and the superior ionization and sampling efficiency obtained using a nanospray tip, because in nanospray, the higher surface charge density of the initial droplet results in early fissions without extensive evaporation, thereby decreasing the relative concentration of nonvolatile salts. Therefore, smaller and more highly charged droplets result not only in better ESI, but also resist ionization suppression when nonvolatile salts, such as SDS, are present in the CE running buffer. In addition to the better sensitivity, ion signals were also found to be much more stable than the conventional sheath liquid interface. In spite of the clear advantages observed, the signal dropped to $<30\%$ when the concentration of SDS was increased to 35 mM, and to $<10\%$ if the concentration of SDS was increased to 55 mM in the ESI positive-ion mode where the conditions are less demanding.

Capillary gel electrophoresis–mass spectrometry

In CGE it is necessary to use sieving buffers of high concentration to achieve adequate separations; this requirement is the major handicap for on-line CGE–MS coupling, since these sieving buffers contaminate the source and severely decrease the sensitivity of MS detection. In spite of this intrinsic incompatibility, large biomolecules such as DNA, RNA, and proteins are interesting species for CGE, and different studies about the development and application of CGE–MS have been reported. Thus, Garcia and Henion [72] first reported the use of gel-filled capillaries in CGE–MS in 1992. Barry et al. [73] in 1996 used a poly(vinylpyrrolidone) (PVP) matrix (14% (w/v) in 20 mM ammonium acetate at pH 9) for the analysis of short, modified oligonucleotides. Harsch and Vouros [74] in 1998 separated isomeric (acetylamino)fluorene-modified oligonucleotides at the micromolar concentration level using the PVP matrix and CGE-ion trap MS. CGE–MS analysis of oligonucleotides (12- to 20-mer) differing in length by one nucleotide unit, was reported in 2001 by Bayer's group [75]. These authors used a PVA-coated capillary and a buffer made of 20% (w/v) PEG, 20% (v/v) acetonitrile and 20 mM Bis-Tris borate; CE-inlet buffer 20 mM Bis-Tris borate; as sheath flow they employed a solution of 50% acetonitrile and 50% of a 20 mM triethylammoniumacetate aqueous dissolution, detecting the oligonucleotides in the negative mode. Optimizing both the concentration of the sheath flow liquid and the composition of the CE buffer is shown to be fundamental for obtaining a good signal. In a recent study by the same group, different characteristics of the on-line CGE–ESI–MS system for oligonucleotide analysis were studied [76]. Namely, effects of CGE buffer concentration, capillary length, separation and orifice voltage on CGE separation and MS detection of a phosphodiester oligonucleotides mixture were examined. Attention was paid to the influence of the interface, such as geometry of capillary alignment, sheath liquid flow rate and sheath liquid composition on performance of the system. However, although baseline separation

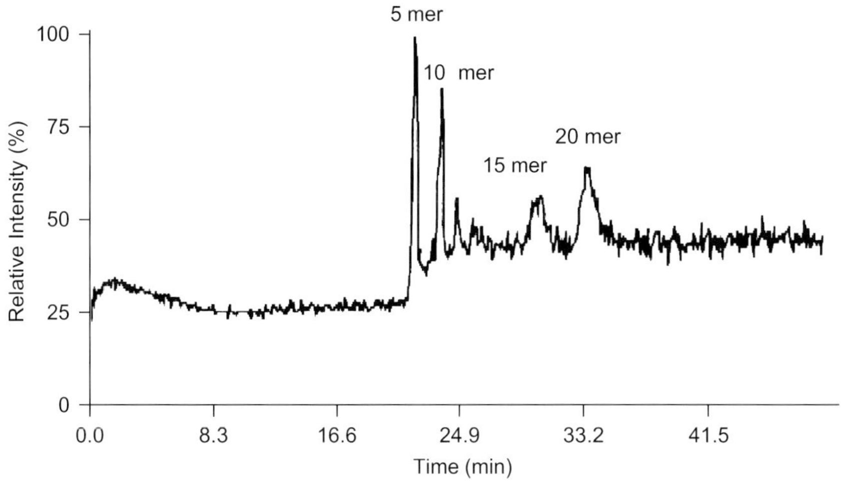

Fig. 9.8. CGE–ESI–MS total ion electropherogram of oligonucleotide mixture: 5-mer (3′-TTTTT-5′), 10-mer (3′-CCCCCCCCCC-5′), 15-mer (3′-AAGAACAGACGAGAA-5′) and 20-mer (3′-CCTCCTCTCTCGCTCCCTCT-5′). Running buffer: 20% (w/v) PEG 35,000, 20% (v/v) acetonitrile and 20 mM BisTris-borate. CE inlet buffer: 20 mM BisTris-borate. Sheath liquid: 2.5 mM triethyl ammonium acetate in acetonitrile–water (1:1, v/v), sheath flow: 15 μl/min. Reprinted from Ref. [76]. Copyright (2003), with permission from Elsevier.

could be achieved of these relatively small oligonucleotides, very noisy electropherograms and MS spectra were obtained, as can be seen in Fig. 9.8. These results demonstrate the difficulty of this type of analysis, which probably explains the few works published thus far on this topic.

Capillary isoelectric focusing–mass spectrometry

Coupling of CIEF with ESI–MS is very attractive for the direct identification of analytes, selectivity enhancement and structure confirmation by MS–MS techniques. Moreover, apart from the information on protein molecular weight (and in some cases structure), CIEF–MS can also provide information about the isoelectric point of these biomolecules. CIEF–MS was first introduced by Lee's group [77–80], which used a coaxial sheath liquid interface to couple both techniques [77–80]. The basic procedure can be described by taking as a reference one of the works of this group (e.g., Ref. [80]). In that work, a group of proteins were first focused using a normal CIEF scheme inside the capillary (the outlet reservoir, containing 20 mM sodium hydroxide as

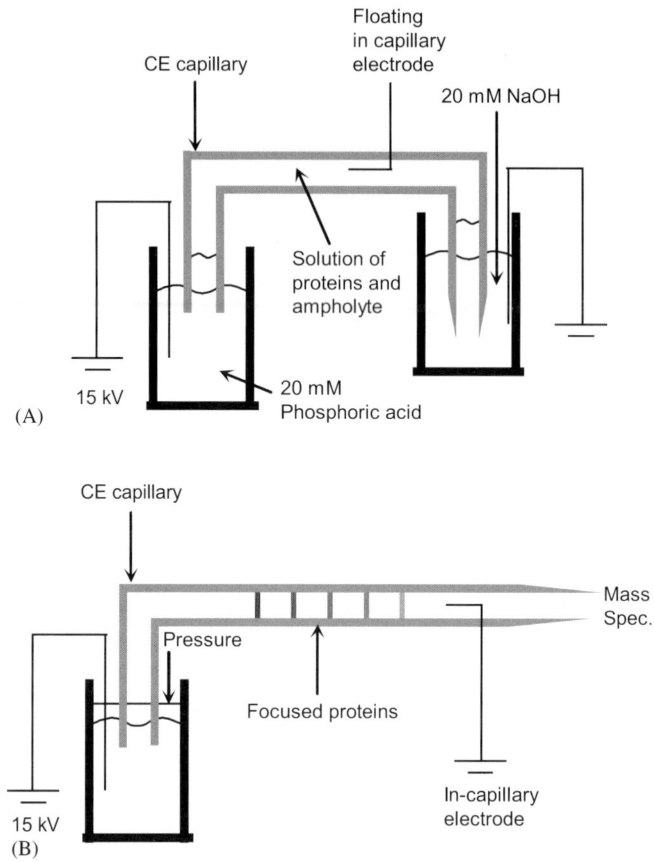

Fig. 9.9. Schematic sheathless CIEF design before (A) and after (B) focusing. From Ref. [9]. Copyright (2002), with permission from Springer-Verlag GmbH & Co. KG.

the catholyte, was located inside the coaxial electrospray housing). Once the focusing was completed, the electric potential was turned off, and the outlet reservoir was removed. The capillary tip was fixed into the ESI–MS instrument about 0.5 mm outside the electrospray needle. The sheath liquid consisted of 50% methanol, 49% water and 1% acetic acid (by vol.) and was delivered. In this way, proteins could be directly identified by CIEF–ESI-MS with a mass accuracy of less than 0.04%. A scheme of the whole process is shown in Fig. 9.9. The integration of CIEF with ESI–MS, as a novel two-dimensional bioanalytical methodology, showed good resolving power and speed. However, the detection sensitivity of CIEF-MS is about 1–2 orders of magnitude lower than that achievable using radiolabeled two-dimensional gel electrophoresis; the presence of carrier ampholyte in ESI caused a marked reduction in the protein ion intensity.

In order to overcome the disadvantage of employing a coaxial liquid sheath flow (i.e., the addition of excess electrolyte to the CIEF effluent significantly decreases analyte sensitivity) different strategies have been developed. Thus, Van der Greef's group introduced a coupled-capillary setup, which incorporated a microdialysis device between a CIEF separation capillary and a transfer capillary, for on-line coupling of CIEF with ESI–MS [81]. Although there was no apparent loss in separation efficiency and resolution of focused protein zones across the microdialysis junction, significant reduction in separation resolution was observed in the reconstructed ion electropherogram owing to analyte diffusion and the absence of a pH gradient for maintaining the focused protein zones in the transfer capillary. Following the same idea (i.e., to skip the use of a coaxial ESI source), a microdialysis junction, based on a microdialysis membrane connecting a separation capillary and a short, sharply tapered microelectrospray emitter capillary was developed by Yang et al. [82] for on-line combination of CIEF with ESI–MS. The microdialysis junction provided the necessary electrical connection across the dialysis membrane for defining the electric fields needed for the CIEF separation and the electrospray process. A microelectrospray emitter produces a stable electrospray of protein analytes without the need for a makeup liquid flow, and eliminates any subsequent sample dilution and reduction in MS sensitivity. The results indicated that at least one order of magnitude improvement in detection limits of a group of standard proteins could be achieved using the microdialysis junction compared with the coaxial-sheath liquid interface [82]. As can be seen in Fig. 9.10, a substantial gain was noted in both signal intensity and signal-to-noise ratio using the microdialysis junction interface. Note that the protein mixture used in Fig. 9.10B was diluted 10-fold relative to the mixture used in Fig. 9.10A.

Recently, Smith and co-workers [57,83–85] have demonstrated the good possibilities of CIEF–Fourier transform ion cyclotron resonance (FTICR)–MS to identify intact proteins on a proteome wide level. Two important factors for successful CIEF analysis include elimination of BGE migration during focusing, and proper mobilization of the focused analyte zones for detection. Although using the CIEF–FTICR coupling, up to 1000 putative proteins in the 2–100 kDa molecular mass range can be detected using 200–300 ng of soluble *Escherichia coli* proteins

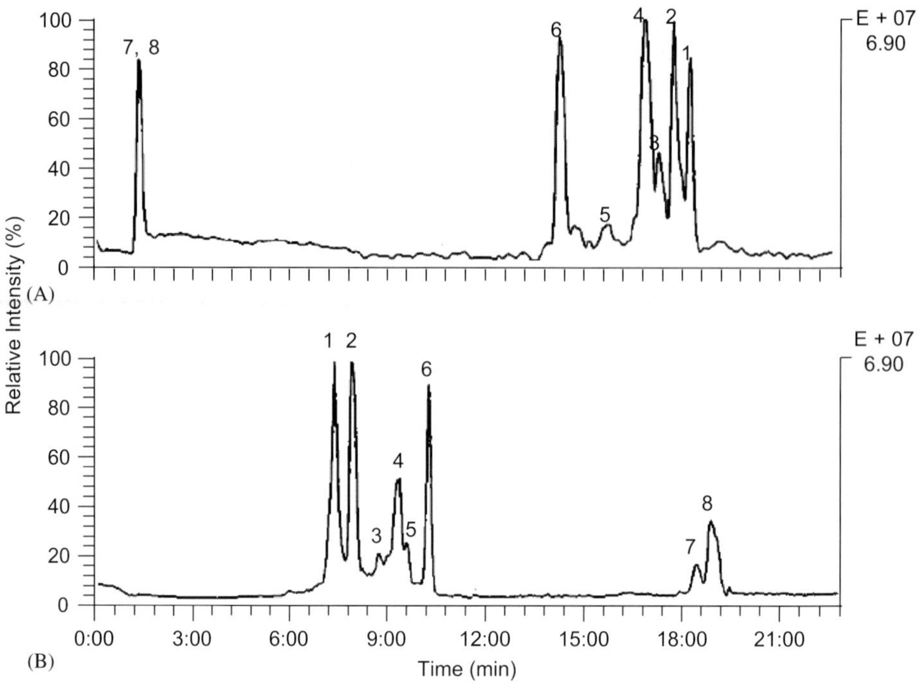

Fig. 9.10. CIEF–ESI–MS separation of model proteins using (A) a coaxial-sheath liquid interface with protein concentration of 0.1 mg/ml (each analyte), and (B) the microdialysis junction with protein concentration of 0.01 mg/ml (each analyte). Model proteins: (1) carbonic anhydrase II (pI 5.4), (2) carbonic anhydrase II (pI 5.9), (3 and 4) carbonic anhydrase I (pI 6.6), (5 and 6) myoglobin (pI 6.8 and 7.2) and (7 and 8) cytochrome *c* (pI 9.6). From Ref. [82]. Copyright (1998), American Chemical Society.

(compared with 0.1–1 mg total protein typically used for 2D-PAGE) in a single 30 min CIEF–FTICR analysis, the high price of the FTICR–MS instrument is an important limitation that has to be taken into account.

Capillary electrochromatography–mass spectrometry

Capillary electrochromatography (CEC), a hybrid between CE and HPLC, has attracted increasing interest over the last 4 years. This interest has been extended to its hyphenated use with MS. As mentioned above, CEC has been the second preferred mode, after CZE, for coupling with MS [86,87]. This can be understood, considering that CEC is an interesting alternative to MEKC for the separation of neutral compounds, since no micelles are used in CEC, which greatly increases its compatibility with MS [88]. Besides, the intrinsic low sample capacity of CZE and the incompatibility of CZE–MS with many separation buffers can be also overcome by using CEC. These advantages can explain the large number of drugs, enantiomers, amino acids, peptides, etc. that have been performed by this company.

The coupling of CEC to MS has been reported in the two modes in which capillary electrochromatography can be carried out, that is, using open tubular-CEC (OT-CEC, where the stationary phase coats the wall of the capillary column typically less than 20 μm i.d.) [89,90] or using the most common packed capillaries (referred to just as CEC, that typically uses 50–100 μm i.d. fused-silica capillaries packed with ~3–10 μm particles of different nature). Besides, CEC can be used in the mode of pressure-assisted, where pressure is additionally applied to the electrical field to increase the analysis speed and to improve the stability of the whole CEC–MS system.

The first reports of CEC–MS using as interface cf-FAB were carried out by Van der Greef's group [91]. In 1993 the same group reported the first coupling of pressure-assisted CEC with ESI–MS using a coaxial sheath flow [92], whose usefulness was further demonstrated and optimized 3 years later by Lord et al. [93] and Lane et al. [94]. In 1995 Bayer's group [95] reported the direct electrospray from a CEC column (i.e., sheathless). In this case the packed capillary, 20 cm long and 100 μm i.d., was packed with 1.5 μm particles and was capped at both ends with two 0.3 cm frits. In this case, supplementary pressure had to be applied to stabilize the spray. In 1998, Warriner et al. [96] reported the first nanospray interface used for coupling reversed- phase/ion-exchange CEC columns with MS which was applied to analyze peptides. This was successful owing to the compatibility of CEC flow rates (100 nl/min) with relatively large orifice nanospray emitters (10 μm). The application of high field strengths led to fast EOF velocities; thus, short retention times were achieved. A more recent application of sheathless nanoelectrospray to CEC of peptides has been reported by Gucek et al. [97], while the coupling between CEC and FTICR MS has demonstrated its usefulness for characterizing complex oligosaccharide pools [98].

A serious problem observed with CEC is bubble formation at the frit at the column end due to degassing, that logically also has place in CEC–MS [99]. This problem has been usually solved by applying pressure at the inlet and outlet of the CEC column. However, in

CEC–ESI–MS, the outlet is usually at the spray tip and, therefore, pressurization is difficult to achieve. As a result, the configuration of both the outlet CEC capillary and the interface used have been indicated as the determining points for obtaining an optimal CEC–MS arrangement. Three main strategies have been described based on the use of: (i) columns with fritted terminations, (ii) columns with tapered ends and (iii) columns with connecting tubing acting as a transfer line. These outlet CEC column designs [87], which can be seen in Fig. 9.11, are typically used together with some of the mentioned ESI interfaces (i.e., sheathless, coaxial sheath flow or liquid junction) for CEC–MS. A deep description of the advantages and drawbacks of the different configurations is beyond the scope of this chapter, but it can be briefly said that tapered capillaries together with a sheathless interface provide the best results in terms of efficiency and sensitivity [87].

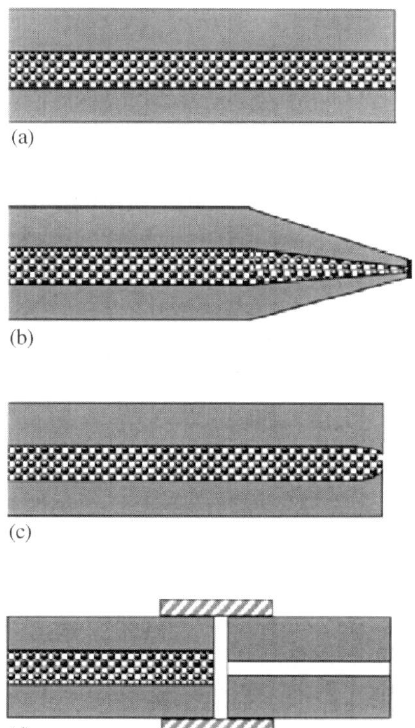

Fig. 9.11. Configuration of column outlet for CEC–ESI–MS. Columns with (A) fritted termination, (B) external taper, (C) internal taper and (D) connecting tube acting as transfer line. Reprinted from Ref. [87]. Copyright (2000), with permission from Elsevier.

However, the experimental difficulties of developing this type of arrangement and its low durability have to be considered. It is also interesting to note that the use of tapered columns, sprayers or supplemental pressure from the inlet appears to reduce bubble formation [100–102]. In the other work, the use of the coaxial sheath flow was compared with the sheathless interface for CEC–MS and, despite the loss in sensitivity (20–40-fold), the sheath flow interface was shown to be superior in terms of ruggedness, and allowed the use of higher electrical fields to achieve faster analysis times [103].

Recently, a newly developed, completely fritless column system has been found to eliminate band broadening, peak distortion and bubble formation associated with frits [100,104]. In this sense, the investigation on new monolithic columns for CEC [105] and new CEC stationary or pseudostationary phases [106] will determine, among other factors, the successful use of CEC–MS in other applications. Also, the use of CEC together with coordination ion spray (CIS–MS, an ionization technique in which charged coordination compounds are formed on-line after the separation and then nebulized) has demonstrated its potential to couple CEC and MS for the analysis of vitamins, estrogens and unsaturated fatty acids [107] as well as for chiral compounds of barbiturates and chlorinated alkyl phenoxypropanoates performed on a permethylated ß-cyclodextrin (CD) stationary phase by pressure-assisted CEC [108]. As an example, the CEC–CIS–MS separation of fenoxaprop ethyl and diclofop methyl enantiomers on a 30 cm, 100 µm i.d. capillary packed with a chiral stationary phase is shown in Fig. 9.12.

An additional problem in CEC–MS (also applicable to other CE modes) is that at the present time, the commercially available instruments have imposed the use of column lengths in the range of 0.5–1 m in length, which is less than ideal in CEC, since the electric field strength across the capillary column is significantly reduced, resulting in low sample linear flow velocities, long elution times and low efficiencies.

On-column transient isotachophoresis

This mode of CE has been mostly combined with MS to enhance the concentration detection limit of the whole analytical procedure (i.e., as on-column preconcentration step [109,110]) and, therefore, will be discussed in Section 9.2.3.4.

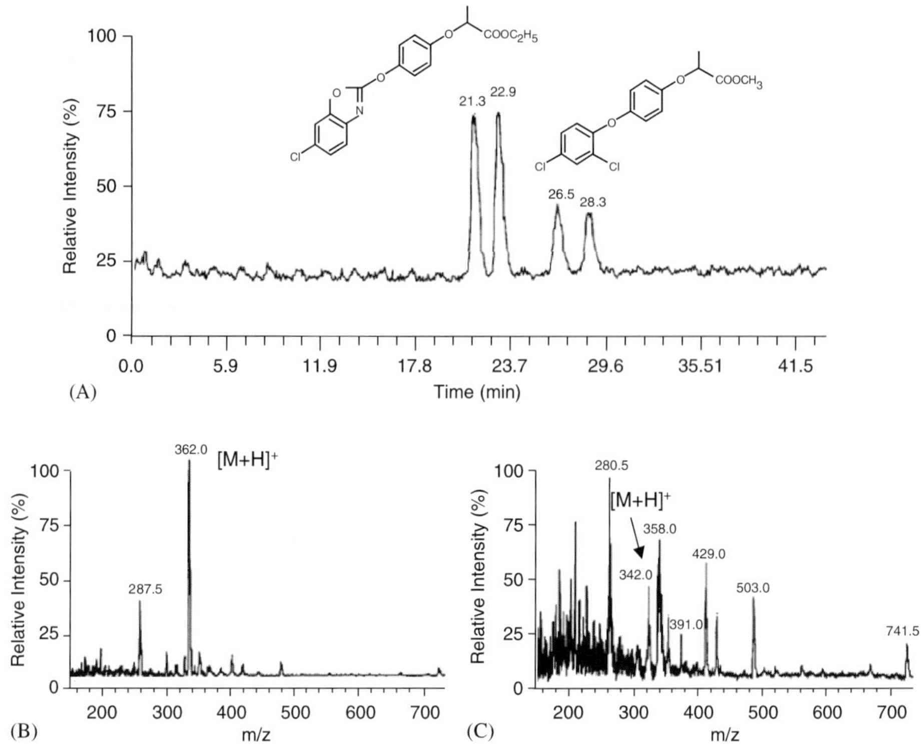

Fig. 9.12. Separation of the enantiomers of fenoxaprop ethyl (21.3 and 22.9 min) and diclofop methyl (26.5 and 28.3 min). Capillary packed with Chira-Dex-silica (30 cm, 100 μm i.d.). Running buffer: 0.5 mM ammonium acetate in water–methanol (40:60, v/v) at pH 6.6. (A) Total ion chromatogram in the positive-ion mode and scanning in the range *m/z* 170–750. Sheath liquid flow rate: 3 μl/min. Corresponding mass spectra of fenoxaprop ethyl (B) and diclofop methyl (C) of the peaks in (A). Reprinted from Ref. [108]. Copyright (2002), with permission from John Wiley & Sons, Inc.–Wiley-VCH Verlag GmbH.

9.2.3.2 Buffers and solvents

Among the different parameters that play a role in determining the compatibility between CE modes and MS, the choice of the separation buffer substances is of paramount importance. Equally important the composition of the additional liquid when a coaxial sheath flow interface is used. As already indicated, CE-UV typically employs aqueous buffers containing nonvolatile compounds, such as borate, phosphateor SDS, which are not compatible with ESI–MS. The use of these compounds can significantly increase the chemical noise during

ESI–MS detection, ruining the sensitivity and, under extreme conditions, clogging the system. Moreover, the use in CE–MS of alkali metal salts would simultaneously lead to the formation of both cationized and protonated analytes, splitting the analyte into several distinct ions and reducing the overall sensitivity. Besides the nature of the buffer, the buffer concentration and its influence on the sensitivity of ESI–MS has also been studied by several authors [6,60,110–112], who have concluded, as a general rule, that volatile acidic buffers of low ionic strength seem to give the best results, although logically, each case will require specific buffer optimization. In this sense, it has to be considered that there are a limited number of ions that can be formed at a given time, in this way, that any ionic compound from the buffer will compete with the analyte in ion formation [6].

As a result, the choice of the buffer (and sheath liquid) composition is a decisive step for obtaining both good CE separation and an adequate MS signal [60]. An example of the deleterious effect that different kind of buffers can have on the MS signal is shown in Fig. 9.13.

An overview of the many CE–MS applications reported up till now shows that it is possible to carry out the CE–MS analysis of practically any kind of compound using buffers (at pH ranging from 2 to 10) composed of different quantities of the following substances: water,

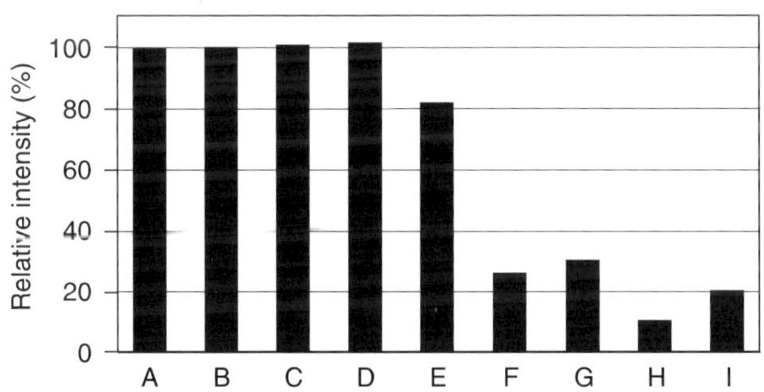

Fig. 9.13. Relative MS signal of 1 μg/ml of mebeverine (monitored at *m*/*z* 430) in various running buffers measured by infusion. Signal obtained in 100 mM formic acid at pH 2.6 was set at 100%. (A) 50 mM formic acid at pH 2.6, (B) 100 mM formic acid at pH 2.4, (C) 25 mM ammonium acetate at pH 6.6, (D) 50 mM ammonium acetate at pH 6.7, (E) 5 mM sodium phosphate at pH 2.5, (F) 25 mM sodium phosphate at pH 2.1, (G) 5 mM sodium phosphate at pH 7.5, (H) 25 mM sodium phosphate at pH 7.5, and (I) 10 mM sodium borate at pH 9.3. Reprinted from Ref. [60]. Copyright (2003), with permission from Elsevier.

ethanol, methanol, acetonitrile, isopropanol, acetic acid, formic acid, aminocaproic acid, ammonium acetate, ammonium bicarbonate, ammonium formate, ammonium hydroxide, triethanolamine and triethylamine. Besides, the use of the so-called nonaqueous media (i.e., buffers containing only organic solvents and a soluble salt) has increased the number of applications of CE–MS. Although the substances used for preparing these nonaqueous buffers are generally the same as those mentioned earlier, the use of other different solvents such as formamide and *N*-methylformamide has also been proposed for nonaqueous-CE–MS [113].

The compounds mentioned above are also used to prepare the sheath liquids when a coaxial-sheath flow interface is employed for coupling CE with MS. In this case, a 1:1 (v/v) hydroorganic solution is usually employed containing some small quantity (less than 1%) of acid or base added depending on whether the ionization mode is positive or negative.

Table 9.3 shows some typical buffers and sheath liquids (when applicable) used in different applications of CZE–MS.

9.2.3.3 Partial filling technique

Involatile compounds such as SDS or CDs can be used together with CE–ESI–MS via this procedure. As a result, PFT is the most widespread approach to couple MEKC or chiral-CE separations with ESI–MS. Originally developed for chiral CE with UV detection to decrease the background absorbance due to the presence of UV-absorbing chiral selectors (e.g., proteins, macrocyclic antibiotics) [128–131], PFT was first applied in the chiral CE–MS application by Jäverfalk et al. [132]. An additional advantage of the use of PFT is that the consumption of a chiral selector (or any compound that interacts with the analytes without interfering with the detection procedure) is extremely low because the inlet CE vial does not contain any chiral selector, thus opening the possibility of exploring exotic and/or expensive chiral selectors for CE–MS [133].

In this technique only a part of the capillary is filled with micelle-containing (or chiral selector-containing) buffer. The analytes separate as they pass the plug of micelles (or chiral selector) and then continue in free-zone electrophoresis until they reach the MS detector. Thereby the surfactants (or chiral selectors) are prevented from reaching the MS at the same time as the analytes. A scheme of this procedure is shown in Fig. 9.14.

TABLE 9.3

Running buffers and sheath liquids typically used in different applications of CZE–MS

Running buffers	Sheath liquids	Applied to	Ion mode	Reference
(i) 50 mM sodium acetate at pH 4.5 (ii) 25 mM ammonium acetate and 1 M acetic acid in acetonitrile–methanol (75:25 v/v) (iii) 75 mM sodium formate and 0.5 M formic acid in acetonitrile–methanol (50:50, v/v)	Methanol–ammonium formate in water (pH 2.5, 200 mM) (50:50, v/v)	Pharmaceuticals	Positive	[114]
(i) 20 mM ammonium acetate in water–acetic acid (99:1, v/v) (ii) 15 mM ammonium acetate in ethanol–acetic acid (98.5:1.5, v/v) (iii) 10 mM ammonium acetate in water–methanol–acetic acid (59.5:40:0.5, by vol.) (iv) 10 mM ammonium acetate in methanol–acetic acid (99.5:0.5, v/v)	Sheathless	Catcholamines	Positive	[115]
0.1 M formic acid	Methanol–water (80:20, v/v) containing 1% acetic acid	Histones	Positive	[116]
Formic acid–ammonium formate at pH 3.5 with 2% of methanol	Identical to separation buffer	Fungicide residues	Positive	[117]
2 mM acetic acid–ammonium acetate in acetonitrile–water (50:50, v/v) at pH 5.2	Sheathless	Fatty acids and prostaglandins	Negative	[34]
4 mM pyromellitic acid, 4 mM naphtalene disulfonate, 2 mM diethylene triamine in water–methanol (80:20, v/v).	Identical to separation buffer	Carboxilic acids: succinic, maleic, malonic, and glutaric acids	Negative	[118]
30 mM (+)-(18-crown-6)-2,3,11,12-tetracarboxylic acid	Sheathless	Enantiomers of amino acids	Positive	[119]
30 mM formic acid buffer at pH 3.20	2-Propanol-5 mM formic acid, pH 3.20 (adjusted with ammonium hydroxide, 4%, w/v)	Heparin depolymerized disaccharides	Negative	[120]
(i) 30 mM ammonium formate/formic acid at pH 2.79 (ii) 300 mM ammonium formate/formic acid at pH 2.23	Methanol–water (50:50, v/v) containing 0.05% formic acid	Peptides	Positive	[121]
0.9 M Ammonium formate/formic acid at pH 2	Methanol–water (50:50, v/v) containing 0.05% formic acid	Peptides	Positive	[122]
0.2 M Formic acid, 7 mM NH_3 in water–acetonitrile (90:10, v/v)	2-Propanol–water (50:50, v/v)	Peptides	Positive	[48]
75 mM ammonium acetate/acetic acid at pH 5.5	Methanol–water (50:50, v/v) containing 0.05% acetic acid	Basic proteins and food analysis	Positive	[123]
30 mM ammonium acetate–methanol 90:10, at pH 10 (adjusted with ammonium hydroxide)	2-Propanol–water (25:75, v/v))	Phenolic compounds	Negative	[124]
40 mM ammonium acetate, 15% acetonitrile (pH 9.5)	2-Propanol–water (80:20 v /v) containing 0.1%triethylamine	Flavonoids	Negative	[125]
40 mM ammonium acetate at pH 8.5	2-Propanol–water (50:50, v/v)	Alkaloids	Positive	[126]
1 M acetic acid, 20 mM ammonium acetate in methanol–acetonitrile (87.5:12.5, v/v)	Methanol–acetonitrile, (87.5:12.5, v/v)	Synthetic polymers	Positive	[127]
Acetonitrile containing 25 mM ammonium acetate and 1 M acetic acid	2-Propanol–water (80:20, v/v) containing 0.5%formic acid	Fluoxetine and related compounds	Positive	[38]

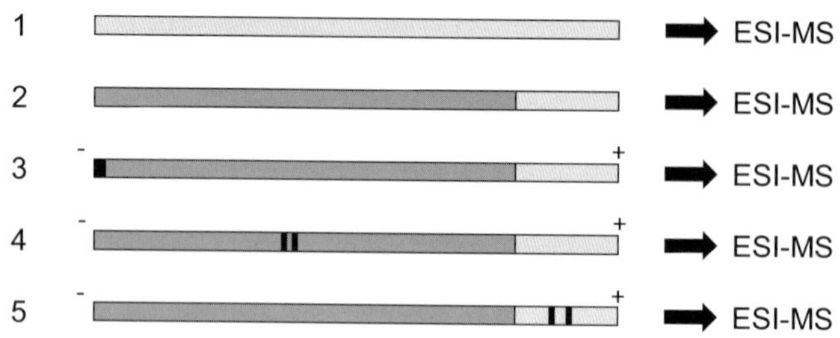

Fig. 9.14. Schematic chiral separation using the PFT. Steps: (1) Capillary filling with running buffer without chiral selector, (2) partial capillary filling with running buffer with chiral selector, (3) injection of analytes, (4) enantiomeric separation of analytes through the capillary zone containing chiral selector, and (5) migration of enantiomers to detector through the capillary zone without chiral selector.

The many examples of the use of PFT together with MS known so far demonstrate that this technique is useful for MEKC–MS and chiral CE–MS analysis of drugs, metabolites, natural products, agrochemicals, etc. Moreover, many combinations are available that even further broaden the application of this procedure. For instance, PFT has been applied in CE–MS together with nonaqueous media [134], neutral and charged CDs [135,136], crown ethers [137], ammonium, lithium and sodium salts of dodecyl sulfate [138], etc. As an example, Fig. 9.15 shows the chiral CE–MS separation of different amphetamines using the PFT together with a running buffer containing 40 mM ammonium formate and 24 mg/ml of 2-hydroxypropyl-ß-CD at pH 3. In this case, the use of a polyvinyl alcohol-coated capillary eliminated the EOF during the separation and allowed the filling of 90% of the capillary with the chiral buffer.

In spite of the large number of advantages and applications reported on the use of PFT in CE–MS, this procedure also has serious drawbacks that have to be considered [67]: (i) The selectivity is altered since MEKC (or chiral-CE) and CZE are combined. (ii) The efficiency is decreased since an extra band-broadening mechanism occurs at the micelle (or chiral selector) zone buffer boundary. (iii) The migration time window in MEKC is smaller since it is determined by the effective mobility of the surfactant monomers and not by the effective mobility of the micelles [139]. (iv) Total separation volume is smaller as a result of the relatively short surfactant (or chiral) zone [139]. (v) Optimization of PFT–CE–MS analysis is tedious since it incorporates many

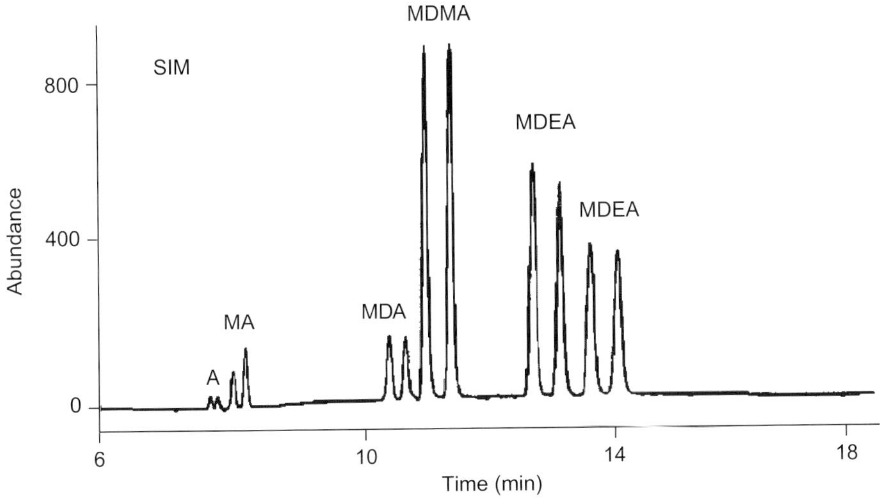

Fig. 9.15. CE–ESI-MS enantioseparation of different amphetamines. CE conditions: PVA-coated capillary (55 cm l_t, 50 μm i.d.), running buffer: 40 mM ammonium formate at pH 3 in the presence of 24 mg/ml HP-β-CD, partial filling of the capillary, 90%, pressure injection: 50 mbar for 6 s, voltage: 30 kV, temperature: 25°C. ESI–MS conditions: SIM positive-ion mode (6 ions), capillary voltage: 4.5 kV, fragmentor: 70 V, drying gas: N_2, temperature: 150°C, flow rate: 2 l/min, nebulizer pressure: 0 psi, sheath liquid composition: 0.5% formic acid in water–isopropanol (20:80, v/v) at a flow rate of 2 μl/min. Reprinted from Ref. [135]. Copyright (2001), with permission from John Wiley & Sons, Inc.–Wiley-VCH Verlag GmbH.

variables, including the length and concentration of the zone that contains the surfactant (or chiral selector) that affects the efficiency, resolution and sensitivity obtained [137,140]. In general, these drawbacks result in a different selectivity, a lower resolution and a lower peak capacity than in normal MEKC or normal chiral CE.

9.2.3.4 Improving sensitivity in CE–MS

The limited surface area and inner volume (~1–2 μl) of the capillaries typically used in CE (and CE–MS) lead to limited loading capacity, resulting in poor concentration limits of detection (usually in the micromolar range, although under determined conditions this limit can be considerably reduced). As indicated by Tomer [8] one straightforward way to increase sensitivity in CE–MS would be to increase the sensitivity of the MS instrument or to increase the efficiency of analyte transfer to the gas phase and ionization. However, it is not probable that in the near-term these instrumental improvements will lead (in an affordable way) to the increase of sensitivity needed for obtaining adequate detection limits using standard CE–MS instruments.

This low sensitivity has proven to be a major limitation of CE–MS in the analysis of many samples, and has certainly presented a significant barrier to the routine application of CE–MS. In order to overcome this important limitation, different approaches have been worked out for trying to increase the amount of sample that can be loaded onto the column without significant detriment to the separation efficiency [141,142]. These can be summarized in the use of (i) CIEF (described in Section 9.2.3.1), (ii) on-column transient isotachophoresis (tCITP), (iii) preconcentration devices (using membrane concentrators, packing the inlet of the CE capillary with C_{18} or using affinity packings) and (iv) other different arrangements that, in general, are combinations of the other three.

Transient-capillary isotachophoresis

The first applications of capillary isotachophoresis (CITP) together with MS detection were described in the late 1980s [143,144], demonstrating the possibilities of this approach. In one of these works [144], however, it could be seen that analytes that were at a very low level in the sample were concentrated during the CITP process into very narrow bands, giving rise to poor mass spectra. This problem was solved by using a dual-column CITP–CZE [144]. In 1993, tCITP was described as an alternative means of concentrating analytes prior to MS detection [109]. In tCITP, more than one-third of the capillary is filled with the sample solution and, as in ITP, analytes are first stacked in bands that have the same concentration as the leading electrolyte, and next separated in a normal free-zone electrophoretic process. In this way, the problem linked to the too narrow bands arriving the MS instrument was solved and therefore, increases in sensitivity up to two orders of magnitude could be obtained [145–149], allowing the injection of sample volumes as large as 0.9 μl [149].

Although recently the use of CITP together with MS has been shown to provide a 100-fold sensitivity improvement in the determination of aromatic sulfonates [150], the use of this preconcentration procedure seems to have decreased significantly in the last few years. One of the main reasons behind this is that in tCITP, the varying nature and volume of leading buffer and terminal (or separation) buffer can significantly alter analyte peak shape, separation efficiency and migration time; therefore, many experiments are usually needed to optimize the separation performance for each class of analytes.

Preconcentration devices

Basically, these devices are employed as a tool for both preconcentration and cleanup of samples. To do this, two basic designs based on the use of chromatographic packing material or membrane preconcentration devices have been developed. Usually, the CE capillary is connected to a transfer capillary by means of a short piece of PTFE sleeve (ca. 1–10 mm) containing either the chromatographic stationary phase (typically C_{18}) or the membrane (typically made of styrene divinylbenzene). Sample is loaded through the transfer capillary by pressure and desalted by washing with the separation buffer. The analytes are then preferentially desorbed with a small plug of an elution buffer usually containing some organic solvent.

Some examples of the multiple applications of this type of device are given in Table 9.4. It has to be noted that many of these applications have been made by the same group, which probably explains why it is difficult for other laboratories to carry out this procedure.

TABLE 9.4

Examples of preconcentration devices used in CE/MS and CE/MS/MS [142]

	Analytes	Reference
Small organic molecules	Haloperidol standards and urine metabolites	[151–154]
	3-Phenylamino-1,2-propanediol/3-phenylamino-L-alanine standards and rat hepatocyte incubation	[155,156]
Peptides	Peptide standards	[151–153,157,158]
	Trypsin digest of α-casein	[159]
	Cell media	[160]
	MHC class I compounds	[44,151,154,158,161–164]
Proteins	Protein standards	[165]
	Aqueous humor	[154,165,166]
	Urine proteins	[154,160]
	Blood dialysate proteins	[151]
	Brain dialysate proteins	[151]
	Cerebrospinal fluid proteins	[151]
	Proteins in tears	[167]
	Renal dialysate proteins	[168]

The selection of the packing material and membranes, as well as the efficacy and stability of the chemistry employed in coupling the affinity ligand to the solid support, is of the utmost importance to enhance efficiency of the system. Thus, the solid phase packing technique significantly impairs the CE separation performance owing to some problems of these preconcentrators such as increased back-pressure, decreased hydrodynamic flow within the separation capillary, relatively large volume of organic solvent needed to elute the analyte of interest, erratic analyte migration times, bubble formation, etc. Although many of these drawbacks can be reduced by the careful selection and design of the preconcentrator, one of their main difficulties, the short life span of an analyte concentrator (i.e., usually between 10 and 35 uses, under the best optimal conditions of fabrication) remains unsolved. Despite the determined efforts by various laboratories to fabricate a reliable analyte concentrator–microreactor device, there is not yet a model system that can be considered ideal. Accordingly, it may take some time before the device is commercially available (earlier attempts at commercialization failed) [141] and, therefore, other new approaches are being investigated.

Other procedures

Several other strategies have been described in the literature to overcome the important drawback that the low sensitivity of CE–MS can pose to this technique. Only in a few cases are these strategies based on entirely new developments, being in general combinations of the different procedures mentioned above or based on the use of stacking strategies [169].

Thus, the off-line combination of SPME and CE–MS has proved to be a powerful and sensitive analytical tool able to detect pesticides in fruit samples at levels ranging from 0.02 to 5 mg/kg [170].

Derforce and co-workers [171] developed a CE–MS procedure in which analytes were dissolved in an electrolyte, the concentration of which was much lower than that of the running buffer. To compensate for the reversal EOF caused by the negative electrical potential applied during injection, a physical pressure in the opposite direction was applied. In this way, when the pressure was too low, the analytes are unable to enter the capillary owing to the EOF. If, on the other hand, the pressure is too high, sample solvent enters into the capillary and the current gradually decreases. Under optimized pressure, the sample is concentrated in the capillary by up to a factor of 100 compared with the normal injection.

An interesting CIEF/ESI–MS setup using a dialysis fibre to connect the IEF capillary with a transfer capillary was described by Van der Greef's group [81]. In this device, only the molecules of the pH-buffer can penetrate the dialysys membrane (having a cut-off of 5 kDa) on account of their size and are removed from the system while the larger protein molecules remain in the capillary, what brings about their cleanup and concentration. The IEF capillary is connected to a transfer capillary, which transports the analyte to the ESI tip.

Recently, an integrated proteome concentration/separation approach for protein and peptide mixtures has been reported [172]. The approach involves on-line combination of CIEF with capillary reversed-phase liquid chromatography (CRPLC) and MS. Upon completion of analyte focusing, the self-sharpening effect greatly restricts analyte diffusion and contributes to analyte stacking in narrowly focused bands with a concentration factor of ca. 240. In addition to analyte focusing, CIEF as the first separation dimension resolves proteins and peptides on the basis of their differences in pI and offers greater resolving power than that achieved in strong CEC. The grouping of two highly resolving and completely orthogonal separation techniques of CIEF and CRPLC, together with analyte focusing and concentration, significantly enhances the dynamic range and sensitivity of conventional MS toward the identification of low-abundance proteins. The CIEF-based multidimensional separation/concentration platform enables the identification of a greater number of yeast soluble proteins than methods presented in the literature needing protein loadings 2–3 orders of magnitude lower [172].

In other work, a multidimensional electrokinetic-based separation/concentration platform coupled with electrospray ionization-Fourier transform ion cyclotron resonance MS (ESI–FTICR–MS) was developed for achieving high-resolution and ultrasensitive analysis of complex protein/peptide mixtures [173]. In this case, a microdialysis junction was employed as the interface for on-line combination of CIEF with tCITP/CZE in an integrated platform. A scheme of this system can be seen in Fig. 9.16. Besides the excellent resolving power afforded by both CIEF and CZE separations, the electrokinetic focusing/stacking effects of CIEF and tCITP greatly enhanced the dynamic range and detection sensitivity of MS for protein identification. The constructed multidimensional separation/concentration platform was demonstrated

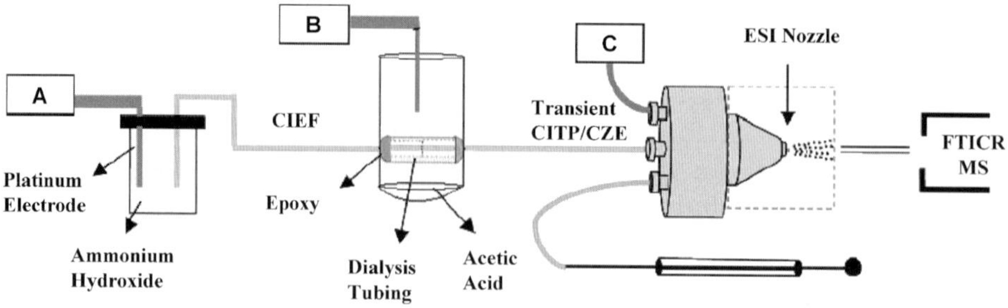

Fig. 9.16. Schematic on-line coupling of CIEF-transient CITP-CZE with ESI-FTICR-MS using a microdialysis junction. Points A, B and C are either connected to ground or a high-voltage power supply (depending on the separation step). Reprinted with permission from Ref. [173]. Copyright (2003), American Chemical Society.

for the analysis of *Shewanella oneidensis* proteome (an organism with implications on bioremediation of environmental pollutants). The electrokinetic-based platform offered an overall peak capacity comparable to those obtained using multidimensional chromatography systems, but with a much shorter run time and no need for column regeneration. Most importantly, a total of 1174 unique proteins, corresponding to 26.5% proteome coverage, were identified from the cytosolic fraction of *S. oneidensis*, while requiring < 500 ng of proteolytic digest loaded in the CIEF capillary. The ultrasensitive capabilities of electrokinetic-based proteome approach were attributed, among other factors, to the concentration effect in CIEF, the electrokinetic stacking of tCITP, the nanoscale peak volume in CZE and the high-sensitivity, high-resolution and high-mass measurement accuracy of FTICR–MS [173]. As an example of the impressive information provided by this system, Fig. 9.17 shows a two-dimensional display of BSA tryptic peptides in one of the CIEF fractions analyzed by transient CITP/CZE/ESI–FTICR–MS. As can be seen, by using this procedure, information about pI and molecular weight of the different fractions analyzed can be obtained.

Cooper and co-workers have developed a miniaturized trypsin membrane reactor housed inside a commonly used capillary fitting enabling rapid and sensitive protein identification by on-line proteolytic digestion and analysis of protein digests using nano-ESI–MS and MALDI-MS. According to the authors, the design and assembly of the capillary fitting-based trypsin membrane reactor were straightforward and highly robust, without the need for expensive fabrication technology

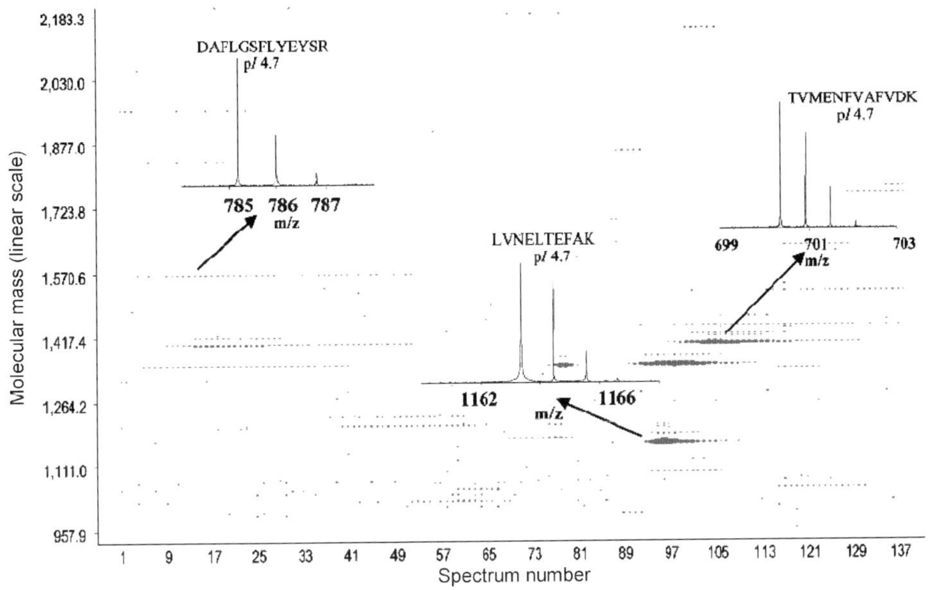

Fig. 9.17. 2D BSA tryptic digest separation in one of the CIEF fractions analyzed by transient CITP-CZE-ESI-FTICR-MS. The spots are identified as three BSA tryptic peptides with similar pI values, and their spectra are shown in the insets. Reprinted with permission from Ref. [173]. Copyright (2003), American Chemical Society.

and procedures. The resultant protein digests could also be further concentrated and resolved using CRPLC or transient CITP/CZE prior to the mass spectrometric analysis in an integrated platform. This nanoscale reaction system enables rapid proteolytic digestion in seconds for a protein concentration of less than 10^{-8} M, consumes very little sample (about 5 fmol) and offers capillary interfaces with various separation and MS techniques [174].

9.2.4 Chip-based CE–MS

The need for high-throughput, mainly for the analysis of proteins, peptides, DNA and chemical libraries in drug development, using extremely small sample amounts has stimulated new developments in microfabricated devices (also called microdevices, chips, microchips, lab-on-a-chip or micro total analysis systems (μTAS)) [175–180]. Microchips are microdevices fabricated on glass, quartz or polymers with standard micromachining techniques, such as chemical etching or photolithography. These devices with a typical length of about 5 cm contain a centrally placed separation channel of 4–11 cm in length and "diameter" of 50–100 μm where the separation takes place. The main advantages of the use of these microdevices can be summarized as: (i) reduced sample requirements, (ii) faster analysis times, (iii) increased automation and (iv) less exposure to lab contaminants. They have multiple applications, such as, sample preparation, solid-phase extraction, tryptic digestion or preconcentration and separation [177–179].

Recently, their use in mass spectrometric applications has shown great possibilities as a result of their flow-rate compatibilities and high throughput [180–182]. Moreover, MS overcomes some of the main limitations linked to the use of laser induced fluorescence detection (LIF) (i.e., MS can be applied to nonfluorescent analytes without derivatization), while it also provides structural information on the analytes. An example of this configuration can be seen in Fig. 9.18, which shows a photograph of a typical chip-MS configuration [175]. There are different procedures to couple chip-ESI–MS providing mass spectral data from continuously flowing samples from these microdevices using liquid junction strategies or spray nozzles with conductive coatings, and they have been summarized in different reviews [11,175]. Briefly, from an engineering point of view, the techniques used to couple microchip to MS using ESI include: flat edge surface [180,183], attachments of ESI emitter tips directly to the exit of the microchips by a double-etching procedure [183,184], hand drilling [185,186] or microfabricated ESI tips [187–189]. Some of these ESI–MS interface techniques have been applied to chip-CE–ESI–MS analysis [183–185,190–192] including quantitative approaches [193], as well as for chip-CIEF–ESI–MS [194,195]. Fig. 9.19 shows one of the first separations that was carried out using chip-CE–ESI–MS [185]. In this case, the analysis of a tryptic digest, using only 150 fmol of protein, was sufficient to allow the protein identification comparing the tryptic fragments to a protein database. The analysis allowed the peptide sequence to be identified with a 45 amol/μl concentration detection limit using chip-MS/MS. In this type of configuration, however, special care has to be taken over their engineering since frequently the interfaces used to couple the microdevice with the MS instrument can impair the CE separation. Apart from the fact that these approaches suffer from difficulties in fabrication, they still require improvements in the limits of detection (LOD) and separation power. Besides, this technique is still far from away to be

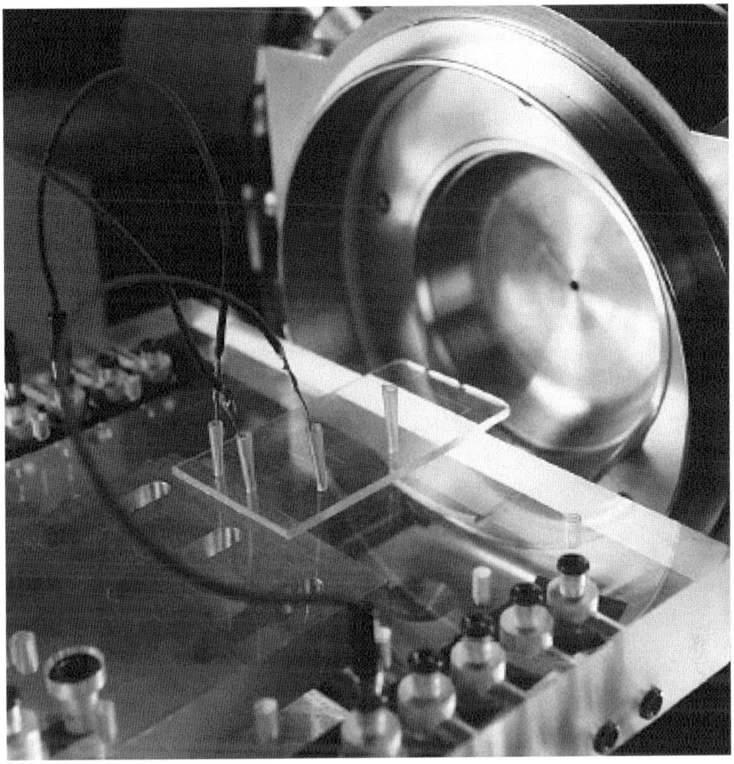

Fig. 9.18. Photograph of a chip-MS configuration, showing the chip support/manipulator, electrical connections, chip with attached capillary, and the sampling orifice of the MS. Reprinted from Ref. [175]. Copyright (2000), with permission from Elsevier.

routinely applied and therefore, most papers deal with the possibility of chip-MS coupling itself, rather than the application.

To increase the chip/ESI–MS throughput, several strategies have been introduced in order to develop the so-called multiplexing. In this case the final goal is to handle multiple samples either simultaneously or sequentially. In an initial design a single chip was used with six different channels and six orifices, the wafer was translated and each channel sequentially sprayed [180]. However, the position of each orifice within the spray interface was shown to be critical and also difficult to optimize it. Since then, different procedures have been developed to overcome these difficulties [181,196–199], such as, for instance, the combined use of a microfabricated CE–ESI–MS with automated sampling from a microwell plate [198], or the use of devices with

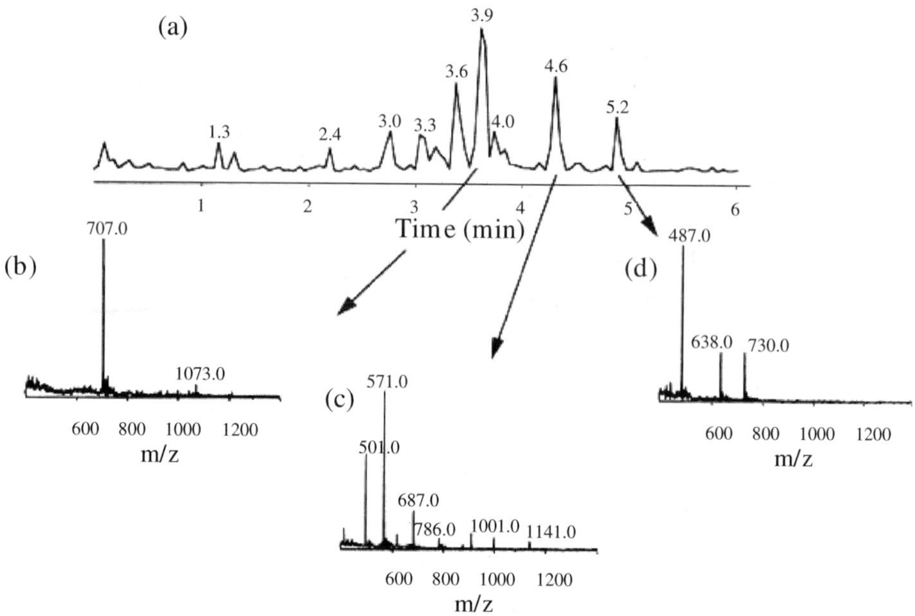

Fig. 9.19. On-chip CE–ESI–MS separation of a tryptic digest of a lectin from *Pisum sativum*. Reprinted with permission from Ref. [185]. Copyright (1999), American Chemical Society.

multiple channels finished in triangular tips for ESI–MS coupling, all made of the same material and integrated in a single microchip [199]. In the latter example, a thin, triangular-shaped thin polymer tip was formed by lithography and etching. A microfluidic channel, 20 μm wide and 10 μm deep, was embossed in a cycloolefin substrate using a silicon master. The triangular tip was aligned with the channel and bonded between the channel plate and a flat plate to create a microfluidic channel with a wicking tip protruding from the end. This structure aided the formation of a stable Taylor cone at the apex of the tip, forming an ESI electrospray ionization source. Because the components are all made by lithographic approaches with high geometrical fidelity, an integrated array system with multiple channels can be formed with the same method and ease as a single channel. This multichannel system was tested in a multiplexed manner and showed reliable operation with no significant cross-contamination between closely spaced channels.

Other investigations in this area focuses on the inclusion within the chip of most common sample preparation protocols together with the development of microdevices made of more affordable, compatible and resistant materials. Integration of all aspects of protein sample handling within a microfuidic device is clearly a goal of many research groups. Thus, as indicated by Guzman and Stubbs [141], microchips represent an evolutionary step toward voltage-driven, robust and versatile miniaturized-separations. With the use of affinity adsorbents on their channels, microchips may revolutionize life science research applications, because they can enable carrying out applications in a massive parallel, multichannel, integrated format. Moreover, the use of selective affinity adsorbents in their microchannels appears particularly promising for enhancing detection capabilities in microchips.

9.3 APPLICATIONS

In this section we do not intend to present an exhaustive review of the different CE–MS applications published up to now (for doing that, readers can check the reviews mentioned in this chapter [5,7–11]), but present, instead an overview of the different possibilities that this analytical technique has demonstrated so far.

9.3.1 Proteins and peptides

Peptides and proteins are biomolecules characterized by the large and complex role they play in living organisms. Study of protein profiling will provide a better understanding of complex cellular regulation, and therefore, provide knowledge about disease mechanisms. Therefore, analytical methods for proteomics studies are in great demand, and in this sense, CE–MS has proved to be a powerful alternative. Since CE–MS development, peptides and proteins are probably the most common species analyzed by this technique. It has been used for determining intact protein mixtures, protein digests, and other natural or synthetic peptides [10]. CE–MS allows a high accurate determination of relative molecular masses of peptides and proteins, and it also, provides useful structural information (e.g., amino acid sequence, sites of post-translational modifications, peptide mapping, noncovalent interactions of peptides and proteins, etc.). The information provided by MS techniques and the development of computer algorithms have permitted the rapid increase of protein and peptide analysis. It is important to consider that *m/z* data from the multiple-charged ions created can be mathematically deconvoluted to calculate the original molecular mass of large ions such as proteins and peptides. In this way, chromatography and electrokinetic separation methods with MS or MS^n are in continuous development.

One of the first problems that has to be faced during protein and peptide CE–MS separations is the possible interaction between positively charged groups and the negatively charged capillary wall. It is normally overcome by using very acidic running buffers or using different coating procedures [200–203].

On-line CIEF–ESI–MS has been used to analyze standard and biological proteins showing, since the pioneering work by Lee and coworkers [77], remarkable possibilities in terms of resolving power, speed and, under given analytical conditions, sensitivity. Thus, hemoglobin variants with an LOD in the range of 10 nM [78], complex glycoforms in bovin serum apotransferrin [204], proteins from recombinant *E. coli* cell lysates [80,205], ovalbumin phosphorylation variants [206] and proteins from cerebrospinal fluid and blood [207] were analyzed by CIEF–MS with a coaxial-sheath flow ESI interface. Advantages of on-line combination of CIEF with coaxial-sheath liquid ESI interface and FTICR–MS, in high resolution, sensitivity and higher mass measurement accuracy than conventional mass spectrometers, were demonstrated on model proteins, human hemoglobin variants (as can be seen in Fig. 9.20) and cell lysates from multiple organisms [208,209]. Characterization of human alcohol dehydrogenase isoenzymes at the attomole level, for a better understanding of human ethanol metabolism, was also conducted by CIEF–ESI–FTICR [210]. Recently, CIEF–ESI–FTICR analysis of intact noncovalent protein complexes was accomplished by Martinović et al. [211]. They found that variations in sheath liquid composition used for ESI ionization affected the stability of protein complexes, making possible the determination of noncovalent complexes or their dissociated subunits with the identical CIEF and different ESI conditions.

CZE–ESI–MS has frequently been used to analyze intact proteins. Thus, CZE–ESI–MS was used to study myoglobin complexes with the noncovalent bound heme group, revealing that pH and organic content on the sheath liquid affected myoglobin complexes stability on the ESI chamber [54]. Other applications of CZE–ESI–MS with a sheath liquid interface, can be found in the literature [10], such as the study of metallothionenin isoforms from different species [212], and the analysis of histones [116] and high mannose-containing glycoproteins

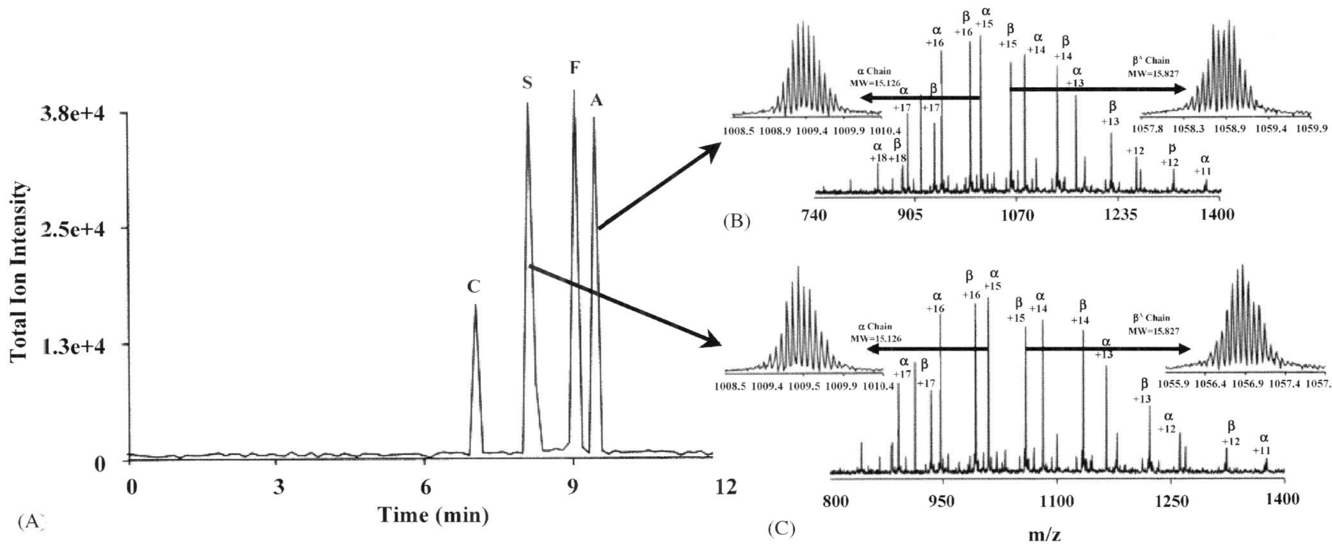

Fig. 9.20. (A) CIEF-ESI-FTICR total ion electropherogram of a mixture of hemoglobin variants C, S, F, and A. (B) Positive ESI mass spectra of hemoglobin A and (C) hemoglobin S. Reprinted with permission from Ref. [208]. Copyright (1998), American Chemical Society.

[213]. Protein analysis of human blood samples, together with model proteins and protein digests, was also performed by CZE–ESI–MS with an in-capillary electrode sheathless interface, obtaining an LOD between 0.6 and 4 fmol for standard proteins [214].

Peptides are involved in numerous biological processes, as neurotransmitters, hormones, immunomodulators, regulators of enzyme activity, antibiotics, toxins, etc. A variety of CE–MS couplings have been used for peptide analysis. Owing to the need for sensitive analysis, a sheathless interface has been used in CE–MS coupling for peptide determination. Different sheathless designs have been used to overcome the reduction in sensitivity due to higher background signal and dilution of the sample produced by coaxial-sheath flow ESI interface. Pressure-assisted CE with a metallized electrospray emitter coupled to a Hadamard transform TOF–MS was used for nine-peptide mixture separation (e.g., bombesin, bradykinin fragment, leucine enkephalin, oxytocin, methionine enkephalin, substance P, luteinizing hormone-releasing hormone, Arg8-vasopressin, bradykinin) with detection limits between 1 and 26 μM (0.1–4.2 pmol) [58]. Standard peptide mixtures (e.g., luteinizing hormone-releasing hormone, leucine enkephalin, renin substrate, angiotensin, β-lipotropin, neurotensin, kassinin, gramicidin S, somatostatin, bradykinin, substance P) and protein digests were analyzed using a capillary gold or silver coating for sheathless CE–MS interfacing [215–220]. Detection limits in the low fmol range were obtained [215,216,218], showing a 10–20-fold signal improvement regarding coaxial-sheath flow interface [215], and identification of glycopeptides was accomplished using low pmol injection of original protein digest [219]. CE–ESI with IT/reTOF (ion trap-reflectron time-of-flight MS) with silver-coated capillary as sheathless interface was used for peptide analysis from hemoglobin (Hb) tryptic digest to study its structural variants [221]. CE–MS with in-capillary platinum electrode sheathless interface was used for tryptic digest of standard peptides and model protein determination [214,222,223]. Fang et al. [30] demonstrated the capabilities of CE–TOF coupling with a sheathless ESI interface using a thin gold wire to maintain CE potential, to analyze mixtures of peptides with a concentration detection limit of 1–2 μM (40–80 fmol), and six peptidic mixture of the sea mollusk *A. californica* at a concentration of 10 μM of each component. Recently, a new and sheathless nanoflow ESI design for CE–IT–MS coupling has been used to obtain highly sensitive analyses of standard bioactive peptides and apomyoglobin digest in the low fmol range [224].

In addition to electrospray interface mode, sensitivity also depends on the preconcentration mode. Several on-line preconcentration systems have been used together with a sheathless interface for CE–MS peptide determination, either with chromatographic packed material or membrane preconcentration devices [33,154,162,225–228], providing in some cases low nanomolar concentration LODs in both model protein digests and biological samples [225,226].

A mixture of angiotensins and a tryptic digest of cytochrome *c* was analyzed by CE–MS with an LOD in the submicromolar range (attomole range), using a liquid junction interface with a replaceable micro-ESI tip enclosed in a subatmosferic chamber [45]. A similar LOD was obtained with a liquid junction for complex peptide mixtures analysis by on-line coupling of a solid-phase extraction device to a CE–MS [229]. In a recent work, a new porous junction has been used for electrical connection in a CE–MS coupling for standard peptide and proteins analysis [47].

Nowadays, it has been demonstrated that coaxial-sheath flow electrospray is a sensitive and robust coupling CE–MS for proteomics. Some examples are given below. Neusüss et al. [48] demonstrated the usefulness of a commercial CE–IT–MS using a coaxial sheath flow for analysis of a standard peptide mixture (e.g., angiotensins I and II, bradykinin, leucine enkephalin) with an LOD in the low nM range (low fmol range), and those from transferring digest, with a simple pH-mediated staking for sample concentration. It has also been demonstrated to be useful for studying the decomposition and enantiomerization of peptides [230] or peptides acting as hormones [231]. Banks et al. [232] demonstrated the benefits of the fast scanning capabilities of TOF mass detector on-line with CE in rapid peptide and protein separations. They reached the separation of five standard peptides in 100–200 s, generating narrow peaks (1–2 s) with an LOD of 8 fmol. CE–ESI–TOF–MS with a coaxial sheath flow was also used for determination of patterns of peptides and proteins present in human urine and hemodialysis fluids for the diagnosis of kidney disease [233,234].

In any characterization of protein digests by CE–MS, there still are two important limitations: (i) the buffer choice (i.e., the buffer must provide an adequate CE separation without ruining the MS signal), and (ii) the frequent generation of "unexpected" peptidic fragments during the enzymatic protein hydrolysis. Recently we have published a new approach [122] to help to solve these difficulties, namely, a theoretical

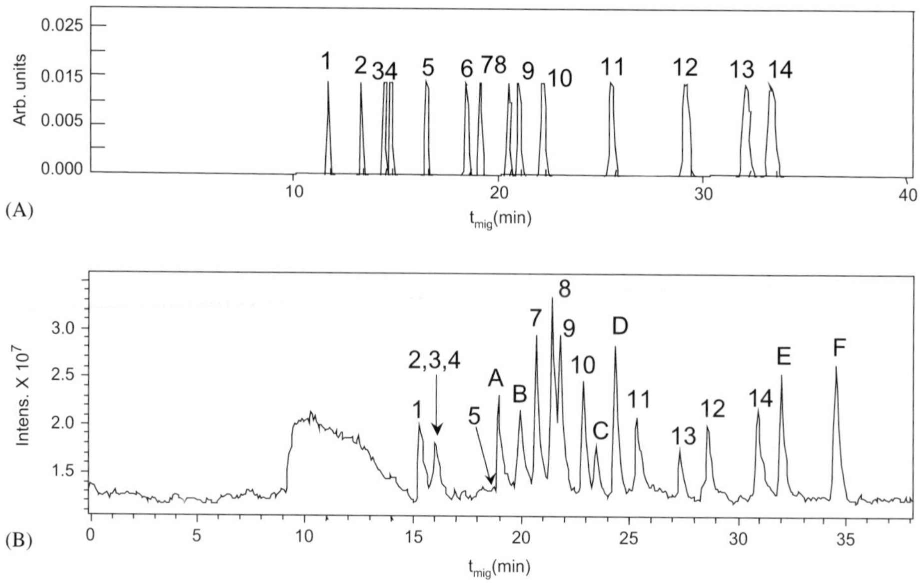

Fig. 9.21. (A) Predicted CE–ESI–MS electropherogram of cytochrome *c* trypsin digest. (B) Experimental CE–ESI–MS total ion electropherogram of cytochrome *c* trypsin digest. CE conditions: fused-silica capillary (87 cm l_t, 50 μm i.d.), running buffer: 0.9 M ammonium formate/formic acid at pH 2, voltage: 20 kV, temperature: 25 °C, hydrodynamic injection: 0.5 psi for 45 s. ESI–MS conditions: positive-ion mode, drying gas: N_2 at a flow of 8 l/min and a temperature of 120°C, nebulizer pressure: 4 psi, sheath liquid: methanol–water (50:50, v/v) at a flow rate of 4 μl/min. Reprinted from Ref. [122]. Copyright (2003), Reprinted with permission from John Wiley & Sons, Inc.–Wiley-VCH Verlag GmbH.

model that relates the electrophoretic behaviour of peptides to their sequence. The effectiveness of this procedure was demonstrated by the fast attainment of good CE–ESI–MS conditions for analyzing the peptides obtained from an enzymatic protein hydrolysate in a single run. This strategy was demonstrated to provide useful information for helping to characterize "unexpected" fragments from protein digests. An example of the usefulness of this procedure is given in Fig. 9.21, in which the theoretical and experimental CE–MS electropherograms are given for a group of peptides obtained from a digested protein. The peaks indicated with numbers correspond to the expected fragments from the digestion of cytochrome *c* with trypsin while the peaks marked with letters indicate the appearance of "unexpected" fragments. As can be seen, using the simulation program, it was possible to find a buffer capable of providing a good separation of this complex mixture. Moreover, by comparing both electropherograms, it can be seen that the simulation program also gave a good estimation of the expected order of migration of the different peptides, which made the assignment of the different compounds.

The tCITP focusing step has been used for the preconcentration of peptide samples analyzed by CE–ESI–MS. Using this concentration step, different neurotransmitters and neuropeptides in brain tissue have been analyzed by CE–ESI–MS, and detection limits between 5 and 20 nM (1–3 fmol) have been obtained [148]. ITP–CE–MS was used for analysis of a peptidomimetric drug and endogenous peptides (substance P and calcitonin gene-related peptide) in brain stem extract, obtaining detection limits of 6 nM (3 fmol) [149]. Bateman et al. [235] evaluated different adsorption preconcentration systems coupled to CE–ESI–MS for the analysis of tryptic peptides and glycopeptides at concentrations of 50 nM.

CE–ESI–MS with a coaxial-sheath flow interface, has also been used to study the stability of goserelin and luteinizing hormone-releasing hormone analogs [236], characterization of glycopeptides and glycoproteins [237], and study of recombinant (C125S mutated interleukin) and naturally modified (bovine β-casein) proteins through their proteolytic fragments separation and tandem MS analysis [238].

The PFT was used to separate closely related peptides (neurotensin and angiotensin II derivatives) using MEKC with a nonionic surfactant, coupled to MS with a coaxial-sheath flow interface [239].

With regard to chip-based CE–MS, several works can be found in the literature, in which proteins and peptides have been used as model compounds for microfabricated devices coupling with MS detection [183–185,190,191,194,195,198]. In Fig. 9.22, a chip-CE–ESI–MS separation of peptides from 100 ppm of cytochrome *c* trypsin digest is shown to take place in about 40 s.

9.3.2 Drugs and metabolites

CE has become an alternative to LC for the monitoring of drugs. With the introduction of electrospray MS, the coupling of CE and MS has opened the way to widely applied drug monitoring [240]. Moreover, CE–MS enables quantitative analysis of various drugs and their metabolites down to the low μg–ng/ml level. Although the coupling of CE to MS is today far from a routine application in drug monitoring, CE–MS

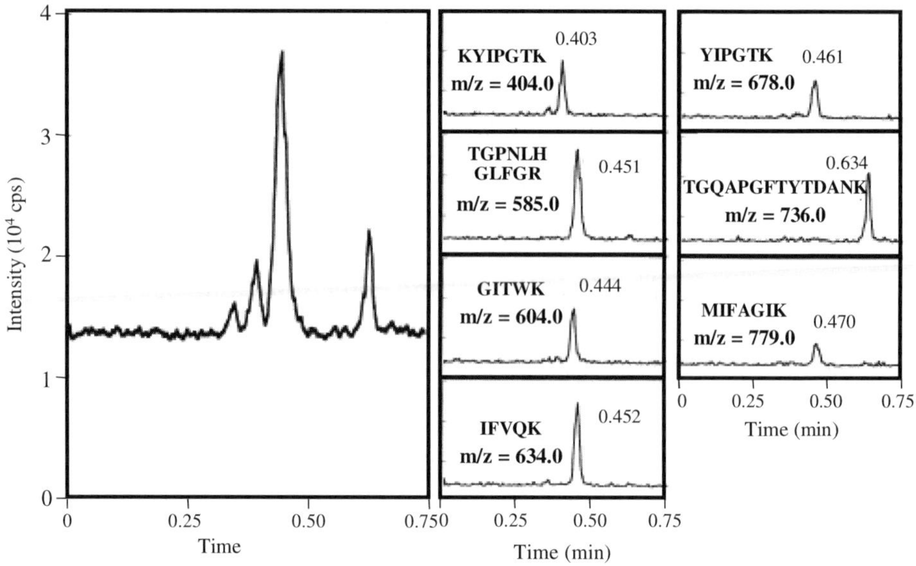

Fig. 9.22. Analysis of cytochrome *c* trypsin digest on a chip-CE–MS system. Running buffer: 50 mM acetic acid-ammonium acetate buffer (pH 5.7) containing 30% (v/v) acetonitrile, injection time: 0.5 s, electric field strength: 400 V/cm. (A) total ion electropherogram, and (B) selected ion electropherograms. Reprinted from Ref. [191]. Copyright (2003), with permission from Elsevier.

is finding an increased application in clinical research. For instance, many disorders cause the accumulation of specific metabolites in biological fluids, and their detection can help to monitor disease progression, effects of drugs, etc. Besides, structural information from mass spectrum and fragment patterns is a very useful tool for metabolites identification. Some interesting applications reported in pharmacological and clinical research using CE–ESI–MS are described below.

CZE–ESI–MS has been used for the analysis of opioids and their glucuronic acids in urine for unambiguous drug abuse determination [169], and for separation of methylenedioxyamphetamine-related designer drugs [241]. Carnitine and acylcarnitines were analyzed in biological fluids as markers of metabolic disorders, obtaining limits of quantitation around 0.1–1 μM [242]. A new CE–MS method for morphine and amphetamine determination with LODs of 30 and 50 ng/ml, respectively, was also developed, and it was further used for detection of morphine in human urine [243]. Quaternary ammonium drugs, abusively used in equine sports, were determined at 1 ng/ml in horse urine [244]. Other pharmaceuticals, such as nonopioid analgesics and some of their metabolites [245,246], and ebrotidine and its metabolites [247], were also analyzed by CE–MS in urine samples. Recently, a method for analysis of pharmaceutical formulation containing lidocaine has been validated by CZE–ESI–MS [248].

MEKC–MS is an important analytical tool that has also been used for drug and metabolite analysis. Direct on-line coupling of MEKC to ESI–MS using the surfactant SDS has been used to analyze several pharmaceuticals (e.g., acetaminophen, caffeine, guaifenesin, phenacetin) [69], mebeverine and related compounds with LOD about 1 μg/ml [60], and determination of synthetic chemical drugs (as presented in Fig. 9.23) used as adulterants in Chinese medicines [70]. A high background signal from SDS [69], and suppression of MS signal as function of SDS concentration were observed [60,70]. Furthermore, MEKC has been used in combination with PFT for standard pharmaceutical separation to prevent the entrance of the surfactants into the MS system [63,139]. Caffeine and ethenzamide separation by PFT–MEKC–MS was demonstrated, and a study of the influence of several parameters on PFT–MEKC system was also made [139]. MEKC–ESI–MS separation of mepenzolate and pipenzolate was made using a system with a coupled capillary and the possibility of voltage switching and buffer renewal, developed to avoid entrance of micelles into the ESI interface [65]. MS-compatible surfactants such as perfluorooctanilic acid (PFOA) for the direct MEKC–ESI–MS coupling were used for the separation of terbutaline and related compounds [67]. The use of a low concentration of high-molecular-mass surfactant butyl acrylate-butyl methacrylate acid copolymer sodium salt (BBMA) was used for detection of caffeine, noscapine, azelastine, dibucaine, eperisone, isopropylantipyrine, propanolol and temepidium bromide. However, owing to signal decrease with increase BBMA concentration, PFT was used for the analysis of caffeine and its metabolites, which needed a higher BBMA concentration for their separation [63].

Nonaqueous CE (NACE) has also been coupled to MS in pharmaceutical and clinical separation areas. Several pharmaceuticals and metabolites were studied by Veuthey's group [38,113,249,250]: fluoxetine and its related compounds [38], amphetamines and related compounds analyzed in spiked urine samples [249], venlafaxine and its phase I metabolites [250] and β-blockers [113]. Tamoxifen and its metabolites analyzed by NACE–ESI–MS, with a low concentration of SDS dissolved in the running buffer have also been reported [68].

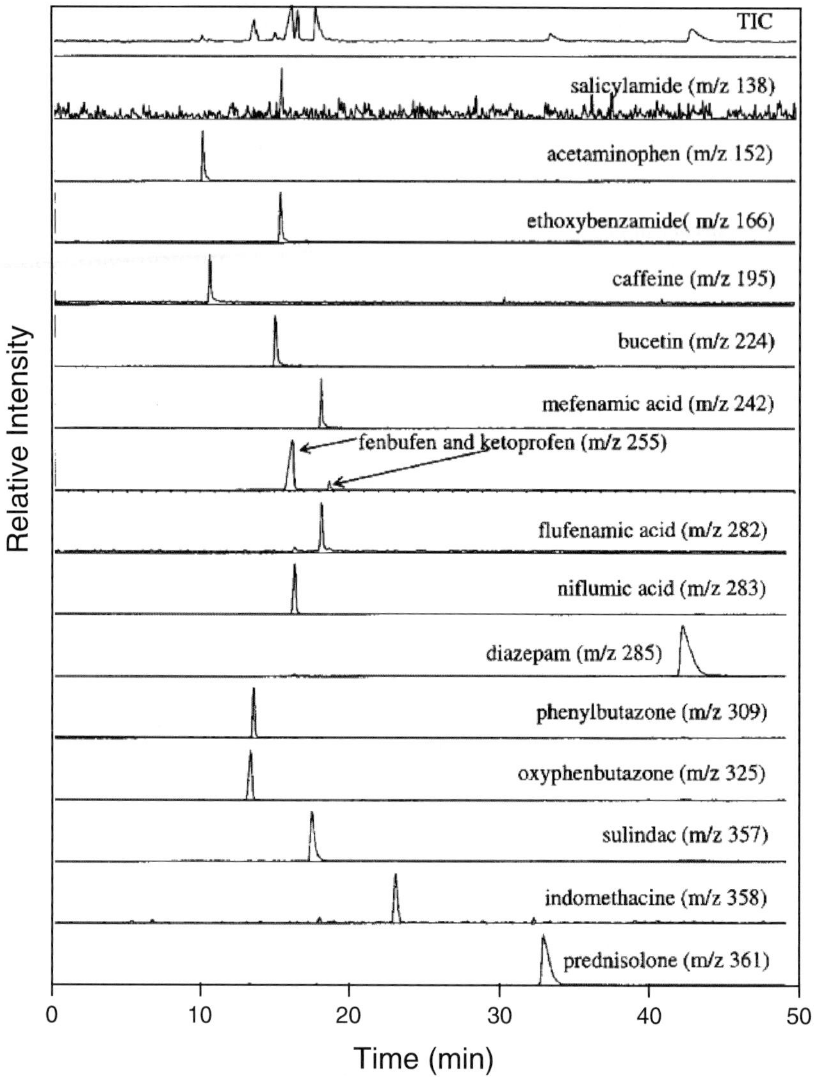

Fig. 9.23. MEKC/ESIc–MS electropherograms of 16 chemical drugs. CE conditions: fused-silica capillary (90 cm total length (l_t), 50 µm i.d.), running buffer: 40 mM ammonium acetate, 20 mM SDS, at pH 9, voltage: 24 kV. ESI–MS conditions: positive-ion mode, capillary voltage: 4 kV, temperature: 200°C, sheath liquid composition: water–methanol–formic acid (30:70:0.2, by vol.) at a flow rate of 3 µl/min. From Ref. [70]. Copyright (2001), with permission from John Wiley & Sons Limited.

Mixtures of benzodiazepines, corticosteroids and thiazide diuretic drugs [102], flunitrazepam and its metabolites [88] were analyzed by CEC–ESI–MS. An example of the application of CEC–MS coupling to drug analysis is shown in Fig. 9.24.

Coupled chip to MS was used for separation of berverine, palmatine, carnitine and some of its acyl analogs in less than 1 min in standard mixtures and in fortified human urine at concentrations between 35 and 124 mM [190,192].

Enantiomeric separation is of special importance in pharmaceutical and clinical science, since enantiomers may have different pharmacological or toxicological properties. The development of efficient and sensitive methods for pharmacokinetic studies and control of enantiomeric purity of drugs are of major interest (a more exhaustive description on chiral-CE can be found in chapter 13). Although there are various strategies to undertake chiral separation by CE, addition of a chiral selector to the running buffer (direct chiral separation mode) is generally used. Among all chiral selectors available, CDs are the most frequently used selectors. However, these compounds are essentially involatile and, therefore, their use in CE–ESI–MS brings about a significant loss of electrospray efficiency and MS sensitivity [251]. Modifications of this conventional chiral CE separation are being investigated to minimize or prevent the introduction of chiral selectors into the ESI–MS section. Some strategies have been used to avoid interface contamination, such as, for instance, the multiple voltage switching technique, the use of counter-migration principle between analyte and chiral selector, and the PFT. Furthermore, the use of compatible CE–MS chiral selectors has also been described. A detailed revision of the different procedures developed up to now to carry out chiral CE–MS has been done by Shamsi [133], and some representative applications of these procedures are given below.

Enantiomers of terbutaline and ephedrine were separated using heptakis(2,6-di-*O*-methyl)-β-CD (DM-β-CD) without any modification to prevent the entrance of chiral selectors into the ESI interface [252]. Following the same procedure, chiral analysis of the terbutaline, ketamine and propanolol pharmaceuticals, with an LOD of about 10 μM, was reached by using DM-β-CD and hydroxypropil-β-CD (HP-β-CD) containing water/methanol buffers [251]. In this work [251], the suppression effect of the CDs on the analyte signal was also evaluated, yielding similar results for both CDs, and as can be seen in Fig. 9.25, a quantity of 5 mM of these CDs is enough to originate a 6-fold ESI–MS

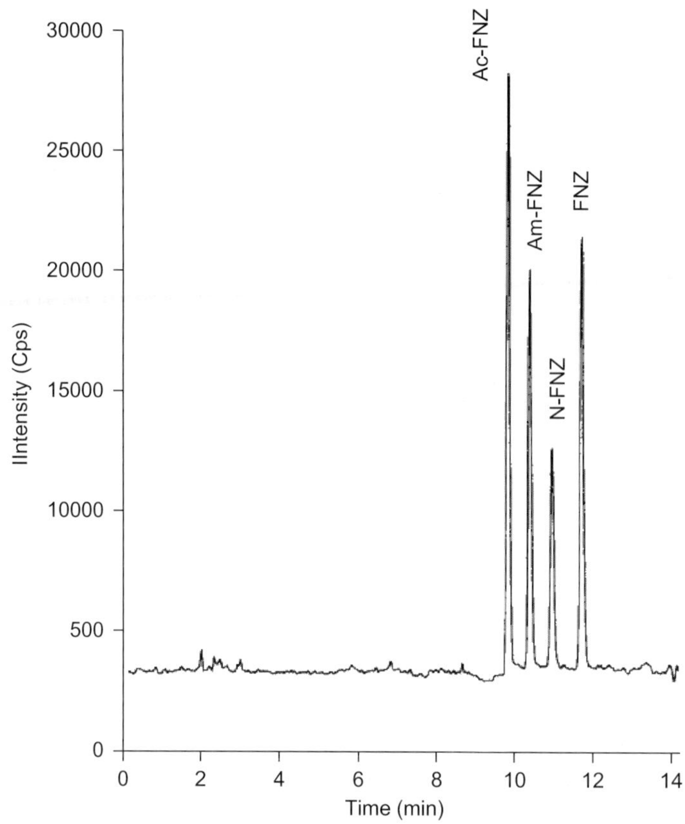

Fig. 9.24. Separation of flunitrazepam (FNZ) and its major metabolites by CEC–MS. CEC conditions: CEC column (Hypersil C_{18} capillary (33.5 cm, 100 μm i.d.), packed length 25 cm) connected with a void FS capillary (28 cm, 50 μm i.d.), mobile phase: formic acid/ammonium hydroxide (pH 8)/methanol (40:60, v/v), 10 mM total ionic strength, voltage: 30 kV, temperature: 20°C, electrokinetic injection: 10 kV for 4 s of 1 μg/ml of each benzodiazepine. ESI–MS conditions: SIM positive-ion mode (4 ions), capillary voltage: 5 kV, fragmentor: 70 V, drying gas: N_2 at a flow of 5 l/min and a temperature of 150 °C, nebulizer pressure: 2 psi, sheath liquid: water–methanol (50:50, v/v) with 0.5% formic acid at a flow rate of 5 μl/min. Reprinted from Ref. [88]. Copyright (2002), with permission from John Wiley & Sons, Inc.–Wiley-VCH Verlag GmbH.

signal supression. A system developed to avoid introduction of DM-β-CD into the ESI ion source was used for the separation of ropivacaine enantiomers, using a coupled capillary with possibility of voltage switching during separation, which was subdivided into different steps [253].

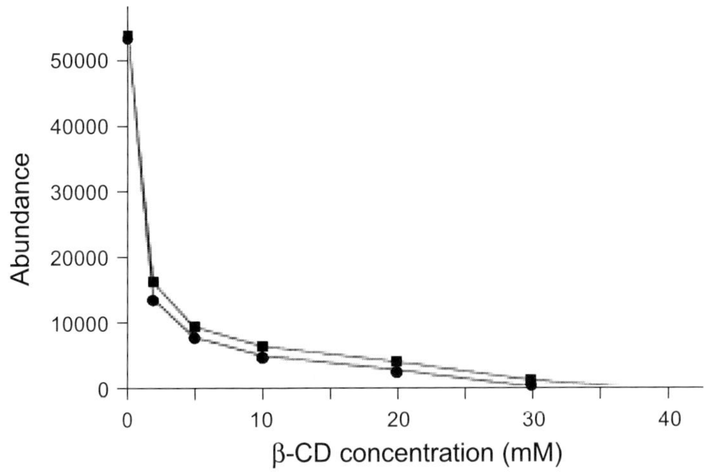

Fig. 9.25. Dependence of ESI–MS signal of protonated propanolol (10^{-4} M) on the concentration of (■) DM-β-CD and (•) HP-β-CD in operating CE buffer consisting of 0.8 M acetic acid and 5 mM ammonium acetate in methanol–water (80:20, v/v). Reprinted from Ref. [251]. Copyright (1998), with permission from Elsevier.

Counter-migration principle has been used in the analysis of chiral arylpropionic acids such as ketoprofen, naproxen, ibuprofen, carprofen and etodolac, using vancomycin as chiral selector [254]. Using the same counter-migration principle, to prevent the entrance of selectors into the ion source, charged CDs, such as carboxymethyl ether β-CD (CM-B-CD), sulfobutyl ether β-CD (SBE-B-CD) and 2-hydroxypropyltrimethylammonium salt of β-CD (TMA-B-CD), were used to separate etilefrine, mianserine, dimethidene and tropic acid enantiomers [136].

The potential of CE–ESI–MS using PTF in chiral pharmaceutical and metabolite separation has been demonstrated in several works using diverse chiral selectors such as: (i) neutral CDs [132,135, 255–257]; (ii) negatively charged CDs [135,258,259]; (iii) positively charged CDs [255]; (iv) macrocyclic antibiotics [255]; (v) chiral crown ethers [137]. A racemic mixture of bupivacaine and ropivacaine was separated using methyl-β-CD. The method permitted the detection of an impurity at a percentage of 0.25% of (*R*)-ropivacaine in (S)-ropivacaine [132]. DM-β-CD was used for determination of about 1 and 2% enantiomeric impurity in a commercial *R*- and *S*-camphorsulfonic acid, respectively [255], and clenbuterol enantiomers in aqueous solution and in plasma samples, with an LOD of 0.15 μg/ml (3 fmol) [256]. HP-β-CD was used for chiral separation of methadone enantiomers using experimental designs for PTF-CE–ESI–MS method development [257], and for chiral separation of six chiral amphetamines (Fig. 9.15). [135]. In the same work, negatively charged CDs (CD-β-CD and sulfated-β-CD) were employed for enantiomeric separation of methadone, venlafaxine, tropane alkaloids, and some of their metabolites. Analysis of methadone and one of its phase I metabolites in a patient serum sample is also shown. Enantiomeric tramadol and its phase I metabolites were separated in the presence of SBE-β-CD, and were also analyzed in plasma [258]. Separation of the anaesthetic drugs, bupivacaine, mepivacaine, ketamine and prilocaine were reached using sulfated-β-CD as chiral selector [259]. Positively charged chiral selectors such as quaternary ammonium-β-CD (QA-β-CD) and the protein avidin were used for chiral separation of tropic acid and arylpropionic acids (e.g., ibuprofen, ketoprofen and warfarin) respectively [255]. Quiral crown ether (+)-(18-crown-6)-2,3,11,12-tetracarboxilic acid has also been used with PFT for enantiomeric separation of α-amino-ε-caprolactan, cycloserine and 3-aminopyrrolidine [137].

Enantiomeric and nonenantiomeric separation of underivatized amino acids has also been described. Standard mixtures of protein amino acids were analyzed by CZE–ESI–MS using a coaxial-sheath flow interface, with an LOD in the low-μM range. The methods were applied to biological (and also food samples) analysis [260,261]. Recently, a sheathless interface coupling has been used in CE–MS for the determination of 20 standard amino acids with an LOD of 1–14 fmol. The method was applied to identification of phenylketonurea and tyrosinemia diseases in blood samples [119]. The use of (+)-(18-crown-6)-2,3,11,12-tetracarboxylic acid (18-C-6-TCA) as running buffer was used for the chiral separation of 11 amino acids, as shown in Fig. 9.26, providing also high separation efficiency of 20 L-amino acids, with LODs in the nM range, owing to the good ionization efficiencies of the amino acids–18-C-6-TCA complexes [262].

9.3.3 Pesticides

Nowadays, the use of pesticides is a widespread practice in agriculture to increase crop yield, to control weed growth, and insect plagues, etc. Monitoring of pesticides in vegetable foods, air, soil and water is an important task owing to their high toxicity and environmental pollutant power. Although many analytical procedures are based on LC and GC separation methods, it has been demonstrated that CE can,

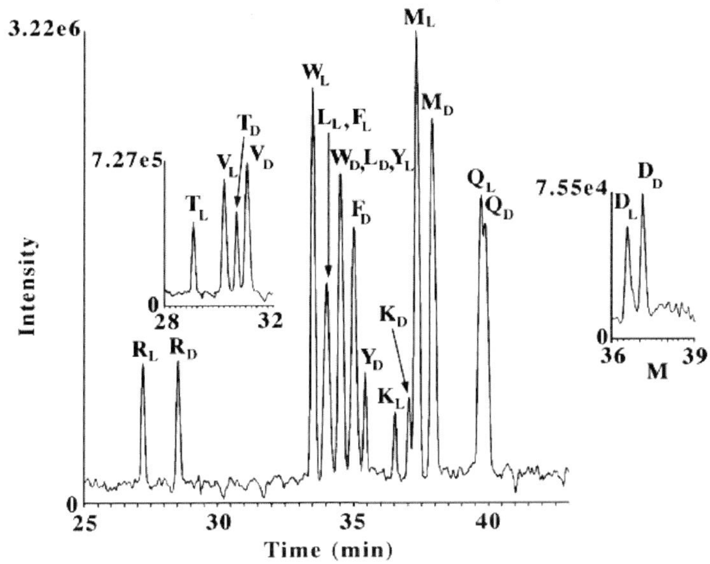

Fig. 9.26. Selected ion electropherogram of the separation of 11 amino acid enantiomers (500 fmol of each amino acid) using 30 mM 18–C-6-TCA solution as the BGE/chiral selector reagent and a fused-silica capillary (130 cm l_t, 20 μm i.d.). Analysis was carried out with 5 psi of inlet pressure to maintain stable ESI due to a lowered EOF under 30 mM 18-C-6-TCA conditions. Reprinted with permission from Ref. [119]. Copyright (2003), American Chemical Society.

in many cases, be a faster and more economical procedure to analyze pesticides. Nearly all pesticide analyses reported by CE–ESI–MS have been done using a coaxial-sheath liquid interface.

Quaternary ammonium herbicides (quats) are a very important commercial and environmental pesticide group. CE–ESI–MS was used for analysis of the non-UV-absorbing herbicides chlormequat and choline chloride. The method was useful for determination of purity of a formulation sample of chlomequat [263]. Among quaternary ammonium herbicides, the nonselective contact herbicides, paraquat and diquat, have been studied by CE–ESI–MS in several works. A mixture of five quaternary ammonium herbicides (paraquat, diquat, difenzoquat, chlormequat and mepiquat) were analyzed in spiked mineral water samples by CZE–ESI–MS [49,264], obtaining an LOD between 5 and 10 ng/ml [49]. Fig. 9.27 shows the electropherogram as well as MS and MS/MS spectra obtained from a highly contaminated irrigation water sample analyzed using this method [49].

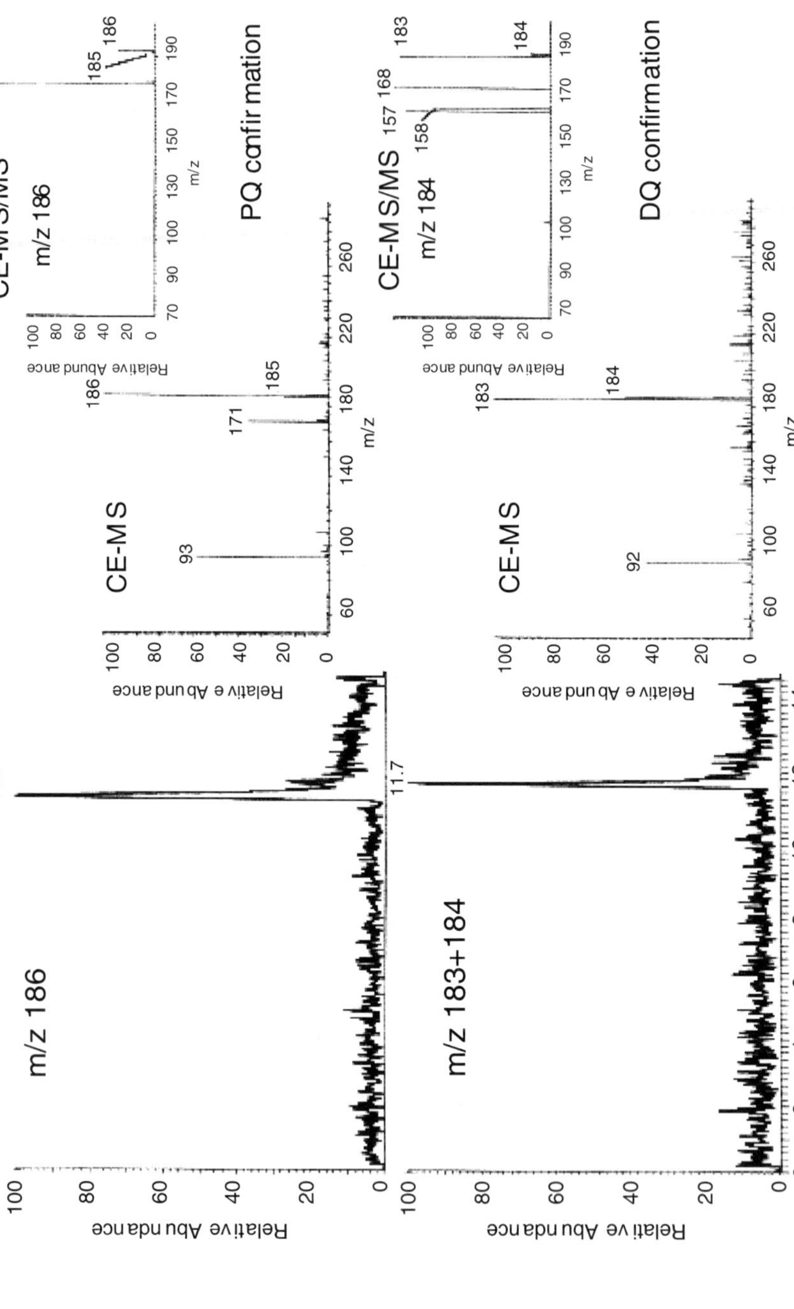
m/z 186
m/z 183+184
Time (min)
Relative Abundance
CE-MS
CE-MS/MS
m/z 186
m/z 184
PQ confirmation
DQ confirmation
m/z

Phenolic herbicides are also well-recognized environmental pollutants. Analysis of phenolic chlorinated herbicides and related compounds has also been carried out by CE–ESI–MS. A mixture of four of these herbicides was determined, and the method was applied to analyze the impurities from an MCPP (2-(2-methyl-4-chlorophenoxy) propionic acid) formulation [265]. A group of 16 chlorinated phenolic herbicides and related compounds were determined in spiked drinking water at a concentration of 5 μg/ml. Estimated LODs were between 8 and 250 ng/ml (4–110 pg) [266]. Otsuka et al. [267] have demonstrated the stereoselective separation of the chlorinated (dichloprop and fenoprop) and non-chlorinated (mecoprop) phenoxy acid herbicides using TM-β-CD without a further mechanism to avoid entrance of the selector into the ESI interface [267]. A comparison between CEC–ESI–MS and CEC–CIS–MS (CEC-coordination ion spray-MS) using permethyl-β-CD-silica as a chiral stationary phase was made using the chlorinated phenoxy acids, fenoxaprop and diclofop (as well as barbiturates) as model compounds. CIS–MS detection permitted an easier mass spectrum interpretation owing to the high affinity of the chiral herbicides for some of the metal complexes that improve MS detection [108].

Triazine derivatives are a very important selective herbicides group, used for the control of weeds in agricultural crops. Eight triazine herbicides (prometryne, terbutryn, prometon, ametryne, simetryn, propazine, atrazine and simazine) were analyzed by CZE–ESI–MS with reverse EOF using a dynamically coated capillary [268]. MEKC was used in combination with PFT for the separation of a triazine herbicide mixture (prometryne, ametryne, propazine and atrazine) as model system [269]. In a later work, MEKC was also combined with ESI–MS for chlorotriazine herbicides (atrazine, desethylatrazine, desisopropylatrazine and simazine) separation. In this case, adjusting the running buffer pH produced an anodically migrating micelles SDS,

---

Fig. 9.27. Electropherogram of a contaminated irrigation water sample and CE–MS and CE–MS-MS spectra of the identified compounds paraquat (PQ) and diquat (DQ). CE conditions: fused-silica capillary (80 cm l_t, 50 μm i.d.), running buffer: 200 mM formic acid/ammonium acetate at pH 3 containing 50% methanol, voltage: 25 kV with 35 kPa of inlet pressure during separation, temperature: 25°C, hydrodynamic injection: 35 kPa for 10 s. ESI-MS conditions: SIM positive-ion mode, capillary voltage: 4 kV, drying gas: N_2 at a flow of 13.5 l/h and a temperature of 250°C, sheath liquid: methanol-acetic acid (9:1, v/v) at a flow rate of 3 μl/min. Reprinted from Ref. [49]. Copyright (2002), with permission from Elsevier.

preventing in this way, the entrance of surfactant into the ESI interface [64].

The fungicides thiabendazole and procymidone were analyzed in fruits and vegetables by CZE–ESI–MS. Extraction of these compounds from samples were made using a C_8 solid-phase column. LODs were 0.1 and 0.01 μg/ml for procymidone and thiabendazole, respectively; sensitivity levels 10 times lower than the maximum residue limits authorized in some of the analyzed fruits and vegetables. Fig. 9.28 shows an application of this CE–ESI–MS method in real samples obtained from local markets [117]. Recently, acidic pesticides (acifluorfen, haloxyfop, picloram, *o*-phenylphenol and ioxynil) have been determined in fruits by the same group using an off-line solid-phase microextraction and CE–ESI–MS. LODs were between 0.5 and 10 μg/ml; these LOD values were lower than those obtained by CE-UV [170].

9.3.4 Other compounds

Different physiologically and ecologically dangerous compounds such as those described above have also been analyzed by CE–ESI–MS. 3-Quinuclinidol and four alkyl derivatives at 0.1–0.05 ppm [270], phenolic compounds in water from olive oil manufacturers with LODs between 1 and 386 pg [271] and aromatic sulfonates used in industrial and domestic processes in water samples [150] have been analyzed by CE–ESI–MS. Other toxins such as tetramethylammonium (a harmful substance founded in sea snails) [272], and paralytic shellfish poisoning (a group of related toxins founded in contaminated shellfish) [146], were also analyzed by CZE–ESI–MS.

Natural compounds from plants, with a great variety of properties such as flavonoids [125], isoflavonoids [273] and alkaloids [274,275] have also been analyzed with CE–ESI–MS coupling. In Fig. 9.29, an example of CE–ESI–MS analysis of naringenin in a phytomedicine from herbal drugs is shown.

Low-molecular-weight carboxylic acids (e.g., malonic acid, succinic acid, maleic acid and glutaric acid) were analyzed by CE–ESI–MS with LODs between 1 and 10 ppm (60–600 pg) [118].

Oligonucleotides and their derivatives have acquired increasing importance in biology and biomedicine area. A few examples of oligonucleotide separations by CE-–ESI–MS can be founded in the bibliography. As commented in Section 9.2.3.1, CGE and ESI–MS hyphenation is a very difficult task owing to the high buffer concentration

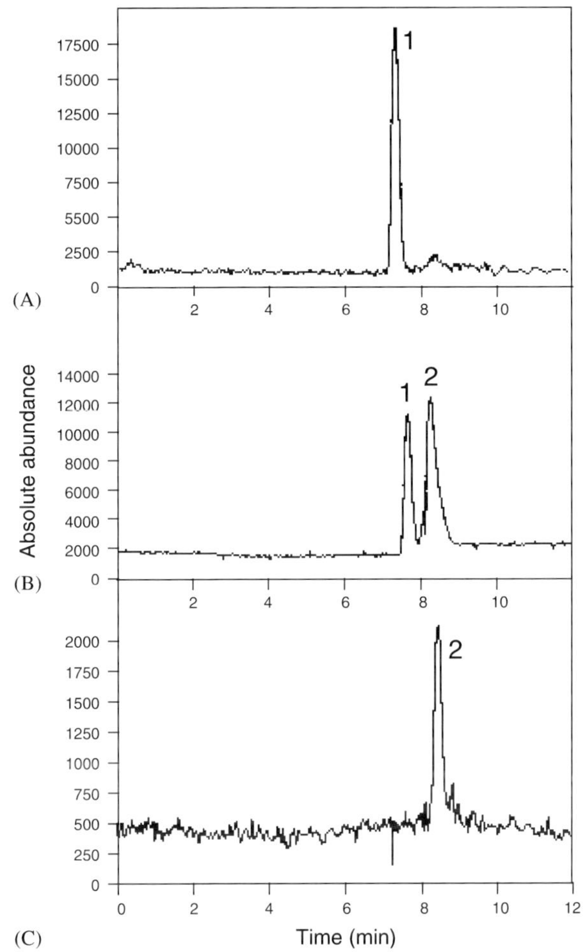

Fig. 9.28. CE–ESI–MS electropherograms of (A) orange that contains thiabendazole at 0.95 mg/kg, (B) pear that contains thiabendazole and procymidone at 0.42 and 0.81 mg/kg respectively, and (C) strawberry sample that contains procymidone at 0.2 mg/kg. CE conditions: fused-silica capillary (150 cm l_t, 75 μm i.d.), running buffer: 12 mM ammonium fomate/20 mM formic acid at pH 3.5 with 2% methanol, voltage: 30 kV, temperature: 25°C, hydrodynamic injection: 0.5 psi for 15 s. ESI–MS conditions: SIM positive-ion mode, capillary voltage: 4 kV, drying gas: N_2 at a flow of 11 l/min and a temperature of 150°C, nebulizer pressure: 40 psi, sheath liquid: similar composition as running buffer at a flow rate of 13 μl/min. Reprinted from Ref. [117]. Copyright (2002), with permission from Elsevier.

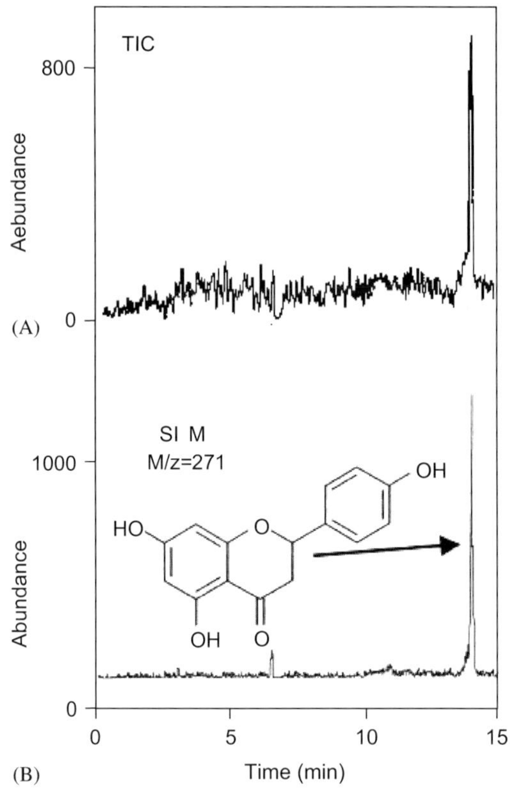

Fig. 9.29. CE–ESI–MS of naringenin in a phytomedicine prepared from five different herbal drugs. (A) Total ion current, (B) selected ion trace. CE conditions: fused silica capillary (60 cm l_t, 50 μm i.d.), running buffer: 40 mM ammonium acetate at pH 9.5 and 15% acetonitrile, voltage: 25 kV. ESI–MS conditions: negative-ion mode, sheath liquid composition: in propanol–water–triethylamine (80:20:0.1, by vol.) with 2 mM ammonium acetate at a flow rate of 4 μl/min. Reprinted from Ref. [125]. Copyright (2002), with permission from John Wiley & Sons, Inc.–Wiley-VCH Verlag GmbH.

used in CGE, among other considerations. Interfacing CGE using a PVP matrix with ESI–MS, was used for the analysis of synthetic oligonucleotides of 4, 5 and 6 bases in length, in concentrations of about 100 nM [172], modified and unmodified oligonucleotides (see Fig. 9.30) [74] and a four mixture of isomeric oligonucleotides [74]. Phosphodiester and phosphorothioates oligonucleotides with 12–15-mer and 12–20-mer, respectively, were analyzed using a PEG gel [75]. Cisplatin, one of the most widely used anticancer agents, and the search for

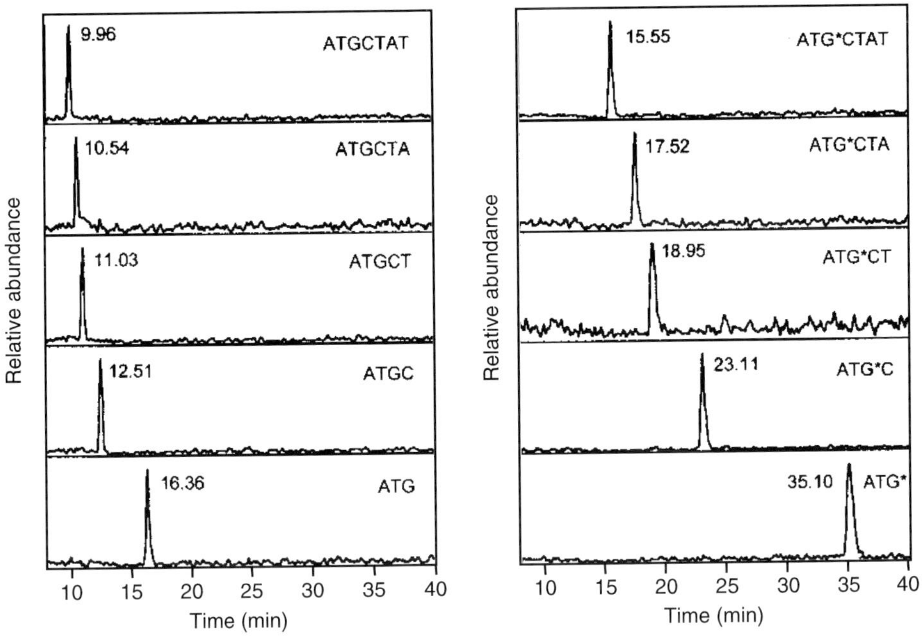

Fig. 9.30. CE separation in aqueous PVP solution. Field strength 360 V/cm. MS on-line detection. Monitored mass data traces: $(ATGCTAT)^{2-}$, *m/z* 1047; $(ATGCTA)^{2-}$, *m/z* 895; $(ATGCT)^{-}$, *m/z* 1477; $(ATGC)^{-}$, *m/z* 1173; $(ATG)^{-}$, *m/z* 884; $(ATG^{*}CTAT)^{2-}$, *m/z* 1157; $(ATG^{*}CTA)^{2-}$, *m/z* 1005; $(ATG^{*}CT)^{2-}$, *m/z* 849; $(ATG^{*}C)^{-}$, *m/z* 1394 and $(ATG^{*})^{-}$, *m/z* 1105. Reprinted with permission from Ref. [74]. Copyright (1998), American Chemical Society.

DNA adducts has been, and remains, a research area of its own. CZE–ESI–MS seems to be an interesting, less expensive and less complicated alternative to current complicated techniques [276].

Suitability of CE with ESI–MS detection has also been demonstrated for carbohydrate analysis (see Fig. 9.31) [277]. A standard mixture of carbohydrates with LODs of about 0.5–3 mg/ml and application to different wine samples [277], oligosaccharides from enzymatic heparin digestion [278], and derivatized and underivatized glycans [50], were analyzed by CZE–ESI–MS. CEC–ESI–MS was used for the analysis of complex oligosaccharides mixtures using hydrophilic monolithic columns [105], and FT–MS for high mass resolution [98]. PFT–MEKC–ESI–MS has also been used for the analysis of iridoid glycosides in plant samples from different species [279].

Recently, our group has demonstrated the great possibilities of CE–ESI–MS for the analysis of synthetic polymers [127]. Thus, Fig. 9.32

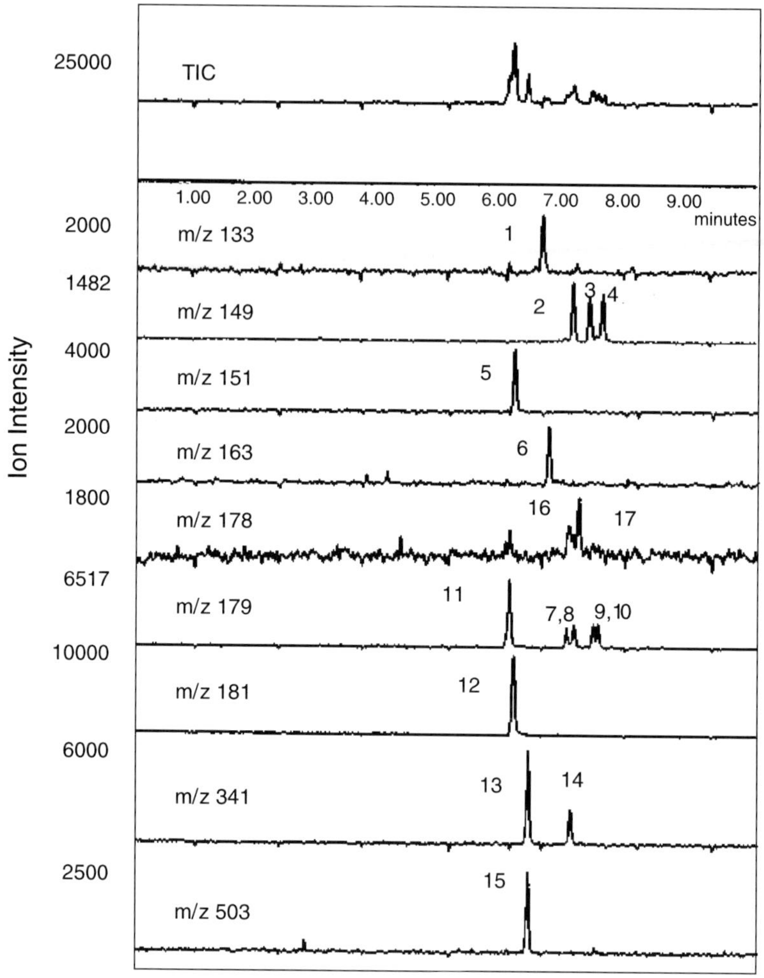

Fig. 9.31. CE–MS TIC- and SIM-electropherograms for a standard mixture containing 17 carbohydrates at a concentration level of 25 mg/ml each. Running buffer: 300 mM DEA, voltage: 20 kV, injection: 9 s at 50 mbar. Sheath liquid: 2-propanol–water (80:20, v/v) with 0.25% DEA at a flow rate of 4 μl/min. Peaks: 1, deoxyribose; 2, arabinose; 3, ribose; 4, xylose; 5, xylitol; 6, fucose; 7, galactose; 8, glucose; 9, manose; 10, fructose; 11, inositol; 12, mannitol; 13, sucrose; 14, maltose; 15 raffinose; 16 glucosamine; 17, galactosamine. Reprinted from Ref. [277]. Copyright (2001), with permission from John Wiley & Sons, Inc.–Wiley-VCH Verlag GmbH.

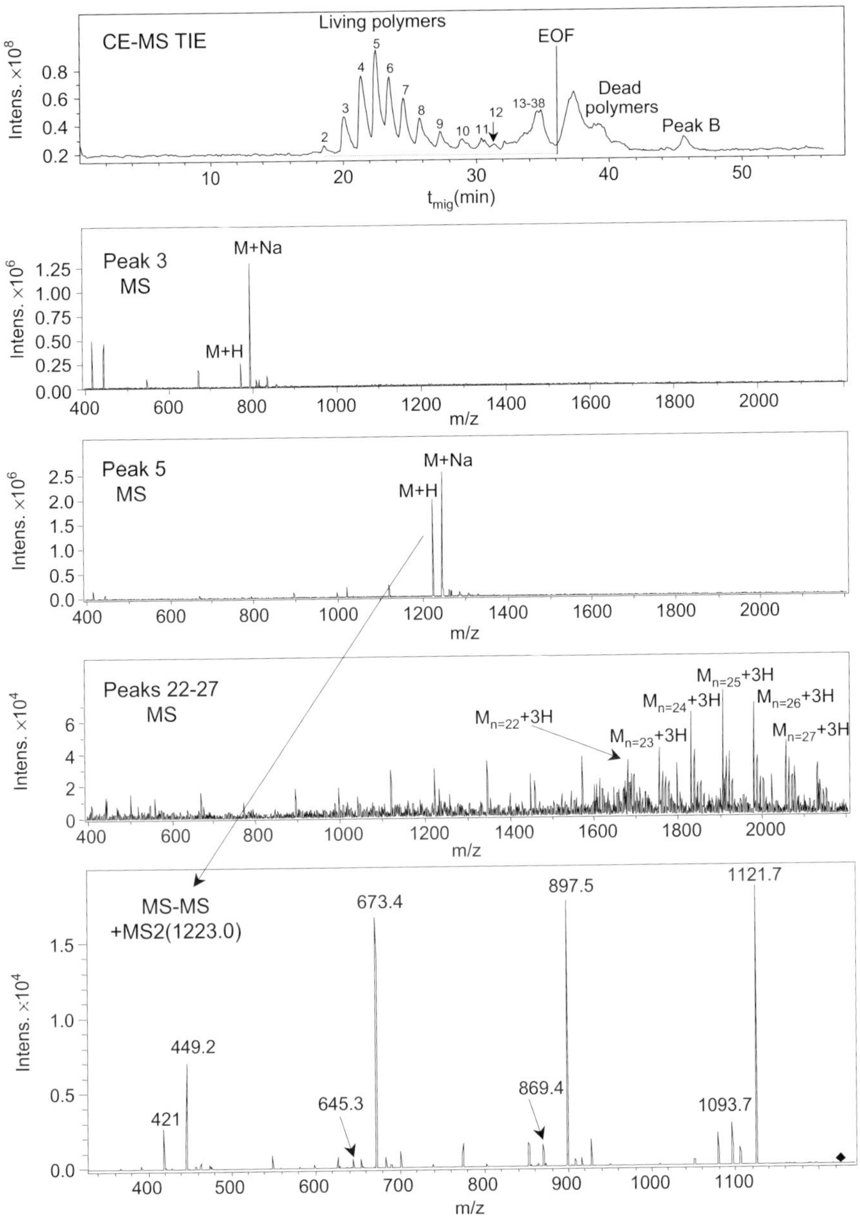

Fig. 9.32. CE–MS total ion electropherogram, MS and MS–MS spectra of some living polymers of poly(N_ε-trifluoroacetyl-L-lysine). CE conditions: fused-silica capillary (87 cm l_t, 50 μm i.d.), running buffer: 1 M acetic acid, 20 mM ammonium acetate in methanol/acetonitrile, (87.5:12.5, v/v), voltage: 25 kV, temperature: 25°C. ESI–MS conditions: positive-ion mode, drying gas: N_2 at a flow of 4 l/min and a temperature of 300°C, nebulizer pressure: 3 psi, sheath liquid: methanol-acetonitrile (87.5:12.5, v/v) at a flow rate of 5 μl/min. Redrawn from Ref. [127]. Copyright (2004), American Chemical Society.

shows the CE–MS total ion electropherogram (TIE), MS and MS–MS spectra of some living polymers of poly(N_ε-trifluoroacetyl-L-lysine) obtained using nonaqueous CE conditions. Under these conditions, the molecular mass characterization of this polymer up to a degree of polymerization of 38 could be established. Moreover, it was demonstrated that nonaqueous-CE–MS can provide interesting information on the chemical structure of the polymer end groups and other final by-products of the polymerization reaction.

9.4 FUTURE OUTLOOK

It has been recognized by many authors that at the moment, the limited concentration sensitivity of CE–MS prevents its use for routine analysis. Its intrinsic advantages of low solvent and sample consumption, speed of analysis and high efficiency led many analysts to expect a wide acceptance of CE–MS by analytical laboratories. However, these expectations have not been fully realized, and in practice, the situation regarding the use and introduction of CE–MS instruments in the laboratory has not advanced much. Although, as described in this chapter, interesting applications and developments have been reported, we cannot yet talk in terms of a widespread acceptance of this coupling. In our opinion, before that wide acceptance can occur, the main problem that needs to be solved is the sensitivity problem; CE–MS practitioners will then be able to take advantage of the full analytical potential of this coupling. Thus, some instrumental and methodological improvements have to be carried out before CE–MS can be considered a routine technique: (i) ideally, the sensitivity of the more affordable mass spectrometers must be enhanced (or the cost of the more sensitive analyzers, such as FT–ICTR, reduced considerably); (ii) a higher development and application of preconcentration techniques in conjunction with CE–MS is desirable, since it will enable the detection of very minor components in complex mixtures; (iii) although the ESI interface based on a coaxial sheath design is widely used, it would be interesting to additionally consider with a commercial/universal design based on a sheathless interface, which could reduce the chemical background noise while improving detection limits; and (iv) the intrinsic incompatibility of several CE modes (CIEF, and mostly MEKC and CGE) with the ionization sources has to be worked out more extensively; the development of new procedures that can overcome or at least reduce such incompatibility would largely increase the number of applications of this technique. Similarly, the study of new column and interface designs and new stationary or pseudostationary phases should increase the compatibility between CEC and MS and, as a consequence, their widespread acceptance. This increase can also be imagined for affinity CE-MS, a procedure that can become very useful for studying in-depth molecular interactions and which is expected to be a very useful tool for combinatorial libraries and drug discovery as well as for drug metabolism studies [141,280]. On the other hand, it seems clear that the research on high-throughput analysis of chemical and biological compounds using CE-on-chip will continue to be a hot topic in the near future although, at the moment, chip-CE–MS is at its infancy.

ACKNOWLEDGMENTS

C.S. thanks Consejería de Educación y Cultura (Comunidad de Madrid) for a fellowship. The authors acknowledge to CICYT, AGL2002-04621-C02-02 for financial support.

REFERENCES

1 J.A. Olivares, N.T. Nguyen, C.R. Yonker and R.D. Smith, *Anal. Chem.*, 59 (1987) 1230–1232.
2 R.D. Smith, J.A. Olivares, N.T. Nguyen and H.R. Udseth, *Anal. Chem.*, 60 (1988) 436–441.
3 R.D. Smith, C.J. Barinaga and H.R. Udseth, *Anal. Chem.*, 60 (1988) 1948–1952.
4 J.C. Severs and R.D. Smith. In: R.B. Cole (Ed.), *Electrospray Ionization Mass Spectrometry*, Wiley, New York, 1997.
5 J.F. Banks, *Electrophoresis*, 18 (1997) 2255–2266.
6 K.B. Tomer, L.J. Deterding and C.E. Parker. In: M.G. Khaledi (Ed.), *High Performance Capillary Electrophoresis Theory, Techniques and Applications*, Wiley, New York, 1998.
7 A. Brocke, G. Nicholson and E. Bayer, *Electrophoresis*, 22 (2001) 1251–1266.
8 K.B. Tomer, *Chem. Rev.*, 101 (2001) 297–328.
9 M. Moini, *Anal. Bioanal. Chem.*, 373 (2002) 466–480.

10 P. Schmitt-Kopplin and M. Frommberger, *Electrophoresis*, 24 (2003) 3837–3867.
11 E. Gelpi, An update on recent developments, *J. Mass Spectrom.*, 37 (2002) 241–253.
12 M.A. Moseley, L.J. Deterding, K.B. Tomer and J.B. Jorgenson, *Rapid Commun. Mass Spectrom.*, 3 (1989) 87–93.
13 J. Preisler, P. Hu, T. Rejtar and B.L. Karger, *Anal. Chem.*, 72 (2000) 4785–4795.
14 J. Preisler, P. Hu, T. Rejtar, E. Moskovets and B.L. Karger, *Anal. Chem.*, 74 (2002) 17–25.
15 S. Takeda, Y. Tanaka, M. Yamane, Z. Siroma, S-I. Wakida, K. Otsuka and S. Terabe, *J. Chromatogr. A*, 924 (2001) 415–420.
16 Y. Tanaka, K. Otsuka and S. Terabe, *J. Pharm. Biomed. Anal.*, 30 (2003) 1889–1895.
17 M. Dole, L.L. Mack, R.L. Hines, R.C. Mobley, L.D. Ferguson and M.B. Alice, *J. Chem. Phys.*, 49 (1968) 2240–2249.
18 M. Yamashita and J.B. Fenn, *J. Phys. Chem.*, 88 (1984) 4451–4459.
19 M. Yamashita and J.B. Fenn, *J. Phys. Chem.*, 88 (1984) 4671–4675.
20 J.M. Ding and P. Vouros, *Anal. Chem.*, 71 (1999) 378A–385A.
21 J. Cai and J. Henion, *J. Chromatogr. A*, 703 (1995) 667–692.
22 D. Figeys and R. Aebersold, *Electrophoresis*, 19 (1998) 885–892.
23 R.S. Ramsey and S.A. McLuckey, *J. Microcol. Sep.*, 7 (1995) 461–469.
24 J.H. Wahl, D.C. Gale and R.D. Smith, *J. Chromatogr.*, 659 (1994) 217–222.
25 S. Nilsson and K.E. Markides, *Rapid Commun. Mass Spectrom.*, 14 (2000) 6–11.
26 Y.Z. Chang, Y.R. Chen and G.R. Her, *Anal. Chem.*, 73 (2001) 5083–5087.
27 L. Bendahl, S.H. Hansen and J. Olsen, *Rapid Commun. Mass Spectrom.*, 16 (2002) 2333–2340.
28 X. Zhu, S. Thiam, B.C. Valle and I.M. Warner, *Anal. Chem.*, 74 (2002) 5405–5409.
29 P. Cao and M. Moini, *J. Am. Soc. Mass Spectrom.*, 8 (1997) 561–564.
30 L. Fang, R. Zhang, E.R. Williams and R.N. Zare, *Anal. Chem.*, 66 (1994) 3696–3701.
31 M. Mazereeuw, A.J.P. Hofte, U.R. Tjaden and J. Van der Greef, *Rapid Commun. Mass Spectrom.*, 11 (1997) 981–986.
32 J.C. Hannis and D.C. Muddiman, *Rapid Commun. Mass Spectrom.*, 12 (1998) 443–448.
33 M.B. Barroso and A.P. De Jong, *J. Am. Soc. Mass Spectrom.*, 10 (1999) 1271–1278.
34 M.A. Petersson, G. Hulthe and E. Fogelqvist, *J. Chromatogr. A*, 854 (1999) 141–154.
35 M. Moini, *Anal. Chem.*, 73 (2001) 3497–3501.

36 K. Huikko, T. Kotiaho and R. Kostiainen, *Rapid Commun. Mass Spectrom.*, 16 (2002) 1562–1568.
37 G.A. Ross, *LC-GC Europe*, 1 (2001) 2–6.
38 S. Cherkaoui and L.C. Veuthey, *Electrophoresis*, 23 (2002) 442–448.
39 F. Foret, T.J. Thompson, P. Vouros, B.L. Karger, P. Gebauer and P. Bocek, *Anal. Chem.*, 66 (1994) 4450–4458.
40 E.D. Lee, W. Muck, T.R. Covey and J.D. Henion, *Biomed. Environ. Mass Spectrom.*, 18 (1989) 844–850.
41 S. Pleasance, P. Thibault and J. Kelly, *J. Chromatogr.*, 591 (1992) 325–339.
42 J.C. Severs, A.C. Harms and R.D. Smith, *Rapid Commun. Mass Spectrom.*, 10 (1996) 1175–1178.
43 T. Wachs, R.L. Sheppard and J. Henion, *J. Chromatogr. B*, 685 (1996) 335–342.
44 R.E. Settlage, P.S. Russo, J. Shabanowitz and D.F. Hunt, *J. Microcol. Sep.*, 10 (1998) 281–285.
45 F. Foret, H. Zhou, E. Gangi and B.L. Karger, *Electrophoresis*, 21 (2000) 1363–1370.
46 M. Jussila, K. Sinervo, S.P. Porras and M.L. Riekkola, *Electrophoresis*, 21 (2000) 3311–3317.
47 J.T. Whitt and M. Moini, *Anal. Chem.*, 75 (2003) 2188–2191.
48 C. Neususs, M. Pelzing and M. Macht, *Electrophoresis*, 23 (2002) 3149–3159.
49 O. Nuñez, E. Moyano and M.T. Galcerán, *J. Chromatogr. A*, 974 (2002) 243–255.
50 L.A. Gennaro, J. Delaney, P. Vouros, D.J. Harvey and B. Domon, *Rapid Commun. Mass Spectrom.*, 16 (2002) 192–200.
51 S.A. McLuckey and J.M. Wells, *Chem. Rev.*, 101 (2001) 571–606.
52 C.H. Sin, E.D. Lee and M.L. Lee, *Anal. Chem.*, 63 (1991) 2897–2900.
53 J.G. Boyle and C.M. Whitehouse, *Anal. Chem.*, 64 (1992) 2084–2089.
54 G. Brenner-Weiss, F. Kirschhofer, B. Kuhl, M. Nusser and U. Obst., *J. Chromatogr. A*, 1009 (2003) 147–153.
55 M. Wetterhall, M. Palmblad, P. Hkansson, K.E. Markides and J. Bergquist, *J. Proteome Res.*, 1 (2002) 361–366.
56 Y. Shen, N. Tolić, R. Zhao, L. Paša-Tolić, L. Li, S.J. Berger, R. Harkewics, G.A. Anderson, M.E. Belov and R.D. Smith, *Anal. Chem.*, 73 (2001) 3011–3021.
57 S. Martinovic, T.D. Veenstra, G.A. Anderson, L. Paša-Tolić and R.D. Smith, *J. Mass Spectrom.*, 37 (2002) 99–107.
58 F.M. Fernandez, J.M. Vadillo, J.R. Kimmel, M. Wetterhall, K. Markides, N. Rodriguez and R.N. Zare, *Anal. Chem.*, 74 (2002) 1611–1617.
59 J. Varghese and R.B. Cole, *J. Chromatogr. A*, 652 (1993) 369–376.

60 G.W. Somsen, R. Mol and G.J. De Jong, *J. Chromatogr. A*, 1000 (2003) 953–961.
61 K. Isoo, K. Otsuka and S. Terabe, *Electrophoresis*, 22 (2001) 3426–3432.
62 W. Nashabeh, K.F. Greve, D. Kirby, F. Foret, B.L. Karger, D.H. Reifsnyder and S.E. Builder, *Anal. Chem.*, 66 (1994) 2148–2154.
63 H. Ozaki and S. Terabe, *J. Chromatogr. A*, 794 (1998) 317–325.
64 L.Y. Yang, A.K. Harrata and C.S. Lee, *Anal. Chem.*, 69 (1997) 1820–1826.
65 M.H. Lamoree, U.R. Tjaden and J. Van der Greef, *J. Chromatogr. A*, 712 (1995) 219–225.
66 Y. Ishihama, H. Katayama and N. Asakawa, *Anal. Biochem.*, 287 (2000) 45–54.
67 P. Petersson, M. Jornten-Karlsson and M. Stalebro, *Electrophoresis*, 24 (2003) 999–1007.
68 W. Lu, G.K. Poon, P.L. Carmichael and R.B. Cole, *Anal. Chem.*, 68 (1996) 668–674.
69 Y. Tanaka, Y. Kishimoto, K. Otsuka and S. Terabe, *J. Chromatogr. A*, 817 (1998) 49–57.
70 H.L. Cheng, M.C. Tseng, P.L. Tsai and G.R.A. Her, *Rapid Commun. Mass Spectrom.*, 15 (2001) 1473–1480.
71 Y.R. Chen, M.C. Tseng, Y.Z. Chang and G.R.A. Her, *Anal. Chem.*, 75 (2003) 503–508.
72 F. Garcia and J.D. Henion, *Anal. Chem.*, 64 (1992) 985–990.
73 J.P. Barry, J. Muth, S-J. Law, B.L. Karger and P. Vouros, *J. Chromatogr. A*, 732 (1996) 159–166.
74 A. Harsch and P. Vouros, *Anal. Chem.*, 70 (1998) 3021–3027.
75 T. Freudemann, A. Von Brocke and E. Bayer, *Anal.Chem.*, 73 (2003) 2587–2593.
76 A. Brocke, T. Freudemann and E. Bayer, *J. Chromatogr. A*, 991 (2003) 129–141.
77 Q. Tang, A.K. Harrata and C.S. Lee, *Anal. Chem.*, 67 (1995) 3515–3519.
78 Q. Tang, A.K. Harrata and C.S. Lee, *Anal. Chem.*, 68 (1996) 2482–2487.
79 Q. Tang, A.K. Harrata and C.S. Lee, *J. Mass Spectrom.*, 31 (1996) 1284–1290.
80 Q. Tang, A.K. Harrata and C.S. Lee, *Anal. Chem.*, 69 (1997) 3177–3182.
81 M.H. Lamoree, U.R. Tjaden and J. Van der Greef, *J. Chromatogr. A*, 777 (1997) 31–39.
82 L. Yang, C.S. Lee, S.A. Hofstadler and R.D. Smith, *Anal. Chem.*, 70 (1998) 4945–4950.
83 P.K. Jensen, L. Paša-Tolić, G.A. Anderson, J.A. Horner, M.S. Lipton, J.E. Bruce and R.D. Smith, *Anal. Chem.*, 71 (1999) 2076–2084.

84 L. Paša-Tolić, P.K. Jensen, G.A. Anderson, M.S. Lipton, K.K. Peden, S. Martinović, N. Tolić, J.E. Bruce and R.D. Smith, *J. Am. Chem. Soc.*, 121 (1999) 7949–7950.

85 T.D. Veenstra, S. Martinović, G.A. Anderson, L.J. Paša-Tolić and R.D. Smith, *J. Am. Soc. Mass Spectrom.*, 11 (2000) 78–82.

86 V. Spikmans, N.W. Smith, M.G. Tucker, R. Horsten and M. Mazereeuw, *LC-GC Europe*, 13 (2000) 486–494.

87 G. Choudhary, A. Apffel, H. Yin and W. Hancock, *J. Chromatogr. A*, 887 (2000) 85–101.

88 X. Cahours, S. Cherkaoui, G. Rozing and J-L. Veuthey, *Electrophoresis*, 23 (2002) 2320–2326.

89 J.T. Wu, P. Huang, M.X. Li, M.G. Qian and D.M. Lubman, *Anal. Chem.*, 69 (1997) 320–326.

90 V. Schurig and S. Mayer, *J. Biochem. Biophys. Methods*, 48 (2001) 117–141.

91 E.R. Verheij, U.R. Tjaden, W.A.M. Niessen and J. Van der Greef, *J Chromatogr*, 554 (1991) 339–349.

92 M. Hugener, A.P. Tinke, W.M.A. Niessen, U.R. Tjaden and J. Van der Greef, *J. Chromatogr.*, 647 (1993) 375–385.

93 G.A. Lord, D.B. Gordon, L.W. Tetler and C.M. Carr, *J. Chromatogr. A*, 700 (1995) 27–33.

94 S.J. Lane, R. Boughtflower, C. Paterson and T. Underwood, *Rapid Commun. Mass Spectrom.*, 9 (1995) 1283–1287.

95 K. Schmeer, B. Behnke and E. Bayer, *Anal. Chem.*, 67 (1995) 3656–3658.

96 R.N. Warriner, A.S. Craze, D.E. Games and S.J. Lane, *Rapid Commun. Mass Spectrom.*, 12 (1998) 1143–1149.

97 M. Gucek, M. Gaspari, K. Walhagen, R.J. Vreeken, E.R. Verheij and J. Van der Greef, *Rapid Commun. Mass Spectrom.*, 14 (2000) 1448–1454.

98 A.H. Que, Y. Mechref, Y. Huang, J.A. Taraszka, D.E. Clemmer and M.V. Novotny, *Anal. Chem.*, 75 (2003) 1684–1690.

99 K. Walhagen, M. Gaspari, U.R. Tjaden, G.P. Rozing and J. Van der Greef, *Rapid Commun. Mass Spectrom.*, 15 (2001) 878–883.

100 G.A. Lord, D.B. Gordon, P. Myers and B.W. King, *J. Chromatogr. A*, 768 (1997) 9–16.

101 G. Choudhary, C. Horvath and J.F. Banks, *J. Chromatogr A*, 828 (1998) 469–480.

102 M.R. Taylor and P. Teale, *J. Chromatogr. A*, 768 (1997) 89–95.

103 M. Gaspari, M. Gucek, K. Walhagen, R.J. Vreeken, E.R. Verheij, U.R. Tjaden and J. Van der Greef, *J. Microcol. Sep.*, 13 (2001) 243–249.

104 E. Rapp and E. Bayer, *J. Chromatogr. A*, 887 (2000) 367–378.

105 A.H. Que and M.V. Novotny, *Anal. Chem.*, 74 (2002) 5184–5191.

106 P. Viberg, M. Jornten-Karlsson, P. Petersson, P. Spegel and S. Nilsson, *Anal. Chem.*, 74 (2002) 4595–4601.
107 C. Rentel, P. Gfrörer and E. Bayer, *Electrophoresis*, 20 (1999) 2329–2336.
108 A. Brocke, D. Wistuba, P. Gfrörer, M. Stahl, V. Schurig and E. Bayer, *Electrophoresis*, 23 (2002) 2963–2972.
109 T.J. Thompson, F. Foret, P. Vouros and B.L. Karger, *Anal. Chem.*, 65 (1993) 900–906.
110 M.A. Moseley, J. Shabanowitz, D.F. Hunt and K.B. Tomer, *J. Am. Soc. Mass Spectrom.*, 3 (1992) 289–300.
111 J.H. Wahl and R.D. Smith, *J. Capillary Electrophor.*, 1 (1994) 62–71.
112 J.R. Perkins, C.E. Parker and K.B. Tomer, *J. Am. Soc. Mass Spectrom.*, 3 (1992) 139–149.
113 L. Geiser, S. Cherkaoui and J-L. Veuthey, *J. Chromatogr. A*, 979 (2002) 389–398.
114 J. Senior, D. Rolland, D. Tolson, S. Chantzis and V. De Biasi, *J. Pharm. Biomed. Anal.*, 22 (2000) 413–421.
115 K. Vuorensola, H. Siren, R. Kostiainen and T. Kotiaho, *J. Chromatogr. A*, 979 (2002) 179–189.
116 C. Aguilar, A.J.P. Hofte, U.R. Tjaden and J. Van der Greef, *J. Chromatogr. A*, 926 (2001) 57–67.
117 R. Rodríguez, Y. Picó, G. Font and J. Mañes, *J. Chromatogr. A*, 949 (2002) 359–366.
118 S.K. Johnson, L.L. Houk, D.C. Johnson and R.S. Houk, *Anal. Chim. Acta*, 389 (1999) 1–8.
119 C.L. Schultz and M. Moini, *Anal. Chem.*, 75 (2003) 1508–1513.
120 V. Ruiz-Calero, E. Moyano, L. Puignou and M.T. Galcerán, *J. Chromatogr. A*, 914 (2001) 277–291.
121 C. Simó, P. López Soto-Yárritu and A. Cifuentes, *Electrophoresis*, 23 (2002) 2288–2295.
122 C. Simó and A. Cifuentes, *Electrophoresis*, 24 (2003) 834–842.
123 C. Simó, C. Elvira, N. González, J. San Román, C. Barbas and A. Cifuentes, *Electrophoresis*, 25 (2004) 2056–2064.
124 F. Lafont, M.A. Aramendia, I. García, V. Borau, C. Jiménez, J.M. Marinas and F.J. Urbano, *Rapid. Commun. Mass Spectrom.*, 13 (1999) 562–567.
125 C.W. Huck, G. Stecher, W. Ahrer, W.M. Stoggl, W. Buchberger and G.K. Bonn, *J. Sep. Sci.*, 25 (2002) 904–908.
126 L. Mateus, S. Cherkaoui, P. Christen and J-L. Veuthey, *Electrophoresis*, 20 (1999) 3402–3409.
127 C. Simó, H. Cottet, W. Vayaboury, O. Giani, M. Pelzing and A. Cifuentes, *Anal. Chem.*, 76 (2004) 335–344.
128 S. Fanali and C. Desiderio, *J. High Resol. Chromatogr.*, 19 (1996) 322–330.
129 T.J. Ward, Dann III and A.P. Brown, *Chirality*, 8 (1996) 77–83.

130 S. Fanali, C. Desiderio and J. Aturki, *J. Chromatogr. A*, 772 (1997) 185–194.
131 Y. Tanaka and S. Terabe, *J. Chromatogr. A*, 694 (1995) 277–284.
132 E.M. Jäverfalk, A. Amini, D. Westerlund and P.E. Andrén, *J. Mass Spectrom.*, 33 (1998) 183–186.
133 S.A. Shamsi, *Electrophoresis*, 23 (2002) 4036–4051.
134 F. Menzinger, Ph. Schmitt-Kopplin, M. Frommberger, D. Freitag and A. Kettrup, *Fresenius J. Anal. Chem.*, 371 (2001) 25–34.
135 S. Cherkaoui, S. Rudaz, E. Varesio and J-L. Veuthey, *Electrophoresis*, 22 (2001) 3308–3315.
136 G. Schulte, S. Heitmeier, B. Chankvetadze and G. Blaschke, *J. Chromatogr. A*, 800 (1998) 77–82.
137 Y. Tanaka, K. Otsuka and S. Terabe, *J. Chromatogr. A*, 875 (2000) 323–330.
138 J. Suomi, S.K. Wiedmer, M. Jussila and M.L. Riekkola, *J. Chromatogr. A*, 970 (2002) 287–296.
139 P.G. Muijselaar, K. Otsuba and S. Terabe, *J. Chromatogr. A*, 802 (1998) 3–15.
140 S. Grard, P. Morin, M. Dreux and J.P. Ribet, *J. Chromatogr. A*, 926 (2001) 3–10.
141 N.A. Guzman and R.J. Stubbs, *Electrophoresis*, 22 (2001) 3602–3628.
142 Q. Yang, A.J. Tomlinson and S. Naylor, *Anal. Chem. News Features*, 71 (1999) 183A–189A.
143 R.D. Smith, J.A. Loo, C.J. Barinaga, C.G. Edmonds and H.R. Udseth, *J. Chromatogr.*, 480 (1989) 211–232.
144 A.P. Tinke, J.J. Reinhoud, W.M.A. Niessen, U.R. Tjiaden and J. Van der Greef, *Rapid Commun. Mass Spectrom.*, 6 (1992) 560–563.
145 S. Auriola, P. Thibault, I. Sadovskaya and E. Altmann, *Electrophoresis*, 19 (1998) 2665–2676.
146 S.J. Locke and P. Thibault, *Anal. Chem.*, 66 (1994) 3436–3446.
147 J. Gysler, B. Helk, S. Ddambacher, U.R. Tjaden and J. Van der Greef, *Pharm. Res.*, 16 (1999) 695–701.
148 E.M. Jäverfalk-Hoyes, U. Bondesson, D. Westerlund and P.E. Andrén, *Electrophoresis*, 20 (1999) 1527–1532.
149 M. Larsson and E.S.M. Lutz, *Electrophoresis*, 21 (2000) 2859–2865.
150 M.J. Cugat, C. Aguilar, R.M. Marce, F. Borrull and M. Calull, *Electrophoresis*, 23 (2002) 2279–2287.
151 A.J. Tomlinson, N.A. Guzman and S. Naylor, *J. Cap. Elect.*, 2 (1995) 247–266.
152 A.J. Tomlinson, L.M. Benson, W.D. Braddock, R.P. Oda and S. Naylor, *J. High Resol. Chromatogr.*, 78 (1995) 381–383.
153 A.J. Tomlinson and S. Naylor, *J. Cap. Elect.*, 2 (1995) 225–233.

154 A.J. Tomlinson, L.M. Benson, S.C. Jameson, D.H. Johnson and S. Naylor, *J. Am. Soc. Mass Spectrom.*, 8 (1997) 15–24.
155 L.M. Benson, A.J. Tomlinson, A.N. Mayeno, G.J. Gleich, D. Wells and S. Naylor, *J. High Resol. Chromatogr.*, 19 (1996) 291–294.
156 S. Naylor, L.M. Benson, B.L. Williamson, A.J. Tomlinson and G.J. Gleich. In: N.J. Gooderham (Ed.), *Drug Metabolism: Towards the Next Millennium* IOS Press, Amsterdam, 1998.
157 A.J. Tomlinson and S. Naylor, *J. Liq. Chromatogr.*, 18 (1995) 3591.
158 S. Naylor, Q. Ji, K.L. Johnson, A.J. Tomlinson, W.C. Kieper and S.C. Jameson, *Electrophoresis*, 19 (1998) 2207–2212.
159 K.R. Jonscher and J.R. Yates, *Proceedings of the 45th ASMS Conference on MS and Allied Topics*, Palm Springs, CA, 1997 p 1192.
160 A.J. Tomlinson, L.M. Benson, R.P. Oda, W.D. Braddock, B.L. Riggs, J.A. Katzmann and S. Naylor, *J. Cap. Elect.*, 2 (1995) 97–104.
161 A.J. Tomlinson, L.M. Benson, S.C. Jameson and S. Naylor, *Electrophoresis*, 17 (1996) 1801–1807.
162 A.J. Tomlinson, S.C. Jameson and S. Naylor, *J. Chromatogr. A*, 744 (1996) 273–278.
163 K.A. Hogquist, A.J. Tomlinson, W.C. Kieper, M.A. McGargill, M.C. Hart, S. Naylor and S.C. Jameson, *Immunity*, 6 (1997) 389–399.
164 A.J. Tomlinson, R. Gallagher, P.J. Derrick, G.W. Butcher, S.J. Powis and S. Naylor, *Proceedings of the 44th ASMS Conference on MS and Allied Topics*, Portland, OR, 1996 p 942.
165 E. Rohde, A.J. Tomlinson, D.H. Johnson and S. Naylor, *J. Chromatogr. B*, 713 (1998) 301–311.
166 E. Rohde, A.J. Tomlinson, D.H. Johnson and S. Naylor, *Electrophoresis*, 19 (1998) 2361–2370.
167 F. Li, A.J. Tomlinson, Q. Ji, F.L. Neely and S. Naylor, *Proceedings of the 45th ASMS Conference on MS and Allied Topics*, Palm Springs, CA, 1997 p 1290.
168 S. Naylor, A.J. Tomlinson, J.M. Londowski and R. Kumar, *Proceedings of the 44th ASMS Conference on MS and Allied Topics*, Portland, OR, 1996 p 934.
169 A.B. Wey and W. Thormann, *J. Chromatogr. A*, 924 (2001) 507–518.
170 R. Rodríguez, J. Mañes and Y. Picó, *Anal. Chem.*, 75 (2003) 452–459.
171 D.L.D. Derforce, J. Raymackers, L. Meheus, F. Van Wijnendaele, A. De Leenheer and E.G. Van den Eeckhout, *Anal. Chem.*, 70 (1998) 3060–3068.
172 J. Chen, B.M. Balgley, D.L. De Voe and C.S. Lee, *Anal. Chem.*, 75 (2003) 3145–3152.
173 D. Mohan, L. Paša-Tolić, C.D. Masselon, N. Tolic, B. Bogdanov, K.K. Hixson, R.D. Smith and C.S. Lee, *Anal. Chem.*, 75 (2003) 4432–4440.
174 J.W. Cooper, J. Chen, Y. Li and C.S. Lee, *Anal. Chem.*, 75 (2003) 1067–1074.

175 R.D. Oleschunk and D.J. Harrison, *Trends Anal. Chem.*, 19 (2000) 379–388.
176 B. Grass, G. Weber, A. Neyer, M. Schilling and R. Hergenröder, *Spectrochim. Acta*, 57 (2002) 1575–1583.
177 P.A. Auroux, D. Iossifidis, D.R. Reyes and A. Manz, *Anal. Chem.*, 74 (2002) 2637–2652.
178 G.J.M. Bruin, *Electrophoresis*, 21 (2000) 3931–3951.
179 N. Lion, J.O. Gellon, H. Jensen and H.H. Girault, *J. Chromatogr. A*, 1003 (2003) 11–19.
180 Q. Xue, F. Foret, Y.M. Dunayevski, P.M. Zavracky, N.E. McGruer and B.L. Karger, *Anal. Chem.*, 69 (1997) 426–430.
181 R.S. Ramsey and J.M. Ramsey, *Anal. Chem.*, 69 (1997) 1174–1178.
182 D. Figeys, Y. Ning and R. Aebersold, *Anal. Chem.*, 69 (1997) 3153–3160.
183 B. Zhang, H. Liu, F. Foret and B.L. Karger, *Anal. Chem.*, 71 (1999) 3258–3264.
184 B. Zhang, F. Foret and B.L. Karger, *Anal. Chem.*, 72 (2000) 1015–1022.
185 J. Li, P. Thibault, N.H. Brings, C.D. Skinner, C. Wang, C.L. Coyler and J.D. Harrison, *Anal. Chem.*, 71 (1999) 3036–3045.
186 P. Thibault, N.H. Brings, C.D. Skinner, C. Wang, C.L. Coyler and J.D. Harrison, *Anal. Chem.*, 71 (1999) 3292–3296.
187 L. Licklider, X. Wang, A. Desai, Y. Tai and T. Lee, *Anal. Chem.*, 72 (2000) 367–375.
188 A. Schultz, T.N. Corso, S.J. Prosser and S. Zhang, *Anal. Chem.*, 72 (2000) 4058–4063.
189 M. Svedberg, A. Pettersson, S. Nilsson, J. Bergquist, L. Nyholm, F. Nikolajeff and K. Markides, *Anal.Chem.*, 75 (2003) 3934–3940.
190 T. Wachs and J. Henion, *Anal. Chem.*, 73 (2001) 632–638.
191 Y. Tachibana, K. Otsuka, S. Terabe, A. Arai, K. Suzuki and S. Nakamura, *J. Chromatogr. A*, 1011 (2003) 181–192.
192 Y. Deng, J. Henion, J. Li, P. Thibault, C. Wang and D.J. Harrison, *Anal. Chem.*, 73 (2001) 639–646.
193 Y. Deng, H. Zhang and J. Henion, *Anal. Chem.*, 73 (2001) 1432–1439.
194 J. Wen, Y. Lin, F. Xiang, D.W. Matson, H.R. Udseth and R.D. Smith, *Electrophoresis*, 21 (2000) 191–197.
195 A. Chartogne, U.R. Tjaden and J. Van der Greef, *Rapid Commun. Mass Spectrom.*, 14 (2000) 1269–1274.
196 D. Figeys, S.P. Gygi, G. Mckinnon and R. Aebersold, *Anal. Chem.*, 70 (1998) 3728–3734.
197 Z.H. Fan and D.J. Harrison, *Anal. Chem.*, 66 (1994) 177–184.
198 B. Zhang, F. Foret and B.L. Karger, *Anal. Chem.*, 73 (2001) 2675–2681.
199 J. Kameoka, R. Orth, B. Ilic, D. Czaplewski, T. Wachs and H.G. Craighead, *Anal. Chem.*, 74 (2002) 5897–5901.

200 H. Hu and N.J. Dovichi, *Anal. Chem.*, 74 (2002) 2833–2850.
201 C.K. Larive, S.M. Lunte, M. Zhong, M.D. Perkins, G.S. Wilson, G. Gokulrangan, T. Williams, F. Afroz, C. Schöneich, T.S. Derrick, C.R. Middaugh and S. Bogdanowich-Knipp, *Anal. Chem.*, 71 (1999) 389R–423R.
202 M. Takashi, *Electrophoresis*, 20 (1999) 3116–3121.
203 K. Vaclav, *Electrophoresis*, 20 (1999) 3084–3105.
204 L. Yang, Q. Tang, A.K. Harrata and C.S. Lee, *Anal. Biochem.*, 243 (1996) 140–149.
205 C-X. Zhang, F. Xiang, L. Paša-Tolić, G.A. Anderson, T.D. Veenstra and R.D. Smith, *Anal. Chem.*, 72 (2000) 1462–1468.
206 J. Wei, L. Yang, A.K. Harrata and C.S. Lee, *Electrophoresis*, 19 (1998) 2356–2360.
207 N.J. Clarke and S. Naylor, *Biomed. Chromatogr.*, 16 (2002) 287–297.
208 L. Yang, C.S. Lee, S.A. Hofstadler, L. Paša-Tolić and R.D. Smith, *Anal. Chem.*, 70 (1998) 3235–3241.
209 P.K. Jensen, L. Paša-Tolić, K.K. Peden, S. Martinović, M.S. Lipton, G.A. Anderson, N. Tolić, K-K. Wong and R.D. Smith, *Electrophoresis*, 21 (2000) 1372–1380.
210 S. Martinović, L. Paša-Tolić, C. Masselon, P.K. Jensen, C.L. Stone and R.D. Smith, *Electrophoresis*, 21 (2000) 2368–2375.
211 S. Martinović, S.J. Berger, L. Paša-Tolić and R.D. Smith, *Anal. Chem.*, 72 (2000) 5356–5360.
212 C.B. Knudsen, I. Bjørnsdottir, O. Jøns and S.H. Hansen, *Anal. Biochem.*, 265 (1998) 167–175.
213 B. Yeung, T.J. Porter and J.E. Vath, *Anal. Chem.*, 69 (1997) 2510–2516.
214 P. Cao and M. Moini, *J. Am. Soc. Mass Spectrom.*, 9 (1998) 1081–1088.
215 J.F. Kelly, L. Ramaley and P. Thibault, *Anal. Chem.*, 69 (1997) 51–60.
216 D. Figeys, I. Van Oostveen, A. Ducret and R. Aebersold, *Anal. Chem.*, 68 (1996) 1822–1828.
217 M.E. McComb and H. Perreault, *Electrophoresis*, 21 (2000) 1354–1362.
218 J-T. Wu, M.G. Qian, M.X. Li, L. Liu and D.M. Lubman, *Anal. Chem.*, 68 (1996) 3388–3396.
219 K.P. Bateman, R.L. White, M. Yaguchi and P. Thibault, *J. Chromatogr. A*, 794 (1998) 327–344.
220 M.E. McComb, A.N. Krutchinsky, W. Ens, K.G. Standing and H. Perreault, *J. Chromatogr. A*, 800 (1998) 1–11.
221 M.X. Li, J-T. Wu, L. Liu and D.M. Lubman, *Rapid Commun. Mass Spectrom.*, 11 (1997) 99–108.
222 M. Moini, P. Cao and A.J. Bard, *Anal. Chem.*, 71 (1997) 1658–1661.
223 P. Cao and M. Moini, *Rapid Commun. Mass Spectrom.*, 12 (1998) 864–870.
224 G.M. Janini, T.P. Conrads, K.L. Wilkens, H.J. Issaq and T.D. Veenstra, *Anal. Chem.*, 75 (2003) 1615–1619.

225 G.M. Janini, M. Zhou, L-R. Yu, J. Blonder, M. Gignac, T.P. Conrads, H.J. Issaq and T.D. Veenstra, *Anal. Chem.*, 75 (2003) 5984–5993.

226 J.C.M. Waterval, G. Hommels, P. Bestebreurtje, C. Versluis, A.J.R. Heck, A. Bult, H. Lingeman and W.J.M. Underberg, *Electrophoresis*, 22 (2001) 2709–2716.

227 C.J. Herring and J. Qin, *Rapid Commun. Mass Spectrom.*, 13 (1999) 1–7.

228 J.C.M. Waterval, P. Bestebreurtje, H. Lingeman, C. Versluis, A.J.R. Heck, A. Bult and W.J.M. Underberg, *Electrophoresis*, 22 (2001) 2701–2708.

229 D. Figeys, A. Ducret and R. Aebersold, *J. Chromatogr. A*, 763 (1997) 295–306.

230 S. De Boni, C. Neusüß, M. Pelzing and G.K.E. Scriba, *Electrophoresis*, 24 (2003) 874–882.

231 V. Sanz-Nebot, F. Benavente, E. Balaguer and J. Barbosa, *Electrophoresis*, 24 (2003) 883–891.

232 J.F. Banks Jr. and T. Dresch, *Anal. Chem.*, 68 (1996) 1480–1485.

233 T. Kaisera, A. Hermannb, J.T. Kielsteinb, S. Wittkea, Bartela, R. Krebsa, F. Hausadela, M. Hillmanna, I. Golovkoa, P. Koestera, H. Hallerb, E.M. Weissingera, D. Fliserb and H. Mischak, *J. Chromatogr. A*, 1013 (2003) 157–171.

234 S. Wittkea, D. Fliserb, M. Haubitzb, S. Bartela, R. Krebsa, F. Hausadela, M. Hillmanna, I. Golovkoa, P. Koestera, H. Hallerb, T. Kaisera, H. Mischaka and E.M. Weissinger, *J. Chromatogr. A*, 1013 (2003) 173–181.

235 K.P. Bateman, R.L. White and P. Thibault, *J. Mass Spectrom.*, 33 (1998) 1109–1123.

236 M.A. Hoitink, E. Hop, J.H. Beijnen, A. Bult, J.J. Kettenes-van den Bosch and W.J.M. Underberg, *J. Chromatogr. A*, 776 (1997) 319–327.

237 J.F. Kelly, S.J. Locke, L. Ramaley and P. Thibault, *J. Chromatogr. A*, 720 (1996) 409–427.

238 T. Liu, X-X. Shao, R. Zeng and Q-C. Xia, *J. Chromatogr. A*, 855 (1999) 695–707.

239 K. Koezuka, H. Ozaki, N. Matsubara and S. Terabe, *J. Chromatogr. B*, 689 (1997) 3–11.

240 G. Guetens, G. De Boeck, M.S. Highley, M. Wood, R.A.A. Maes, A.A.M. Eggermont, A. Hanauske, E.A. De Bruijn and U.R. Tjaden, *J. Chromatogr. A*, 976 (2002) 239–247.

241 H-J. Gaus, Z.Z. Gögüs, K. Schmeer, B. Behnke, K-A. Kovar and E. Bayer, *J. Chromatogr. A*, 735 (1996) 221–226.

242 K. Heinig and J. Henion, *J. Chromatogr. B*, 735 (1999) 171–188.

243 J.L. Tsai, W-S. Wu and H-H. Lee, *Electrophoresis*, 21 (2000) 1580–1586.

244 F.P.W. Tang, G.N.W. Leung and T.S.M. Wan, *Electrophoresis*, 22 (2001) 2201–2209.

245 S. Heitmeier and G. Blaschke, *J. Chromatogr. B*, 721 (1999) 109–125.
246 S. Heitmeier and G. Blaschke, *J. Chromatogr. B*, 721 (1999) 93–108.
247 S. Sentellas, L. Puignou, E. Moyano and M.T. Galcerán, *J. Chromatogr. A*, 888 (2000) 281–292.
248 L. Geiser, S. Rudaz and J-L. Veuthey, *Electrophoresis*, 24 (2003) 3049–3056.
249 L. Geiser, S. Cherkaoui and J-L. Veuthey, *J. Chromatogr. A*, 895 (2000) 111–121.
250 S. Cherkaoui, S. Rudaz and J-L. Veuthey, *Electrophoresis*, 22 (2001) 491–496.
251 W. Lu and R.B. Cole, *J. Chromatogr. B*, 714 (1998) 69–75.
252 R.L. Sheppard, X. Tong, J. Cai and J.D. Henion, *Anal. Chem.*, 67 (1995) 2054–2058.
253 M.H. Lamoree, A.F.H. Sprang, U.R. Tjaden and J. Van der Greef, *J. Chromatogr. A*, 742 (1996) 235–242.
254 S. Fanali, D. Desiderio, G. Schulte, S. Heitmeier, D. Strickmann, B. Chankvetadze and G. Blaschke, *J. Chromatogr. A*, 800 (1998) 69–76.
255 Y. Tanaka, Y. Kishimoto and S. Terabe, *J. Chromatogr. A*, 802 (1998) 83–88.
256 B. Toussaint, M. Palmer, P. Chiap, P. Hubert and J. Crommen, *Electrophoresis*, 22 (2001) 1363–1372.
257 S. Rudaz, S. Cherkaoui, J-Y. Gauvrit, P. Lanteri and J-L. Veuthey, *Electrophoresis*,, 22 (2001) 3316–3326.
258 S. Rudaz, S. Cherkaoui, P. Dayer, S. Fanali and J-L. Veuthey, *J. Chromatogr. A*, 868 (2000) 295–303.
259 S. Cherkaoui and J-L. Veuthey, *J. Pharm. Biomed. Anal.*, 27 (2002) 615–626.
260 T. Soga and D.N. Heiger, *Anal. Chem.*, 72 (2000) 1236–1241.
261 A. Martin-Girardeau and M-F. Renou-Gonnord, *J. Chromatogr. B*, 742 (2000) 163–171.
262 M. Moini, C.L. Schultz and H. Mahmood, *Anal. Chem.*, 75 (2003) 6282–6287.
263 D. Wycherley, M.E. Rose, K. Giles, T.M. Hutton and D.A. Rimmer, *J. Chromatogr. A*, 734 (1996) 339–349.
264 E. Moyano, D.E. Games and M.T. Galcerán, *Rapid Commun. Mass Spectrom.*, 10 (1996) 1379–1385.
265 M.W.F. Nielen, *J. Chromatogr. A*, 712 (1995) 269–284.
266 X. Song and W.L. Budde, *J. Chromatogr. A*, 829 (1998) 327–340.
267 K. Otsuka, C.J. Smith, J. Grainger, J.R. Barr, D.G. Patterson Jr., N. Tanaka and S. Terabe, *J. Chromatogr. A*, 817 (1998) 75–81.
268 C-Y. Tsai, Y-R. Chen and G-R. Her, *J. Chromatogr. A*, 813 (1998) 379–386.
269 W.M. Nelson, Q. Tang, K. Harrata and C.S. Lee, *J. Chromatogr. A*, 749 (1996) 219–226.

270 P. Bednář, K. Lemr, P. Barták, J. Ševčík, J. Hlaváč, J. Stýskala, I. Wiedermannová and Z. Stránský, *J. Mass Spectrom.*, 37 (2002) 1213–1218.
271 F. Lafont, M.A. Aramendia, I. García, V. Borau, C. Jiménez, J.M. Marinas and F.J. Urbano, *Rapid Commun. Mass Spectrom.*, 13 (1999) 562–567.
272 J-Y. Zhao, P. Thibault, T. Tazawa and M.A. Quilliam, *J. Chromatogr. A*, 781 (1997) 555–564.
273 M.A. Aramendia, I. Garcia, F. Lafont and J.M. Marinas, *J. Chromatogr. A*, 77 (1995) 327–333.
274 M. Unger, D. Stöckigt, D. Belder and J. Stöckigt, *J. Chromatogr. A*, 767 (1997) 263–276.
275 G. Bianco, P. Schmitt-Kopplin, A. Crescenzi, S. Comes, A. Kettrup and T.R.I. Cataldi, *Anal. Bioanal. Chem.*, 375 (2003) 799–804.
276 U. Warnke, J. Gysler, B. Hofte, U.R. Tjaden, J. Van der Greef, C. Kloft, W. Schunack and U. Jaehde, *Electrophoresis*, 22 (2001) 97–103.
277 C.W. Kampfl and W. Buchberger, *Electrophoresis*, 22 (2001) 2737–2742.
278 S. Duteil, P. Gareil, S. Girault, A. Mallet, C. Feve and L. Siret, *Rapid Commun. Mass Spectrom.*, 13 (1999) 1889–1898.
279 J. Suomi, S.K. Wiedmer, M. Jussila and M-L. Riekkola, *Electrophoresis*, 22 (2001) 2580–2587.
280 M.A. Kelly, T.J. McLellan and P.J. Rosner, *Anal. Chem.*, 74 (2002) 1–9.

Chapter 10

Inductively coupled plasma-mass spectrometry for specific detection in capillary electrophoresis

Gloria Álvarez Llamas, Mª del Rosario Fernández de la Campa, Elisa Blanco González and Alfredo Sanz Medel

10.1 THE INDUCTIVELY COUPLED PLASMA-MASS SPECTROMETRY (ICP-MS) AS A SPECIFIC ATOMIC DETECTOR

Since its introduction in the 1980s, inductively coupled plasma-mass spectrometry (ICP-MS) has become one of the most powerful techniques for elemental analysis and is used worldwide. Its potential has already been demonstrated in an extremely wide range of applications in many different fields, and it exhibits unique features in terms of sensitivity, specificity, multi-elemental capacity, linear dynamic range, possibility of measuring isotopic ratios and easy modification of the sample introduction system [1].

10.1.1 ICP-MS fundamentals

Although different types of mass analysers have been reported in the literature, the quadrupole type (Q) is the most commonly used. Unfortunately, many spectral interferences (particularly the overlapping of analyte with polyatomic MS signals) occur in such Q-mass analysers. The more general approach to overcome such spectral interferences is to use double focussing (DF), also called sector field (SF), ICP-MS instruments owing to their much higher "resolving power", R (e.g., R up to 10,000 versus the typical $R = 300$ in quadrupoles).

However, working in high-resolution mode means a consequent decrease in sensitivity, and that may compromise the determination of a

Comprehensive Analytical Chemistry XLV
M.L. Marina, A. Ríos and M. Valcárcel (Eds)
Volume XLV ISSN: 0166-526X DOI: 10.1016/S0166-526X(05)45010-2

target element present at low concentrations. Recently, the incorporation of collision cells or dynamic reaction cells into ICP-(Q)MS instruments has been found to enable suppression of the signals of unwanted ions (i.e., resolution of analyte signals from polyatomics) by means of selective reactions with an appropriate gas to destroy interfering ions. On the other hand, the two types of mass analysers described (Q and DF) are sequential devices suffering from important limitations in those applications demanding simultaneous detection of the ions. Good examples of this are multi-isotope and multi-elemental determination in fast transient signals where acceptable results can only be achieved at the expense of precision. Time-of-flight (TOF) mass analysers allow the acquisition of a full mass spectra in a much shorter period of time (30,000 spectra per second) than in sequential systems, so these instruments should be ideal for isotopic ratio measurements or multi-elemental analyses in transient signals [2]. In terms of isotope ratios precision attainable, TOF mass analysers are recommended when measuring transient signals faster than 8 s (full-width at half-maximum intensity), or if the number of isotopes to be measured is above 15 [3]. Unfortunately, detection limits observed with the (TOF) instrument were 5–20 times worse than those by using the quadrupole. Thus, the reduced sensitivity of TOF mass analysers may be an important drawback when coupled to capillary electrophoresis (CE), particularly considering the low sample volumes typically injected.

In the present chapter, particular attention will be paid to the quadrupole mass analyser because ICP-(Q)MS has been most frequently used as a specific detector after CE separations. The main components of an ICP-MS (quadrupole type) can be seen in Fig. 10.1.

Samples to be measured by ICP-MS are generally in liquid state. Thus, the most common sample introduction system is via nebulisation. As can be seen in Fig. 10.1, liquid samples are delivered to the nebuliser by means of a peristaltic pump (or by self-aspiration with pneumatic devices), leading to aerosol formation. Larger droplets are eliminated in a spray chamber usually cooled at 2 °C, after which analytes are transported to the ICP torch. The torch, whose position can be finely adjusted in the three axes, consists of three quartz concentric tubes, and the sample aerosol is transported through the central torch channel by means of an adequate nebuliser gas flow (usually Ar). An auxiliary Ar gas flow goes through the intermediate torch channel (usually at 1 l/min) in order to locate the plasma basis with respect to the torch. A much higher Ar gas flow (typically 14–16 l/min) is supplied through the

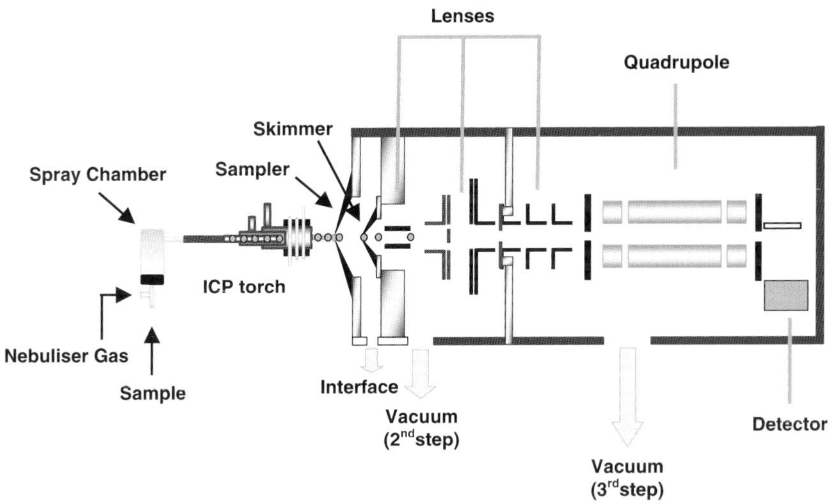

Fig. 10.1. Schematic view of an ICP-MS quadrupole type.

external torch tubing, which is surrounded by an induction coil connected to a radio frequency (RF) generator. The magnetic field generated there by the RF energy induces an electric current in the gas stream (eddy currents) when seeded with electrons (e.g., by means of a Tesla spark). Argon ions are then formed, and their amount is continuously increased by collisions of free electrons with neutral Ar atoms up to an equilibrium situation where a steady-state plasma is formed.

Once the ICP is formed, the sample aerosol is introduced into the plasma (via nebulisation) where it is desolvated, vaporised, atomised and ionised, and sample-formed ions are vacuum-extracted towards the mass analyser through the sampler cone. Sample introduction, plasma generation and all the processes leading to sample ionisation take place at atmospheric pressure. However, the mass analyser zone is under high vacuum (i.e., 10^{-4}–10^{-5} Pa). Therefore, an ICP-MS interface to accommodate such large pressure difference is needed. To do so, a first step intermediate vacuum (i.e., $3-5 \times 10^2$ Pa) in the interface between sampler and skimmer cones is secured by means of a rotary pump. After passing through the skimmer cone, ions are pushed by differential pressure towards a set of lenses held at different potentials to focus them in a linear beam (ion optics) towards the mass analyser. Extraction and Einzel lenses accelerate and focus the ions, respectively, while Omega lenses modify ion paths in order to direct photons to an exit lens, where they collide. Once in the mass analyser (quadrupole), ions are separated in response to their mass-to-charge ratio in the high-vacuum zone achieved by means of a turbomolecular pump. The quadrupole, as the name suggests, consists of four parallel, cylindrical poles (or rods) arranged symmetrically. The voltage connections to the rods are such that opposite rods have the same polarity, while the adjacent rods have opposite polarity, and the voltage applied has two components: a direct current component, U, and an RF component, $V_0(\cos \omega t)$, where ω is the frequency of the RF voltage. The beam of ions coming from the source, on entering this electric field in the quadrupole, will oscillate in such a way that only ions of a certain mass (m/z, really) at a given RF voltage will proceed through the whole quadrupole assembly and reach the detector. Under these conditions, all other masses will have unstable trajectories in the quadrupole (they will be lost by absorption on the rods of the quadrupole). In this way, mass separation is achieved and the quadrupole behaves as a mass filter. In order to produce a mass spectrum, U and V_0 are varied (while the ratio U/V_0 is kept essentially constant), and in a millisecond time frame, a full spectrum (scanning from mass 1 to 250) can be obtained.

10.1.2 ICP-MS as detection system in hyphenated techniques

Elements play important roles in biological and environmental processes. Traditionally, estimates of total elemental concentrations in a certain sample have been required in order to assess their functions. However, today it is recognised worldwide that the distribution of elements among the different elemental species present in a given sample will determine their actual toxicity, essentiality, bioavailability, metabolism and transport. Therefore, the "speciation analysis" concept, involving those activities under taken to identify or measure the quantities of one or more individual chemical species in a sample, has been actively pursued in the last decade. The distribution of an element among defined chemical species in a system is then termed "element speciation" [4].

In speciation analysis, a powerful analytical separation technique, able to differentiate among the species under study, is typically coupled on-line to a highly sensitive and specific detector (which makes it possible to determine the elements present in each of the species previously separated). In fact, hyphenated techniques have developed into the most powerful tool for carrying out speciation studies in the most varied fields, including proteomics or metabolomics, and in bio-inorganic, clinical, pharmaceutical and environmental fields [5–9].

Considering the low levels at which trace elements are naturally present in most biological and environmental matrices, ICP-MS is now established as the most powerful detection choice for speciation purposes owing to its atomic specificity, multi-elemental character, wide linear dynamic range and capacity to provide isotopic abundance information [10–15].

The coupling of liquid chromatography (LC) to ICP-MS has been widely documented in the literature [16–18]. Typical flows emerging from the chromatographic column (0.5–2 ml/min) match those required for conventional nebulisers quite well. Therefore, the LC–ICP-MS interface design is very simple, consisting of a polyetheretherketone (PEEK) or Teflon tube directly connecting the exit of the LC column to the entrance of the nebuliser. As conventional nebulisers for ICP suffer from a low transport efficiency (which does not usually exceed 2%), μLC is becoming an increasingly used alternative to be coupled to ICP-MS because sample flows are highly reduced compared with those used in conventional LC [19]. In this way, transport efficiencies are substantially increased.

In any case, the main limitation for this type of coupling arises from the composition of the mobile phase. High organic solvent contents, common in reversed-phase HPLC, will cause plasma instability, high reflected powers and even a real risk of extinguishing the plasma. Carbon deposits may also take place in the sampler and skimmer cones, degrading ion transmission from the plasma to the mass analyser and resulting in reduced sensitivity. In the same way, mobile phases with high salinity contents may clog the solution channel in the nebuliser and cause salt deposits in the ICP-MS cones with a similar effect. Consequently, it may be the case in reversed-phase or ionic exchange LC that mobile phases chosen to fulfil the requirements needed in terms of separation may not be adequate when using ICP-MS as the online detection system. The interested reader can refer to chapter 4, entitled "Liquid Chromatographic Techniques for Trace Element Speciation Analysis" in [12] (pp. 81–121) for a review of this topic.

Coupling of gas chromatography (GC) to an ICP-MS detector has unique advantages over liquid-phase separation techniques. The direct introduction of sample species as a vapour results in practically 100% transport efficiency. Also, better atomisation and ionisation of analyte species is accomplished, as no energy from the plasma needs to be used in desolvation and volatilisation processes. In contrast to HPLC, however, the analytes often have to be derivatised in order to form volatile compounds. Besides, the GC–ICP interface itself is more complicated than in a LC-ICP-MS coupling, as it needs to ensure the stability of species in gas phase and secure their separation achieved in the GC column throughout their way from the exit of the GC column to the plasma. A recent review compiles the state of the art of this important coupled technique, including interfacing considerations and applications in environmental and biological samples [20].

To a lesser extent, supercritical fluid chromatography (SFC) has also been coupled to ICP-MS [10,21]. Here, the mobile phase needs to be changed from a supercritical state used for the separation to a gas phase, which can be easily handled by the ICP-MS. Up to now, however, very limited applications have been published on this topic.

10.2 INTERFACING CE TO ICP-MS: POTENTIALS AND LIMITATIONS

CE now constitutes a powerful alternative to the more established liquid chromatographic techniques. This is so because of its higher separation efficiency, lower amounts of sample and reagents consumed and shorter analysis times. Its use may be of particular interest in applications where interactions of analytes within the chromatographic stationary phase may take place, making it difficult to preserve species stability (which is very important in speciation analysis). As only a few nanolitres are typically injected into the capillary, CE may also be the separation technique of choice in research studies where samples are available only in limited amounts (i.e., studies of biological tissues and fluids). However, it is well known that the main limitation of CE is its comparative lack of sensitivity, resulting in concentration-based detection limits typically two orders of magnitude higher than those obtained by using HPLC separations.

Keeping all these considerations in mind, the main advantages derived from the CE–ICP-MS couplings are as follows: (i) complex mixtures may be resolved under mild separation conditions (securing original speciation of samples); (ii) in most cases, previous derivatisation stages are not needed; (iii) owing to the nl/min flows, optimum CE separation conditions can be easily extended to ICP-MS detection, without problems derived from incompatibilities between liquid mobile phase composition and plasma stability, if the interface is adequate; of course, high volume samples are not required; and (iv) elements and isotopes in any of the species present in the sample may be identified on-line and reliably quantified at the same time, owing to the specificity, multi-elemental capability and extremely high sensitivity of the ICP-MS.

In any case, the main limitation in CE–ICP-MS work is the interface itself, as documented by the great number of attempts to build up an efficient interface (which highlights the difficulty of the task). A comprehensive review, which includes practical considerations and applications of CE–ICP-MS in speciation studies has been recently published [22].

In order to clarify the different ways of possible coupling of a CE separation to an ICP-MS detector, an extended view of the state of the art of this hybrid technique, its background, applications and trends are revisited in the present chapter. The main aspects to be discussed in the following sections will deal with: (i) interface design via nebulisation; (ii) interface design via volatile species generation; and (iii) other recent approaches to improve sensitivity/detection limits achievable by CE–ICP-MS.

10.3 CE–ICP-MS COUPLING VIA NEBULISATION

As detailed in section 10.1, the sample introduction system in the common configuration of a commercial ICP-MS instrument consists of a nebuliser and a spray chamber, where a fine aerosol of the sample solution is generated and directly transported to the plasma. Most samples to be analysed are in the liquid state and so they can be directly nebulised. As LC–ICP-MS couplings consist of a direct connection between the chromatographic column and the nebulisation system, without any further modification, many attempts have been made to adopt the same paradigm for CE–ICP-MS couplings. Unfortunately, the interface in this latter case is much more critical than in LC–ICP-MS, and it is important to consider all the aspects involved in such interface design.

10.3.1 The interface: key to success

The term "interface" refers to the instrumental part of the coupling that allows the transport of analytes from the outlet end of the CE capillary to the ICP-MS detector. The challenge of creating such an interface requires that the following conditions to be fulfilled:

(i) The electrical connection has to be maintained, despite the impossibility of keeping the CE capillary outlet immersed in the cathodic reservoir (as typically occurs in a CE–UV system). For this purpose, a "make-up" flow of an electrolyte solution (also named "sheath flow") is additionally introduced through the interface, wetting the capillary end in a constant mode and closing the electrical circuit. Alternatives to the make-up solution consist of coating the outside of the CE capillary with conducting silver paint [23] or inserting a Pt wire, used as an electrode, through the exit of the CE capillary for electrical contact with the electrolyte solution [24].

(ii) The very low flows typically emerging from the CE capillary (0.1–0.9 μl/min) should match those required for the nebuliser under ICP-MS optimum operation conditions (typically 100–1000 μl/min). To match them, the addition of a make-up solution through the interface is the common choice.

(iii) The separation resolution achieved in the capillary (as shown by UV detection) should be preserved in the CE–ICP-MS coupling. Particularly, special care has to be taken when using pneumatic concentric nebulisers as, owing to their geometry, a suction effect may take place at the tip of the nebuliser. In consequence, a laminar flow may be induced inside the CE capillary, and so the separation previously achieved may be totally degraded. Further considerations on this topic will be discussed later on.

(iv) The interface should guarantee high transport efficiency, otherwise the high sensitivity that characterises ICP-MS may be extremely reduced. This is of critical importance here owing to the typical low injection volumes in CE separations.

Figure 10.2 shows the schematics of six typical interface designs reported in the literature, as an illustration.

As stated before, the most common way of maintaining the electrical connection at the end of the CE capillary, when coupled to ICP-MS detection, is by means of a sheath or make-up flow. That solution is introduced through the interface by means of a pumping device or in self-aspiration mode (for this, a suction effect needs to be originated at the nebuliser tip). The main body of the interface design consists of a T-piece or a cross-connector to which a Pt wire or stainless-steel tubing is fixed in order to close the electrical circuit. Different configurations for such basic arrangements can be seen in Fig. 10.2A, B and C. Particular attention has to be paid to the interface designed by Schaumlöffel and Prange [25], which is schematically presented in Fig. 10.2D, because it has been patented [26] and is commercially available. In this case, a low dead volume spray chamber is incorporated

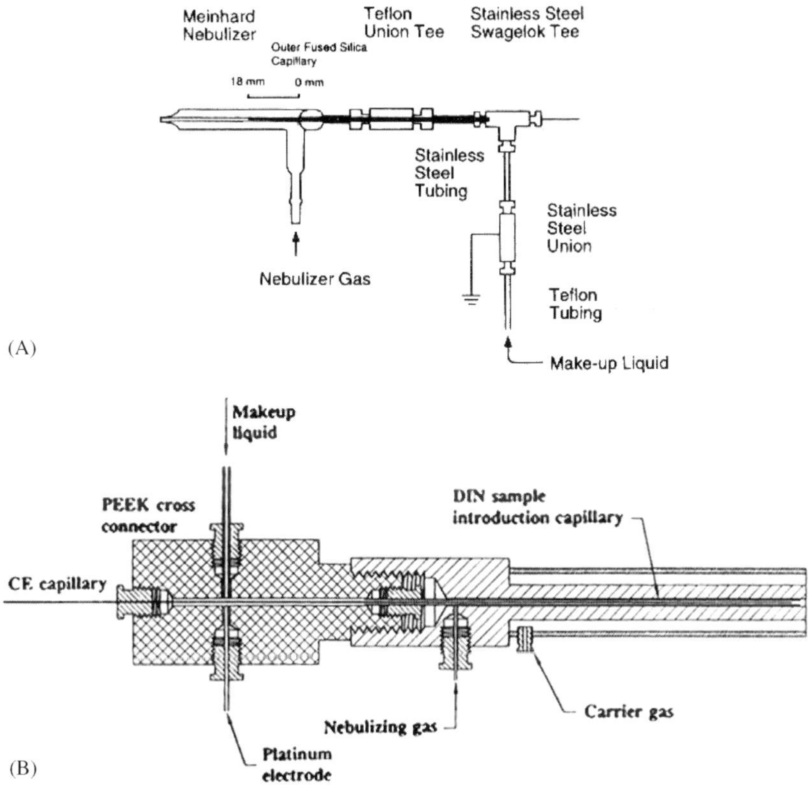

(A)

(B)

Fig. 10.2. Selected CE–ICP-MS interfaces reported in the literature. (A) Reproduced with permission from Ref. [30]. Copyright (1995), American Chemical Society; (B) Reproduced with permission from Ref. [69]. Copyright (1995), American Chemical Society; (C) Reproduced with permission from Ref. [29]. Copyright (1996), American Chemical Society; (D) Reproduced from Ref. [25] with permission from Springer; (E) Reproduced with permission from Ref. [23]. Copyright (1995), American Chemical Society; (F) Reproduced from Ref. [37]. Copyright (2000), with permission from Elsevier.

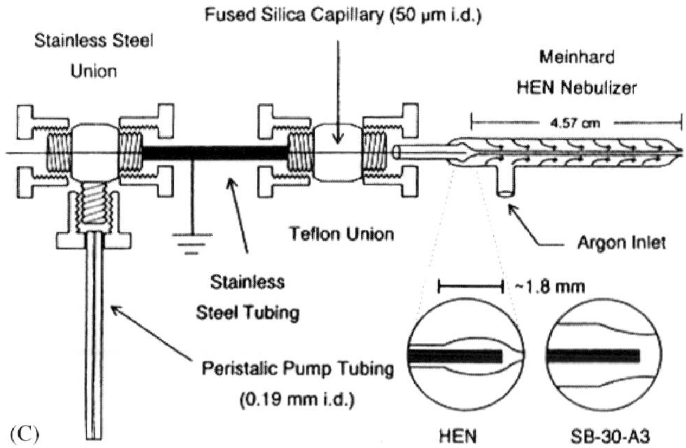

(C)

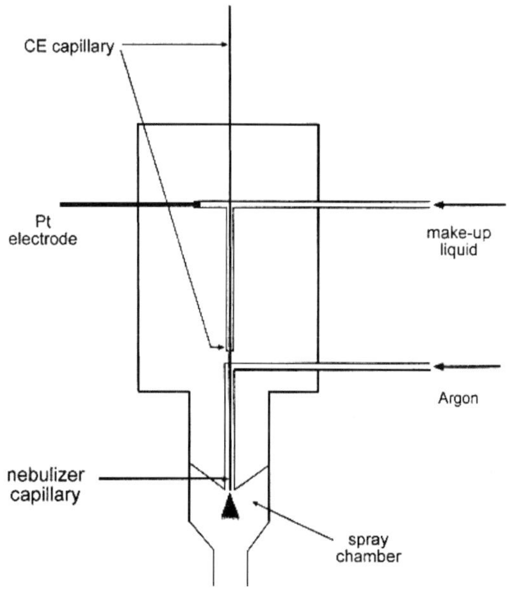

(D)

Fig. 10.2. Continued.

and all dimensions are minimised in order to allow self-aspiration of the nebuliser at very low flows (2–12 μl/min). In this way, any laminar flow (which would degrade the resolution achieved in the CE separation) is avoided. Figures 10.2E and F show less popular interfaces where the sheath or make-up solution is not needed.

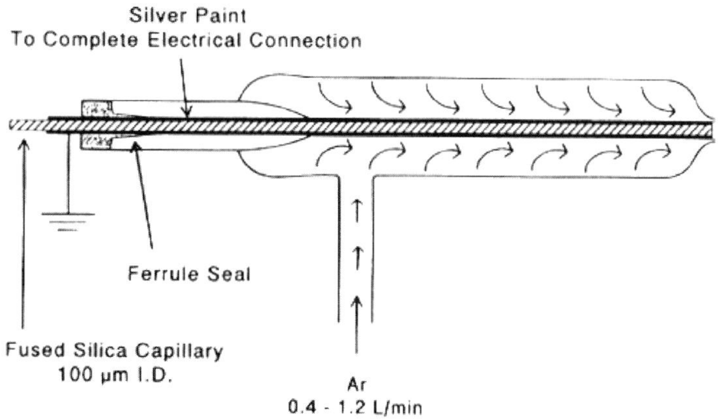

(E)

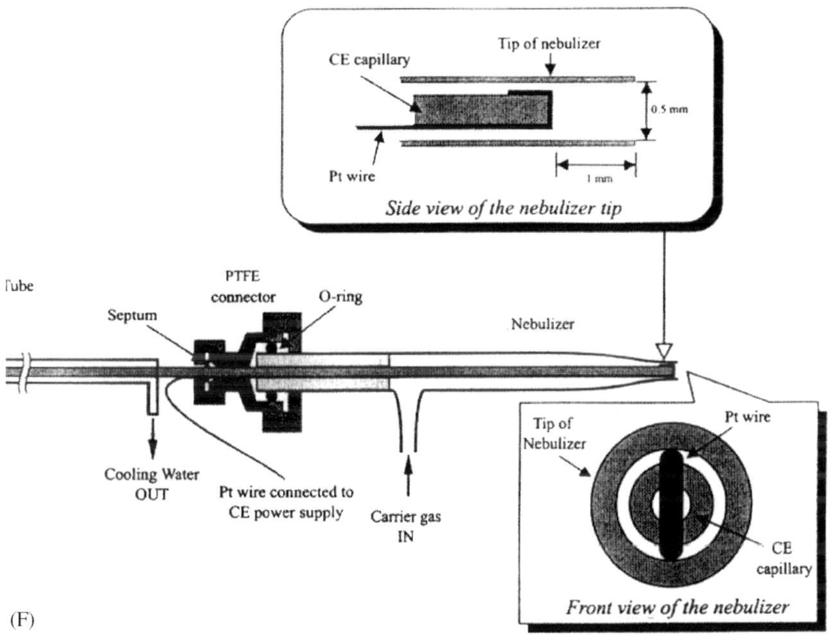

(F)

Fig. 10.2. Continued.

10.3.2 Considerations about sensitivity/detection limits in CE–ICP-MS

In order to reach the sensitivity/detection limits desired, parameters affecting peak shape, aerosol characteristics and transport efficiency through the interface should be carefully controlled. In general, it can be stated that the narrower the peaks, the lower are the detection limits, and the better is the resolution (possible suction effects must be controlled and dead volumes should be reduced to a minimum). On the other hand, nebuliser design and spray chamber type have a direct influence on primary, secondary and tertiary aerosols formed, and consequently, on analyte transport efficiency as well. Olesik explains in detail two general misconceptions to be avoided [27]: first, the transport efficiency provided by a nebuliser is not as dependent on the nebuliser itself as it is on the solution flow delivered to it. As a general rule, lower flows are desirable to increase analyte transport efficiency, the main goal being not the production of a fine primary aerosol but a reduction of losses of droplets of all sizes through secondary and tertiary aerosol formation (by droplet–droplet collisions and coagulation in the spray chamber) [28]. So it is not exactly right to state that a conventional concentric nebuliser provides, as such, lower transport efficiency than a microconcentric one. The fact is that the latter is specially designed to adopt lower solution flows, owing to which the resulting sample transport efficiency is higher. If operated at the same low flow, differences between both nebulisers would be diminished (although other additional problems would arise for the conventional nebuliser, i.e., the difficulty of controlling the laminar flow induced by a suction effect).

The second misconception is directly related to the first and refers to a possible "dilution" effect of the analyte concentration by the sheath solution. In detection systems such as ICP-MS, the signal obtained is directly dependent on the total mass of the analyte reaching the detector per unit time. Consequently, as long as this factor and the analyte transport efficiency are maintained, the sheath flow should not have any effect on detection limits. The fact is that the higher the make-up solution flow, the lower is the ICP-MS signal observed; in this case, however, the sheath solution may cause a decrease in sensitivity that is due to a decrease in transport efficiency, rather than an apparent dilution effect [27].

10.3.3 The importance of the selected nebuliser in CE–ICP-MS couplings

Since 1995, a great number of CE–ICP-MS interfaces have been proposed in the literature [22] for the specific detection of many elements, present in traces or ultratraces, in different species and samples.

The nebuliser used is a key feature of the design. It is well known that in pneumatic nebulisers, fragmentation of the solution flow into

Pneumatic Nebulisers	
Meinhard (conventional)	
CF	HECFMN
Babington	OCN
HEN	CEI-100
MCN	DIN
MicroMist	DIHEN

Ultrasonic Nebulisers
USN

Fig. 10.3. Typical nebulisers used in CE–ICP-MS couplings (CF, cross-flow; HECFMN, high-efficiency cross-flow micronebuliser; OCN, oscillating capillary nebuliser; HEN, high-efficiency nebuliser; MCN, microconcentric nebuliser; DIN, direct injection nebuliser; DIHEN, direct injection high-efficiency nebuliser; USN, ultrasonic nebuliser).

small droplets occurs by impact with a gas stream. Depending on the geometry of both solution and gas delivery channels, pneumatic nebulisers can be classified into concentric and cross-flow nebulisers. A schematic diagram showing the acronyms of nebulisers most commonly employed in CE–ICP-MS couplings is presented in Fig. 10.3.

As can be seen, a lot of designs have already been proposed. Thus, the main features of CE–ICP-MS interfaces reported based on different nebulisers will be discussed.

10.3.3.1 Concentric nebulisers

The type of nebuliser most commonly used in CE–ICP-MS couplings is the concentric one. Its main feature is a coaxial disposition of the channels through which the solution containing the analytes and the nebuliser gas stream flow. At the nebuliser annular tip, both flows join and fragmentation of the liquid takes place, leading to the formation of a primary aerosol. Usually, the gas channel experiences a narrowing at the end of the nebuliser, and so, the velocity of the gas flow may increase at that point while pressure decreases (Venturi effect). This pressure drop may induce a laminar flow in the capillary, causing the characteristic flat flow profile of the electroosmotic flow (EOF) to turn into parabolic flow (see Fig. 10.4) when coupled to CE. This "suction effect" of the nebuliser may induce in this way an important band broadening inside the CE capillary, and the achieved separation is likely to be degraded (see lower part in Fig. 10.4).

The laminar flow induced by suction effect depends not only on the nebuliser itself but also on the CE capillary dimensions (internal diameter and length), position of CE capillary end inside the nebuliser and the nebuliser gas flow rate used. Although suction effect is generally viewed as a negative effect that should be avoided, several authors have pointed out some of its possible benefits, i.e., its capacity to separate cationic, neutral and anionic substances in one run, shorter migration times and additional resolution when forcing a laminar flow away from the detector [29]. Owing to the atomic (specific) detection provided by ICP-MS, the overlapping of signals from different elements as a result of laminar flow is not a problem. In such cases, the suction effect only implies a desirable reduction of analysis time.

Several approaches have been proposed to eliminate or compensate for the laminar flow. As referred to above, a sheath solution may be additionally introduced through the interface (i.e., by a T-piece or cross-connector, see Fig. 10.2), optimising its flow rate or the container position. Another option to apply certain pressure to the CE capillary inlet and away from the detector [30]. By reducing the nebuliser gas flow, the magnitude of the laminar flow may be decreased. Making an inlet porous frit inside the capillary, able to permit EOF through it but avoiding the influence of the laminar flow on the separation, has also been reported [31].

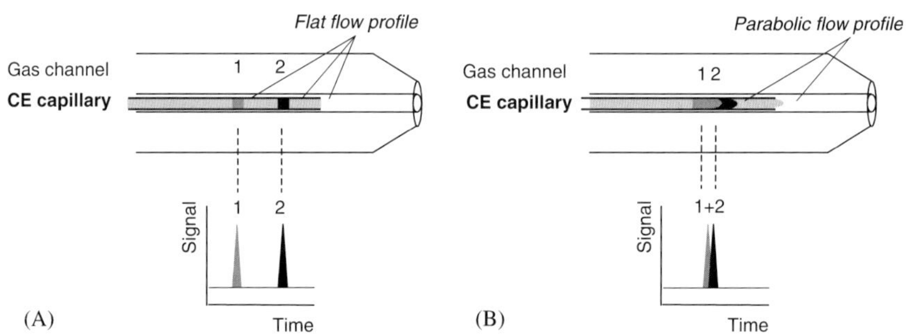

Fig. 10.4. Schematic view of two analytes (1 and 2) in the CE capillary introduced inside a concentric nebuliser. (A) A possible suction effect is totally controlled, resulting in absence of laminar flow. (B) A laminar flow with parabolic profile is generated inside the CE capillary affecting resolution of analytes.

The Meinhard nebuliser is the most common nebuliser in direct nebulisation, and is also the first nebuliser incorporated in CE–ICP-MS interfaces. Its native geometry is shown in Fig. 10.5, which serves as a model for further modifications (according to needs and authors).

In 1995, Olesik et al. [23] designed a pioneer interface by externally coating the last 4–5 cm of the CE capillary with silver paint in order to close the electrical circuit (see Fig. 10.2E). The separation capillary was fixed inside the central solution channel of the Meinhard nebuliser and a suction effect was then reported (the resulting EOF was 0.051 μl/min, while a natural aspiration flow at 2.0 μl/min occurred). Other phenomena such as band broadening, together with the influence of nebuliser gas flow rate or injected volume on peak shape, were also studied.

Lu et al. [30] investigated in detail the effects of CE capillary position and flow rates of sheath solution and nebuliser gas. The resolution achieved with UV detection for a mixture of two main isoforms of metallothionein (MT1 and MT2) and ferritine was degraded when the CE–ICP-MS interface was attached. A schematic view of the interface can be seen in Fig. 10.2A. Different positions of the CE capillary inside the central channel of the nebuliser were assayed, and it was found that the further inside the capillary was placed, the stronger was the suction effect, shortening migration times and worsening resolution. The sheath solution flow also decreased owing to a reduction in the free annular space of the nebuliser channel when the CE capillary was inserted towards the nebuliser tip, leading to an increase in the signal. It is therefore clear that there is a need to reach a compromise between sensitivity and resolution. The nebuliser gas flow had the same effect, and higher signals were obtained by increasing gas flow,

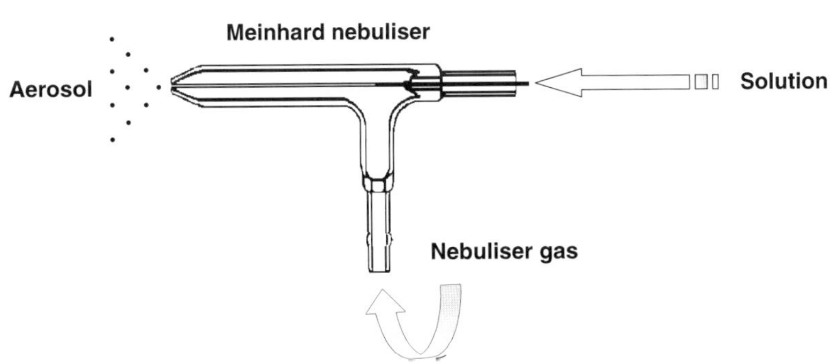

Fig. 10.5. Aerosol formation in a Meinhard nebuliser.

but laminar flow affecting CE separation also increased. The use of longer capillaries and the application of pressure at the inlet end and away from the detector contributed to improve resolution.

Silva da Rocha et al. [32] carried out mercury speciation studies with an interface also based on a Meinhard nebuliser and a T-piece through which an additional sheath or make-up solution flow was incorporated. A Pt wire completed the electrical connection. Two ICP-MS were compared, a quadrupole and a DF instrument. Optimisation of sheath and nebuliser gas flows resulted in opposite trends. Signals increased with higher gas flows, and a significant change in migration times could be observed. Under optimum conditions, cysteine complex of ethyl mercury, methyl mercury and inorganic mercury could be resolved in <10 min (despite being negatively charged). In this case, it seems to be that laminar flow contributed to shorten analysis times.

With a similar interface, Costa-Fernández et al. [33] coupled CE to a ICP-(TOF)MS, proving the capacity of such a detection system for measurements in CE transient signals. The application of high negative voltage and certain pressure at CE capillary inlet end during separation and towards the detector, resulted in narrower peaks and shorter analysis times, compromising resolution in some cases. Besides, mass spectral separation of metal species that were not electrophoretically resolved is possible owing to the specificity and multi-elemental capability of ICP-(TOF)MS was also reported.

Apart from those interfaces based on a conventional Meinhard nebuliser, some modifications or new concentric designs have appeared in the literature.

Michalke and Schramel [34] proposed a homemade nebuliser, paying special attention to the interface design to overcome problems derived from the suction effect. CE capillary position inside the nebuliser was carefully controlled and a sheath solution flow was introduced at 4 μl/h. Two ways of evaluating the presence of laminar flow were described: the first one consists of filling the CE capillary with buffer solution and measuring the current value after application of voltage. Then, the capillary inlet is kept in open air and the nebuliser gas is switched on for a period of time. If a suction effect was created through the interface, an air bubble would be introduced into the capillary, and when tipping again the capillary inlet in buffer solution and applying voltage, the current would be zero. The second procedure consists of maintaining the capillary inlet immersed in the sample solution while the nebuliser gas is switched on. After 1 min, the CE capillary is washed with water solution, and in case of laminar flow, an analyte signal is observed. A "two-step procedure" has also been proposed: a separation and a detection stage. Analytes are separated into discrete bands along the first part of the CE capillary under an electric field. After a while, the current is stopped and the zones are pushed by pressure of buffer towards the detector. Resolution is maintained and migration times become shorter.

In a further work, a modified spray chamber was incorporated where an additional Ar–H_2 gas stream is introduced. The main purpose of the new design was avoiding aerosol condensation at the chamber walls, which resulted in a marginal improvement of detection limits by a factor of 1.5 [35].

Kirlew and Caruso [36] tried to extend the promising analytical features of an oscillating capillary nebuliser (OCN) by its use in CE–ICP-MS couplings. Basically, an OCN consists of two concentric flexible fused-silica tubes, through which the gas and the sheath solution flow externally and internally, respectively, and a CE capillary, which is placed inside the nebuliser constituting the central tube. Owing to the flexibility of the silica tubing, an oscillation takes place, contributing to the formation of small aerosol droplets. However, when adapting the original design to the interface, the resulting signals were 21% of those attainable by means of a conventional concentric nebuliser. Further experiments with a two-tube OCN were carried out with UV detection while also evaluating the influence that grounding the nebuliser at the CE capillary had on the separation. The purpose was to investigate the capability of a sol-gel frit created at the inlet end of the CE capillary to avoid the suction effect [31].

As mentioned before, Deng and Chan [37] designed a novel interface (see Fig. 10.2F) where the sheath solution flow is avoided. Instead, a Pt electrode is inserted [24] or wrapped around the CE capillary [37], and studies were carried out with ICP-OES (Optical Emission Spectrometry) detection. The higher the gas nebuliser flow, the stronger the suction effect, and thus low flow values were recommended. Studies on the influence of CE capillary internal diameter on ICP-MS signals were also presented.

All studies detailed so far point out the difficulty of controlling the laminar flow in the capillary in order to preserve resolution using concentric nebulisation devices. Of course, if species to be separated are far away from each other in the time scale using UV detection, they may be clearly resolved with interfaces based on conventional concentric nebulisers, with the additional advantage of shorter analysis times. Moreover, the multi-elemental character of the ICP-MS results in their use to solve speciation problems where species containing different metals or metalloids have to be detected. Additional resolution (to that obtained by the separation) can be achieved by adequate mass resolution. However, most CE–ICP-MS applications would require the development of improved interfaces, aiming at avoiding or at least controlling as for as possible, the laminar flow typically generated by the basic nebulisation process.

10.3.3.2 Cross-flow nebulisers

An alternative to concentric nebulisers are cross-flow ones (designed with the sample solution flowing orthogonal to the nebuliser gas capillary). Although the relative position of the two capillary tips should be carefully optimised owing to their influence on signal intensity, they are strongly recommended when working with high salinity samples or even slurries, as they are less prone to clogging than concentric devices.

Despite their quite different geometry, the existence of a suction effect in CE–ICP-MS interfaces based on commercial cross-flow nebulisers has been reported. Majidi and Miller-Ihli [38] evaluated the performance of a standard cross-flow nebuliser with a commercial Scott spray chamber and a homemade concentric nebuliser with a low-volume cyclonic one. Interestingly, a suction effect was only observed when using the cross-flow-based interface, and in that case, detection limits for cadmium bound to metallothionein were 10 times lower. Resolutions were not compared owing to the different lengths of CE capillaries used.

Different comparative studies show that low sample consumption nebulisers perform better than standard cross-flow ones in CE–ICP-MS couplings (both in terms of sensitivity and LODs). However, cross-flow devices offer higher ruggedness and better resolution due to a weaker suction effect than microconcentric nebulisers (MCNs) [39,40]. A multicapillary design was proposed in order to improve the analytical characteristics of a conventional cross-flow-based interface, resulting in a signal increase proportional to the number of capillaries assembled [41]. In an attempt to combine the main features of MCNs with those typically attached to cross-flow devices, Li et al. [42] designed a high-efficiency cross-flow micronebuliser (HECFMN), which could be successfully adapted to CE–ICP-MS coupling. The reduced dimensions of such nebuliser capillaries allowed a stable operation of the interface at liquid flow rates as low as 5 µl/min, avoiding any suction effect that might have degraded the separation, and also improving aerosol characteristics and analyte transport efficiency to the plasma.

With a different disposition of tubings transporting gas and liquid flows, Babington type nebulisers can be also considered cross-flow nebulisers. Figure 10.6 shows a schematic view of a standard cross-flow nebuliser with a right-angle configuration (A) and a typical Babington-type nebuliser (B). As a main differential feature, Babington-type nebulisers do not generate any suction effect (both flows are apart). Therefore, CE-UV profiles could be preserved with ICP-MS detection [43,44] using such devices. Sensitivity, however, may be a problem with such nebulisation in CE–ICP-MS.

10.3.3.3 Ultrasonic nebulisers

A nonpneumatic nebuliser, with a completely different mechanism of aerosol formation, is the ultrasonic nebuliser. In this case, the liquid flow is directed towards the surface of a piezoelectric transducer, which oscillates at high frequencies provided by an ultrasonic generator. As a consequence, the nebuliser gas flow does not have a direct influence on aerosol formation. Analyte transport efficiencies are generally higher (20%) than in pneumatic devices, but worse precision and longer washout times are commonly associated with ultrasonic nebulisers. In fact, their applications for CE–ICP-MS couplings are very scarce. Lu and Barnes [45] pointed out the absence of a suction effect, but a

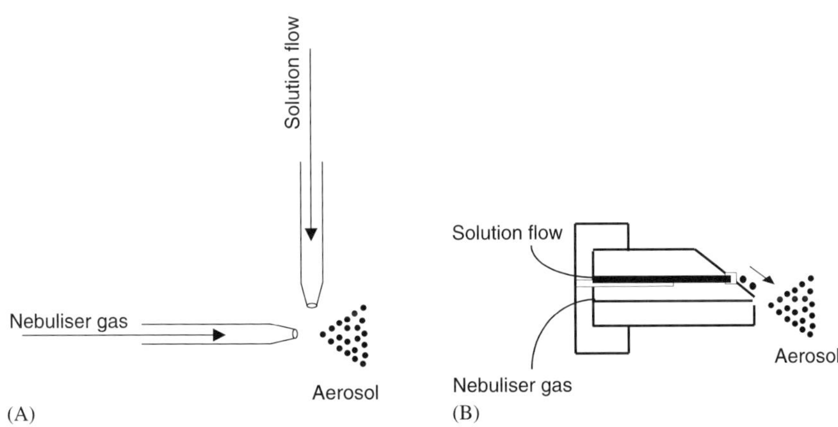

Fig. 10.6. Representation of aerosol formation in two different types of "cross-flow" nebulisers: (A) standard and (B) V-groove Babington.

back-pressure in the CE capillary altering the separation process may take place owing to transducer oscillation. A very precise localisation of the capillary at a certain distance from the oscillating surface turned out to be mandatory. Then, signals that were higher by a factor of 3.6–8.4 could be obtained. However, signal-to-noise ratios were only improved by a factor of 1.3–2.8 [45]. Kirlew et al. [46] studied in some depth the performance of two low-cost homemade ultrasonic nebulisers, as compared with a commercial one, evaluating different parameters affecting resolution and sensitivity in CE–ICP-MS speciation measurements.

10.3.3.4 Micronebulisers

One of the main reasons that prompted the development of nebuliser designs with reduced dimensions was the need to deal with problems where low amounts of sample are available. On the other hand, the particular requirements of microflow separation techniques, where the flow to be coupled to ICP-MS detection is much lower than in conventional applications, also promoted new developments in micronebulisers to cope with the main limitations unveiled in the CE–or μLC–ICP-MS couplings that were investigated.

Considering that micronebulisers are generally based on pneumatic mechanisms of aerosol formation, in order to better understand the main procedures for the construction of such reduced-dimension devices, perhaps it may be of interest to start with a discussion of the processes affecting aerosol formation in pneumatic nebulisers.

Several studies have aimed to investigate aerosol characteristics generated under different experimental conditions in conventional pneumatic nebulisers (liquid uptake rate, gas flow rate, type of solvents and gases employed and nebuliser dimensions). Drop-size distributions at different locations inside the spray chamber and contained in primary, secondary and tertiary aerosols were investigated in order to define the main processes affecting analyte transport efficiency. In general, an increase in the mean drop size of the primary aerosol was found to result in a reduction of analyte mass transport and therefore in sensitivity. Therefore, the production of a fine primary aerosol should be aimed at. The cross-sectional area of the annular gas outlet seems to have a direct influence on droplet size, and the smaller the area, higher is the gas velocity; thus higher kinetic energy is available to interact with the liquid flow, resulting in more droplets of smaller diameter [47]. With a lower influence, a similar trend was observed for the liquid cross-sectional area. Besides, liquid–gas interaction efficiency tends to be higher in concentric designs than in cross-flow nebulisers [48].

On the other hand, as the sample uptake rate increases, so does the transport rate, but the opposite trend is observed for aerosol transport efficiency. As more primary aerosol droplets of all sizes form, the volume of primary aerosol contained in drops with lower diameters also increases with sample uptake, so other processes affecting secondary and tertiary aerosols were investigated in order to understand the decrease in transport efficiency. This phenomenon can be attributed to the more severe losses of droplets of all sizes while they travel from the nebuliser exit to the plasma, and so, decreasing liquid sample uptake in order to reduce aerosol densities numbers is also recommended. In the same way, high gas flow rates resulted in fine primary aerosols and reduced loss of analytes during secondary and tertiary processes [28].

From all basic studies it appears that reduced dimensions in micronebulisers result in a more efficient gas–liquid interaction, resulting in higher analyte transport efficiencies compared with those achieved with (bigger) conventional pneumatic nebulisers. Thus, the use of micronebulisers at low flows (80–85 μl/min) may result in detection limits similar to or better than those obtained with conventional nebulisers (operating at 1–2 ml/min) [49,50].

Several micronebuliser designs have been described in the literature. In Table 10.1, the main characteristics and differences found among them are summarised.

Of particular interest is a recent study concerning the general performance of micronebulisers [51]. As shown throughout this chapter, one of the main features to be controlled in a CE–ICP-MS coupling is the laminar flow created inside the CE capillary as a consequence of possible nebuliser suction effect. Thus, the influence of the nebuliser type used is clear, but this latter study points out how its magnitude may vary when using different units of a certain nebuliser model. Three units of a MicroMist nebuliser were investigated, and particular attention was paid to the influence of the nebuliser gas flow and the internal diameter of the sample tubing on the sheath solution uptake by self-aspiration. The uptake rate found for a given nebuliser gas flow was different from the one provided by the manufacturer and also different for the three investigated units. Besides, a marked decrease in such solution flow could be particularly observed for one of the MicroMist nebulisers at high gas flows, which authors attributed to a

TABLE 10.1.

Main characteristics found among the different micronebulisers reported in the literature

Characteristics	Effect	Ref.
Material of construction (glass, PVDF, PEEK)	If opaque, the adjustment of CE capillary end inside the nebuliser is more difficult	[58]
Disposition of tubing, transporting gas and liquid flows (concentric or cross-flow)	Efficiency of gas–liquid interaction	[48]
Location of liquid and gas outlets related one to each other	Efficiency of gas–liquid interaction	[85]
Gas outlet cross-sectional area	May determine operating pressure and amount of energy available for nebulisation. Aerosol characteristics	[47,48]
Liquid outlet cross-sectional area	Interaction between gas and liquid flows. Aerosol characteristics	[48]

back-pressure created in the system due to a turbulence phenomena at the nebuliser tip. On the other hand, the wider the sample uptake tubing the higher the self-aspirated solution flow, although the relationship was not proportional. These observations are of relevance when coupling the CE to an ICP-MS by means of micronebulisers.

There is no doubt by now that micronebuliser-based interfaces are today the most successful of CE–ICP-MS couplings. Therefore, we will review the designs and applications reported so far.

Interfaces with micronebulisers and spray chamber

The high efficiency nebuliser (HEN; from Meinhard Glass Products, Santa Ana, CA, USA) is one of the first reported micronebulisers for CE–ICP-MS couplings. With a concentric geometry similar to that of the conventional Meinhard, one of its main advantages is its ability, owing to its reduced dimensions, to prevent laminar flow inside the CE capillary by means of using lower sheath solution flows [29], apart from providing improved detection limits.

Sutton et al. [52] pointed out the importance of the nebuliser choice to improve ICP-MS detection limits in CE. The use of the HEN instead of a conventional Meinhard resulted in narrower peaks, a less noisy baseline and lower detection limits for a series of lanthanides.

Another micronebuliser of extended use in this coupling is the one named MCN from CETAC Technologies (Omaha, NE, USA). Taylor et al. [53] carried out the separation of the main isoforms of metallothionein by a CE–ICP-MS interface based on a MCN nebuliser. In order to maintain the resolution achieved with UV detection, the laminar flow had to be avoided by resorting to low sheath solution flows. However, they found that the optimum flow required for such a purpose was higher than the one providing the higher sensitivity. Consequently, the authors proposed the application of pressure at the inlet end of the CE capillary and away from the detector in order to counterbalance the suction effect. In this way, resolution was preserved without compromising sensitivity.

Day et al. [54] suggested the introduction of the sheath solution by self-aspiration of the nebuliser, pointing out the simplicity of the methodology and the avoidance of pulsing and variations in the flow typically occurring with pumping devices. Care was taken to adjust the sheath solution level so that migration times were coincident with those obtained by UV detection, which reflected a perfect control of the suction effect. Emphasising the ease with which the micronebuliser could be clogged, the authors also pointed out that such clogging was to be avoided in order to preserve a constant self-aspirated flow. Lee and Jiang [55] reported optimum conditions for separation of methyl, ethyl and inorganic mercury by CE–ICP-MS based on an MCN nebuliser.

A comparison between MCN (with a cyclonic spray chamber) and cross-flow (with a Scott type one) was reported by Tu et al. [39] While the latter device presented lower suction effect and higher ruggedness, the MCN allowed lower detection limits by 10–20 times for mercury speciation. If sufficient resolution is achieved, the authors pointed out the possibility of not completely eliminating laminar flow in order to reduce analysis times. In a similar comparison, Baker and Miller-Ihli [40] studied in detail the influence of both, the nebuliser gas flow and the sheath solution flow rates, on the performance of MCN and cross-flow nebulisers. Again, the higher suction effect observed in the micronebuliser may degrade resolution, although its smaller dimensions allowed compensating for this with low sheath solution flows, resulting in improved aerosol production and transport. Detection limits for cadmium in metallothioneins were 10 times lower with MCN, but the frequent clogging of the nebuliser may be an important drawback in real-life analysis.

The use of a MicroMist nebuliser (Glass Expansion, West Melbourne Vic, Australia) in CE–ICP-MS couplings has also been reported in the literature. Similar to the HEN, the gas and liquid flow channels are concentric, and the CE capillary position inside the nebuliser can be finely adjusted by direct observation. With a MicroMist-based interface, chiral separation of DL-selenomethionine could be achieved by CE–ICP-MS [56].

MicroMist analytical performance in terms of instrumental set-up and figures of merit was compared with a HEN for metal speciation studies in metallothioneins using quadrupole and DF mass analysers. Although easier to operate, mainly on account of the higher gas pressures required for the HEN, separation profiles previously achieved with UV detection could be maintained with both nebulisers. Resolution and detection limits for Cd-MTs achieved with both interfaces were very similar in a quadrupole ICP-MS. However, the MicroMist interface was more robust and user-friendly in a DF instrument, and lower detection limits were observed [57].

Arsenic speciation studies were carried out by CE with ICP-MS detection in order to gain enough sensitivity to detect species of interest in environmental samples (detection limits down to 1 μg/l). For such a purpose, two different nebulisers were evaluated and compared: MicroMist and MCN. Although the MCN nebuliser seemed to be less sensitive to matrix effects, the MicroMist led to higher transport efficiency and was less prone to blockage. Besides, as it is made of glass, locating the capillary end inside the nebuliser is much easier than in a MCN (which is made of polyvinilidenefluoride (PVDF) and PEEK). The influence of both nebuliser gas and sheath solution on laminar flow was also investigated [58].

As a way to control induced laminar flow inside the CE capillary using concentric devices, B'Hymer et al. [59] proposed placing a sol–gel frit at the inlet end. The research was carried out with a MicroMist-based interface and a homemade spray chamber, resulting in a response that was at least three-times higher than that achieved with an MCN in previous studies. The authors observed a similar separation profile for a mixture of MT I and ferritine when (i) a frit was created inside the CE capillary and the sheath solution reservoir was maintained 10 cm below the interface; and (ii) the reservoir was positioned 25 cm above it when a free CE capillary was used. This fact proved the advantages of frits for avoiding suction effects, although peak efficiencies decreased and the use of electrokinetic injection was mandatory.

Kuczewski et al. [60] compared an MCN nebuliser (with a typical self-aspirated flow of 0.1 ml/min) with a MicroMist nebuliser FM02, whose aspirated uptake at nominal pressure was 0.2 ml/min (instead of the 0.1 ml/min of FM01). After separating different oxidation states of plutonium and neptunium, authors found up to two times higher efficiency of aerosol formation with the MCN, although the MicroMist interface was preferred owing to its better stability. No significant differences were found when comparing a standard Scott-type spray chamber and a Cinnabar small-volume cyclonic one in terms of resolution, stability and precision of the signal.

Perhaps the CE–ICP-MS interface more widely reported in the literature (since it is commercially available) is the CEI-100 from CETAC. Apart from a typical micronebuliser, its design includes a low-volume spray chamber (5 ml) with no waste exit, so that the entire aerosol generated at the outlet end of the nebuliser is directly introduced into the plasma [26]. A schematic view of the interface can be seen in Fig. 10.2D. The nebuliser capillary was changed to a narrower id. resulting in flow rates, in the self-aspiration mode, between 2 and 12 μl/min. A Pt electrode was located inside one of the arms of a cross piece through which the CE capillary and the sheath solution were introduced. Owing to its reduced dimensions, a possible suction effect was minimised, thereby avoiding a laminar flow inside the CE capillary that may have degraded the separation. As shown in Fig. 10.7, migration times of a test sample containing Rb turned out to be independent of the nebuliser gas flow rate (increasing gas flow rates are represented in small letters in Fig. 10.7).

Metal speciation studies in metallothioneins [61–64], speciation analysis of nickel in hyperaccumulating trees [65] and determination of nucleotides [66,67] and sulphur-containing amino acids [68] are some of the applications reported in the literature using the CEI-100 interface.

Interfaces with just a micronebuliser (without spray chamber)

A nebuliser must produce droplets $<10\,\mu m$ in diameter (1) for the aerosol transport efficiency to be high, and (2) for the desolvation,

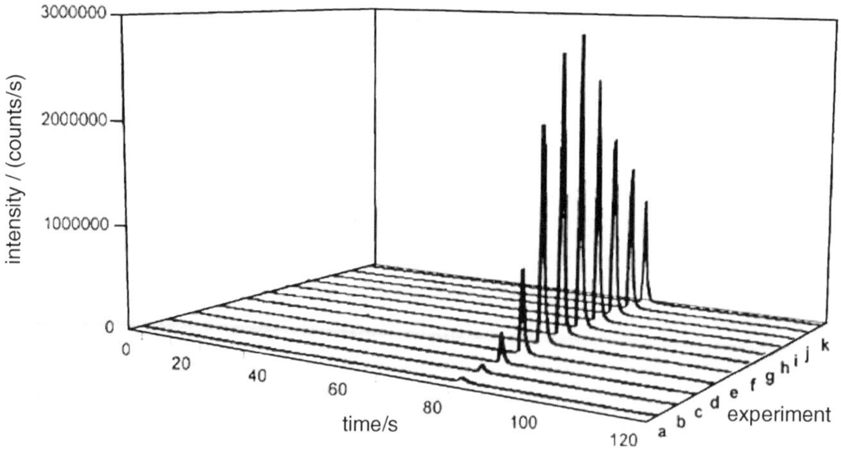

Fig. 10.7. Influence of nebuliser gas flow rate on Rb-signal intensity and migration time in CE–ICP-MS coupling by means of CEI-100 interface. Reproduced from Ref. [25], with permission from Springer.

volatilisation and atomisation processes to be rapid in the ICP. However, conventional nebulisers generate polydispersive aerosols with droplet diameters up to 100 μm. For this reason, a spray chamber is traditionally required in order to eliminate larger droplets and so enhance precision and reduce interferences. However, this means an important loss of analytes on their way to the plasma.

As stated before, the design of HENs with reduced dimensions resulted in finer aerosols when compared with conventional devices, so different authors concentrated their efforts on the construction of interfaces based on such micronebulisers without a spray chamber. First attempts were made by means of a commercial direct injection nebuliser or DIN, with the interface shown in Figure 10.2B. Mixtures containing alkali, alkaline earth, heavy metal and selenium ions, together with inorganic and organic arsenic species could be separated without any noticeable suction effect inside the CE capillary [69].

Tangen et al. [70] designed an MCN for direct introduction of the formed aerosol into the ICP. The interface body consisted of a cross-piece through which a sheath solution flow is pumped at a very low flow (i.e., 2.5 μl/min). Several inner diameters of nebuliser gas and liquid flow channels were investigated as well as their relative positions. Performance of the interface was tested to separate alkaline, alkaline earth and chromium ions. Further studies were carried out in an MCN nebuliser whose original inner capillary was substituted by a narrower one. A CE capillary of 20 μm id. had to be used in order to avoid significant suction effects [71].

A homemade sample introduction system without a spray chamber was also constructed by Bendahl et al. [72] and compared to the commercial "direct injection high efficiency nebuliser" (DIHEN), in terms of sensitivity, precision and detection limits. This micronebuliser has been described in detail [50]. In the new design, the sample capillary id. and wall thickness, and also the gas annulus area were reduced compared to the DIHEN, resulting in a sensitivity one order of magnitude higher. Besides, at 10 μl/min sample flow the new design provided detection limits similar to those obtained with the DIHEN at 85 μl/min. Selenium speciation was carried out with such an interface, and four selenium compounds could be baseline resolved without being affected by a significant suction effect.

In conclusion, since 1995, a great amount of knowledge has been gained about factors affecting aerosol formation and transport efficiency in nebulisation-based CE–ICP-MS interfaces. Once the basis for maintaining separation profiles typically achieved with UV detection was set, the low concentrations of analytes in biological or environmental samples have prompted the development of new designs aiming at even lower detection limits. In fact, the superiority of the ICP-MS detection in terms of sensitivity and detection limits over UV detection has been proved for the direct determination of such elements. Also, the superiority of micronebulisers is well established now. However, to preserve the favourable ICP-MS analytical figures of merit when coupled to CE, proper selection and optimisation of the interface are mandatory.

10.4 CE–ICP-MS COUPLING VIA VOLATILE SPECIES GENERATION

The better performance of micronebulisers compared with conventional devices derives from a higher nebulisation efficiency, which leads to an increased transport efficiency of the sample as a whole to the plasma. This fact shows the influence of the sample introduction stage on the development of hybrid techniques. In speciation analysis most of the work carried out is unielemental (i.e., it focuses on different species of a single given element). Thus, it is reasonable to extend this idea of improving the transport efficiency to just the desired element, rather than to the sample as a whole. A well-known approach to improve analyte transport in atomic spectrometry is the formation of their volatile species. Therefore, it is not surprising that conversion of analyte species into volatile forms has been investigated by several authors in CE–ICP-MS couplings.

10.4.1 Advantages of volatile species generation in CE–ICP couplings

The advantages of such analyte introduction systems for the CE–ICP couplings can be summarised as follows: (a) analyte transport efficiencies are about 90–100% (ultrasonic nebulisation is 20–30%); (b) the energy of the plasma is directly invested in atomisation and excitation of the sample, but not in desolvation or vaporisation processes; (c) analytes are isolated from a complex matrix, removing potentially interfering species that remain in the solution; (d) without a nebuliser, problems derived from a high salinity or viscosity of the samples and solvents are avoided; (e) micronebulisers in the interface make it possible to greatly improve detection limits, but their high cost, fragility and frequent blockage are important limitations that could be overcome for a given analyte by means of volatile species generation (VSG) systems Thus, analytes emerging from the CE capillary could be subjected on-line to VSG before ICP-MS detection.

Of course, this methodology can only be applied to those elements forming volatile species, and consequently, the general multi-elemental character of the ICP-MS detector may be lost.

10.4.2 Interface designs based on VSG

Despite the advantages pointed out above, CE–ICP-MS interfaces incorporating an on-line VSG system as sample introduction device are scarce.

Magnuson et al. [73] described, in a pioneering work, a CE–VSG-ICP-MS interface for arsenic speciation. The outlet end of the CE capillary is introduced into a cross connector and the nanoflow, emerging from it, is implemented by a side arm of the cross with an additional make-up buffer pumped through a T-piece, which is also connected to a platinum ground wire. The total effluent is delivered first to a three-way manifold where it mixes with HCl, and then to a second one through which $NaBH_4$ is introduced (both reagents are needed in order to form analyte volatile species). The authors have pointed out the necessity of including an isolating peristaltic pump in the instrumental set-up, placed between the cross exit and the HCl mixing tee. In that way, a back-pressure created inside the CE capillary by the high amounts of hydrogen generated is prevented. The analyte volatile species migrate across a membrane separator and are directly transported to the plasma by an Ar gas stream.

A very similar CE–VSG-ICP-MS interface was then employed for speciation analysis of both arsenic and selenium [74]. Maintaining the isolating pump, the platinum ground wire is here directly connected to the cross piece, through which concentrated HCl (in order to reduce Se^{VI} to Se^{IV}) is directly introduced joining the CE effluent. Then, a make-up flow of water is incorporated in a three-way manifold, by means of the isolating pump, which is placed just before the $NaBH_4$ mixing tee. In this way, a direct contact of the concentrated HCl solution with the peristaltic pump tubing is avoided (this solution is introduced in the flow now by the isolating pump).

Both studies detailed above are lacking a direct comparison of the figures of merit achieved by this novel sample introduction scheme and those attained by conventional nebulisation. Silva da Rocha et al. [75] investigated a different CE–ICP-MS interface, based on VSG, to be applied to mercury speciation. It was described evaluated and compared to a Meinhard nebuliser-based interface using both a quadrupole and a DF mass analyser. The CE capillary was introduced through a T-piece (grounding tee) and inserted into a second one (mixing tee). A Pt wire is placed in the side arm of the grounding tee, and the HCl solution is continuously pumped through it. By the side arm of the mixing tee, the $NaBH_4$ is incorporated and the mixed effluent is transported to a gas–liquid separator. Volatile species are then transported by an Ar gas flow into the separator. This Ar flow, on its way to the plasma, merges with a second Ar gas flow incorporated in the system through an additional tee. The most salient feature of this design compared with the one described before is its simplicity, because it avoids an additional expensive isolating pump (possibly owing to the lower amounts of hydrogen generated, as a consequence of the relatively low concentrations of $NaBH_4$ used for mercury cold vapour generation). Detection limits were lower with the VSG interface than with a Meinhard nebuliser by at least four times in the quadrupole system and two times using a DF mass analyser. However, repeatability of migration

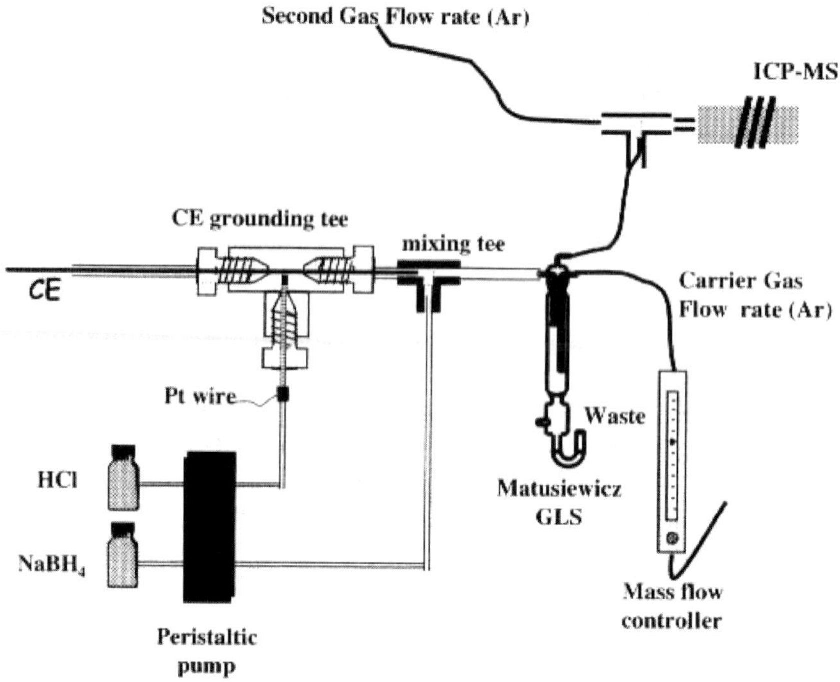

Fig. 10.8. Schematic diagram of a typical CE–VSG–ICP-MS system. Reproduced from Ref. [75], with permission from The Royal Society of Chemistry.

times was degraded compared with the results obtained by nebulisation. A schematic diagram of this simple CE–VSG–ICP-MS system is shown in Fig. 10.8.

10.5 STRATEGIES TO IMPROVE SENSITIVITY IN CE–ICP-MS COUPLING

Throughout the present chapter, apart from delineating general operation principles and mechanisms involved in analyte delivery from the outlet end of the CE capillary to the ICP, a comprehensive review of the main contributions reported so far has been included. However, it may be of interest for future research to consider here those particular applications involving strategies to improve the sensitivity of CE–ICP-MS hybrid technique.

First of all, the use of highly sensitive ICP-MS instrumentation, i.e., DF mass analysers used at low-resolution settings, can provide detection limits that are better than those obtained by using a quadrupole instrument [32,57,76].

Preconcentration strategies such as those based on stacking CE techniques should follow [77]: different methodologies were evaluated in longer CE capillaries (about 150 cm length) based on a two-step procedure. This consists of applying hydraulic pressure after 10–15 min of electrophoretic separation in order to push the bands of analytes towards the detector once they were separated. Detection limits for $PtCl_6^{2-}$ could be improved almost three orders of magnitude using an isotachophoresis technique. If additional resolution is required, as is the case for the separation of arsenic species, a pH stacking methodology may be advisable. By placing the sample in the middle of the CE capillary between a leading electrolyte and a terminating one, the pH inside the CE capillary experiences a successive decrease, resulting in changes of the analytes' mobilities depending on their p*K* values. Capillary isoelectric focussing may result in improved resolution, as demonstrated for selenium speciation analysis. The total length of the CE capillary can be filled with sample, and so a direct repercussion on detection limits achievable is clear.

Another stacking methodology proposed in the literature for CE separations is "large volume sample stacking" (LVSS). Although highly extended with UV detection, only two applications with ICP-MS have been reported so far. Basically, it involves the introduction of a long plug of sample, dissolved in a lower concentration buffer or in water, in a capillary containing a higher concentration support buffer. When applying a reverse polarity, the result is a rapid migration of the analytes towards the detector end, until they cross the concentration boundary between the sample plug and the separation buffer. They experience a decrease in electrophoretic mobility owing to the lower field in this region, forming a narrow band in the buffer zone. At the same time, and before the analyte separation process takes place, the sample matrix is pumped out from the injection end of the capillary. This methodology resulted in a six times improvement in detection limits for mercury species by CE–ICP-MS coupling based on a conventional Meinhard nebuliser [32]. Further studies allowed reducing detection limits by one order of magnitude when this technique of LVSS was applied to metal speciation studies in metallothioneins using a HEN-based interface [78]. In that case, particular attention had to be paid to avoiding clogging of the micronebuliser during the stacking process. With optimum conditions previously obtained with UV detection [79], speciation studies in high salinity real-life samples, such as eel liver cytosolic extracts, were possible in control and Cd-exposed groups. Fig. 10.9 illustrates the excellent electropherograms obtained for Cd speciation in a mixture of rabbit liver metallothioneins (MT1 and MT2) by CE–ICP-(Q)MS. As can be seen in Figs. 10.9A and B, separation resolution could be preserved when applying the LVSS technique, and a 10 times improvement in Cd signals was obtained compared with the normal procedure. Figure 10.9C shows the electropherogram corresponding to a 10-times diluted mixture of MT1 and MT2 by LVSS, resulting in signals similar to those obtained with the normal procedure for the concentrated sample.

10.6 SOME SELECTED CONTRIBUTIONS AND APPLICATIONS

In view of more recently published contributions to the field, it may be stated that in contrast to 8 years ago research on this topic is no longer exclusively focused on interface design. The considerable efforts dedicated to investigate and overcome the main technical difficulties arising when interfacing the capillary to the nebuliser have paved the way for interesting applications of CE–ICP-MS techniques.

Thus, first contributions to the field [23,30,69] were followed by reports on practical considerations (laminar flow existence, preservation of UV resolution and improvement of detection limits) and comparative studies among different nebuliser designs [29,40,44,52,53]. Homemade devices [25,42,72] and particular strategies developed to improve the attainable sensitivity have been of great value [32,78].

One novel approach to overcome main interferences caused by polyatomic ions in ICP-MS is the use of collision or reaction cells. The published contributions involving CE separations coupled to these advanced ICP-MS detectors reflect, once again, the state of development achieved by the hybrid technique under review. In this way, element-specific analysis of DNA nucleotides via phosphorus measurements [67] and determination of sulphur-containing amino acids [68] have been described.

Last, but not least, particular attention deserves to be paid to those contributions focused on the miniaturisation of the whole system. It may be of interest to point out the coupling of a microfluidic chip-based CE system to a MicroMist nebuliser for chromium, copper and arsenic speciation [80]. Similarly, the CE–ICP-OES interface based on movable

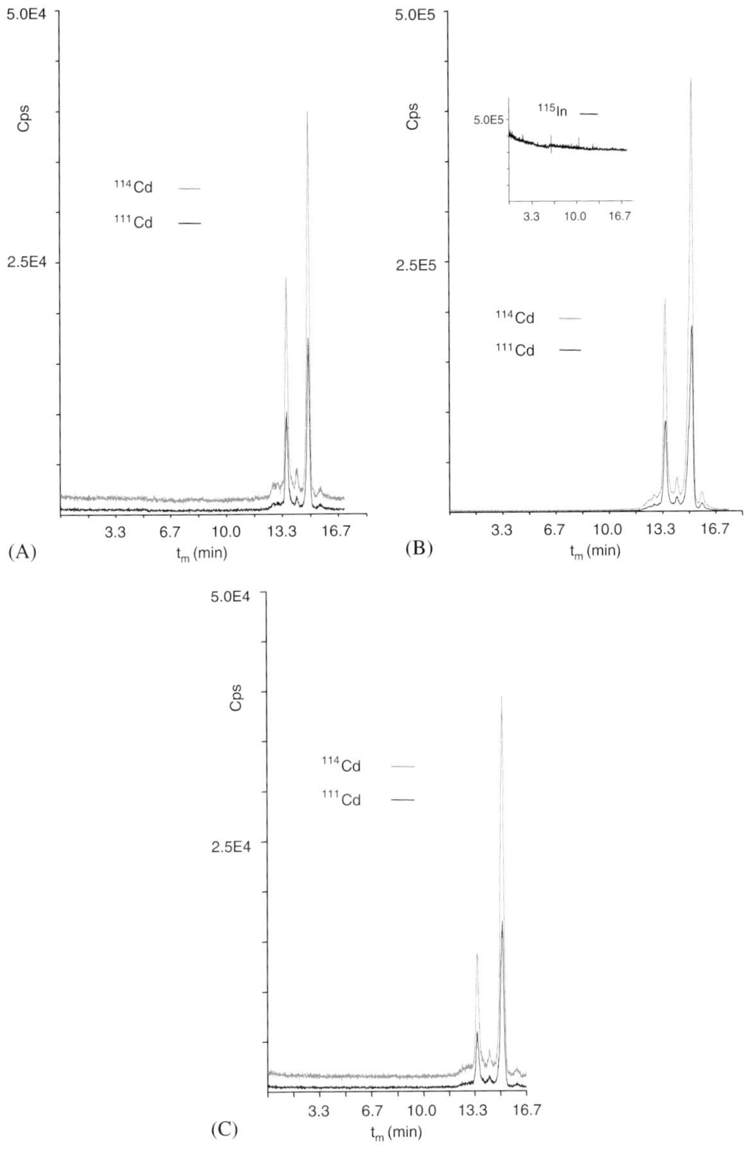

Fig. 10.9. Electropherograms obtained for Cd speciation. (A) 4 µg/ml Cd in MT1 and in MT2 analysed with the normal procedure (sample dissolved in water); (B) 4 µg/ml Cd in each isoform analysed using the LVSS technique; (C) 0.4 µg/ml Cd in each isoform analysed as in (B) Reproduced from Ref. [78], with permission from The Royal Society of Chemistry.

reduction bed hydride generation is another contribution to minimisation of reagent consumption, which was applied, in this case, to arsenic speciation studies [81]. Such aspects clearly underscore the advantages of CE from the "Green Chemistry" point of view.

10.7 CONCLUSIONS AND FUTURE PROSPECTS

In brief, the analytical potential of CE–ICP-MS hybrid techniques is clearly evident at present. The combination of the high-resolution power that characterises CE separations with the specificity and extremely high sensitivity provided by the ICP-MS detector guarantees a fruitful coupling. The aim of this chapter was to show that this hybrid technique is now a complementary approach to HPLC–ICP-MS for speciation analysis, an alternative of high value, particularly when low sample volumes are available or interactions with a chromatographic material have to be avoided in order to preserve the integrity of species [65,82–84]. Moreover, as CE–MS has become a well-established tool for Proteomics [85], it can be envisaged that many of such applications can be extended to "hetero-atom-tagged" experiments [15] where CE–ICP-MS techniques may play an important instrumental role.

REFERENCES

1 A. Montaser, *Inductively coupled plasma mass spectrometry*, Wiley-VCH, New York, 1998.
2 N.H. Bings, J.M. Costa-Fernández, J.P. Guzowski Jr., A.M. Leach and G.M. Hieftje, *Spectrochim. Acta B*, 55 (2000) 767.
3 M. Vázquez Peláez, J.M. Costa-Fernández and A. Sanz-Medel, *J. Anal. At. Spectrom.*, 17 (2002) 950.
4 D.M. Templeton, F. Ariese, R. Cornelis, L.G. Danielsson, H. Muntau, H.P. van Leeuwen and R. Lobinski, *Pure Appl. Chem.*, 72 (2000) 1453.
5 A. Sanz-Medel, *Spectrochim. Acta B*, 53 (1998) 197.
6 R. Lobinski, *Appl. Spectrosc.*, 51 (1997) 260A.
7 J. Szpunar, *Analyst*, 125 (2000) 963.
8 K.G. Heumann, *Anal. Bioanal. Chem.*, 373 (2002) 323.
9 J. Szpunar, R. Lobinski and A. Prange, *Appl. Spectrosc.*, 57 (2003) 102A.
10 K. Sutton, R.M.C. Sutton and J.A. Caruso, *J. Chromatogr. A*, 789 (1997) 85.
11 F. Vanhaecke and L. Moens, *Fresenius J. Anal. Chem.*, 364 (1999) 440.

12 K. Ackley, K.L. Sutton and J.A. Caruso, In: J.A. Caruso, K.L. Sutton and K.L. Ackley (Eds.), *Elemental Speciation New Approaches for Trace Element Analysis*, Elsevier, Amsterdam, 2000.
13 C.A. Ponce de León, M. Montes-Bayón and J.A. Caruso, *J. Chromatogr. A*, 974 (2002) 1.
14 C.N. Ferrarello, M.R. Fernández de la Campa and A. Sanz-Medel, *Anal. Bioanal. Chem.*, 373 (2002) 412.
15 A. Sanz-Medel, M. Montes-Bayón and M.L. Fernández Sánchez, *Anal. Bioanal. Chem.*, 377 (2003) 236.
16 B. Michalke, *Trends Anal. Chem.*, 21 (2002) 142.
17 B. Michalke, *Trends Anal. Chem.*, 21 (2002) 154.
18 M. Montes-Bayón, K. DeNicola and J.A. Caruso, *J. Chromatogr. A*, 1000 (2003) 457.
19 B.W. Acon, J.A. McLean and A. Montaser, *J. Anal. At. Spectrom.*, 16 (2001) 852.
20 B. Bouyssiere, J. Szpunar and R. Lobinski, *Spectrochim. Acta B*, 57 (2002) 805.
21 N.P. Vela and J.A. Caruso, *J. Biochem. Biophys. Methods*, 43 (2000) 45.
22 S.S. Kannamkumarath, K. Wrobel, K. Wrobel, C. B'Hymer and J.A. Caruso, *J. Chromatogr. A*, 975 (2002) 245.
23 J.W. Olesik, J.A. Kinzer and S.V. Olesik, *Anal. Chem.*, 67 (1995) 1.
24 Y.Y. Chan and W.T. Chan, *J. Chromatogr. A*, 853 (1999) 141.
25 D. Schaumlöffel and A. Prange, *Fresenius J. Anal. Chem.*, 364 (1999) 452.
26 D. Schaumlöffel and A. Prange, Patent No. 198 41 288.6 (1998).
27 J.W. Olesik. In: J.A. Caruso, K.L. Sutton and K.L. Ackley, (Eds.), *Elemental Speciation New Approaches for Trace Element Analysis*, Elsevier, Amsterdam, 2000.
28 J.W. Olesik and L.C. Bates, *Spectrochim. Acta*, 50B (1995) 285.
29 J.A. Kinzer, J.W. Olesik and S.V. Olesik, *Anal. Chem.*, 68 (1996) 3250.
30 Q. Lu, S.M. Bird and R.M. Barnes, *Anal. Chem.*, 67 (1995) 2949.
31 C. B'Hymer, R.M.C. Sutton, K.L. Sutton and J.A. Caruso, *Anal. Commun.*, 36 (1999) 349.
32 M. Silva da Rocha, A.B. Soldado, E. Blanco-González and A. Sanz-Medel, *J. Anal. At. Spectrom.*, 15 (2000) 513.
33 J.M. Costa-Fernández, N.H. Bings, A.M. Leach and G.M. Hieftje, *J. Anal. At. Spectrom.*, 15 (2000) 1063.
34 B. Michalke and P. Schramel, *Fresenius J. Anal. Chem.*, 357 (1997) 594.
35 B. Michalke and P. Schramel, *Electrophoresis*, 19 (1998) 270.
36 P.W. Kirlew and J.A. Caruso, *Appl. Spectrosc.*, 52 (1998) 770.
37 B. Deng and W.T. Chan, *J. Chromatogr. A*, 891 (2000) 139.
38 V. Majidi and N.J. Miller-Ihli, *Analyst*, 123 (1998) 803.
39 Q. Tu, J. Qvarnström and W. Frech, *Analyst*, 125 (2000) 705.
40 S.A. Baker and N.J. Miller-Ihli, *Appl. Spectrosc.*, 53 (1999) 471.

41 V. Majidi, J. Qvarnström, Q. Tu, W. Frech and Y. Thomassen, *J. Anal. At. Spectrom.*, 14 (1999) 1933.
42 J. Li, T. Umemura, T. Odake and K.I. Tsunoda, *Anal. Chem.*, 73 (2001) 5992.
43 G. Álvarez-Llamas, M.R. Fernández de la Campa and A. Sanz-Medel, *Anal. Chim. Acta*, 448 (2001) 105.
44 C. Casiot, O.F.X. Donard and M. Potin-Gautier, *Spectrochim. Acta B*, 57 (2002) 173.
45 Q. Lu and R.M. Barnes, *Microchem. J.*, 54 (1996) 129.
46 P.W. Kirlew, M.T.M. Castellano and J.A. Caruso, *Spectrochim. Acta B*, 53 (1998) 221.
47 A. Canals, J. Wagner, R.F. Browner and V. Hernandis, *Spectrochim. Acta*, 43B (1988) 1321.
48 A. Canals, V. Hernandis and R.F. Browner, *Spectrochim. Acta*, 45B (1990) 591.
49 S.H. Nam, J.S. Lim and A. Montaser, *J. Anal. At. Spectrom.*, 9 (1994) 1357.
50 J.A. McLean, H. Zhang and A. Montaser, *Anal. Chem.*, 70 (1998) 1012.
51 E.G. Yanes and N.J. Miller-Ihli, *Spectrochim. Acta B*, 58 (2003) 949.
52 K.L. Sutton, C. B'Hymer and J.A. Caruso, *J. Anal. At. Spectrom.*, 13 (1998) 885.
53 K.A. Taylor, B.L. Sharp, D.J. Lewis and H.M. Crews, *J. Anal. At. Spectrom.*, 13 (1998) 1095.
54 J.A. Day, K.L. Sutton, R.S. Soman and J.A. Caruso, *Analyst*, 125 (2000) 819.
55 T.H. Lee and S.J. Jiang, *Anal. Chim. Acta*, 413 (2000) 197.
56 J.A. Day, S.S. Kannamkumarath, E.G. Yanes, M. Montes-Bayón and J.A. Caruso, *J. Anal. At. Spectrom.*, 17 (2002) 27.
57 G. Álvarez-Llamas, M.R. Fernández de la Campa, M.L. Fernández Sánchez and A. Sanz-Medel, *J. Anal. At. Spectrom.*, 17 (2002) 655.
58 M. Van Holderbeke, Y. Zhao, F. Vanhaecke, L. Moens, R. Dams and P. Sandra, *J. Anal. At. Spectrom.*, 14 (1999) 229.
59 C. B'Hymer, J.A. Day and J.A. Caruso, *Appl. Spectrosc.*, 54 (2000) 1040.
60 B. Kuczewski, C.M. Marquardt, A. Seibert, H. Geckeis, J.V. Kratz and N. Trautmann, *Anal. Chem.*, 75 (2003) 6769.
61 A. Prange, D. Schaumlöffel, P. Brätter, A.N. Richarz and C. Wolf, *J. Fresenius Anal. Chem.*, 371 (2001) 764.
62 D. Pröfrock, A. Prange, D. Schaumlöffel and W. Ruck, *Spectrochim. Acta B*, 58 (2003) 1403.
63 K. Polec, J. Szpunar, O. Palacios, P. González-Duarte, Silvia Atrian and R. Lobinski, *J. Anal. At. Spectrom.*, 16 (2001) 567.
64 K. Polec-Pawlak, D. Schaumlöffel, J. Szpunar, A. Prange and R. Lobinski, *J. Anal. At. Spectrom.*, 17 (2002) 908.

65 D. Schaumlöffel, L. Ouerdane, B. Bouyssiere and R. Lobinski, *J. Anal. At. Spectrom.*, 18 (2003) 120.
66 C.F. Yeh and S.J. Jiang, *Analyst*, 127 (2002) 1324.
67 D. Pröfrock, P. Leonhard and A. Prange, *J. Anal. At. Spectrom.*, 18 (2003) 708.
68 C.F. Yeh, S.J. Jiang and T.S. Hsi, *Anal. Chim. Acta*, 502 (2004) 57.
69 Y. Liu, V. Lopez-Avila, J.J. Zhu, D.R. Wiederin and W.F. Beckert, *Anal. Chem.*, 67 (1995) 2020.
70 A. Tangen, W. Lund, B. Josefsson and H. Borg, *J. Chromatogr. A*, 826 (1998) 87.
71 A. Tangen and W. Lund, *J. Chromatogr. A*, 891 (2000) 129.
72 L. Bendahl, B. Gammelgaard, O. Jons, O. Farver and S.H. Hansen, *J. Anal. At. Spectrom.*, 16 (2001) 38.
73 M.L. Magnuson, J.T. Creed and C.A. Brockhoff, *J. Anal. At. Spectrom.*, 12 (1997) 689.
74 M.L. Magnuson, J.T. Creed and C.A. Brockhoff, *Analyst*, 122 (1997) 1057.
75 M. Silva da Rocha, A.B. Soldado, E. Blanco and A. Sanz-Medel, *J. Anal. At. Spectrom.*, 16 (2001) 951.
76 J.A. Day, J.A. Caruso, J.S. Becker and H.J. Dietze, *J. Anal. At. Spectrom.*, 15 (2000) 1343.
77 B. Michalke and P. Schramel, *Analysis*, 26 (1998) M51.
78 G. Álvarez-Llamas, M.R. Fernández de la Campa and A. Sanz-Medel, *J. Anal. At. Spectrom.*, 18 (2003) 460.
79 G. Álvarez-Llamas, A. Rodríguez-Cea, M.R. Fernández de la Campa and A. Sanz-Medel, *Anal. Chim. Acta*, 486 (2003) 183.
80 Q.J. Song, G.M. Greenway and T. McCreedy, *J. Anal. At. Spectrom.*, 19 (2004) 883.
81 X.D. Tian, Z.X. Zhuang, B. Chen and X.R. Wang, *Analyst*, 123 (1998) 899.
82 V. Vacchina, S. Mari, P. Czernic, L. Marquès, K. Pianelli, D. Schaumlöffel, M. Lebrun and R. Lonbinski, *Anal. Chem.*, 75 (2003) 2740.
83 K. Polec-Pawlak, P. Zambenedetti, J. Szpunar, R. Lobinski and P. Zatta, *J. Anal. At. Spectrom.*, 19 (2004) 41.
84 B. Gammelgaard and L. Bendahl, *J. Anal. At. Spectrom.*, 19 (2004) 135.
85 M. Moini, *Anal. Bioanal. Chem.*, 373 (2002) 466.

Chapter 11

Vibrational spectroscopic detection in capillary electrophoresis (CE)

Stephan Kulka and Bernhard Lendl

11.1 INTRODUCTION

Vibrational spectroscopy embraces infrared (IR) and Raman spectroscopy. Both techniques provide direct, molecular-specific information by measuring vibrational and rotational transitions in molecules. For measurement in condensed phase as is required for detection in capillary electrophoresis (CE), only vibrational transitions are observed. According to Hooke's law, the frequencies of the observed fundamental vibrations depend on the strength and the masses of the atoms involved and are characteristic for a given functional group or structural subunit. As a consequence, it is possible to determine the presence of these vibrations in the molecule under investigation if bands are observed at the corresponding wavenumbers (wavelengths) in the recorded spectra. Owing to the sensitivity of the vibrational transitions that involve large parts of the molecule (fingerprint region), IR and Raman spectroscopy are considered powerful techniques to distinguish between structurally similar molecules and isomers. When considering the whole vibrational spectrum, it is therefore possible to confirm the identity of a given molecule where appropriate reference spectra are available against which a visual or computer-aided comparison can be made. Reference spectra recorded under comparable conditions are important because the exact position and intensities of the bands in IR and Raman spectra are different for neat (powders, liquids) or dissolved analyte spectra. Furthermore, they will depend on the solvent and pH value, because different hydrogen bonding as well as dissociation may occur, which strongly alter the appearance of the obtained spectra. The sensitivity of vibrational spectra in inter- and intramolecular interactions can also be used advantageously for the

Comprehensive Analytical Chemistry XLV
M.L. Marina, A. Ríos and M. Valcárcel (Eds)
Volume XLV ISSN: 0166-526X DOI: 10.1016/S0166-526X(05)45011-4

determination of secondary structures of proteins as well as for differentiation of diastereomers. In addition to providing qualitative information on the separated compounds, IR and Raman spectroscopy also make it possible to obtain quantitative information on the concentration of the measured analytes to be obtained.

11.2 IR SPECTROSCOPY

Whereas both IR and Raman spectroscopy provide information on the vibrational transitions in molecules, their physical fundamentals are different. To give rise to bands in the IR spectra it is required that the dipole moment of the part of the molecule involved in the vibration changes during vibration. In principle, IR spectroscopy can be divided into three different ranges: near-infrared (NIR, 10000–4000 cm^{-1}), mid-infrared (MIR, 4000–400 cm^{-1}) and far-infrared (FIR, 400–100 cm^{-1}). In NIR, overtones and combinations of vibrations can be found. Molar absorptivities in the NIR range are much smaller than in the MIR spectral region, where the fundamental vibrations are observed. Furthermore, the structural information gained from NIR is less detailed than from MIR. Therefore NIR, while it is advantageously used in modern processes in analytical chemistry, has found no applications as a detection technique in CE so far. FIR spectra give information about long-range intermolecular interactions. This part of IR spectrometry is currently experiencing an important development, as new powerful light sources and detectors are being introduced [1–3]. However, no application of FIR in CE has been reported yet.

MIR spectroscopy in CE is an absorption technique, and uses thermal, broad-band light sources with an interferometer to record the transmitted light. For quantitative analysis, Beer's law can be used. The hyphenation of MIR spectroscopy to CE is difficult owing to the low amount of sample available for measurement in CE. Therefore, high-quality Fourier transform (FT) spectrometers are required to achieve a successful hyphenation. However, miniature, powerful and room-temperature-operated quantum cascade lasers may soon be an alternative for sensitive, functional, group-specific detection in CE.

11.2.1 On-line MIR spectroscopic detection in CE

The advantage of On-line detection lies mainly in the real-time acquisition of data, allowing the whole electrophoretic peak to be recorded over time. However, to achieve on-line MIR detection, important experimental difficulties have to be overcome. Fused-silica capillaries are not transparent in the MIR spectral region, and therefore, transmission measurements through the capillaries such as the ones made in the UV–visible spectral region, are not possible in the MIR region. In addition, the buffers employed in CE also show strong absorptions in the MIR spectral region. For aqueous solutions, the optical path must be in the low micrometer region (typically between 8 and 25 μm) to allow measurements in the information-rich fingerprint region. However, in the case of such short optical path lengths, part of the spectral region will be blocked in regions of strong solvent absorption. To record the spectra of the eluting analytes, the spectrum of the buffer must be subtracted from the spectra recorded during the separation. This can be achieved efficiently by taking the spectrum of the buffer as the background spectrum. As measurements are made on-line, there are, in contrast to off-line detection, no limitations concerning the employed buffer system. However, as spectra are recorded in solution spectral reference libraries, which usually contain spectra measured in solid form cannot be used.

Kölhed et al. [4] first showed that the use of a micromachined IR-transparent flow cell (CE cell) made out of calcium fluoride (Fig. 11.1) and inserted between two pieces of standard fused-silica capillaries makes on-line MIR detection in capillary zone electrophoresis possible. This CE cell is placed in a holder and standard capillaries are attached on both sides and sealed off with o-rings (Fig. 11.2).

When checked with a standard UV detector, it was shown that this cell did not influence the separations. In this study, adenosine, guanosine and adenosine-5′-monophosphate (AMP) were separated

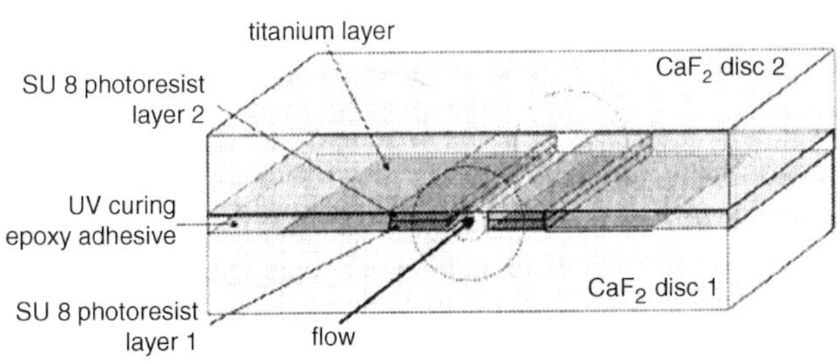

Fig. 11.1. Schematic view of the CE–FT-IR cell (with scaling altered to show details of key features). Reprinted with permission from Ref. [4]. Copyright (2002), American Chemical Society.

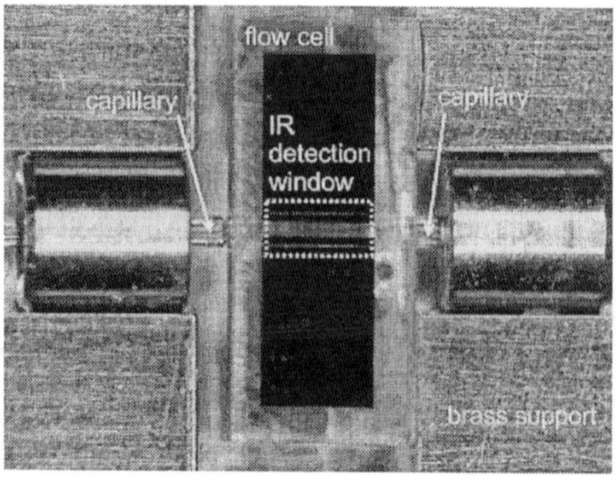

Fig. 11.2. Microscopic view of the CE–FT-IR cell with connected capillaries housed in its supporting block. Reprinted with permission from Ref. [4]. Copyright (2002), American Chemical Society.

and analyzed (Fig. 11.3). A linear calibration was made ranging from 0.8 to 8 mM for adenosine and AMP, and from 0.64 to 3.2 mM for guanosine, leading to correlation coefficients higher than 0.99.

The same technology was applied to a Micellar electrokinetic chromatography (MEKC) [5] separation, where neutral analytes can be separated as well. In this work, paracetamol, caffeine, *p*-nitrobenzyl alcohol and *m*- and *p*-nitrophenol were studied. It was shown that with a background electrolyte containing 40 mM of sodium dodecyl sulfate (SDS) in addition to the borax buffer, Fourier-transform infrared (FTIR) detection still achieved useful results. In this application, MIR spectra for substance identification could also be recorded. Linear calibration for all analytes has been shown to cover the low mM concentration range (from 1.08 to 1.44 mM). Thus, on-line FTIR detection for different modes of CE might be of great interest in applications where not only the quantification, but also the identification of the analyte play a role.

Hinsmann et al. [6] applied the already discussed technology with the CE cell to nonaqueous CE, expanding the application range of CE to charged hydrophobic compounds. On-line FT-IR was applied to study the separation and direct identification of (*R*,*S*)-3,5-dinitrobenzoyl leucine (DNB-Leu) enantiomers using *o*-(tert-butyl carbamoyl) quinine (tBuCQN) as a chiral selector. The background

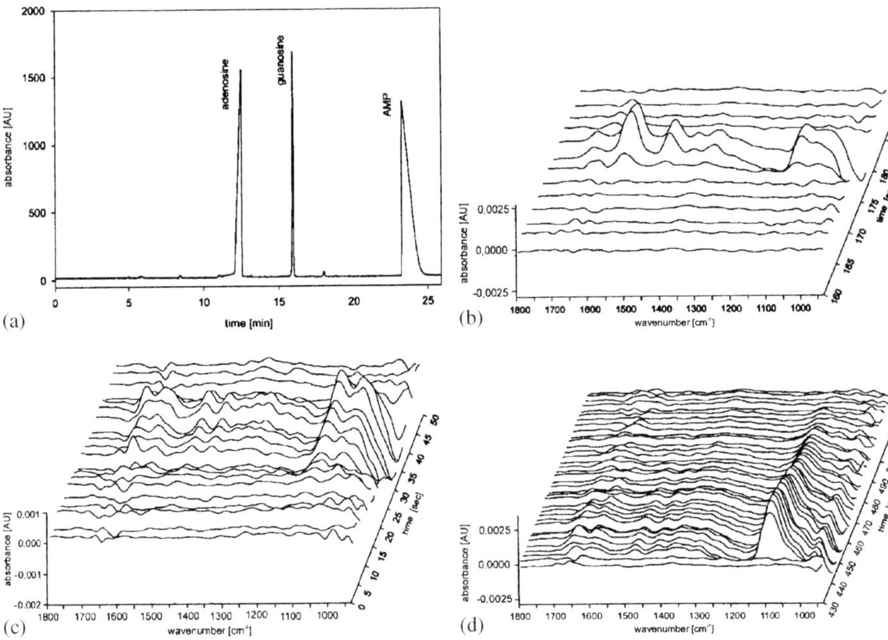

Fig. 11.3. (a) Electropherogram with conventional UV detection at 254 nm showing a separation of adenosine, guanosine, and AMP. CE conditions: 15 mM Borax buffer at pH 10, hydrodynamic injection 20 s at 27 cm height, 250 V/cm, and 60 cm to UV detection. (b–d) 3D stack plots of infrared spectra from a separation of adenosine (b), guanosine (c), and AMP (d). Reprinted with permission from Ref. [4]. Copyright (2002), American Chemical Society.

electrolyte was a mixture of ethanol and methanol containing some modifiers and the chiral selector. It could be shown that the formed diastereomers could be directly distinguished by FTIR spectroscopy (Fig. 11.4). In this paper, however, no quantitative results were reported.

Recently, research efforts have been directed toward developing MIR detection for chip-based separation systems [7,8]. For this purpose, IR-compatible chip materials such as CaF_2 need to be used. Because of the reduced overall dimensions of chip-based separations, microscopic detection is mandatory. High-quality IR spectra close to the diffraction limit (sample spot 10–20 μm in diameter) can be obtained using a synchrotron radiation source coupled to an FTIR microscope. Kulka et al. [7] carried out the first IR detection in CE-IR chips using synchrotron powered IR microscopy as a detection system.

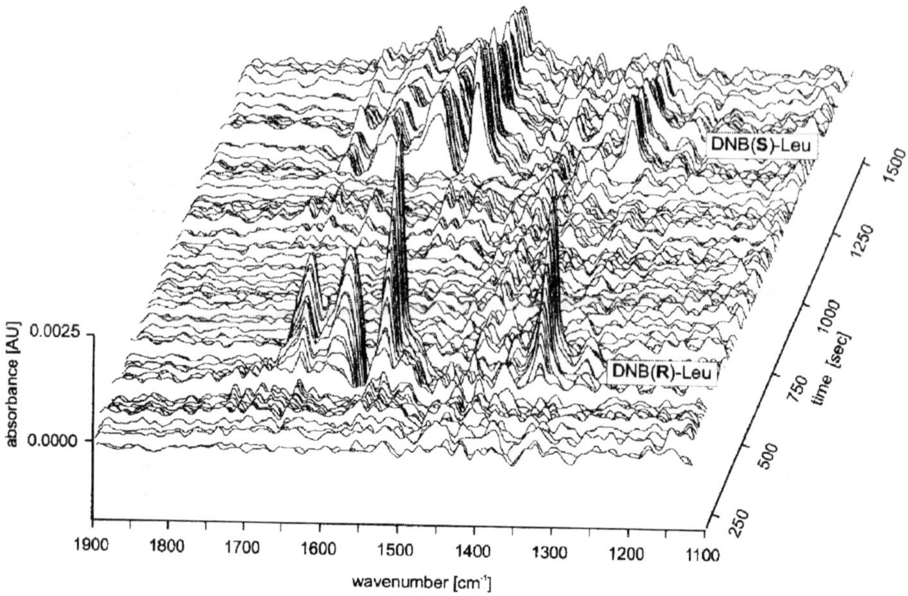

Fig. 11.4. 3D IR stack plot recorded during a CE–FT-IR separation run. Reproduced from Ref. [6], with permission from the Society of Applied Spectroscopy.

These experiments showed that the secondary structure of proteins can be determined on-chip (Fig. 11.5).

In a different work using a conventional thermal light source and etched channels on a CaF_2 chip, Pan et al. [8] showed that spectra from the produced channels could be recorded with a conventional IR microscope.

Another approach was adopted by Patterson et al. [9], who used attenuated total reflectance (ATR) infrared microspectroscopy as a detection technique for CE. The tip of the capillary was placed about 1 μm above the Ge ATR crystal where the spectra were taken. A plastic reservoir around the ATR prevented evaporation of the sample and allowed for the grounding of the Pt wire (Fig. 11.6).

The analytes studied were citrate and nitrate as well as succinylcholine chloride and sodium salicylate, and acetone was used as a neutral marker. It was shown that these analytes could be separated using capillaries of different diameters and concentrations to study the effect of dead volume and sample concentration on the ATR detection. As the time to collect a spectrum was quite low, the signal-to-noise ratio would

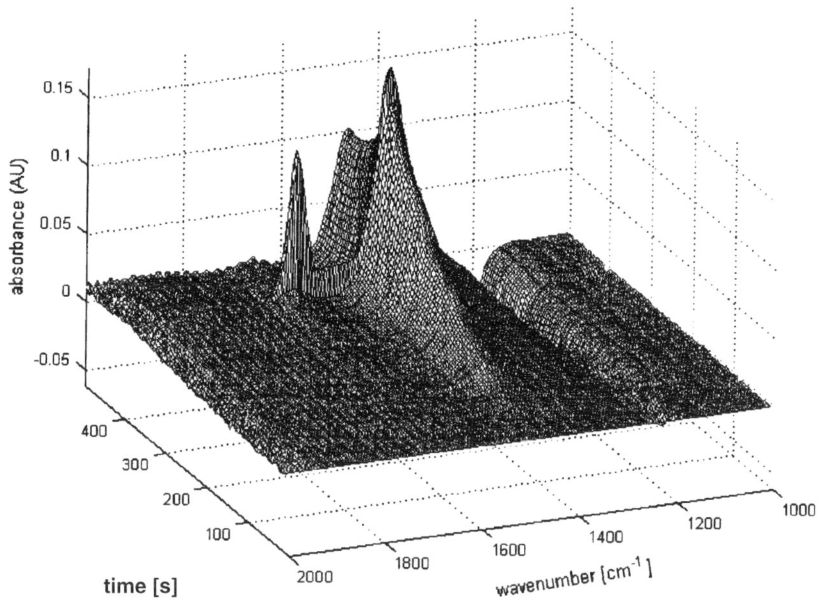

Fig. 11.5. 3D plot of the recording obtained upon loading the CE-IR-chip (design B) with 2 g/l myoglobin followed by electrophoretic movement of the analyte through the detection window (~450 s). Reproduced from Ref. [7], with permission from Springer.

be improved if faster detectors would be used as pointed out in the outlook. However, the spectra obtained were equivalent to the ones taken in a conventional ATR measurement. The advantage of this set-up is the simplicity of the design and the minimal broadening of the eluted analytes. The limit of detection achieved in an injection of 820 pl of succinyl chloride was 1.3 mM.

11.2.2 FTIR spectroscopic detection in CE after solvent elimination

Solvent elimination techniques aim to deposit the separated analytes on a target surface and to measure them after the solvent has evaporated. By this method, spectral contributions from the solute molecules will be eliminated. As the analytes are measured in dry state, available spectral libraries can, in principle, be applied to these spectra. Furthermore, as spectrum acquisition is decoupled from the time scale of separation, signal averaging over a prolonged period of time can be carried out on a given analyte deposit if required. Todebush et al. [10]

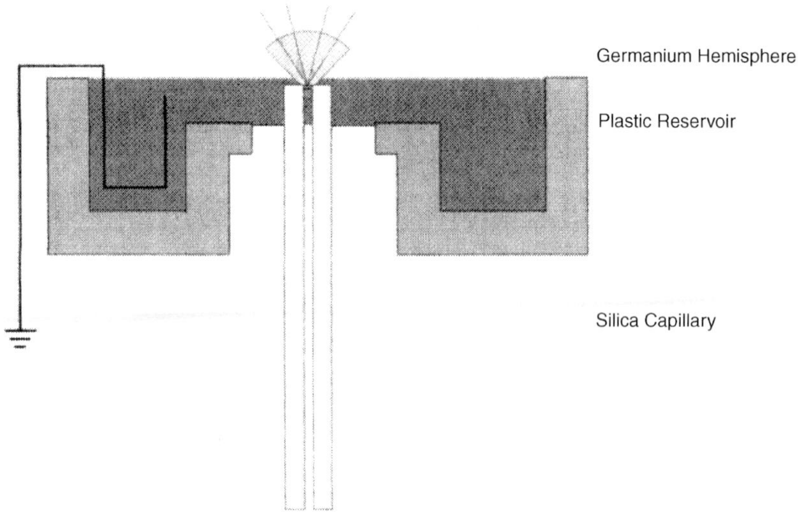

Fig. 11.6. Diagram of CE–ATR-FT-IR interface showing Ge ATR crystal, reservoir, capillary and platinum grounding wire. A representative analyte plug about to exit the capillary is shown. Top of the reservoir was open and could be filled with the desired electrolyte solution using a pipet (not to scale). Reprinted with permission from Ref. [9]. Copyright (2004), American Chemical Society.

introduced this approach using a nebulizer interface made of stainless steel. The elute from the capillary was deposited onto a calcium fluoride or zinc selenide window that was moved to prevent mixing of the deposited analytes (Fig. 11.7).

A typical deposit was round in shape and had a diameter between 200 and 400 μm. The spots were dried and analyzed off-line with an FTIR microscope (Fig. 11.8).

In this work the analytes studied were *p*-aminobenzoic acid, acetylsalicylic acid, sodium benzoate, *N*-acetyl-D-glucosamine (GlcNAc), nicotinamide, caffeine and salicylic acid in concentrations varying from 5.0×10^{-2} to 1×10^{-3} M. A separation was run with caffeine, salicylic acid, *p*-aminobenzoic acid and sodium benzoate, while the other analytes were used to compare the performance of the hyphenated system with conventional IR microscopy measurements of manually deposited analytes. These results showed that the spectra obtained were not influenced by the presence of the background electrolyte used during the separation if it was volatile enough to be removed through the nebulizer interface.

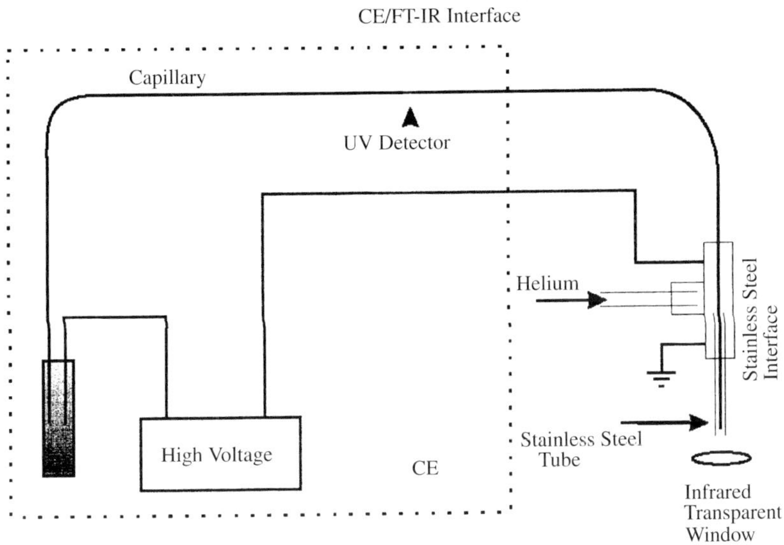

Fig. 11.7. Metal nebulizer CE/FT-IR interface. The ground potential of the high-voltage supply is connected to the metal nebulizer. Reprinted with permission from Ref. [10]. Copyright (2003), American Chemical Society.

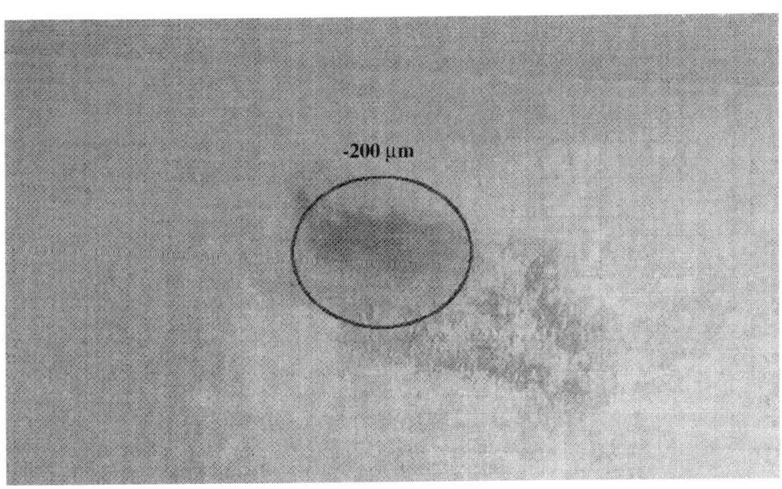

Fig. 11.8. CE/FT-IR deposit image. Sample, *p*-aminobenzoic acid, electrolyte: ammonium acetate; deposit size: ~200 μm. Reprinted with permission from Ref. [10]. Copyright (2003), American Chemical Society.

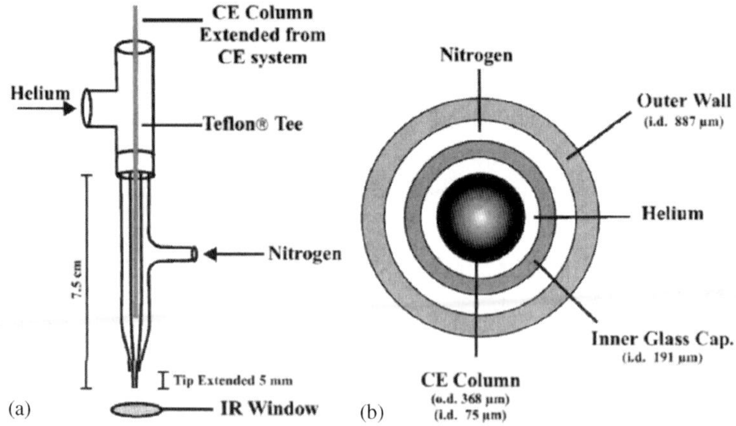

Fig. 11.9. (a) A schematic representation of the interface with the PTFE tee and microconcentric glass nebulizer. Helium and nitrogen ports are open to the capillary and shell, respectively. The resulting aerosol is deposited directly onto a calcium fluoride or zinc selenide IR window. (b) Cross-section of interface tip showing the layers of capillary, helium, nitrogen and glass. Reprinted from Ref. [11]. Copyright (2004), with permission from Elsevier.

As the deposits with the metal nebulizer exhibited a significant amount of splatter patterns and interference patterns, Jarman et al. [11] introduced a glass nebulizer interface to overcome these problems (Fig. 11.9).

It could be shown that the new design of the nebulizer with an appropriate orifice allowed the uniform deposition of the analyte. The analytes used to test the stability and separation ability of the interface were caffeine, salicylic acid and GlcNAc, with final concentrations of 5.0×10^{-3} M.

In contrast to on-line MIR detection, no quantitative results have been reported for off-line MIR detection so far.

11.3 RAMAN SPECTROSCOPY

For a vibration to be active in Raman spectroscopy, the polarizability of the bonds involved must change during vibration. In contrast to MIR spectrometry, a monochromatic light source situated in the UV-NIR spectral region is employed, and the inelastically scattered light is detected. The energies of the vibrational transitions are then calculated as the difference between the laser frequency and the frequency of the detected light. The absolute wavelength of the detected light is shifted toward the excitation wavelength and is in general found in the Vis-NIR spectral regions. Raman spectroscopy has experienced a dramatic development in recent years. FT-Raman instruments are used when excitation in the NIR spectral region (mostly 1064 nm) is undertaken, whereas dispersive instruments are used when UV and Vis laser are used to induce Raman scattering [12,13]. The main problem of Raman spectroscopy remains the low probability of the Raman scattering process (10^{-10} when compared with Rayleigh scattering), and therefore, its inherently low concentration sensitivity. When combined with confocal microscopy, however, Raman spectroscopy, can reach surprising sensitivities as seen, for example, in single-cell imaging [14,15]. To increase concentration sensitivities, two different approaches can be used. One is the surface-enhanced Raman scattering (SERS), where small deposits of noble metals, mostly Ag, are used as substrates for the analytes; the other is Resonance Raman scattering (RRS), which makes use of the possibility of enhancing the signal of analytes absorbing in the wavelength of the exciting light. These effects can generate different enhancements of the signal ranging from 10^2 to 10^6, depending on the experimental circumstances. An important advantage of Raman compared with FTIR spectroscopy (MIR) for detection in CE is that shorter wavelengths that need to be measured in Raman spectroscopy, as in these spectral regions the fused-silica capillaries are transparent. Therefore, on-capillary measurements are possible, in contrast to MIR detection. A further advantage of Raman compared with MIR detection is the fact that although water shows only weak Raman scattering, it is a strong IR absorber. Therefore, it is understandable that so far, more work has been done on the coupling of Raman spectroscopy and CE.

11.3.1 Raman scattering in CE

The first work coupling CE and Raman detection was reported by Chen and Morris [16],who used optical fibers to collect Raman scattered light directly from a capillary. The capillaries were aligned and focused onto the entrance slit of a spectrometer fitted with a photomultiplier tube (PMT) (Fig. 11.10).

The analytes used were methyl orange and methyl red at 1×10^{-4} M, which were detected on the base of the intense Raman band at $1410\,cm^{-1}$ using an excitation wavelength of 442 nm. As a high spectral

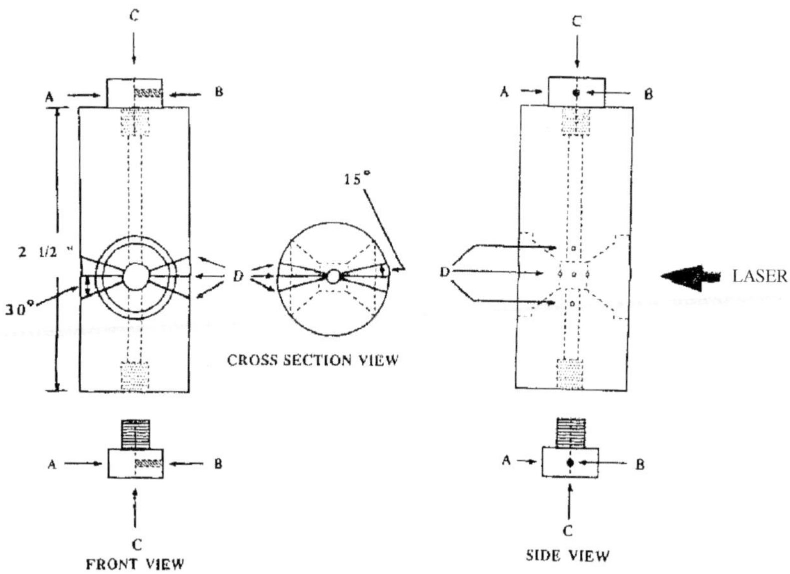

Fig. 11.10. Holder for electrophoresis capillary. Front, cross section, and side views. (A) Threaded cap; (B) set screw; (C) bore for capillary; (D) bore for optical fiber. Reproduced from Ref. [16], with permission from the Society of Applied Spectroscopy.

resolution was not necessary in this application, it was set to 30 cm^{-1} to increase the signal-to-noise ratio.

In a slightly modified system, a charge-coupled device (CCD) and lenses were employed, and the collection efficiency was thereby enhanced. The spectral resolution could be set to 12 cm^{-1}, again using methyl orange and methyl red as test analytes [17]. In this work, pre-resonance Raman spectra were taken with an integration time of 5 s to achieve detection limits as low as 6×10^{-7} M. Further increase of integration time lowered the detection limit but was incompatible with the fast separation process; therefore, the voltage was reduced when the sample zone arrived at the detector window to allow integration times until 40 s and a limit of detection (LOD) of 1×10^{-7} M.

The intensity of the Raman signal can be increased if the concentration of the sample zone in the capillary is directly increased. One way to achieve this goal is sample stacking, which depends on the conductivity differences between the dilute sample zone and the higher concentrated background electrolyte. Therefore, the electric field is higher in the dilute sample zone than in the background electrolyte. Consequently, the sample ions move faster to the boundary formed with the background electrolyte where they slow down and stack up. This preconcentration method was used by Kowalchyk et al. [18] in their work where a 1×10^{-5} M mixture of nitrate and perchlorate was separated and detected. They used nitrate and perchlorate as model analytes at an excitation wavelength of 532 nm (Fig. 11.11).

As the analyte concentrations in the steady state in Isotachophoresis (ITP) can be between 0.01 and 0.03 M when using electrolytes with a concentration of 0.1 M, it is the preferred separation and preconcentration method for Raman detection [19].

Analytes in a matrix containing a high concentration of background electrolyte cannot be preconcentrated via sample stacking methods.

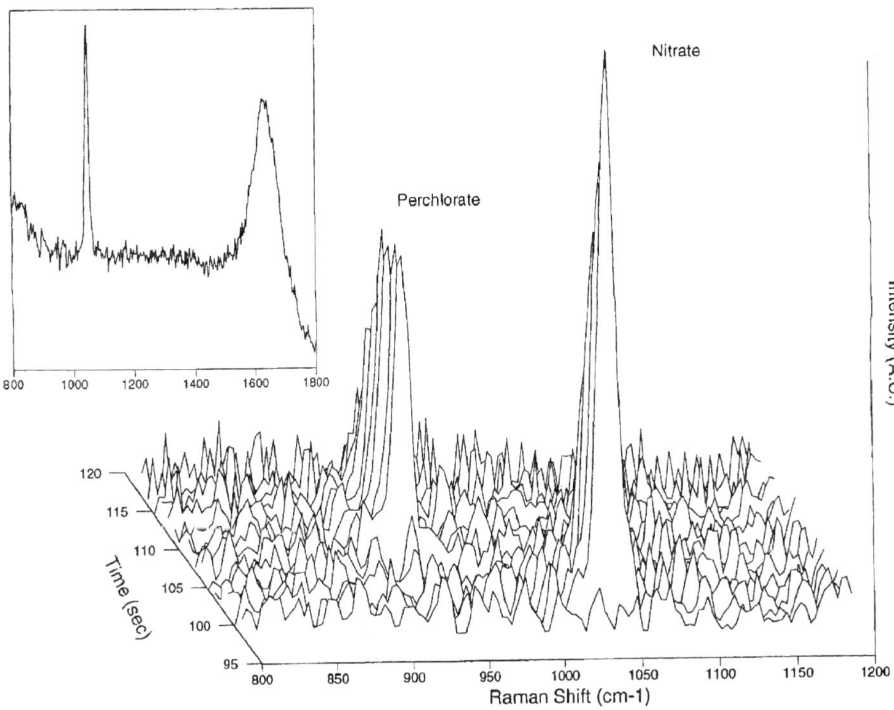

Fig. 11.11. One-second integration Raman spectra of nitrate/perchlorate separation. Sample concentrations 5×10^{-4} M in each anion; injection time 15 s at –2 kV. A 3% T coated 75-μm-i.d. capillary was used. The spectra have been flat fielded and ratioed to the background electrolyte spectrum. A 5-point Savitsky–Golay quadratic smooth was employed. The inset is the raw spectrum at 100 s. Reproduced from Ref. [18], with permission from the Society of Applied Spectroscopy.

ITP requires a leading electrolyte (LE) of higher mobility and a trailing electrolyte (TE) of lower mobility than any of the ions of interest. This technique is also called the "moving boundary" technique, as all zones move at the same velocity. This velocity occurs since the electric field is self-adjusting to maintain a constant velocity, as velocity equals mobility. Walker et al. [20] employed ITP with Raman microscopic detection for the first time, separating adenosine 5′-triphosphate, adenosine 5′-diphosphate and adenosine 5′-monophosphate in phosphate buffer. The exciting laser was an Nd-YAG laser at 532 nm (Fig. 11.12). Limits of detection in the range of 5×10^{-6} M were achieved for the studied analytes.

As Raman spectroscopy deals with absolute wavelengths in the UV-Vis-NIR spectral region where high-performance optics are available, fiber-optic probes have also been used for detection in CE [21]. In this work, Walker et al. again separated adenosine 5′-triphosphate, cytidine 5′-monophosphate, guanosine 5′-monophosphate and uridine 5′-monophosphate. The fiber-optic head is much smaller than a microscope making it possible to couple it inside the safety housing of the CE

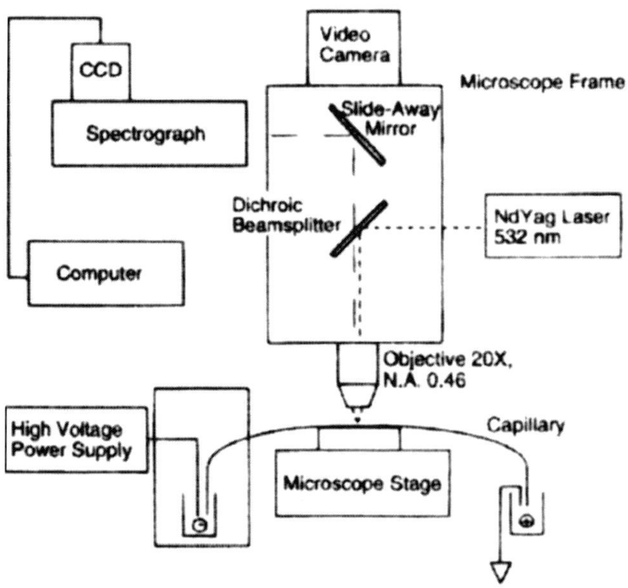

Fig. 11.12. Block diagram of the capillary isotachophoresis Raman spectroscopy system. Component details in text. Reprinted with permission from Ref. [20]. Copyright (1995), American Chemical Society.

system. Owing to coupling and insertion losses, the signal intensity of the fiber-probe was only about 28% of the signal gained with the conventional Raman microscope. However, spectra of the ribonucleotides with an initial concentration of 2×10^{-5} M could be obtained with a similar quality as in the previous work [20], although the absolute signals were lower (Fig. 11.13).

Walker et al. [22] furthermore applied statistical methods to the results of an isotachophoretic separation of paraquat and diquat detected by a Raman microscope, again using 532 nm as excitation wavelength. The detection limit could be lowered substantially when using discorrelation against the leading electrolyte Raman spectrum with a t-test ($t = 5$, significant difference with 99% confidence level) and factor analysis to extract spectra of analytes during the electrophoresis run. As the factor analysis is a form of signal averaging, good spectra could be achieved with a low starting concentration of about 10^{-7} M. Paraquat could be detected at initial concentrations of 180 ppb visually and 90 ppb using statistical methods (Fig. 11.14).

Raman microscopes can be employed for probing small volumes down to a few femtoliters. The microchip is directly coupled to the Raman microprobe. In the field of electrophoresis the miniaturization is advantageous as the speed of the separation increases and the heat dissipation due to a high surface-to-volume ratio is very efficient, and is

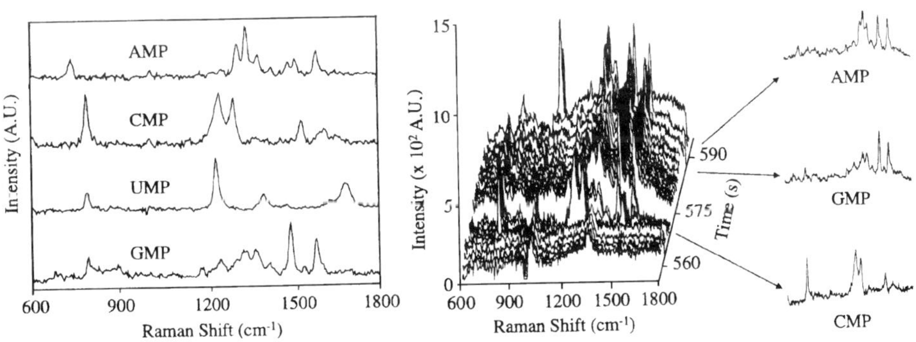

Fig. 11.13. Fiber-optic Raman spectra of AMP, CMP, UMP and GMP at a starting contained in a 50-mm i.d.- capillary. The spectra were during ITP, at 1-s integrations, 2 W 532 nm excitation (left).On-capillary Raman isotachopherogram of AMP, CMP and GMP at a starting concentration of 3.3×10^{-5} M. Leading electrolyte, 0.1 M Na_2SO_4; trailing electrolyte, 0.1 M MOPS. Sample introduction 120 s at 200 V cm; separation at 200 and 66 V cm; 2 W 532 nm excitation (right). Reprinted from Ref. [21]. Copyright (1998), with permission from Elsevier.

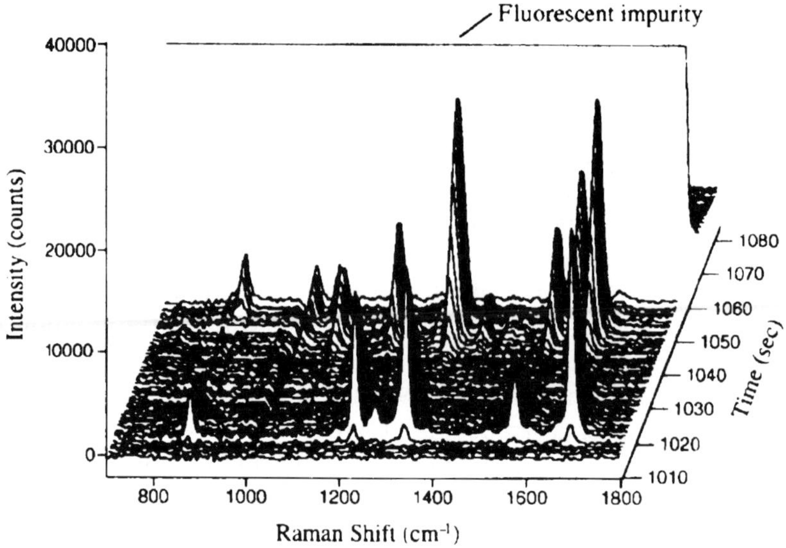

Fig. 11.14. On-capillary Raman isotachopherogram spectra of 1.5×10^{-5} M paraquat and diquat in deionized water. Leading electrolyte, 0.1 M Na_2SO_4; trailing electrolyte, 0.1 M Tris. Sample introduction, 30 s at +4.0 kV; 400-mW, 532 nm excitation. Reproduced from Ref. [22], with permission from the Society of Applied Spectroscopy.

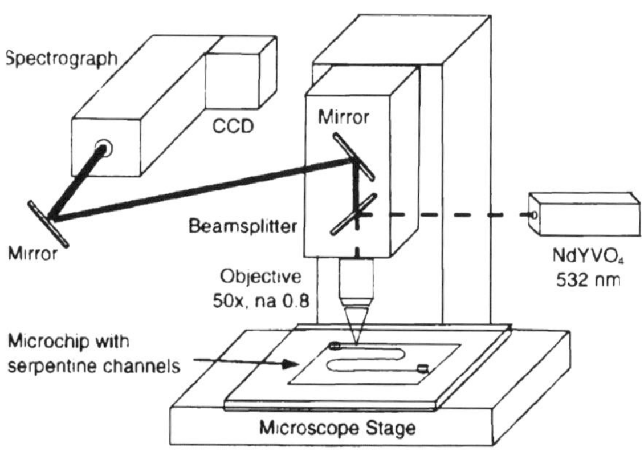

Fig. 11.15. Diagram of Raman spectroscopy/microchip isotachophoresis system. Component details in text. Reprinted with permission from Ref. [23]. Copyright (1998), American Chemical Society.

particularly useful for ITP as highly-concentrated buffers are used. Walker et al. [23], using the same analytes as in the previous work [22], carried out the separation in a serpentine designed chip with a total length of 21 cm. The interface between the Raman microscope and the microchip was straightforward as the microchip was directly attached to the microscope stage using clips, making this interface even simpler than for discrete capillaries (Fig. 11.15).

The detected concentrations of paraquat and diquat in this work were 2.3×10^{-7} M with no statistical treatment employed in this case.

11.3.2 Surface-enhanced Raman scattering in CE

As already mentioned in the introduction, SERS is an important method to increase the signal intensities obtained in Raman spectroscopy. Therefore, methods to use this enhancing factor in the coupling of CE and Raman spectroscopy were developed. Most of them are off-line methods, as the eluents are deposited onto different SERS substrates and are detected subsequently.

However, one on-line method where silver colloids were added to the background electrolyte was developed [24], which is also the first work that employed SERS as a detection method for CE. The analytes used were riboflavin and the very SERS-active rhodamine 6G; the detection limits were 1×10^{-6} M for riboflavin and 10^{-9} M for rhodamine (S/N = 10) (Fig. 11.16) when a microprobe spectrometer operated in a confocal mode with an excitation wavelength of 515 or 488 nm was used. The silver colloidal solution was prepared as described by Lee and Meisel [25], but it could be shown that the silver colloid had only a little influence on the separation performance when compared with a CE with laser-induced fluorescence detection (CE–LIF). Different concentrations of silver colloid in the background electrolyte were checked for their influence of the S/N of the SERS spectra; however the particle size of the colloids was not optimized. The only obvious disadvantage of this work lies in the necessity of cleaning the capillary after each run, as the spectra degraded because of decreased resolution, and increased background fluorescence owing to an assumed gradual buildup of silver particles on the walls of the capillary.

He et al. [26] were the first to couple CE and Raman spectroscopy off-line. The interface used consisted of the CE system, a computer-controlled translation stage and a fiber-optic Raman microprobe with a He–Ne laser (632.8 nm) as excitation source (Fig. 11.17).

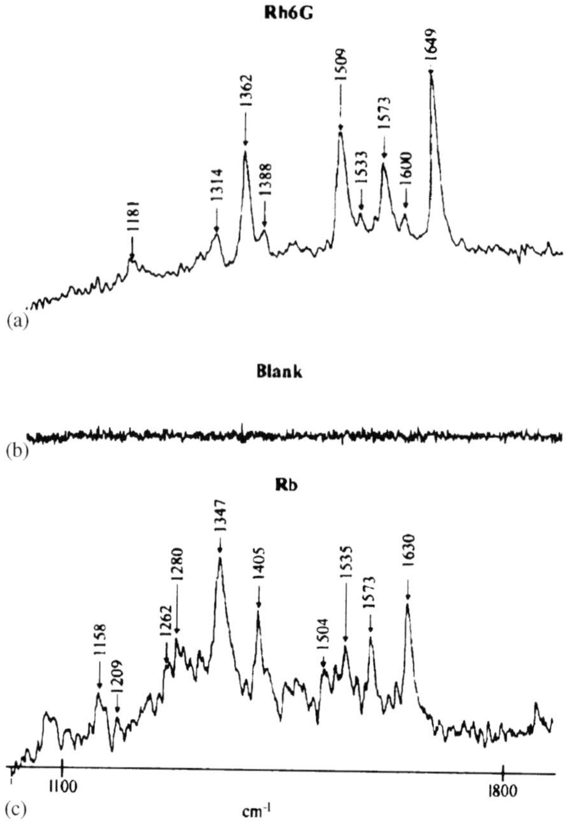

Fig. 11.16. CE-SERS spectra: (A) SERS spectrum of 1.10^{-8} M rhodamine 6G injected with an elution time of 9 min; (B) blank spectrum between the two analytes of interest at a time of 11 min; (C) SERS spectrum of 5×10^{-6} M Rb injected with an elution time of 13 min. CE conditions were same as before. SERS conditions were 514.5 nm, 200 mW, and 1 s acquisition time. Reprinted with permission from Ref. [24]. Copyright (2000), American Chemical Society.

For confirmation of the separation achieved by the CE–SERS system, standard on-line UV detection was used. The retention times obtained corresponded well. The authors evaluated four different substrates for deposition of the eluate, namely, Au-LI Ag, Ag colloid, sputtered Ag and Au-LI Ag-Ag film substrates with Au-LI Ag-Ag showing the strongest enhancements. The Au-LI substrate was prepared using a colloidal Au solution (~17 nM) and re-immersing the colloidal Au into a mixture of equal volumes of LI Ag initiator and enhancer solutions (Nanoprobes, Inc.). The sputtered substrates were

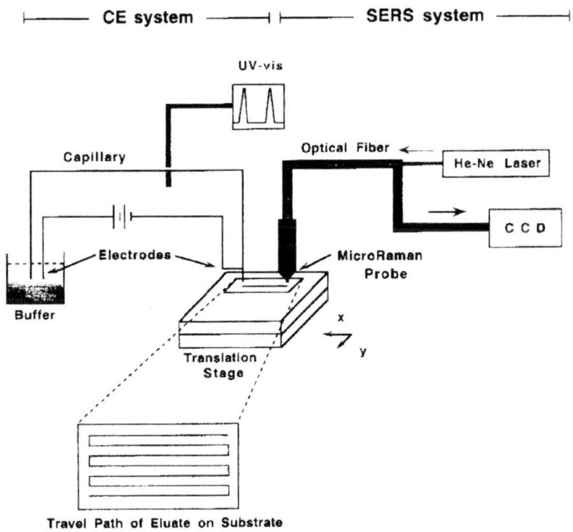

Fig. 11.17. Diagram of CE/SERS system using an optical fiber for probing the deposited sample. Component details in text. Reprinted with permission from Ref. [26]. Copyright (2000), American Chemical Society.

prepared using an Ar plasma sputterer, while the Ag colloid was prepared by reduction with ethylene diaminetetraaceticacid (EDTA) (Fig. 11.18).

The analytes used were *trans*-1,2-bis(4-pyridyl)ethylene and *N*,*N*-dimethyl-4-nitrosoaniline. The CE–SERS system was also applied to the amino acids tyrosine and tryptophan and also to the environmental pollutants chlorophenol and dichlorophenol. As this work was mainly intended as a feasibility study, no quantification was done in this work.

DeVault and Sepaniak [27] explored another approach for deposition as they used electrofilament deposition onto a moving SERS substrate using an Ar-ion laser (514 nm) as excitation source. The on-column separation was checked by LIF. The analytes used were benzyloxyresufin, riboflavin and resorufin (all 10^{-5} M), all compounds of biological significance. In this work the preparation of the SERS substrates was carefully evaluated and optimized, as were the separation conditions (Fig. 11.19).

The possibility of optimizing the separation and the detection step in an off-line experiment separately is one of the main advantages of such an approach. Also, the use of a self-assembled monolayer (SAM) formed out of dodecanethiol to enhance the SERS signal by selective absorption was studied for resorufin. Using SAM, the sample deposited was better

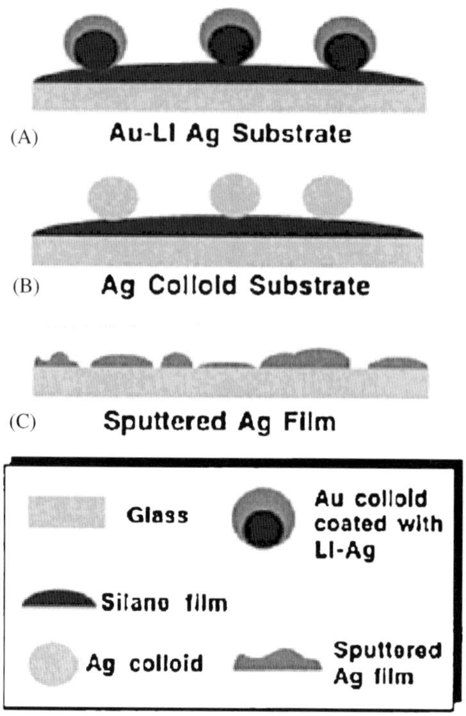

Fig. 11.18. Schematic views of the different used SERS substrates. Reprinted with permission from Ref. [26]. Copyright (2000), American Chemical Society.

spatially focused and the detection limit could therefore be lowered by a factor of 10, resulting in a limit of detection of 10^{-6} M (Fig. 11.20).

In the approach of Seifar et al. [28] thin-layer chromatography (TLC) plates were used as substrates followed by application of the silver sol after the deposition (Fig. 11.21).

The analytes used in this experiment were three different dyes, Acid Orange 7, Food Yellow 3 and Food Red 1. The wavelength of the excitation source, an Ar-ion laser, was 514.5 nm. The eluate of the CE run was immediately absorbed on moving the TLC plates so that no dispersion of the spot was observed. The detectable amount deposited on the TLC plate was 2–3 pmol (about 1 ng) (Fig. 11.22).

The possibility of using resonance Raman spectroscopy (RRS) to enhance the signal was first employed by Chen and Morris [17] using a laser line of 532 nm. In a more recent work by Dijkstra et al. [29] the possibilities of CE-UV-RRS were explored. The employment of excitation light in the UV range makes the application of RRS more widely

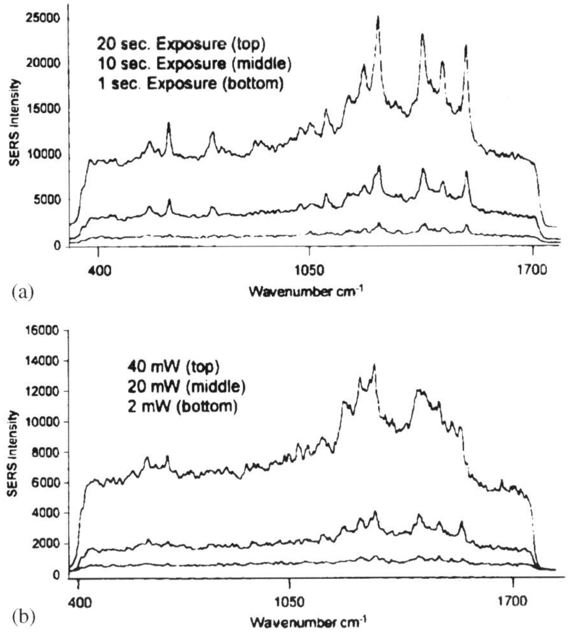

Fig. 11.19. SERS spectra of rhodamine 6G showing the effects of (a) increasing the diode-array exposure time and (b) increasing the laser power. A solution of 10^{-8} M rhodamine 6G diluted in CE buffer was EF deposited onto the SERS substrate. Experimental conditions are described before. Reproduced from Ref. [27], with permission from Wiley-VCH-Verlag GmbH.

usable as more compounds show absorption in the UV rather than in the visible region. In this work two excitation wavelengths (244 and 257 nm) were used that were generated by a frequency-doubled Ar-ion laser. The analytes studied were aromatic sulfonic acids, nucleotides and substituted pyrenes (Fig. 11.23).

The detection was carried out on-line and, therefore, experiments were carried out to improve the detectability of the analyte by increasing the volume of the injection and sample stacking. It could be shown that the overall sensitivity was increased, and only some peak-broadening could be observed. The limits of detection for the nucleotides were calculated and found to range from 25 μM for thymidine-5-monophosphate disodium salt (TMP) to 190 μM for uridine-5′-monophosphate disodium salt (UMP).

As both SERS and RRS are methods enhancing the signal-to-noise ratio, the combination of these techniques is promising to yield even better results. The mechanism of SERRS is somewhat different from

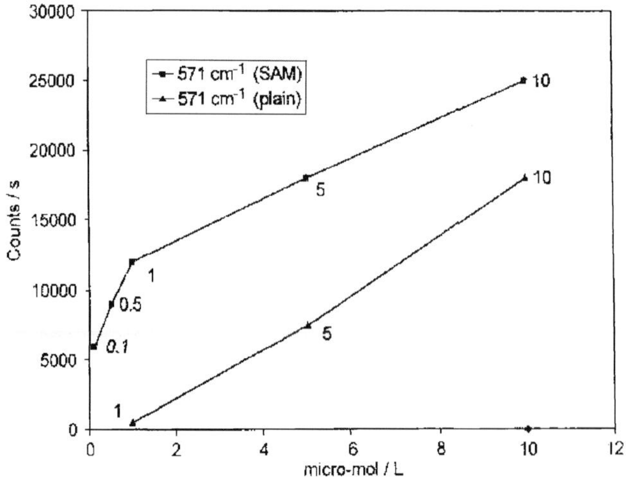

Fig. 11.20. Calibration curves for resorufin EF deposited onto plain SERS substrate and onto dodecane-SERS substrate. The resorufin Raman band at 571 cm^{-1} was used for calibration. Experimental conditions are described before. Reproduced from Ref. [27], with permission from Wiley-VCH-Verlag GmbH.

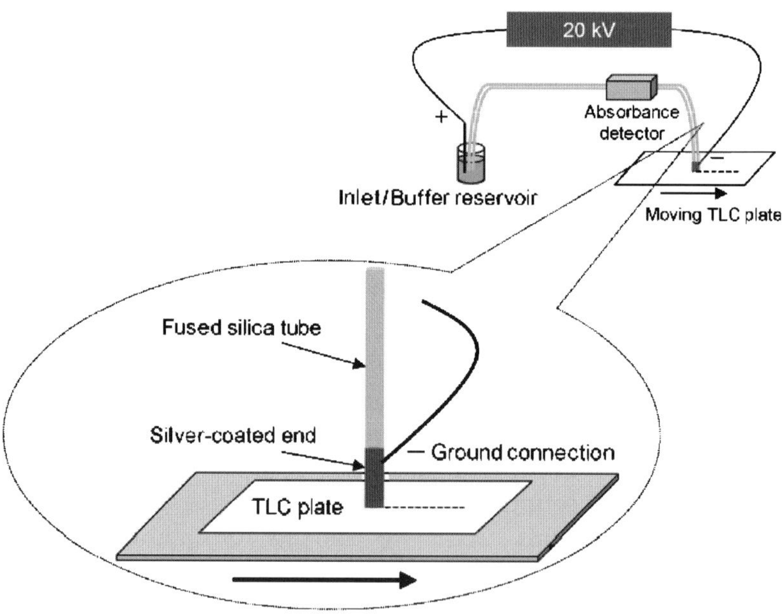

Fig. 11.21. Schematic of the modified CE setup. Reproduced from Ref. [28], with permission from Wiley-VCH Verlag GmbH.

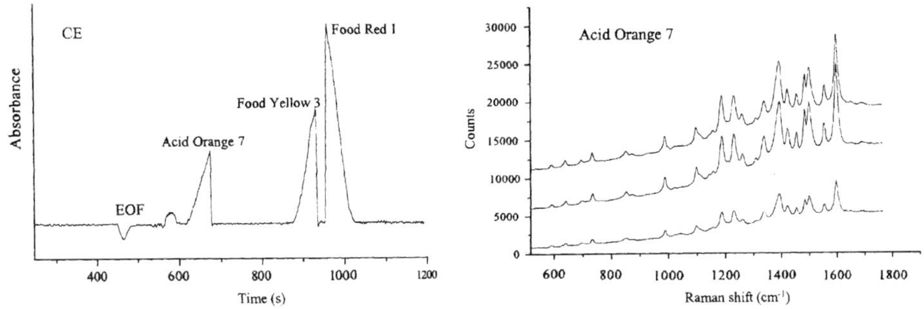

Fig. 11.22. Typical electropherogram of a mixture of Acid Orange 7 (0.2 mM), Food Yellow 3 (0.2 mM) and Food Red 1 (0.4 mM) with absorbance detection at 500 nm. After deposition on a silica TLC plate the SERRS spectra of three successive CE runs of each dye were recorded using nitric acid as additive. The translation speed of the silica plate was 3.4 mm min^{-1} until elution of Food Yellow 3 and, then, increased to 6.1 mm min^{-1}(only one Acid Orange 7 is shown). Reproduced from Ref. [28], with permission from Wiley-VCH Verlag GmbH.

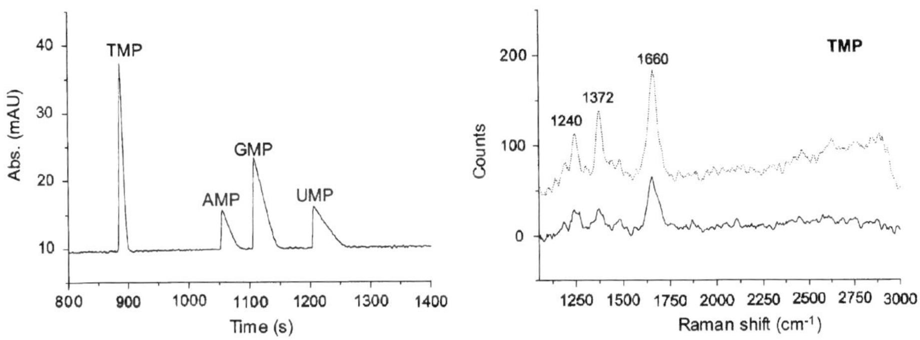

Fig. 11.23. CE-UV electropherogram ($\lambda = 260$ nm) of a mixture of 0.5 mg/ml of each TMP, AMP, GMP and UMP, using an injection volume of 23 nl (left). Background-corrected UV-RRS spectra acquired of TMP (right). Excitation wavelength: 244 nm (lower spectrum) or 257 nm (higher spectrum). Exposure time: 2 s (only TMP is shown here). Reprinted with permission from Ref. [29]. Copyright (2003), American Chemical Society.

SERS, as SERRS is less sensitive to changes in molecular orientation on the substrate. Dijkstra et al. [30] studied different substrates in off-line SER(R)S experiments. Again, the procedure was similar as before [28]: the eluate was deposited onto the substrates, dried and detected,

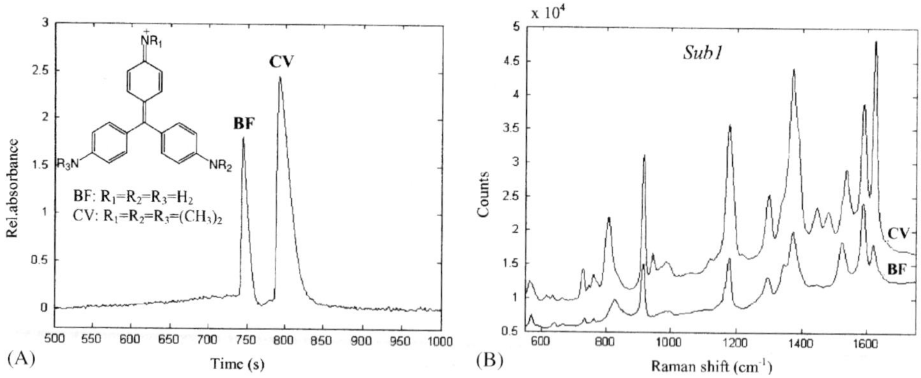

Fig. 11.24. At-line CE–SERRS using different substrates. (A) Electropherogram of a solution containing 100 μg/ml of each BF and CV, recorded with an absorbance detector at 560 nm. Injection volume: 13 nl. The CE effluent was deposited on the substrates at a table speed of 6.1 mm/min. SERRS spectra were recorded on (B) Sub1 after deposition of the whole CE run. Exposure time for all substrates: 80 × 2 s (here only one substrate is shown). Reprinted from Ref. [30], Copyright (2004), with permission from Elsevier.

and the excitation wavelength was 514.5 nm. The substrates studied were an etched silver foil, a vapor-deposited silver film, a silver-oxalate-precoated silica TLC plate and a silica TLC plate, to which colloid and poly(L-lysine) were added after deposition. As in this work, the effect of SERS and SERRS was studied using *trans*-1,2-bis(4-pyridyl)ethylene (BPE) as a model compound for SERS, and crystal violet (CV) for SERRS, respectively. The limits of detection were 50 μM for BPE and 0.5 μM for CV, respectively (Fig. 11.24).

Except the silver-oxalate-coated TLC plate, all substrates used showed good results, while in the case of the TLC plate the background signals increased with time, obviously owing to a drying process attributable to amorphous carbon.

REFERENCES

1 M. Abo-Bakr, K. Feikes, K. Holldack, P. Kuske, W.B. Peatman, U. Schade, G. Wustefeld and H.-W. Hübers, *Phys. Rev. Lett.*, 90 (2003) 094801-1-094801/4.
2 P.F. Taday, I.V. Bradley, D.D. Arnone and M. Pepper, *J. Pharm. Sci.*, 92 (2002) 831–838.

3 L.M. Miller, G.D. Smith and G.L. Carr, *J. Biol. Phys.*, 29 (2003) 219–230.
4 M. Kölhed, P. Hinsmann, P. Svasek, J. Frank, B. Karlberg and B. Lendl, *Anal. Chem.*, 74 (2002) 3843–3848.
5 M. Kölhed, P. Hinsmann, B. Lendl and B. Karlberg, *Electrophoresis*, 24 (2003) 687–692.
6 P. Hinsmann, L. Arce, P. Svasek, M. Lämmerhofer and B. Lendl, *Appl. Spectrosc.*, 58 (2004) 662–666.
7 S. Kulka, N. Kaun, J. Baena Rodriguez, J. Frank, P. Svasek, D. Moss, M.J. Vellekop and B. Lendl, *J. Anal. Bioanal. Chem.*, 378 (2004) 1735–1740.
8 T. Pan, R.T. Kelly, M.C. Asplund and A.T. Woolley, *J. Chromatogr. A*, 1027 (2004) 231–235.
9 B.M. Patterson, N.D. Danielson and A.J. Sommer, *Anal. Chem.*, 76 (2004) 3826–3832.
10 R.A. Todebush, L.-T. He and J.A. de Haseth, *Anal. Chem.*, 75 (2003) 1393–1399.
11 J.L. Jarman, R.A. Todebush and J.A. de Haseth, *J. Chromatogr. A*, 976 (2002) 19–26.
12 J.M. Chalmers and P.R. Griffiths, *Handbook of Vibrational spectroscopy*, Wiley, Chicester, 2002.
13 L.A. Lyon, C.D. Keating, A.P. Fox, B.E. Baker, L. He, S.R. Nicewarner, S.P. Mulvaney and M.J. Natan, *Anal. Chem.*, 70 (1998) 341R–61R.
14 P. Rösch, M. Schmitt, W. Kiefer and J. Popp, *J. M. Struct.* (2003) 661–662 363–369.
15 K.C. Schuster, I. Reese, E. Urlaub, J.R. Gapes and B. Lendl, *Anal. Chem.*, 72 (2000) 5529–5534.
16 C.-Y. Chen and M.D. Morris, *Appl. Spectrosc.*, 42 (1988) 515–518.
17 C.-Y. Chen and M.D. Morris, *J. Chromatogr.*, 540 (1991) 355–363.
18 W.K. Kowalchyk, P.A. Walker III and M.D. Morris, *Appl. Spectrosc.*, 49 (1995) 1183–1188.
19 M.D. Morris. In: M. Chalmers John (Ed.), *Handbook of Vibrational spectroscopy*, Wiley, Chicester, 2002.
20 P.A. Walker III, W.K. Kowalchyk and M.D. Morris, *Anal. Chem.*, 67 (1995) 4255–4260.
21 P.A. Walker III and M.D. Morris, *J. Chromatogr. A*, 805 (1998) 269–275.
22 P.A. Walker III, J.M. Shaver and M.D. Morris, *Appl. Spectrosc.*, 51 (1997) 1394–1399.
23 P.A. Walker III, M.D. Morris, M.A. Burns and B.N. Johnson, *Anal. Chem.*, 70 (1998) 3766–3769.
24 W.F. Nirode, G.L. DeVault, M.J. Sepaniak and R.O. Cole, *Anal. Chem.*, 72 (2000) 1866–1871.
25 P.C. Lee and D.J. Meisel, *J. Phys. Chem.*, 86 (1982) 3391–3395.
26 L. He, M.J. Natan and C.D. Keating, *Anal. Chem.*, 72 (2000) 5348–5355.

27 G.L. DeVault and M.J. Sepaniak, *Electrophoresis*, 22 (2001) 2211–2303.
28 R.M. Seifar, R.J. Dijkstra, A. Gerssen, F. Ariese, U.A. Th. Brinkman and C. Gooijer, *J. Sep. Sci.*, 25 (2002) 814–818.
29 R.J. Dijkstra, E.V. Efremov, F. Ariese, U.A. Th. Brinkman and C. Gooijer, *Anal. Chem.*, 75 (2003) 5697–5702.
30 R.J. Dijkstra, A. Gerssen, E.V. Efremov, F. Ariese, U.A. Th. Brinkman and C. Gooijer, *Analytica Chimica Acta*, 508 (2004) 127–134.

Chapter 12

Coupling nuclear magnetic resonance to capillary electrophoresis

Dimuthu A. Jayawickrama and Jonathan V. Sweedler

12.1 INTRODUCTION

As described in earlier chapters, a number of detection methods have been successfully coupled with Capillary Electrophoresis (CE). The merger of the high separation efficiency of CE and the information-rich detection of Nuclear Magnetic Resonance (NMR) spectroscopy creates a unique class of on-line analytical techniques. NMR is a commonly used spectroscopic technique in both academic research and industrial applications. The analytical capabilities of NMR are unparalleled when compared with many other spectroscopic techniques, for a number of reasons. For example, NMR is an indispensable tool for structure elucidation of small to large organic molecules (proteins/peptides) [1], in drug metabolism/pharmacokinetics and synthetic chemistry/natural product chemistry [2]. Alone or in conjunction with mass spectrometry (MS), NMR can determine the structures of the majority of small molecules. The ability to detect nuclear spins greater than spin quantum number zero allows the application of NMR to over 120 isotopes. Besides structure elucidation, NMR is a noninvasive technique that can probe equilibrium chemical kinetics, binding and molecular interactions [3]. The diagnostic capabilities of NMR allow it to measure pH [4,5] and temperature [6–8]. Bulk (viscosity) and molecular properties (diffusion coefficients) can also be extracted with NMR [9]. However, a major limitation of NMR is its inherently poor sensitivity, which restricts it to applications with analytes in the micromolar range or higher. To achieve acceptable levels of signal to noise (*S*/*N*) in static NMR, long observation times or a large number of scans are required. However, this approach is not entirely practical for on-line NMR

Comprehensive Analytical Chemistry XLV
M.L. Marina, A. Ríos and M. Valcárcel (Eds)
Volume XLV ISSN: 0166-526X DOI: 10.1016/S0166-526X(05)45012-6

detection, particularly in the case of liquid chromatography (LC) or CE. In CE, a typical analyte band moves across the detection in few seconds, which limits the NMR observation time, often preventing the acquisition of the highest quality NMR spectra. A number of techniques to improve CE–NMR detection are described in the following sections.

12.1.1 NMR sensitivity improvements

What restricts the application of NMR for CE? As shown in Table 12.1, the limit of detection (LOD) of NMR with capillary separation is several orders of magnitude poorer than other selected detection methods. The ongoing research on NMR sensitivity improvement focuses on the use of NMR data processing to improve signals. However, it is necessary to balance the *S/N* improvement with the tradeoff in spectral resolution. A good example is the application of apodization functions prior to Fourier transformation (FT). In ^{13}C-NMR acquisition, proton decoupling is used to record high *S/N* ^{13}C spectra. However, the resulting spectra lack proton–carbon coupling information. Other approaches to increase the NMR sensitivity use hardware improvements. As a historical example, the use of pulsed radio frequencies (RF) (Fourier transform) instead of continuous wave (CW) spectrometers was a major step toward acquiring a large number of scans of the same sample in a relatively short period, thus increasing *S/N*. Another well-known method to improve NMR sensitivity is to use a higher magnetic field, magnets as sensitivity increases with 7/4th to the power of the magnetic field. Recently, GHz magnets have been introduced, but their current cost prevents their use in most facilities.

TABLE 12.1

LODs for common analytical methods used with capillary separations

Method	LOD (mol)
Fluorescence	10^{-18}–10^{-23}
Mass spectrometry	10^{-13}–10^{-21}
Electrochemical	10^{-15}–10^{-19}
Radiochemical	10^{-14}–10^{-19}
UV–Vis absorbance	10^{-13}–10^{-16}
NMR	10^{-9}–10^{-11}

Reprinted with permission from Ref. [40] Copyright 1999, American Chemical Society.

Besides spectrometer hardware, there are several sample- and probe-dependent approaches to improve performance, including nuclear polarization techniques such as spin polarization-induced nuclear overhauser enhancement (SPINOE) [10,11] and dynamic nuclear polarization (DNP) [12,13]. sensitivity improvements of 50-fold for ^{1}H and 70-fold for ^{13}C have been reported with SPINOE [14]. The SPINOE applications are limited to nonaqueous systems in which Xe is soluble. DNP, which exploits saturating electron spins (electron radicals) coupled to nuclear spins to increase NMR sensitivity, has been employed as an NMR detector in LC [15]. The potential of this approach is exciting and we expect that it will have a powerful effect on hyphenated NMR separation techniques in the future. New pulse sequences based on magnetic field gradients can also improve NMR signals [16,17].

The NMR probe specifications and performances are important factors affecting the quality of the NMR data recorded. The NMR probe includes a RF coil that excites the sample and detects the weak signals generated by the precessing NMR-active nuclei. NMR probes with cryogenically cooled RF coils provide an ~4-fold sensitivity enhancement [18–21]. The requirement for a cryogenically cooled coil causes either a large gap between the NMR coil and the sample or a large thermal gradient between the NMR coil and a liquid sample, making it difficult to interface smaller superconducting coils to nanoliter volume liquid samples as required for CE–NMR. We certainly expect this technology to continue to advance.

NMR microcoil probes are another method to improve NMR mass sensitivity. To date, the solenoidal microcoil probe is the most mass-sensitive probe available for small-volume samples. The NMR microcoil was first introduced to study tissue surfaces. However, its large spectral linewidths hindered widespread applications. The first high-resolution microcoil NMR probe was reported in 1995 with a linewidth of 1 Hz [22]. A microfabricated NMR coil was also introduced in 1997 [23]. Advances in miniaturization may further decrease NMR sensitivity. At this time, NMR microcoil probes size-matched to the volumes of the capillaries used for CE have become the most common method for interfacing CE and NMR.

12.1.2 CE and NMR Hyphenation

The Hyphenation of NMR to electrophoresis (E–NMR) was first introduced by Johnson and He in 1989 [24] to study the electrophoresis process. Although this work was not conducted at the capillary level and did not involve a separation either, it demonstrated the capability of combining these two techniques to measure electrophoretic mobilities and diffusion coefficients in mixed solutions. This technique is now used for the physiochemical characterization of complex chemical systems [25–27]. The first electrophoresis separation coupled to NMR used a solenoidal NMR probe in a 300-MHz magnet [28,29]. This original work described the issues involved with CE–NMR Hyphenation, such as electrophoretic and current-induced effects on the appearance and *S/N* of NMR spectra. CE–NMR applications using both solenoidal microcoils [28,29] and saddle coils [30–32] have since demonstrated the capabilities and the drawbacks of CE–NMR. An integrated system to perform CE–NMR, capillary electrochromatography CEC–NMR and capillary column LC–NMR was introduced by Bayer, Albert and their groups in 1998, using a saddle-type NMR coil [30]. One published application included the analysis of paracetamol metabolites in human urine [31,32]. A unique application using a two-solenoidal-coil NMR probe demonstrates simultaneous CE measurements and multidimensional NMR acquisition [33]. In addition to separations, CE–NMR has also been employed as a diagnostic tool to measure intracapillary temperature, pH, sample injection performance and sample plug profile, as described below.

12.2 CE–NMR HYPHENATION

Coupling CE to NMR can be performed without any major modifications to the CE or NMR spectrometer. All the magnetic materials should be kept 2–3 m away from the magnet, except with the shielded NMR, where the distance can be much less. Because the platinum electrodes typically used in CE are not magnetic, the inlet and outlet buffers can be located inside or next to the magnet as required by the application. The design of the NMR probe is critical for recording high-quality CE–NMR data. An overview of the important characteristics of the Hyphenation of CE to NMR spectroscopy follows.

12.2.1 The size-matched NMR probe

As mentioned earlier, the NMR probe (RF coil) is a vital component in NMR instrumentation. It has been shown theoretically [34] and verified experimentally [22] that the miniaturization of the NMR coil to accommodate the sample can provide highly mass-sensitive NMR detectors. Several figures of merit are worth defining when comparing NMR probe performance. The mass sensitivity (S_m) and concentration sensitivity (S_c) are defined as the minimum amount and concentration of a sample as defined by the following:

$$S_m = \frac{S/N}{\text{mol}\, t^{1/2}} \tag{12.1}$$

$$S_c = \frac{S/N}{Ct^{1/2}} \tag{12.2}$$

where S/N is the signal-to-noise ratio of the peak of interest, mol the number of moles within the NMR-active volume (V_{obs}), C the sample concentration based on peak of interest and t the NMR acquisition time. Another useful figure of merit when assessing probe performance, especially when using NMR as an on-line detector, is the normalized limit of detection (nLOD) for concentration (Eq. (12.3)) and mass (Eq. (12.4)). These equations take into account acquisition time and assume a S/N of 3 for the detection limit criteria:

$$\text{nLOD}_c = \frac{3Ct^{1/2}}{S/N} \tag{12.3}$$

$$\text{nLOD}_m = \frac{3\ \text{mol}\, t^{1/2}}{S/N} \tag{12.4}$$

The NMR observed factor (f_0) is defined as the ratio of the probe volume, V_{obs}, and total sample volume, V_{tot}. In theory, the maximum sensitivity can be achieved by bringing f_0 to unity. However, magnetic susceptibility effects caused by different solvents (with different magnetic susceptibilities) in the probe can severely impair NMR sensitivity and spectral resolution. Therefore, NMR sample volumes often extend beyond the probe regions. As a result, a typical 5-mm NMR probe requires 750 µl of sample although the V_{obs}~220 µl. This is also true at the small scale.

Probe miniaturization to analyze smaller samples has resulted in a number of modified probes that are suitable to couple to capillary separation techniques. Although a number of different geometries have been adapted for NMR coils, we discuss only two categories of probes, the Helmholtz (saddle type) and the solenoidal coil based probes (Figs. 12.1(A) and (B)), as these have been used as on-line detectors for CE.

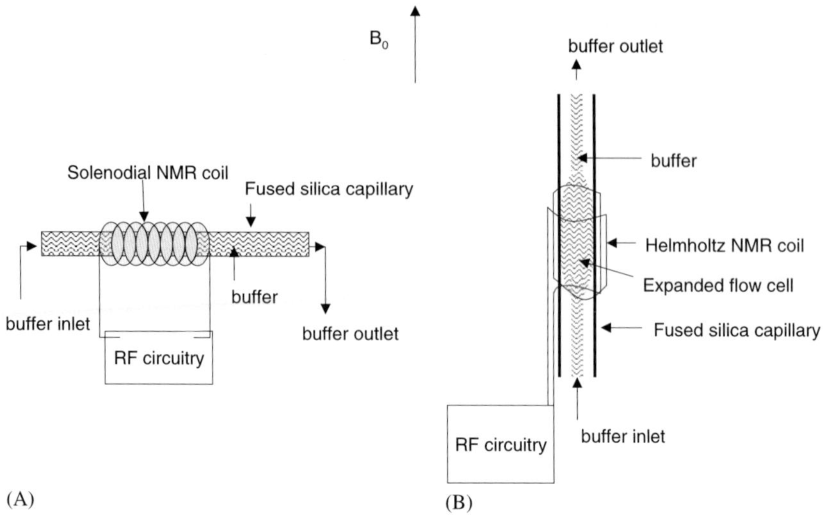

Fig. 12.1. (A) Solenoidal RF coil directly wrapped around a fused silica capillary placed orthogonal to B_0. (B) Helmholtz coil with an expanded flow cell is placed in parallel to B_0.

The introduction of a saddle coil to house 1.7-mm-diameter sample containers was a key step in NMR in probe miniaturization [35,36]. A saddle-type inverse coil for 60 µl NMR-active volumes (required volume 140 µl) was another major development [37]. The smallest commercially available saddle coil requires a 5 µl total sample volume with a 2.5 µl active volume [38]. Most importantly, this probe achieves a five-fold *S*/*N* enhancement compared with full-sized saddle coil probes.

Solenoidal microcoils have been designed to cover volumes of a few nanoliters to 10 µl with coil length typically between 1 and 3 mm, and are perhaps the easiest small-volume probes to fabricate. The *S*/*N* per unit volume for solenoidal coils with a diameter greater than 100 µm is inversely proportional to coil diameter, whereas the *S*/*N* per unit volume of coils with a diameter smaller than 100 µm varies with the square root of coil diameter. As S_m is dependent on *S*/*N* per unit volume, the S_m improves as the coil diameter decreases for a microcoil with a fixed length-to-diameter ratio [39]. Therefore, theoretically, a size-matched solenoidal coil provides a several-fold increase in sensitivity. Overall, solenoidal sensitivity enhancement is due to these two effects: the geometry of the solenoid allows a 2-fold increase in performance, and the second enhancement is caused by the ability to size-match the probe to the nanoliter volume range. In practice, solenoidal coils provide up to ~20-fold higher mass sensitivity than a 5-mm standard saddle-type NMR coil [22].

Which probe should be used? To answer this question, one needs to consider what type of sample one has. Table 12.2 compares the mass and concentration sensitivities of sucrose using a standard saddle coil (Varian 5 mm, V_{obs}~222 µl) and a 1-mm-long microcoil probe ($V_{obs} = 5$ nl) [40]. As is obvious from the table, the concentration sensitivity of the large-volume saddle-type coil is superior to that of the microcoil. On the other hand, the mass sensitivity achieved with the microcoil is greater than that of the saddle coil. If the amount of available sample is small and can be dissolved in a volume similar to V_{obs}, then it is beneficial to use a microcoil probe. The advantages of the increased mass sensitivity allow a shorter NMR acquisition time. In one study, 100-fold less time (or a 10-fold increased sensitivity) was reported with an NMR microcoil in comparison with a commercial nanoprobe [41]. If the sample of interest has a low concentration and cannot be concentrated because of solubility or other issues, saddle coils are more useful than microcoil probes.

The previous discussion has been general and applies to most applications. There are a number of advantages of using solenoidal coils for CE. A solenoidal coil can be fabricated easily by wrapping a piece of Cu wire around a capillary that houses the sample. This Cu wire acts as the RF transmitter/receiver, whereas the capillary serves as the sample holder and probe holder. The NMR active volume can be manipulated either by changing the number of turns in the coil or by changing the capillary diameter. The miniature nature of the coil permits more than one coil to be within the homogeneous region of the magnetic field to achieve high NMR throughput [42–44]. Recently, an eight-microcoil

TABLE 12.2

Performance comparisons of NMR probes

Figures of merit	Varian 5-mm probe ($V_{obs} = 222\,\mu l$)	microcoil probe ($V_{obs} = 5$ nl)
S_c ($S/N \cdot mM^{-1} s^{-1/2}$)	30	0.028
S_m ($S/N \cdot \mu mol^{-1} s^{-1/2}$)	134	5580
$nLOD_c$ ($mM\,s^{1/2} \cdot S/N^{-1}$)	0.10	110
$nLOD_m$ ($nmol\,s^{1/2} \cdot S/N^{-1}$)	22	0.54

Reprinted with permission from Ref. [40]. Copyright 1999, American Chemical Society.

probe has been used to analyze multiple samples [45]. One significant issue with CE–NMR is the electrophoretic current-induced magnetic field. This effect can be severe with the microcoil probe but has no effect with saddle coils because of their orientation in the magnetic field. CE is more than just a separation technique as it concentrates individual components while being separated. This is an added advantage when using inherently less concentration-sensitive NMR as an on-line detector.

12.2.2 Interfacing CE to NMR

The direct coupling of conventional and commercially available NMR spectrometers to CE has been a challenge. One reason is that the micron-scale (nanoliter volume) requirements imposed by the CE capillary do not match the millimeter dimensions of commercially available NMR probes. Because of this, the adaptation of a commercial NMR system is common with custom NMR probes. A typical CE–NMR interface schematic is shown in Fig. 12.2(A). Three main configurations have been reported for CE–NMR [28,30,46]. The most easily adopted configuration is the one in which the inlet and outlet buffers are outside the magnet, as shown in Fig. 12.2(A). However this lengthens the capillary and prolongs the separation time. A second configuration involves the CE–NMR instrumentation with the inlet and outlet buffers inside the magnet bore, as shown in Fig. 12.2(B). However more space is required to house both inlet and outlet buffer vials and therefore this approach becomes most useful with wider bore (i.e., 89-mm-diameter) magnets. A third hybrid configuration has also been used for CE–NMR measurements (Fig.12.2(B)).

When using a saddle coil, the CE capillary is threaded through the coil and allows easy exchange of the capillary. The solenoidal coil is fabricated directly on the separation capillary or can be fabricated around a polyimide capillary sleeve (the so-called sleeve probe), which can hold both the coil and the separation capillary [47]. Directly wrapping the wire around the CE capillary provides the highest filling factor. The sleeve probe technology permits the exchange of a capillary as required by a particular application even though the filling factor is lower than that of directly wrapped coils. Because of the magnetic susceptibility differences, the typical NMR spectral linewidth reaches 1–2 Hz with sleeve probes. On the plus side, the sleeve probe technology extends the probe's lifetime, as the probe is not vulnerable to capillary breakage.

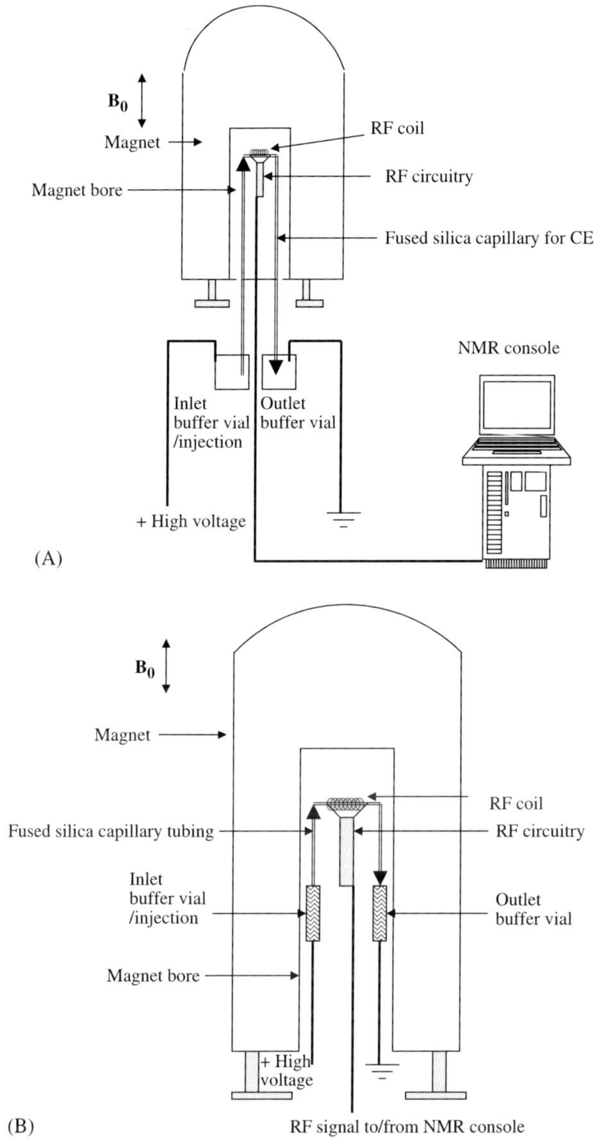

Fig. 12.2. Instrumental schematic for CE–NMR: (A) both inlet and outlet buffer vials kept outside the magnet bore; (B) both inlet and outlet buffer vials kept inside the magnet bore; and (C) the inlet buffer vial is kept outside the magnet bore while the outlet buffer vial is kept inside the magnet bore. While the solenoidal coil is shown, saddle coils are also possible.

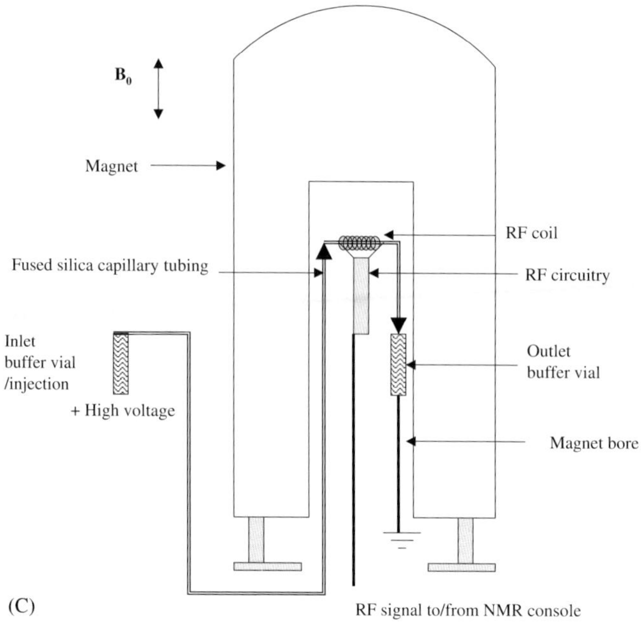

Fig. 12.2. Continued

Typical CE analyte peak volumes are in the low-nanoliter-volume range. CE–NMR experiments can be performed with NMR observe volumes as low as 5 nL [46]. However, further miniaturization of the observe volume reduces the NMR sensitivity. The difficulty of fabrication has precluded the use of smaller saddle-type coils to date. The smallest saddle coil probe reported so far has a 2.5 μl NMR observe volume, but has not been used for CE–NMR experiments [38]. Standard saddle type coils can be modified for CE–NMR by using insets in the larger volume saddle coils. A range of effective volumes from 250 to 400 nl has been reported for CE–NMR using a saddle coil as an on-line detector [30–32].

12.2.3 CE–NMR measurements

The NMR acquisition in CE–NMR is similar to that in LC–NMR in many aspects except for scale. However, the micrometer dimensions of the CE capillary require smaller volumes of buffer compared with the mobile-phase volumes used in standard LC. For example, LC–NMR typically requires hundreds of milliliters of mobile phase (including the reservoirs), which restricts the routine use of deuterated solvents because of the expense of the solvent. However, the total solvent volume required for CE is generally less than a few milliliters (including the buffer vials), permitting the inexpensive use of deuterated buffers in routine CE–NMR.

Critical parameters, governing the CE–NMR performance are the electrophoretic current-induced magnetic field and thermal effects. As mentioned in Section 12.2.1, the most mass-sensitive solenoidal NMR microcoil is formed with the coil axis perpendicular to B_0. A CE current-induced secondary magnetic field is created in this coil configuration that affects the spectral line shape. Equation (12.5) describes the relationship between the electrophoretic current-induced magnetic field, B_i and the radial distance from the center of the capillary, r: [40]

$$B_\mathrm{i} = \frac{\mu_0 i r}{2\pi R^2} \tag{12.5}$$

where μ_0 is the permeability constant, i the electrophoretic current and R the capillary internal diameter. The induced field gradient perturbs the uniform magnetic field surrounding the sample and deteriorates the S/N and the NMR spectral features. Figure 12.3 demonstrates induced field effect on the methyl triplet of triethylamine (TEA) [48].

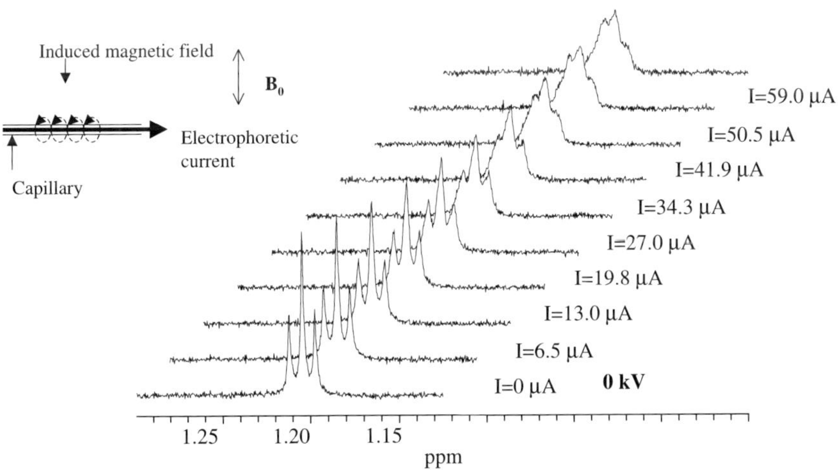

Fig. 12.3. Current-induced magnetic field effect on CE–NMR spectra of the TEA methyl peak in 1 M borate buffer with increasing applied voltage of 0.0–9.0 kV in increments of 1.0 kV. Inset: directions of electrophoretic current and induced magnetic field. (Reprinted with permission from Ref. [48]. Copyright 2002, American Chemical Society.)

The NMR signal broadens with increasing electrophoretic current. The scalar coupling of the triplet is beyond recognition at currents higher than 42 μA. The linewidth increases from 1.5 to 15 Hz and the corresponding *S/N* decrease is about 87%. The shimming procedures adopted in NMR to restore magnetic field homogeneity do not effectively eliminate these induced field effects. A straightforward answer is to record NMR spectra when the flow is stopped. In this method the applied voltage is terminated as the peak maximum reaches the NMR coil. NMR data are then acquired in the absence of electrophoretic flow. Although this permits the recording of high-resolution NMR, stopped-flow can lead to reduced separation efficiency. Nevertheless, stopped-flow approaches remain the only method available to acquire time-consuming but more informative multidimensional NMR.

How can one minimize the induced field-gradient effects in CE–NMR using solenoidal probes? NMR data acquisition under quiescent conditions has been described [46]. In this method, the separation voltage is periodically interrupted and NMR data are acquired for 1 min for every 15 s of applied voltage. However, this periodic voltage stop can still lead to reduced separation efficiency and poorer peak resolution. The recent introduction of a dual-coil NMR probe eliminates induced field effects. Figure 12.4 illustrates dual-coil CE–NMR instrumentation. In this design, two sleeve probe coils are fabricated to facilitate the use of two outlet capillaries. The two outlet capillaries are

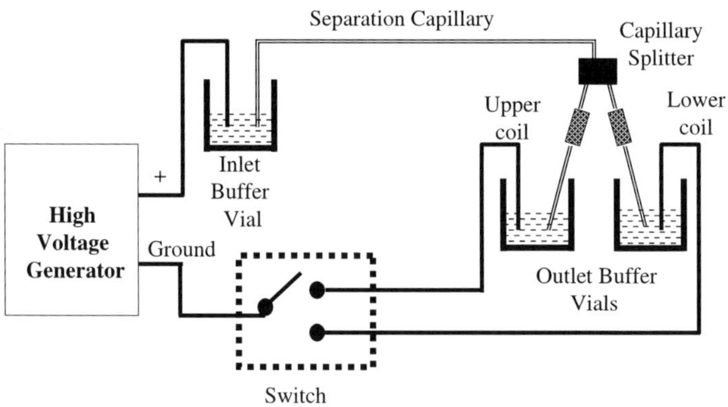

Fig. 12.4. Instrumental schematic for dual-coil continuous CE–NMR with two-microcoil NMR detection showing the arrangement of the separation capillary, the two outlet capillaries and the two NMR detection coils. (Reprinted with permission from Ref. [48]. Copyright 2002, American Chemical Society.)

connected to a single separation capillary through a capillary splitter. Figure 12.5 demonstrates a series of NMR spectra recorded while alternating electrophoretic flow between the two outlets. The CE–NMR spectra were recorded at one outlet capillary while the second capillary was at the separation voltage. The NMR data were acquired from the capillary with no applied voltage. Thus, the local magnetic gradient field effects observed in Fig. 12.3 were completely eliminated, and shigh-resolution NMR spectra with good *S*/*N* were recorded. The current-induced magnetic field effects can also be minimized, but not eliminated, by post processing [49].

The best sensitivity with a saddle coil is achieved with an axis parallel to B_0. In this configuration, the CE current does not change the lineshape. Therefore, NMR data are acquired continuously without

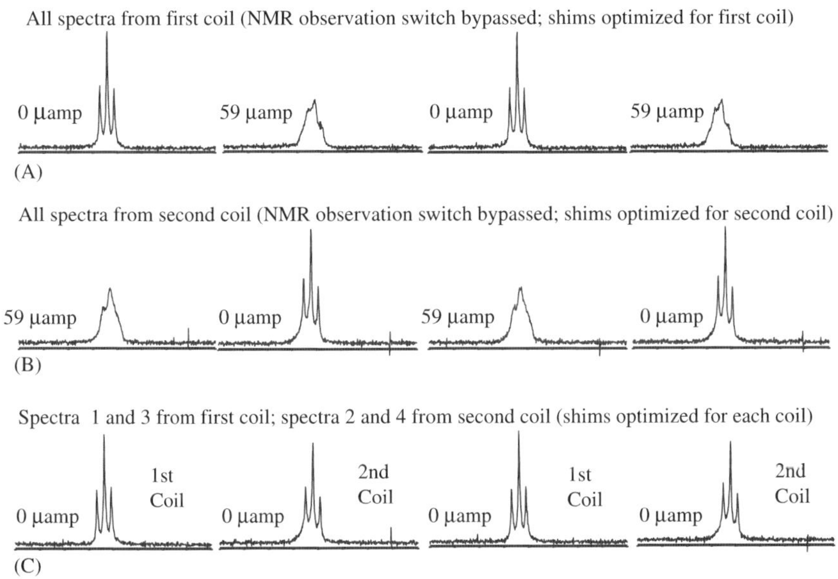

Fig. 12.5. Arrays of two-microcoil CE–NMR spectra of the methyl peak of triethylamine in 1 M borate buffer. Spectra acquired during alternation of electrophoresis flow between two outlet capillaries. (A) All spectra acquired from upper coil (shim settings optimized for upper coil; NMR observation switch bypassed). (B) All spectra acquired with lower coil (shim settings optimized for lower coil; NMR observation switch bypassed). (C) NMR spectra acquired from microcoil on outlet capillary without electrophoretic flow (shim settings optimized for active coil; NMR observation switch in-line). (Reprinted with permission from Ref. [48]. Copyright 2001, American Chemical Society.)

interrupting the applied voltage and special instrumentation modifications. The work by Bayer and Albert groups demonstrates the absence of induced field effects on NMR spectra (Fig. 12.6) [30]. The slight change of chemical shift for some resonances is due to Joule heating at higher voltages.

Continuous NMR data acquisition with CE–NMR is also affected by the different migration rates of sample bands. In CE, the migration time of an analyte is dictated by its electrophoretic mobility. As a result, the NMR detection time (sample residence time) varies from one analyte to another. Residence time can affect the signal intensity and linewidth. Therefore, corrections may need to be adopted when interpreting data quantitatively. The different migration times can change

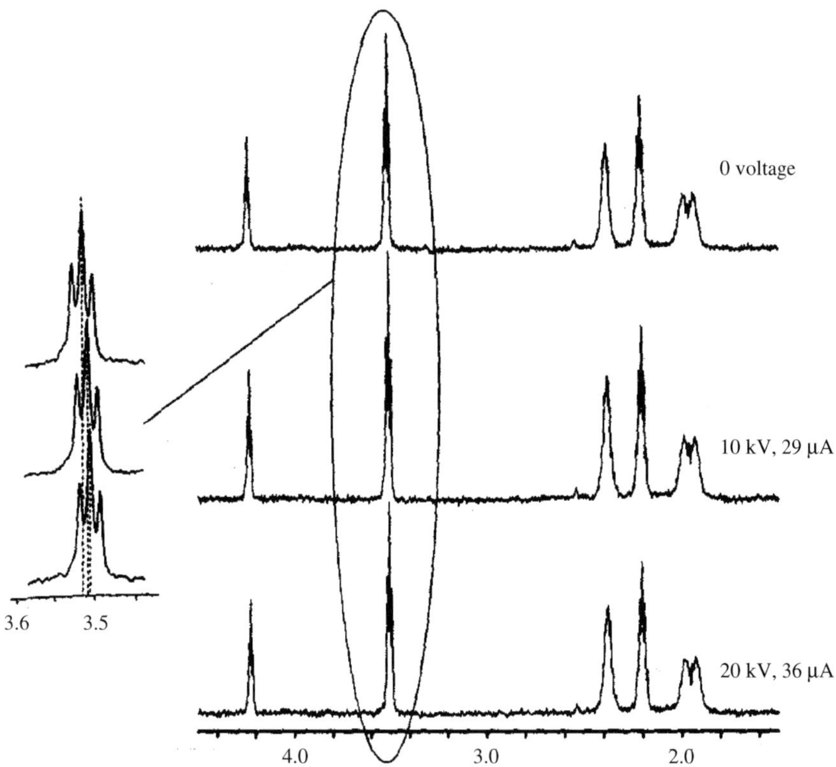

Fig. 12.6. Static 600-MHz ^{1}H-NMR spectra of lysine under CZE conditions: (1) without voltage, (2) with 10 kV (29 μA) and (3) with 20 kV (36 μA). Inset shows slight change in chemical shift due to temperature. (Reprinted with permission from Ref. [30]. Copyright 1998, American Chemical Society.)

the effective relaxation times. The sample residence time, τ, of an analyte in a continuous-flow experiment can be related to the NMR active volume, V_{obs}, and the flow rate, F: [50]

$$\tau = \frac{V_{obs}}{F} \tag{12.6}$$

For a fixed V_{obs}, a decrease of flow rates by 50% increases the NMR detection time by twofold. The effective relaxation time (T_n) of flowing spins is related to τ.

$$\frac{1}{T_n} = \sum \frac{1}{T_1} + \frac{1}{\tau} \tag{12.7}$$

With flow, system spin-lattice relaxation time (T_1) and spin-spin relaxation time (T_2) are reduced according to the following equations:

$$\frac{1}{T_{1\ \mathrm{flow}}} = \frac{1}{T_{1\ \mathrm{static}}} + \frac{1}{\tau} \tag{12.8}$$

$$\frac{1}{T_{2\ \mathrm{flow}}} = \frac{1}{T_{2\ \mathrm{static}}} + \frac{1}{\tau} \tag{12.9}$$

where $T_{1\,\mathrm{flow}}$ and $T_{2\,\mathrm{flow}}$ are spin–lattice and spin–spin relaxation times of resonance of interest in a flowing system, respectively. The $T_{1\,\mathrm{static}}$ and $T_{2\,\mathrm{static}}$ are spin–lattice and spin–spin relaxation times of the same resonance under static conditions. In static systems, NMR data acquisition and repetition time can be optimized to attain the highest NMR *S/N* by choosing a repetition time of between three and five T_1 relaxation times. As shown in Eq. (12.8), the $T_{1\,\mathrm{flow}}$ (effective T_1) is reduced in flowing systems and this allows the recording of NMR spectra with faster repetition times. Therefore, NMR acquisition parameters must be optimized in continuous-flow CE to achieve highest possible *S/N*.

12.3 APPLICATIONS

A wide range of applications has been reported with CE–NMR, including drug metabolite analyses, amino acid and small-molecule separations and analgesic separations. A recent study extends the capabilities of CE–NMR by adopting cyclic continuous Capillary Electrophoresis [33], and capillary isotachophoresis (cITP), a modified version of CE, have both been applied to achieve high concentration sensitivity [47]. The ability of NMR to observe the physical and chemical environment around an analyte can be used to explore otherwise undetected electrophoretic events [46,51,52]. This section describes both the separation and diagnostic capabilities of CE–NMR.

12.3.1 CZE–NMR

The first CE–NMR experiments used capillary zone electrophoresis (CZE) [29]. The instrumentation had both an outlet and the inlet within the magnet bore and a solenoidal coil for NMR detection. Three different sizes of flow cells with NMR-active volume ranging from 5 to 200 nl were tested in this study. The first reported separation involves separation of a mixture of amino acids—arginine, cysteine, and glycine—at high millimolar levels [28]. The detection was made with a microcoil in the stopped-flow mode, with an LOD of 50 ng. However a relatively high volume of 20 nl was injected in order to acquire NMR data in 16 s, resulting in the low separation efficiency of 5000. As described in Section 12.2.3, the use of solenoidal coil as a detector in parallel to B_0 is limited by CE current-induced magnetic field gradients. This problem can be eliminated by the periodic stopped-flow technique [46]. A mixture of arginine and TEA was analyzed using this method. In this study, the LOD for arginine and TEA were reported to be 57 ng (330 pmol) and 9 ng (88 pmol) with field-amplified stacking. Both sample stacking and stopped-flow facilitated a 2–4-fold increase in the NMR sensitivity without loss of separation efficiency. For example, the CE separation efficiency reported for arginine was 50,000 with the periodic stopped flow.

A number of CE–NMR studies have been reported with saddle-type coils. A continuous-flow CE–NMR experiment reported the separation of a mixture of lysine and histidine [30] with an LOD of 336 ng (2.3 nmol) for lysine. The analysis of biological fluids is a challenge because of their complex nature. CE–NMR with a saddle coil has been used to separate and detect paracetamol metabolites in human urine [31,32]. The CE–NMR electropherogram in Fig. 12.7 illustrates this separation [31]. Several major metabolites, paracetamol glucuronide (I), paracetamol sulfate conjugates (II), and an endogenous material (hippurate) (III) were detected. The analysis was performed with an 85-μm-i.d. capillary with an NMR observe volume of 400 nl at 600 MHz. The LOD reported for this study is 10 ng. The structures of these compounds were determined with the assistance of NMR chemical shifts. This study demonstrates the power of CE–NMR to separate and identify unknown components on-flow.

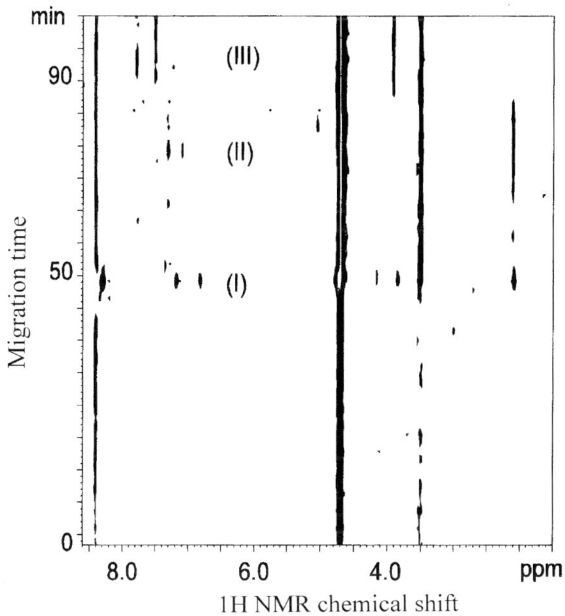

Fig. 12.7. On-flow contour plot of the 600 MHz ^{1}H-NMR detected CE separation of the human urine extract. (I) paracetamol glucuronide (II) paracetamol sulfate conjugates, and (III) endogenous material (hippurate). (From Ref. [31], reproduced by permission of the Royal Society of Chemistry).

The buffer pH is one parameter that can be used to change the separation conditions. Separation of a mixture of caffeine and aspartame has been performed with continuous-flow CE–NMR, using a saddle-coil-based probe [53]. The separation using glycine at basic pH results in spectral overlap of several components and in glycine resonances. This has been resolved by switching to a formate buffer at pH 5. However, the formate buffer increased the migration time of aspartame, and reversed the migration and reduced the separation efficiency. As demonstrated in this study, some experimental conditions are a compromise between separation efficiency and NMR spectral resolution.

CE in a closed loop, or continuous-CE (also known as cyclic CE), is one approach to increase the separation efficiency [54–56]. In this method, CE is continued in a capillary loop by switching the applied voltage appropriately to achieve the desired separation. Recently, a novel arrangement of capillaries has been proposed to separate and isolate analytes in CE–NMR [33]. This method adopts an independently operable two-loop-five-junction capillary configuration with two NMR coils as shown in Fig. 12.8(A). A schematic of the complete cyclic CE–NMR is shown in Fig. 12.8(B). The two-loop capillary CE system is located inside the magnet bore. The dual-coil probe has two independently operable 1-mm-long solenoidal NMR coils with $V_{obs}{\sim}5$ nl. In this example, a sample mixture of alanine (Ala), threonine (Thr) and valine (Val) was injected and separated. The two-dimensional CE–NMR electropherogram in Fig. 12.9 demonstrates the separation of amino acids after one cycle in loop 1 detected by coil A. Ala was then transferred to loop 2 to record more time-consuming 2D-NMRs under a stopped-flow condition. While acquiring a 2D-NMR of Ala, the mixture of Val and Thr was further interrogated in loop 1. The separation of Val and Thr is illustrated in Fig. 12.10. (These two bands were then located in loops 1 and 2 to obtain 2D-NMR in the absence of the applied voltage.) Because of the low electrophoretic current (7–8 μA) only minimum current-induced magnetic effects are observable; thus, scalar coupling patterns are recognizable in many instances (see Fig. 12.10). The separation efficiency obtained for Ala is ~3100. However, better separation efficiencies were recorded for Val (13,500) and Thr (15,500). This work demonstrates the simultaneous and continuous electrophoretic separation in one loop and NMR data acquisition in the second loop. In addition to the separation, total structure elucidation of molecules is feasible with two-loop continuous CE–NMR.

12.3.2 Sample concentration methods

NMR sensitivity enhancement in CE is vital to improve the concentration sensitivity of the mass-sensitive solenoidal microcoils. One approach involves CE sample stacking. Although the first CE–NMR sample stacking is rather modest in comparison with lower concentration sample stacking [46], this method demonstrates how to use the increased mass sensitivity of the smallest microcoils to improve NMR concentration sensitivity. In this study, field-amplified sample stacking is used in CE–NMR to improve concentration sensitivity. This work demonstrates concentration enhancement by 2- to 3-fold for arginine and 4.3-fold for TEA, and 55,000 separation efficiency.

cITP is another method to increase concentration sensitivity. The sample components concentrate between the leading electrolyte and trailing electrolyte. The concentrations or the stacking capability can

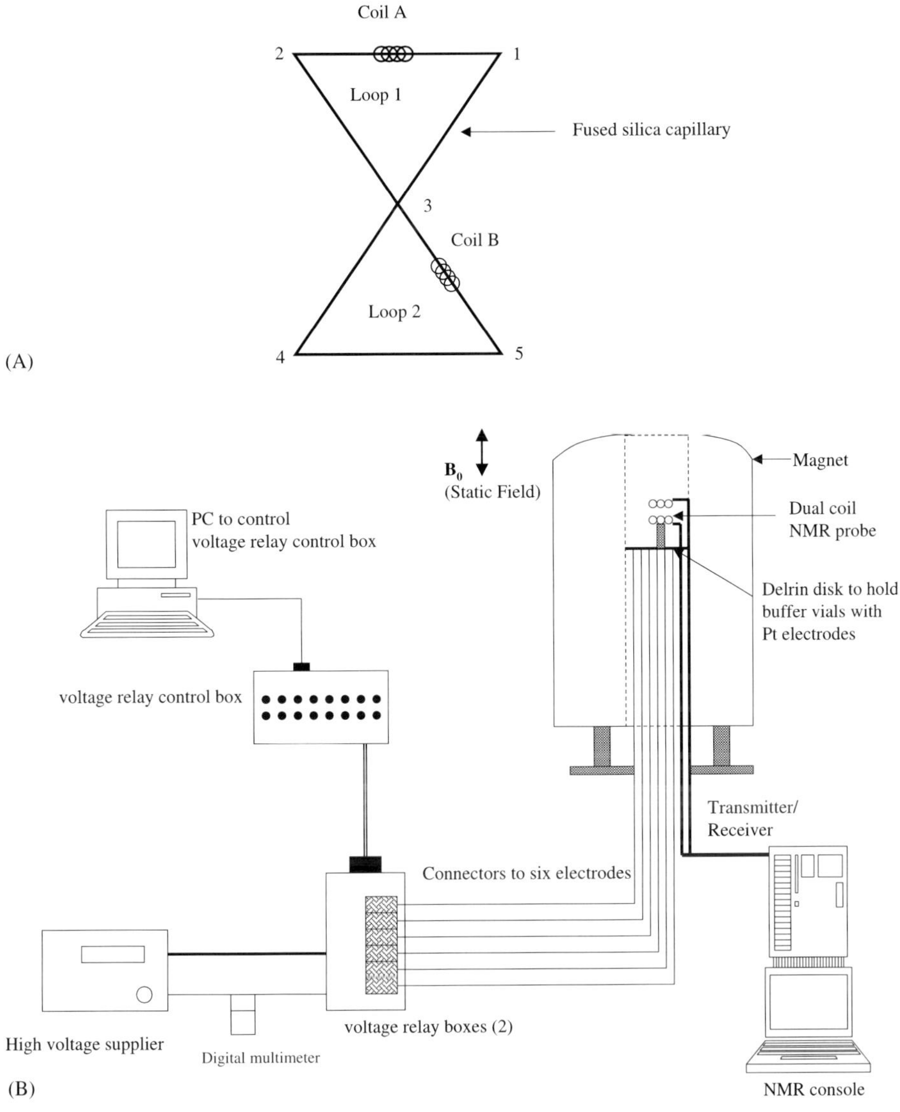

Fig. 12.8. (A) Schematic of the two-loop-five-junction cyclic CE system showing the two NMR microcoils located in the two loops. (B) Detailed instrumental arrangement used for cyclic CE–NMR in the NMR probe illustrating the controlled circuitry used. (Reprinted with permission from Ref. [33]. Copyright 2004, American Chemical Society.)

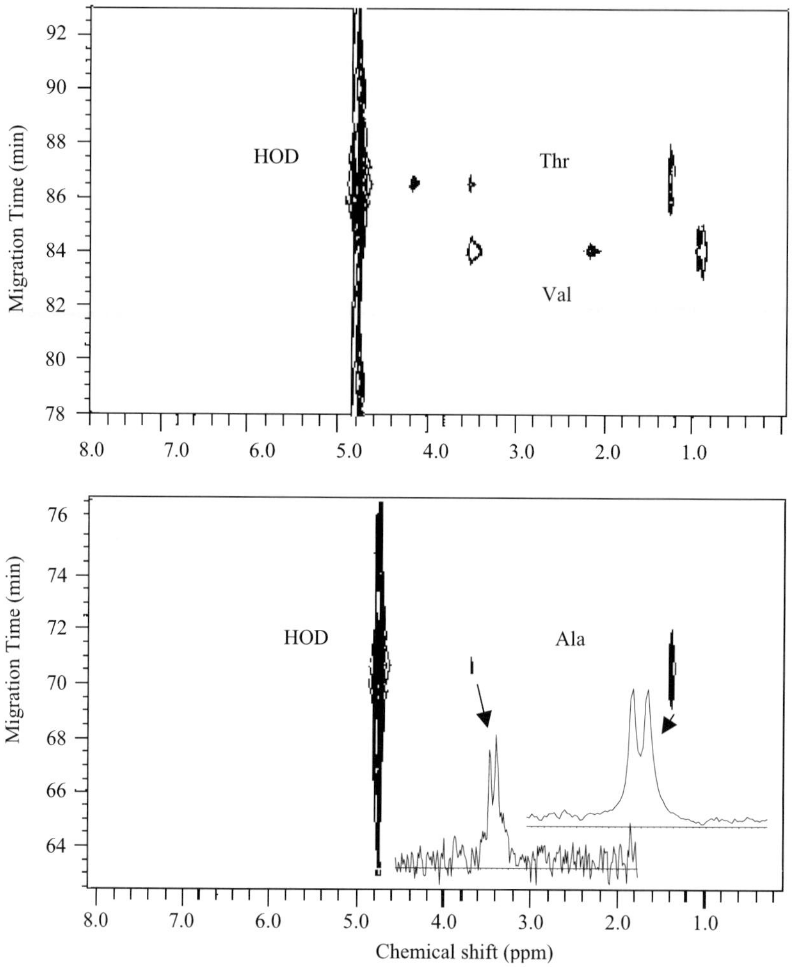

Fig. 12.9. Two dimensional CE–NMR electropherogram for the separation of Ala, Thr, and Val after one cycle. One-dimensional NMR spectra at the peak maximum are in the inset. (Reprinted with permission from Ref. [33]. Copyright 2004, American Chemical Society.)

be changed by adjusting the leading electrolyte concentration. In cITP–NMR, microliter volume injection is reduced to nanoliter volumes during the separation and presented to the detector. The first cITP–NMR measurement demonstrated a concentration sensitivity enhancement by 1000-fold [47]. A 200 μM tetraethylammonium

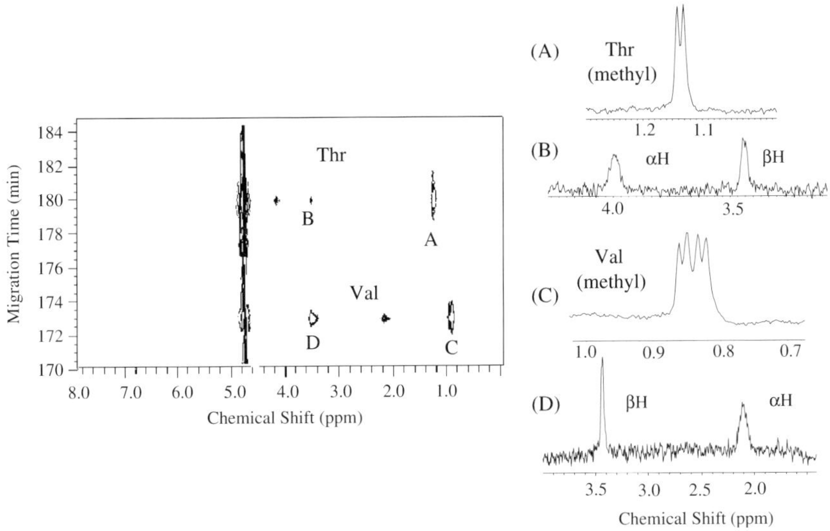

Fig. 12.10. Two-dimensional CE–NMR electropherogram for the separation of Thr and Val after the second cycle. One-dimensional NMR spectra from the peak maximum are shown in the inset. (Reprinted with permission from Ref. [33]. Copyright 2004, American Chemical Society.)

bromide (TEAB) sample was concentrated on-flow, and achieved a concentration of 20 mM. Figure 12.11(A) shows CE-^{1}H-NMR of TEAB without sample stacking. The same spectrum under cITP conditions is shown in Fig. 12.11(B). Enhancing the concentration by 100 also increases the observation efficiency from 0.5% to 50%. As a result, high *S/N* NMR data can be recorded in a shorter time period. For example the inset in Fig. 12.11(B) illustrates a COSY spectrum acquired in 22 min under stopped-flow conditions using a 30-nL V_{obs} solenoidal coil. Theoretically, the increase in concentration brought about by cITP permits the recoding of the COSY spectrum 10,000 times faster than without sample stacking. The concentration sensitivity enhancement reported in this initial cITP-NMR work is superior to other capillary-scale NMR measurement.

The capabilities of cITP-NMR have been further explored with trace amount analysis of the beta blocker, atenolol, used for cardiovascular diseases [57]. This work employed a specially designed dual solenoidal coil (Fig. 12.12). The first coil acted as a scout coil so that selected analyte bands could be stopped at the second coil to acquire 2D-NMR

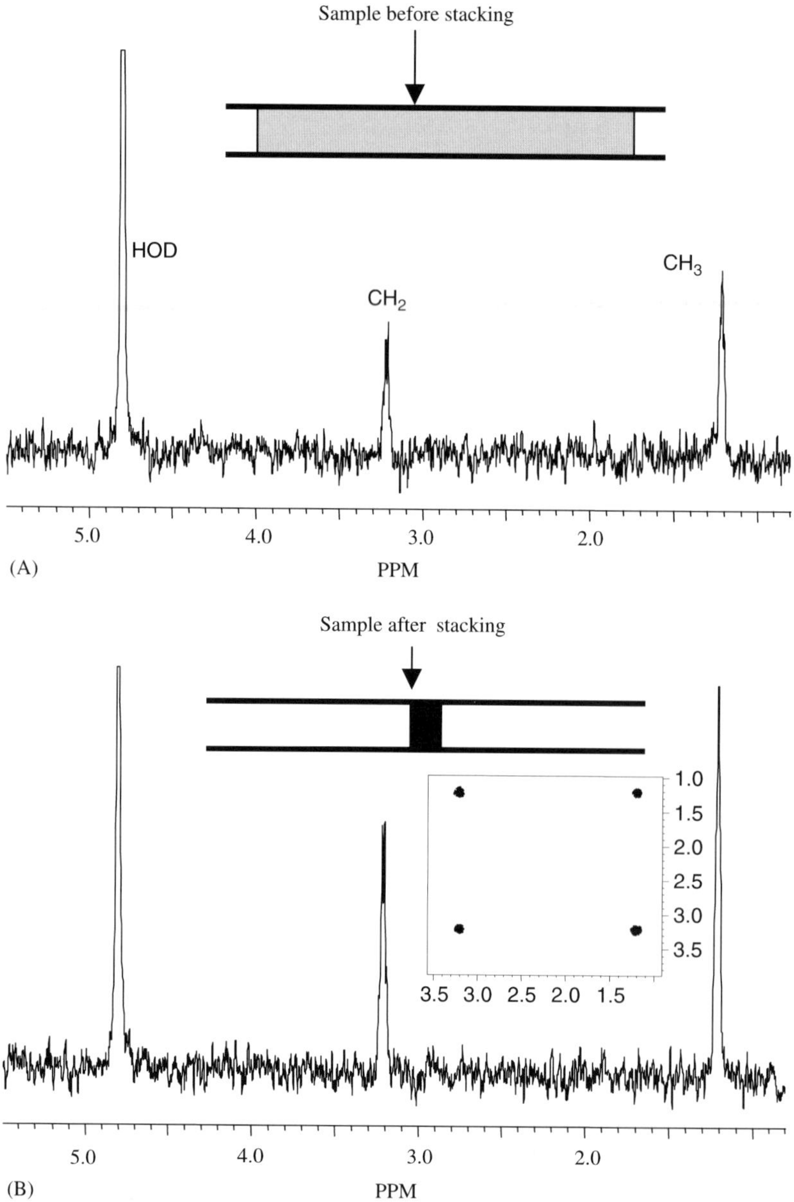

Fig. 12.11. ^{1}H-NMR spectra of (A) capillary filled with 5 mM 200μm TEAB without sample stacking (*S/N* of peak at 1.2 ppm = 13) and (B) 8 μl of TEAB injected; cITP stacked (*S/N* of peak at 1.2 ppm = 30). Inset shows COSY spectrum of cITP stacked TEAB obtained in 22 min. (Reprinted with permission from Ref. [47]. Copyright 2001, American Chemical Society.)

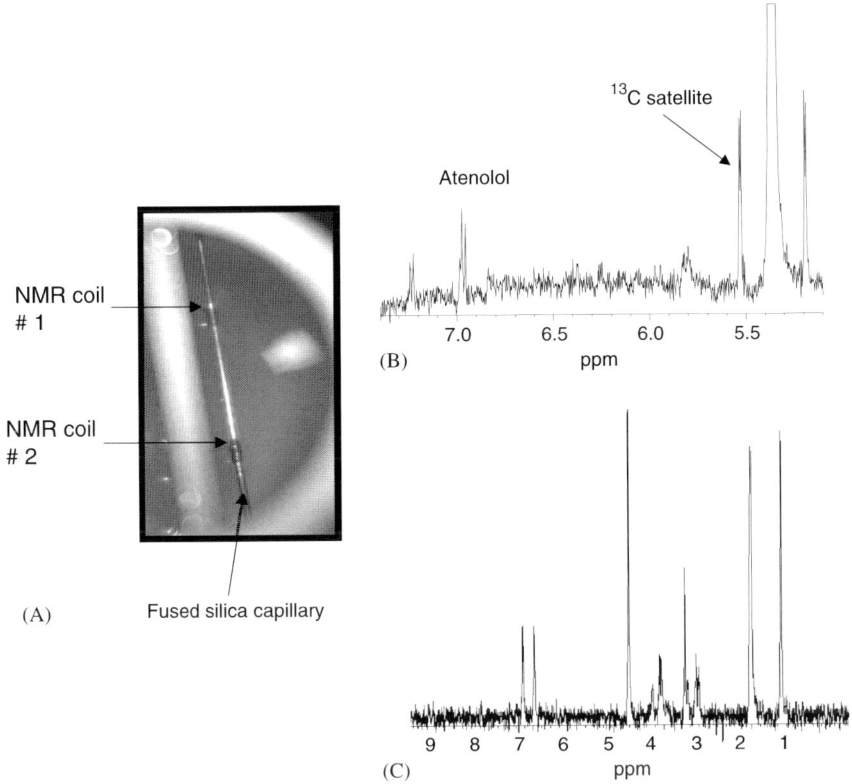

Fig. 12.12. (A) Dual serial microcoil NMR probe arrangement used for cITP–NMR. (B) 5 mm static ^{1}H-NMR spectrum of expanded area showing ^{13}C satellite peaks of anomeric sucrose peak and atenolol aromatic peaks of a 200 μM atenolol and 200 mM sucrose sample. (C) On-flow cITP–NMR spectrum depicting the atenolol sample band at peak maximum (sample: 200 μM atenolol and 200 mM sucrose). (Reprinted with permission from Ref. [57]. Copyright 2002, American Chemical Society.)

data. In this study 1.9 nmol (200 μM) atenolol was successfully analyzed in the presence of 200 mM sucrose in acetate buffer. Figure 12.12(B) shows a ^{1}H spectrum obtained for the atenolol/sucrose sample with a 5 mm NMR probe under static condition. The atenolol peaks are fairly weak and observed below the ^{13}C satellites peaks. The cITP-separated and stacked NMR spectrum is shown in Fig. 12.12(C). The cITP–NMR spectrum has a higher *S/N*, and the estimated concentration is ~40 mM. This study exhibits the power of cITP–NMR to separate and detect trace (0.1%) levels of charged molecules in the presence of

1000-fold excess neutral molecules. By carefully selecting experimental conditions and fine-tuning buffer pH, more complex mixtures can be analyzed using cITP-NMR.

Knowledge of the chiral composition of drugs can be a key to understanding their activity and fate. Owing to structural similarities, chiral separations can be a challenging task. CE is acknowledged as a powerful chiral separation technique because of its high separation efficiency. Chiral separation in CE can be achieved in the presence of chiral selector molecules such as cyclodextrin (CD) [58], micelles [59] and other specially designed selector molecules [60]. In general, the effective electrophoretic mobility",4,0,4?>electrophoretic mobility (μ_e), of each stereoisomer is altered by forming diastereoisomers. NMR is a powerful spectroscopic technique to discriminate optical isomers using chiral shift molecules [61]. NMR parameters such as chemical shift, linewidth, relaxation time and diffusion coefficient can be used to evaluate the extent of the chiral separation. Information gathered with NMR can reveal structure-interaction-relationships between the chiral molecule and the selector molecule to expose the fundamental aspects of chiral discrimination and separation.

An on-line cITP-NMR method has been reported for chiral separation of a 2 nmol mixture of alprenolol [62]. This study demonstrated the capability of cITP–NMR to concentrate and separate *S*- and *R*-alprenolol in acetate buffer with α-CD and β-sulfated-CD as chiral selector molecules. The cITP–NMR instrumentation was similar to the first cITP–NMR work, which used a 500 MHz magnet and a single solenoidal coil of 30 nl V_{obs}. Figures 12.13(A) and (B) illustrate cITP–NMR separation of S- and R-alprenolol. The magnetic susceptibility mismatch of two separating bands marks the boundaries of the two separating bands. For example, the separation of sample band *S*-alprenolol from the leading electrolyte and the R-alprenolol band are marked by spectra at 59 min and at 62 min respectively. More importantly, the chemical shift changes, coupling pattern and line broadening (Fig. 12.14(A) and (B)) exemplify the power of cITP–NMR to identify *S*- and *R*-alprenolol. The concentration enhancement for the *S*- isomer is ~200 and for the *R*-isomer ~220 times, which is equivalent to having 84% of the injected *R*-alprenolol and 76% of the injected *S*-alprenolol within the NMR active coil. The power of on-flow cITP–NMR to concentrate, separate and obtain structural information were demonstrated in this study.

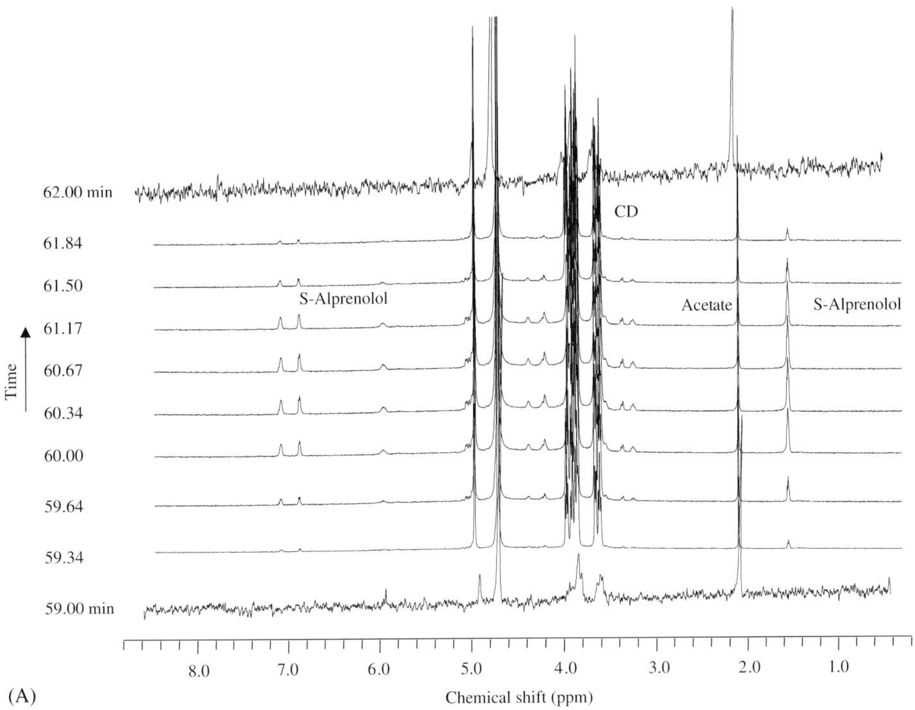

Fig. 12.13. Microcoil ^{1}H-NMR spectra. (A) On-flow cITP-NMR spectra of S-alprenolol band as a function of run time. The estimated concentration at the peak maximum 25 mM. (B) On-flow cITP-NMR spectra of R-alprenolol band as a function of run time. The estimated concentration at the peak maximum 28 mM. The band boundaries are marked with low *S/N* spectra due to magnetic susceptibility mismatches. (Reproduced with permission from Ref. [62]. Copyright 2004, Springer-Verlag.)

12.3.3 Diagnostic capabilities of CE/cITP–NMR

NMR is a well-known and powerful diagnostic tool. The presence of protonated solvents in many CE buffers allows NMR to follow events occurring in the buffer. The diagnostic capabilities of NMR have been exploited in a number of electrophoretic NMR studies. The first such diagnostic work describes the fate of a gravimetrically injected H_2O plug [46]. The H_2O plug exhibits a leading parabolic-type concentration profile and tailing flat profile. The flat profile has been attributed to voltage effects.

The Joule heat generated as the electrophoretic current passes through the buffer in CE can limit the separation efficiency and be

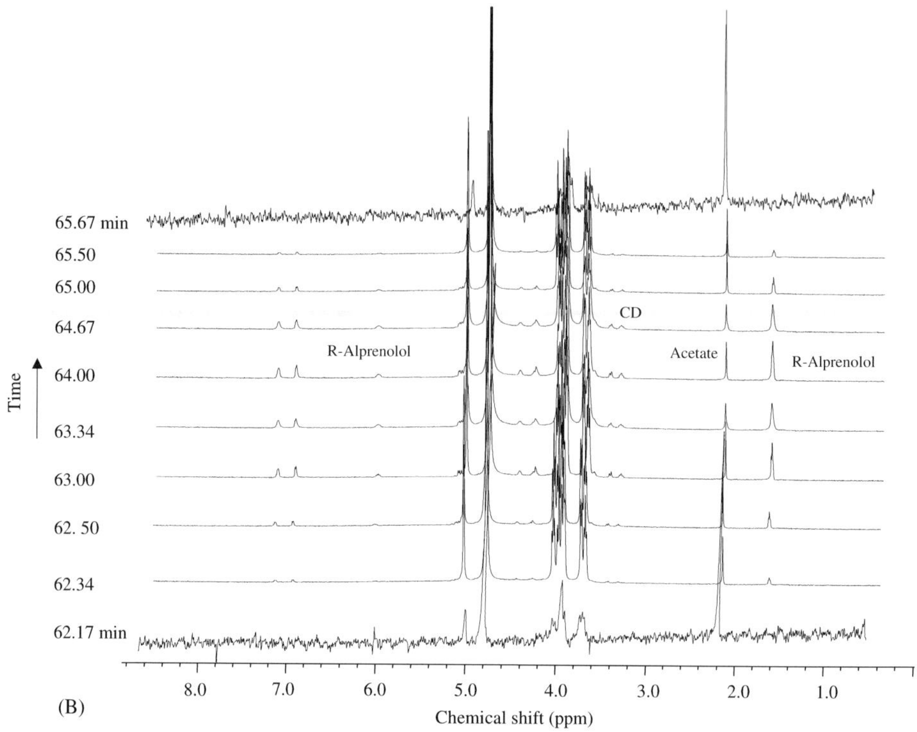

Fig. 12.13. Continued

detrimental for thermally liable samples. Variation of the temperature of the buffer solution can also change pH, peak shapes, migration times, reproducibility, etc. However, the temperature change can also be beneficial in certain CE measurements. Temperature-induced protein conformational changes [63] and temperature-regulated DNA separation [64] are two typical applications of temperature manipulation in CE. Properly regulated intracapillary temperature is critical for achieving good CE resolution and reproducible migration times. NMR thermometry is a valuable technique to measure the temperature of a system where such a probe is not available. The ^{1}H frequency of water, which has a linear dependency [52] is commonly used in NMR to determine the temperature. NMR thermometry to monitor temperature changes in CE has been reported using nanoliter volume NMR coils [52]. CE–NMR has been performed with the inlet and outlet buffer inside the magnet with natural air convection. The intracapillary temperature determined on the basis of ^{1}H proton frequency is shown in

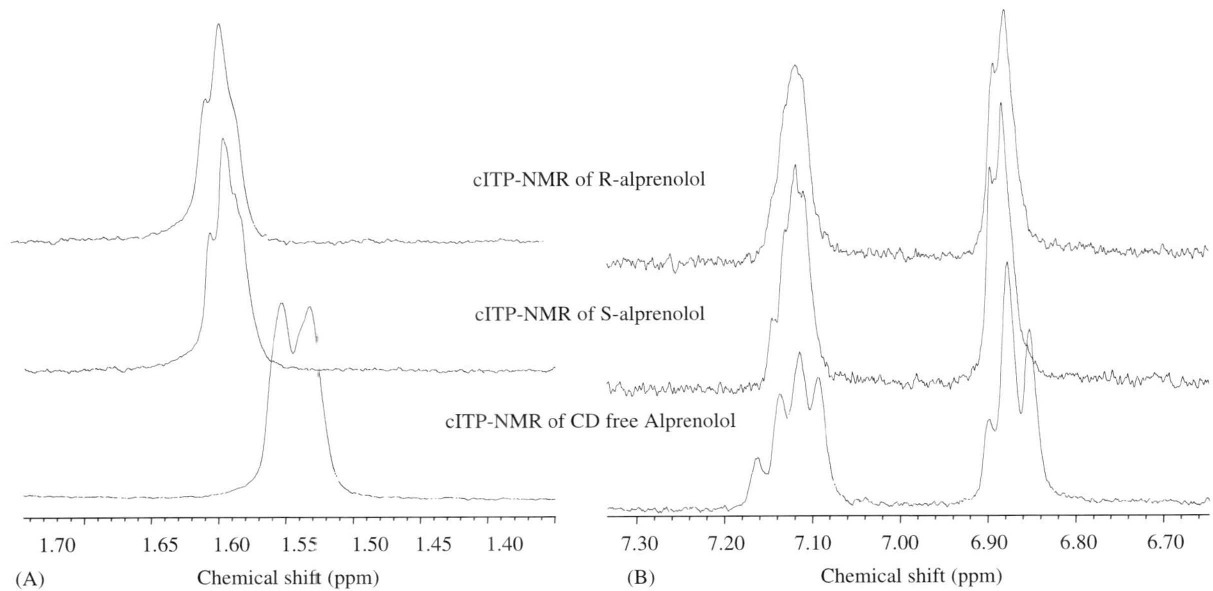

Fig. 12.14. The cITP–NMR spectra of CD-free alprenolol, *S*-alrepnolol and *R*-alprenolol: (A) the methyl region and (B) the aromatic region. The cITP-NMR spectrum of CD-free alprenolol was acquired under the same cITP conditions in the absence of CD in the buffer. (Reproduced with permission from Ref. [62]. Copyright 2004, Springer-Verlag.)

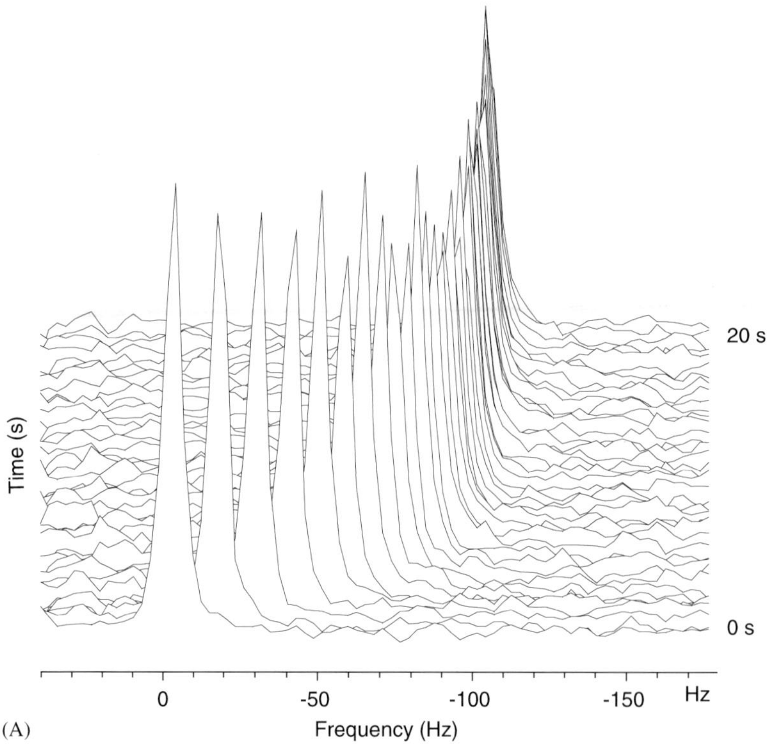

20 s
Time (s)
0 s
0
-50
-100
-150
Hz
Frequency (Hz)

(A)

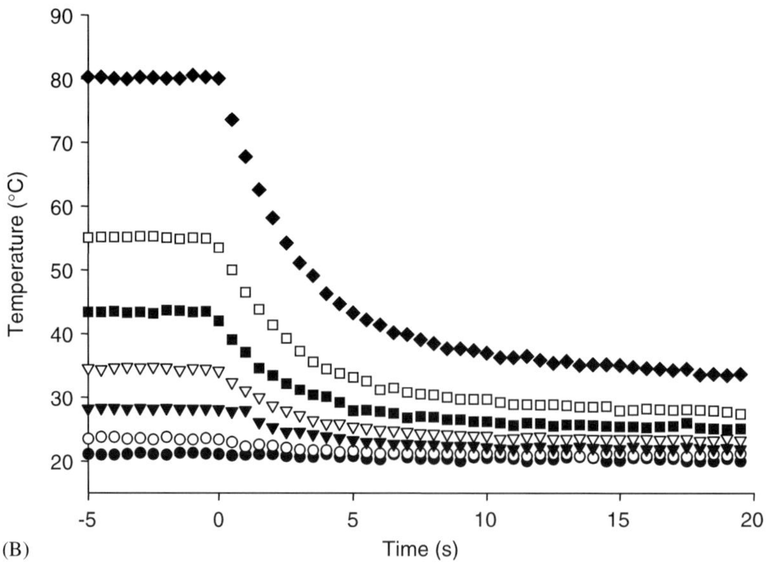

Temperature (°C)
90
80
70
60
50
40
30
20
-5
0
5
10
15
20
Time (s)

(B)

Fig. 12.15(A). This figure denotes the H_2O peak frequency changes during the first 20-s time interval after applying 12 kV. The measurements were performed under natural air convection using a solenoidal coil placed in parallel with the B_0 field. Therefore, no current-induced NMR peak deterioration is observed. The linear least-squares analysis of H_2O chemical shift data to thermocouple reading yields a regression (r^2) of 0.995 or better. The correlation between frequency shift and temperature change remains constant and is ideal to monitor intracapillary temperature. Figure 12.15(B) illustrates the cooling curves determined with NMR thermometry. Importantly, the time taken to reach the ambient temperature after applying 15 kV is longer than any other lower applied voltages. This explains the importance of a capillary wash not only to regenerate the capillary wall chemistry, but also to achieve equilibrium temperature. One interesting investigation is the temperature of analyte bands. The temperature of a 1 mM NaCl plug injected into a 50 mM borate buffer is shown in Fig. 12.16. This temperature behavior is similar to change in current of the same CE–NMR experiment. As illustrated in Fig. 12.16, the temperature of a NaCl plug is 20 °C higher than the buffer temperature. This is because 1 mM NaCl has lower conductivity (0.13 mS/cm) than 50 mM borate buffer (3.69 mS/cm). In this study, the temperature has been recorded with subsecond temporal resolution and 1 mm spatial resolution.

NMR detection in cITP proves to be a valuable investigative tool to reveal otherwise undetectable events. The buffer pH change has been monitored with cITP–NMR [51]. The acetate chemical shift change at the interface of trailing electrolyte and sample band is shown in Fig. 12.17(A). The trailing electrolyte pH is more acidic than that of the sample. However, the split in the acetate peak confirms the presence of a sharp boundary between the analyte band and the trailing electrolyte.

Fig. 12.15. (A) Stacked plot of 40 ^{1}H-NMR spectra of H_2O collected during the initial 20 s after voltage was applied to the capillary. The acquisition of the first spectrum coincides with the application of 12 kV across 38-cm-long 100-μm-i.d/360-μm-o.d. fused silica capillary filled with 50 mM borate buffer in H_2O. (B) Temperature as a function for 38-cm-long 50-μm-i.d/360-μm-o.d. fused silica capillary filled with 50 mM phosphate buffer in H_2O using a vertical solenoidal coil. Time zero corresponds to discontinuation of a voltage which had been applied for the previous 5 min. ◆15 kV, □12 kV, ■10 kV, ∇8 kV, ▼6 kV, ○4 kV, •2 kV.(Reproduced with the permission from Ref. [52]. Copyright 2000, American Chemical Society.)

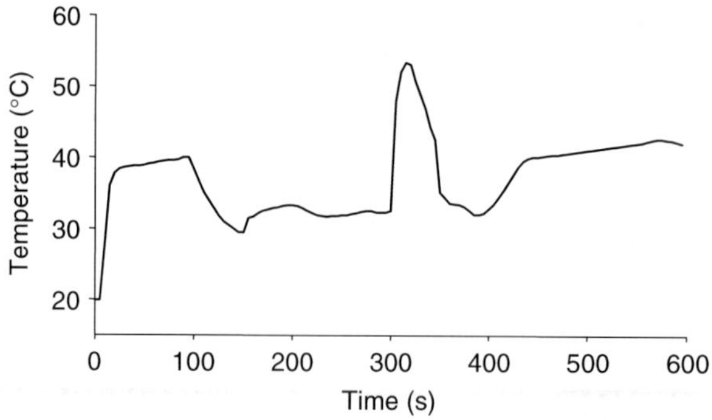

Fig. 12.16. Injection of a plug of 1 mM NaCl in H_2O with the electrolyte consisting of 50 mM borate buffer in H_2O using a solenoidal coil in vertical position. (Reprinted with permission from Ref. [52]. Copyright 2000, American Chemical Society.)

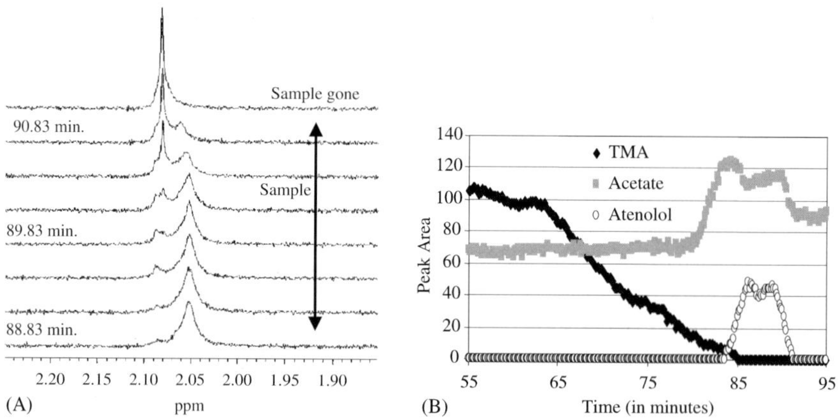

Fig. 12.17. (A) Progression of cITP-NMR spectra displaying acetate chemical shift during passage of interface between focused sample band and trailing electrolyte (TE) through the NMR detector. (B) TMA, acetate and atenolol peak areas as a function of run time (min) in cITP-NMR. (Reprinted with permission from Ref. [51]. Copyright 2002, American Chemical Society.)

As defined by the Kohlrausch regulating function, once the steady state is reached, the concentration of the individual electrolytes bands should remain constant [65]. However, this cITP–NMR reports some interesting behavior of electrolytes. As exemplified in Fig. 12.17(B), the leading electrolyte, tetramethylammonium (TMA) cation shows a steady decline and almost disappears with the appearance of the analyte peak, atenolol. If the cITP system is in the steady state, the TMA signal should remain constant until the atenolol peak appears. This work demonstrates the capability of cITP–NMR to improve our understanding of the dynamic processes otherwise undetected in cITP.

12.4 CONCLUSIONS AND FUTURE DIRECTIONS

As demonstrated in this chapter, CE–NMR is becoming a powerful hyphenated technique. Sample amounts as small as picomoles have been successfully analyzed. The induced current is a major obstruction to obtain high-resolution NMR data with solenoidal coil-based NMR probes. Recent work by Shapira et al. [66,67] shows the possibilities of recording NMR with a single scan under an inhomogeneous magnetic field. This approach could be useful to record NMR in the presence of an electrophoretic current-induced magnetic field. The multiple coil approach to CE–NMR is an exciting and still-evolving area of research. This will increase the throughput of CE–NMR using a single magnet. For example, a four-coil probe is operable as four independent CE–NMR instruments. However, even with the most mass sensitive microcoil probes, the sensitivity of NMR is still below that of other spectroscopic methods. Stated plainly, the promise of CE–NMR is high, but improvements in sensitivity are needed if CE–NMR is to fulfill its high promise.

How can the sensitivity be improved? The multiple coil probes can improve the *S/N* by carefully co-adding analyte signals from each NMR coil. An attractive, alternative technique to improve the NMR sensitivity is DNP and chemically induced dynamic nuclear polarization (CIDNP). The applications of DNP, to a certain degree, have been demonstrated to enhance ^{13}C signals in flow NMR [15]. However, the application of CIDNP is yet to be realized in CE–NMR. Nuclear polarization techniques are probably the most likely approach to detect less-sensitive NMR nuclei. The high separation efficiency of CE and unique detection capability of NMR are ideal to analyze complex biological mixtures. Obviously, increases in field strength and higher temperature superconducting materials will both allow improvements in performance.

A major area of interest is pattern recognition in metabonomics and proteomics. The CE–multiple coil NMR probes can replace more time-consuming off-line analytical methods to analyze a large number of samples. A single scan of flow 2D NMR measurement has been reported for LC–NMR. This may be easily adopted to record on-line multidimensional NMR in CE. Overall, we certainly expect the improvements in the performance of this new hyphenated technique to expand the range of applications requiring the separating ability of CE and the chemical information content of NMR.

REFERENCES

1 S.M.V. Freund and M. Bycroft, *Curr. Innov. Mol. Bio.*, 2 (1995) 155–169.

2 C. Marchioro, S. Davalli, S. Provera and G.H. Raza, *Rec. Res. Dev. Org. Bioorg. Chem.*, 4 (2001) 151–164.

3 Y. Ito, *RIKEN Rev.*, 24 (1999) 59–61.

4 W.R. Adam, A.P. Koretsky and M.W. Weiner, *Am. J. Physiol.*, 251 (1986) F904–F910.

5 G.K. Radda, J.J.H. Ackerman, P. Bore, P. Sehr, G.G. Wong, B.D. Ross, Y. Green, S. Bartlett and M. Lowry, *Int. J. Biochem.*, 12 (1980) 277–281.

6 C.S. Zuo, A. Mahmood and A.D. Sherry, *J. Magn. Reson.*, 151 (2001) 101–106.

7 D. Hechtfischer and G. Schuster, *Temperature*, 6 (1992) 107–111.

8 O. Avenel, P. Piejus and E. Varoquaux, *IEEE T. Instrum. Meas.*, 19 (1970) 408–411.

9 Q. He and C.S. Johnson Jr., *J. Magn. Reson.*, 81 (1989) 435–439.

10 H.W. Long, H.C. Gaede, J. Shore, L. Reven, C.R. Bowers, J. Kritzenberger, T. Pietrass, A. Pines, P. Tang and J.A. Reimer, *J. Am. Chem. Soc.*, 115 (1993) 8491–8492.

11 G. Navon, Y.Q. Song, T. Room, S. Appelt, R.E. Taylor and A. Pines, *Science*, 271 (1996) 1848–1851.

12 W. Mueller-Warmuth and K. Meise-Gresch, *Adv. Magn. Reson.*, 11 (1983) 1–45.

13 H.C. Dorn, J. Gu, D.S. Bethune, R.D. Johnson and C.S. Yannoni, *Chem. Phys. Lett.*, 203 (1993) 549–554.

14 R.J. Fitzgerald, K.L. Sauer and W. Happer, *Chem. Phys. Lett.*, 284 (1998) 87–92.

15 S. Stevenson and H.C. Dorn, *Anal. Chem.*, 66 (1994) 2993–2999.

16 Y. Sakamoto and T. Takamura-Enya, *Polycycl. Aromat. Comp.*, 22 (2002) 441–450.

17 T. Parella, F. Sanchez-Ferrando and A. Virgili, *J. Magn. Reson.*, 126 (1997) 274–277.

18 G.E. Martin, *Cryogenic NMR Probe. Appl.*, 9 (2002) 33–35.

19 H.C. Keun, O. Beckonert, J.L. Griffin, C. Richter, D. Moskau, J.C. Lindon and J.K. Nicholson, *Anal. Chem.*, 74 (2002) 4588–4593.

20 P.F. Flynn, D.L. Mattiello, H.D.W. Hill and A.J. Wand, *J. Am. Chem. Soc.*, 122 (2000) 4823–4824.
21 D. Moskau, C. Ritcher, H. Kovaks, M. Salzmann, L. Baselgia, M. Haeberli, D. Marek and O. Schett, *Spec. Anal.*, 32 (2003) 39–41.
22 D.L. Olson, T.L. Peck, A.G. Webb, R.L. Magin and J.V. Sweedler, *Science*, 270 (1995) 1967–1970.
23 J.A. Rogers, R.J. Jackman, G.M. Whitesides, D.L. Olson and J.V. Sweedler, *Appl. Phys. Lett.*, 70 (1997) 2464–2466.
24 C.S. Johnson and Q. He, *Adv. Magn. Reson.*, 13 (1989) 131–159.
25 U. Bohme and U. Scheler, *Colloid. Surf. A*, 222 (2003) 35–40.
26 E. Li and Q. He, *J. Magn. Reson.*, 156 (2002) 181–186.
27 Q. He and Z. Wei, *J. Magn. Reson.*, 150 (2001) 126–131.
28 N. Wu, T.L. Peck, A.G. Webb, R.L. Magin and J.V. Sweedler, *Anal. Chem.*, 66 (1994) 3849–3857.
29 N. Wu, T.L. Peck, A.G. Webb, R.L. Magin and J.V. Sweedler, *J. Am. Chem. Soc.*, 116 (1994) 7929–7930.
30 K. Pusecker, J. Schewitz, P. Gfroerer, L.-H. Tseng, K. Albert and E. Bayer, *Anal. Chem.*, 70 (1998) 3280–3285.
31 K. Pusecker, J. Schewitz, P. Gfrorer, L.-H. Tseng, K. Albert, E. Bayer, I.D. Wilson, N.J. Bailey, G.B. Scarfe, J.K. Nicholson and J.C. Lindon, *Anal. Commun.*, 35 (1998) 213–215.
32 J. Schewitz, P. Gfrorer, K. Pusecker, L.-H. Tseng, K. Albert, E. Bayer, I.D. Wilson, N.J. Bailey, G.B. Scarfe, J.K. Nicholson and J.C. Lindon, *Analyst*, 123 (1998) 2835–2837.
33 D.A. Jayawickrama and J.V. Sweedler, *Anal. Chem.*, 76 (2004) 4894–4900.
34 D.I. Hoult and R.E. Richards, *J. Magn. Reson.*, 24 (1976) 71–85.
35 J.N. Shoolery and E.W. Southwick, *J. Agric. Food Chem.*, 27 (1979) 1400–1402.
36 J.N. Shoolery and R.E. Majors, *Am. Lab.*, 9 (1977) 51–61.
37 R.C. Crouch and G.E. Martin, *Magn. Reson. Chem.*, 30 (1992) S66.
38 G. Schlotterbeck, A. Ross, R. Hochstrasser, H. Senn, T. Kuhn, D. Marek and O. Schett, *Anal. Chem.*, 74 (2002) 4464–4471.
39 A.G. Webb, *Prog. Nucl. Mag. Res. Sp.*, 31 (1997) 1–42.
40 M.E. Lacey, R. Subramanian, D.L. Olson, A.G. Webb and J.V. Sweedler, *Chem. Rev.*, 99 (1999) 3133–3152.
41 D.L. Olson, M.E. Lacey and J.V. Sweedler, *Anal. Chem.*, 70 (1998) 645–650.
42 Y. Li, A.M. Wolters, P.V. Malawey, J.V. Sweedler and A.G. Webb, *Anal. Chem.*, 71 (1999) 4815–4820.
43 E. MacNamara, T. Hou, G. Fisher, S. Williams and D. Raftery, *Anal. Chim. Acta*, 397 (1999) 9–16.
44 M.A. MacNaughtan, T. Hou, J. Xu and D. Raftery, *Anal. Chem.*, 75 (2003) 5116–5123.

45 H. Wang, L. Ciobanu, A.H. Edison and A.G. Webb, *J. Magn. Reson.*, 170 (2004) 206–212.
46 D.L. Olson, M.E. Lacey, A.G. Webb and J.V. Sweedler, *Anal. Chem.*, 71 (1999) 3070–3076.
47 R.A. Kautz, M.E. Lacey, A.M. Wolters, F. Foret, A.G. Webb, B.L. Karger and J.V. Sweedler, *J. Am. Chem. Soc.*, 123 (2001) 3159–3160.
48 A.M. Wolters, D.A. Jayawickrama, A.G. Webb and J.V. Sweedler, *Anal. Chem.*, 74 (2002) 5550–5555.
49 Y. Li, M.E. Lacey, J.V. Sweedler and A.G. Webb, *J. Magn. Reson.*, 162 (2003) 133–140.
50 A.G. Webb, J.V. Sweedler and D. Raftery. In: K. Albert (Ed.), *On-line LC-NMR and Related Techniques*, Wiley, Chichester, 2002, pp. 259–277.
51 A.M. Wolters, D.A. Jayawickrama, C.K. Larive and J.V. Sweedler, *Anal. Chem.*, 74 (2002) 4191–4197.
52 M.E. Lacey, A.G. Webb and J.V. Sweedler, *Anal. Chem.*, 72 (2000) 4991–4998.
53 P. Gfroerer, J. Schewitz, K. Pusecker, L.-H. Tseng, K. Albert and E. Bayer, *Electrophoresis*, 20 (1999) 3–8.
54 J.C.T. Eijkel, A. Van-den-Berg and A. Manz, *Electrophoresis*, 25 (2004) 243–252.
55 J.G. Zhao and J.W. Jorgenson, *J. Microco. Sep.*, 11 (1999) 439–449.
56 J.G. Zhao, T. Hooker and J.W. Jorgenson, *J. Microco. Sep.*, 11 (1999) 431–437.
57 A.M. Wolters, D.A. Jayawickrama, C.K. Larive and J.V. Sweedler, *Anal. Chem.*, 74 (2002) 2306–2313.
58 T. Inou and J.P. Chang, *J. Liq. Chromatogr. R. T.*, 26 (2003) 2351–2367.
59 S.A.A. Rizvi and S.A. Shamsi, *Electrophoresis*, 24 (2003) 2514–2526.
60 T. Ivanyi, K. Pal, I. Lazar and D.L. Massart, *J. Chromatogr. A*, 1028 (2004) 325–332.
61 J. Belleney, C. Bui and F.J. Carriere, *Magn. Reson. Chem.*, 28 (1990) 606–611.
62 D.A. Jayawickrama and J.V. Sweedler, *Anal. Bioanal. Chem.*, 378 (2004) 1528–1535.
63 V.J. Hilser, G.D. Worosolla and E. Freire, *Anal. Biochem.*, 208 (1993) 125–131.
64 H.J. Issaq, H. Xu, K.C. Chan and M.C. Dean, *J. Chromatogr. B*, 738 (2000) 243–248.
65 B.J. Wanders and F.M. Everaerts, *Handbook Capillary Electrop.*, CRC Press, Boca Raton, FL, 1994.
66 B. Shapira and L. Frydman, *J. Am. Chem. Soc.*, 126 (2004) 7184–7185.
67 B. Shapira, A. Karton, D. Aronzon and L. Frydman, *J. Am. Chem. Soc.*, 126 (2004) 1262–1265.

Chapter 13

Chiral analysis by capillary electrophoresis

Carmen García-Ruiz and Maria Luisa Marina

13.1 INTRODUCTION

Compounds made up of the same atoms, bonded by the same sequence of bonds, but possessing different three-dimensional structures (configurations) are called stereoisomers. Two types of stereoisomers are enantiomers and diastereomers. Whereas enantiomers are pairs of stereoisomers related as an object to its mirror image, diastereomers do not bear a mirror-image relationship to each other. A 1:1 mixture of both enantiomers of a chiral compound forms a racemate or racemic mixture. There are different types of chiral compounds (some examples are illustrated in Table 13.1) [1]:

(i) Chiral compounds with one or more than one asymmetrical center. They have one or more than one atom joined to four different groups. These compounds have 2^n stereoisomers (n is the number of asymmetric centers).
(ii) Chiral compounds with axial chirality. They have a biphenyl structure with at least two voluminous groups in ortho position that hinder the free rotation on the simple central bond. These chiral stereoisomers are called atropisomers.
(iii) Asymmetrical cyclic compounds, characterized by the absence of planes, axes or centers of symmetry.

The enantiomers of a chiral compound have the same chemical properties, except for their reaction with chiral substances, and they have the same physical properties, except for the direction in which each one rotates the polarized light. In fact, according to the direction of rotation of the plane of polarized light, they are designated as d or (+)

Comprehensive Analytical Chemistry XLV
M.L. Marina, A. Ríos and M. Valcárcel (Eds)
Volume XLV ISSN: 0166-526X DOI: 10.1016/S0166-526X(05)45013-8

TABLE 13.1

Examples of different types of chiral compounds

Chiral compounds with one or more than one asymmetric center (i)		Chiral compounds with axial chirality (ii)	Asymmetric cyclic compounds (iii)
Alanine ($n = 1$)	Tartaric acid ($n = 2$)	2,2′,3,4′,5′,6-Hexachlorobiphenyl (PCB 149)	α-Hexaclorocyclohexane (α-HCH)
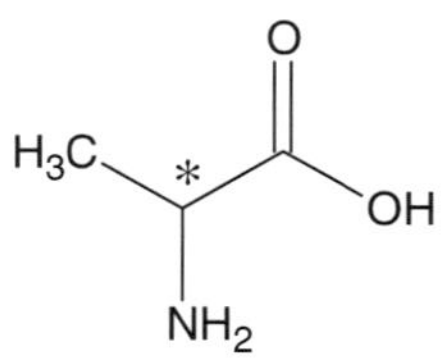	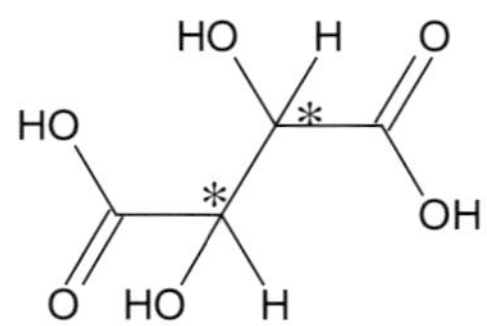	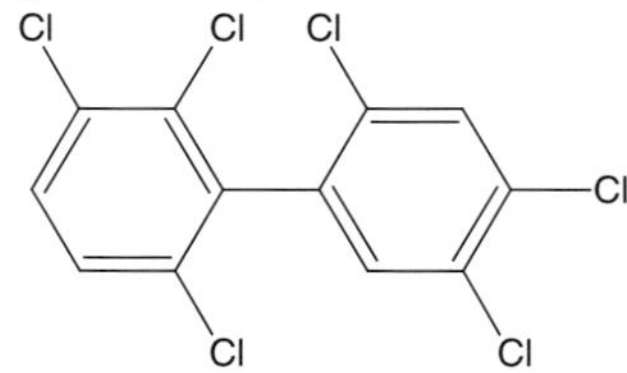	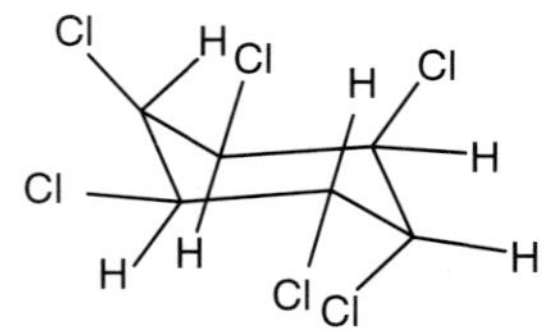

(dextrogiro) or l or (−) (levogiro). This nomenclature is not related to the molecular configuration of the molecule. The two nomenclatures that pertain to the molecular configuration are the d/l system (only applicable to biological molecules with one asymmetrical atom, e.g., amino acids) or *R*/*S* system (systematic nomenclature applicable to molecules with one or more asymmetrical atom) [2,3].

Nature is chiral because it mainly uses one of the two enantiomers of a chiral compound. That is why living material has amino acids, and therefore peptides, enzymes and other proteins, in only one of the mirror image forms. Carbohydrates and nucleic acids such as DNA and RNA are other examples. Thus, receptors of cell machinery, for example, enzymes, are chiral, preferring to bind to one of the enantiomers of a chiral compound [4]. This is why most biological processes have a high degree of enantioselectivity: each enantiomer may have a different biological activity.

When a drug is administered as a racemic mixture, one enantiomer may have pharmacological effects while the other could have no or few effects, or it could show some undesired side effects. A good case in point of one enantiomer showing an undesired activity is the well-known case of thalidomide. It was prescribed to pregnant women in the early 1960s as a cure for morning sickness. However, this drug was found to be responsible for a large number of very severe birth malformations and even fatalities of babies. Later, it was discovered that these problems were caused by one of the thalidomide enantiomers, *R*-thalidomide, which was very toxic. Nowadays, a high percentage of chiral drugs are commercialized as pure enantiomers, and the control of their enantiomeric purity is essential in such cases [1,4].

In the case of chiral pesticides, the use of racemic mixtures when only one of the enantiomers is biologically active increases environmental pollution with respect to the utilization of the active enantiomer alone. This fact has generated interest in the preparation of formulations of pure enantiomers in the agricultural industry. For instance, the case of phenoxy acid herbicides can be cited: only the *R*-enantiomers of these compounds are active. On other occasions, each enantiomer has a different activity and the enantiomeric ratio is controlled to produce a certain effect. As an example, the activity of uniconazole as plant growth regulator and diniconazole as fungicide is given by their enantiomeric proportion because the *R*-enantiomer shows stronger fungicidal activity than the *S*-enantiomer, which has higher plant growth-regulating activity. Since uniconazole has a higher plant growth regulating activity than diniconazole but is less active as a fungicide, products containing a high proportion of the *R*- and *S*-enantiomer for diniconazole and uniconazole, respectively, have been developed as a high-activity fungicide and an effective plant growth regulator [5]. Finally, other compounds of environmental interest are persistent pollutants. Since the toxicity of each enantiomer may be different, the real toxicity of samples depends on the individual content of each enantiomer. In addition, it is known that the degradation of these compounds by microorganisms may be enantioselective producing a selective enrichment of one of the enantiomers in the environment [1].

In the case of food samples, the separation of enantiomers may provide information about possible adulterations in foods and beverages. As an example, the addition of inexpensive synthetic amino acids to mask water dilution in beverages can be detected through chiral analysis of selected amino acids due to the presence of d- and l-forms in the racemic synthetic amino acids and only the l-forms in the natural amino acids. However, it is also possible to detect the presence of d-amino acids in fermented products or treated products. This is because fermentation processes or certain treatments can produce different degrees of racemization of l-amino acids in their d-forms. Thus, the enantioseparation of certain chiral compounds in foods may be used to control and monitor fermentation processes and products, or to evaluate and identify the treatment, age and storage effects. In addition, chiral separations allow the analysis of chiral metabolites of many chiral and prochiral constituents of foods and beverages [6].

All of this shows that there are many reasons to discriminate between the enantiomers of a chiral compound and to study them separately.

Enantiomeric discrimination can be done mainly using enantioselective detectors or separation techniques (see Fig. 13.1). Also, it is important to note that the use of miniaturized systems such as sensors [7] or microchips [8] are an emerging trend in chiral analysis. Available enantioselective detectors, i.e., optical rotatory dispersion and circular dichroism, rely on the differential interaction of enantiomers with circularly polarized light and they have a limited sensitivity [9]. Therefore, the discrimination of enantiomers is usually performed by using separation techniques. In addition to classical methods, where the formation of diastereomeric pairs using chiral reagents followed by repeated crystallization or the use of stereoselective enzymes are

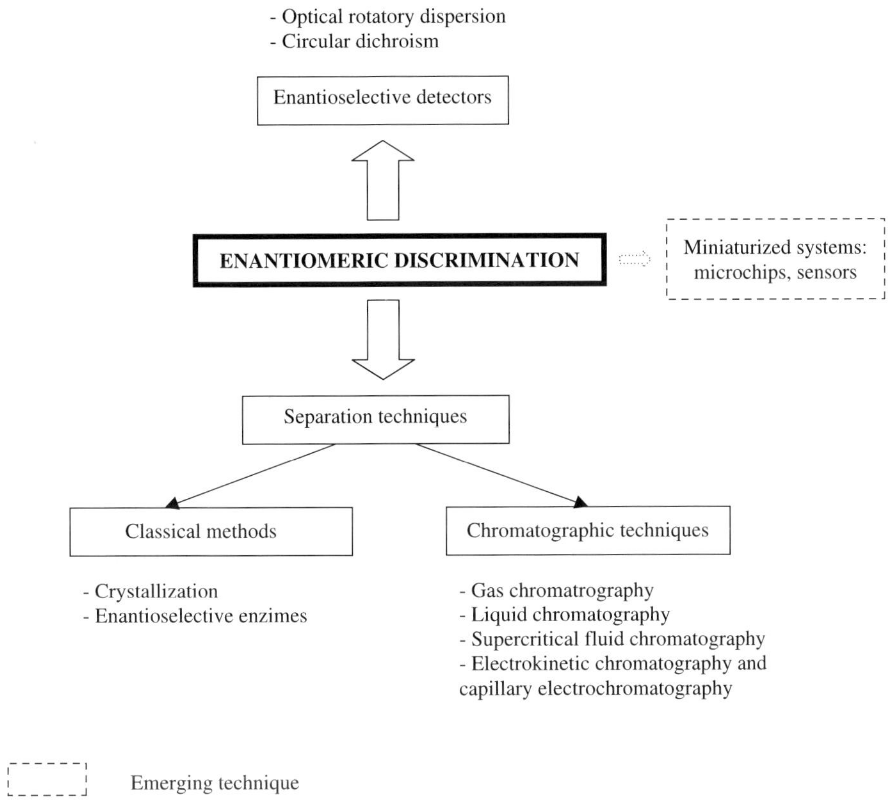

Fig. 13.1. Alternatives to discriminate between the enantiomers of a chiral compound.

performed, chromatographic techniques are especially relevant in chiral analysis [10–12]. In this regard, it is interesting to note that gas chromatography (GC), high-performance liquid chromatography (HPLC) and supercritical fluid chromatography (SFC) have been widely used in order to achieve this kind of separations. There are two types of separation methods for resolving enantiomers: (i) *indirect chiral separations*, where enantiomers are derivatized into diastereomers using an optically pure reagent, enabling their separation in an achiral environment because of their different chemical properties; and (ii) *direct chiral separations*, which are based on the formation of labile diastereomers with chiral selectors or chiral stationary phases present in the separation technique. In this case, the presence of a chiral environment is absolutely necessary. The disadvantages of the indirect separation of enantiomers, such as time-consuming derivatization reactions and expensive chemical and optical pure reagents, make that most of the applications in chiral analysis are performed by direct separation methods.

In addition to the use of the above-mentioned chromatographic techniques, a continuous increase in the interest for capillary electrophoresis (CE) concerning the separation of enantiomers has been observed in the recent years. In fact, this separation technique provides fast and efficient separations of chiral compounds, requires low consumption of sample, especially important in biological applications where only limited amounts of sample are available and needs a minimum amount of chiral selectors, which are quite expensive. In addition, chiral CE has higher flexibility than other chromatographic techniques because the selector–selectand interactions can be intensified by using a large excess as well as a combination of chiral selectors. The importance of enantioseparations in the pharmaceutical and biomedical field has prompted that several big pharmaceutical companies consider CE a technique of choice for enantioseparations. Therefore, nowadays CE not only complements but also competes with gas or liquid chromatographic techniques in chiral analysis [13,14].

The main goals of this chapter are: (i) to describe the separation modes mainly employed to achieve chiral analyses by CE with emphasis on the selection of the chiral selector, a key factor for a successful enantioseparation; (ii) to point out the sensitivity requirements in chiral analysis and to describe the preconcentration strategies and alternative detection systems to on-line UV detection used in chiral analysis by CE; and (iii) to show the most recent applications in pharmaceutical and biomedical analysis, the main application field, and to provide an overview of other applications of chiral CE including environmental and food analysis. Finally, the future perspectives of CE in chiral analysis will be described taking into account recent developments in this field.

13.2 SEPARATION MODES IN CE FOR CHIRAL ANALYSIS

An enantiomeric separation in CE is not based on an electrophoretic mechanism because the electrophoretic mobilities of the enantiomers of a chiral compound are equal and nonselective. Actually, the enantioselective recognition of the enantiomers of a chiral compound is due to their different interaction with a chiral selector and therefore caused by a chromatographic mechanism. This basic concept has been clearly developed and defended by Chankvetadze [15,16] and Chankvetadze and Blaschke [17], who established that the chiral separation principle is absolutely the same in the enantiomeric separation of a charged chiral analyte with a neutral chiral selector and in that of an uncharged chiral analyte with a charged chiral selector. This separation principle relies on the different partition of enantiomers between the bulk solution and the chiral pseudophase. However, originally and also in a few recent papers, chiral separations with neutral chiral selectors were considered within the capillary zone electrophoresis (CZE) format, and only those produced by charged chiral selectors were included within the electrokinetic chromatography (EKC) separation mode. In this chapter, according to Chankvetadze, all chiral separations in CE will be considered within the EKC mode regardless of the chiral selector charge. On the other hand, although much less used, capillary electrochromatography (CEC) and non-aqueous CE (NACE) are also employed for chiral analysis. CEC has recently become a tool for enantioseparation that is accessible to many laboratories, and is now entering the field of practical applications. However, the use of NACE for enantiomeric separations is mainly focused on the study of different separation selectivities concerning those obtained in aqueous media.

13.2.1 Electrokinetic chromatography

The separation principle of EKC has been described in chapter 2. In chiral EKC, the chiral selector acts as chiral pseudophase and interacts differently with each enantiomer. A large number of chiral selectors are currently available; especially noteworthy are cyclodextrins (CDs), chiral crown ethers, chiral surfactants, macrocyclic antibiotics, proteins, polysaccharides and ligand-exchange type selectors [17–21]. A brief description of the characteristics and usefulness of these chiral selectors in EKC and the basic separation mechanism involved will be presented here.

13.2.1.1 Cyclodextrins

CDs are the most widely used chiral selectors in CE. According to the literature, 86% of the papers published till January 2004 mention the use of CDs as chiral selectors (see Fig. 13.2).

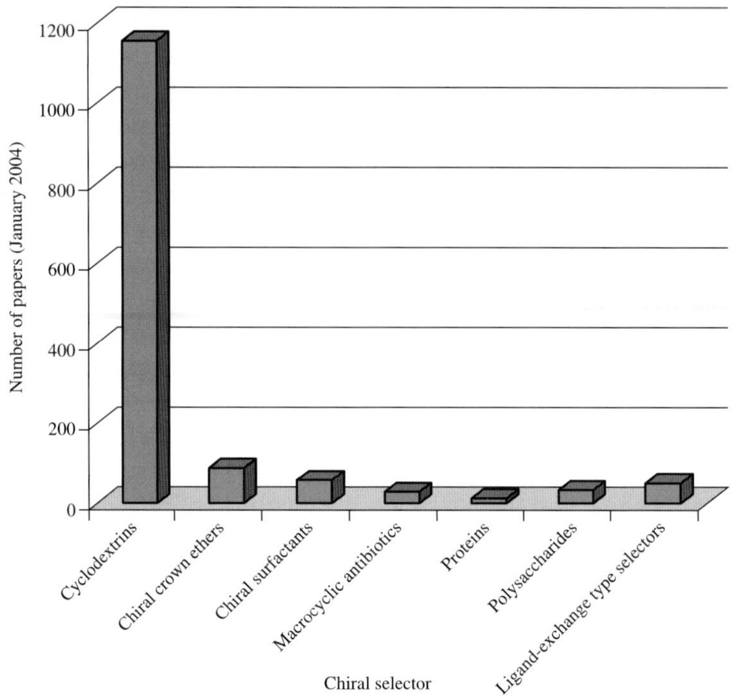

Fig. 13.2. Number of papers published using the different chiral selectors described in EKC.

Naturally occurring α-, β- and γ-CD are cyclic oligosaccharides containing 6, 7 and 8 glucopyranose units, respectively. They have the shape of a truncated cone with a relatively hydrophobic inside and a hydrophilic outside owing to the presence of hydroxyl groups. In addition to the naturally occurring CDs (also called native CDs), there are a large amount and variety of CD derivatives commercially available (a desired CD can be requested from companies such as Cyclolab) [20].

The most widely accepted mechanism for the enantioresolution of a chiral analyte with CDs involves the inclusion of the analyte into the cavity of the CD and the establishment of secondary interactions with the hydroxyl group on the CD rim, i.e., the formation of inclusion or host–guest complexes. Although inclusion does not ensure enantioresolution, it is a necessary condition, so the size and shape of the selector and selectand should be considered. Hence small molecules (one or two rings) fit better into small-cavity CDs (α- or β-CD), whereas only large molecules (3 or 4 rings) can be included into the largest cavity CD

(γ-CD) [18,21,22]. However, it is important to take into consideration that this is not a general rule because the complete inclusion of the analyte into the cavity of CD is not always a necessary prerequisite for enantioseparations; a partial inclusion or external intermolecular interaction may also be sufficient [17].

From the naturally occurring CDs, β-CD has the highest enantioresolution power. However, the extent of chiral discrimination exhibited by native CDs is quite modest, which can be ascribed to the inherent symmetry of these CDs. Derivatization of CDs has made it possible to enhance the enantioselectivity achieved by native CDs and to increase the number of applications [23]. In addition, β-CD has limited solubility in water, but this fact can be overcome using CD derivatives of neutral or ionic nature. Special attention has been paid to charged CD derivatives because they have helped to enantioresolve neutral chiral analytes [18,24]. Table 13.2 summarizes the structure and characteristics of the neutral and ionic CDs most widely used until now [25]. The neutral CD derivatives 2-hydroxypropyl-β-CD (HP-β-CD), heptakis(2,6-di-*O*-methyl)-β-CD also called dimethyl-β-CD (DM-β-CD) and heptakis(2,3,6-tri-*O*-methyl)-β-CD, also called trimethyl-β-CD (TM-β-CD), have been widely used in the pharmaceutical, biomedical and environmental fields. They have shown higher solubility in water and additional enantioselectivity than the native β-CD; however they can only enantioseparate charged analytes. In order to separate the enantiomers of neutral and charged analytes, charged CDs have been successfully employed although they increase the current of the final buffer solution. The anionic CD derivatives most widely used until now are carboxymethylated-β-CD (CM-β-CD), sulfated-β-CD (sulfated-β-CD) and sulfobutylether-β-CD (SBE-β-CD). These CD derivatives have shown high enantioselectivity for a large number of cationic and neutral compounds and even for some anionic compounds. The cationic CD derivatives more employed are 2-hydroxy-propyl-trimethylammonium-β-CD (QA-β-CD) and 6-monodeoxy-6-monoamino-β-CD (β-CD-NH_2). Of all these CD derivatives, which are commercially available, anionic CDs have been used more widely than cationic CD derivatives. Table 13.2 also shows that these CDs are single-isomer or randomly substituted derivatives. The single-isomer CDs are synthesized to yield only a single molecular species, whereas the randomly substituted CDs potentially possess several molecular structures with different degrees of substitution (an average degree of substitution is then usually given). The use of randomly substituted CDs shows two main disadvantages: (i) since they are

TABLE 13.2

Characteristics and structure of the commercial CDs most widely used as chiral selectors [25]

Cyclodextrin	Nature	Structure	Molecular weight (d.s.)	pK	Solubility in water at 25 °C (g/100 ml)
α-Cyclodextrin (α-CD)	Native CD	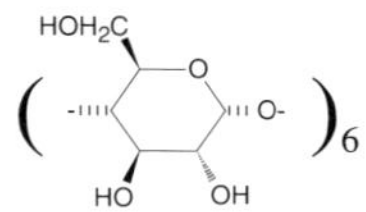	972	12.33	14.5
β-Cyclodextrin (β-CD)	Native CD	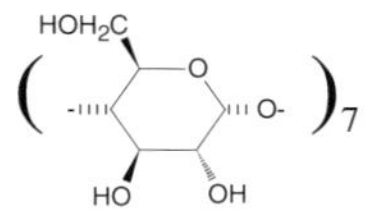	1135	12.20	1.87
γ-Cyclodextrin (γ-CD)	Native CD	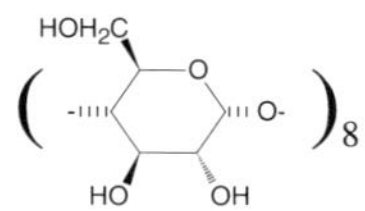	1297	12.08	23.3
2-Hydroxypropyl-β-cyclodextrin (HP-β-CD)	Neutral CD derivative (randomly substituted)	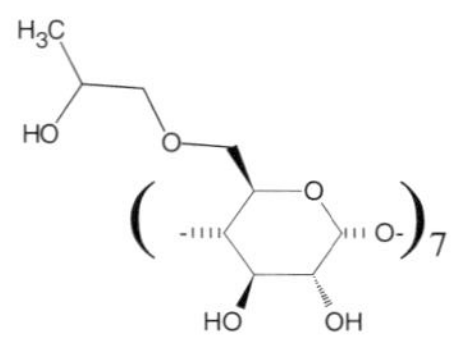	~1380 (d.s. ~4)	> 12	60

Heptakis(2,6-Di-*O*-methyl-β-cyclodextrin (dimethyl-β-cyclodextrin, DM-β-CD)	Neutral CD derivative (single-isomer)	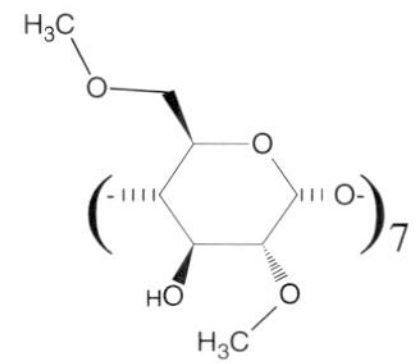	1331	> 12	> 33
Heptakis(2,3,6-tri-*O*-methyl)-β-cyclodextrin (trimethyl-β-cyclodextrin, TM-β-CD)	Neutral CD derivative (single-isomer)	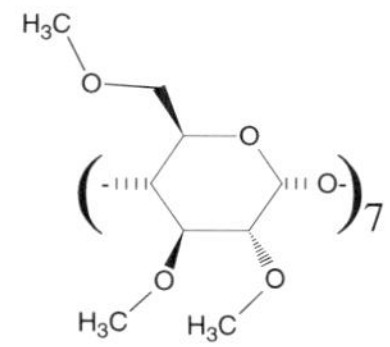	1429	> 12	> 33
Carboxymethyl-β-cyclodextrin (CM-β-CD)	Anionic CD derivative (randomly substituted)	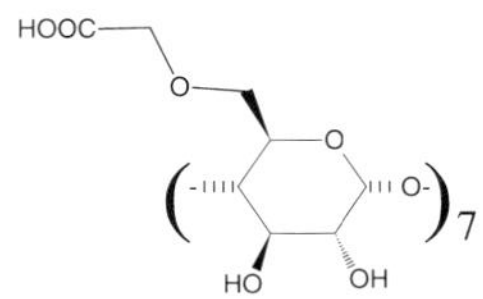	~1309 (d.s. ~3)	< 4	> 100
Sulfated-β-cyclodextrin (sulfated-β-CD)	Anionic CD derivative (randomly substituted)	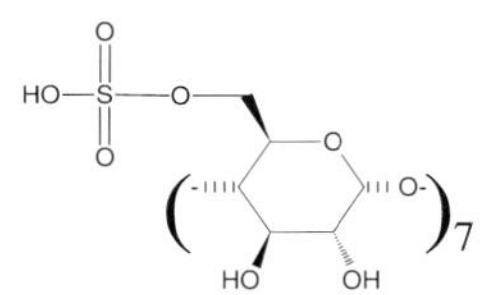	~2359 (d.s. ~12)	~2	> 33

continued

TABLE 13.2 (*continued*)

Cyclodextrin	Nature	Structure	Molecular weight (d.s.)	p*K*	Solubility in water at 25 °C (g/100 ml)
Sulfobutylether-β-cyclodextrin (SBE-β-CD)	Anionic CD derivative (randomly substituted)	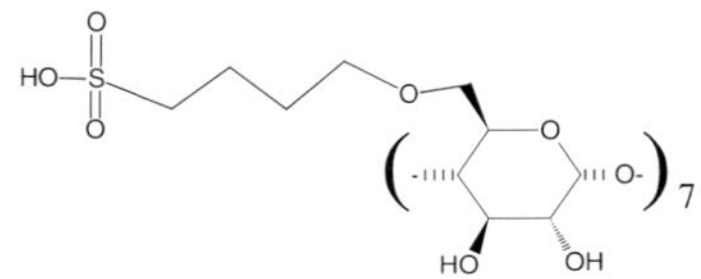	~2160 (d.s. ~7)	<3	80
Quaternary ammonium-β-cyclodextrin (QA-β-CD)	Cationic CD derivative (randomly substituted)	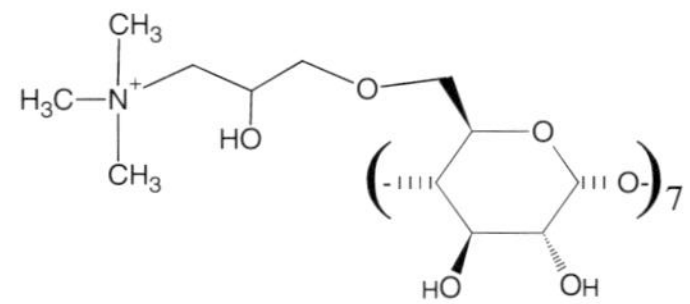	~1669 (d.s. ~3.5)	~8	>100
6-monodeoxy-6-monoamino-β-cyclodextrin (β-CD-NH_2)	Cationic CD derivative (single-isomer)	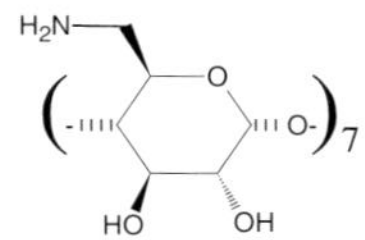	1170	8.2	>100

d.s., degree of substitution.

mixtures of many isomeric forms differing in the degree of substitution and in the position of the constituents, high variability in the selectivity obtained from different commercial suppliers and even from batch to batch for the same supplier can be observed; and (ii) it is not possible to establish the robustness of a certain chiral method and complete its validation. As a result, it can be emphasized that a recent trend in chiral analysis by CE is the use and development of new single-isomer CD derivatives [18,19,26].

Sometimes, the use of mixtures of CDs is necessary in order to achieve the separation of certain enantiomers [27–29]. Although mixtures of native CDs with neutral CD derivatives have been used, most of the systems combine two CDs (usually called dual CD systems), mixing an ionic CD with a neutral one. This approach has been very effective for the enantioseparation of neutral compounds and the separation of the enantiomers of several chiral compounds in mixtures. In several cases, the ionic CD acts as carrier of the neutral analyte, and it is the neutral CD that shows chiral recognition. In this case, the separation mechanism is similar to that producing an enantioseparation in micellar EKC with CDs (CD-MEKC), where the anionic achiral micelles act as carriers of the neutral analyte and the neutral CD performs the enantiomeric discrimination.

When CDs are used as chiral selectors, the choice of the appropriate CD is crucial to achieve the desired enantiomeric separation. Unfortunately, it is not possible to predict a successful enantiomeric separation on the basis of the chemical structure of the analyte and the chiral selector. Recently, Evans and Stalcup [30] have published an interesting review showing the broad spectrum of chiral compounds that can be enantioseparated using sulfated CDs. In addition, comprehensive strategies for chiral separations using these chiral selectors are provided. In general terms, Fig. 13.3 shows a scheme to develop a chiral method by CE as proposed by the authors. It is recommended to begin using as chiral selectors the CDs indicated in Table 13.2 on the basis of their demonstrated wide range of application and feasibility. Other factors to be considered for the achievement of an enantiomeric separation are: the pH of the buffer solution, CD concentration and temperature. It is important to note that the pH and the appropriate buffer solution should be chosen as a function of the pK_a of the analytes and the charged CD. In addition, for a certain compound, the enantiomeric resolution changes with the CD concentration, generally reaching a maximum value that has to be determined. The influence of the

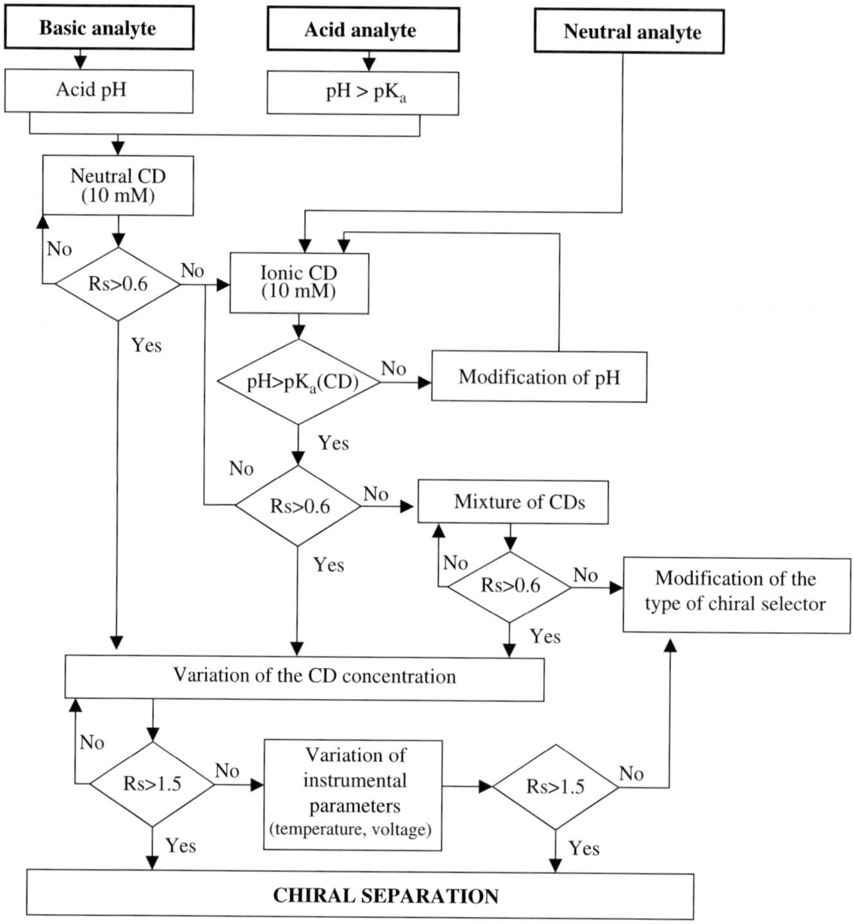

Fig. 13.3. Scheme for the development of a chiral separation by CE.

temperature is another important factor to take into consideration. It has been observed that, in general, the enantioresolution increases when the temperature decreases [13,31]. Finally, if CDs are not the appropriate chiral selectors, then the use of other chiral selector should be tested. However, this scheme requires the study of the effect of a selected parameter while keeping the other parameters constant, which is a time-consuming task. Chemometric experimental designs (Plackett-Burman, central composite or factorial designs) can be used to reduce the number of experiments required for optimization and to consider the possible interdependence of parameters [17]. The different mathematical models describing the dependence of chiral selectivity on the buffer pH and the CD concentration not only allow the optimization of such parameters, but also seem to be useful to predict the separation selectivity in certain conditions [17,32,33]. However, any of these designs or models are useful in the most important step in the development of a chiral method by CE: the choice of the most appropriate chiral selector.

13.2.1.2 Chiral crown ethers

Chiral crown ethers are macrocyclic polyethers, and only (+)-(18-crown-6)-2,3,11,12-tetracarboxylic acid ($18C6H_4$) has proved to be effective as chiral selector for CE. It is a charged chiral selector with the following pK_a values: 2.1, 2.8, 4.3 and 4.9, and it does not interfere with UV detection [19].

Crown ethers form inclusion complexes with alkali, alkaline-earth and primary ammonium cations. The main interactions are assumed to be the formation of hydrogen bonds between the three hydrogens attached to the nitrogen of the analyte (primary amine) and the dipoles of the oxygens of the macrocyclic ether (see its structure in Table 13.3). As carboxylic groups of the crown ether are perpendicular to the plane of the macrocyclic ring, different diastereomeric inclusion complexes with each enantiomer are formed in a tripod-like arrangement [18,22,34].

Separation of amino acids and other analytes bearing a primary amine group has been carried out at low pH values in Tris or triethylamine/citric acid buffers without alkali or alkaline-earth ions, which exhibit a high affinity to the crown ether and would diminish enantioresolution. In fact, it is important to consider the chemical structure of the analyte before using a crown ether as chiral selector. The analyte has to contain a primary amine group, because second or tertiary amines cannot provide the type of complexation required for chiral recognition. The enantioresolution increases when the distance between the amine functional group and the asymmetrical center is short. Although the existence of bulky substituents on the asymmetric center of the chiral analyte improves the enantioseparation, if the substitution is thorough, the enantioseparation can disappear for steric hindering [18].

It is interesting to comment on the synergic effect reported by Armstrong et al. [35] when a crown ether and a CD were used for the

TABLE 13.3

Structure, molecular weight and applicability of some other chiral selectors used in CE [25]

Chiral selector type	Name	Structure	Molecular weight	Applicability
Crown ether	(+)-(18-crown-6)-2,3,11,12-Tetracarboxylic acid ($18C6H_4$)	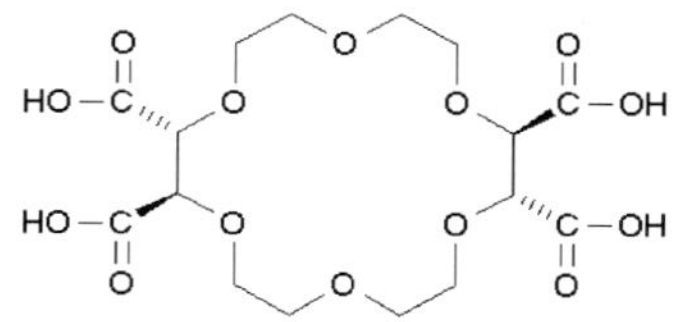	440.4	Restricted to compounds containing primary amine groups
Bile salts (natural surfactants)	Sodium cholate (SC)	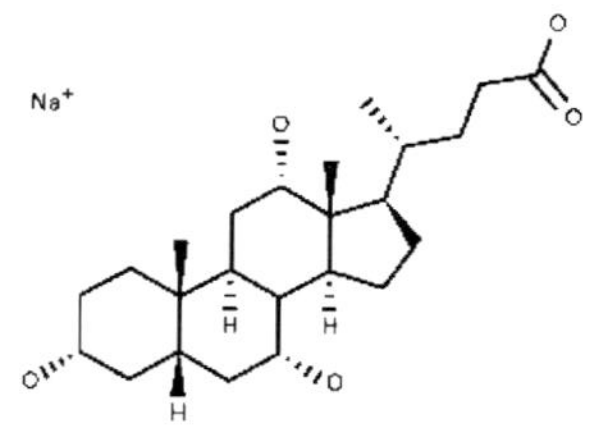	430.6	Appropriate to analytes with rigid structure of fused rings
	Sodium deoxycholate (SDC)	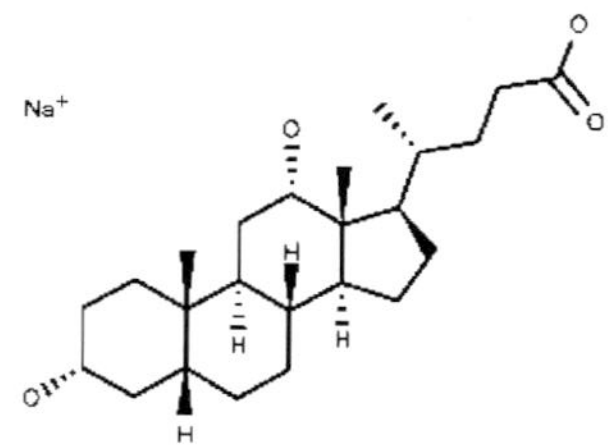	414.6	

Bile salts (natural surfactants)	Sodium taurocholate (STC)	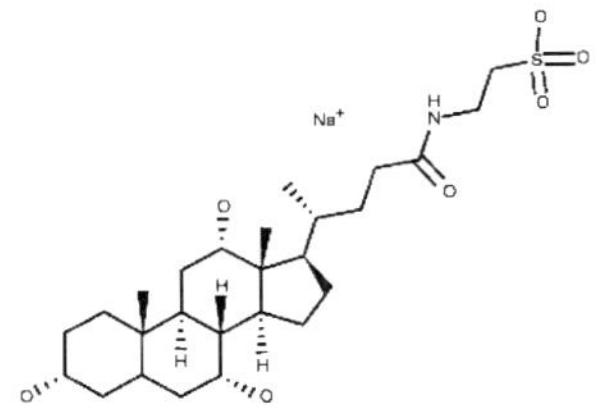	537.7	
	Sodium taurodeoxycholate (STDC)	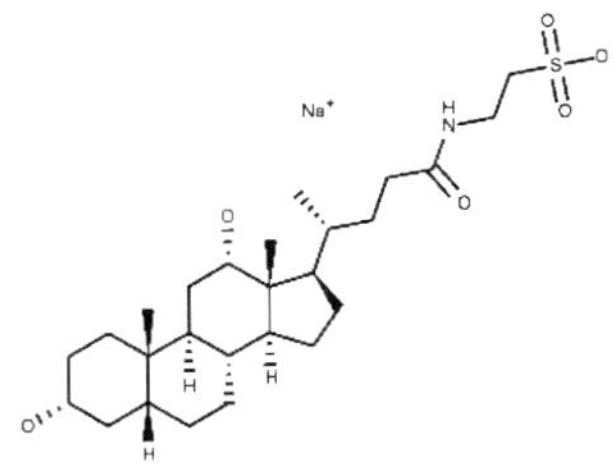	521.7	
Polymeric surfactants	Polysodium *N*-undecanoyl-l-leucylvalinate (poly-l-SULV)	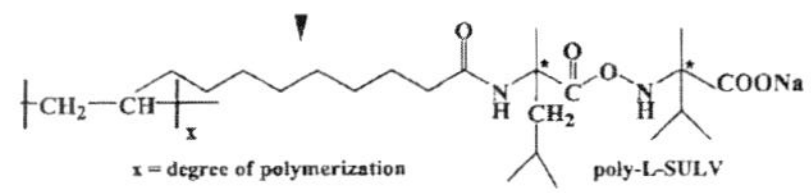	14,650(±625)	Adequate for the enantiomeric separation of cationic and neutral analytes. Anionic analytes are difficult to resolve. Compatible with MS detection
Macrocyclic antibiotics	Ristocetin A	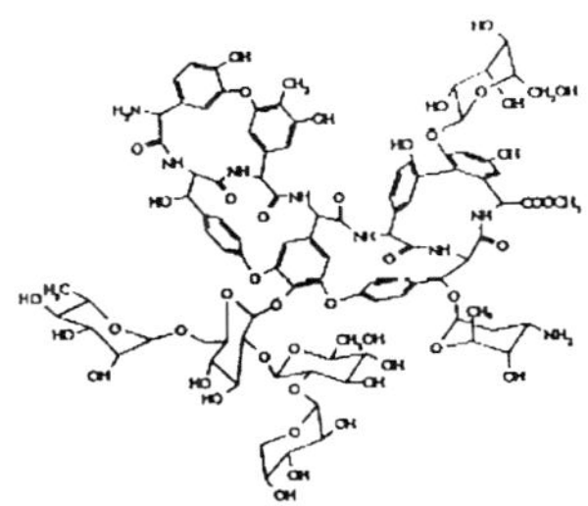	2066	Problems of light absorption and adsorption to the capillary walls. Adequate for neutral and anionic compounds of a wide range of structures

continued

TABLE 13.3 (*continued*)

Chiral selector type	Name	Structure	Molecular weight	Applicability
	Vancomycin	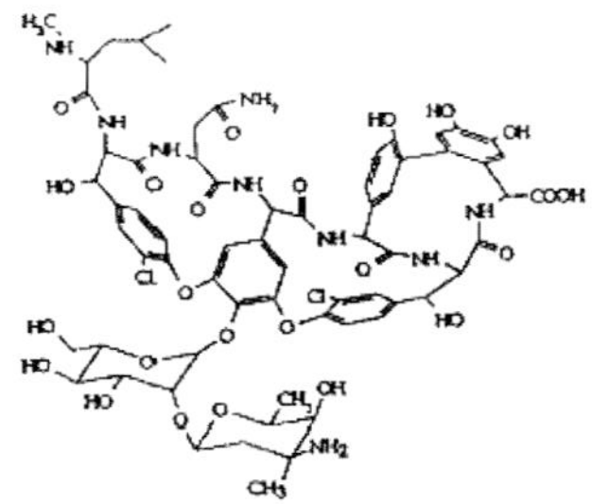	1449	Problems of light absorption and adsorption to the capillary walls. Adequate for neutral and anionic compounds of a wide range of structures
Macrocyclic antibiotics	Teicoplanin	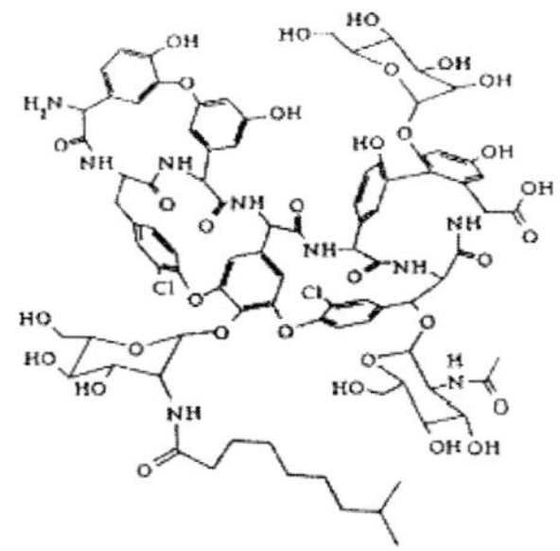	1877	
Proteins	Serum albumin	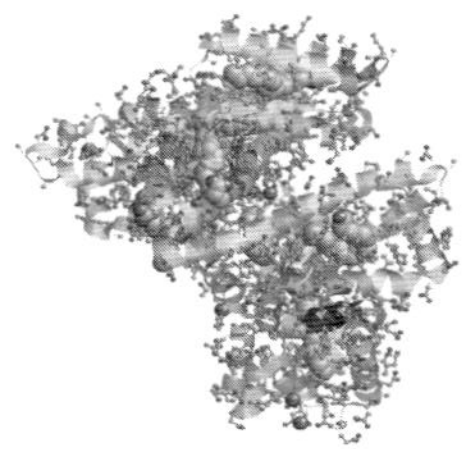	—	Problems of light absorption and adsorption to the capillary walls. Appropriate for neutral and ionic compounds

Polysaccharides	Dextrin	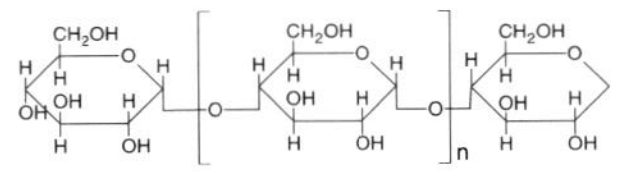	—	Appropriate for anionic and cationic compounds
	Dextran	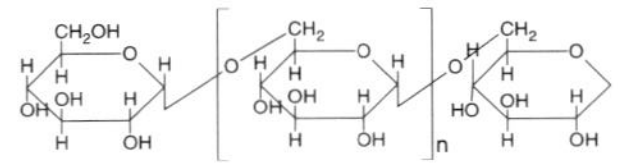	—	
	Chondroitin	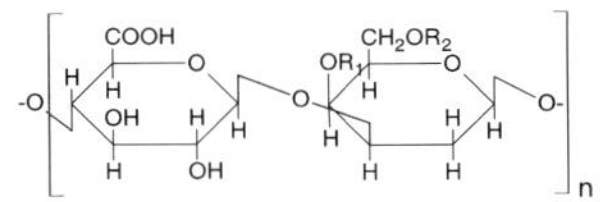	30,000–50,000	Adequate for neutral and charged analytes
	Heparin	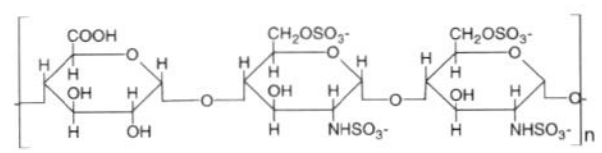	6000–30,000	

enantioseparation of organic racemates containing a primary amine functional group.

13.2.1.3 Chiral surfactants

There are two types of chiral surfactants: naturally occurring detergents and synthetic chiral surfactants. The natural chiral surfactants used in chiral CE can be classified into bile salts, digitonins and saponins families. Among them, bile salts were the first chiral surfactants employed [36] and are also the most popular. They form helical micelles with a reversed micelle conformation that defines the mechanism of enantioseparation [22]. Of the four different bile salts (see Table 13.3) sodium taurodeoxycholate has shown the highest enantioresolution power. Enantiodiscrimination with bile salts appears to be especially favorable in analytes with a rigid structure of fused rings. In this respect, bile salts are complementary to CDs, because this type of structures is unlikely to be resolved with CDs, owing to their decreased inclusion capacity [18].

Synthetic chiral surfactants, (also called polymeric surfactants) can derived from natural sugars: alkyl-glucoside and steroidal glucoside type surfactants [37,38] or aminoacids: N-(n-alkanolyl) or N-(n-alkyloxycarbonyl)-amino acids chiral surfactants [37–39]. These pseudophases seem to provide better mass transfer and increased rigidity and stability than conventional micelles [39]. A promising type of synthetic chiral surfactants are polymerized surfactants, also known as micelle polymers, which have shown some advantages (zero critical micelle concentration (CMC), the structure of the polymer remains constant with changes in the analytical conditions, i.e., the addition of organic solvents, and they are compatible with mass spectrometry (MS) detection) over micelles of low-molecular-weight surfactants, which are equilibrium self-assemblies [37,40]. As an example, the micelle polymer poly(sodium *N*-undecanoyl-l-valinate (poly l-SUV) has been successfully employed as chiral selector in micellar EKC with MS detection [40]. Although synthetic chiral surfactants have shown novel and promising results [41,42], the problem is that they are at an incipient stage and most of them are not commercially available.

13.2.1.4 Macrocyclic antibiotics

Macrocyclic antibiotics used in CE are mainly of two types: glycopeptides, which are especially efficient for the separation of neutral and anionic compounds, and ansamycins, which are particularly adaptable for the enantioresolution of cationic analytes [18,43–47]. Ansamycins have lower enantiodiscrimination power than glycopeptides. Among glycopeptides, ristocetin A, vancomycin and teicoplanin (see Table 13.3) are the most effective as chiral selectors [43]. Separation of enantiomers occurs when they interact with the chiral environment provided by the macrocyclic antibiotics (which are macromolecules with multiple asymmetrical centers) forming diastereoisomers with different stability constants and thus different mobilities in CE. Although the precise mechanism by which these chiral selectors provide enantiorecognition has not yet been unequivocally established, there is evidence that it involves inclusion into hydrophobic cavities, dipole–dipole interactions, hydrogen bonding and electrostatic or π–π interactions. Macrocyclic antibiotics present several drawbacks when they are added to the background electrolyte (BGE): (i) they have a strong adsorption to the inner wall of fused-silica capillaries, a problem that can be overcome by using coated capillaries or adding additives to the buffer solution; (ii) they show low chemical stability (they can be degraded at high temperatures and acid or basic pH), a problem that can be avoided by selecting appropriate experimental conditions; and (iii) they absorb in the low-UV range causing low sensitivity, a problem that can be solved using the partial filling technique (see chapter 9) combined with a counter-current mode (in this case, the electroosmotic flow (EOF) must be suppressed or controlled to prevent the selector from being carried to the detection window. Selector and analyte are used under conditions where they have opposite charges, and the selector migrates away from the detector window, which will be reached only by the analytes [48]).

13.2.1.5 Proteins

Proteins used in CE include albumins, glycoproteins, enzymes and other proteins such as casein, human serum transferrin or ovotransferrin [49]. Serum albumins (see Table 13.3), which are plasma proteins, are the proteins most frequently used for chiral separations in CE. Depending on the pH selected, proteins can be positively or negatively charged and can be used for the enantioseparation of neutral and charged compounds. When proteins are dissolved in the running buffer, they work as a pseudostationary phase, and enantiomeric separations are based on the differential affinity of the proteins to individual enantiomers, which is why this separation mode is called *affinity capillary electrophoresis* (ACE) [50]. When protein solutions are used in CE, two practical precautions must be taken: (i) avoiding the interaction of proteins with the wall of the capillary, by means of additives or coated capillaries and (ii) preventing the use of high concentrations (usual concentrations are lower than 100 μM) due to the background UV absorption in short-wavelength regions of protein solutions. The latter drawback can be overcome by using the partial filling technique explained in chapter 9, based on its usefulness for the coupling of CE to MS.

13.2.1.6 Polysaccharides

There are electrically neutral polysaccharides, appropriate for the enantioseparation of basic and acidic analytes, and charged polysaccharides, effective for the enantioresolution of neutral as well as charged compounds [51]. The enantioseparation mechanism of these selectors involves a variety of interactions and hence enables the enantioresolution of widely different structures.

Among neutral polysaccharides, dextrins and dextrans (see their basic structures in Table 13.3) have been used in chiral CE, although dextrins have shown more enantioresolution power than dextrans [52]. The helical structure of dextrins might be responsible for chiral recognition [22].

Most of the charged polysaccharides used in CE so far are naturally occurring, such as heparin or chondroitin sulfates (for their basic structures, see Table 13.3). Since they have a wide molecular mass range, their enantioselectivity depends on molecular mass distribution and the different ratios of the unit component. Chondroitin sulfate C has shown higher enantioresolution power than heparin for the enantioseparation of basic drugs [53]. A new trend in chiral CE is the use of synthetic or modified polysaccharides since, as in the case of CD derivatives, small modifications or the introduction of some residues to polysaccharides may result in a change of enantioselectivity [52,53].

13.2.1.7 Ligand-exchange type selectors

Enantioresolution by ligand-exchange complexes relies on the formation of a multi-component chelate complex consisting of a central cation [basically Cu(II), but also Ni(II) or Zn(II)], a chiral bifunctional ligand (usually l-amino acids) and the analyte enantiomers. The enantioseparation mechanism is based on the formation of ternary mixed-metal complexes of different stability between the complex formed by the metal and the chiral ligand (Me(l-Lig)) with the analyte enantiomers (S-A and R-A) according to the following equilibria:

$$Me(\textsc{l}-Lig)_2 + S - A \rightleftharpoons Me(\textsc{l}-Lig)S - A + \textsc{l} - Lig$$

$$Me(\textsc{l}-Lig)_2 + R - A \rightleftharpoons Me(\textsc{l}-Lig)R - A + \textsc{l} - Lig$$

These selectors can enantioresolve analytes with functional groups capable of forming the complexes, such as amino acids, dipeptides, amino alcohols and hydroxyl acids [18,54–56].

From a practical point of view, it is important to take into account that these selectors may cause two problems: (i) low detection sensitivity or rough baselines arising from the high UV absorption of the buffer solution and (ii) poor resolution caused by slow ligand-exchange kinetics [18].

13.2.2 Capillary electrochromatography

CEC is a hybrid technique of CE and HPLC (see chapter 2). In CEC the driving force is the EOF, which provides plug-like profiles giving higher efficiencies than HPLC. Enantioseparations are possible in three different modes: (i) in capillaries packed with chiral stationary phases, (ii) in wall-coated open tubular capillaries or (iii) in capillaries packed with achiral stationary phases in combination with BGEs where the chiral selector is added [10,17,57].

The most intensively developing mode in chiral CEC is the use of capillaries packed with chiral stationary phases. Many of the chiral stationary phases which have earlier proved to be useful for HPLC enantioseparations have been adapted for chiral CEC, such as CD and their derivatives [58,59], macrocyclic antibiotics [60], proteins [49], polysaccharides [61–63] and ligand-exchange-type selectors [64].

The use of monolithic columns (also called polymeric continuous beds) represents a current trend in chiral CEC [65–70]. In this case, the preparation of the monolithic phase is by in situ polymerization in the column instead of packing on the basis of silica gel [10]. The use of these columns may also be considered a different working mode in CEC [71,72].

Most of the studies on chiral CEC report the development of novel chiral stationary phases and capillary columns as well as the transfer of HPLC methods to chiral CEC, showing particular separations of model chiral compounds. In addition, CEC has become a tool for enantioseparation more accessible to laboratories because different approaches for the preparation of enantioselective columns have been reported in the literature in detail. Several reviews on chiral CEC have been published recently and show the progress, trends and applicability of CEC on enantioseparations [71–76]. Although fundamental aspects and new chiral stationary phases must be studied, it is a fact that nowadays, chiral CEC is entering into the field of practical applications, thereby showing its potential and possible limitations [77–87]. As an example, Fig. 13.4 shows the separation of a mixture of the enantiomers of warfarin and coumachlor with different percentages of organic modifier, ionic strength and pH in the mobile phase. It can be observed that the enantioresolution increases when the percentage of acetonitrile decreases, the concentration of ammonium acetate increases and the pH decreases. Thus, 70% acetonitrile containing 5 mM ammonium acetate at pH 4 was selected for a good enantioresolution and separation of the enantiomers of both compounds. This mobile-phase composition was used for the quantitative chiral assay of human plasma samples. As illustrated in Fig. 13.5A, for a plasma sample spiked with racemic warfarin (and using racemic coumachlor as internal standard) it was possible to detect the single enantiomers with a limit of detection (LOD) as low as 25 ng/ml. Fig. 13.5B also reveals that 1% of the minor enantiomer can be easily detected by this CEC method. In addition, the hyphenation of chiral CEC with electrospray ionization mass spectrometry (ESI-MS), although nowadays very challenging, seems to be very promising because it combines sensitivity with high specificity and selectivity. Since with this detection technique the use of chiral selectors needed in chiral CE is limited, the fact that the chiral selector can be immobilized in the CEC column is an advantage of this CE mode [87].

13.2.3 Non-aqueous capillary electrophoresis

In NACE the aqueous buffer employed in CE is replaced by an organic solvent (mainly formaldehyde or methanol) containing an electrolyte. In this way, additional selectivities other than those obtained in aqueous CE systems can be obtained, and the analysis of compounds with poor solubility in water can be easily performed [88]. For chiral separations, CDs, crown ethers and ion-pairing compounds have been used until now, as shown in Table 13.4.

The native β-CD in a BGE in formamide or *N*-methylformamide has been used for the separation of the enantiomers of pharmaceutical

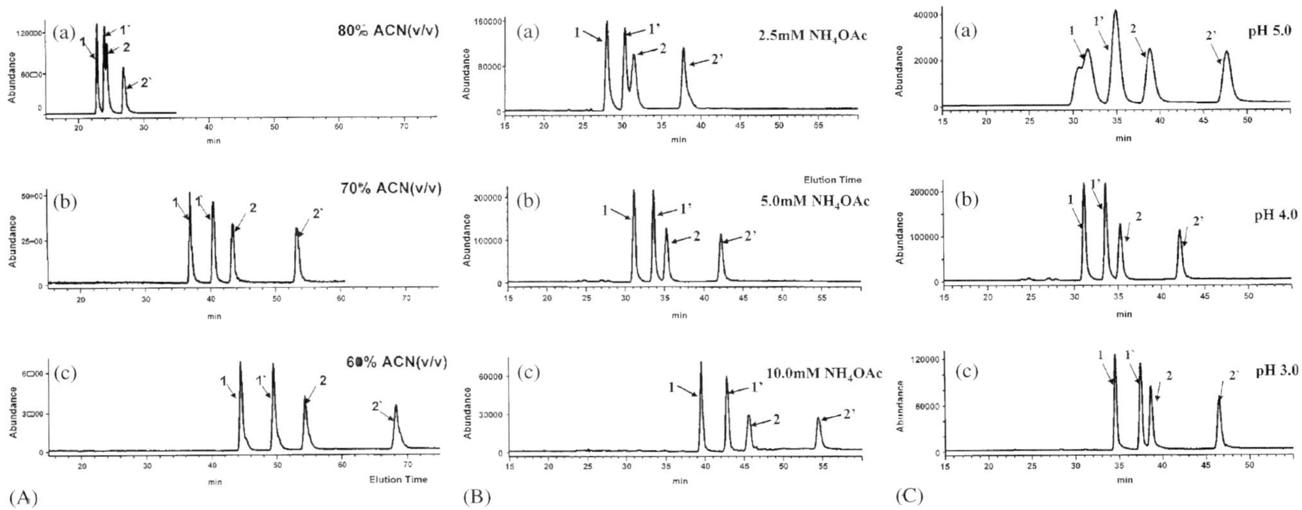

Fig. 13.4. Electropherograms showing the effects of (A) percentage of acetonitrile (ACN, v/v) in 10 mM NH_4OAc at pH 3; (B) mobile-phase ionic strength in 70% ACN at pH 3 and (C) mobile-phase pH in 70% ACN containing 5.0 mM NH_4OAc. Peak identification: 1, (*R*)-warfarin; 1′, (*S*)-warfarin, 2, (*R*)-coumachlor, 2′, (*S*)-coumachlor. CEC conditions: 60-cm-long, 75-μm-i.d. tapered capillary packed with 5-μm (3*R*,4*S*)Whelk-O 1 CSP; run voltage, 30 kV (3 kV/s); injection at 6 kV for 8 s. ESI-MS conditions: SIM negative ion mode; capillary voltage, −2500 V; fragmentor voltage, 80 V; drying gas flow rate and temperature, 5 l/min and 150°C, respectively; nebulizer pressure, 4 psi; sheath liquid, 5 mM NH_4OAc in CH_3OH/H_2O (50:50, v/v) at 5 μl/min. Reprinted with permission from Ref. [87]. Copyright (2003), American Chemical Society.

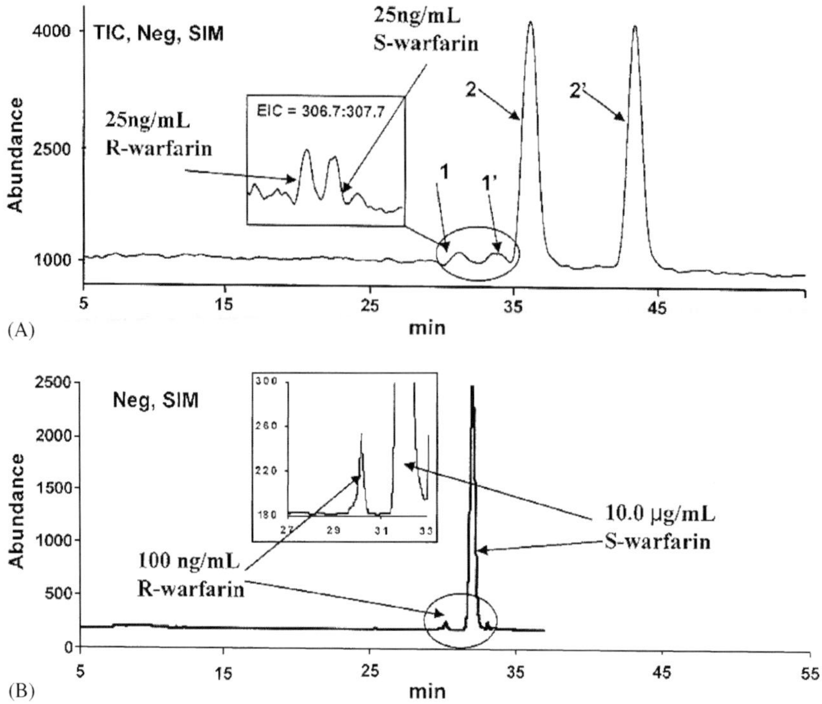

Fig. 13.5. Electropherograms of (A) (±)-warfarin at LOD and (B) a plasma sample spiked with 1% of the minor enantiomer ((*R*)-warfarin) along with 99% major enantiomer ((*S*)-warfarin). Conditions: mobile phase: 70% ACN containing 5.0 mM NH_4OAc at pH 4.0. Other conditions as in Fig. 13.4. Peak identification as in Fig. 13.4. Reprinted with permission from Ref. [87]. Copyright (2003), American Chemical Society.

amines and dansyl-amino acids [89,90]. Charged CD derivatives of anionic and cationic nature have also been used. The anionic CDs heptakis-(2,3-dimethyl-6-sulfato)-β-CD, heptakis-(2,3-diacetyl-6-sulfato)-β-CD, octakis-(2,3-dimethyl-6-sulfo)-γ-CD and octakis-(2,3-diacetyl-6-sulfato)-γ-CD have been used as good chiral selectors for a wide range of basic pharmaceuticals, using acidic BGEs in methanol [91–95]. The cationic QA-β-CD in an acetic acid BGE in formamide has been used for the enantiomeric separation of acid analytes [96].

The crown ether $18C6H_4$ in a tetra-*n*-butyl ammonium perchlorate electrolyte in formamide has been a good chiral selector for the enantiomeric separation of aromatic amines, amino acids and amino alcohols [97].

TABLE 13.4

Experimental conditions employed to achieve enantiomeric separations by NACE

Analytes	Chiral selector (conc.)	BGE	Detection	Reference
Drugs (trimipramine, thioridazine, mianserin, nefopam, primaquine, propiomazine, trihexylphenidyl, trimeprazine, chlophedianol, chlorcyclizine and ethopropazine)	β-CD (100 mM)	150 mM citric acid+100 mM Tris in formamide	UV at 254 nm	[89]
Dansyl-amino acids (Pro, Ala, Ser, Lys, Asn, Leu, Nval, Asp, Met, Thr and Val)	β-CD (100 mM)	10 mM NaCl in *N*-methylformamide	UV at 254 nm	[90]
Weak base drugs (epinephrine, isoproterenol, metaproterenol, oxyphencyclimine and propranolol)	Heptakis-(2,3-dimethyl-6-sulfato)-β-CD (12–40 mM)	25 mM phosphoric acid+12.5 mM NaOH in methanol	UV at 214 nm	[91,92]
Weak base drugs (cipofibrate, fenoprofen, ibuprofen and other 12 basic drugs)	Heptakis-(2,3-diacetyl-6-sulfato)-β-CD (10–40 mM)	50 mM dichloroacetic +25 mM triethylamine in methanol	UV at 214 nm	[93]
Weak base drugs (bupropion, oxprenolol and other 32 basic drugs)	Octakis-(2,3-dimethyl-6-sulfo)-γ-CD (0–40 mM)	25 mM phosphoric acid+12.5 mM NaOH in methanol	UV at 214 nm	[94]
Weak base drugs (atenolol, labetalol, chlorpheniramine, propafenone, tetrahydrozoline, tolperisone, tetrahydrozoline and other 19 basic drugs)	Octakis-(2,3-diacetyl-6-sulfato)-γ-CD (2.5–30 mM)	25 mM phosphoric acid+12.5 mM NaOH in methanol	UV at 214 nm	[95]
Non-stereoidal anti-inflamatory drugs (profens); 1,1′-binaphthyl-2,2′-diyl-hydrogen phosphate, *N*-[1-(1-naphthyl)ethyl]phthalamic acid; derivatized amino acids (Leu, Met, Nleu, Nval, Phe, Ser, Thr, Try and Val)	QA-β-CD (20 mM)	20 mM ammonium acetate+1% acetic acid in formamide	UV at 254 nm	[96]
Aromatic amines (1-naphthylethylamine and 1-phenylethylamine); amino acids (Phe and Tryp); amino alcohols (DOPA, norephedrine, noradrenaline and 2-amino-1,2-diphenylethanol)	18C6H$_4$ (10 mM)	100 mM Tetra-*n*-butyl ammonium perchlorate in formamide	UV at 260/280/300 nm	[97]
Basic drugs (atenolol, bisoprolol, bunitrolol, metoprolol, pindolol, propranolol, salbutamol, ephedrine, epinephrine, atropine, prometazine, bupivacaine, trimipraine, flecainide and mexiletine)	(+)- or (−)-Camphorsulfonate (30 mM)	0.2 mM Tween 20, 1 M acetic acid in methanol/ acetonitrile	UV at 214 nm	[98]

continued

TABLE 13.4 (*continued*)

Analytes	Chiral selector (conc.)	BGE	Detection	Reference
Benzoyl, 3,5-dinitrobenzoyl and 3,5-dinitrobenzyloxycarbonyl amino acid derivatives (Leu, Phg, αMeLeu, NmeLeu, tLeu, Phe, βPhen, Pro, α-amino butyric acid, β-amino-butyric acid and pipecolinic acid)	Tert-butyl carbamoylated quinine (10 mM)	12.5 mM Ammonia + 100 mM octanoic acid in ethanol/methanol (60:40)	UV at 214 nm	[99–101]
Benzoyl-3,5-dinitrobenzoyl and 3,5-dinitrobenzyloxycarbonyl amino acid derivatives (Leu, Phg, αMeLeu, NmeLeu, tLeu, Phe, βPhen, Pro, α-amino butyric acid, β-amino-butyric acid and pipecolinic acid)	1-Adamantyl carbamoylated quinine (10 mM)	12.5 mM Ammonia + 100 mM octanoic acid in ethanol/methanol (60:40)	UV at 214 nm	[102]
Benzoyl-3,5-dinitrobenzoyl and 3,5-dinitrobenzyloxycarbonyl amino acid derivatives (Leu, Phg, αMeLeu, NmeLeu, tLeu, Phe, βPhen, Pro, α-amino butyric acid, β-amino-butyric acid and pipecolinic acid)	trans-1,4-Cyclohexylene-bis(carbamoylated-11-dodecylthio-dihydroquinine (10 mM)	12.5 mM Ammonia + 100 mM octanoic acid in ethanol/methanol (60:40)	UV at 214 nm	[103]
Amines (pronethalol, labetalol, bambuterol, metoprolol and pronethanol)	(−)-2,3:4,6-Di-*O*-isopropylidene-2-keto-l-gulonic acid (100 mM)	40 mM NaOH in methanol	UV at 214 nm	[104]
Cationic and amphoteric compounds (cinchona alkaloid derivatives and other structurally related basic compounds like mefloquine)	(+)- or (−)-3,5-Dinitrobenzoyl-leucine (10 mM)	12.5 mM tetraethylammonium + 100 mM octanoic acid in ethanol/methanol (60:40)	UV at 216/330 nm	[105]

In NACE, ionic compounds capable of forming ion-pair complexes of different mobility with the enantiomers of a chiral compound have been used as often as CDs (see Table 13.4). This is because in NACE the formation of ion pairs is favored in comparison with aqueous CE. The enantiomerically pure (+) or (−)-camphorsulfonate was the fist chiral counter-ion reported for the enantioseparation of basic drugs in a non-aqueous medium [98]. Figure 13.6 shows the enantiomeric separation of racemic and non-racemic metoprolol (spiked with *S*-metoprolol) in (+)-*S*-camphorsulfonate with Tween 20 and acetic acid in methanol/acetonitrile. Also, different quinine derivatives in ethanol/methanol BGEs have been used for the enantiomeric separation of benzoyl, 3,5-dinitrobenzoyl and 3,5-dinitrobenzyloxycarbonyl amino acid derivatives [99–103]. Other chiral counter-ions used in chiral NACE have been: (−)-2,3:4,6-di-*O*-isopropylidene-2-keto-l-gulonic acid and the pure enantiomers of 3,5-dinitrobenzoyl-leucine (see Table 13.4). The first has been used for the separation of the enantiomers of different amines [104], whereas the pure enantiomers of the N-derivatized amino acid have been employed for the enantiomeric separation of basic and amphoteric compounds [105].

13.3 IMPROVING THE DETECTION SENSITIVITY FOR CHIRAL ANALYSIS BY CE

For some applications, high detection sensitivity in chiral analysis is required. The analysis of limited amounts of low concentrations of biological samples, the determination of enantiomeric purities or the analysis of environmental samples can be mentioned as examples. In these cases, it is necessary to improve the detection sensitivity achieved by on-line UV detection, which is the most common detection mode used in CE. During the sample preparation [106] it is possible to concentrate the analyte by treating the sample by liquid–liquid extraction, solid-phase extraction, liquid-phase microextraction, filtration, etc., prior to the injection in the CE system. Also, it is possible to use different on-line preconcentration techniques, as described in detail in chapter 3. However, the use of alternative detection systems to on-line UV detection has helped to achieve the most spectacular and promising sensitivity enhancements. The combination of preconcentration techniques with alternative detectors makes it possible to achieve the highest sensitivity enhancements (ultratrace levels).

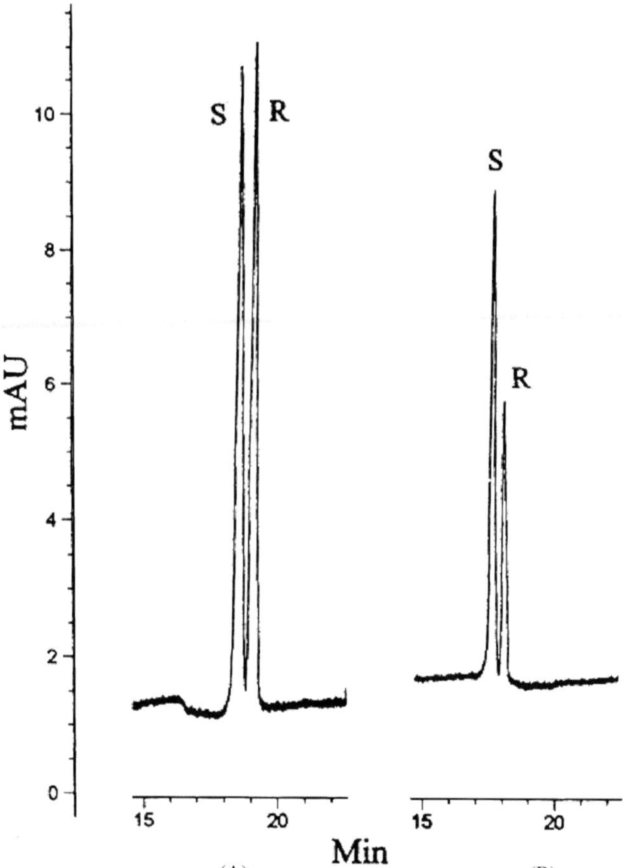

Fig. 13.6. Electropherograms of racemic and non-racemic metoprolol. CE conditions: fused-silica capillary, $l_d = 50\,cm$, $l_t = 64.5\,cm$ with 50 μm i.d.; separation BGE, 30 mM (+)-*S*-camphorsulfonate, 0.2 mM Tween 20, and 1 M acetic acid in methanol/acetonitrile; capillary temperature, 25°C; run voltage, 30 kV; injection at 50 mbar for 3 s. UV detection at 214 nm. Reprinted from Ref. [98]. Copyright (1996), with permission from Elsevier.

The usefulness of off-line sample treatment prior to CE analysis will be widely illustrated in section 13.4, which is devoted to the applications of CE in chiral analysis. Therefore, only an updated overview on the applications of on-line preconcentration techniques as well as the use of alternative detection systems to on-line UV detection in the field of chiral CE is briefly presented here. Furthermore, these two last alternatives present an additional interest when developing automatized processes.

13.3.1 On-line preconcentration techniques employed in chiral analysis by CE

Owing to the interest of on-line systems as a means to ease the automatization of processes, a strong trend in analytical chemistry, on-line preconcentration of enantiomers in CE is of great interest. Surprisingly, only a few works dealing with on-line preconcentration in chiral analysis by CE have been published. This seems to be due to the more elaborate approach required (such as the use of a zone of the capillary filled with packing material or the coupling of two capillaries to perform a preconcentration by isotachophoresis (ITP)) and the difficulty of keeping the enantioresolution of a chiral analyte after an on-line preconcentration based on special injection techniques (stacking or sweeping). Table 13.5 shows the main applications performed in chiral CE using on-line preconcentration techniques. Thus, the calcium channel blocker Verapamil has been determined in human plasma after ultrafiltration and an on-line preconcentration using a capillary zone filled with packing material, obtaining an LOD of 5×10^{-9} M [107]. Likewise, the detection of $\sim 10^{-8}$ M of the enantiomers of tryptophan and norleucine labeled in urine and complex ionic matrices has been performed by coupling an ITP system to a CE system [108,109].

The enhancement of the sensitivity in the determination of enantiomers separated by CE employing special injection techniques such as stacking or sweeping techniques is not well established, although a few works have been published on this subject. Thus, the detection of drugs in plasma samples after on-line preconcentration by stacking has been reported [110,111] to achieve LODs in the 10^{-8}–10^{-9} M range. However, only preliminary results on the stacking and sweeping preconcentration of a chiral fungicide (LOD$\sim 10^{-5}$ M) as well as the acetonitrile stacking of naphthyl and dansylated amino acid enantiomers have been reported recently [112,113]. To illustrate the latter, Fig. 13.7 shows the effect of the sample matrix on the stacking and separation of *R*/*S*-1,1′-binaphthyl diyl hydrogen phosphate and *R*/*S*-1,1′-bi-2-naphthol. It can be observed that only sample matrices containing acetronitrile make it possible to maintain the enantioresolution of the chiral compounds tested.

TABLE 13.5

Analytes, detection conditions, LODs and applications performed using on-line preconcentration in chiral CE

On-line preconcentration	Analyte	Detection	LOD (M)	Application/comment	Reference
Preconcentration zone filled with packing material	Verapamil	UV at 200 nm	5×10^{-9}	Determination of the unbound concentration of Verapamil enantiomers in human plasma	[107]
ITP	l-tryptophan	UV at 200 nm	$\sim 10^{-8}$	Detection and quantitation of the analyte in urine and complex ionic matrices	[108]
ITP	l-tryptophan and 2,4-dinitrophenyl-labeled norleucine	UV at 200 nm	$\sim 10^{-8}$	Detection of the analytes in urine, complex ionic matrices, and matrices with mixtures of the enantiomers at significantly different concentrations	[109]
Stacking	Isoproterenol	Amperometric	$\sim 3 \times 10^{-8}$	Analysis of plasma to establish the pharmacokinetic of this catecholamine	[110]
Stacking (FESI)	Adrenoreceptor antagonist	UV at 200 nm	$\sim 4 \times 10^{-9}$	Analysis of plasma samples	[111]
Stacking (SRMM)	Triadimenol	UV at 200 nm	$\sim 10^{-5}$	Preliminary study of the on-line preconcentration of the fungicide	[112]
Sweeping	Triadimenol	UV at 200 nm	$\sim 10^{-5}$	Preliminary study of the on-line preconcentration of the fungicide	[112]
Stacking	Naphthyl enantiomers and dansylated amino acid	UV at 220 nm	—	Preliminary study of the on-line preconcentration by stacking with acetonitrile	[113]

FESI, field-enhanced sample injection; SRMM, stacking with reverse migrating micelles.

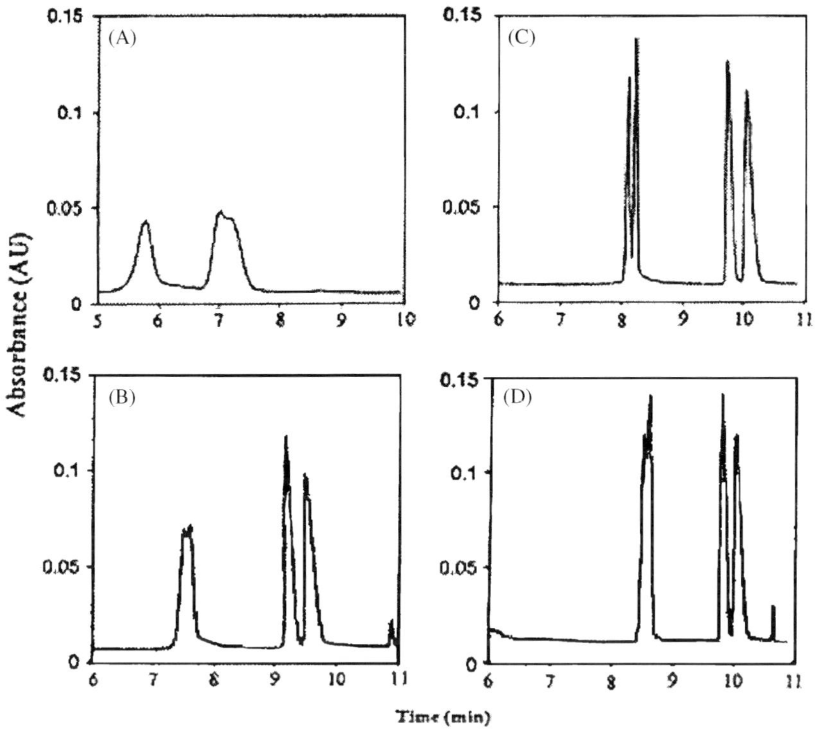

Fig. 13.7. Electropherograms showing the effect of the sample matrix on the stacking and separation of *R*/*S*-1,1′-binaphthyl diyl hydrogen phosphate and *R*/*S*-1,1′-bi-2-naphthol (2×10^{-2} mg/ml each enantiomer). Sample dissolved in (A) distilled deionized water, (B) 1% NaCl, (C) 3% NaCl, (D) acetonitrile–3% NaCl mixture (2:1 v/v). CE conditions: capillary, $l_d = 40$ cm, $l_t = 50.5$ cm with 50 μm i.d.; separation buffer: 50 mM sodium cholate, 10 mM Na_2HPO_4/6 mM $Na_2B_4O_7$ (pH 9.0) and 20% acetonitrile; run voltage, 25 kV; sample injection size = 9% capillary volume. UV detection at 220 nm. Reprinted from Ref. [113]. Copyright (2003), with permission from John Wiley & Sons, Inc.–Wiley-VCH Verlag GmbH.

13.3.2 Alternative detection systems to on-line UV detection employed in chiral analysis by CE

In on-line UV detection (LODs~10^{-5} M are usually obtained), a low enhancement of the sensitivity (~3–10 times) can be observed by using special designs of detection windows (bubble and Z-cells) (see chapter 5) [114]. The main problem of these designs is the decrease in enantioresolution observed when the separated bands of the enantiomers cross the special detection window. In addition, Z-cells may cause problems of buffer leak and current instability, which explains their limited use.

To overcome the relatively poor concentration detection sensitivity associated with on-line UV detection, the more effective option is the choice of alternative detection systems such as other optical detectors, MS detectors or electrochemical detectors.

The *alternative optical detectors* to UV detection developed and used until now in chiral analysis by CE are laser-induced fluorescence (LIF) and phosphorescence detection. Although *LIF detection* is limited to compounds with native fluorescence or derivatized with a fluorophore, it provides low LODs and additional selectivity. Table 13.6 shows the most recent applications of chiral CE with LIF detection. Very favorable LODs ranging from 0.4×10^{-9} to 10^{-8} M have been obtained for the detection of labeled amino acids [115–120]. A high sensitivity in the detection of chiral amino acids in orange juices has been reported. Figure 13.8 reveals the existence of d-Asp in an orange concentrate from Brazil, where only l-amino acids were expected [120]. The detection of about 10^{-7} M of phenprocoumon [121] and tramadol and its metabolite [122] in urine samples has been performed with good selectivity. Much lower detection limits ($\sim 10^{-9}$ M) have been obtained for phenoxy acid herbicides labeled with 7-aminonaphthalene-1,3-disulfonic acid (ANDSA) [123]. Recently, *phosphorescence detection*, which is a novel detection technique in CE, has been used in chiral CE. Although the applicability range of quenched phosphorescence detection for neutral molecules seems to be fairly wide, only one application has been published so far. LODs (10^{-7} M) more than three orders of magnitude higher than those obtained by UV detection have been achieved for the neutral camphorquinone [124]. An interesting example where this molecule is enantioselectively degraded by yeast is illustrated in chapter 7.

The hyphenation of chiral CE with *MS detection* is increasing in recent years as reflected in literature (see the literature referenced in Table 13.6, where most of the applications of chiral CE with MS detection have been collected). This detection technique is universal, selective and makes it possible obtain structural information. However, it presents an important problem associated with the introduction of non-volatile chiral selectors into the mass spectrometer detector, which affects the sensitivity and stability of the mass spectrometer. Although there are some papers where the chiral selector is introduced directly into the mass spectrometer, producing satisfactory results [125–127],

TABLE 13.6

Analytes, detection conditions, LODs and applications performed using alternative detection systems to on-line UV detection in chiral CE

Analyte	Detection	LOD (M)	Application/comment	Reference
Fluoxetine and norfluoxetine	UV at 195 nm (zeta-shaped cell)	$\sim 3 \times 10^{-6}$	Analysis of clinical serum and plasma samples of patients under depression therapy	[114]
CBI-amino acids (Thr, Asp, Ile, Tyr, Phe)	LIF: $\lambda_{exe} = 442$ nm, $\lambda_{em} = 490$ nm	$\sim 0.4 \times 10^{-9}$	Detection and separation of the derivatized amino acids	[115]
CBI-amino acids (Ala, Glu, Val, Phe, Tyr, Trp)	Diode LIF: $\lambda_{exc} = 635$ nm	$\sim 2 \times 10^{-8}$	Separation and detection of analytes by CD-capillary gel electrophoresis	[116]
CBI-amino acids (Tyr, Ile, Asp, Met, Trp, Phe)	LIF: $\lambda_{exc} = 457$ nm	3×10^{-8}	Assessment of enantiomeric purity for hydrolyzed synthetic peptides	[117]
CBI-tryptophan	LIF: $\lambda_{exc} = 457$ nm	3×10^{-8}	Analysis of biological samples	[118]
CBI-aspartic acid	LIF: $\lambda_{exc} = 457.9$ nm, $\lambda_{em} = 495$ nm	$\sim 10^{-8}$	Study of the occurrence and postnatal changes of d-Asp in rat brain	[119]
FITC-amino acids (Pro, Asp, Ser, Asn, Glu, Ala, Arg)	LIF: $\lambda_{exc} = 488$ nm, $\lambda_{em} = 520$ nm	$\sim 3 \times 10^{-9}$	Separation, detection and quantitation of amino acids in orange juices and orange concentrates	[120]
Phenprocoumon	LIF: $\lambda_{exc} = 325$ nm, $\lambda_{em} = 405$ nm	$\sim 7 \times 10^{-7}$	Analysis of urine samples of a patient treated with the analyte and other drugs	[121]
Tramadol and *O*-demethyl tramadol glucuronide	LIF $\lambda_{exc} = 257$ nm	$\sim 2 \times 10^{-7}$	Direct assay of tramadol in human urine and quantitation of its metabolite	[122]
ANDSA-phenoxy acid herbicides (silvex, mecoprop, dichlorprop, 2,4-CPAA, 2,4,5-CPAA, PPA, 2-CPPA, 3-CPPA and 4-CPPA)	LIF: $\lambda_{exc} = 325$ nm, $\lambda_{em} = 420$ nm	$\sim 10^{-9}$	Detection and chiral and achiral separation of phenoxy acid herbicides derivatives	[123]
Camphorquinone	Quenched phosphorescence	$\sim 7 \times 10^{-7}$	Detection, enantioseparation and monitoring the stereoselective biodegradation of the analyte by yeast	[124]
Terbutaline and ephedrine	ISP-MS	$\sim 10^{-7}$	Direct coupling of CE to ISP-MS. Detection of the enantiomers in a spiked urine sample	[125]
Methylenedioxy-amphetamines	ISP-MS	$\sim 2 \times 10^{-5}$	Direct coupling of CE to ISP-MS. Separation of the enantiomers and racemic amphetamines	[126]
Dichorprop, fenoprop and mecoprop	ESI-MS	—	Preliminary study examining the direct coupling of CE with ESI-MS	[127]

continued

TABLE 13.6 *(continued)*

Analyte	Detection	LOD (M)	Application/comment	Reference
Ropivacaine	ESI-MS	—	Special set-up with a coupled capillary system with the possibility of voltage switching. Separation and detection of the enantiomers of this basic drug	[128]
Anionic drugs (carprofen, flurbiprofen, ketoprofen, naproxen, etodolac and its metabolites, ibuprofen and its metabolites)	ESI-MS	—	Preliminary study showing the enantioseparation and detection of drugs and their metabolites by CE–MS	[129]
Terbutaline, ketamine and propranolol	ESI-MS	$\sim10^{-4}$	Investigation of the conditions for the enantioseparation of these drugs obtaining an adequate sensitivity by CE–MS	[130]
Basic drugs (etilefrine, mianserine, dimethindene and chlorpheniramine), tropic acid	ESI-MS	—	Enantioseparation by CE–MS using the counter-migration principle of analyte and chiral selector	[131]
Camphorsulfonic acid, tropic acid, ibuprofen, ketoprofen and warfarin	ISP-MS	$\sim10^{-5}$	Enantioseparation by CE–MS using the partial filling technique	[132]
Bupivacaine and ropivacaine	ESI-MS	$\sim10^{-4}$	Enantioseparation of these anaesthetics by CE–MS using the partial filling technique	[133]
Tramadol and its main phase I metabolites	ESI-MS	$<5\times10^{-7}$	Enantioseparation, detection and stereoselective analysis of plasma by CE–MS using the partial filling technique	[134]
3-Aminopyrrolidine, α-amino-ε-caprolactam and cycloserine	ESI-MS	$<5\times10^{-4}$	Enantioseparation of the primary amines by CE–MS using the partial filling technique	[135]
Adrenoreceptor antagonist	MS-MS	$\sim2\times10^{-8}$	Study of the effect of operating parameters on the enantioseparation and sensitivity of the analyte by CE–MS-MS using the partial filling technique	[136]
Clenbuterol	MS	$\sim10^{-6}$	Enantioseparation of the analyte by CE–MS using the partial filling technique. Analysis of plasma samples	[137]

Amphetamines, methadone, venlafaxine and tropane alkaloids	ESI-MS	$\sim 10^{-8}$	Enantioseparation of drugs and their metabolites by CE–MS using the partial filling technique. Determination of methadone and its metabolites in serum samples	[138]
Bupivacaine, mepivacaine, ketamine and prilocaine	ESI-MS	$< 10^{-4}$	Enantioseparation of these anesthetic drugs by CE–MS using the counter-current and partial filling technique	[139]
1,1′-binaphtol	ESI-MS	$\sim 10^{-5}$	Optimization of the enantioseparation of the analyte by MEKC–MS using a micelle polymer	[140]
Amino acids and neurotransmitters	ESI-MS	$\sim 10^{-9}$	Enantioseparation of non-derivatized amino acids and neurotransmitters by CE–MS using a crown ether as chiral selector and buffer. Analysis of lysed red blood cells	[141]
Amphetamines	ESI-MS	$\sim 2 \times 10^{-7}$	Analysis of urine samples of amphetamine addicts	[142]
Three basic drugs	ESI-MS-MS	$\sim 10^{-5}(10^{-8})$	Study of the variables for the enantioseparation and detection of drugs by CE–MS with a homemade interface. Detection and quantitation of its enantiomers *in vivo*	[143]
Amphetamines-type stimulants	ESI-MS-MS	$\sim 10^{-3}$	Enantioselective identification of the analytes using reversed-polarity CE–MS. Detection and identification of the impurities of l-ephedrine and d-pseudoephedrine	[144]
Warfarin and coumachlor	ESI-MS	$\sim 10^{-7}$	Enantioseparation and detection of the structurally similar enantiomers. Assay of warfarin in human plasma samples	[87]
Isoproterenol	Amperometric	$\sim 3 \times 10^{-8}$	Analysis of plasma to establish the pharmacokinetic of this catecholamine	[110]
Amine derivatives [*threo*-2-amino-1-(4-nitrophenyl)-1,3-propanediol and *threo*-2-(dimethylamino)-1-(4-nitrophenyl)-1,3-propanediol]	Amperometric	$< 10^{-5}$	Determination of the enantiomeric purity of laboratory synthetic batches	[146]

CBI-amino acids, cyanobenzoisoindole-labeled amino acids after derivatization with naphthalene-2,3-dicarboxaldehyde; FITC-amino acids, fluorescein isothiocyanate labeled amino acids; ANDSA-phenoxy acids herbicides, 7-aminonaphthalene-1,3-disulfonic acid-labeled phenoxy acid herbicides; ISP, ion spray; ESI, electrospray; 2,4-CPAA, (2,4-dichlorophenoxy) acetic acid; 2,4,5-CPAA, (2,4,5-trichlorophenoxy)acetic acid; PPA, 2-phenoxypropionic acid; 2-CPPA, 2-(2-chlorophenoxy)propionic acid; 3-CPPA, 2-(3-chlorophenoxy)propionic acid; 4-CPPA, 2-(4-chlorophenoxy)propionic acid.

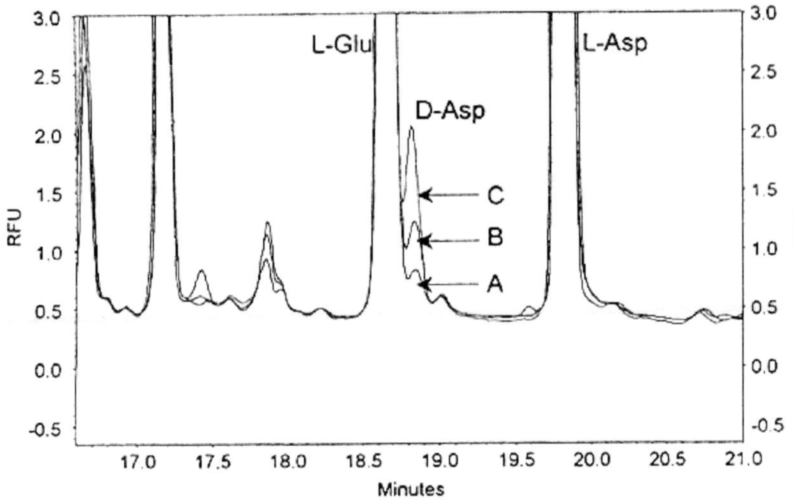

Fig. 13.8. Electropherograms of (A) an orange concentrate from Brazil (C1), (B) C1 plus 1.5×10^{-6} M d-Asp; and (C) C1 plus 4.4×10^{-6} M d-Asp. CE conditions: capillary, $l_d = 40$ cm, $l_t = 50$ cm with 50 μm i.d.; separation buffer, 100 mM sodium tetraborate, 30 mM SDS at pH 9.4 with 20 mM β-CD; capillary temperature, 15°C; run voltage, 20 kV; injection at 0.5 psi for 5 s. LIF detection at 488 nm (excitation wavelength) and 520 nm (emission wavelength). Reprinted with permission from Ref. [120]. Copyright (2002), American Chemical Society.

most of the papers published to date avoid the introduction of the chiral selector into the mass spectrometer [87,128–144].

The partial filling technique or the counter-migration technique (based on the significantly higher counter-current electrophoretic mobility of charged chiral selectors that do not enter into the mass spectrometer) are the two techniques used until now to preclude the introduction of the non-volatile chiral selectors into the mass spectrometer. In chiral analysis by CE with MS detection, the partial filling technique has been the most widely used (see Table 13.6), although the combination of counter-current migration and partial filling technique seems to provide very successful results. With MS detection, LODs ranging from 10^{-9} to 10^{-3} M (see Table 13.6) have been obtained. Although possible, nowadays it is not easy to improve the sensitivity of MS with respect to UV detection for highly absorbent compounds. There are two main reasons: the low volume injected in the CE system (nl) and the dilution produced by the sheath liquid required for the

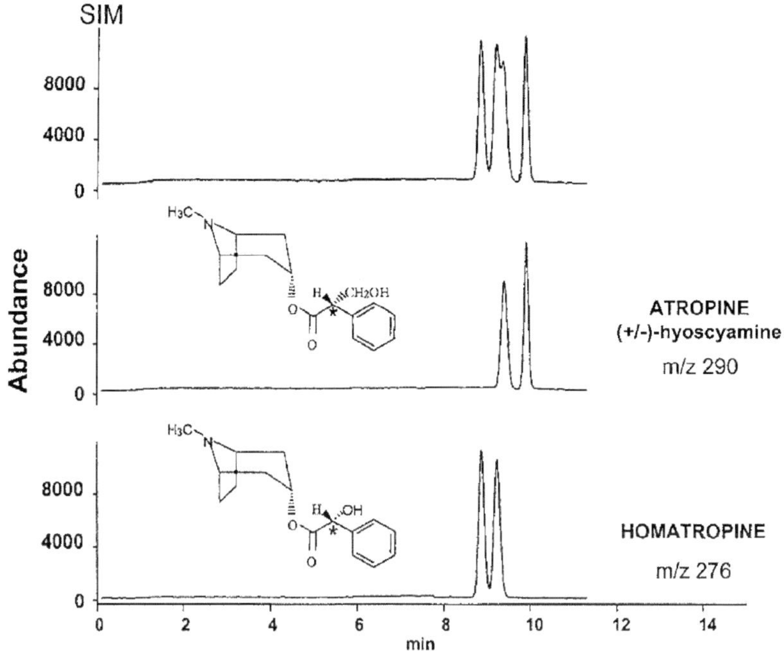

Fig. 13.9. Electropherograms showing the enantiomeric separation of a standard solution of atropine and homatropine. CE conditions: PVA-coated capillary, $l_d = 61.5$ cm, $l_t = 70$ cm with 50 µm i.d.; separation buffer, 30 mM ammonium acetate at pH 7.0 with 2 mg/ml sulfated β-CD; run voltage, 25 kV; temperature, 25°C; injection at 50 mbar for 10 s. ESI-MS conditions: SIM positive-ion mode; capillary voltage, 4000 V; fragmentor voltage, 120 V; drying gas flow rate and temperature, 6 l/min and 150°C, respectively; nebulizer pressure, 15 psi; sheath liquid, 0.5% formic acid in water/isopropanol (20:80, v/v) at 3 µl/min. Reprinted from the Ref. [138]. Copyright (2001), with permission from John Wiley & Sons, Inc.–Wiley-VCH Verlag GmbH.

nebulization. A review on the modes and applications of chiral CE–MS has been published recently [145]. As an example, Fig. 13.9 displays the chiral separation of atropine and homatropine enantiomers by CE–ESI-MS. For these compounds, LODs were in the ppb level, which correspond to an improvement of about 1000 in comparison with UV detection. Interestingly, this important gain in sensitivity is mainly attributed to the high proton affinity of tropane alkaloids as well as their low molar absorptivity [138]. Table 13.6 also shows that the range of applications of CE–MS in chiral analysis has increased during the last few years. Thus, the chiral analysis of biological samples

[137,138,141–143], the identification of drugs and its metabolites [138] or the determination of enantiomeric impurities of drugs [144] have been performed using MS as detection system. It is also important to emphasize that the direct coupling of CEC to MS is very promising and has been recently applied to detect a chiral drug in biological fluids [87].

Electrochemical detection is the detection technique less used in CE and also in chiral CE. Although it provides higher selectivity and sensitivity than classical UV detection, its applicability is limited to electroactive compounds. Table 13.6 groups the applications performed in chiral CE using amperometric detection. LODs ranging from 3×10^{-8} to 10^{-5} M have been reported. The pharmacokinetic of isoproterenol has been established analyzing plasma samples [110] and the enantiomeric purity of laboratory synthetic batches has been determined for amine derivatives [146]. An interesting example is illustrated in Fig. 13.10, which shows the amperometric detection of a 1:1 standard mixture of the d and l enantiomers of (A) *threo*-2-amino-1-(4-nitrophenyl)-1,3-propanediol and (B) *threo*-2-(dimethylamino)-1-(4-nitrophenyl)-1,3-propanediol using a Cu disk electrode of 220 μm and a working potential of 675 mV [146].

13.4 APPLICATIONS OF CHIRAL ANALYSIS BY CE

From the first enantiomeric separation by CE reported in 1985 [147], the use of CE in chiral separations has grown constantly as shown by the exponential number of published papers and the increasing use of CE in the industry with this aim, especially in the pharmaceutical sector. Currently, there are more than 1000 publications on chiral CE that describe its fundamental principles, selectivity manipulation, developments and applications. There are special issues of international journals [14,148–151] and even a book [13] devoted to chiral analysis by CE and/or CEC. A brief overview of the different applications of CE in chiral analysis will be presented.

13.4.1 Pharmaceutical and biomedical analysis

Pharmaceutical and biomedical analysis is the main application field in chiral CE. This is why only the most recent applications of CE in chiral analysis will be presented here. Some review articles will be also cited in order to provide additional information on the CE applications

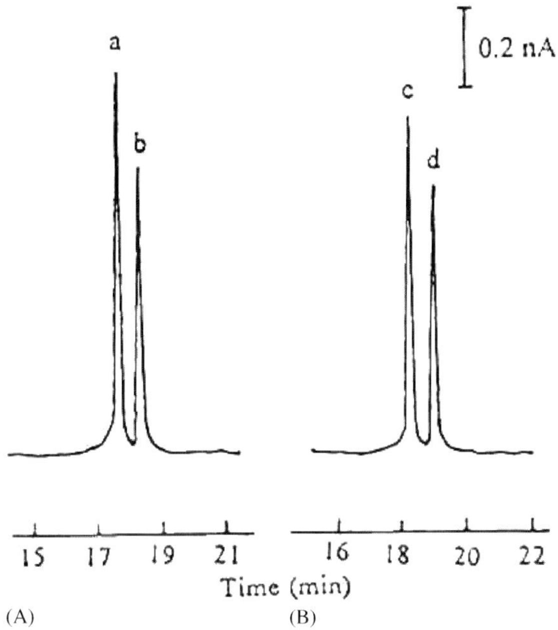

Fig. 13.10. Electropherograms showing the enantiomeric separation of a 1:1 standard mixture of the d and l enantiomers of (A) *threo*-2-amino-1-(4-nitrophenyl)-1,3-propanediol and (B) *threo*-2-(dimethylamino)-1-(4-nitrophenyl)-1,3-propanediol. CE conditions: fused-silica capillary, $l_t = 80\,cm$ with $25\,\mu m$ i.d.; separation buffer, 0.10 M NaOH containing 12 mM β-CD; run voltage, 20 kV; temperature, 30°C; injection at 20 kV for 10 s. Amperometric detection using a Cu disk electrode of 220 µm and a working potential of 675 mV. Reprinted with permission from Ref. [146]. Copyright (1998), American Chemical Society.

published until now. These applications may be divided into the following categories: (i) enantiomeric separation of drugs, (ii) determination of the enantiomeric purity of drugs, (iii) quantitative chiral analysis of pharmaceutical preparations and biological samples, (iv) validation of chiral methods for the analysis of drugs and (v) monitoring the stereoselectivity of metabolic processes. These topics are briefly discussed below with reference to the most recent literature.

13.4.1.1 Enantiomeric separation of drugs

Many applications and developments in chiral CE involving drugs have been published in recent years. We do not intend to provide an overview of all the drugs enantioseparated by CE. This information may be acquired from some previous reviews [14–17,21,22,152–160] to which the reader can refer. Most of the papers related to the enantiomeric separation of drugs compare the ability of different chiral selectors to perform the separation of a particular chiral compound or study the possibility of separating different racemic analytes with a particular chiral selector. In addition, some studies on theoretical aspects have been developed in order to understand the interactions between the analytes and the chiral selector. Table 13.7 shows some recent chiral separations of different types of drugs. It can be observed that CDs are the favorite chiral selectors. In fact, they have enabled the development of fast enantiomeric separations such as those shown in Fig. 13.11. The sympathomimethic drugs norepinephrine and epinephrine, and the β-agonists clenbuterol, terbutaline and salbutamol (for which no baseline resolution was obtained) were enantiomerically separated in analysis times ranging from 2 to 3 min using the neutral CD permethylated-β-CD (PM-β-CD, with a degree of substitution (d.s.) $\sim$12–13) in acetate buffer at pH 5.

Several new chiral selectors have been introduced in chiral CE during the last few years. Thus, the micelle polymer polysodium *N*-undecanoyl-l-leucylvalinate (poly l-SULV) has provided good enantioseparations for a large number of neutral, anionic and cationic drugs [165]. Also, new CD derivatives such as heptakis(2-*N*,*N*-dimethylcarbamoyl)-β-CD [164], hydroxyethyl-β-CD [165], highly sulfonated-β-CD [143] and phosphated γ-CD [114] have been introduced and employed. In addition, dual CD systems have been used, providing additional selectivity and promising results for the separation of the enantiomers of related compounds in mixtures [114,122,143,178].

Another aspect of recent interest is the study of the migration order of the enantiomers [161]. These studies intend to provide a better understanding of chiral CE as well as its more effective application. A review showing the most important aspects of enantiomer migration order in CE has been published recently [181].

Finally, Table 13.7 shows that UV detection is the detection method preferred to achieve the enantiomeric separation of standard compounds of pharmaceutical interest.

13.4.1.2 Determination of the enantiomeric purity of drugs

The analysis of the stereochemical purity of compounds is of critical importance in chiral drug synthesis and development as well as the quality control of drug substances. While the enantiomeric purity, or

TABLE 13.7

Recent applications of chiral CE in pharmaceutical and biomedical analysis

Application	Analyte	Sample	Sample preparation	Chiral selector (conc.)	BGE	Detection	Reference
Enantiomeric separation	Labetadol and nadolol	Standard mixtures	DI	Sulfated β-CD (10 g/l)	Phosphate (pH 5–6)	UV at 200 nm	[161]
Enantiomeric separation	Ketoprofen, ibuprofen and fenoprofen	Standard mixtures	DI	TM-β-CD (50 mM)	Phosphate (pH 5)-tetraethyl-ammonium	UV at 253 nm	[162]
Enantiomeric separation	Glutethimide	Standard	DI	β-CD/γ-CD/SBE-β-CD/CM-β-CD	Phosphate (pH 5)	UV at 210 nm	[163]
Enantiomeric separation	Acid (ketoprofen and others) and basic drugs (mianserin, cyclopentolat, fluoxetine, nisoxetine, verapamil, aminoglutethimide, dimethindene, doxapram hydroxyzine and meclozine)	Standard	DI	Heptakis(2-*N*,*N*-dimethylcarbamoyl)-β-CD (36 mM)	Phosphate (pH 7/3)	UV at 220 nm	[164]
Enantiomeric separation	58 Compounds (binaphthyl, paveroline, and coumarinic derivatives; benzodiazepinones; barbiturates; PTH-amino acids among others)	Standard mixtures	DI	Poly l-SULV (5–50 mM)	Phosphate/borate (pH 7–9)	UV at 220 nm	[165]
Enantiomeric separation	*N*-propionyl-6,7-dimethoxy-2-aminotetralin	Standard	DI	Hydroxyethyl-β-CD (90 mM)-STDC (150 mM)	Borate (pH 9)	UV at 214 nm	[166]
Enantiomeric separation—validation for enantiomeric purity determination	M3 antagonist	Standard mixtures of enantiomers	Filtration	Highly sulfated β-CD (2%)	Phosphate (pH 2.5)–acetonitrile (1%)	UV at 200 nm	[167]
Enantiomeric separation—validation for enantiomeric purity determination	Melagatran and ximelagatran	Standard mixtures of enantiomers	DI	DM-β-CD (30 mM)	Phosphate (pH 1.8)–methanol (20%)	UV at 200 nm	[168]
Enantiomeric separation—validation for enantiomeric purity determination	Tetrahydro-naphthalenic derivatives	Standard mixtures of enantiomers	DI	Highly sulfated β-CD (2.5%)	Phosphate (pH 2.5)	UV at 210 nm	[169]
Enantiomeric separation—validation for enantiomeric purity determination	Amine derivatives [*threo*-2-amino-1-(4-nitrophenyl)-1,3-propanediol; *threo*-2-(dimethylamino)-1-(4-nitrophenyl)-1,3-propanediol]	Laboratory synthetic batches	DI	β-CD (12 mM)	NaOH (0.1 M)	Amperometric	[146]
Enantiomeric separation—validation for enantiomeric purity determination and analysis of degradation products	Cizolirtine and its degradation products	Standards and degraded solutions (pH, oxidation)	DI	HP-β-CD (60 mM)	Tetraborate (pH 9.2)–SDS–butanol (5%)	UV at 205 nm	[170]
Validation for enantiomeric purity determination	Ketoprofen	Pharmaceutical preparation	DI	TM-β-CD (75 mM)	Acetate (pH 5)	UV at 254 nm	[171]
Enantiomeric separation—validation for quantitative analysis of pharmaceutical preparations	Terbutaline, epinephrine, norepinephrine, clenbuterol and salbutamol	Pharmaceutical preparation	DI	Permethyl-β-CD (30 mM)	Acetate (pH 5)	UV at 230 nm	[172]

continued

TABLE 13.7 *(continued)*

Application	Analyte	Sample	Sample preparation	Chiral selector (conc.)	BGE	Detection	Reference
Validation for quantitative analysis of pharmaceutical preparations	Salbutamol	Pharmaceutical preparations	DI	CM-β-CD (13.1 mM)	Acetate (pH 5)	UV at 230 nm	[173]
Enantiomeric separation—quantitative analysis of pharmaceutical preparations	Amphetamines-type stimulants (norpseudoephedrine, ephedrine, pseudoephedrine, norephdrine, amphetamine, methamphetamine, methylenedioxyamphetamine, methylenedioxymethamphetamine and methylenedioxyethylamphetamine)	Illicit methanphetamine seizures	Filtration	Highly sulfated-γ-CD (2.5 mM)	Formate (pH 3.1)	ESI-MS-MS	[145]
Enantiomeric separation—quantitative analysis of biological samples	Warfarin and coumachlor	Human plasma	SPE	CEC packed with (3R,4S)-Whelk-O1 phase	Acetonitrile (70% v/v) with NH_4OAc (pH 4)	ESI-MS	[87]
Quantitative analysis of biological samples	Verapamil	Human plasma	Ultrafiltration	TM-β-CD (25 mM)	Run buffer (pH 2.5)	UV at 200 nm	[107]
Quantitative analysis of biological samples	Isoproterenol	Human plasma	Microdialysis	M-β-CD (mM)	Lithium acetate (pH 4.75)	Amperometric	[110]
Enantiomeric separation—validation for quantitative analysis of biological samples	Disopyramide and mono-*N*-dealkyl-disopyramide	Human plasma	Sample clean-up/LLE	Sulfated β-CD (0.2%)	Acetate (pH 5)	UV at 214 nm	[175]
Enantiomeric separation—validation for quantitative analysis of biological samples	Praziquantel and *trans*-4-hydroxy-praziquantel	Human plasma	LLE	Sulfated β-CD (2%)–SDC (20 mM)	Borate (pH 10)	UV at 214 nm	[176]
Enantiomeric separation—validation for quantitative analysis of biological samples	Citalopram and desmethyl-citalopram	Human plasma	LPME	Sulfated-β-CD (1%)	Phosphate (pH 2.5)–acetonitrile (12%)	UV at 200 nm	[179]
Quantitative analysis of biological samples	Adrenoreceptor antagonist	Plasma	SPE	HP-β-CD (3.5 mM)	Formic acid–ammonia (pH 4)	UV at 200 nm	[111]
Quantitative analysis of biological samples	Fluoxetine and norfluoxetine	Plasma and serum	LLE	DM-β-CD (0.5 mg/ml)+Phosphated-γ-CD (0.6 mg/ml)	Phosphate (pH 2.5)	UV at 195 nm	[114]
Enantiomeric separation—quantitative analysis of biological samples	3-Aminopyrrolidine, α-amino-ε-caprolactam, cycloserine	Plasma	LLE	$18C6H_4$ (2–5 mM)	Ammonium formate (pH 4)	ESI-MS	[135]
Enantiomeric separation—quantitative analysis of biological samples	Amphetamines/methadone, venlafaxine/tropane alkaloids	Plasma	SPE	HP-β-CD (24 mg/ml)/CM-β-CD (1–2 mg/ml)/sulfated-β-CD (2 mg/ml)	Ammonium acetate (pH 3–7)	ESI-MS	[139]
Quantitative analysis of biological samples	Phenprocoumon	Urine	DI	α-CD (2 mg/ml)	Phosphate-TEA (pH 5.4)	LIF: λ_{exc} = 325 nm, λ_{em} = 405 nm	[121]
Quantitative analysis of biological samples	Tramadol and *O*-demethyl tramadol glucuronide	Urine	DI	CM-β-CD (40 mg/ml)+M-β-CD (22.5 mg/ml)	Borax (pH 10.0)	LIF: λ_{exc} = 257 nm	[122]

Quantitative analysis of biological samples	Amphetamines	Urine	SPE	β-CD (3 mM)+DM-β-CD (1–2 mg/ml)	Formic acid (pH 2.2)	ESI-MS	[142]
Enantiomeric separation—quantitative analysis of biological samples	Ofloxacin	Urine	DI	Sulfated β-CD (0.35 mM)	Phosphate (pH 2.5)	UV at 291 nm	[177]
Enantiomeric separation—quantitative analysis of biological samples	l-tryptophan and/or 2,4-dinitrophenyl-labeled norleucine	Urine and complex ionic matrices	ITP	α-CD (80 mM)	Borate (pH 9)	UV at 200 nm	[108,109]
Enantiomeric separation—quantitative analysis of biological samples	CBI-tryptophan	Urine, cerebrospinal fluid, rat brain tissue, aplysia ganglia	Centrifugation	HP-γ-CD (15 mM)	Borate (pH 9)–SDS	LIF: $\lambda_{exc} = 457$ nm	[118]
Enantiomeric purity—Quantitative analysis of biological samples	CBI-amino acids (Tyr, Ile, Asp, Met, Trp and Phe)	Rat brain	Hydrolysis	β-CD (20 mM)	Borate (pH 9)–SDS-methanol (15%)	LIF: $\lambda_{exc} = 457$ nm	[117]
Quantitative analysis of biological samples	CBI-aspartic acid	Rat brain	Homogenization and centrifugation	β-CD (20 mM)	Borate (pH 9)–SDS–methanol	LIF: $\lambda_{exc} =$ 458 nm, $\lambda_{em} = 495$ nm	[119]
Enantiomeric separation—quantitative analysis of biological samples	Non-derivatized amino acids (including Asp, Gln, Trp); neurotransmitters (dopamine, serotonin and norepinephrine)	Lysed red blood cells	DI	18C6H$_4$ (30 mM)	18C6H$_4$	ESI-MS	[142]
Enantiomeric separation—quantitative analysis of biological samples	Three basic drugs	*In vivo* samples	LLE	Highly sulfonated-β-CD (0.1–0.7%)	Acetic acid–ammonium acetate	ESI-MS-MS	[143]
Enantiomeric separation—validation for enantiomeric purity determination	Naproxen and methyl naproxen	Enzyme-catalyzed real sample	DI	Sulfated β-CD (3.5%)+HP-β-CD (2%)	Tris-phosphate (pH 5.5)	UV at 210 nm	[178]
Enantiomeric separation—quantitative analysis of biological samples monitoring the stereoselectivity of metabolic processes	Ibuprofen	Human plasma	LLE	Sulfated β-CD (2%)	Phosphoric acid-tetraethylammonium (pH 2.6)	UV at 220 nm	[174]
Enantiomeric separation—validation for quantitative analysis of biological samples monitoring the stereoselectivity of metabolic processes	Ketoprofen	Human serum	LLE	TM-β-CD (50 mM)	Phosphate-triethanolamine (pH 5)	UV at 253 nm	[180]

DI, direct injection of the appropriate diluted solution; SPE, solid-phase extraction; LLE, liquid–liquid extraction; LPME, liquid-phase microextraction; PTH-amino acids, phenylthiohydantoin-labeled amino acids; other abbreviations as in Table 13.6.

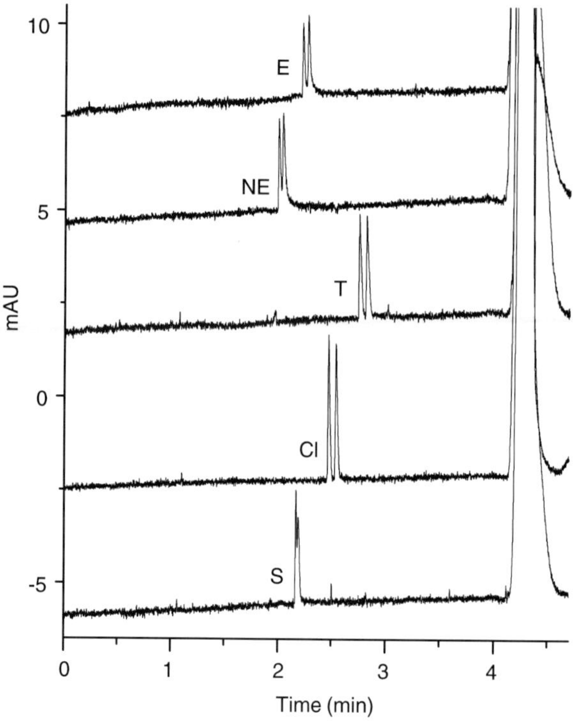

Fig. 13.11. Electropherograms showing the enantiomeric separation of racemic mixtures of epinephrine (E), norepinephrine (NE), terbutaline (T), clenbuterol (Cl) and salbutamol (S). CE conditions: fused-silica capillary, l_d = 25 cm, l_t = 33.5 cm with 50 μm i.d.; separation buffer, 25 mM acetate (pH 5) containing 30 mM permethyl-β-CD; run voltage, 20 kV; temperature, 15°C; injection at 30 mbar for 2 s sample followed by 30 mbar for 2 s buffer. UV detection at 230 nm. Reprinted from Ref. [172]. Copyright (2001), with permission from John Wiley & Sons, Inc.–Wiley-VCH Verlag GmbH.

enantiomeric excess (ee), is often measured in chemical synthesis, an enantiomeric impurity (ei) is usually quantified in analytical chemistry.

The ee in percentage is calculated by the following equation:

$$ee = \frac{R - S}{R + S} \times 100$$

where R and S are the concentration or peak area of the enantiomers. Logically, an ee value of 100 stands for an enantiomerically pure substance and 0 for a racemic mixture.

The enantiomeric impurity in percentage of the S-enantiomer in R-enantiomer, for example, can be calculated by

$$\mathrm{ei} = \frac{S}{R+S} \times 100$$

where ei gives the percentage of one enantiomer in the mixture.

Nowadays, a high percentage of chiral drugs are commercialized as pure enantiomers, and the determination of their purity is essential. The International Conference on Harmonization (ICH) guidelines on impurities (Topics Q3A and B) [182] exclude enantiomeric impurities from their scope, but the same guiding principle is expected to apply as for achiral drugs. It defines certain thresholds for the content of impurities above which they should be identified and quantified. These thresholds have recently been revised (February 2002 and 2003) establishing that for drug substances where the maximum daily dose (MDD) is 2 g/day or below, impurities must be reported if they are present above 0.05%, identified if above 0.10% and quantified if above 0.15%. For these requirements the sensitivity of detection has to be taken into consideration [184–186]. The most recent applications of CE to the determination of the enantiomeric purity of drug substances are mentioned in Table 13.7. The development and validation of a chiral method to determine the enantiomeric purity of an M3 antagonist [167], the new drug melagatran and its prodrug (ximelagatran) [168], tetrahydronaphthalenic derivatives [169] and cizolirtine and its degradation products [170] have been performed in standard solutions. On the other hand, the determination of the enantiomeric impurity of pharmaceutical preparations subjected to a stability test [171], and of laboratory synthetic batches has also been achieved [146]. It is interesting to observe that most of the applications grouped in Table 13.7 related to the determination of the enantiomeric purity of drugs have been performed using UV detection. An interesting example has been reported by Blanco et al. [171]. They have detected as little as 0.04% of (R)-ketoprofen in (S)-ketoprofen using UV detection at 254 nm. Figure 13.12 shows the presence of the (R)-enantiomer in an oral solution containing the active enantiomer alone ((S)-ketoprofen) after 3 months at 30°C and 60% relative humidity analyzed during the course of a stability study. Results obtained revealed the presence of the enantiomeric impurity at levels from 0.04% to 0.10% in different batches of 24 samples of the pure enantiomer. On the other hand, the enantiomeric purity remained constant with time.

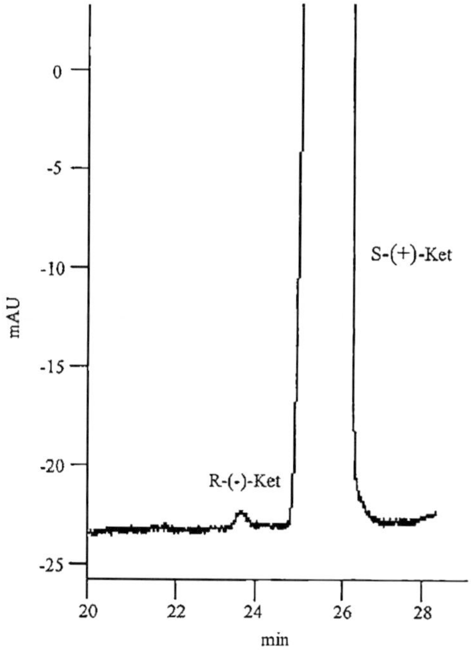

Fig. 13.12. Electropherogram showing the R impurity contained in an oral solution marked as pure (*S*)-ketoprofen after 3 months at 30°C and 60% relative humidity. CE conditions: fused-silica capillary, $l_d = 56$ cm, $l_t = 64.5$ cm with 50 μm i.d.; separation buffer, 60 mM acetic acid at pH 5 with 75 mM HP-β-CD; capillary temperature, 35°C; run voltage, 20 kV; injection at 50 mbar for 5 s. UV detection at 198 nm. Reprinted from Ref. [171]. With permission from Springer-Verlag.

13.4.1.3 Quantitative chiral analysis of pharmaceutical preparations and biological samples

The papers dealing with the chiral analysis of pharmaceutical preparations by CE have mostly focused on the development of quantitative methods of analysis and their validation [172,173] or to the determination of the enantiomeric purity of its active component [171,146] as shown in Table 13.7, which groups some recent examples of the analysis of pharmaceutical formulations or synthetic batches. As an example, Fig. 13.13 shows the electropherograms corresponding to a standard solution of salbutamol (0.045 mg/ml) and to five dilutions in dimethylsulfoxide of different pharmaceutical preparations containing 0.045 mg/ml of salbutamol sulfate. Although the absence of interfering peaks and similar enantiomeric resolutions for all samples was

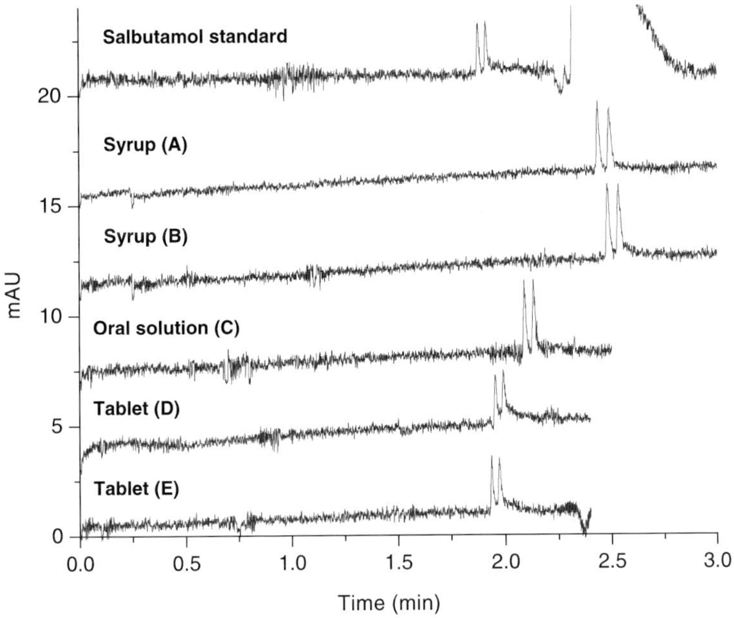

Fig. 13.13. Electropherograms showing the separation of salbutamol enantiomers in a standard solution (0.045 mg/ml) and in dilutions of five different pharmaceutical preparations (0.045 mg/ml in salbutamol). CE conditions: fused-silica capillary, $l_d = 25$ cm, $l_t = 33.5$ cm with 50 μm i.d.; separation buffer, 25 mM acetate at pH 5 with 13.1 mg/ml CM-β-CD; run voltage, 20 kV; temperature, 25 °C; injection at 30 mbar for 2 s sample followed by 30 mbar for 2 s buffer. UV detection at 230 nm. Reprinted from the Ref. [173]. Copyright (2003), with permission from John Wiley & Sons, Inc.–Wiley-VCH Verlag GmbH.

observed, the migration times were affected owing to the effect of matrix components on the EOF [173].

However, one of the most interesting applications of chiral CE is the analysis of drugs in biological matrices. Some recent applications in the analysis of biological samples are also included in Table 13.7. Other applications in this field not included in this table can be found in a recent review devoted to the enantioselective determination of drugs in body fluids by CE [157]. Table 13.7 shows that drugs have been analyzed mainly in plasma [87,107,110,111,114,135,139,174–176,179] and urine [121,122,143,177] although they have also been detected in *in vivo* samples [143] and enzyme catalyzed real samples [178]. The analysis of other analytes such as amino acids has been achieved in different biological fluids such as plasma, urine, cerebrospinal fluid, rat brain tissue, aplysia ganglia or lysed red blood cells [108,109,117–119,142]. An interesting example is the study of postnatal changes of aspartic acid (Asp) in rat brain. Figure 13.14 shows the electropherograms obtained for brain samples from a rat 1 day before birth and for a rat 90 days after birth. The peaks of d- and l-Asp (CBI-labeled amino acids) were well resolved with no interference from other amino acids or endogenous components in this biological sample matrix. It can be observed that LIF detection was employed in this work providing appropriate sensitivity for the analysis of small mass/volume samples. Owing to the relatively low sensitivity of UV detection in CE for drugs or amino acids in biological matrices, LIF [117–119,121,122], MS [87,135,139,142,143,145] or amperometric detection [110] have generally been used to overcome this problem. These detection systems also provide additional selectivity compared with UV detection, yielding successful results in the analysis of complex biological samples. In addition, the analysis of biological samples usually requires the use of off-line solid-phase extraction or liquid–liquid extraction to avoid matrix interferences, as shown in Table 13.7.

13.4.1.4 Validation of chiral methods for the analysis of drugs

Nowadays, in order to meet the requirements of the pharmaceutical industry, the validation of a chiral method for the analysis of drugs must follow the ICH guidelines on analytical validation (Topics Q2A and B) [182]. The guidelines define the parameters to be used for different types of tests and recommend acceptable methods of validation. Requirements for the validation of analytical methods apply equally to chiral and non-chiral substances and must be addressed in drug development. The characteristics usually determined are the precision (repeatability, intermediate precision and reproducibility), accuracy, LODs and quantitation, linearity, linear range and robustness [182,183]. Table 13.7 shows the chiral methods validated most recently. It can be observed that most of the chiral methods validated are applied in the determination of the enantiomeric purity of different samples. Although most of these samples are standard mixtures [167–170], others such as laboratory synthetic batches [146], degraded drug solutions [170], pharmaceutical preparations [171] and an enzyme-catalyzed sample [178] have also been analyzed. Enantioselective methods have also been validated for the quantitative analysis of pharmaceutical preparations [172,173] and biological samples [175,176,179,180].

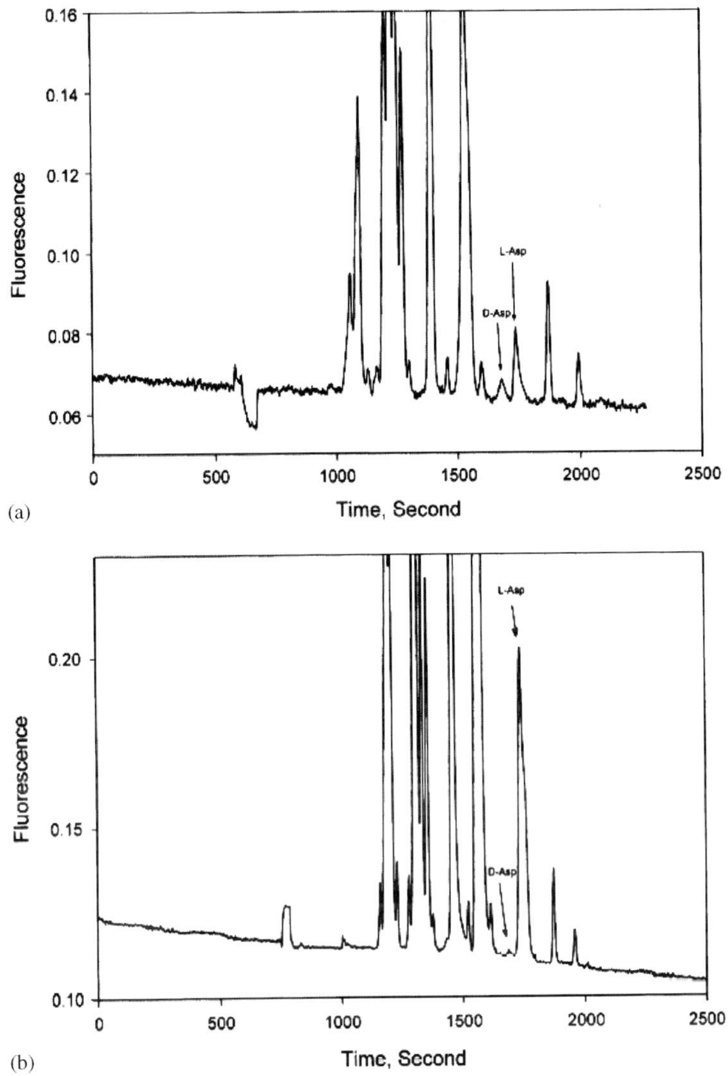

Fig. 13.14. Electropherograms corresponding to (a) a brain sample from a rat 1 day before birth, and (b) a brain sample from a rat 90 days after birth. CE conditions: fused-silica capillary, $l_d = 50$ cm, $l_t = 60.2$ cm with 75 μm i.d.; separation buffer, 50 mM sodium borate at pH 9.0 with 50 mM SDS, 20 mM β-CD and 15% methanol; run voltage, 15 kV; temperature, 25 °C; injection at a height differential of 20 cm for 20 s. LIF detection with $\lambda_{ex} = 457$ nm and $\lambda_{em} \geqslant 495$ nm. Reprinted from Ref. [119]. Copyright (2001), with permission from Elsevier.

13.4.1.5 Monitoring the stereoselectivity of metabolic processes

Another important application of chiral CE is to monitor the stereoselectivity of a metabolic process when a certain chiral drug is administered as a racemate or pure enantiomer to healthy volunteers and patients. This is included within the pharmacokinetic studies required for drugs. Since a recent review summarizes the pharmacokinetic applications of CE [187], only some examples are included in Table 13.7 in order to reflect the most recent works in this field. Thus, ibuprofen [174] or ketoprofen [180] have been stereospecifically analyzed in human plasma with the aim of studying their pharmacokinetics. The electropherograms corresponding to the enantioselective analysis of ibuprofen in a human plasma sample spiked with 10.0 μg/ml of each enantiomer, and in a plasma sample from a healthy volunteer collected 3.0 and 7.0 h after the administration of a 600 mg oral dose of racemic ibuprofen are shown in Fig. 13.15. It can be observed that ibuprofen enantiomers are clearly separated without interferences from endogenous compounds from the matrix, which were eliminated by liquid–liquid extraction. Peaks corresponding to the two enantiomers of fenoprofen can also be observed in the electropherograms because this compound was used as an internal standard owing to its structural similarity with ibuprofen. It is very interesting to observe the stereoselective disposition of ibuprofen in the plasma of the volunteer because the levels of the (+)-enantiomer were higher than those of the (−)-enantiomer [174]. These different pharmacokinetic properties of the enantiomers of this non-stereoidal anti-inflammatory drug were mainly attributed to the unidirectional inversion of the pharmacologically inactive (−)-enantiomer to the active one (the (+)-(*S*)-ibuprofen) in *in vivo* systems [187].

13.4.2 Environmental analysis

The application of chiral CE in environmental analysis is quite limited compared with its application in pharmaceutical and biomedical analysis. The different applications of chiral CE in environmental analysis may be divided into the following: (i) enantiomeric separation of compounds of environmental interest, (ii) determination of the enantiomeric purity of agricultural formulations, (iii) quantitative chiral analysis of environmental samples, and (iv) monitoring of the stereoselectivity of metabolic and (bio)degradation processes.

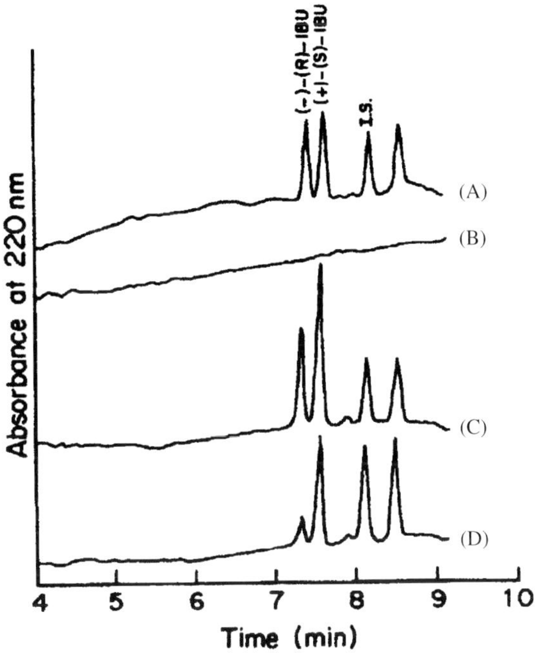

Fig. 13.15. Electropherograms for the enantioselective analysis of ibuprofen in (A) a human plasma sample spiked with 10.0 μg/ml of each enantiomer, (B) a blank human plasma sample, and a plasma sample from a healthy volunteer collected (C) 3.0 and (D) 7.0 h after receiving a 600 mg oral dose of racemic ibuprofen. CE conditions: fused-silica capillary, $l_d = 30$ cm, $l_t = 36$ cm with 50 μm i.d.; separation buffer, 100 mM phosphoric acid/triethanolamine at pH 2.6 with 2.0% (w/v) sulfated β-CD; run voltage, −15 kV; temperature, 25°C; injection at 5.5 kPa for 4 s. UV detection at 220 nm. Reprinted from Ref. [174]. Copyright (2002), with permission from John Wiley & Sons, Inc.–Wiley-VCH Verlag GmbH.

13.4.2.1 Enantiomeric separation of compounds of environmental interest

The chiral separation of compounds of environmental interest by CE has been reviewed in several articles [188,189] and also in a chapter of an encyclopaedia [190]. Compounds enantiomerically separated by CE include herbicides, fungicides and persistent organic pollutants. Table 13.8 groups the applications developed on the enantioseparation of different compounds of environmental interest by CE. It is shown that most of the applications are devoted in developing of analytical methodologies to achieve the enantiomeric separation of the

TABLE 13.8

Applications of chiral CE in environmental analysis

Application	Analyte	Sample	Sample preparation	Chiral selector (conc.)	BGE	Detection	Reference
Enantiomeric separation	Phenoxy acid herbicides (silvex, mecoprop, dichlorprop, PPA, 2-CPPA, 3-CPPA, 4-CPPA)	Standard mixtures	DI	OG or NG (10–150 mM)	Phosphate (pH 6.5)	UV at 230 nm	[191]
Enantiomeric separation	Phenoxy acid herbicides (silvex, mecoprop, dichlorprop, 2,4-CPAA, 2,4,5-CPAA, PPA, 2-CPPA, 3-CPPA, 4-CPPA)	Standard mixtures	DI	TM-β-CD	Acetate (pH 4.5)	UV at 230 nm	[192]
Enantiomeric separation	Phenoxy acid herbicides (PPA, 3-CPPA, 4-CPPA, 2,4-CPPA, 4-CPMPA, 2,4,5-CPPA)	Standard mixtures	DI	HP-β-CD (15 mM)	Formate (pH 5)	UV at 220 nm	[193]
Enantiomeric separation	Phenoxy acid herbicides (mecoprop, dichlorprop)	Standard mixtures	DI	Ethylcarbonate-β-CD (~9 mM)	Phosphate/citric acid (pH 5)	UV at 214 nm	[194]
Enantiomeric separation	Phenoxy acid herbicides (dichlorprop, fenoprop, mecoprop)	Standard mixtures	DI	TM-β-CD (20 mM)	Ammonium acetate (pH 4.6)	ESI-MS	[127]
Enantiomeric separation	ANDSA-phenoxy acid herbicides (silvex, mecoprop, dichlorprop, 2,4-CPAA, 2,4,5-CPAA, PPA, 2-CPPA, 3-CPPA and 4-CPPA)	Standard mixtures	DI	β-CD (5 mM) + TM-β-CD (30 mM)	Phosphate/borate (pH 5)	LIF: λ_{exc} = 325 nm, λ_{em} = 420 nm	[123]
Enantiomeric separation	Aryloxyphenoxy-propionic herbicides (fluaziafop, halossifop, fenoxaprop) and flamprop	Standard mixtures	DI	1-Allylterguride (25 mM)	β-Alanine/acetate (pH 5.3)-MeOH	UV at 230 nm	[195]
Enantiomeric separation	Imidazolinone (imazaquin and imazamethabenz) and aryloxyphenoxy propionic herbicides (diclofop)	Standard mixtures	DI	DM-β-CD or TM-β-CD (10 mM)	Acetate (pH 4.6)	UV at 214 nm	[196]

Enantiomeric separation	Organophosphorus (ruelene, isofenphos, dialifos, fenamifos and malathion) phenoxy acid methyl esters (fenoprop, mecoprop and dichlorpor-methyl ester), organochlorine herbicides (p,p′-DDT, p,p′-DDD, o,p′-DDT, o,p′-DDD, p,p′-DDE and o,p′-DDE) and metolachlor	Standard mixtures	DI	HP-β-CD (60 mM)/γ-CD (20–40 mM)	Borate (pH 9)–SDS–MeOH (0–20%)/AcN (20%)	UV at 200 nm	[197]
Enantiomeric separation	Uniconazole and diniconazole	Standard mixtures	DI	γ-CD (50 mM)	Borate (pH 9)–SDS–urea–2-methyl-2-propanol (5%)	UV at 220 nm	[198,199]
Enantiomeric separation	PCBs (45, 84, 88, 91, 95, 132, 136, 139, 149, 171, 183, 196)	Standard mixtures	DI	γ-CD (50 mM)	CHES (pH 10)–SDS–urea	UV at 235 nm	[200]
Enantiomeric separation	PCBs (88, 131, 132, 135, 136, 139, 144, 171, 175, 183, 196)	Standard mixtures	DI	γ-CD (40 mM)	Borate (pH 9)–SDS–urea	UV at 230 nm	[201]
Enantiomeric separation	PCBs (45, 84, 88, 91, 95, 132, 136, 139, 149, 171, 183, 196)	Standard mixtures	DI	γ-CD (22 mM)+β-CD (73 mM)	CHES (pH 10)–SDS–urea	UV at 235 nm	[202]
Enantiomeric separation	PCBs (84 95, 176)	Standard	DI	SC (150 mM)	CHES (pH 10)–urea	UV at 235 nm	[203]
Enantiomeric separation	PCBs (45, 84, 88, 91, 95, 131, 135, 136, 139, 144, 175, 176)	Standard mixtures	DI	SC (50 mM)+γ-CD (50 mM)	CHES (pH 9)–SDS–urea	UV at 235 nm	[204]
Enantiomeric separation	PCBs (45, 88, 91, 95, 131, 132, 136, 139, 144, 149, 171, 176)	Standard mixtures	DI	β-CD (10 mM)+CM-β-CD (20 mM)	MES (pH 6.5)	UV at 230 nm	[205]
Enantiomeric separation—enantiomeric purity determination	Arylalanine herbicides (flamprop isopropyl)	Standards and enantiomerically pure commercial preparation	DI	SBE-β-CD (20 mg/ml)	Borate (pH 9)–MeOH	UV at 205 nm	[206]
Enantiomeric separation—enantiomeric purity determination	Uniconazole and diniconazole	Standard mixtures	DI	CM-γ-CD (5 mM)	Phosphate (pH 6.5)	UV at 220 nm	[5]

continued

TABLE 13.8 *(continued)*

Application	Analyte	Sample	Sample preparation	Chiral selector (conc.)	BGE	Detection	Reference
Enantiomeric separation—quantitative analysis of environmental samples	Chlorophenoxy acid herbicides (2,4-CPAA, 2,4,5-CPAA, 2,4-CPPA, 2,4,5-CPPA, 2,4-CPBA and 2-CPMBA)	Spiked lake water	SPE	α-CD (4 mM) + β-CD (1 mM)	Phosphate (pH 5.6)	UV at 200 nm	[207]
Enantiomeric separation—quantitative analysis of environmental samples	Free acid herbicides (Haloxyfop, fluazifop, fenoxaprop and flamprop free acids); phenoxy acid herbicides (diclofop, mecoprop, dichlorprop, fenoprop and PPA)	Spiked soil	SLE/LLE	Vancomycin (6 mM)	Boric/acetic/phosphoric (pH 5)	UV at 230 nm	[208]
Enantiomeric separation—quantitative analysis of environmental samples	Aryloxypropionic (mecoprop and fenoprop); aryloxyphenoxy-propionic acidic herbicides (fluazifop and haloxyfop)	Spiked ground and river water	SPE	Vancomycin + γ-CD	Boric/acetic/phosphoric (pH 5)	UV at 230 nm	[209]
Monitoring the stereoselectivity of biodegradation processes	PCBs (45, 88, 91, 95, 136, 144, 149, 176)	Culture medium	LLE	β-CD (10 mM) + CM-β-CD (20 mM)	MES (pH 6.5)	UV at 230 nm	[210]
Enantiomeric separation—monitoring the stereoselectivity of biodegradation processes	Thiobencarb sulfoxide	Spiked soil culture medium	SLE/LLE	HP-γ-CD (60 mM)	Phosphate/borate (pH 8.5)–SDS	UV at 220 nm	[211]

PCBs, polychlorinated biphenyls; SDS, sodium dodecyl sulfate; CHES, 2-[*N*-cyclohexylamino]ethanesulfonic acid; MES, 2-morpholinoethanesulfonic acid; 2,4-CPPA, 2-(2,4-dichlorophenoxy)propionic acid, 2,4-CPBA, 2-(2,4-dichlorophenoxy)butyric acid; 4-CPMPA, 2-(4-chlorophenoxy)-2-methylpropionic acid; 4-CPMBA, 2-(4-chlorophenoxy)-2-methylbutyric acid; 2,4,5-CPPA, 2-(2,4-dichlorophenoxy)propionic acid, SLE, solid–liquid extraction; other abbreviations as in Table 13.7.

compounds in standard solutions. Thus, the separation of the enantiomers of different herbicides (phenoxy acids [123,127, 191–194,208], aryloxyphenoxypropionic acids [195,196,209], imidazolinone derivatives [196], organophosphorus herbicides, DDT congeners and methyl esters of phenoxy acids [197], arylalanine herbicides [206], chlorophenoxy acids [207] and free acid herbicides [208]), fungicides such as uniconazole and diniconazole [5,198,199], and persistent pollutants such as polychlorinated biphenyls (PCBs) [200–205] has been achieved. Although it can be observed that CDs are the chiral selectors most widely used, other chiral selectors such as the alkylglucoside surfactants *n*-octyl-β-d-glucopyranoside (OG) and *n*-nonyl-β-d-glucopyranoside (NG) [201], the ergot alkaloide 1-allylterguride [195], the bile salt SC [203,204] or the macrocyclic antibiotic vancomycin [208,209] have also been used for the separation of the above-mentioned compounds. As an example, Fig. 13.16 shows the separation of a mixture of seven phenoxy acid herbicides using OG as chiral selector. It can be observed that the enantiomeric separation of only six of the seven phenoxy acid herbicides studied was achieved. In addition, the chiral resolution was possible only at surfactant concentrations above the CMC (OG has a CMC of 25 mM), which means that the presence of the chiral surfactant in the micellar form is critical for chiral recognition. It can be observed that the enantioresolution increases with an increase in the concentration of OG [191].

For the separation of multicomponent mixtures of chiral compounds of environmental interest, dual CD systems [123,205,207] or mixtures of micellar systems with CDs [197–200,204,211] or without CDs [191,203] have proved to be very appropriate.

13.4.2.2 Determination of the enantiomeric purity of agricultural formulations

Contrary to pharmaceutical industry where there are regulatory guidelines to control chiral drugs in order to justify the stereoisomeric forms selected for marketing and where many chiral drugs are synthesized nowadays as pure enantiomers, any regulations are applied to chiral pesticides being produced and used as racemic mixtures in most cases. As stated in the introduction of this chapter, this practice causes unnecessary pollution when only one enantiomer of the agrochemical is active. In this case, the use of formulations of pure enantiomers is desirable. In this regard, it is interesting to note that there are a few pesticides that are synthesized as pure enantiomers. This is the case of

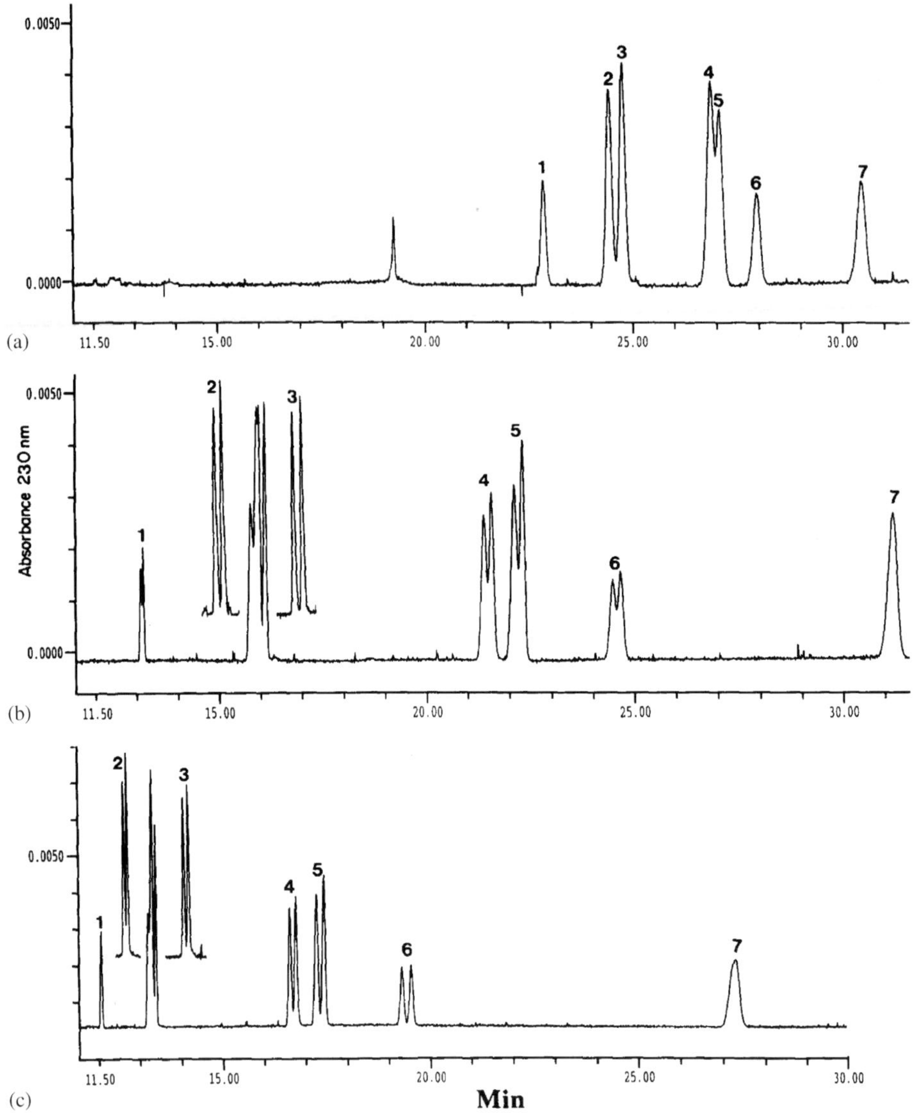

Fig. 13.16. Electropherograms of phenoxy acid herbicides in phosphate buffer at pH 6.5 containing: (a) 10 mM OG, (b) 60 mM OG, and (c) 150 mM OG. CE conditions: fused-silica capillary, $l_d = 50$ cm, $l_t = 57$ cm with 50 μm i.d.; separation buffer, 200 mM sodium phosphate at pH 6.5 with different concentrations of OG; run voltage, 20 kV; temperature, 15 °C; injection at 3.5 kPa for various lengths of time. UV detection at 230 nm. Peaks: 1, silvex; 2, 2-(2,4-dichlorophenoxy)propionic acid; 3, 2-(4-chloro-2-methylphenoxy)propionic acid; 4, 2-(4-chlorophenoxy)propionic acid; 5, 2-(3-chlorophenoxy)propionic acid; 6, 2-(2-chlorophenoxy)propionic acid, 7, 2-phenoxypropionic acid. Reprinted from Ref. [191]. Copyright (1997), with permission from Elsevier.

the herbicide flamprop isopropyl, the commercial preparation of which containing the enantiomerically pure herbicide has been analyzed with a chiral CE method using an anionic CD as chiral selector [206]. On the other hand, some pesticides are commercialized with a given proportion of enantiomers in order to produce a certain activity. For example, the compounds uniconazole and diniconazole are an effective plant growth regulator, and a high-activity fungicide, respectively. Formulations are prepared with different ratios of the *R*- and the *S*-enantiomers, which have different activities. Figure 13.17 shows the enantiomeric separation of uniconazole and diniconazole in a mixture of both fungicides by a fast chiral CE method using a low concentration of the negatively charged CM-γ-CD as chiral selector. This separation enabled the estimation of the enantiomeric ratio for both compounds [5].

13.4.2.3 Quantitative chiral analysis of environmental samples

Chiral analysis of environmental samples by CE is a clear challenge. In fact, the findings reported by the articles published in the literature related to the chiral separation of compounds of environmental interest have not been applied to the analysis of real environmental samples. Thus, spiked samples have only been analyzed generally. However, both separation and analysis clearly demonstrated the high potential of CE in this field.

Table 13.8 groups applications of chiral CE to the analysis of pesticides in water [207,209] and soil [208,211] spiked samples. Samples were previously treated with SPE [207,209] or LLE [208,211] to (i) eliminate the interferences from the sample matrix, and (ii) preconcentrate the analytes contained in the sample. As an example, Fig. 13.18 shows the enantiomeric separation of four herbicides in a river water sample spiked with the herbicide standards before treatment with SPE. The baseline profile was ascribed to a slight but detectable contamination of the samples by humic matter. Sample pretreatment enabled the concentration of the analytes, improving detection sensitivity, and minimized sample contamination by humic acids, which usually interfere strongly with the peaks of interest [208].

13.4.2.4 Monitoring the stereoselectivity of metabolic or (bio)degradation processes

Although biological activity of chiral agrochemicals generally resides in only one of the enantiomers, as stated above, they are usually employed as racemic mixtures. When these racemic mixtures are applied, they

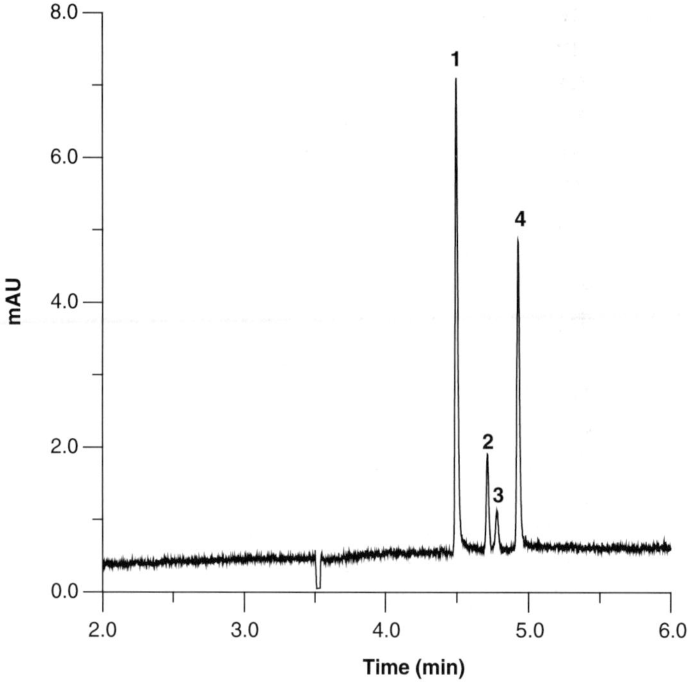

Fig. 13.17. Electropherograms showing the enantiomeric separation of uniconazole (peaks 1 and 2) and diniconazole (peaks 3 and 4). CE conditions: fused-silica capillary, $l_d = 50\,cm$, $l_t = 58.5\,cm$ with 50 μm i.d.; separation buffer, 50 mM phosphate buffer at pH 6.5 containing 5 mM CM-γ-CD; run voltage, 20 kV; temperature, 50 °C; injection at 50 mbar for 2 s. UV detection at 220 nm. Reprinted from Ref. [5]. Copyright (2000), with permission from John Wiley & Sons, Inc.–Wiley-VCH Verlag GmbH.

are often degraded in the environment at different rates, these degradation processes exhibiting a high degree of stereoselectivity.

The racemic haloxifop ethoxyethyl ester herbicide contained in a spiked soil sample was hydrolyzed stereoselectively to the haloxyfop free acid metabolite, as the production of this metabolite was not racemic. In fact, the analysis by chiral CE using vancomycin as chiral selector of an extract of the soil sample after 72 h at room temperature revealed that the metabolite contained a mixture of *R*- and *S*-enantiomers, where the *R* form was about 72% (peak areas ratio) [208]. Another example of the application of chiral CE to the monitoring of the stereoselectivity of degradation processes is the development

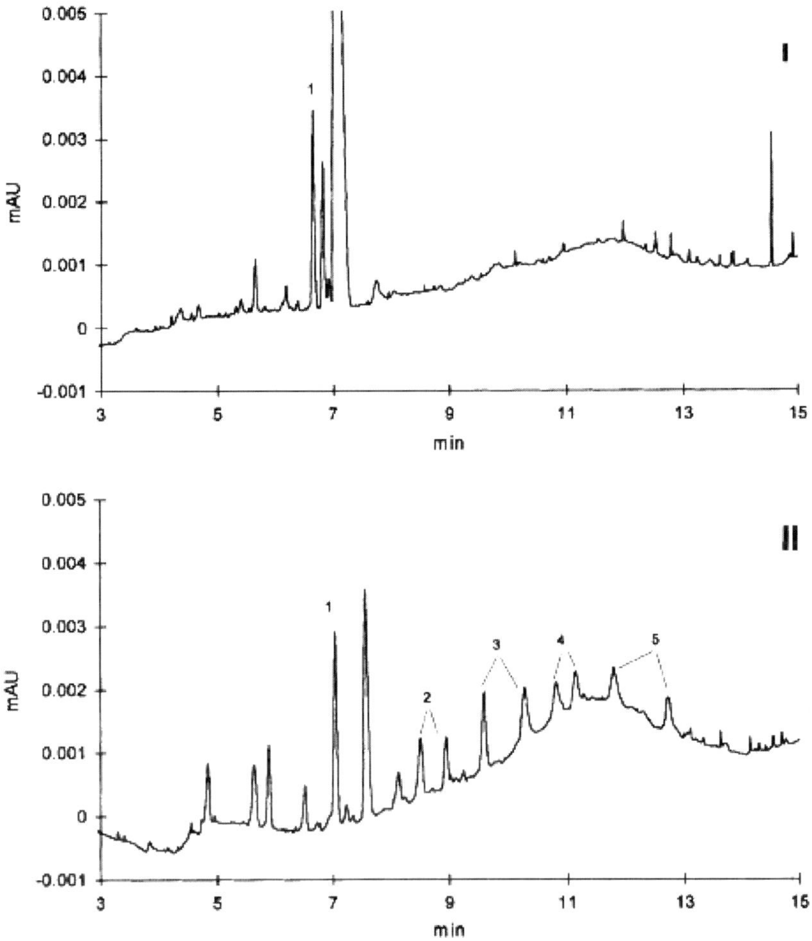

Fig. 13.18. Electropherograms obtained for river water extracts of (I) a blank sample and (II) a river water sample spiked before extraction with: mecoprop (2×10^{-9} M) and fenoprop, fluazifop, haloxyfop (1.5×10^{-9} M, single-isomers). CE conditions: polyacrylamide-coated fused-silica capillary, $l_d = 33$ cm, $l_t = 37.5$ cm with 50 μm i.d.; separation buffer, 75 mM Britton-Robinson (boric/acetic/phosphoric) at pH 5 with 10 mM γ-CD; partial filling with 8 mM vancomycin in 10 mM γ-CD containing running buffer (34.5 kPa for 34 s); run voltage, 15 kV; temperature, 25°C; injection at 34.5 kPa for 4 s. UV detection at 205 nm. Peaks: 1, l-β-phenyllactic acid as internal standard; 2, mecoprop; 3, fenoprop; 4, fluazifop; 5, haloxyfop. Reprinted from Ref. [209]. Copyright (1999), with permission from John Wiley & Sons, Inc.–Wiley-VCH Verlag GmbH.

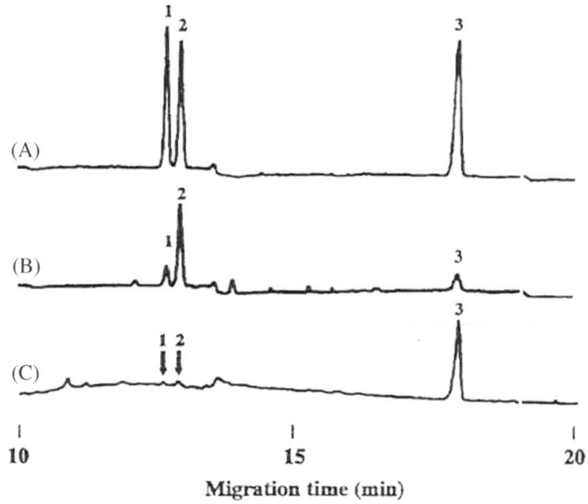

Fig. 13.19. Electropherograms obtained for thiobencarb sulfoxide enantiomers (peaks 1 and 2) and the herbicide thiobencarb (peak 3) for (A) standard solution, (B) thiobencarb sulfoxide produced by rat liver S9 catalysis of thiobencarb, and (C) thiobencarb sulfoxide produced in thiobencarb-spiked soil. CE conditions: fused-silica capillary, $l_d = 56$ cm, $l_t = 64.5$ cm with 75 μm i.d.; separation buffer, 20 mM phosphate–5 mM borate at pH 8.5 containing 100 mM SDS and 60 mM HP-γ-CD; capillary temperature, 20°C; run voltage, 20 kV; injection at 50 mbar for 2 s. UV detection at 220 nm. Reprinted from Ref. [211]. Copyright (2002), with permission from John Wiley & Sons, Inc.–Wiley-VCH Verlag GmbH.

of a chiral CE method that employed CD-modified MEKC to monitor the degradation of the herbicide thiobencarb to thiobencarb sulfoxide by *S*-oxygenation in soil samples (see Fig. 13.19). The enantiomeric ratio between (+) and (−) thiobencarb sulfoxide was 30:70, showing the stereoselectivity of the degradation of this herbicide [211].

The enantiomeric separation of the above-mentioned thiobencarb sulfoxide by CE also enabled the study of the stereoselectivity of the metabolism of thiobencarb herbicide when treated with rat liver microsomal fractions. In this case, an enantiomeric ratio of 15:85 between (+) and (−) thiobencarb sulfoxide was obtained, showing the stereoselectivity of this metabolic process (Fig. 13.19) [211].

On the other hand, biodegradation of pollutants may be stereoselective, which implies a selective enrichment of one of the enantiomers in the environment. Controlling the degradation of pollutants is very interesting in bioremediation studies, which uses living organisms to degrade hazardous organic pollutants to environmentally safe levels in samples (soils, subsurface materials, waters, sludges and residues). In these studies, a similar degradation of both enantiomers is desirable. There is only one published article where the stereoselectivity of biodegradation of chiral PCBs was monitored using a chiral CE method consisting of a dual CD system. Figure 13.20 shows the peaks corresponding to the enantiomers of PCBs 149 and 95 after their degradation during different incubation times by bacteria isolated from a polluted soil. It can be observed that in this case the microorganism produced an effective but non-stereoselective biodegradation of PCBs [210].

13.4.3 Food analysis

The number of works dealing with chiral CE methods in food analysis is less than those dealing with the chiral analysis of pharmaceutical and clinical samples. The first review of chiral electromigration methods in food analysis was published by Simó et al. [212]. Other applications in the determination of amino acids in food and agricultural analysis were reported in a review devoted to the recent advances in amino acid analysis by CE [213]. To date, the main applications in this field are four: (i) enantiomeric separation of food components, (ii) determination of the enantiomeric purity of food components, (iii) chiral analysis of food samples and (iv) monitoring the stereoselectivity of food processing.

13.4.3.1 Enantiomeric separation of food components

Some recent applications of chiral CE in food analysis are illustrated in Table 13.9. The enantiomeric separation of food components has been mainly achieved using CDs as chiral selectors [215–218], although mixtures of CDs with micellar systems [120,214,222,223], a macrocyclic antibiotic (vancomycin) [221] and a copper (II) sulfate/d-quinic acid mixture have also been employed. HP-β-CD is a good chiral selector to achieve the enantiomeric separation of acids (lactic or panthothenic acids) or food colorants, whereas a CD with a wider inner cavity (γ-CD) has been useful to separate the enantiomers of flavanone-7-*O*-glycosides. Chiral selectors such as ligand-exchange complexes [220] or macrocyclic antibiotics [221] have enabled the separation of the enantiomers of other acids such as tartaric, aspartamic or glutamic acid. Mixtures of SDS micelles with β-CD in the absence or presence of bile salts such as TDC have been useful for separating the enantiomers of derivatized amino acids [120,214]. Concerning detection, although

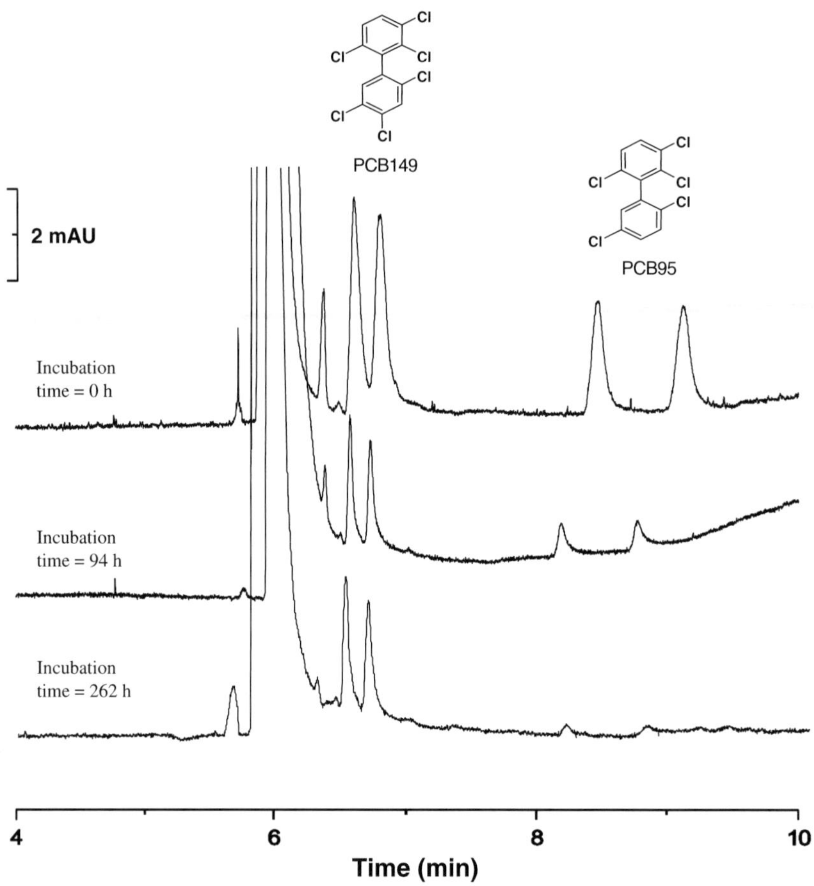

Fig. 13.20. Electropherograms obtained during the biodegradation of a mixture of PCBs 149 and 95 at three different incubation times. CE conditions: fused-silica capillary, $l_d = 50$ cm, $l_t = 58.5$ cm with 50 μm i.d.; separation buffer, 50 mM MES (pH 6.5), 2 M urea, 20 mM CM-γ-CD and 10 mM β-CD; capillary temperature, 45°C; run voltage, 20 kV; injection at 50 mbar for 1 s. UV detection at 230 nm. Reprinted from Ref. [210]. Copyright (2002), with permission from John Wiley & Sons, Inc.–Wiley-VCH Verlag GmbH.

UV absorbance detection is the main detection system used until now (see Table 13.9), alternative detection systems such as LIF have allowed additional selectivity and increased sensitivity [120]. Also, indirect UV detection has been employed. As an example, the chiral analysis of the UV non-absorbing aspartic and glutamic acid enantiomers using vancomycin as chiral selector is shown in Fig. 13.21. The enantiomeric separation of both amino acids and their detection at

TABLE 13.9

Applications of chiral CE in food analysis

Application	Analyte	Sample	Sample preparation	Chiral selector (conc.)	BGE	Detection	Reference
Enantiomeric separation—enantiomeric purity determination	CBI-selenomethyonine and CBI-selenoethyonine	Commercial sample	DI	TDC (50 mM)–β-CD (50 mM)	Phosphate/borate (pH 7)-SDS	UV at 230 nm	[214]
Enantiomeric separation—quantitative analysis of food samples	Synthetic food colorants	Ice cream bars and fruit soda drinks	DI	HP-β-CD (5 mM)	Borax (pH 9.5)	UV at 200 nm	[215]
Enantiomeric separation—quantitative analysis of food samples	Pantothenic acid	Soft drink	DI	HP-β-CD (60 mM)	Phosphate (pH 7.0)–methanol (10%)	UV at 200 nm	[216]
Enantiomeric separation—quantitative analysis of food samples	Lactic acid	Yogurts and beverages (wine, sake, beer and soft drink)	DI	HP-β-CD (240 mM)	Phosphate (pH 6)	UV at 200 nm	[217]
Enantiomeric separation—quantitative analysis of food samples	Flavanone-7-*O*-glycosides	Lemon juice	DI	γ-CD (5 mM)	Borate (pH 10.0)	UV at 290 nm	[218]
Enantiomeric separation—quantitative analysis of food samples	Maleic hydrazide	Potatoes, onions	SLE/SPE	SC (40 mM)	Phosphate (pH 7)	UV at 220 nm	[219]
Enantiomeric separation—quantitative analysis of food samples	Tartaric acid	Grape juices, wines, soft drinks, sakes, jams, candies and pickles	Centrifugation	copper(II) sulfate (1 mM)/d-quinic acid (10 mM)	(pH 5)	UV at 250 nm	[220]
Enantiomeric separation—quantitative analysis of food samples	Aspartamic and glutamic acids	Beer/teeth dentine	SPE	Vancomycin (10 mM)	Sorbic acid/ histidine	Indirect UV at 238 nm and 254 nm	[221]
Enantiomeric separation—quantitative analysis of food samples monitoring the stereoselectivity of food processing	FITC-amino acids (Pro, Asp, Ser, Asn, Glu, Ala, Arg)	Orange juices and orange concentrates	Thermal treatment and centrifugation	β-CD (20 mM)	Tetraborate (pH 9.4)-SDS	LIF: λ_{exc} = 488 nm, λ_{em} = 520 nm	[119]
Enantiomeric separation—monitoring the stereoselectivity of food processing	Vinclozolin	Wine	SPE and fractionation with HPLC	γ-CD (50 mM)	Phosphate/borate (pH 8.5)–SDS	UV at 203 nm	[222]
Enantiomeric separation—monitoring the stereoselectivity of food processing	Imazalil	Orange	LLE/SPE	HP-β-CD (4 mM)	Phosphate (pH 3)–SDS	UV at 200 nm	[223]

Other abbreviations as in Tables 13.7 and 13.8.

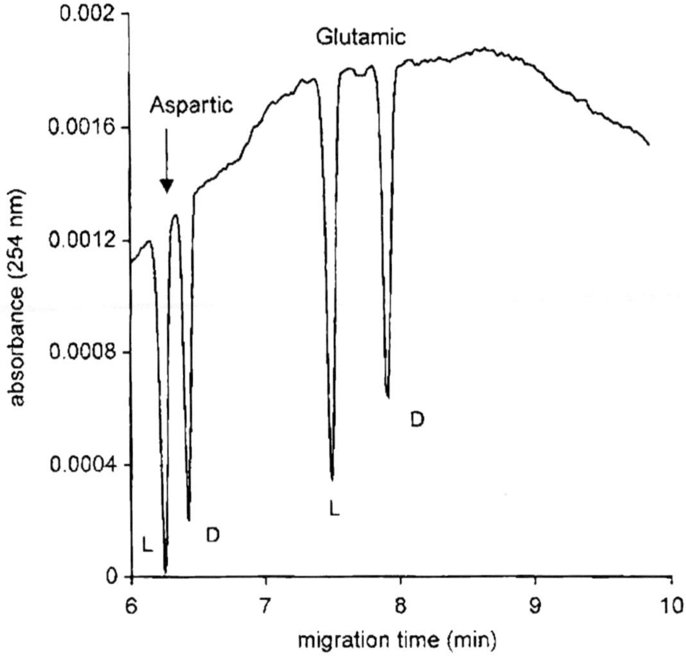

Fig. 13.21. Electropherograms showing the simultaneous enantiomeric separation of aspartic and glutamic acids. CE conditions: polyacrylamide-coated capillary, $l_d = 30.5\,\text{cm}$, $l_t = 35\,\text{cm}$ with 50 µm i.d.; separation buffer, 10 mM sorbic acid/histidine at pH 5 with 10 mM vancomycin (partial filling at 130 psi s); run voltage, −10 kV; temperature, 25°C; injection at 10 psi for 1 s. Indirect UV detection at 254 nm. Reprinted from [221]. Copyright (2001), with permission of John Wiley & Sons, Inc.–Wiley-VCH Verlag GmbH.

254 nm was achieved using the absorbing co-ion sorbic acid (10 mM) in the separation buffer [221].

13.4.3.2 Determination of the enantiomeric purity of food components
The determination of the enantiomeric purity of chiral compounds is a subject of great interest in food analysis. In fact, as stated in the introduction of this chapter, the separation of the enantiomers of certain food components may provide information about possible adulterations in foods and beverages.

The enantiomer-selective analysis of amino acids can provide information concerning adulteration and quality of juices. Thus, high quality orange juices only contain l-amino acids whereas juices of inferior quality also contain some d-amino acids (possible addition of inexpensive racemic amino acids). The enantiomeric separation of seven FITC-labeled amino acids (Pro, Asp, Ser, Asn, Glu, Ala and Arg) was achieved by chiral CE using LIF detection. The method was applied to the determination of d-Ala, d-Asp, d-Arg and d-Glu in orange juices and orange concentrates of different geographical origins. The detection of d-Asp in an orange juice has been shown in Fig. 13.8 as an example of sensitive detection [120].

The development of a CD-modified MEKC method for the separation of CBI-labeled selenoaminoacids selenomethionine and selenoethionine with UV detection was applied to the determination of the enantiomeric purity of selenomethionine in a commercial sample sold as "pure" l-selenomethionine. Figure 13.22 shows the electropherogram obtained for the CBI-selenomethionine derivative contained in the commercial sample. As can be seen, an impurity of about 6% of the d-enantiomer was present in the l-selenomethionine sample [214].

13.4.3.3 *Chiral analysis of food samples*

As opposed to the enantiomeric separation of drugs or compounds of environmental interest, all the enantiomeric separations of food components have been applied to the analysis of food samples, as shown in

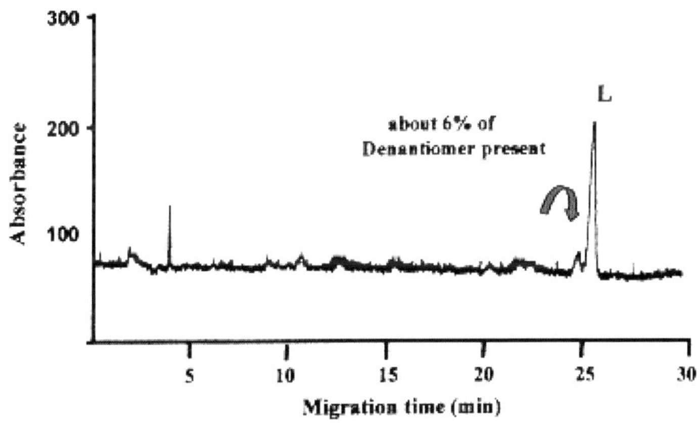

Fig. 13.22. Electropherogram of a commercial sample of "pure" l-selenomethionine (as CBI derivative). CE conditions: fused-silica capillary, l_d = 50 cm, l_t = 67 cm with 50 μm i.d.; separation buffer, 10 mM boric acid/30 mM phosphate at pH 7 with 30 mM β-CD, 50 mM TDC, and 50 mM SDS; run voltage, 12 kV; room temperature; injection at 70 mbar for 0.6 s. UV detection at 230 nm. Reprinted from Ref. [214]. Copyright (2000), with permission from Elsevier.

Table 13.9. Food samples such as beverages (orange/lemon/grape juices, beer, wine, shakes and soft drinks), fruit (oranges), yogurts, candies, jams, ice cream, pickles, potatoes and onions have been analyzed.

With respect to the analytes determined, it should be emphasized that the chiral separation of amino acids and peptides is of special interest in food analysis. Their enantioseparation has been described in detail in some reviews [213,224,225]. Table 13.9 shows that the analysis of acids (pantothenic, lactic, tartaric, aspartamic and glutamic) [216,217,220,221], food colorants [215], flavonoids [218] as well as the pesticides maleic hydrazide [219], viclozolin [222] and imazalil [223] in different food samples has also been recently reported. As an example, Fig. 13.23 shows the peaks of both enantiomers of lactic acid in yogurt and shake [217]. It can be observed that the ratios between d- and l-lactic acid varied, which was attributed to the properties of the lactic bacteria that took part in the lactic acid fermentation.

The treatment of the sample prior to the injection in the CE system may be necessary in food analysis in order to prevent matrix interferences. In fact, although for some samples the direct injection of the appropriate sample dilutions in the CE system was possible [214–218], physical treatments (centrifugation or thermal treatment) [220,120] or

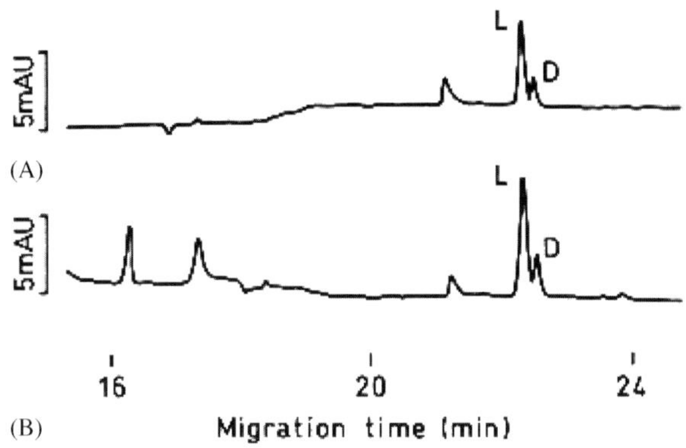

Fig. 13.23. Electropherograms showing the enantiomeric separation of l- and d-lactic acid in (A) yogurt diluted 200-fold and (B) shake diluted 10-fold. CE conditions: polyvinylalcohol-coated bubble cell capillary, $l_d = 50$ cm, $l_t = 58.5$ cm with 50 μm i.d.; separation buffer, 90 mM phosphate buffer at pH = 6.0 with 240 mM HP-β-CD; capillary temperature, 16°C; run voltage, −30 kV; injection at 50 mbar for 200 s. UV detection at 200 nm. Reprinted from Ref. [217]. Copyright (2000), with permission from Elsevier.

extraction procedures (LLE or SPE) [219,221–223] have also been performed. Also, as stated above, the use of alternative detection systems to UV detection allows the increase in sensitivity required to assess adulterations or to detect minor components in food samples [120].

13.4.3.4 Monitoring the stereoselectivity of food processing

Monitoring the stereoselectivity of food processing may furnish information about the treatments applied during the preparation of foods where racemization or degradation may occur (fermentation, thermal treatment, etc.). Although the literature shows that there have been very few applications of chiral CE methods with this aim, the high potential of chiral CE promises an important increase in the number of applications in this field in the near future. As an example, the detection of d-amino acids in orange juice can be indicative of a possible microbial spoilage. Also, these d-amino acids have been proposed as possible molecular markers of thermal treatment of juices [120].

However, food processing is not only stereoselective for food components. Also, the degradation of pesticides in food samples has proved to be enantioselective. Chiral CE methods have been developed to study the stereoselectivity of the degradation of residues of fungicides in food samples. As shown in Table 13.9, vinclozolin and imazalil in wine and orange samples have been studied [222,223]. Vinclozolin is a dicarboximide fungicide (3-(3,5-dichlorophenyl)-5-methyl-5-vinyloxazolidine-2,4-dione) that has been widely used in Europe to protect fruits, vegetables, ornamental plants and turf grasses. The enantiomeric separation of vinclozolin by CE under the experimental conditions indicated in Table 9 has enabled the determination of the peak area ratio of (+)- and (−) enantiomers in wine samples. A ratio of 2:3 was obtained (not racemic), suggesting that the degradation rates during the wine-making process were different for the two enantiomers (see Fig. 13.24) [222]. Likewise, imazalil (1-[2-(2,4-dichlorophenyl)-2-(2-propenyloxy)ethyl]-1H-imidazole can control a wide range of fungal diseases in fruits and vegetables, being widely employed as post-harvest fungicide. Under the conditions specified in Table 13.9, chiral CE has enabled the enantiomeric separation of imazalil enantiomers with a resolution close to 6. The investigation of the presence of imazalil enantiomers in orange samples revealed that this fungicide was present in most samples analyzed. In some of these orange samples, the level of both imazalil enantiomers was the same, but in the other

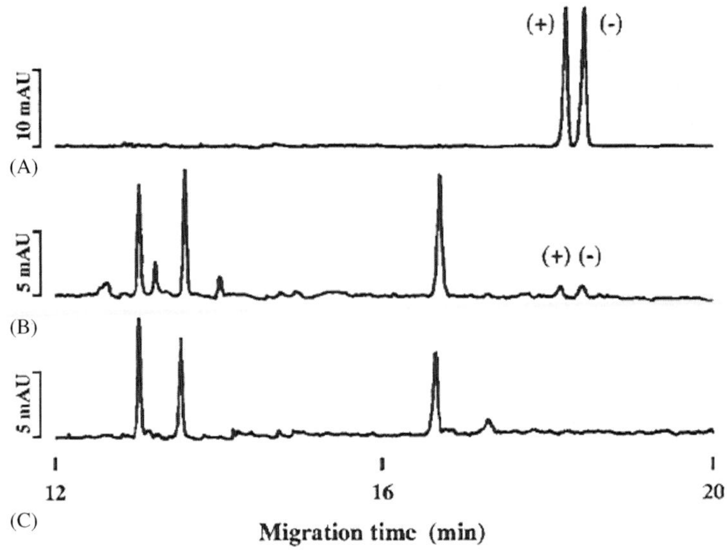

Fig. 13.24. Electropherograms of (A) racemic vinclozolin and (B, C) red wines. CE conditions: fused-silica capillary, $l_d = 56$ cm, $l_t = 64.5$ cm with 75 μm i.d.; separation buffer, 20 mM phosphate–5 mM borate at pH 8.5 with 100 mM SDS and 50 mM γ-CD; capillary temperature, 20°C; run voltage, 20 kV; injection at 50 mbar for 2 s. UV detection at 203 nm. Reprinted from Ref. [223]. Copyright (2003), with permission from American Chemical Society.

samples, the level of (−)-imazalil was found to be lower than that of (+)-imazalil, suggesting that (−)-imazalil degraded more quickly than (+)-imazalil in oranges.

13.4.4 Other applications

In process chemistry, the separation of the enantiomers of chiral compounds also plays an important role in controlling the enantioselectivity of a given stereoselective synthetic process. The importance of enantioselective or asymmetric synthesis as a tool for obtaining enantiomerically pure compounds has grown dramatically in the last two decades as a result of the increasing use of pure enantiomers in different fields. In fact, in 2001 the Nobel Laureates in Chemistry contributed to the development of chiral catalysts for hydrogenation (William Knowles and Ryoji Noyori) and oxidation (Barry Sharpless) reactions in order to produce only one of the enantiomeric forms of a chiral compound [4]. The separation of the enantiomers for subsequent determination of the enantiomeric purity is very important to study the stereoselectivity of this kind of reactions.

A first example reported on the application of chiral CE in process chemistry was the monitoring of the chiral resolution of β-lactam pieprazinyl amide intermediate in aqueous and non-aqueous (mother liquors) samples where mostly the *S*-enantiomer was present. The degradation of products as well as the optical purity was assessed in this work in order to control an enantioselective synthetic process for β-lactam pieprazinyl amide in the presence of a chiral acid. A phosphate buffer at pH 2.5 with 5 mM β-CD as chiral selector was used [226].

On the other hand, chiral CE has been applied to study the asymmetric epoxidation of cinnamyl alcohol with titanium (IV) alkoxide compounds as catalysts in order to evaluate their catalytic activity and the stereoselectivity of the epoxidation processes. A 20 mM succinylated β-CD dissolved in a 10 mM borate buffer at pH 10 was employed. Figure 13.25 shows the electropherogram corresponding to the simultaneous separation of (2*S*, 3*S*)-(−)-3-phenylglycidol and (2*R*, 3*R*)-(+)-3-phenylglycidol enantiomers in the presence of an excess of cinnamyl alcohol in the sample obtained by asymmetric epoxidation of cinnamyl alcohol with $[Ti(O^iPr)_2(OGly)_2]$ as catalyst [227].

13.5 CHIRAL ANALYSIS BY MICROCHIP ELECTROPHORESIS

Enantiomeric separations on micromachined electrophoretic devices have been reported recently. Until now, only a few original articles and a review published in 2003 have focused on the description of developments in chiral analysis by microchip electrophoresis. Some reasons to perform chiral separations with microchips, which provide a format that is more miniaturized than classical CE, are (i) improved separation performance, (ii) reduced instrument size and reagent consumption and (iii) enhanced separation speed and sample throughput [8].

Hutt et al. [228] published the first article on chiral microchip electrophoresis. They explored the feasibility of using microfabricated electrophoresis devices to analyze life signs in extraterrestrial environments. In this work, enantiomeric ratios of amino acids extracted from the Murchison meteorite was determined. Figure 13.26 shows the electropherograms of FITC-labeled amino acid extracts of samples taken from the interior and the exterior of the Murchison meteorite.

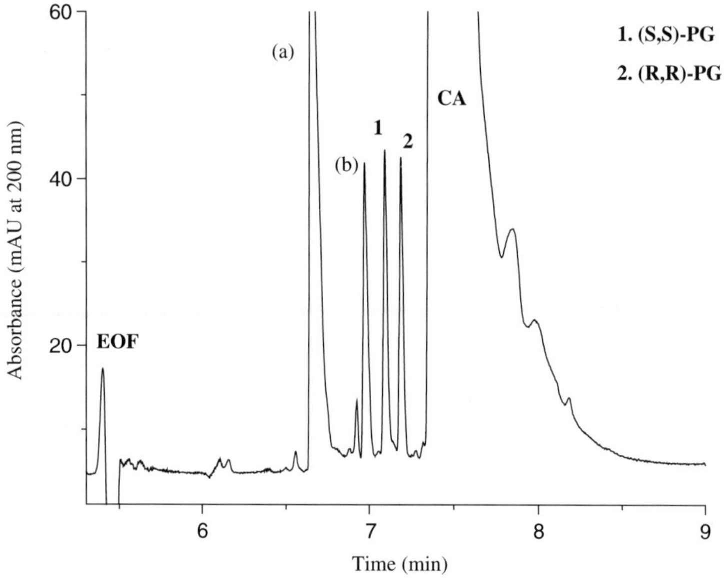

Fig. 13.25. Electropherograms of the separation of (2*S*, 3*S*)-(−)-3-phenylglycidol [(*S*,*S*)-PG] and (2*R*, 3*R*)-(+)-3-phenylglycidol [(*R*,*R*)-PG] enantiomers present in the sample obtained by asymmetric epoxidation of cinnamyl alcohol (CA) with $[Ti(O^iPr)_2(OGly)_2]$ as catalyst. CE conditions: fused-silica capillary, $l_d = 50$ cm, $l_t = 58.5$ cm with 50 μm i.d.; separation buffer, 10 mM borate buffer with 20 mM succinylated-β-CD (pH 10.0); capillary temperature, 15°C; run voltage, 30 kV; injection at 50 mbar for 2 s. UV detection at 200 nm. Reprinted from Ref. [227]. Copyright (2004), with permission from Elsevier.

The fast chiral separations of a variety of basic and acidic compounds have been carried out on microfluidic quartz chips by Ludwig et al. [229]. A microchip electrophoresis instrument equipped with a linear imaging UV-detector was used for the enantiomeric separation of 19 compounds using sulfated CDs as chiral selectors. Figure 13.27 illustrates the successful separation of a mixture of 3 chiral drugs in a single run in less than 11 s utilizing a separation length of only 12 mm.Obviously, these results show that microchip electrophoresis has a great potential for fast chiral analysis. Finally, it is important to consider that high-speed separations are also needed in high-throughput screening systems, which are demanded in combinatorial techniques for the development of new drugs and asymmetric catalysts where thousands of potential catalysts can be generated per day and should be tested [8].

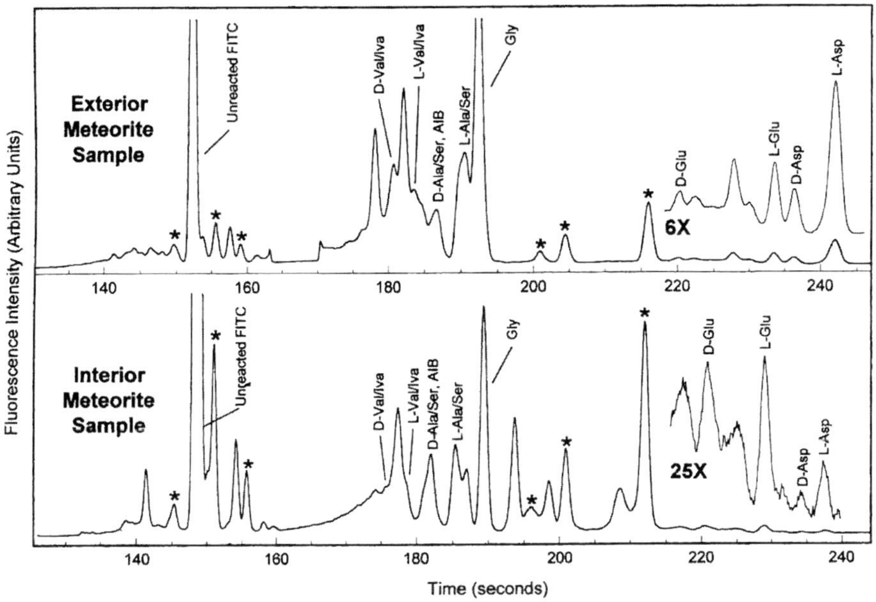

Fig. 13.26. Electropherograms of FITC-labeled amino acid extracts of samples taken from the interior and exterior of the Murchison meteorite. The labeled bands were identified by co-injection of amino acid standards (all protein amino acid enantiomers) or by retention times (Gly, AIB and d/l Iva). The insets present magnifications of the acidic regions of the electropherograms. Microchip electrophoresis device with a separation channel of 19 cm; separation buffer, 10 mM carbonate buffer with 12 mM SDS, and 5 mM γ-CD (pH 10.0). Fluorescence detection. Reprinted with permission from Ref. [228]. Copyright (1999). American Chemical Society.

13.6 FUTURE PERSPECTIVES

Nowadays, chiral CE has shown its potential application in the pharmaceutical, biomedical, food and environmental fields. CE is especially useful in the pharmaceutical sector, where several companies have selected CE for enantioseparations. Although more applications in the pharmaceutical and biomedical fields will be developed in the near future, it is clear that chiral CE will also be used for the analysis of food components, agrochemicals and pollutants to overcome the actual drawbacks posed by the low sensitivity achieved by on-line UV detection, which is the most frequent detection system used in CE.

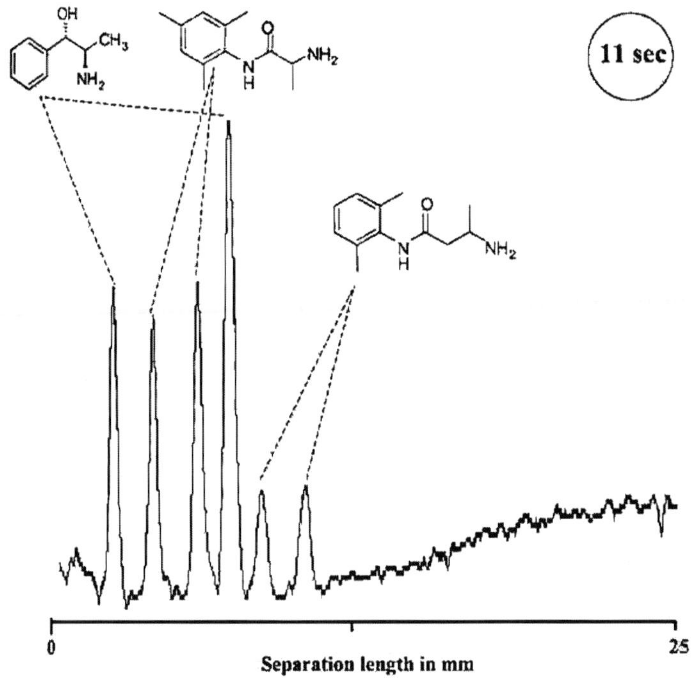

Fig. 13.27. Electropherograms showing the simultaneous chiral separation of three basic drugs in 11 s. Microchip electrophoresis device with a separation chanel of 12 mm; separation buffer, 25 mM triethylammonium phosphate buffer with 5% highly sulfated-γ-CD (pH 2.5). UV detection at 200 nm. Reprinted from Ref. [229]. Copyright (2003), with permission from John Wiley & Sons, Inc.–Wiley-VCH Verlag GmbH.

The coupling of chiral CE with MS has been growing in the recent years and is entering the field of practical applications. It is expected that new chiral selectors more suitable for this hyphenation will be developed in the future. A clear example is the use of new polymeric micelles, which have proved very successful as chiral selectors and useful for CE–ESI-MS [141,165]. Crown ethers have shown high ionization efficiency when introduced in the mass spectrometer [142]; however, its enantioselectivity is limited to chiral compounds containing primary amino groups. With regard to the use of CDs, which are the favorite chiral selectors in CE, the combination of counter-current migration and partial filling techniques in order to avoid the introduction of non-volatile CDs into the mass spectrometer seems to be very successful and promising.

For quantitative measurements, the validation of the chiral method developed is needed. It is important to take into account that with randomly substituted CD derivatives, which have proved to be very good chiral selectors, the evaluation of the robustness of the method is not possible owing to the variability from one commercial supplier to another. To overcome this problem, the development and use of new single-isomer CD derivatives have been promoted [164,230]. These single-isomer derivatives can provide a better understanding of the chiral recognition mechanism.

Although chiral CEC has been developed during the last decade, the current status of this separation technique has shown its potential in chiral analysis. However, a better knowledge of the separation mechanisms is expected, as well as a greater availability of the columns (most of them prepared in the laboratories) and a considerable increment of the number of applications in different fields. One of the most promising topics in CEC is its coupling with MS. In fact, a recent paper has presented CEC–ESI-MS as a very powerful technique for the assay of enantiomers in human fluids [87].

Different chiral selectors employed in chiral CE have been described throughout this chapter. New chiral selectors have recently been developed and more will be developed in the near future. However, the main obstacle to achieve a chiral separation has not yet been solved: selecting the right chiral selector is not easy, and exhaustive screening of different chiral selectors is usually required. In addition, a chiral separation is also affected by other experimental parameters such as the concentration of the chiral selector, pH of the buffer, temperature, characteristics of the capillary, etc. The use of chemometric models to decrease the number of experiments needed to achieve an enantioseparation may be very useful. Although models for the prediction of the enantioseparation selectivity as a function of certain conditions have been developed [17,32,33], more efforts are required to aid in the selection of the appropriate chiral selector, the key factor to achieve the separation of enantiomers.

Finally, microchip electrophoresis is a promising new technique for the separation of enantiomers and aids in performing chiral separations in seconds on tiny micromachined devices. A recent review provides an overview of the original works published until now on this topic with emphasis on the approaches to improve detection and resolution with these devices [8]. Microchip electrophoresis is a very interesting and challenging trend in chiral analysis; therefore, its development and application to solve real problems is expected to grow in the future.

ACKNOWLEDGMENTS

The authors thank the Ministry of Science and Technology (Spain) for the research project BQU2003-03638. C. García-Ruiz thanks the same institution for her research contract from the *Ramón y Cajal* program (RYC-2003-001).

REFERENCES

1 R. Kallenborn and H. Hühnerfuss, *Chiral Environmental Pollutants. Trace Analysis and Ecotoxicology*, Springer, Berlin, 2001.
2 R.T. Morrison and R.N. Boyd, *Química Orgánica*, Addison-Wesley Iberoamericana, Delaware, 1990.
3 D.L. Nelson and M.M. Cox, *Lehninger Principles of Biochemistry*, 3rd ed., Worth Publishers, New York, 2000.
4 Nobel prize in Chemistry 2001 [on-line] http://www.nobel.se/chemistry/laureates/2001/ [Checked on June 2004].
5 Y. Martín-Biosca, C. García-Ruiz and M.L. Marina, *Electrophoresis*, 21 (2000) 3240–3248.
6 D.W. Armstrong, C.D. Chang and W.Y. Li, *J. Agric. Food Chem.*, 38 (1990) 1674–1677.
7 R.-I. Stefan, J.F. van Staden and H. Aboul-Enein, *Comb. Chem. High Thr. Scr.*, 3 (2000) 445–454.
8 D. Belder and M. Ludwig, *Electrophoresis*, 24 (2003) 2422–2430.
9 T.J. Edkins and D.R. Bobbitt, *Anal. Chem.*, 73 (2001) 488A–496A.
10 G. Gübitz and M.G. Schmid, *Biopharm. Drug Dispos.*, 22 (2001) 291–336.
11 G. Subramanian (Ed.), *Chiral Separation Techniques. A Practical Approach*. Wiley-VCH, Weinheim, 2001.
12 N.M. Maier, P. Franco and W. Lindner, *J. Chromatogr. A*, 906 (2001) 3–33.
13 B. Chankvetadze, *Capillary Electrophoresis in Chiral Analysis*, Wiley, Chichester, UK, 1997.
14 B. Chankvetadze and S. Fanali (Eds.), Enantioseparations, *Electrophoresis*. 24, 2407–2732, Wiley-VCH, Weinheim, 2003.
15 B. Chankvetadze, *J. Chromatogr. A*, 792 (1997) 269–295.
16 B. Chankvetadze, *Trend. Anal. Chem.*, 18 (1999) 485–498.
17 B. Chankvetadze and G. Blaschke, *J. Chromatogr. A*, 906 (2001) 309–363.

18 M. Blanco and I. Valverde, *Trend. Anal. Chem.*, 22 (2003) 428–439.
19 A. Rizzi, *Electrophoresis*, 22 (2001) 3079–3106.
20 J.J. Szejtli, *Chem. Rev.*, 98 (1998) 1743–1753.
21 S. Fanali, *J. Chromatogr. A*, 792 (1997) 227–267.
22 G. Gübitz and M.G. Schmid, *J. Chromatogr. A*, 792 (1997) 179–225.
23 C.J. Easton and S.F. Lincoln, *Chem. Soc. Rev.*, 25 (1996) 163.
24 Z. Juvancz and J. Szeijtli, *Trend. Anal. Chem.*, 21 (2002) 379–387.
25 Sigmaaldrich [on-line] http://www.sigmaaldrich.com [Checked on September 2004].
26 A. Amini, *Electrophoresis*, 22 (2001) 3107–3130.
27 I.S. Lurie, *J. Chromatogr. A*, 792 (1997) 297–307.
28 M. Fillet, P. Hubert and J. Crommen, *J. Chromatogr. A*, 875 (2000) 123–134.
29 F. Lelievre, P. Gareil, Y. Bahaddi and H. Galons, *Anal. Chem.*, 69 (1997) 393–401.
30 C.E. Evans and A.M. Stalcup, *Chirality*, 15 (2003) 709–723.
31 S. Fanali, *J. Chromatogr. A*, 875 (2000) 89–122.
32 B.A. Williams and G. Vigh, *J. Chromatogr. A*, 777 (1997) 295–309.
33 W. Yang, A. Yu, X. Yu and H. Chen, *Electrophoresis*, 22 (2001) 2025–2031.
34 R. Kuhn, *Electrophoresis*, 20 (1999) 2605–2613.
35 D.W. Armstrong, L.W. Chang and S.S.C. Chang, *J. Chromatogr. A*, 793 (1998) 115–134.
36 S. Terabe, M. Shibata and Y. Miyashita, *J. Chromatogr.*, 480 (1989) 403–411.
37 K. Otsuka and S. Terabe, *J. Chromatogr. A*, 875 (2000) 163–178.
38 Z. El Rassi, *J. Chromatogr. A*, 875 (2000) 207–233.
39 H.H. Yarabe, E. Billiot and I.M. Warner, *J. Chromatogr. A*, 875 (2000) 179–206.
40 C.P. Palmer, *Electrophoresis*, 23 (2002) 3993–4004.
41 E. Billiot, J. Macossay, S. Thibodeaux, S.A. Shamsi and I.M. Warner, *Anal. Chem.*, 70 (1998) 1375–1381.
42 J.K. Rugutt, H.H. Yarabe, S.A. Shamsi, D.R. Billodeaux, F.R. Fronczek and I.M. Warner, *Anal. Chem.*, 72 (2000) 3887–3895.
43 T.J. Ward and A.B. Farris III, *J. Chromatogr. A*, 906 (2001) 73–89.
44 H.Y. Aboul-Enein and I. Ali, *Chromatographia*, 52 (2000) 679–691.
45 C. Desiderio and S. Fanali, *J. Chromatogr. A*, 807 (1998) 37–56.
46 D.W. Armstrong and U.B. Nair, *Electrophoresis*, 18 (1997) 2331–2342.
47 H. Fangmin and H. He, *Chinese J. Anal. Chem.*, 30 (2002) 621–626.
48 T.M. Oswald and T.J. Ward, *Chirality*, 11 (1999) 663–668.
49 Y. Tanaka and S. Terabe, *J. Biochem. Biophys. Methods*, 48 (2001) 103–116.

50 J. Haginaka, *J. Chromatogr. A*, 875 (2000) 235–254.
51 H. Nishi and Y. Kuwahara, *J. Biochem. Biophys. Methods*, 48 (2002) 89–102.
52 H. Nishi and Y. Kuwahara, *J. Pharm. Biomed. Anal.*, 27 (2001) 577–585.
53 H. Nishi, K. Nakamura, H. Nakai and T. Sato, *Anal. Chem.*, 67 (1995) 2334–2341.
54 M.G. Schmid, N. Grobuschek, O. Lecnik and G. Gübitz, *J. Biochem. Biophys. Method.*, 48 (2001) 143–154.
55 G. Gübitz and M.G. Schmid, *Electrophoresis*, 21 (2000) 4112–4135.
56 O. Lecnik, M.G. Schmid, A. Presser and G. Gübitz, *Electrophoresis*, 23 (2002) 3006–3012.
57 G. Gübitz and M.G. Schmid, *Enantiomer*, 5 (2000) 5–11.
58 V. Schurig and D. Wistuba, *Electrophoresis*, 20 (1999) 2313–2328.
59 D. Wistuba, K. Cabrera and V. Schurig, *Electrophoresis*, 22 (2001) 2600–2605.
60 A.S. Carter-Finch and N.W. Smith, *J. Chromatogr. A*, 848 (1999) 375–385.
61 M. Girod, B. Chankvetadze and G. Blaschke, *J. Chromatogr. A*, 887 (2000) 439–455.
62 S. Mayer, X. Briand and E. Francotte, *J. Chromatogr. A*, 875 (2000) 331–339.
63 D. Mangelings, N. Hardies, M. Maftouh, C. Suteu, D.L. Massart and Y.V. Heyden, *Electrophoresis*, 24 (2003) 2567–2576.
64 M. Lämmerhofer, E. Tobler and W. Lindner, *J. Chromatogr. A*, 887 (2000) 421–437.
65 M. Lämmerhofer, E.C. Peters, C. Yu, F. Svec, J.M.J. Frechet and W. Lindner, *Anal. Chem.*, 72 (2000) 4614–4622.
66 D. Wistuba and V. Schurig, *Electrophoresis*, 21 (2000) 3152–3159.
67 M. Quaglia, E. De Lorenzi, C. Sulitzky, G. Massolini and B. Sellergren, *Analyst*, 126 (2001) 1495–1498.
68 Z.L. Chen and T. Hobo, *Electrophoresis*, 22 (2001) 3339–3346.
69 J.W. Kang, D. Wistuba and V. Schurig, *Electrophoresis*, 23 (2002) 1116–1120.
70 Z.L. Chen, H. Ozawa, K. Uchiyama and T. Hobo, *Electrophoresis*, 24 (2003) 2550–2558.
71 J.W. Kang, D. Wistuba and V. Schurig, *Electrophoresis*, 23 (2002) 4005–4021.
72 D. Wistuba and V. Schurig, *Electrophoresis*, 21 (2000) 4136–4158.
73 C. Fujimoto, *Anal. Sci.*, 18 (2002) 19–25.
74 S. Fanali, P. Catarcini, G. Blaschke and B. Chankvetadze, *Electrophoresis*, 22 (2001) 3131–3151.
75 M. Lämmerhofer, F. Svec, J.M.J. Frechet and W. Lindner, *Trend. Anal. Chem.*, 19 (2000) 676–698.

76 A. Dermaux and P. Sandra, *Electrophoresis*, 20 (1999) 3027–3065.
77 W. Bicker, D. Hebenstreit, M. Lämmerhofer and W. Lindner, *Electrophoresis*, 24 (2003) 2532–2542.
78 E. Tobler, M. Lämmerhofer, G. Mancini and W. Lindner, *Chirality*, 13 (2001) 641–647.
79 B. Chankvetadze, I. Kartozia, C. Yamamoto, Y. Okamoto and G. Blaschke, *J. Pharm. Biomed. Anal.*, 30 (2003) 1897–1906.
80 M. Meyring, B. Chankvetadze and G. Blaschke, *J. Chromatogr. A*, 876 (2000) 157–167.
81 M. Meyring, D. Strickmann, B. Chankvetadze, G. Blaschke, C. Desiderio and S. Fanali, *J. Chromatogr. B*, 723 (1999) 255–264.
82 C. Desiderio, S. Rudaz, J.L. Veuthey, M.A. Raggi and S. Fanali, *J. Sep. Sci.*, 25 (2002) 1291–1296.
83 S. Fanali, S. Rudaz, J.L. Veuthey and C. Desiderio, *J. Chromatogr. A*, 919 (2001) 195–203.
84 K. Kawamura, K. Otsuka and S. Terabe, *J. Chromatogr. A*, 924 (2001) 251–257.
85 Z. Liu, K. Otsuka and S. Terabe, *Electrophoresis*, 22 (2001) 3791–3797.
86 E. Carlsson, H. Wikstrom and P.K. Owens, *Chromatographia*, 53 (2001) 419–424.
87 J. Zheng and S.A. Shamsi, *Anal. Chem.*, 75 (2003) 6295–6305.
88 M. Fillet, A.C. Servais and J. Crommen, *Electrophoresis*, 24 (2003) 1499–1507.
89 F. Wang and M.G. Khaledi, *Anal. Chem.*, 68 (1996) 3460–3467.
90 I.E. Valkó, H. Sirén and M.-L. Riekkola, *J. Chromatogr. A*, 737 (1996) 263–272.
91 H. Cai and G. Vigh, *J. Pharm. Biomed. Anal.*, 18 (1998) 615–621.
92 M. Tacker, P. Glukhovskiy, H. Cai and G. Vigh, *Electrophoresis*, 20 (1999) 2794–2798.
93 J.B. Vincent and G. Vigh, *J. Chromatogr. A*, 816 (1998) 233–241.
94 M.B. Busby, O. Maldonado and G. Vigh, *Electrophoresis*, 23 (2002) 456–461.
95 W. Zhu and G. Vigh, *J. Chromatogr. A*, 892 (2000) 499–507.
96 F. Wang and M.G. Khaledi, *J. Chromatogr. A*, 817 (1998) 121–128.
97 Y. Mori, K. Ueno and T. Umeda, *J. Chromatogr. A*, 757 (1997) 328–332.
98 I. Bjørnsdottir, S.H. Hansen and S. Terabe, *J. Chromatogr. A*, 745 (1996) 37–44.
99 V. Piette, M. Fillet, W. Lindner and J. Crommen, *J. Chromatogr. A*, 875 (2000) 353–360.
100 V. Piette, M. Lammerhofer, W. Lindner and J. Crommen, *Chirality*, 11 (1999) 622–630.
101 V. Piette, M. Lammerhofer, W. Lindner and J. Crommen, *J. Chromatogr. A*, 987 (2003) 421–427.

102 V. Piette, W. Lindner and J. Crommen, *J. Chromatogr. A*, 894 (2000) 63–71.
103 V. Piette, W. Lindner and J. Crommen, *J. Chromatogr. A*, 948 (2002) 295–302.
104 Y. Carlsson, M. Hedeland, U. Bondesson and C. Pettersson, *J. Chromatogr. A*, 922 (2001) 303–311.
105 E. Zarbl, M. Lämmerhofer, P. Franco, M. Petracs and W. Lindner, *Electrophoresis*, 22 (2001) 3297–3307.
106 H. Kataoka, *Trend. Anal. Chem.*, 22 (2003) 232–244.
107 J.Y. He, A. Shibukawa, M. Zeng, S. Amane, T. Sawada and T. Nakagawa, *Anal. Sci.*, 12 (1996) 177–181.
108 M. Danková, D. Kaniansky, S. Fanali and F. Ivanyi, *J. Chromatogr. A*, 838 (1999) 31–43.
109 S. Fanali, C. Desiderio, E. Olvecka, D. Kaniansky, M. Vojtek and A. Ferancova, *J. High Resol. Chromatogr.*, 23 (2000) 531–538.
110 M.E. Hadwiger, S.R. Torchia, S. Park, M.E. Biggin and C.E. Lunte, *J. Chromatogr. B*, 681 (1996) 241–249.
111 S. Grard, P. Morin and J.P. Ribet, *Electrophoresis*, 23 (2002) 2399–2407.
112 K. Otsuka, M. Matsumura, J.-B. Kim and S. Terabe, *J. Pharm. Biomed Anal.*, 30 (2003) 1861–1867.
113 T.M.H. Choy, W.H. Chan, A.W.M. Lee and C.W. Huie, *Electrophoresis*, 24 (2003) 3116–3123.
114 C. Desiderio, S. Rudaz, M.A. Raggi and S. Fanali, *Electrophoresis*, 20 (1999) 3432–3438.
115 T. Ueda, F. Kitamura, R. Mitchell, T. Metcalf, T. Kuwana and A. Nakamoto, *Anal. Chem.*, 63 (1991) 2979–2981.
116 T. Kaneta, H. Shiba and T. Imasaka, *J. Chromatogr. A*, 805 (1998) 295–300.
117 Y.M. Liu and S.L. Zhao, *LC GC*, 19 (2001) 414–420.
118 S.L. Zhao and Y.M. Liu, *Electrophoresis*, 22 (2001) 2769–2774.
119 S.L. Zhao, Y.Z. Feng, M.H. LeBlanc and Y.M. Liu, *J. Chromatogr. B*, 762 (2001) 97–101.
120 C. Simó, C. Barbas and A. Cifuentes, *J. Agric. Food Chem.*, 50 (2002) 5288–5293.
121 B. Chankvetadze, N. Burjanadze and G. Blaschke, *Electrophoresis*, 22 (2001) 3281–3285.
122 U.B. Soetebeer, M.O. Schierenberg, H. Schulz, P. Andresen and G. Blaschke, *J. Chromatogr. B*, 765 (2001) 3–13.
123 Y. Mechref and Z. El Rassi, *Anal. Chem.*, 68 (1996) 1771–1777.
124 C. García-Ruiz, M. Siderius, F. Ariese and C. Gooijer, *Anal. Chem.*, 76 (2004) 399–403.

125 R.L. Sheppard, X.C. Tong, J.Y. Cai and J.D. Henion, *Anal. Chem.*, 67 (1995) 2054–2058.
126 H.J. Gaus, Z.Z. Gogus, K. Schmeer, B. Behnke, K.A. Kovar and E. Bayer, *J. Chromatogr. A*, 735 (1996) 221–226.
127 K. Otsuka, C.J. Smith, J. Grainger, J.R. Barr, D.G. Patterson, N. Tanaka and S. Terabe, *J. Chromatogr. A*, 817 (1998) 75–81.
128 M.H. Lamoree, A.F.H. Sprang, U.R. Tjaden and J. vanderGreef, *J. Chromatogr. A*, 742 (1996) 235–242.
129 S. Fanali, C. Desiderio, G. Schulte, S. Heitmeier, D. Strickmann, B. Chankvetadze and G. Blaschke, *J. Chromatogr. A*, 800 (1998) 69–76.
130 W.Z. Lu and R.B. Cole, *J. Chromatogr. B*, 714 (1998) 69–75.
131 G. Schulte, S. Heitmeier, B. Chankvetadze and G. Blaschke, *J. Chromatogr. A*, 800 (1998) 77–82.
132 Y. Tanaka, Y. Kishimoto and S. Terabe, *J. Chromatogr. A*, 802 (1998) 83–88.
133 E.M. Javerfalk, A. Amini, D. Westerlund and P.E. Andren, *J. Mass Spectrom.*, 33 (1998) 183–186.
134 S. Rudaz, S. Cherkaoui, P. Dayer, S. Fanali and J.L. Veuthey, *J. Chromatogr. A*, 868 (2000) 295–303.
135 Y. Tanaka, K. Otsuka and S. Terabe, *J. Chromatogr. A*, 875 (2000) 323–330.
136 S. Grard, P. Morin, M. Dreux and J.P. Ribet, *J. Chromatogr. A*, 926 (2001) 3–10.
137 B. Toussaint, M. Palmer, P. Chiap, P. Hubert and J. Crommen, *Electrophoresis*, 22 (2001) 1363–1372.
138 S. Cherkaoui, S. Rudaz, E. Varesio and J.L. Veuthey, *Electrophoresis*, 22 (2001) 3308–3315.
139 S. Cherkaoui and J.L. Veuthey, *J. Pharm. Biomed. Anal.*, 27 (2002) 615–626.
140 S.A. Shamsi, *Anal. Chem.*, 73 (2001) 5103–5108.
141 M. Moini, C.L. Schultz and H. Mahmood, *Anal. Chem.*, 75 (2003) 6282–6287.
142 R. Iio, S. Chinaka, S. Tanaka, N. Takayama and K. Hayakawa, *Analyst*, 128 (2003) 646–650.
143 E.K. Kindt, S. Kurzyniec, S.C. Wang, G. Kilby and D.T. Rossi, *J. Pharm. Biomed. Anal.*, 31 (2003) 893–904.
144 Y.T. Iwata, T. Kanamori, Y. Ohmae, K. Tsujikawa, H. Inoue and T. Kishi, *Electrophoresis*, 24 (2003) 1770–1776.
145 A.A. Shamsi, *Electrophoresis*, 23 (2002) 4036–4051.
146 X.M. Fang, F.Y. Gong and Y.Z. Fang, *Anal. Chem.*, 70 (1998) 4030–4035.
147 E. Gassmann, J.E. Kuo and R.N. Zare, *Science*, 230 (1985) 813–814.
148 S. Fanali (Ed.), CE Enantioseparations. *Electrophoresis*. 20, 2577–2800, Wiley-VCH, Weinheim, 1999.

149 S. Fanali and B. Chankvetadze, (Ed.), CE/CEC Enantioseparations. *Electrophoresis*. 22, 3077–3354, Wiley-VCH, Weinheim, 2001.
150 H. Nishi and S. Terabe (Ed.), Enantiomer separations by CE. *J. Chromatogr. A*. 875, 1–471, Elsevier, Amsterdam, 2000.
151 S.A.C. Wren (Ed.), The separation of enantiomers by CE. *Chromatographia*. 54, S5–S93, Vieweg, Wiesbaden, 2001.
152 G.K.E. Scriba, *Electrophoresis*, 24 (2003) 2409–2421.
153 G.K.E. Scriba, *J. Pharm. Biomed. Anal.*, 27 (2002) 373–399.
154 B. Chankvetadze, *J. Sep. Sci.*, 24 (2001) 691–705.
155 A. Amini, *Electrophoresis*, 22 (2001) 3107–3130.
156 G. Blaschke and B. Chankvetadze, *J. Chromatogr. A*, 875 (2000) 3–25.
157 S. Zaugg and W. Thormann, *J. Chromatogr. A*, 875 (2000) 27–41.
158 B. Koppenhoefer, X.F. Zhu, A. Jakob, S. Wuerthner and B.C. Lin, *J. Chromatogr. A*, 875 (2000) 135–161.
159 M. Fillet, Ph. Hubert and J. Crommen, *J. Chromatogr. A*, 875 (2000) 123–134.
160 K.D. Altria, M.A. Kelly and B.J. Clark, *Trend. Anal. Chem.*, 17 (1998) 214–226.
161 S.L. Tamisier-Karolak, M.A. Stenger and A. Bommart, *Electrophoresis*, 20 (1999) 2656–2663.
162 M. Blanco, J. Coello, H. Iturriaga, S. Maspoch and C. Pérez-Maseda, *J. Chromatogr. A*, 793 (1998) 165–175.
163 A.M. Abushoffa, N. Burjanadze, G. Blaschke, J. Crommen and B. Chankvetadze, *J. Sep. Sci.*, 25 (2002) 10–16.
164 T. Christians and U. Holzgrabe, *J. Chromatogr. A*, 911 (2001) 249–257.
165 S.A. Shamsi, B.C. Valle, F. Billiot and I.M. Warner, *Anal. Chem.*, 75 (2003) 379–387.
166 P. Castelnovo and C. Albanesi, *Electrophoresis*, 18 (1997) 996–1001.
167 S. Song, L. Zhou, R. Thompson, M. Yang, D. Ellison and J.M. Wyvratt, *J. Chromatogr. A*, 959 (2002) 299–308.
168 H. Wilkström and P.K. Owens, *J. Sep. Sci.*, 25 (2002) 1167–1174.
169 C. Foulon, J.F. Goossens, E. Fourmaintraux, J.P. Bonte and C. Vaccher, *Electrophoresis*, 23 (2002) 1121–1128.
170 A. Gómez-Gomar, E. Ortega, C. Calvet, R. Mercé and J. Frigola, *J. Chromatogr. A*, 950 (2002) 257–270.
171 M. Blanco, J.M. González, E. Torras and I. Valverde, *Anal. Bioanal. Chem.*, 375 (2003) 157–163.
172 C. García-Ruiz and M.L. Marina, *Electrophoresis*, 22 (2001) 3191–3197.
173 E. Ekiert, C. García-Ruiz, M.A. García and M.L. Marina, *Electrophoresis*, 24 (2003) 2680–2686.
174 V.A.P. Jabor, V.L. Lanchote and P.S. Bonato, *Electrophoresis*, 23 (2002) 3041–3047.

175 V.A.P. Jabor, V.L. Lanchote and P.L. Bonato, *Electrophoresis*, 22 (2001) 1406–1412.
176 V.A.P. Jabor and P.S. Bonato, *Electrophoresis*, 22 (2001) 1399–1405.
177 T. de Boer, R. Mol, R.A. de Zeeuw, G.J. de Jong and K. Ensing, *Electrophoresis*, 22 (2001) 1413–1418.
178 H.J. Lu, Z.Q. Ruan, J.W. Kang and Q.Y. Ou, *Anal. Lett.*, 34 (2001) 1657–1668.
179 S. Andersen, T.G. Halvorsen, S. Pedersen-Bjergaard, K.E. Rasmussen, L. Tanum and H. Refsum, *J. Pharm. Biomed. Anal.*, 33 (2003) 263–273.
180 F.K. Glowka, *J. Pharm. Biomed. Anal.*, 30 (2002) 1035–1045.
181 B. Chankvetadze, *Electrophoresis*, 23 (2002) 4022–4035.
182 International Conference on Harmonisation [on-line] http://www.ich.org [Checked on March 2004].
183 H. Wätzig, M. Degenhardt and A. Kunkel, *Electrophoresis*, 19 (1998) 2695–2752.
184 U. Schmitt, S.K. Branch and U. Holzgrabe, *J. Sep. Sci.*, 25 (2002) 959–974.
185 L.G. Blomberg and H. Wan, *Electrophoresis*, 21 (2000) 1940–1952.
186 M.L. Riekkola and S.K. Wiedmer, *Process Contr. Qual.*, 10 (1997) 169–180.
187 C.C. Lin, Y.T. Li and S.H. Chen, *Electrophoresis*, 24 (2003) 4106–4115.
188 A.K. Malik and W. Faubel, *Crit. Rev. Anal. Chem.*, 31 (2001) 223–279.
189 D.T. Eash and R.J. Bushway, *J. Chromatogr. A*, 880 (2000) 281–294.
190 C. García-Ruiz, A.L. Crego and M.L. Marina, Organic analysis in environmental samples by capillary electrophoresis. In: R.A. Meyers (Ed.), *Encyclopedia of Analytical Chemistry*. Wiley, Chichester, 2000, pp. 3002–3035.
191 Y. Mechref and Z. El Rassi, *J. Chromatogr. A*, 757 (1997) 263–273.
192 A.W. Garrison, P. Schmitt and A. Kettrup, *J. Chromatogr. A*, 688 (1994) 317–327.
193 Y. Martín Biosca, C. García-Ruiz and M.L. Marina, *Electrophoresis*, 22 (2001) 3216–3225.
194 O. Zerbinati, F. Trotta and C. Giovannoli, *J. Chromatogr. A*, 875 (2000) 423–430.
195 B.A. Ingelse, J.C. Reijenga, M. Flieger and F.M. Everaerts, *J. Chromatogr. A*, 791 (1997) 339–342.
196 K.V. Penmetsa, R.B. Leidy and D. Shea, *J. Chromatogr. A*, 790 (1997) 225–234.
197 Ph. Schmitt, A.W. Garrison, D. Freitag and A. Kettrup, *J. Chromatogr. A*, 792 (1997) 419–429.
198 R. Furuta and T. Doi, *J. Chromatogr. A*, 676 (1994) 431–436.
199 R. Furuta and T. Doi, *Electrophoresis*, 15 (1994) 1322–1325.

200 M.L. Marina, I. Benito, J.C. Díez-Masa and M.J. González, *Chromatographia*, 42 (1996) 269–272.
201 J. Grainger, Z. Liu, K. Marano, J. Barr and D.G. Patterson Jr., *Organohalogen Comp. (Dioxin' 97)*, 31 (1997) 286–291.
202 M.L. Marina, I. Benito, J.C. Díez-Masa and M.J. González, *J. Chromatogr. A*, 752 (1996) 265–270.
203 A.L. Crego, M.J. González and M.L. Marina, *Electrophoresis*, 19 (1998) 2113–2118.
204 A.L. Crego, M.A. García and M.L. Marina, *J. Microcol. Sep.*, 12 (2000) 33–40.
205 C. García-Ruiz, Y. Martín-Biosca, A.L. Crego and M.L. Marina, *J. Chromatogr. A*, 910 (2001) 157–164.
206 C. Desiderio, C.M. Polcaro and S. Fanali, *Electrophoresis*, 18 (1997) 227–234.
207 Y.Z. Hsieh and H.Y. Huang, *J. Chromatogr. A*, 745 (1996) 217–223.
208 C. Desiderio, C.M. Polcaro, P. Padiglioni and S. Fanali, *J. Chromatogr. A*, 781 (1997) 503–513.
209 C.M. Polcaro, C. Marra, C. Desiderio and S. Fanali, *Electrophoresis*, 20 (1999) 2420–2424.
210 C. García-Ruiz, R. Andrés, J.L. Valera, F. Laborda and M.L. Marina, *J. Sep. Sci.*, 25 (2002) 17–22.
211 S. Kodama, A. Yamamato, A. Matsunaga, K. Okamura, R. Kizu and K. Hayakawa, *J. Sep. Sci.*, 25 (2002) 1055–1062.
212 C. Simó, C. Barbas and A. Cifuentes, *Electrophoresis*, 24 (2003) 2431–2441.
213 V. Poinsot, C. Bayle and F. Couderc, *Electrophoresis*, 24 (2003) 4047–4062.
214 A. Pérez Méndez, E. Blanco González and A. Sanz-Medel, *Anal. Chim. Acta*, 416 (2000) 1–7.
215 K.L. Kuo, H.Y. Huang and Y.Z. Hsieh, *Chromatographia*, 47 (1998) 249–256.
216 S. Kodama, A. Yamamoto and A. Matsunaga, *J. Chromatogr. A*, 811 (1998) 269–273.
217 S. Kodama, A. Yamamoto, A. Matsunaga, T. Soga and K. Minoura, *J. Chromatogr. A*, 875 (2000) 371–377.
218 N. Gel-Moreto, R. Streich and R. Galensa, *J. Chromatogr. A*, 925 (2001) 279–289.
219 D.T. Kubilius and R.J. Bushway, *J. Liq. Chrom. Rel. Tech.*, 22 (1999) 593–601.
220 S. Kodama, A. Yamamoto, A. Matsunaga and K. Hayakawa, *J. Chromatogr. A*, 932 (2001) 139–143.
221 P. Bednar, Z. Aturki, Z. Stranski and S. Fanali, *Electrophoresis*, 22 (2001) 2129–2135.

222 S. Kodama, A. Yamamoto, Y. Saitoh, A. Matsunaga, K. Okamura, R. Kizu and K. Hayakawa, *J. Agric. Food Chem.*, 50 (2002) 1312–1317.
223 S. Kodama, A. Yamamoto, T. Ohura, A. Matsunaga and T. Kanbe, *J. Agric. Food Chem.*, 51 (2003) 6128–6131.
224 G.K.E. Scriba, *Electrophoresis*, 24 (2003) 4063–4077.
225 H. Wan and L.G. Blomberg, *J. Chromatogr. A*, 875 (2000) 43–88.
226 C. Silverman, *J. Cap. Elec.*, 4 (1997) 181–187.
227 S. Morante-Zarcero, A.L. Crego, I. Sierra, M. Fajardo and M.L. Marina, *Electrophoresis*, 25 (2004) 2745–2754.
228 L.D. Hutt, D.P. Glavin, J.L. Bada and R.A. Mathies, *Anal. Chem.*, 71 (1999) 4000–4006.
229 M. Ludwig, F. Kohler and D. Belder, *Electrophoresis*, 24 (2003) 3233–3238.
230 W.H. Zhu and G. Vigh, *Anal. Chem.*, 72 (2000) 310–317.

Chapter 14

Electrochemical detection in capillary electrophoresis on microchips

Antonio Javier Blasco and Alberto Escarpa

14.1 INTRODUCTION

At the time of writing this chapter, the first decade after the revolutionary introduction of the micro total analysis system (μ-TAS) concept by A. Manz and co-workers in 1990 [1] is already over. The μ-TAS concept developed from the modification of the total analysis system (TAS) approach by downsizing and integrating its multiple steps (injection, reaction, separation, detection) onto a single device, yielding a sensor-like system with fast response time, low sample consumption, on-site operation and high stability [1,2]. As an extension of the μ-TAS concept, these microsystems are now called "lab-on-a-chip" to account for the fact that not only analytical tasks but also other chemical functions, such as synthesis, can be performed on such a device [3]. Today, both terms are used indiscriminately.

Since the initial description of the μ-TAS concept, and owing to the unique solutions it can offer, there has been a burst of interest in the development of analytical systems utilizing microchip formats [4,5]. The advantages of a μ-TAS can be summarized as follows: faster analysis (scale decreasing), portability (decentralized chemical analysis), reliability (fractions of microlitres of sample and reagents are moved around with very high accuracy), reduction of sample and reagent consumption (decreasing cost), as well as reduction of wastes (friendly to the environment), automation of chemical analysis, high-throughput screening (possibility of processing samples in parallel) and the possibility of carrying out *on line* analysis [4,5].

Capillary electrophoresis on microchips (CE-microchips) was one of the earliest examples of a μ-TAS system [6–8]. In the μ-TAS field,

Comprehensive Analytical Chemistry XLV
M.L. Marina, A. Ríos and M. Valcárcel (Eds)
Volume XLV ISSN: 0166-526X DOI: 10.1016/S0166-526X(05)45014-X

CE-microchips has undoubtedly become the most rapidly advancing and fruitful branch so far. The reason for this dominance of electrophoresis separations over other separation techniques lies partially in the inherent simplicity of fabrication and operation combined with unique features with respect to separation speed, sample injection and reagent consumption. CE-microchips employ channels etched into a planar substrate, which are based upon microfabrication techniques developed in the semiconductor industry. Early applications of CE-microchips employed glass or quartz for chip substrates because they are optically transparent and exhibit electroosmotic flow (EOF) properties similar to those of fused silica [9–11]. Since then, a number of other materials, including polymer/plastics, have been investigated [10–13]. Microchannels are fabricated in microchips using photolithography or micromolding to form channels for sample injection, capillary electrophoresis separation, and analyte detection. Once all solutions, including those of the samples, are loaded, the samples are typically transferred electrokinetically into an injector region. Then their components are separated by application of a high voltage, and afterwards detected with a suitable detection system.

Moreover, CE-microchips, as μ-TAS itself, has the potential to simultaneously assay hundreds of samples in a matter of minutes or less. The rapid analysis combined with massively parallel analysis arrays should yield an ultra-high throughput. These microchips typically consume only picolitres of sample and they may potentially be prepared on board for a complete integration of sample preparation, separation and detection. Also, a key advantage to the use of these planar devices/substrates is ease of manufacture and the possibility of adapting microfabrication techniques for mass production. Therefore, these features make CE-microchips an attractive technology for the next generation of capillary electrophoresis instrumentation [9–11].

On the other hand, most applications of CE-microchips have used laser-induced fluorescence (LIF) detection [14] because the laser can easily be focused directly onto the small channel, and also because low detection limits are attainable. However, CE-microchips with LIF detection has some disadvantages. Most compounds do not possess a fluorescent functionality. For this reason, LIF detection usually requires a suitable fluorophore. These fluorophores are often very bulky and usually eliminate at least one charged group. This can cause the analytes of interest to exhibit similar electrophoretic mobilities; therefore, a longer separation channel or postcolumn derivatization scheme may be needed to effectively use LIF detection. While offering high sensitivity and powerful information, LIF requires a bulky off-chip control instrumentation that greatly compromises the benefits of miniaturization and portability. Meanwhile, it is important to emphasize that the success of fluorescence detection is partly due to its high sensitivity but also due to the facts that the analysis of biochemically relevant species is of great interest and that amino acids and related species can be derivatized readily with a fluorescent tag (e.g. with fluorescein isothiocyanate). DNA may also be easily labelled with intercalators. For other classes of analytes, however, the technique is not as easily implemented and direct fluorescence detection cannot be considered a universal approach. It is therefore not surprising that, in order to extend the use of microfluidic separation devices beyond the now established biochemical applications, much attention is being given to the development of other detection methods based, for example, on electrochemical techniques, where the chemical signal is directly converted into the electrical domain via simple electrodes. As a consequence, the overall system can be more compact and the goal of miniaturization can be driven further.

Electrochemistry offers great promise for such microsystems, with features that include remarkable sensitivity (approaching that of fluorescence), inherent miniaturization of both the detector and control instrumentation, responses non-dependent on the optical path length or sample turbidity, low cost, low power requirements and high compatibility with advanced micromachining and microfabrication technologies [14–18]. Moreover, complete self-contained microsystems may be designed through an on-chip integration of the potentiostatic circuit. While the sensitivity of optical detection schemes is usually compromised by the use of thin channels, reducing the size of the working electrode commonly enhances the performance of these technologies (owing to a more effective contact between the flowing solution and the electrode substrate). A remarkably high sensitivity can thus be attained in connection to low-dead volume devices.

Electrochemical detection has often been considered incompatible with CE because the combination of the high voltages applied in the electrophoretic separation and sensitive electrodes has been seen as a conflict. However, it has been found that if the design of the detector cell is appropriate, the separation voltage does not interfere with the electrochemical measurement. Electrochemical methods have therefore recently found wider acceptation for conventional capillary electrophoresis, and much of the know-how gained has been transferred to the microchip format [14].

The numerous papers on capillary electrophoresis in microfabricated channels with electrochemical detection that have been published since the first *proof-of-concept* demonstrate the enormous analytical potential of these miniaturized separation devices. Excellent review articles about CE microchips with electrochemical detection (ED) have been published in the last 4 years [15–18] discussing the most recent advances in different electrochemical detection modalities, as well as the strengths and weaknesses of these approaches. The importance of the matter has also been reflected in the fact that the first published paper in *Lab-on-a-chip Journal* (2001) was a review devoted to the recent developments in detection methods, especially ED, for microfabricated analytical devices [14].

Therefore, the aim of this chapter is to provide comprehensive directions on the fundamental aspects as well as the recent advances involving this exciting technology: CE-microchips with ED. These comprehensive directions (and their localization into μ-TAS concept)

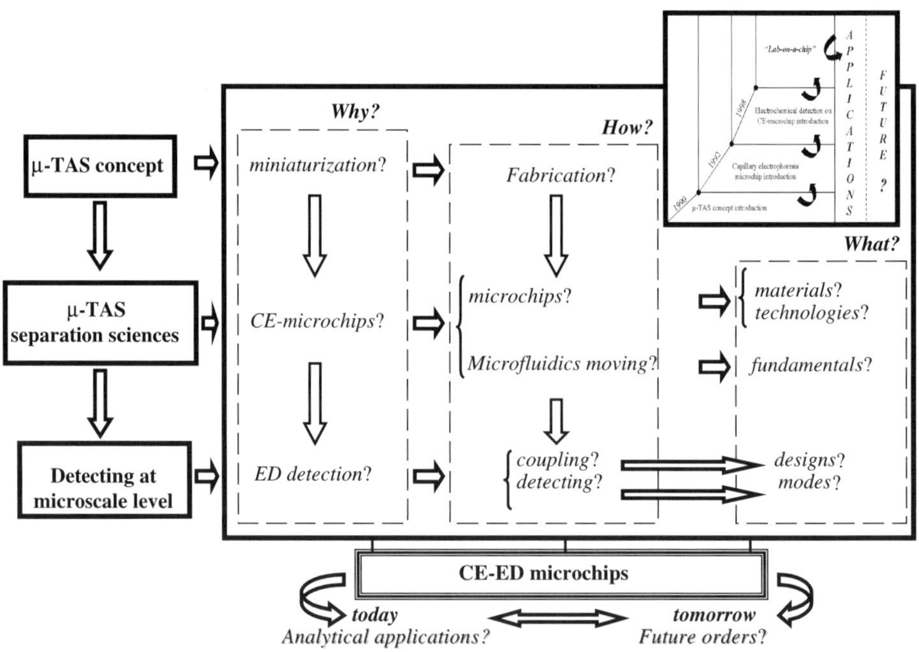

Fig. 14.1. Comprehensive directions of CE Microchips with ED: localization of the CE–ED microchips into the μ-TAS evolution (inset).

are schematized in Fig. 14.1, which summarizes the reasons and modes of CE–ED microchips. Fundamental questions such as "why miniaturization?", "why CE microchips?" and "why ED?" have been answered in this introduction. In the following sections we will try to give responses to other related questions such as "how to make?" and "what to use?". Therefore, a brief summary on micromachining/microfluidic handling outlines in microchip technology will be given. Detection configurations/design and detection modes will be studied in detail. At the end of the chapter, selected analytical applications and the main future prospects will be described.

To make a more careful study, the authors invite the readers to refer chapter 8 of this book, selected reviews written by relevant scientists [10,11,14–18] and the excellent published books [3,19] on this subject.

14.2 PRINCIPLES AND DESIGNS OF CAPILLARY ELECTROPHORESIS WITH ELECTROCHEMICAL DETECTION MICROSYSTEMS

The typical layout of a CE-microchip (both a simple cross-T injector (A), and a twin-T injector (B)) is depicted in Fig. 14.2. It has a network

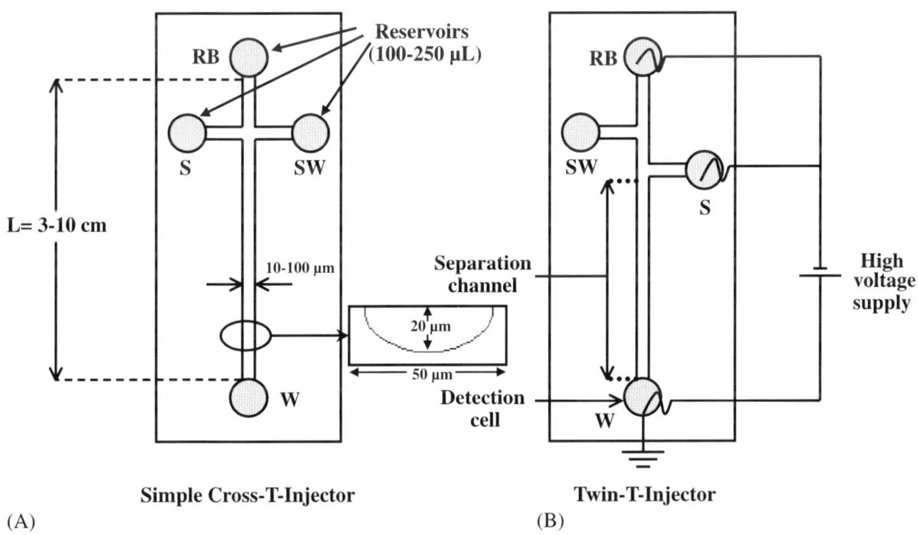

Fig. 14.2. Simple cross and twin-T chip layouts: (A) dimensions, (B) connections (RB, running buffer; S, sample; SW, sample waste and W, main detection waste).

of channels with widths varying from 10 to 100 μm with typical straight separation channels between 3 and 10 cm. In a typical set-up, buffer solutions are introduced on sample (S), running buffer (RB), and waste (W) reservoirs. The reservoir volume is often defined by their capacities (100–250 μl). Electric fields are applied to the reservoirs by high-voltage power supplies (1–5 kV), and platinum electrodes are placed in the reservoirs for injection and separation because samples are normally electrokinetically driven into the network channels. The electrochemical detector consists of a potentiostat, normally using a three-electrode system. Reference electrodes are usually Ag/AgCl, and platinum wires are used as counter electrodes.

14.2.1 Material and microfabrication techniques

An accurate and detailed description of microfabrication techniques is obviously beyond the scope of this chapter, but the readers can find more information in a number of excellent books [3,19]. However, since one of the most attractive features of ED on microchips is its compatibility with microfabrication technologies, it should be interesting to give a brief overview of these microfabrication techniques in order to better understand the designs and principles that control the CE–ED microdevices.

Microchips used in CE are mainly fabricated using various glass substrates [3,10], from inexpensive soda lime glass to high-quality quartz. Glass substrates are the most common because of their good optical properties (optically transparent, which allows optical detection and visual inspection), resistance to many chemicals, well-understood surface characteristics (exhibit an EOF) close to fused silica $\approx 9.5 \times 10^{-4}\,cm^2/V\,s$ at $pH > 9$), dielectric properties (possibility of working with the high voltages used in CE), and well-developed microfabrication methods (adapted from the silicon microfabrication industry). Other advantages of glass are its hardness, high thermal stability and biocompatibility (wide range of applications: DNA separations, enzyme/immunoassays, cell biology).

Various polymer materials have been used to fabricate microchips for CE separations, the most used being poly-methyl-methacrylate (PMMA) as an example of a thermoplastic material (because a wide range of microfabrication methods are available for this compound), and poly dimethylsylosane (PDMS) as an example of an elastomer [3,12]. PDMS has been widely discussed owing to several characteristics: (i) optical transparency; it is suitable for optical detection down to 320 nm; (ii) it cures at low temperatures and moulding can easily be replicated through the process of prototyping, master formation, and soft lithography; (iii) it can seal reversible to itself and other materials by van der Waals contact with clear smooth surface at room temperature; (iv) its surface chemistry can be controlled to form EOF using a plasma technique [12].

Polymers also show good resistance to chemical treatment and good biocompatibility. It is possible to find a polymer that has the optical qualities desired for a given application (i.e. PDMS is transparent to the UV region of the electromagnetic spectrum while the most thermoplastic polymers, such as PMMA and polycarbonate (PC), are transparent in the visible region). The large EOF in glass is partly due to the high surface charges it has because of the deprotonation of silanol groups. However, many polymers do not contain ionizable functional groups and, thus, they would be expected to produce much smaller EOFs, being in clear disadvantage. Indeed, under controlled microfabrication techniques, it is possible to generate EOF very acceptable for PMMA of $1.2 \times 10^{-4}\,cm^2/V\,s$. Although native PDMS exhibits a very small EOF value, subjecting it to oxidation in a plasma discharge changes the surface characteristics. $OSi(CH_3)_2O$ groups are transformed to $OSi(OH)_{4-n}$ producing SiOH groups that lead to surfaces that can be readily deprotonated as glass or fused silica [20]. Also, polymer microchips are of increasing interest because their potentially low manufacturing costs may allow them to be disposable. The combination of low-cost polymer microfabrication with the low cost of electrochemical detectors should be drawn as the true disposable "lab-on-a-chip".

Next, the most important aspects involved in both glass and polymer microfabrication technologies will be described.

14.2.1.1 Glass micromachining

Structures on glass substrates are usually generated using standard photolithographic technologies [3,19,21–24]. Fig. 14.3A presents an example diagram of such a procedure.

First, a layer of photoresist is spin-coated on top of the sacrificial mask layer (Cr/Al) (Fig. 14.3A: deposition). Photoresist is a polymer that becomes soluble (positive) or insoluble (negative) in developer solutions after exposure to light (Fig. 14.3A: exposition light). In the next step, the photoresist spun on the microchip is exposed in the region defined by a photomask, typically using an aligner. The photomask is a

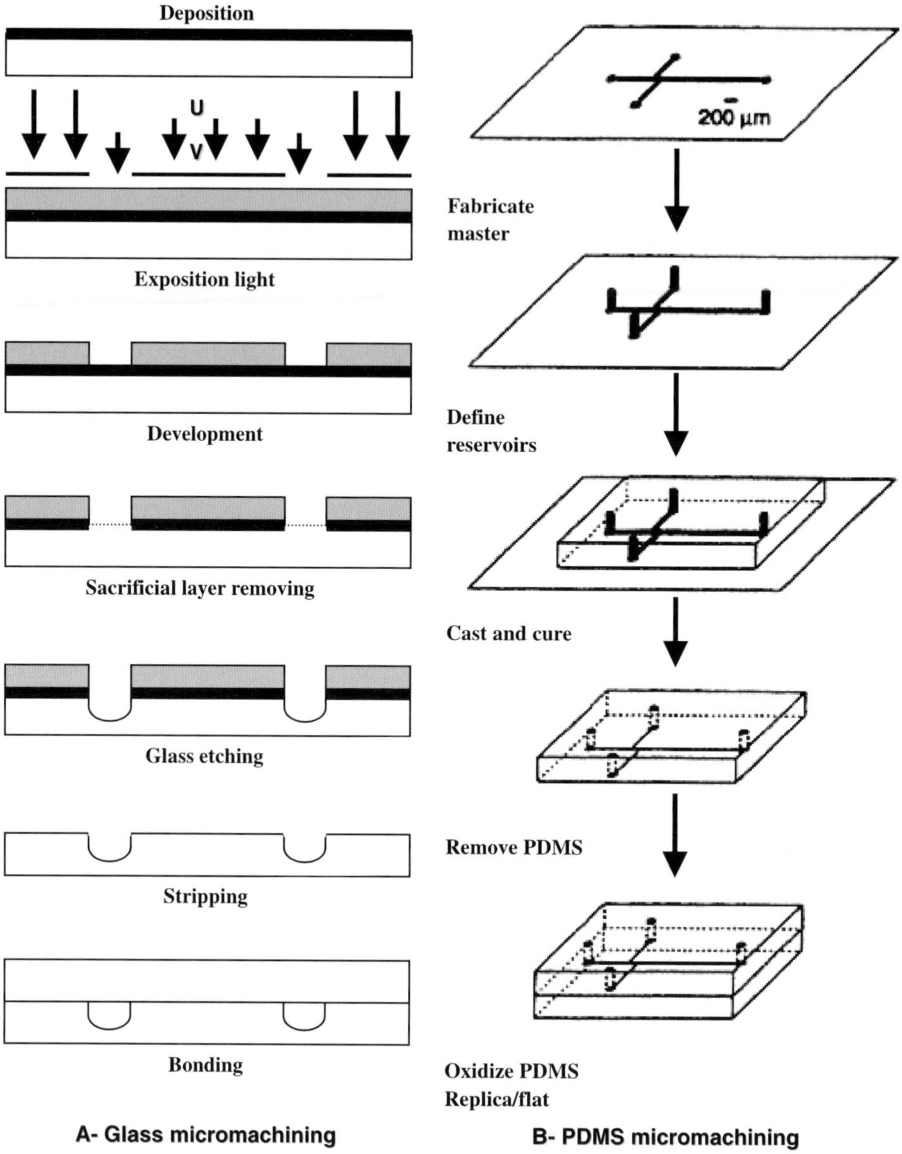

Fig. 14.3. (A) Schematic processing sequence for fabricating microchannels on a glass bottom substrate. Reprinted from Ref. [3]. Copyright (2004), with permission from John Wiley & Sons, Inc.–Wiley-VCH Verlag GmbH. (B) Schematic diagram illustrating how microfabricated structures are made in PDMS. Reprinted with permission from Ref. [27]. Copyright (1998), American Chemical Society.

plate with a user-designed pattern that is transparent, while the background is opaque (or vice versa) to the exposition light. After the microchip is baked to harden unexposed resist, the exposed photoresist is dissolved with a developer solution (Fig. 14.3A: development). The sacrificial mask layer (Cr/Al) of the exposed region is removed using the appropriate etchants (Fig. 14.3A: sacrificial layer removing). During this time, the sacrificial layer underneath the unexposed photoresist remains intact.

After pattern transfer and development, the portions of the microchip that are to be etched have been unmasked and are now ready for chemical etching (Fig. 14.3A: glass etching). HF is used as the primary etchant and can be prepared in various solutions, including HF/NH_4F, HF/HNO_3 and concentrated HF. The etching rate of HF for glass is readily controllable if the temperature is controlled. Following etching of the microchannels, the photoresist and sacrificial mask layer are stripped (Fig. 14.3A: stripping), and the access holes drilled (not shown in figure). The access holes can be drilled on the etched substrate or on another blank glass wafer. When holes are drilled on the etched substrate, aligning two substrates for bonding is much easier. Finally, the substrate is bonded to another piece of substrate to form a finished microchip (Fig. 14.3A: bonding) [3,10].

14.2.1.2 Polymer micromachining

The two major ways of machining polymers are replication from a master (moulding methods) and direct machining [3,19]. Replication methods often produce a microstructure by allowing a polymer workpiece to form an inverse copy of a mould. These methods are: hot embossing [10,25], injection moulding [10,26] and casting [10,19,27]. The formation of microchannels and other structures using *moulding* methods generally involves two primary steps: (i) moulder (also known as master) fabrication, and (ii) channel pattern transfer from that moulder to polymer substrates. In both methods, a very soft or even liquid form of polymer is poured or pressed into a mould, after which the material is hardened and removed from the mould. On the other hand, direct machining methods remove small amounts of polymer in places where the microstructures (microchannels, reservoirs) should be located. This type of micromachining is called laser ablation [10,28].

Hot embossing is the process of pressing a mould into heated, softened thermoplastic, followed by cooling, thereby producing an inverted replicate of the mould. The first step in hot embossing consists of heating the mould and the polymer to glass transition temperature. Once the polymer begins to soften it takes the form and the shape of the mould. The mould and the polymer are then cooled below the glass transition temperature to harden the polymer, and afterwards the polymer is removed from the mould. Embossing can take several minutes per device and can be a useful tool in rapid prototyping devices [25].

In the injection moulding technique the molten polymer material is injected into the mould. For microinjection moulding, the mould is heated to the softening temperature of the polymer to prevent the injected polymer material from hardening too early [29]. After injection of molten polymer into the mould, this is slowly cooled, so that the polymer hardens and then, the polymer microstructure can be removed from the mould. Injection moulding allows very high-throughput production with low production costs (moulded devices are released every 5–10 s) [26].

Casting uses a chemical process to harden the polymer [19]. Two components, a base and a hardener or cure, are mixed just prior to use. Immediately upon mixing, the chemical curing process starts. After some time, this process results in hardening of the polymer (at atmospheric pressure and temperature). The liquid mixture is poured into the mould, and as the polymer sets it takes the shape of the mould. The polymer structure can then be removed from the mould. This technique is very popular with elastomers such as PDMS because they form hermetic, reversible seals and smooth surfaces like glass or silicon by adhesion. Casting is the simplest among these three moulding processes, but requires contact with the mould for minutes or hours. A schematic representation of this type of replica moulding is shown in Fig. 14.3B [27]. Channels in PDMS are easily formed by replica moulding, which is simply the generation of a negative replica of the master in PDMS by pouring the prepolymer over the master and subsequent curing at atmospheric pressure at somewhat elevated temperatures. In the last step, the PDMS replica containing the channel network was sealed irreversibly with a second flat slab of PDMS. In the example shown, this is due to oxidation of both surfaces in an oxygen plasma discharge. Oxidation results in a stronger seal than with untreated PDMS, and it seems to be easier to fill oxidized PDMS channels with liquids; as we explained before, higher cathodic EOFs were generated, which is consistent with a more negatively charged surface of PDMS after oxidation.

On the other hand, laser micromachining (such as photoablation) is a direct machining method based on the removal of polymer material by using intense UV or infrared radiation provided by a laser.

The photoablation process involves absorption of a short-wavelength laser pulse to break covalent bonds in long-chain polymer molecules with production of a shock wave that ejects decomposed polymer fragments [3,10,28]. Many commercially available polymers can be photoablated, including PC, PMMA, polystyrene, nitrocellulose and poly (tetrafluoroethylene) [12,28]. The laser energy can be specially patterned using a mask with the subsequent generation of microcavities and channels in various geometries or by controlling the position of the laser with x–y stages. The resulting structures are generally characterized as having little thermal damage, straight vertical walls and well-defined depth [30]. However, laser ablation does not lend itself well to mass production.

14.2.2 Fluids moving: principles of injection

A very important fluidic handling function is the ability to dispense very well-defined and small volumes of solutions with reproducibility. Especially, the injection of a well-defined, reproducible injection plug into the separation channel is of paramount importance to find the right trade-off between separation efficiency and detection sensitivity. These small volumes are very important in many analytical applications.

The integrated injectors are usually either cross-channel pieces, formed by orthogonally intersecting the separation channel with a channel connecting the sample to waste (T-injector) (Fig. 14.2A) or twin-T injectors (Fig. 14.2B), where the two arms of the sample to waste channel are offset to form a large injector region. Integrated sample injectors permit volume-defined electrokinetic sample injection of short injection plugs with reproducibility [11].

The type of injection employed depends on the sample matrix, precision required and the number of voltage sources available [15]. Three different established injection methodologies have been employed with CE–ED microchips: unpinched [15,31–33], pinched [3,15,32,34] and gated injections [3,15,35]. In each of these approaches the detection reservoir is held at constant ground.

The easiest injection procedure is the unpinched approach, which is shown in Fig. 14.4A (in general, it is the common strategy used)

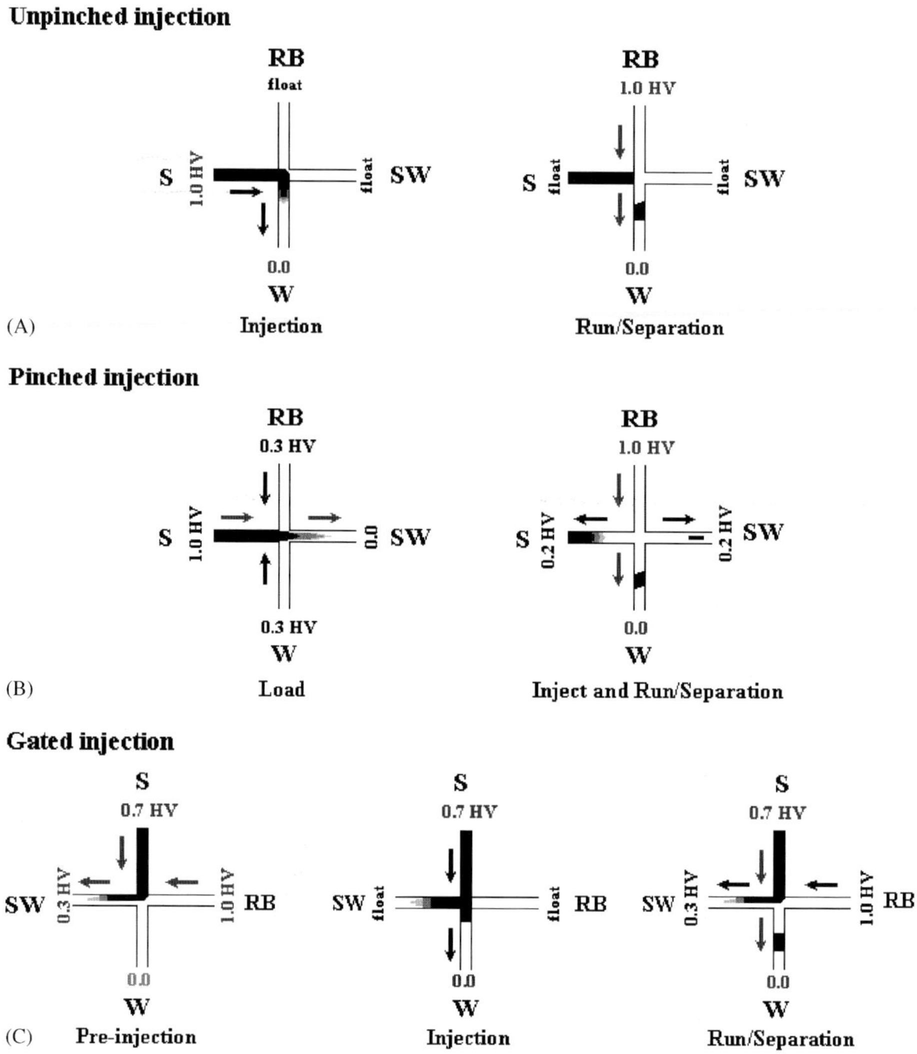

Fig. 14.4. Injection protocols: (A) unpinched, (B) pinched and (C) gated. Reprinted from Ref. [3]. Copyright (2004), with permission from John Wiley & Sons, Inc.–Wiley-VCH Verlag GmbH.

[15,31–33]. This principle can be carried out using just a single power supply. A high voltage is applied to the sample reservoir for a short time with the electrochemical reservoir held at ground (*injection*). Sample is introduced directly into the separation channel by electrokinetic injection. After the injection is completed, the high voltage is switched back to the buffer reservoir and the separation is initiated (*run/separation*). This approach does not use pushback voltages to keep sample in the main separation channel, and can result in irreproducible injections and large plugs.

Pinched injections require voltage control of each reservoir (Fig. 14.4B) [3,15,32,34]. The sample reservoir is placed at the end of one of the side arms, and prior to injection, the sample is continuously pumped across the intersection to the waste reservoir (*load*). Additionally, the voltages are arranged so that there is also flow from the buffer reservoir and the reservoir at the end of the main channel (electrochemical reservoir) towards the intersection and onto the sample waste reservoir. This is to avoid premature bleeding of sample into the main channel. By adjusting the voltages (flow rates), one can "pinch" the sample stream more or less on its passage through the intersection (*inject and run/separation*). For this reason, this technique is often called "pinched injection".

Fig. 14.4C shows the principle of gated injection [3,15,35], which also requires voltage control over each reservoir but with other configurations. Before injection (*preinjection*), the main waste reservoir at the end of the separation channel (electrochemical reservoir) is set to ground. Thus, it functions as the anode to which the EOF is directed. All other reservoirs are at higher positive potentials. The topmost reservoir contains the sample, and to ensure pumping out of the reservoir, the potential is set to a large value. For the same reason, the sample waste reservoir is set to a relatively low voltage to receive the sample flow. The remaining reservoir is filled with a buffer and set at a slightly higher voltage than the sample reservoir. This allows flow out of this reservoir, to keep sample from entering the main channel by flowing right across the junction and into the sample waste reservoir. At the same time, this reservoir also provides a flow of fresh buffer into the main channel. The simplest way to facilitate injection is to remove the power supplies at the sample waste and the buffer reservoir from the electric circuit (*injection step*). Voltage is still applied at the other reservoirs, allowing sample to flow into the main channel. The injection step is finished as soon as the original voltage distribution is established again, thereby cutting off the sample flow into the main channel and releasing a plug of sample flowing to the anode (*run and separation steps*). Because the injected volume is mainly determined by the timing of the injection sequence and sample can enter the main channel for as long as this electrokinetic valve is open, the injection procedure is called "gated injection". In this scheme, the injection is biased, meaning that during the time of injection, positively charged ions are injected to a larger extent than neutral species, which are also injected to a larger extent than negatively charged ions. This fact needs to be taken into consideration when designing a method for quantitative analysis involving one or more gated injection steps.

The main differences between them are that pinched injection volume is predefined and fixed, and that if sufficient time is given to the loading step, then the composition of the sample solution present at the intersection is the same as the composition of the original sample solution, and consequently there is no electrokinetic bias upon injection. Unpinched injection becomes the easiest but less reproducible scheme, although several reports include good relative standard deviations (less than 5%) [32,33].

14.2.3 Detection implementation

The detector design should ensure well-defined mass transport, minimal band broadening and electrical isolation (decoupling) from the high separation voltage (typically 1–5 kV). The latter is attributed to the fact that the current associated with the high separation voltage is usually several orders of magnitude larger than that measured at the electrochemical detector. High sensitivity, selectivity (via the applied potential and electrode material), simple handling, long-term stability and rigidity are additional requirements. To reach these requirements, different and very creative approaches have recently been published in the most relevant literature [15,16,18].

Two main approaches have been proposed. Taking into account the relative position between both working electrode and separation channel [18], the configurations can be classified as: *end-channel*, *in-channel* and *off-channel* detection (Fig. 14.5A). In *end-channel* detection (both externally mounted and internally mounted), the electrode is placed just outside of the separation channel. For *in-channel* detection, the electrode is placed directly in the separation channel, and *off-channel* detection involves grounding the separation voltage by means of a decoupler before it reaches the detector.

On the other hand, taking into account the relative position between both working electrode and flow direction [16] (Fig. 14.5B), we can find three configurations: *flow by*, where the direction of flow is parallel to the electrode surface (means using two plates); *flow onto*, where the electrode surface is normal to the flow direction; and *flow through*,

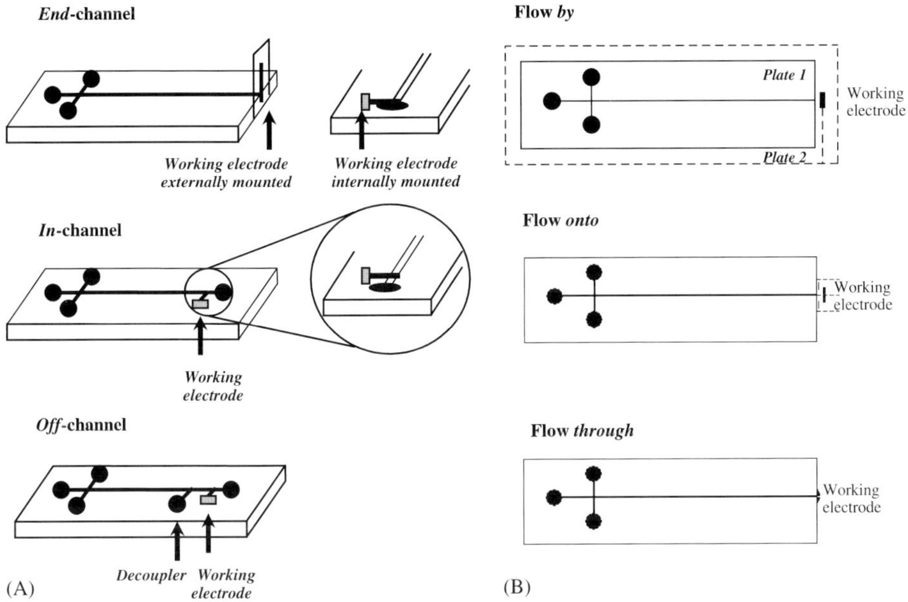

Fig. 14.5. (A) Common configurations of electrochemical detectors for CE microchips, based on different capillary/working-electrode arrangements: *end-channel* (external or internal), *in-channel*, and *off-channel*. Reprinted from Ref. [16]. Copyright (2002), with permission from John Wiley & Sons, Inc.–Wiley-VCH Verlag GmbH. (B) Common configurations of electrochemical detectors for CE microchips, based on different position of the electrode relative to the flow direction: flow by (parallel using two plates); flow onto (with the surface normal to the flow direction); flow through (with the detector placed directly on the channel exit). Reprinted from Ref. [18]. Copyright (2002), with permission from Elsevier.

where the electrode is placed directly on the channel exit. As can be observed, in all cases, the common characteristic is the arrangement of the working electrode outside of the separation channel because such placing results in self-isolation from the high separation potential, owing to the dramatic drop of the potential across the capillary (as explained before: to a negligible value at its outlet). All of them should be understood as *end-channel* detection protocols.

14.2.3.1 End-channel detection

End-channel detection involves the alignment of the electrode at the end of the channel (tens of micrometres), and it means that the electrode is outside the channel. Separation voltage has a minimal influence on the potential applied in electrochemical detectors because most of the voltage is dropped across the channel. However, the separation voltage can cause a small but significant shift in the potential of the working electrode [16,18]. Therefore, to determine the appropriate detection potential for a given analyte, it is necessary to obtain a hydrodynamic voltammogram for the compound of interest under the exact separation conditions that will be employed [36]. The main advantage of this design is that no decoupler is necessary, and that the whole system is simpler and rugged because the channel is all in one piece. However, the main disadvantage is a loss of separation efficiency due to the relatively short length of the channel and the large distance between the end of the channel and the working electrode. This separation distance is also crucial for the noise signal obtained and it can lead to a complete loss of the analytical current.

Next, the most relevant and representative CE–ED microchips *end-channel* configurations will be described. Figs. 14.6–14.9 show selected CE–ED microchips *end-channel* designs. In these figures, the microsystem designs are clearly depicted and representative electropherograms are also enclosed—as representative examples—to show their analytical performance.

Mathies' group [34] described one of the firsts CE–ED microchip; the design of this configuration is showed in Fig. 14.6. This excellent work shows the integration of microfabricated CE on glass substrates with electrochemical detection in only one microdevice. Photolithographic placement of the working electrode just outside the exit of the electrophoresis channel provided high-sensitivity electrochemical detection with minimal interference from the high-voltage separation. In this approach, *flow by* was the configuration used in relation to flow direction-detector position. The suitability of the microsystem was illustrated giving examples of separations of neurotransmitters in less than 100 s with a good resolution (Fig. 14.6D), and detection limits of 3.7 μM for dopamine as well as with DNA analysis.

Wang and co-workers [32] (Fig. 14.7) have reported a very interesting glass CE–ED microchip system that utilizes a single working electrode using end-channel configuration. This configuration has been described by the coupling of a single carbon ink screen-printed working electrode (flow-onto configuration), and a sputtered gold electrode (not shown in the figure) [31] (flow-through configuration). A Plexiglas holder was designed to hold the microchip and to integrate

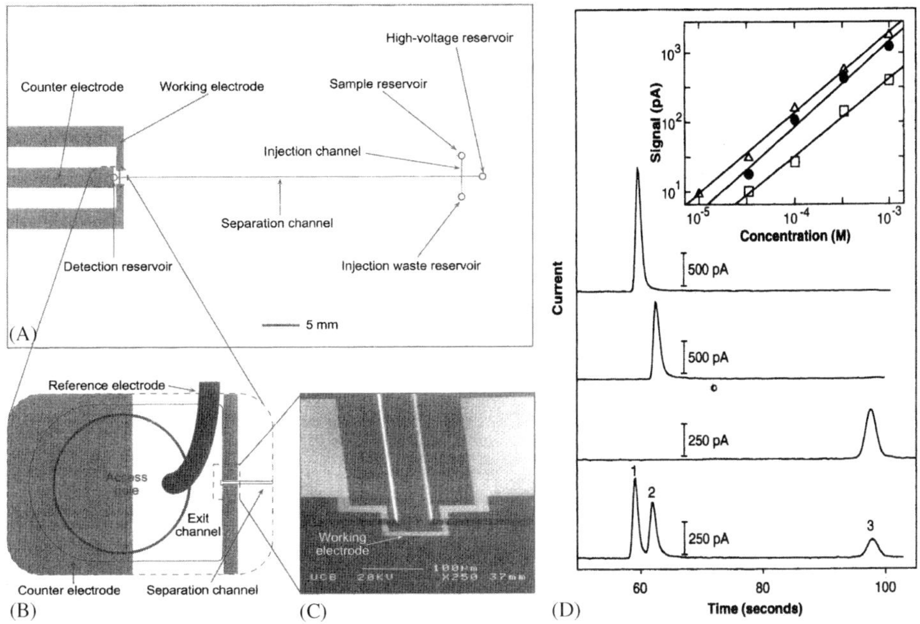

Fig. 14.6. *End-channel* configuration of electrophoresis microchip with integrated ED cell: (A) Full chip view. (B) Expanded view of the integrated electrochemical detector. (C) Scanning electron micrograph of the detection region showing the location of the working electrode in the exit channel. (D) Electrophoretic analysis of neurotransmitters: (a) separation of 1 mM dopamine; (b) separation of 1 mM epinephrine; (c) separation of 1 mM catechol; (d) separation of 330 M dopamine (1), epinephrine (2), and catechol (3). Reprinted with permission from Ref. [34]. Copyright (1998), American Chemical Society.

the electrochemical detector and reservoirs. The easy replacement and alignment of the screen-printed working electrode is one of the most attractive features, as well as the adaptation to other complex microchip formats using the same electrochemical cell detection/Plexiglas set-up. The reliability of the new set-up was widely demonstrated using model analytes (dopamine and catechol). Variables influencing the separation efficiency and amperometric response, including the carbon ink used (Fig. 14.7), channel-electrode spacing, separation voltage and detection potential were optimized. Good detection limits (submicromolar range) were also obtained. A fast nitroaromatic explosives electrophorctic separation (less than 4 min) was also obtained using this configuration. Wang's group [37] has also reported the suitability of coupling SPE with PMMA microchips (totally disposable "lab-on-a-chip")

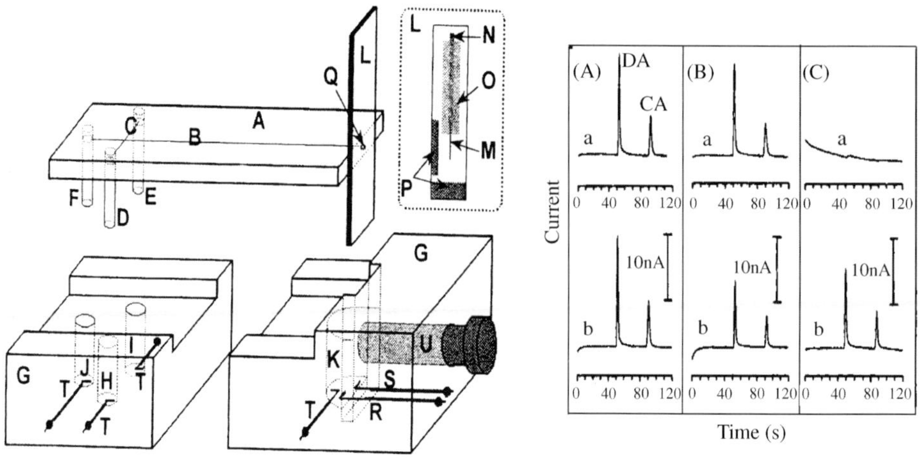

Fig. 14.7. *End-channel* configuration of electrophoresis microchip with integrated ED cell: chip, chip holder and electrochemical cell (left). Influence of different carbon inks on the detection of 100 μM dopamine (DA) and catechol (CA) (right). Reprinted with permission from Ref. [32]. Copyright (1999), American Chemical Society.

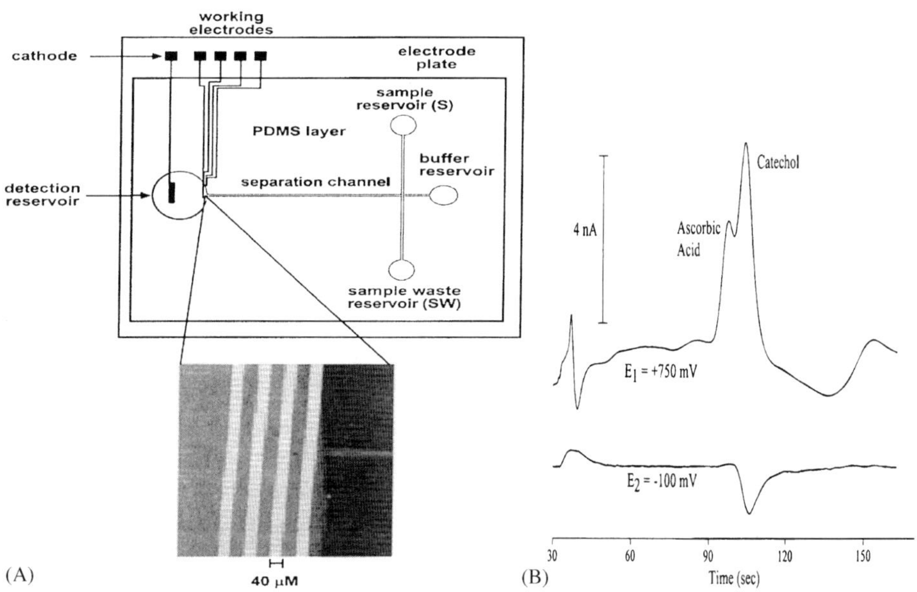

Fig. 14.8. *End-channel* configuration of electrophoresis microchip with integrated ED cell. (A) Full chip view. (B) Dual-electrode detection of an unresolved mixture of 100 μM ascorbic acid and catechol. Reprinted with permission from Ref. [33]. Copyright (2000), American Chemical Society.

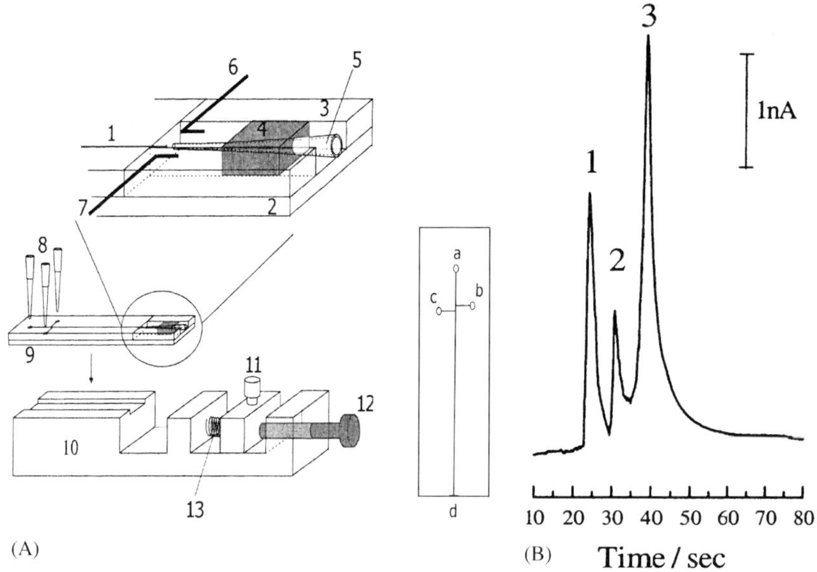

Fig. 14.9. *End-channel* configuration of electrophoresis microchip with integrated ED cell: (A) chip, chip holder and electrode alignment. (B) Electropherogram for 0.1 mM (1) dopamine, (2) 5-HT, (3) epinephrine. Reprinted with permission from Ref. [39]. Copyright (2002), American Chemical Society.

(design is not shown). This group has also described recently a new *end-channel* detection configuration that utilizes a thick film carbon working electrode and a PDMS microchip [38] (design is not shown). In this design, the working electrode is integrated onto a glass base plate while a PDMS chip contains the separation channels. The two plates are bonded together so that the separation channel ends before coming into contact with the working electrode. The separation channel depth and electrode height are similar, creating a flow-onto/flow-by hybrid arrangement. The limits of detection (LODs) obtained for model analytes, such as epinephrine and catechol, by using this *end-channel* configuration were lower than those obtained in glass chips and PDMS-based carbon electrodes. These improvements were attributed to effective isolation of the working electrode from the high separation voltage (noise decreasing) and to the enhanced mass transport of analyte due to the combined flow-onto/flow by flux.

Lunte's group [33] has reported an excellent configuration on the development of a PDMS microchip electrophoresis system employing electrochemical detection (Fig. 14.8). For the first time, the described device consists of a top layer of PDMS containing the separation and injection channels, and a bottom glass layer onto which four gold detection electrodes have been deposited. It was also the first report of dual-electrode electrochemical detection in a PDMS microchip format. The performance of the chip was evaluated using catechol as a test compound (detection limits were 4 μM). Also, the benefit of the dual detection in terms of detection selectivity was illustrated in the detection of an unresolved mixture (ascorbic acid and catechol) by applying suitable detection potentials (Fig. 14.8).

Recently, a novel CE microchip system with ED, using *end-channel* detection with a replaceable microelectrode has been proposed [39]. The *end-channel* electrochemical detection on the CE microchip was achieved by fixing the working electrode (carbon fibre, Pt, Au) through a guide tube on the end of the separation channel (Fig. 14.9). The alignment of the electrode was carried out accurately and reproducibly allowing a low noise and good reproducibility with detection limits for dopamine of 2.4×10^{-7} M. An electrophoretic separation of three neurotransmitters was carried out in less than 50 s (Fig. 14.9). Also, the interest of this configuration is that this kind of electrode can be fabricated in general laboratories and can be replaced quickly with electrodes of different materials.

When PDMS microchips are used, it is much more difficult to accurately maintain the location of the microchannel at the micrometre level each time for PDMS microchip than for the glass one. Recently, a three-dimensional adjustor has been reported [40]. Also, very recently to overcome the limitations of *end-channel* detection systems, a microfabricated CE chip containing an integrated sheath-flow electrochemical detector has been proposed [41]. In this approach, the sheath-flow channels join the end of the separation channel from each side, and gravity-driven flow carries the analytes to electrochemical detectors placed at different working distances from the separation channel exit.

Table 14.1 summarizes the features of these designs, and the characteristics of the chip and working electrode used, injection protocols, detector configuration and analytical performance are shown. Also, a brief comment is enclosed in order to know the relevant "key" in each design.

The two general strategies for integration of electrodes into CE-microchips have been *end-channel* and microfabrication. *End-channel* configuration allows electrode cleaning and use of chemically modified

TABLE 14.1

Features of the main CE–ED microchip with amperometric detection: end- and in-channel designs

Characteristics/materials: Chip	Characteristics/materials: Working electrode	Injection characteristics	Detector configuration: Electrode/channel	Detector configuration: Electrode/flow	Analytical performance			Comments	Ref. (year)
						Theoretical plates /m	LODs	Microfabrication Integrated	
Glass $L = 50$ mm ($46\,\mu m \times 8\,\mu m$)	Sputtered Pt	Pinched injection T-injector (simple cross)	End-channel ($30\,\mu m$)	Flow by	Dopamine	21,000	$3.7\,\mu M$		
					Epinephrine	17,000	$6.5\,\mu M$		
					Catechol	12,000	$12\,\mu M$	Chip+electrodes (first CE–DE microchip)	[34] (1998)
					Linear range: 10–100 μM				
					Analysis time: 100 s				
						Theoretical plates	LODs		
Glass $L_{ef} = 75$ mm ($50\,\mu m \times 20\,\mu m$)	Thick-film (Screen-printed) Carbon	Unpinched injection T-injector (double T)	End-channel ($60\,\mu m$)	Flow onto	Dopamine	4,400	3.8×10^{-7} M	First SPE configuration	[32] (1999)
					Catechol	4,100	7.8×10^{-7} M	Explosives separation Response in less than 4 min	
					Linear range: 20–100 μM				
					Signal reproducibility: 4.3%				
						Theoretical plates	LODs		
Glass (electrodes) PDMS	Gold (deposited)	Unpinched injection T-injector	End-channel ($15\,\mu m$)	Flow by	Catechol	31,433	$4.0\,\mu M$	Dual detection First CE–ED	[33] (2000)

continued

TABLE 14.1 (*continued*)

Characteristics/materials		Injection	Detector configuration		Analytical performance				Comments	Ref. (year)
Chip	Working electrode	characteristics	Electrode/ channel	Electrode/flow						
(channels) $L_{ef} = 10\,mm$ (25 µm × 50 µm)					Linear range: 10–500 µM				PDMS/glass substrate *Glass: printed electrodes PDMS: etched channels*	
					Signal reproducibility <1.0%					
							Theoretical plates	LODs		
Glass $L_{ef} = 75\,mm$ (50 µm × 20 µm)	Microdisk carbon fibre	Unpinched injection T-injector (double T)	End-channel (70 µm)	Flow onto	Dopamine		4,800	2.4×10^{-7} M	Lowest detection limit for dopamine	
					Separation of dopamine, epinephrine and 5-hydroxy-tryptamine					[39] (2002)
					Configuration comparison		Theoretical plates	LODs		
PDMS $L_{ef} = 35\,mm$ (35 µm × 30 µm)	Carbon fibre	Unpinched injection T-injector	In-channel	Flow by	Catechol	(in-channel)	4,880	4 µM	Electrically isolated potentiostat Comparison with LIF	[42] (2002)
						(end-channel)	1,065			
						Electrode size	Theoretical plates	LODs		
PDMS	Metal microwires (Cu, Au, Pt)	Pinched injection T-injector (double T)	In-channel	Flow through	Dopamine	50 µm	1,338	100 nM	Integration of multiple electrodes	[43] (2004)
						25 µm	1,923	250 nM	Multiple electrode materials	
						Linear range: 0.1–100 µM				

electrodes; however, these designs generally lack portability and the ability to incorporate multiple electrodes. Microfabrication allows the incorporation of multiple electrodes on-chip and is easily made portable; but requires the use of expensive metallization and clean room facilities, and integration of more than one electrode material is challenging. To overcome these drawbacks, *in-channel* and *off-channel* approaches have been introduced.

14.2.3.2 In-channel detection

In-channel strategies involve the placement of the working electrode directly within the separation channel. The analytes migrate over the electrode while still confined to the channel, thus eliminating the band broadening often observed with *end-channel* alignments [18]. Martin and co-workers [42] have reported an excellent work involving the development of an electrically isolated potentiostat, which has made it possible to place the working electrode in the separation channel. *In-channel* configuration versus conventional *end-channel* configuration concept is shown in Fig. 14.10. The miniaturized compact potentiostatic system ($4 \times 9 \times 2\,cm^3$) holds promise for the integration of the entire detection system onto a chip. The *in-channel* configuration helps to eliminate some of the negative separation performance characteristics encountered with the *end-channel* configuration, especially with respect to the alignment of the working electrode at the end of the separation channel as well as with respect to separation efficiencies. In fact, this configuration improved the separation efficiency (Fig. 14.10) (number of plates by a factor of 1.3) and exhibited similar separation performance to those found for LIF in terms of plate height and peak symmetry.

Very recently, a simple and sensitive electrode design for CE–ED microchip has been presented by Liu et al. [43], which is depicted in Fig. 14.11. The new design employs metal microwires as the working electrodes for electrochemical detection. This approach aligns a solid metal microwire through the separation channel, allowing the integration of multiple electrodes and the use of different electrode materials (copper, gold and platinum). Also, the size of the working electrode as well as the relative position of the flow with respect to electrode surface was discussed. As can be observed in the figure, the electrode size has an important influence on the separation efficiency and the detection limits obtained. In fact, the lowest detection limit reported to date for dopamine (100 nM) was achieved without the use of

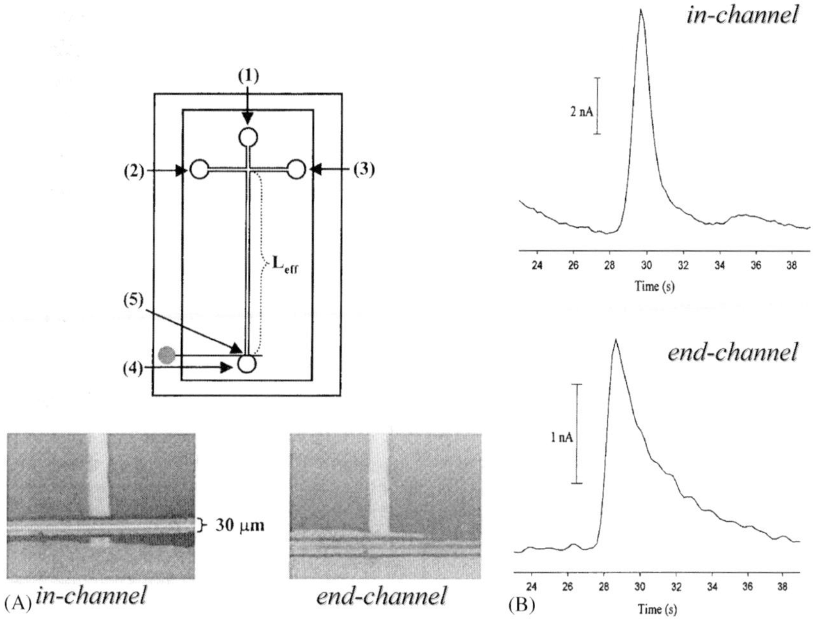

Fig. 14.10. *In-channel* configuration of electrophoresis microchip with integrated ED cell: (A) Chip layout (top) and electrode alignments used in these studies (bottom). (B) Electropherograms for catechol using (a) *in-channel* detection and (b) *end-channel* detection. Reprinted with permission from Ref. [42]. Copyright (2002), American Chemical Society.

a decoupler, as a result of the higher collection efficiency with the new design.

The main features of these *in-channel* designs are also provided in Table 14.1.

14.2.3.3 Off-channel detection

Off-channel detection, has also been employed to overcome the drawbacks of *end-channel* detection. Electrode placement in *off-channel* detection is similar to that in *in-channel* detection, (Fig. 14.5A) but the separation voltage is isolated from the amperometric current through the use of a decoupler. Conceptually speaking, the decoupler effectively shunts the separation voltage to ground, and a field-free region is created where analytes are pushed past the electrode by EOF generated prior to the decoupler. Lunte's group last revision [16] found only two *off-channel* contributions in CE-microchip [44,45]. The first one uses a microchip composed of two different polymers, while the decoupler was

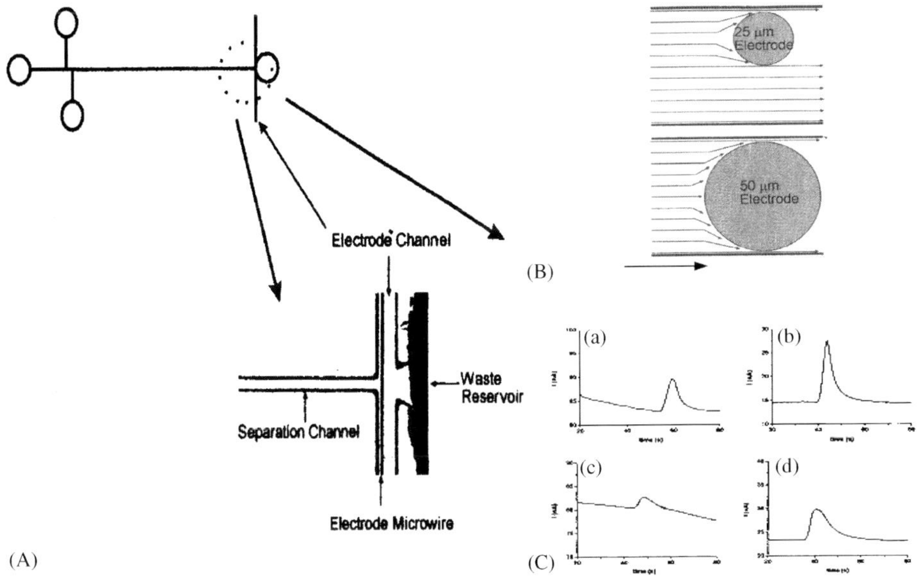

Fig. 14.11. *In-channel* configuration of electrophoresis microchip with integrated ED cell. (A) Schematic of the microchip showing placement of the electrode alignment channel (top). Photograph showing electrode alignment in a completed microchip (bottom). (B) Schematic of flow around the microwire electrode in microchannel. (C) Electropherograms of 100 µM catechol: (a) 25 µm Cu electrode; (b) 25 µm Au electrode; (c) 25 µm Pt electrode; (d) 50 µm Pt electrode. Reprinted with permission from Ref. [43]. Copyright (2004), American Chemical Society.

made creating holes of the top layer. This allowed the placing of the working electrode directly in the separation channel after the decoupler [44]. The other one used a palladium metal film decoupler in a Plexiglas-based microchip with ED [45].

On the other hand, new mixed designs in CE–ED microchips using microfabrication or decoupler strategies have also been recently reported.

Wu et al. [46] have recently described for the first time, the use of a microfabrication technique to simultaneously integrate a three-electrode electrochemical detector and an electric platinized decoupler with an oxygen-plasma-treated PDMS layer containing a CE channel to complete a CE–ED microchip as a whole. Fig. 14.12 shows clearly all the components of this interesting design. This chapter combines the microfabrication advantages with the decoupler concept. At first, the

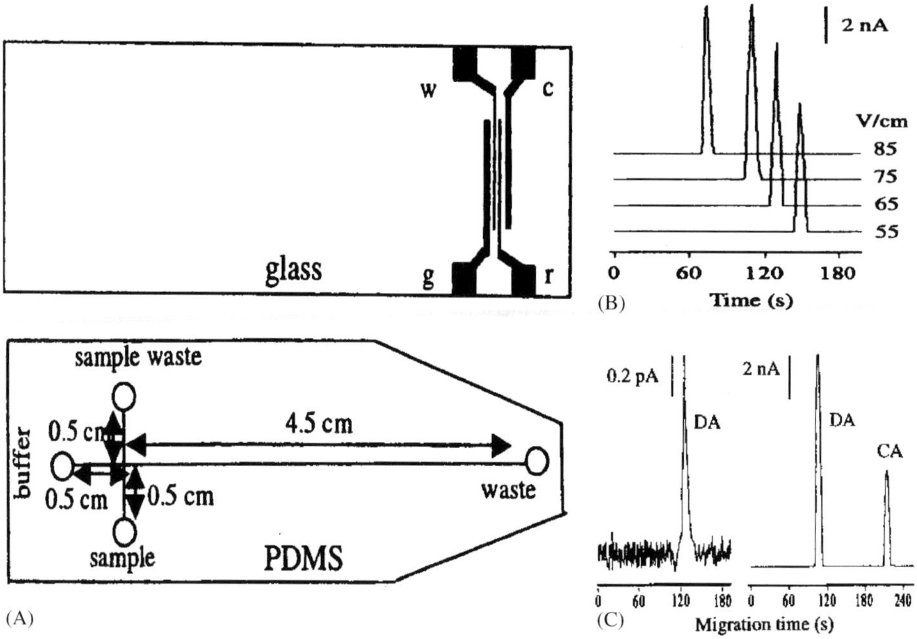

Fig. 14.12. *Off-channel* configuration of electrophoresis microchip with integrated ED cell. (A) Electrode layout on glass (top) and the injection and separation channel made from PDMS (bottom). (B) Electropherograms of 1 mM dopamine performed in different separation strengths of applied electric fields. (C) Electropherograms of 0.125 μM dopamine (DA) in the electric field of 65 V/cm and 1 mM dopamine and 1 mM catechol (CA) in the electric field of 75 V/cm. Reprinted with permission from Ref. [46]. Copyright (2003), American Chemical Society.

material of all electrodes was Au metal. Platinum nanoparticles were electrically deposited on the ground electrode and the reference electrode to be the decoupler and the pseudoreference electrode, separately. The decoupler was located in front of the three-electrode detector to isolate the interference from high-voltage separation/detection potential. Experimentally, the results of baseline offset of EC background current, cyclic voltammetry of ferrocyanide, and analyte (dopamine and catechol) electropherogram demonstrated that the platinized decoupler had sufficient capacity to isolate the interference of the separation electric field (Fig. 14.12).

Very recently, Lai et al. [47] have described an amperometric detector including an *in-channel* dual electrode with a palladium film decoupler. The microchip cleverly integrated a Pd-film decoupler just

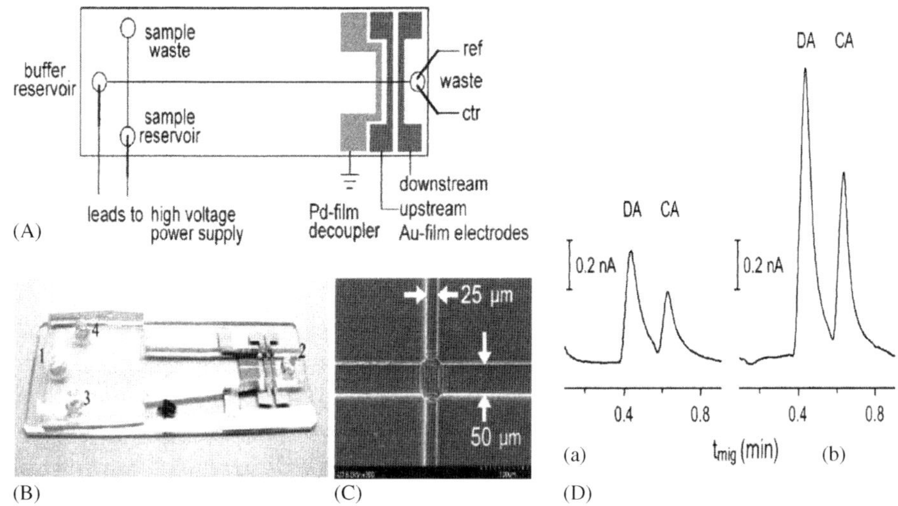

Fig. 14.13. *Off-channel* configuration of electrophoresis microchip with integrated ED. (A) Layout of the electrophoretic chip. (B) A photograph of the microfluidic chip. (C) A SEM micrograph of the PDMS substrate around the cross-region. (D) Electropherograms obtained with detection modes of (a) single electrode and (b) dual-series electrodes. Reprinted from Ref. [47]. Copyright (2004), with permission from Elsevier.

before the two Au working electrodes (Fig. 14.13). The electropherogram showing dopamine and catechol separation in Fig. 14.13 demonstrated that the decoupler had sufficient capacity to isolate the interference of the separation electric field.

A new development of a cellulose acetate decoupler for *on-column* electrochemical detection in CE-microchip has also been described [48]. The decoupler was constructed by aligning a series of 20–30 µm holes through the coverplate of the microchip with the separation channel. Then, a thin film of cellulose acetate was cast within the holes. Detection limits of 25 nM were achieved for dopamine, revealing the excellent isolation of the detection circuit for the separation currents.

Recently, Lunte's group [49] has described the fabrication and evaluation of a fully integrated palladium decoupler/electrochemical detector for CE–ED microchips using a hybrid glass/PDMS device. Both a Pd decoupler and working electrode are deposited onto a glass plate prior to reversibly sealing the glass layer to a PDMS substrate containing the separation channel network. The effect of the Pd decoupler size on the ability to remove hydrogen was also investigated. 500 µm was determined to be the optimum decoupler size and the microsystem was evaluated using dopamine as model analyte.

To date, the work reported has primarily focused on the miniaturization of individual components such as the fluidic circuit or the electrodes rather than a holistic approach aimed at miniaturizing the entire system. In order to create true "lab-on-chip" systems, the electronics as well as the microfluidic circuit and detection scheme need to be miniaturized into a compact system. Very recently, a multiauthor contribution has presented an overview of this approach (total miniaturization) and, in particular, has demonstrated some of the unique advantages of totally microfabricated systems designed for CE–ED [50], offering a multidisciplinary approach to develop a self-contained and transportable CE–ED system that (a) incorporates all necessary electrodes directly on the chip, and (b) utilizes miniaturized supporting electronics designed especially for the purpose of supporting CE–ED microchip designs.

14.2.4 Electrode materials

The detector performance and the success of the CE–ED microsystem are strongly influenced by the material of the working electrode, since that is where the electrochemical reaction of the analyte occurs. The selection of the working electrode depends primarily on the redox behaviour of the target analytes and the background current over the applied potential region. One important aspect is that selectivity in CE–ED microchips can be obtained through a judicious choice of the working electrode material and the applied potential.

On the one hand, carbon, platinum and gold have been used as electrode materials for CE–ED microchips [15,16]. Carbon (including carbon paste or ink and glassy carbon) is the most widely used material for electrodes in CE–ED microchips owing to its minimal fouling and lower overpotential, larger potential window and low background [16]. Lunte's group [51] described the first carbon-based (carbon fibre) dual electrode for a CE-microchip, which used an *end-channel* configuration. This group has also employed carbon paste electrodes in PDMS-based chips for detecting amino acids [52]. Wang's group [32], since the introduction of carbon ink screen-printed electrodes, has reported a large number of possibilities using this type of electrodes [53–56]. The electrodes are printed using patterned stencils. The working electrodes are deposited on a ceramic plate and cured. Contact silver ink is applied next to overlap the carbon. An insulating layer is subsequently printed to cover the carbon–silver junction and is used to define the electrode area ($0.30 \times 2.5\,mm^2$). A large number of analytes have been studied from model analytes (dopamine and catechol), explosives, pollutants and a wide range of analytes of biological and clinical significance (see section 14.3).

Recently, a new technique termed micromoulding of carbon inks for use in microfluidics has been reported [57]. This approach uses PDMS microchannels to define the size of the microelectrode. First, PDMS microchannels of the approximate dimensions desired for the microelectrode are made by soft lithography. The PDMS is then reversibly sealed to a substrate and the microchannels are filled with carbon ink. After a heating step the PDMS mould is removed, leaving a carbon microelectrode with a size slightly smaller than the original PDMS microchannel. The resulting microelectrode (27 μm wide and 6 μm high) can be reversibly sealed to a PDMS-based flow channel.

Platinum electrodes are useful for detecting peroxides, alcohols, aldehydes and hydrazines. Gavin and Ewing [58] described a very interesting electrochemical detector array (of 100 microfabricated electrodes) for parallel spatial resolution of neurotransmitters. Also, Mathies' group [34] has reported the use of a platinum electrode in the catechol measurement employing the design described in section 14.2.3.1.

Gold has also been used as electrode material in CE-microchips. Wang and co-workers [31] have described an *end-channel* configuration using a sputtered gold electrode, gold-coated screen-printed electrodes [37,59–65] and a gold disk electrode [66]. Hilmi and Luong [67] also used a gold electrode in CE-microchips. One important paper was the first description of multiple-electrode electrochemical detection in PDMS-based CE-microchips devices employing gold electrodes [33]. This design was also discussed in section 14.2.3.1.

The power of amperometric detectors can also be enhanced through a deliberate modification of the electrode surface. Electrocatalytic surfaces can be used to accelerate the electron-transfer reactions of species with slow electron-transfer kinetics. Wang's group [68] has reported on a palladium-coated screen-printed carbon electrode for the electrocatalytic detection of various hydrazines. The palladium layer allowed a significant ($>0.3\,V$) lowering of the applied potential and led to sharper and larger peaks as can be observed in Fig. 14.14A. Lunte's group [52] has reported on a cobalt-phthalocyanine (CoPC)-modified carbon-paste

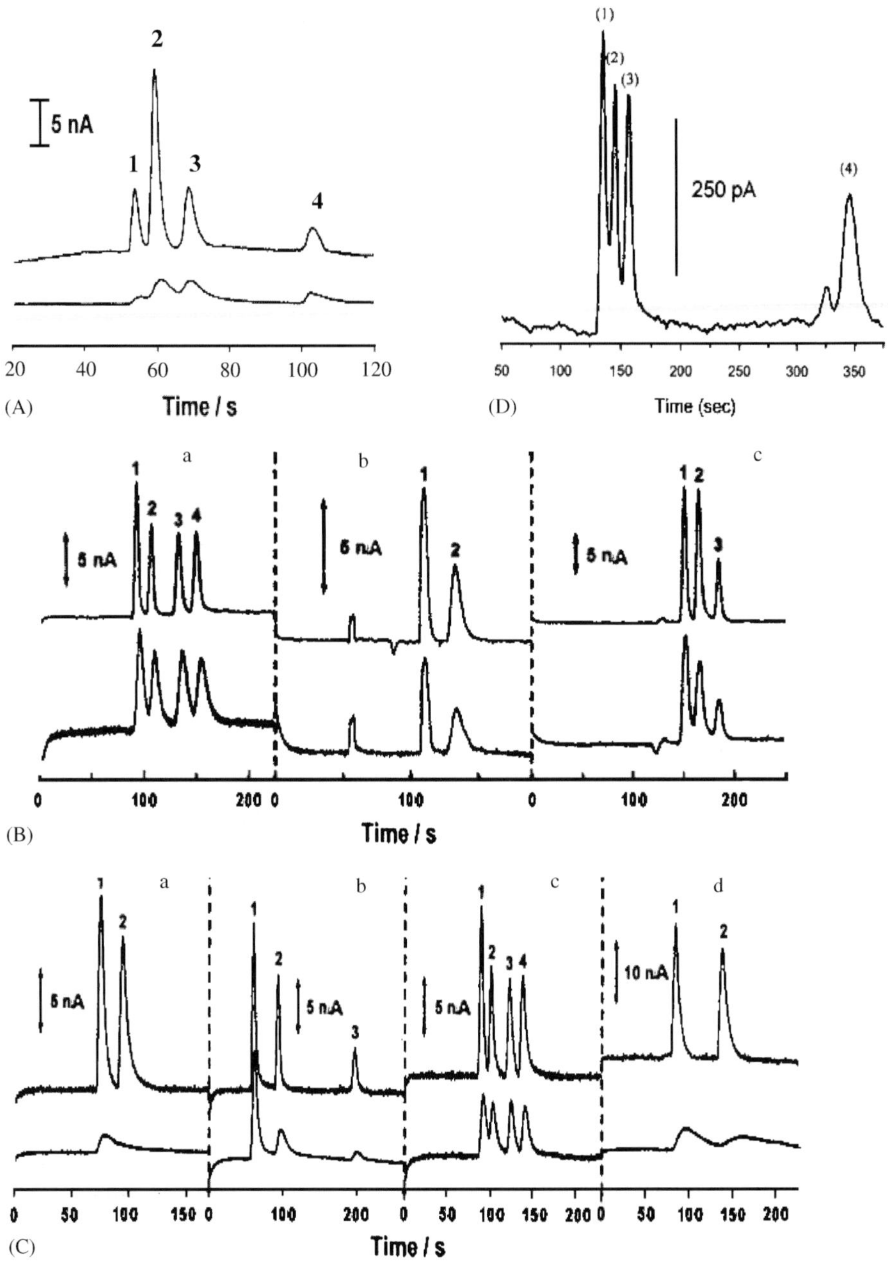

5 nA
1
2
3
4
20 40 60 80 100 120
(A)
Time / s
(1)
(2)
(3)
(4)
250 pA
50 100 150 200 250 300 350
(D)
Time (sec)
a
b
c
5 nA
0 100 200 0 100 0 100 200
(B)
Time / s
a
b
c
d
5 nA
10 nA
0 50 100 150 0 100 200 0 50 100 150 200 0 50 100 150 200
(C)
Time / s

detector for the electrocatalytical detection of thiols. Other attractive routes include selective transport or preferential accumulation in connection with permselective coatings or preconcentrating surfaces, respectively [16].

On the other hand, the most relevant and modern literature reveals how new electrode materials are going to be incorporated into CE–ED microchips technology. Very recently, significant improvements in the performance of a CE-microchip with ED have been observed using a modified carbon-nanotube working electrode [69], as well as boron-doped diamond electrodes [70].

Wang's group [70] has used a chemically vapour-deposited, boron-doped, diamond film band ($0.3 \times 6.0\,mm^2$) electrode for *end-channel* amperometric detection. This material offered enhanced sensitivity, lower noise levels and sharper peaks for several groups of important analytes such as phenols (a), organophosphate nerve agents (b) and nitroaromatic explosives (c), showed in Fig. 14.14B. Detection limits of 70 and 110 ppb were obtained for explosives using this electrode material.

Carbon nanotubes represent an increasingly important group of nanomaterials with unique geometrical, mechanical, electronic and chemical properties. Such properties make single-wall and multiwall carbon nanotubes extremely attractive for the task of ED. The modified carbon-nanotube electrode has allowed amperometric detection at significant low operating potentials and yields substantially enhanced signal-to-noise characteristics [69]. Such advantages have been

Fig. 14.14. (A) Electropherograms for a mixture containing 50 μM hydrazine (1), 100 μM methylhydrazine (2), 300 μM dimethylhydrazine (3), 150 μM phenylhydrazine (4), obtained at the bare (a) and palladium-modified (b) screen-printed electrodes. Reprinted from Ref. [68]. Copyright (2000), with permission from John Wiley & Sons, Inc.–Wiley-VCH Verlag GmbH–. (B) Electropherograms for phenols (a), nerve agents (b) and explosives (c) with screen-printed carbon (bottom) and diamond band (top) electrodes. Reprinted with permission from Ref. [70]. Copyright (2003), American Chemical Society. (C) Electropherograms for hydrazines (a), dopamine, catechol, and ascorbic acid (b), phenols (c), and purines (d) at the bare (bottom) and CNT-modified (top) screen-printed carbon electrodes. Reprinted with permission from Ref. [69]. Copyright (2004), American Chemical Society. (D) Separation of 10 μM dopamine (1), 25 μM (2) epinephrine (2), 25 μM isoprotenerol (3), and 100 μM L-dopa (4). Reprinted with permission from Ref. [71]. Copyright (2003), American Chemical Society.

illustrated in connection with several types of hydrazine (a), dopamine, catechol and ascorbic acid (b), phenols (c) and purines (d) as can be observed in Fig. 14.14C.

The performance of pyrolyzed photoresist carbon films as another attractive working electrode material for ED in CE-microchips has recently been described [71]. This performance is also shown in Fig. 14.14D. The most attractive feature of this material is an exceptionally smooth surface (similar to polished glassy carbon), together with a low oxygen/carbon ratio, which enables low background current levels. The pyrolyzed photoresist carbon films have been introduced as planar carbon electrodes in PDMS-quartz hybrid microchip devices, and their utility has been demonstrated by the separation and detection of various neurotransmitters. This microchip scheme coupled with sinusoidal voltammetry (SV) detection has allowed a detection limit of 100 nM for dopamine.

14.2.5 Modes of detection

Of the three general modes of electrochemical detection–amperometry, potentiometry and conductimetry–amperometry is the most used in the CE-microchip format. More recently, conductimetry has been investigated for microchip-based separations, while potentiometry is not implemented in CE–ED microchips. Both amperometry [16,18] as well as conductimetry [17] coupling in microchip format have recently been revised.

14.2.5.1 Amperometry/voltammetry

Amperometry is the measurement of the resulting current after a constant potential is applied to the working electrode. A reference and auxiliary electrodes are also present. In some cases, only a working and an auxiliary electrode may be employed in a two-electrode configuration [64]. Amperometry has the advantage of generally good detection limits but is restricted to electroactive species. Selectivity is achieved through a judicious choice of the detection potential, with the optimal detection potential normally being determined by hydrodynamic voltammetry. It is the most widely reported electrochemical method for conventional CE, and a number of research groups have recently reported its implementation for microchannel-based separations. The designs described in Section 5.2.3 use amperometry as detection mode, and several will be mentioned later in section 14.3 of this chapter.

Therefore, only the most important amperometric detection advances incorporated into CE–ED microchips will be described below.

SV employs a large amplitude sine wave as the excitation waveform. The resulting faradic current response is non-linear and contains both the signal components at the excitation frequency and the harmonics. The charging current associated with scanning the electrochemical double layer, which is the major background component, is primarily linear, and therefore resides predominantly at the excitation frequency. There is an increase in the discrimination between signal and background at the higher harmonics, resulting in greater sensitivity and selectivity of the amperometric response. Hebert and co-workers [72] have reported an interesting contribution employing SV on microchip detection. Recently, the same authors have reported on a comprehensive direct comparison between constant potential amperometry and SV using microchip platforms [73]. SV was found to be more sensitive (nM) than constant potential amperometry (μM) for the neurotransmitters studied (dopamine, isoproterenol and L-dopa). Applied SV using a pyrolyzed photoresist carbon film has also been described [71].

Pulsed amperometric detection on a hybrid PDMS/glass microchip with a platinum working electrode for the electrochemical detection of glucose, maltose and xylose has been reported [74]. The direct amperometric detection of carbohydrates at noble metal electrodes is characterized by electrode fouling. Carbohydrate detection at a platinum electrode involves adsorption and subsequent anodic dehydrogenation of the molecule at the electrode surface. However, the adsorption of additional molecules quickly becomes limited, as the dehydrogenation product remains adsorbed on the surface of the electrode. As the surface becomes contaminated with the reaction product, the signal rapidly decreases. Pulse amperometry maintains electrode activity via a triple pulse sequence including detection, electrode cleaning and electrode reactivation potentials. Cleaning of the electrode surface via oxidative removal of the adsorbed hydrocarbons occurs at a sufficiently positive potential while regeneration of the electrode surface takes place at a negative potential. Recently, pulse amperometric detection has also allowed the separation and detection of underivatized carbohydrates, amino acids and sulphur-containing antibiotics in CE-microchip format [75,76].

Although the utility of amperometry has widely been demonstrated, a very unique concept about coupling voltammetry on microfluidic chip platforms has been reported [77]. CE-microchips can be understood as ideal platforms for performing microscale voltammetric analysis. The concept deals with the introduction of voltammetric protocols as detection/electrochemical characterization after pumping the analyte using CE-microchips as "injection/sample preparation/liquid manipulation system" because it allows the use of ultra-small detection volumes. This approach is advantageous over nanovial voltammetry, which lacks the sample preparation, liquid-handling and fluid control or manipulation capabilities. But the most attractive feature is that this coupling enhances the power of microchip devices as it adds a new dimension of analytical information and opens up highly sensitive detection schemes. The utility of linear scan, hydrodynamic modulation, rapid square-wave and adsorptive-stripping have been demonstrated using catechol, hydrazine compounds and nickel as model analytes and screen-printed detectors.

14.2.5.2 Conductimetry

Conductimetric detection on microfabricated devices has recently been developed [17,78–88]. Conductimetric detection is a less sensible but universal detection technique that has been applied as a detection mode in CE-microchips, either in the galvanic (a pair of electrodes is placed in the separation channel for liquid impedance measurement) [78,79] or the contactless (no contact between the pair of electrodes and separation channel solution) mode [80–88]. Contactless detection is preferred for three reasons: (i) the electronic circuit is decoupled from the high voltage applied for separation (no direct DC coupling between the electronics and the liquid in the channel), (ii) the formation of glass bubbles at the metal electrodes is prevented, and (iii) electrochemical modification or degradation of the electrode surface is prevented. Both contact and contactless detectors integrated into a microchannel require a physical connection to read out electronics placed inside or even outside the microdevice. These connections, however, should not influence the bonding process or the fluid flow in the microchannel.

Figures 14.15 and 14.16 show two representative designs involving both detection models, galvanic and contactless, respectively. In both strategies, microfabrication also plays an important role in the integration of conductivity detectors.

On the one hand, with respect to the galvanic mode, van den Berg and co-workers [78] developed glass chips for zone electrophoresis and demonstrated the possibility of detecting inorganic ions and fumaric, malic and citric acids with embedded electrodes (see Fig. 14.15). Soper

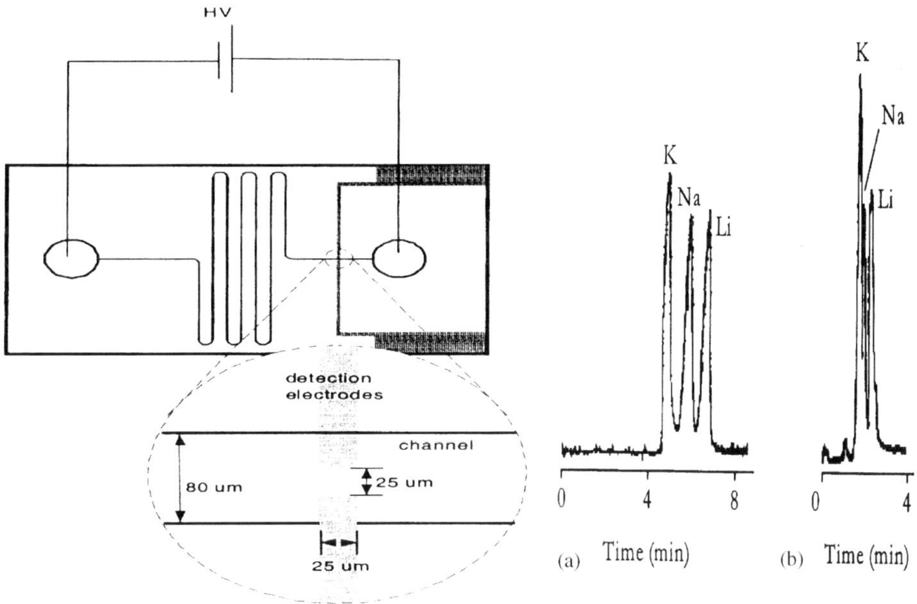

Fig. 14.15. Configuration of electrophoresis microchip with integrated galvanic conductivity detection. (A) Schematic overview of the experimental set-up. (B) (a) Electropherogram of alkali metal ion separations Li^+ in a 6 cm channel. (b) Separation of K^+, Na^+ and Li^+ in a 3 cm channel. Reprinted from Ref. [78]. Copyright (2001), with permission from John Wiley & Sons, Inc.–Wiley-VCH Verlag GmbH.

and co-workers [79] have designed a device in PMMA integrated with Pt wire electrodes, and have shown the separation and sensitive detection of amino acids, proteins and DNA fragments with a built-in conventional contacting conductivity detector using capillary zone electrophoresis (CZE), micellar electrokinctic chromatography (MEKC) and electrochromatography, respectively.

On the other hand, Guijt et al. [80] carried out one of the first and most interesting reports involving contactless conductivity detectors. In this work, a contactless capacitively coupled conductivity detector integrated within a CE microchannel was presented and evaluated. The four-electrode conductivity measurement made use of two outer and two inner electrodes. Capacitive coupling with the liquid inside the channel was obtained by covering the electrodes with a thin dielectric layer of silicon carbide. Other works dealing with CE-microchips with integrated contactless conductivity have also been reported [81,82]. Wang's group [83] has demonstrated contactless detection

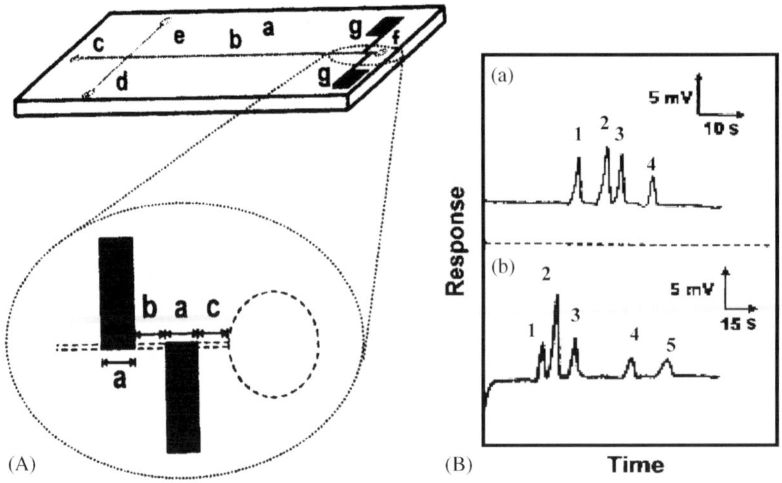

Fig. 14.16. Configuration of electrophoresis microchip with integrated contactless conductivity detection. (A) Microchip (top). Enlarged view of the detector (bottom). (B) Electropherograms showing the separation of cations (a) and anions (b) using the contactless conductivity detection. (a) Mixture containing 450 μM potassium (1), 800 μM barium (2), 650 μM sodium (3), and 550 μM lithium (4). (b) Mixture of chloride (1), sulphate (2), fluoride (3), acetate (4), and phosphate (5), at 800 (1–3) and 1600 μM (4,5). Reprinted with permission from Ref. [83]. Copyright (2002), American Chemical Society.

using another design by placing electrodes on top of a cover plate rather than embedding them in the microdevice. This approach allows a significant simplification of the manufacturing process (see Fig. 14.16). The new contactless conductivity microchip detector is based on placing two planar sensing aluminium film electrodes on the outer side of a PMMA microchip and measuring the impedance of the solution in the separation channel. More recently, the same group has reported on the extension of this concept to dual conductivity/amperometric detection using a glass capillary electrophoresis microchip with nitroaromatic and ionic explosives as tested analytes [84]. The microsystem involved the combination of a contactless conductivity detector with an end-column thick-film amperometric detector. This coupling was made using only a single separation channel and allowed an enhancement of the characterization of both ionic (small cations) and electroactive species (nitroaromatic explosives), as well as the confirmation of peak identity. The same group has described contactless conductivity detection for monitoring separated aliphatic amines in UV-absorbing solvents (dimethylformamide, dimethylacetamide, dimethyl sulphoxide, propylene carbonate media), revealing the suitability of this detector in non-aqueous media [85].

Very recently, this group has also reported an ingenious new movable contactless-conductivity detector [86]. This system relies on positioning the detector at different points along the separation channel by "sliding" the electrode holder. This ingenious concept enables rapid switching between "total" (unresolved solutes) and "individual" (resolved/fingerprints) signal on the basis of placing the detector at the beginning and the end of the separation channel, respectively.

Tanyanyiwa and Hauser [87] have reported an excellent work dealing with a high-voltage approach to capacitively coupled contactless conductivity detection for detection on microfabricated planar electrophoresis devices. These authors demonstrated that better sensitivity was obtained by placing the electrodes into troughs that allow tighter coupling to the separation channel (detection limits for small ions K^+, Na^+ and Mg^{2+} are 0.49, 0.41 and 0.35 μM, respectively). In this work, the detection of heavy metals such as Mn^{2+}, Zn^{2+} and Cr^{3+} was also described. The universal nature of the new design was illustrated by the detection of citric and lactic acids, which are of interest in food and beverage analysis, as well as the detection of three anti-inflammatory drugs.

A multiauthor contribution [88] has recently reported a sophisticated and interesting on-chip, contactless, four-electrode conductivity detection design for CE microdevices. In this work four contactless electrodes were deposited and patterned on the second glass substrate for on-chip conductivity detection. This new configuration improved the analytical performance over the classical two-electrode detection set-up. The novel configuration was illustrated with inorganic ions (K^+, Na^+ and Li^+) and with six organic acids.

Finally, very recently Bai et al. [89] have proposed passive conductivity detection applied to CE-microchips. The separation electrical field is used to generate a potential difference between two electrodes located along the channel. For constant-current electrophoresis, the generated signal is proportional to the resistance of solution passing between electrodes. This principle directly takes advantage of the separation field and the signal is simply measured by a high-impedance voltmeter. The concept has been demonstrated by the separation of three alkali ions on a polymer microchip.

On the other hand, Table 14.2 summarizes the features of the main designs involving both conductivity formats. As can be observed in this

TABLE 14.2

Features of the main CE–ED microchip with conductivity detection: Contact and contactless designs

Chip characteristics	Electrode materials	Design/mode	Characterization of detector			Comments		Ref. (year)
Silicon/glass Serpentine channel $L_{ef} = 3$ cm/6 cm (50 μm × 27 μm)	Pt electrodes 50 μm width	Galvanic contact (perpendicular to flow)		LODs	Interval range	Different microfabrication approaches integrated with a conductivity detector		[78] (2001)
			Na^+	500 μM	1–50 mM	Embedded electrodes		
				LODs	Interval range	Mode	Analytes	
PMMA $L_{ef} = 3$ cm (15 μm × 85 μm)	Pt wires 127 μm diameter 20 μm (separation)	Contact	Alanine	8 nM	15–80 nm	CZE	Aminoacids/ peptides	[79] (2002)
						MEKC	Proteins	
						Electrochromatography (modified C_{18})	Nucleic acids	
Glass (20 μm × 70 μm)	Aluminium+titanium 4 electrodes (2 inner)	Contactless (silicon carbide)		LODs	Interval range			
		Capacitively coupled conductivity detector Electrodes embedded	K^+	20 μM	20 μm–2 mM	Application to two peptides separation Microfabrication		[80,81] (2001/02)
			(K^+, Na^+, Li^+)					
PMMA $L = 60$ mm (50 μm × 50 μm)	Aluminium films planar electrodes (0.8 mm × 10 mm) 10 μm-thick	Contactless (electrodes fixed on corner plate)	Lithium fluoride: 20 μM–7 mM					
			LODs (K^+, Na^+, SO_4^-, Cl^-, F^-)			New and easy design		[83] (2002)
			2.8; 5.8; 5.1; 6.4; 8.1 μM			No embedded electrodes		

table, characteristics of the chip and conductivity electrodes used, characterization of the detector as well as "key" comments for the chosen works are enclosed.

Finally, potentiometric detection on CE-microchips has been employed to a lesser extent. A potentiometric detector based on a miniaturized ion-selective electrode in a micromachined flow-channel has been described by Manz and co-workers [90] for the detection of Ba^{2+}, and proposed for detection in electrophoresis chips. Special membranes must be used for this purpose in order to achieve detection of a range of ions rather than the usually desired selectivity for just a single ion.

14.3 SELECTED ANALYTICAL APPLICATIONS

An important number of analytical applications using ED coupled to the microchip format have been published. A review of all of them is beyond the scope of this chapter (for details, see excellent reviews [9–17]). These applications range from the use of model analytes as catecholamines to sophisticated procedures that do not involve electroactive compounds to enhance the ED power. In fact, a wide range of analytes of biological, clinical and environmental significance has been studied using CE–ED microchips. In this section, selected analytical applications will be briefly described to show the real power as well as the analytical possibilities of these microfluidic platforms.

Next, we will describe the analytical applications involving amperometry as detection mode, and the most representative conductimetry applications. With respect to amperometry, first we will briefly describe applications involving neurotransmitters and related compounds. Second, we will study the most relevant bioanalytical and clinical applications involving different chip layouts and designs. Next, we will show the environmental and other miscellaneous applications involving "green analytical chemistry".

14.3.1 Neurotransmitters and related compounds

Since the electrochemical behaviour of neurotransmitters and related compounds is well known, these analytes have been the most widely studied to illustrate the reliability of the new proposed designs and protocols. Also, neuroactive compounds such as dopamine, epinephrine and nitric oxide are involved in numerous behavioural and physiological processes. Dopamine and epinephrine exhibit peripheral nervous system regulatory effects as well as actions on the central nervous system, such as respiratory stimulation and an increase in psychomotor activity. In addition, changes in physiological levels of dopamine in the brain are associated with schizophrenia and Parkinson disease. Accordingly, catechol [38,42,91,92] and dopamine [38,91–94] are frequently chosen as model analytes because they are electroactive at moderate redox potentials and because dopamine is an important neurotransmitter. An important neuroactive compound that has recently been studied by CE-microchip with ED is nitric oxide (NO) [95]. NO is an important compound in many physiological events, including neurotransmission and inflammation. This compound has a very short half-life and in the presence of oxygen is rapidly oxidized to nitrate and nitrite. Owing to its short half-life, its direct measurement is difficult; therefore, monitoring of nitrate and nitrite has been employed as a useful indicator of NO. This work has reported an indirect method for monitoring NO by determining nitrate and nitrite. This method combines the determination of nitrite by direct amperometric detection following a microchip-based CE separation and conversion of nitrate to nitrite by chemical reduction using Cu-coated Cd granules. Nitrite was separated and detected in 45 s. Recently, an amperometric system with a chemically amplified response for dopamine, noradrenaline and adrenaline has been presented [96]. The principle was the rapid cyclic oxidation of the analytes on the amperometric detector in the presence of the redox active enzyme glucose oxidase in the separation buffer. Detection limits in the range 10^{-7}–10^{-8} M were obtained.

14.3.2 Bioanalytical and clinical

Bioanalytical and clinical analysis are of crucial significance around the world, and many of the unique possibilities of miniaturized systems are very important in this field. Indeed, the "lab-on-a-chip" concept is intimately linked to this area of analytical chemistry.

Wang's group has made tremendous efforts in this area. A wide and interesting range of analytes has been detected using enzymes and antibodies in combination with sophisticated protocols, and this group has developed different strategies involving enzymes and different analytical concepts. Fig. 14.17 shows nice representative applications involving each chip layout: *on-column, pre-* and *post-column* concepts. First, using the "*on-column*" concept, the simultaneous bioassay of glucose, uric acid, ascorbic acid and acetaminophen has been

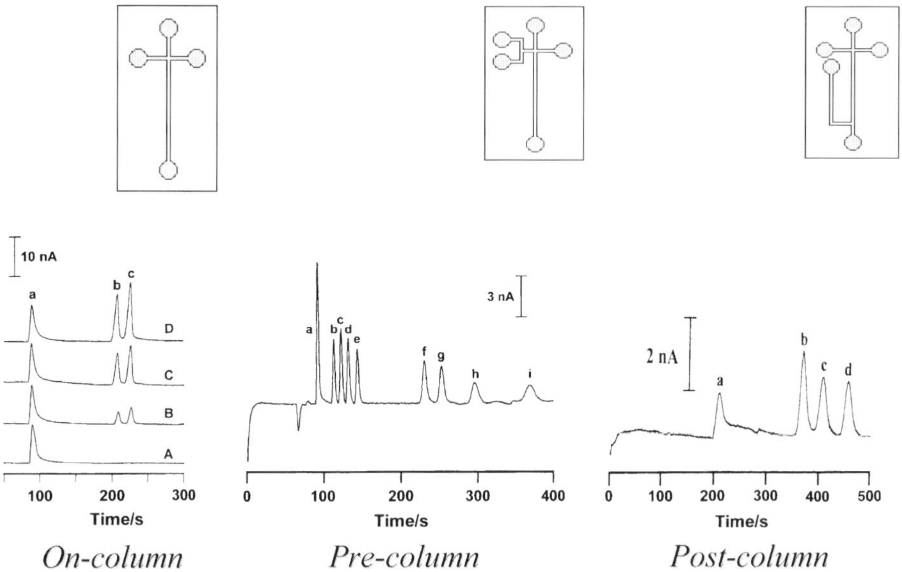

Fig. 14.17. *On-column*: Electropherograms for 1×10^{-3} M glucose (a) solutions containing increasing levels of ascorbic acid (b) and uric acid (c): 2×10^{-4} M (B); 5×10^{-4} M (C); 7×10^{-4} M (D). Reprinted with permission from Ref. [61]. Copyright (2000) American Chemical Society. *Precolumn*: Electropherograms for a mixture containing 1×10^{-4} M (b) histidine, (c) valine, (d) isoleucine, and (e) leucine and 2×10^{-4} M (f) glutamic acid, (g) aspartic acid, (h) arginine, and (i) lysine. Reprinted with permission from Ref. [59]. Copyright (2000), American Chemical Society. *Post-column*: Separation for a mixture of 1.5×10^{-3} M arginine (a), 3×10^{-4} M each of isoleucine (b), alanine (c), and phenylalanine (d). Reprinted from Ref. [62]. Copyright (2002), with permission from John Wiley & Sons, Inc.–Wiley-VCH Verlag GmbH.

demonstrated [61]. The beauty of this strategy consisted in the mixing of the enzyme glucose oxidase with the sample *on-column* liberating hydrogen peroxide after the sample reaction with glucose. Neutral hydrogen peroxide was separated from the other negatively charged analytes as can be observed in the electropherograms shown in Fig. 14.17. By incorporating a precolumn reactor into the chip design, an integrated microfabricated device that performs the on-line derivatization/biocatalytic reactions of multiple enzymes and substrates, the electrophoretic separation of the involved analytes and the electrochemical detection process on a chip platform has been illustrated for the electrophoretic separation of eight amino acids (analysis time, 6 min; LODs down to the 2.5×10^{-6} M level) (Fig. 14.17) [59], as well as for the simultaneous electrochemical measurement of ethanol and glucose [60]. A new biochip strategy for performing postcolumn reaction of a "class" enzyme (amino acid oxidase) on CE-microchip platforms has also been described for the rapid and sensitive measurements of amino acids (arginine, isoleucine, alanine, phenylalanine) in connection with a postcolumn biogeneration of electroactive hydrogen peroxide [62] (Fig. 14.17). The beauty of this work is that the concept can be extended to different target analytes based on the postcolumn reactions of other "class" enzymes, thus improving selectivity and sensitivity in their analytical performance.

Electrochemical enzyme immunoassays are a group of analytical methods with tremendous power owing to the combination of their inherent selectivity (antigen–antibody reaction) with the high sensitivity of the ED coupled with enzyme amplification. A very attractive electrochemical enzyme immunoassay on microchip platforms has been reported using antimouse IgG and mouse IgG, which is shown in Fig. 14.18. The "lab-on-a-chip" protocols integrate precolumn reactions of alkaline phosphatase-labelled antibody with antigen, followed by the electrophoretic separation of the free antibody and antibody–antigen complex. The separation is followed by a postcolumn reaction of the enzyme tracer, with 4-aminophenyl phosphate substrate, and a downstream amperometric detection of the liberated 4-aminophenol product. A remarkable detection limit (1.7×10^{-18} M) was obtained [55]. Recently, the same group has reported microchip-based amperometric immunoassays using redox tracers [63]. In this work, a new protocol based on the use of ferrocene redox label was described. Two reaction formats, based on direct (non-competitive) and competitive modes of operation, were employed for illustrating the use of redox tracers in chip-based electrochemical immunoassays. The direct assays consisted of mixing the ferrocene-tagged antibody and the antigen analyte, a rapid electrophoretic separation of the labelled free antibody and the labelled antigen–antibody complex, and a downstream anodic detection of the ferrocene tracer at a gold-plated carbon screen electrode detector. On the other hand, the competitive assay integrates precolumn reactions of the labelled antigen and the target antigen with the antibody with the electrophoretic separation of the free and bound labelled antigens, along with amperometric detection of the redox tag. The use of a redox tracer offers the advantages of simplified protocols, wider linear ranges, higher stability and higher separation efficiency compared with an analogous use of enzyme tags, as we have described before.

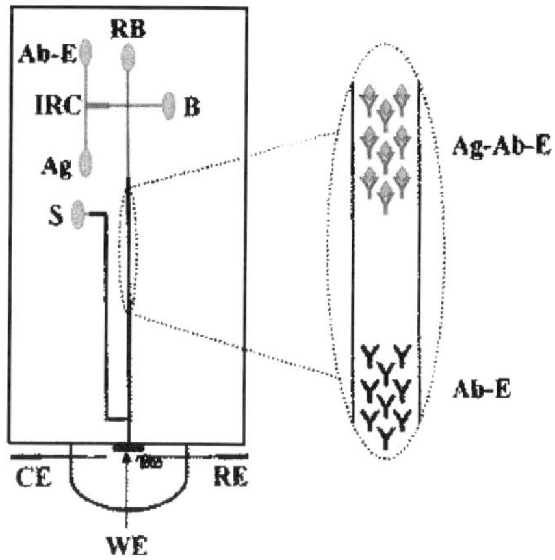

Immunoassay "lab-on-a-chip"

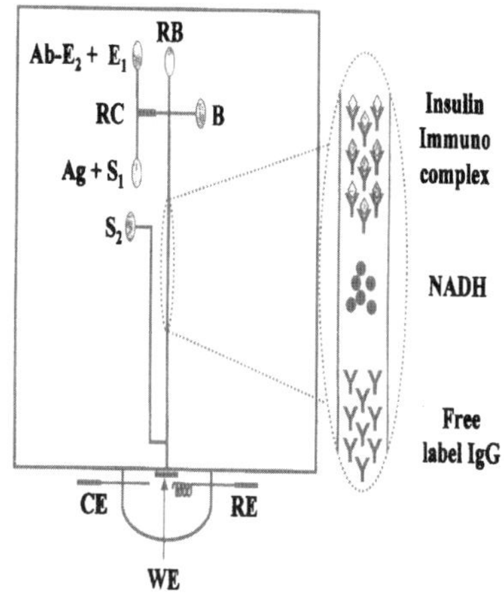

Enzyme + Immunoassay "lab-on-a-chip"

Fig. 14.18. *Immunoassay "lab-on-a-chip"*: RB, running buffer; Ab-E, enzyme-labelled antibody; Ag, antigen; S, substrate; IRC, immunoreaction chamber; RE, reference electrode; CE, counter electrode; WE, working electrode; B, unused reservoir. Reprinted with permission from Ref. [55]. Copyright (2001), American Chemical Society. *Enzyme+ Immunoassay "lab-on-a-chip"*: Schematic of the biochip used (RB) Running buffer, (Ab-E_2) alkaline phosphatase-labelled anti-insulin, (E_1) GDH enzyme, (Ag) insulin, (S_1) glucose, (RC) reaction chamber, (S_2) *p*-NPP, (B) buffer, (WE) working electrode, (CE) counter electrode, and (RE) reference electrode. Reprinted with permission from Ref. [97]. Copyright (2003), American Chemical Society.

Fig. 14.18 also shows an outstanding application, which has recently been reported by Wang and co-workers [97] and deals with the on-chip integration of enzymes and immunoassays. This novel concept was applied to the simultaneous measurement of insulin and glucose. As illustrated in the figure, the implementation of such a dual-bioassay single-channel microchip protocol involves the judicious coupling of multiple enzymatic and immunoreactions through integration of electrophoretic separations with relevant pre- and postcolumn reaction steps. The operation taking place in the biochip includes precolumn reactions of the enzyme-labelled antibody (anti-human insulin) with its antigen (insulin) and the reaction of the glucose-dehydrogenase (GDH) enzyme with its glucose substrate in the presence of NAD^+ cofactor; these reactions are followed by the electrophoretic separation of the free antibody, antibody–antigen complex and NADH product of the glucose enzymatic reaction, and a postcolumn reaction of the alkaline phosphatase (ALP) enzyme tag with its *p*-nitrophenylphosphate (*p*-NPP) substrate. Both the NADH and the *p*-nitrophenol products are monitored at the downstream amperometric detector. Such a coupling of different bioassays on the same microfluidic device reflects the versatility and integration features of such devices.

Representative electropherograms involving recent clinical applications are depicted in Fig. 14.19. Lunte's group has recently reported an interesting work for the detection of homocystein using both conventional and CE-microchip with ED. Homocystein is an amino acid involved in the synthesis and metabolism of the essential amino acid methionine. The separation and detection conditions used for conventional CE were transferred to a microchip format. As can be observed in Fig. 14.19A, both homocystein and reduced glutathione were detected and separated in less than 2 min [98]. A uric acid fast determination (less than 30 s) in urine samples has been also reported using a PDMS/glass microchip with an *off-channel* configuration platinum working electrode [99]. In this work up to six urine samples were evaluated with the microchip device, and the uric acid concentration for each sample was found to be in the expected clinical concentration range (Fig. 14.19B). Wang's group has also recently described a new multianalyte renal marker biochip using a single CE-microchip with a gold-coated thick-film amperometric detector [65]. The four renal markers studied (creatine, creatinine, *p*-aminohippuric acid and uric acid) were readily measured (baseline resolution) within 5 min, and the applicability to urine samples was demonstrated (Fig. 14.19C). Also, CE–ED microchip with a boron-doped diamond electrode has been employed for the fast separation and detection (less than 6 min) of several purines (guanine, hypoxanthine, guanosine, xanthine and uric acid) and purine-containing compounds (nucleosides, nucleotides and oligonucleotides) [100]. This new microchip protocol has opened up a wide range of bioanalytical applications involving assays of this type of compounds.

14.3.3 Environmental

Environmental applications are, on the other hand, one of the most important analytical areas in which CE–ED microchips have a predominant role. The power of simple cross microchips in this area is shown in Fig. 14.20. One of the first applications by Wang's group was related to demands of "green analytical chemistry." Thus, the authors

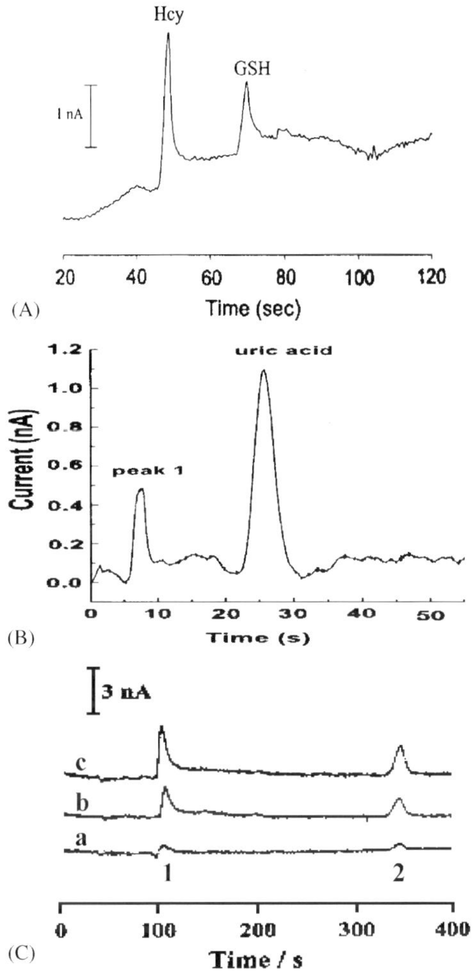

Fig. 14.19. (A) Detection and separation of 200 μM Hcy and GSH. Reprinted from Ref. [98]. Copyright (2002), with permission from John Wiley & Sons, Inc.–Wiley-VCH Verlag GmbH. (B) Electropherogram of uric acid in a urine sample. Reprinted from Ref. [99]. Copyright (2002), with permission from John Wiley & Sons, Inc.–Wiley-VCH Verlag GmbH (C) Assay of renal markers in urine sample. (a) Response for the urine sample; (b) addition of 2.5×10^{-4} M creatinine (1) and 1×10^{-4} M uric acid (2); (c) addition of 2.5×10^{-4} M creatinine (1) and 1×10^{-4} M uric acid (2). Reprinted with permission from Ref. [65]. Copyright (2003), American Chemical Society.

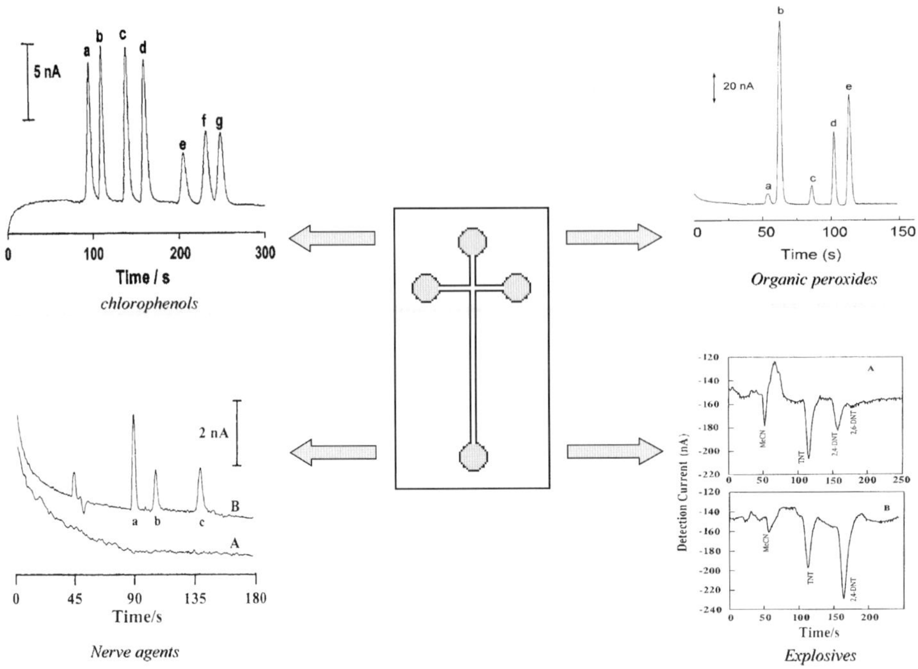

Fig. 14.20. Chlorophenols: separation and detection of phenol and six chlorophenols: (a) 100 μM phenol; (b) 100 μM 2-chlorophenol; (c) 200 μM 2,4-dichlorophenol; (d) 200 μM 2,3-dichlorophenol; (e) 200 μM 2,4,5-trichlorophenol; (f) 200 μM 2,4,6-trichlorophenol; and (g) 200 μM 2,6-dichlorophenol. Reprinted from Ref. [53]. Copyright (2000), with permission from Elsevier. Nerve agents: electropherograms for river water sample before (A) and after (B) the addition of 1.4×10^{-5} M paraoxon (a), 1.5×10^{-5} M methyl parathion (b) and 2.8×10^{-5} M fenitrothion (c). Reprinted with permission from Ref. [54]. Copyright (2001), American Chemical Society. Organic peroxides: electropherograms for mixtures containing oxygen (a), 1.71 mM hydrogen peroxide (b), 4.8 mM peroxyheptanoic acid (c), 0.25 mM peroxipropionic acid (d) and 1.33 mM peroxyacetic acid (e). Reprinted from Ref. [66]. Copyright (2002), with permission from Elsevier. Explosives: analysis of the explosive content in the extract prepared from a contaminated soil sample (A) and in groundwater (B). Reprinted with permission from Ref. [101] Copyright (2000), American Chemical Society.

demonstrated the utility of CE-microchips with thick-film electrodes for the rapid separation (4 min) of seven priority chlorophenolic pollutants down to a concentration level of 1×10^{-6} M [53]. The suitability of CE-microchip with ED has also been demonstrated for detecting toxic organophosphate nerve agents such as paraoxon, methyl parathion, fenitrothion and ethyl parathion in 140 s with applicability to river water samples [54]. A beautiful and fast detection (less than 200 s) of explosives in soil and groundwater samples has also been reported [101]. "Labs-on-a-chip" involving ED devices for *on site* analysis of organic peroxides have also been recently reported. A microfluidic analytical system for the separation and detection of organic peroxides, based on a CE microchip device with an integrated amperometric detector, has been described using the reductive mode at –700 mV. The integrated microsystems offered rapid measurements (less than 130 s) [66]. Fast analysis of thiol-containing degradation of V-type nerve agents (less than 4 min) has been recently described, and its suitability for assays of environmental matrices has also been demonstrated [102]. Representative electropherograms for these examples are clearly shown in Fig. 14.20.

Another environmental-miscellaneous group of relevant analytical applications can be found in the literature. A very attractive strategy using a single-channel chip-based analytical microsystem has allowed the rapid measurement of the total content of relevant compounds (organic explosive or nerve agents), as well as detailed micellar chromatographic identification of each one of them [56]. Fig. 14.21 nicely shows on the right side the strategy schemes employed and on the left side, the electropherograms obtained for both individual and total (less than 1 min) measurements for each compound class studied. This strategy reveals the very unique integration of important modern concepts in Analytical Chemistry, such as screening methods for the measurements of total index in order to detect "alarm" concentrations (vanguard strategies), the separation or confirmation of individual ones (rearguard strategies) and two separation protocols (MEKC and CZE), all into a single microchip. Also, the detection of carbohydrates using copper electrodes and carbon-nanotube/copper composite electrodes has recently been published [103,104].

14.3.4 Conductimetry

On the other hand, as mentioned before, the coupling of conductimetry in CE-microchips was described after that of amperometry. In general, these contributions deal with small inorganic or organic ions in order to demonstrate the reliability of the new proposed designs. However, other interesting applications related to bioanalytical/anti-terrorist areas can be found.

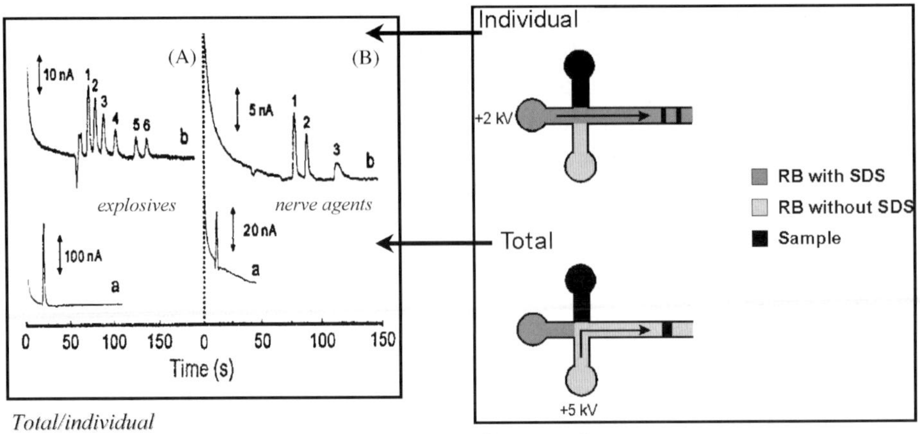

Fig. 14.21. (*Left*) Electropherograms depicting the total (a) and individual (b) assays for mixtures of nitroaromatic explosives (A) and nerve agents (B). Sample mixture A: TNB (1), DNB (2), TNT (3), 2,4-DNT (4), 2-Am-4,6-DNT (5), and 4-Am-2,6-DNT (6) (10 ppm each). Sample mixture B: 10 ppm paraoxon (1), 10 ppm methyl parathion (2), and 40 ppm fenitrothion (3). (*Right*) Schemes of the strategies employed to measure total and individual compounds. Reprinted with permission from Ref. [56]. Copyright (2002), American Chemical Society.

Fig. 14.22. Contact conductivity detection. (A) Free-solution zone electrophoretic separation of 100 μM amino acid mixture consisting of (1) alanine, (2) valine, (3) glutamine, and (4) tryptophan. (B) Free-solution zone electrophoretic separation of a peptide mixture (~0.23 μM total peptide concentration) consisting of (1) bradykinin, (2) bradykinin fragment 1–5, (3) substance P, (4) [Arg^8]-vasopressin, (5) luteinizing hormone, (6) bombesin, (7) leucine enkephalin, (8) methionine enkephalin and (9) oxytocin in an unmodified PMMA microchip. (C) MEKC separation of a protein mixture (1.7 μM total protein concentration with all proteins at similar concentrations within the mixture) consisting of (2) lysozyme, (3) trypsin inhibitor, (4) carbonic anhydrase, (5) ovalbumin, (6) serum albumin, (7) phosphorylase B, (8) β-galactosidase, and (9) myosin detected using indirect, contact conductivity detection. Benzoic acid (1) was added to the mixture as an internal standard. (D) CEC separation of a double-stranded DNA ladder (400 ng/ml) in a C_{18}-modified PMMA device. The ladder consisted of 100, 200, 400, 800, 1200, and 2000 bp fragments. Reprinted with permission from Ref. [79]. Copyright (2002), American Chemical Society.

Galvanic conductimetry detection of small inorganic ions (potassium, lithium, sodium) and the detection of organic acids in the anodic mode (fumaric, malic and citric acids) have also been described [78]. Galloway et al. [79] have shown an impressive range of applications using a PMMA simple cross chip with contact conductivity detection (see Fig. 14.22). In fact, the authors showed a free-solution zone electrophoretic separation

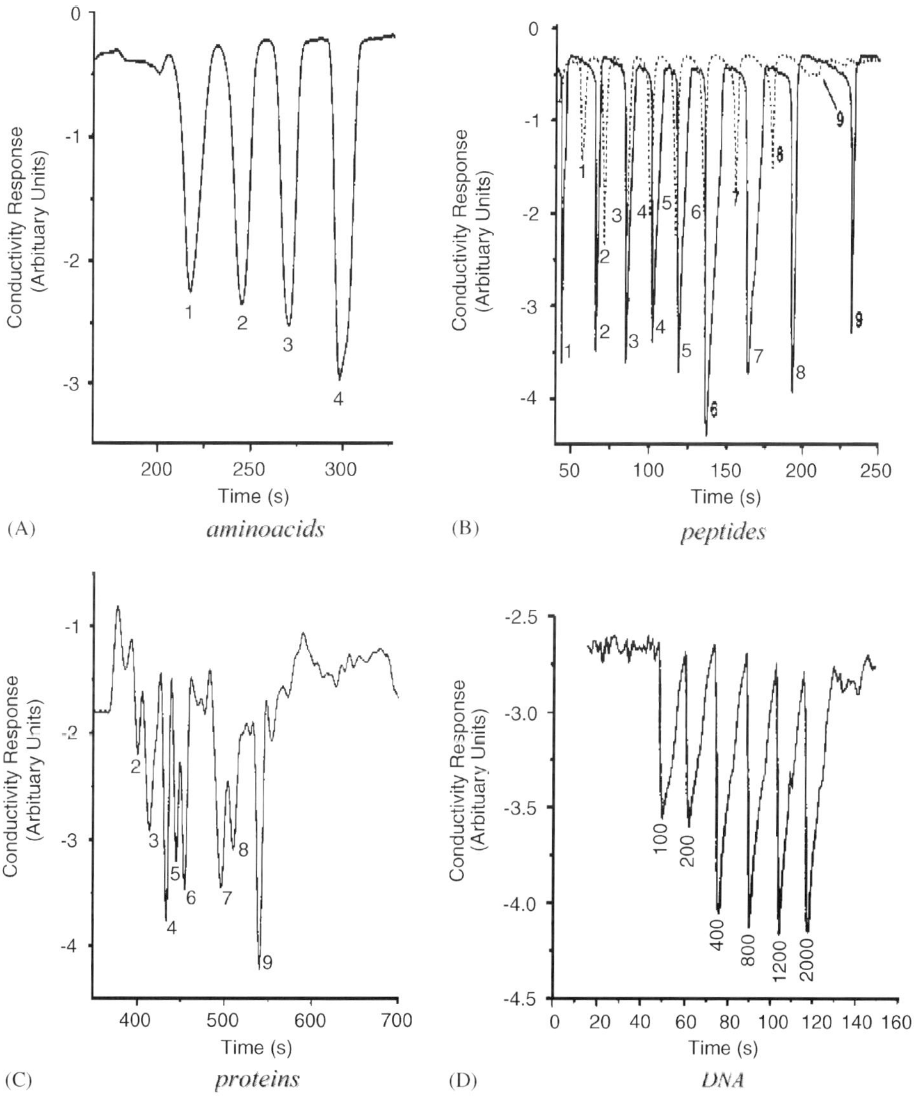

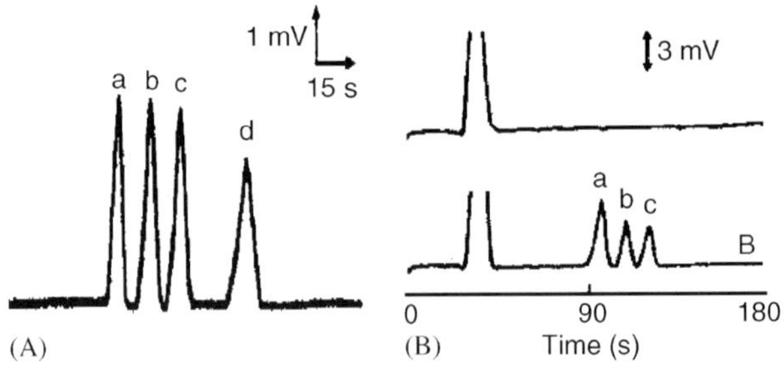

Fig. 14.23. Contactless conductivity detection: (A) electropherogram for a mixture of CWA degradation products: (a) 15 ppm MPA, (b) 20 ppm EMPA, (c) 20 ppm IMPA, and (d) 50 ppm PMPA. (B) Electropherogram for river water (Rio Grande) sample before (top) and after (bottom) the addition of 20 ppm MPA (a), EMPA (b), and IMPA (c). Reprinted with permission from Ref. [105]. Copyright (2002), American Chemical Society.

of four amino acids and nine peptides using indirect detection, a MEKC separation of protein mixtures and a CEC separation of DNA using C_{18}-modified PMMA channels. In another contribution, the application of the four-electrode capacitively coupled conductivity detector in biochemical analysis was also illustrated by a separation of two peptides [80].

The contactless conductivity detector integration in glass microchips has also been described using small inorganic cations as model analytes [81,82]. The elegant coupling of low-cost PMMA chips with the easily constructed contactless conductivity detectors proposed by Pumera et al. [83] was successfully illustrated by applying it to the detection of cations (potassium, lithium, sodium and barium) and anions (chloride, sulphate, fluoride, acetate and phosphate). *p*-, *m*-, *o*-Aminophenol and related explosives were simultaneously detected by dual conductimetry–amperometric detection in order to illustrate their analytical possibilities. Also, in response to recent terrorist activity, Wang's group has reported an interesting conductimetry detection of chemical warfare agents and their degradation products (see Fig. 14.23) [105] as well as explosives [106].

14.4 CONCLUSIONS AND FUTURE DIRECTIONS

The reliability and suitability of CE–ED microchips has been widely demonstrated. It includes the establishment and transference of well-known principles from conventional CE to microchip format, the excellent performance of the integrated microfabricated designs involving separation and detection schemes, excellent analytical performance (fast analysis, good separation efficiency, adequate LODs) and the design of sophisticated protocols and schemes that drive relevant analytical applications.

CE–ED microchips offers a very unique range of analytical possibilities. The combination of the advanced state of CE-microchip (well-known chemistry of glass/polymer materials, microfabrication technologies and the vast knowledge about microfluidic handling) with the great promises of ED (inherent miniaturization, high sensitivity and high compatibility with microfabrication techniques) makes CE–ED microsystems one of the most relevant devices in miniaturization in analytical chemistry, and gives them a prominent role in "lab-on-a-chip"/μ-TAS as a whole.

Different detection designs and configurations have been successfully implemented and applied. Typical problems, such as the coupling between high voltages employed and ED have been overcome. The main strength of CE-microchip coupled to ED schemes lies in the ability to integrate the detector into fully integrated devices. Therefore, since the feasibility of constructing a miniaturized system with the sample preparation, separation and detection units integrated directly onto the CE-microchip with ED has also been described, creating new sophisticated networks in relation to analytical problems should be one of the main future directions. Amperometry and conductimetry can be considered complementary and cover a wide range of analytes. Common to all is the direct signal transduction into the electronic domain, which means that only limited instrumentation is required. This makes the methods particularly attractive for portable and miniaturized devices. Although amperometry is the most widely used detection mode, conductimetry will have a wider grade of acceptation. The combination of the CE–ED microchips alleviates selectivity problems often encountered otherwise.

A wide group of applications (clinical/biological/environmental) has been described using sophisticated routes and protocols. However, the incorporation of new applications to the well-established ones, including clinical diagnosis, forensic security, drug screening, genomics and proteomics and food science is expected.

This is an exciting time for Analytical Chemistry, especially for all areas involving miniaturization and the "lab-on-a-chip" concept. New surprises can be expected from the miniaturization branch in the near future. New microsystems are surely on the way!

ACKNOWLEDGEMENTS

Financial support from Ministry of Science and Technology (Spain) (BQU 2002-01116) is gratefully acknowledged. A. E. acknowledges Dr. Wang, who introduced him to the exciting world of microchips.

REFERENCES

1 A. Manz, N. Graber and H.M. Widmer, *Sens. Actuators*, B1 (1990) 244–248.
2 A. Manz and J.C.T. Eijkel, *Pure Appl. Chem.*, 73 (2001) 1555–1561.
3 O. Geschke, H. Klank and P. Tellesmann (Eds.), *Microsystem Engineering of Lab-on-a-Chip Devices*, Wiley-VCH Verlag & Co. KgaA, Weinheim, 2004.
4 S.C. Jakeway, A.J. de Mello and E.L. Russell, *Fresenius J. Anal. Chem.*, 366 (2000) 525–539.
5 A.J. de Mello, *Anal. Bioanal. Chem.*, 372 (2002) 12–13.
6 A. Manz, D.J. Harrison, E.M.J. Verpoorte, J.C. Fettinger, A. Paulus, H. Lüdi and H.M. Widmer, *J. Chromatogr.*, 593 (1992) 253–258.
7 D.J. Harrison, A. Manz, Z. Fan, H. Lüdi and H.M. Widmer, *Anal. Chem.*, 64 (1992) 1926–1932.
8 S.C. Jacobson, R. Hergenroder, L.B. Koutny and J.M. Ramsey, *Anal. Chem.*, 66 (1994) 1114–1118.
9 C.L. Colyer, T. Tang, N. Chiem and D.J. Harrison, *Electrophoresis*, 28 (1997) 1733–1741.
10 V. Dolník, S. Liu and S. Jovanovich, *Electrophoresis*, 21 (2000) 41–54.
11 G.J.M. Bruin, *Electrophoresis*, 21 (2000) 3931–3951.
12 S.A. Soper, S.M. Ford, S. Qi, R.L. McCarley, K. Kelly and M.C. Murphy, *Anal. Chem.* (2000) 643A–651A.
13 J. Rossier, F. Reymond and P.E. Michel, *Electrophoresis*, 23 (2002) 858–867.
14 M.A. Schwarz and P.C. Hauser, *Lab on a Chip*, 1 (2001) 1–6.
15 N.A. Lacher, K.E. Garrison, R.S. Martín and S.M. Lunte, *Electrophoresis*, 22 (2001) 2526–2536.
16 W.R. Vandaveer IV, S.A. Pasas, R.S. Martín and S.M. Lunte, *Electrophoresis*, 23 (2002) 3667–3677.
17 J. Tanyanyiwa, S. Leuthardt and P.C. Hauser, *Electrophoresis*, 23 (2002) 3659–3666.

18 J. Wang, *Talanta*, 56 (2002) 223–231.
19 M.J. Madou (Ed.), *Fundamentals of Microfabrication (The Science of Miniaturization)*, CRC Press, Boca Raton, FL, 2000.
20 D.C. Duffy, J.C. McDonald, O.J.A. Schueller and G.M. Whitesides, *Anal. Chem.*, 70 (1998) 4974–4984.
21 Z.H. Fan and D.J. Harrison, *Anal. Chem.*, 66 (1994) 177–184.
22 S.C. Jacobson, R. Hergenroder, A.W. Moore and J.M. Ramsey, *Anal. Chem.*, 66 (1994) 4127–4132.
23 L.B. Koutny, D. Schmalzing, T.A. Taylor and M. Fuchs, *Anal. Chem.*, 68 (1996) 18–22.
24 S.C. Jacobson, A.W. Moore and J.M. Ramsey, *Anal. Chem.*, 67 (1995) 2059–2063.
25 L. Martynova, L.E. Locascio, M. Gaitan, G.W. Kramer, R.G. Christensen and W.A. MacCrehan, *Anal. Chem.*, 69 (1997) 4783–4789.
26 R.M. McCormick, R.J. Nelson, M.G. Alonso-Amigo, J. Benvegnu and H.H. Hoopwer, *Anal. Chem.*, 69 (1997) 2626–2630.
27 D.C. Duffy, J.C. McDonald, O.J.A. Schueller and G.M. Whitesides, *Anal. Chem.*, 70 (1998) 4974–4984.
28 M.A. Roberts, J.S. Rossier, P. Bercier and H. Girault, *Anal. Chem.*, 69 (1997) 2035–2042.
29 O. Rötting, W. Röpke, H. Becker and C. Gärtner, *Microsyst. Technol.*, 8 (2002) 32–36.
30 L.G. Reyna and J.R. Sobehart, *J. Appl. Phys.*, 76 (1994) 4367–4371.
31 J. Wang, B. Tian and E. Sahlin, *Anal. Chem.*, 71 (1999) 3901–3904.
32 J. Wang, B. Tian and E. Sahlin, *Anal. Chem.*, 71 (1999) 5436–5440.
33 R.S. Martin, A.J. Gawron, S.M. Lunte and C.S. Henry, *Anal. Chem.*, 72 (2000) 3196–3202.
34 A.T. Woolley, K. Lao, A.N. Glazer and R.A. Mathies, *Anal. Chem.*, 70 (1998) 684–688.
35 Y. Liu, J.C. Fanguy, J.M. Bedsoe and C.S. Henry, *Anal. Chem.*, 72 (2000) 5939–5944.
36 S.R. Wallenborg, L. Nyholm and C.E. Lunte, *Anal. Chem.*, 71 (1999) 544–549.
37 J. Wang, M. Pumera, M.P. Chatrathi, A. Escarpa, R. Konrad, A. Griebel, W. Dörner and H. Löwe, *Electrophoresis*, 23 (2002) 596–601.
38 J. Wang, M. Pumera, M.P. Chatrathi, A. Rodriguez, S. Spillman, R.S. Martín and S.M. Lunte, *Electroanalysis*, 14 (2002) 1251–1255.
39 Y. Zeng, H. Chen, D.-W. Pang, Z.-L. Wang and J.-K. Cheng, *Anal. Chem.*, 74 (2002) 2441–2445.
40 N. Bao, J.-J. Xu, Y.-H. Dou, Y. Cai, H.-Y. Chen and X.-H. Xia, *J. Chromatogr. A*, 1041 (2004) 245–248.
41 P. Ertl, C.A. Emrich, P. Singhal and R. Mathies, *Anal. Chem.*, 76 (2004) 3749–3755.

42 R.S. Martin, K.L. Ratzlaff, B.H. Huynh and S.M. Lunte, *Anal. Chem.*, 74 (2002) 1136–1143.
43 Y. Liu, J.A. Vickers and C.S. Henry, *Anal. Chem.*, 76 (2004) 1513–1517.
44 J.S. Rossier, R. Ferrigno and H.H. Girault, *J. Electroanal. Chem.*, 492 (2000) 15–22.
45 D.-C. Chen, F.-L. Hsu, D.-Z. Zhan and C.-H. Chen, *Anal. Chem.*, 73 (2001) 758–762.
46 C.-C. Wu, R.-G. Wu, J.-G. Huang, Y.-C. Lin and H.-C. Chang, *Anal. Chem.*, 75 (2003) 947–952.
47 C.-C.J. Lai, C.-H. Chen and F.-H. Ko, *J. Chromatogr. A*, 1023 (2004) 143–150.
48 D.M. Osbourn and C.E. Lunte, *Anal. Chem.*, 75 (2003) 2710–2714.
49 N.A. Lacher, S.M. Lunte and R.S. Martin, *Anal. Chem.*, 76 (2004) 2482–2491.
50 R.S. Keynion, T.J. Roussel Jr., M.M. Crain, D.J. Jackson, D.B. Franco, J.F. Naber, K.M. Walsh and R.P. Baldwin, *Anal. Chim. Acta*, 507 (2004) 95–105.
51 A.J. Gawron, R.S. Martin and S.M. Lunte, *Electrophoresis*, 22 (2001) 242–248.
52 R.S. Martin, A.J. Gawron, B.A. Fogarty, F.B. Regan, E. Dempsey and S.M. Lunte, *Analyst*, 126 (2001) 277–280.
53 J. Wang, M.P. Chatrathi and B. Tian, *Anal. Chim. Acta*, 416 (2000) 9–14.
54 J. Wang, M.P. Chatrathi, A. Mulchandani and W. Chen, *Anal. Chem.*, 73 (2001) 1804–1808.
55 J. Wang, A. Ibáñez, M.P. Chatrathi and A. Escarpa, *Anal. Chem.*, 73 (2001) 5323–5327.
56 J. Wang, M. Pumera, M.P. Chatrathi, A. Escarpa and M. Musameh, *Anal. Chem.*, 74 (2002) 1187–1191.
57 M.L. Kovarik, N.J. Torrence, D.M. Spence and R.S. Martin, *The Analyst*, 129 (2004) 400–405.
58 P. Gavin and A.G. Ewing, *J. Am. Chem. Soc.*, 118 (1996) 8932–8936.
59 J. Wang, M.P. Chatrathi and B. Tian, *Anal. Chem.*, 72 (2000) 5774–5778.
60 J. Wang, M.P. Chatrathi and B. Tian, *Anal. Chem.*, 73 (2001) 1296–1300.
61 J. Wang, M.P. Chatrathi, B. Tian and R. Polsky, *Anal. Chem.*, 72 (2000) 2514–2518.
62 J. Wang, M.P. Chatrathi, A. Ibáñez and A. Escarpa, *Electroanalysis*, 14 (2002) 400–404.
63 J. Wang, A. Ibáñez and M.P. Chatrathi, *Electrophoresis*, 23 (2002) 3744–3749.
64 M. Pumera, J. Wang, E. Grushka and R. Polsky, *Anal. Chem.*, 73 (2001) 5625–5628.
65 J. Wang and M.P. Chatrathi, *Anal. Chem.*, 75 (2003) 525–529.

66 J. Wang, A. Escarpa, M. Pumera and J. Feldman, *J. Chromatogr. A*, 952 (2002) 249–254.
67 A. Hilmi and J.H.T. Luong, *Anal. Chem.*, 72 (2000) 4677–4682.
68 J. Wang, M.P. Chatrathi, B. Tian and R. Polsky, *Electroanalysis*, 12 (2000) 691–694.
69 J. Wang, G. Chen, M.P. Chatrathi and M. Musameh, *Anal. Chem.*, 76 (2004) 298–302.
70 J. Wang, G. Chen, M.P. Chatrathi, A. Fujishima, D.A. Tryk and D. Shin, *Anal. Chem.*, 75 (2003) 935–939.
71 N.E. Hebert, B. Snyder, R.L. McCreery, W.G. Kuhr and S.A. Brazill, *Anal. Chem.*, 75 (2003) 4265–4271.
72 N.E. Hebert, W.G. Kuhr and S.A. Brazill, *Electrophoresis*, 23 (2002) 3750–3759.
73 N.E. Hebert, W.G. Kuhr and S.A. Brazill, *Anal. Chem.*, 75 (2003) 3301–3307.
74 J.C. Fanguy and C.S. Henry, *Analyst*, 127 (2002) 1021–1023.
75 C.D. García and C.S. Henry, *Anal. Chem.*, 75 (2003) 4778–4783.
76 C.D. García and C.S. Henry, *Anal. Chim. Acta*, 508 (2004) 1–9.
77 J. Wang, R. Polsky, B. Tian and M.P. Chatrathi, *Anal. Chem.*, 72 (2000) 5285–5289.
78 R.M. Guijt, E. Baltussen, G. van der Steen, R.B.M. Schasfoort, S. Schlautmann, H.A.H. Billiet, J. Frank, G.W.K. van Dedem and A. van den Berg, *Electrophoresis*, 22 (2001) 235–241.
79 M. Galloway, W. Stryjewski, A. Henry, S.M. Ford, S. Llopis, R.L. McCarley and S.A. Soper, *Anal. Chem.*, 74 (2002) 2407–2415.
80 R.M. Guijt, E. Baltussen, G. van der Steen, H. Frank, H. Billiet, T. Schalkhammer, F. Laugere, M. Vellekoop, A. Berthold, L. Sarro and G.W.K. van Demen, *Electrophoresis*, 22 (2001) 2537–2541.
81 A. Berthold, F. Laugere, H. Schellevis, C.R. de Boer, M. Laros, R.M. Guijt, P.M. Sarro and M.J. Vellekoop, *Electrophoresis*, 23 (2002) 3511–3519.
82 J. Lichtenberg, N.F. de Rooij and E. Verpoorte, *Electrophoresis*, 23 (2002) 3769–3780.
83 M. Pumera, J. Wang, F. Opekar, I. Jelínek, J. Feldman, H. Löwe and S. Hardt, *Anal. Chem.*, 74 (2002) 1968–1971.
84 J. Wang and M. Pumera, *Anal. Chem.*, 74 (2002) 5919–5923.
85 J. Wang and M. Pumera, *Anal. Chem.*, 75 (2003) 341–345.
86 J. Wang, G. Chen and A. Muck Jr., *Anal. Chem.*, 75 (2003) 4475–4479.
87 J. Tanyanyiwa and P.C. Hauser, *Anal. Chem.*, 74 (2002) 6378–6382.
88 F. Laugere, R.M. Guijt, J. Bastemeijer, G. van der Steen, A. Berthold, E. Baltussen, P. Sarro, G.W.K. van Dedem, M. Vellekoop and A. Bossche, *Anal. Chem.*, 75 (2003) 306–312.

89 X. Bai, Z. Wu, J. Josserand, H. Jensen, H. Schafer and H.H. Girault, *Anal. Chem.*, 76 (2004) 3126–3131.
90 R. Tranta and A. Manz, *Anal. Chem.*, 72 (2000) 2875.
91 D.P. Manica and A.G. Ewing, *Electrophoresis*, 23 (2002) 3735–3743.
92 U. Backofen, F.M. Matysik and C.E. Lunte, *Anal. Chem.*, 74 (2002) 4054–4059.
93 M.A. Schwarz, B. Galliker, K. Fluri, T. Kappes and P.C. Hauser, *Analyst.*, 126 (2001) 147–151.
94 Y. Zeng, H. Chen, D.W. Pang, Z.L. Wang and J.K. Cheng, *Anal. Chem.*, 74 (2002) 2441–2445.
95 R. Kikura-Hanajiri, R.S. Martin and S.M. Lunte, *Anal. Chem.*, 74 (2002) 6370–6377.
96 M.A. Schwarz, *Electrophoresis*, 25 (2004) 1916–1922.
97 J. Wang, A. Ibáñez and M.P. Chatrathi, *J. Am. Chem. Soc.*, 125 (2003) 8444–8445.
98 S.A. Pasas, N.A. Lacher, M.I. Davies and S.M. Lunte, *Electrophoresis*, 23 (2002) 759–766.
99 J.C. Fanguy and C.S. Henry, *Electrophoresis*, 23 (2002) 767–773.
100 J. Wang, G. Chen, A. Muck, D. Shin and A. Fujishima, *J. Chromatogr. A*, 1022 (2004) 207–212.
101 A. Hilmi and J.H.T. Luong, *Environ. Sci. Technol.*, 34 (2000) 3046–3050.
102 J. Wang, J. Zima, N.S. Lawrence, M.P. Chatrathi, A. Mulchandani and G.E. Collins, *Anal. Chem.*, 76 (2004) 4721–4726.
103 H.-L. Lee and S.-C. Chen, *Talanta*, 64 (2004) 210–216.
104 J. Wang, G. Chen, M. Wang and M.P. Chatrathi, *Analyst.*, 129 (2004) 512–515.
105 J. Wang, M. Pumera, G.E. Collins and A. Mulchandani, *Anal. Chem.*, 74 (2002) 6121–6125.
106 J. Wang, M. Pumera, G.E. Collins, F. Opekar and I. Jelínek, *Analyst.*, 127 (2002) 719–723.

List of Abbreviations

A

ACE Affinity Capillary Electrophoresis
APCI Atmospheric Pressure Chemical Ionization

B

BGE Background Electrolyte

C

CD Cyclodextrin
CE Capillary Electrophoresis
CEC Capillary Electrochromatography
CFS Continuous Flow System
CGE Capillary Gel Electrophoresis
CIEF Capillary Isoelectric Focusing
CITP Capillary Isotachophoresis
CMC Critical Micellar Concentration
CZE Capillary Zone Electrophoresis

E

EKC Electrokinetic Capillary Chromatography
EOF Electroosmitic Flow
ESI-MS Electrospray Ionization Mass Spectrometry

F

FTIR Fourier Transform Infrared Spectroscopy

G

GC Gas Chromatography

H

HPLC High Performance Liquid Chromatography

I

ICP Inductively Coupled Plasma
IR Infrared Spectroscopy

L

LIF Laser-Induced Fluorescence

M

MALDI Matrix-Assisted Laser Desorption/Ionization
MEEKC Microemulsion Electrokinetic Chromatography
MEKC Micellar Electrokinetic Chromatography
MS Mass Spectrometry

N

NACE Non Aqueous Capillary Electrophoresis
NRM Nuclear Magnetic Resonance

P

PAH Polycyclic Aromatic Hydrocarbons
PCB Polycyclic Chlorinated Biphenils
PCR Polymerase Chain Reaction

R

RTPL Room-Temperature Phosphorescence in the Liquid State

S

SDS Sodium Dodecyl Sulphate
SFC Supercritical Fluid Chromatography
SFE Supercritical Fluid Extraction

T

TAS Total Analysis System
tITP Transient Isotachophoresis
TOF-MS Time of Light – Mass Spectrometry

Symbol List

D	Diffusion Coefficient
d_c	Inner Diameter of the Capillary
E	Electric Field
ε	Dielectric Constant
F	Electrostatic Force
F'	Frictional Force
z	Charge
ζ	Zeta Potential
H	Plate Height
η	Viscosity
l	Effective Length of the Capillary
L_{tot}	Length of the Capillary
m	Mass
μ_a	Apparent Mobility
μ_{eo}	Mobility of the EOF
μ_{ep}	Electrophoretic Mobility
N_i	Amount of the Analyte
V	Voltage
v	Volume
v_{eo}	Linear Velocity of the EOF
v_{ep}	Velocity of Migration
v_s	Linear Velocity of Sample Injection
ρ	Density
q	Charge of the Ion
σ	Standard Deviation
r	Radius of the Ion
R'	Capillary Radius
t_i	Injection Time
t_m	Migration Time
t_r	Retention Time
Φ	Quantum Yield
w	Weight

Subject Index